# ANALOG INTEGRATED CIRCUITS

PRENTICE-HALL SERIES IN SOLID STATE PHYSICAL ELECTRONICS

Nick Holonyak, Jr., *Editor*

CHEO Fiber Optics
GREEN Solar Cells: Operating Principles, Technology, and System Applications
HAUS Waves and Fields in Optoelectronics
NUSSBAUM Contemporary Optics for Scientists and Engineers
SOCLOF Analog Integrated Circuits
SOCLOF Applications of Analog Integrated Circuits
STREETMAN Solid State Electronic Devices, 2nd ed.
UMAN Introduction to the Physics of Electronics
VAN DER ZIEL Solid State Physical Electronics, 3rd ed.
VERDEYEN Laser Electronics

# ANALOG INTEGRATED CIRCUITS

**Sidney Soclof**

*California State University*
*Los Angeles*

PRENTICE-HALL, INC., Englewood Cliffs, NJ 07632

*Library of Congress Cataloging in Publication Data*

Soclof, Sidney.
Analog integrated circuits.

(Prentice-Hall series in solid state physical electronics)
Includes bibliographies and index.
1. Linear integrated circuits. I. Title.
II. Series.
TK7874.S64 1985 621.381'73 84–11680
ISBN 0–13–032772–7

Editorial/production supervision and
interior design: *Ellen Denning*
Cover design: *Lundgren Graphics, Ltd.*
Manufacturing buyer: *Anthony Caruso*



Printed in the United States of America

10 9 8 7 6 5 4 3 2 1

ISBN 0-13-032772-7 01

PRENTICE-HALL INTERNATIONAL, INC., *London*
PRENTICE-HALL OF AUSTRALIA PTY. LIMITED, *Sydney*
EDITORA PRENTICE-HALL DO BRASIL, LTDA., *Rio de Janeiro*
PRENTICE-HALL CANADA INC., *Toronto*
PRENTICE-HALL OF INDIA PRIVATE LIMITED, *New Delhi*
PRENTICE-HALL OF JAPAN, INC., *Tokyo*
PRENTICE-HALL OF SOUTHEAST ASIA PTE. LTD., *Singapore*
WHITEHALL BOOKS LIMITED, *Wellington, New Zealand*

# CONTENTS

# PREFACE

This book treats the fabrication, analysis, design, and applications of analog integrated circuits. There is enough material in this book for a two-semester senior or graduate level sequence of courses. It can also be used for a one-semester course. For a one-semester course Chapters 4 through 8 are recommended. If time permits, the first part of Chapter 3 on current sources should also be included as well as Appendices B and C. For a two-semester advanced electronics course sequence covering both analog and digital integrated circuits, this book can be used for the analog part of the course. The material on integrated-circuit fabrication and device characteristics in the first two chapters is common to both analog as well as digital integrated circuits and thus can serve both the analog as well as the digital parts of such a two-course sequence. At the end of every chapter (except Chapter 8) a number of representative problems are given. A computer is useful for many of the problems and a few of the problems *require* the use of a computer; this is indicated by an asterisk preceding the problem. At the end of each chapter is a list of general references for further reading or study. The preparation recommended for this book is a one- or two-semester introductory course in electronic circuit analysis.

In Chapter 1 the basic processes that are used to manufacture various semiconductor devices are presented, with particular emphasis on the processes used for integrated circuits. Included in this chapter is material on crystal growth, wafer preparation, diffusion, ion implantation, oxidation, photolithography, metallization, and chemical-vapor deposition processes including epitaxy.

In Chapter 2 the basic processes that are described in Chapter 1 are combined to produce various semiconductor devices and, in particular, devices for integrated circuits. This chapter discusses the basic characteristics of PN junctions and the advantages of the use of epitaxial layers. The processing sequences used for diodes, bipolar transistors, junction-field-effect transistors, and MOSFETs are described followed by the various techniques used for the isolation of devices on an integrated-circuit chip. There is also a presentation of the fabrication and characteristics of various integrated-circuit devices including NPN and PNP bipolar transistors, JFETs, MOSFETs, diodes, resistors, and capacitors. Other material in Chapter 2 deals with considerations of chip size, circuit complexity and thermal design.

In Chapters 3 and 4 there is a presentation of various electronic circuit configurations common to many types of analog integrated circuits. The first part of Chapter 3 deals with various types of constant-current source circuits that are an important part of many integrated circuits. The second part of this chapter is concerned with low impedance voltage source circuits, and the last part is a study of temperature-compensated voltage reference circuits. The subject matter of Chapter 4 is differential amplifiers. In the first part of this chapter, the design and characteristics of bipolar transistor differential amplifiers is considered. This is followed by a presentation of JFET and MOSFET differential amplifiers. Closely associated with the subject of differential amplifiers is that of active-load circuits which is discussed in the last part of the chapter along with bipolar transistor and FET active loads.

The most widely used type of analog integrated circuit is the operational amplifier, and this is the subject matter of Chapters 5 and 6. The discussion commences in Chapter 5 with an analysis of closed-loop operational-amplifier circuits based on the case of an ideal device. The various nonideal characteristics of operational amplifiers are discussed including the effects of the input offset voltage, input bias current, input and output impedances, common mode gain, and the noise generated both within the operational amplifier and the external resistances. As in any feedback system, the question of system stability is an important one. This is considered in Chapter 5 and the various compensation techniques to insure an adequate margin of stability are also discussed. Later in the chapter, examples of a large number of applications of operational amplifiers are presented. An important category of operational amplifier applications is that of active filters. These applications are discussed at the end of Chapter 5 together with material on switched-capacitor filters.

Whereas Chapter 5 deals mostly with the external circuitry and applications of operational amplifiers, attention in Chapter 6 is turned principally to the internal circuit design of operational amplifiers. The chapter begins with a discussion of the basic configuration of operational amplifiers and a general analysis of the open-loop frequency response, including an analysis of the unity-gain frequency. This is followed by a discussion of the large-signal response characteristics including the slewing rate and the full-power bandwidth. A representative example of an operational-amplifier circuit is then presented and analysed. This analysis includes evaluation of the d-c biasing, the a-c small signal voltage gain of each stage and the entire amplifier, and various other amplifier characteristics including the common-mode rejection ratio, the power supply rejection ratio, the open-loop frequency response, and the equivalent input noise voltage and current. Operational amplifiers that use field-effect transistors in just the input stage or throughout the amplifier are an important type of device and are considered in the latter part of Chapter 6. A number of representative JFET and MOSFET operational amplifier circuits are presented.

The voltage comparator is a device closely related to the operational amplifier and a full discussion of this type of integrated circuit is given in Chapter 7. A consideration of some of the basic characteristics of voltage comparators and how these devices compare to operational amplifiers is thoroughly discussed. Some examples of voltage comparators are then presented. One of the most important characteristics of the voltage comparator is its response time and this characteristic is considered in detail. Various techniques that are used to reduce the response time are also presented. Complementary-symmetry MOSFETs can be used for comparators, especially for high-density, low-power drain applications. The CMOS inverter and CMOS voltage comparators are presented at the end of this chapter.

While the operational amplifier is the preeminant type of analog-integrated circuit, and the voltage comparator is probably the second most widely used analog device, there are many other types of analog integrated circuits. These integrated circuits are the subject of Chapter 8 in which there is a brief presentation of a number of these integrated circuits including voltage regulators, power amplifiers, and video amplifiers. This is followed by a discussion of integrated circuits used in a variety of applications in the areas of communications, control, signal processing, optoelectronics, digital systems, and computer interfacing.

SIDNEY SOCLOF

# ANALOG INTEGRATED CIRCUITS

# 1

# INTEGRATED-CIRCUIT FABRICATION

## 1.1 SILICON DEVICE PROCESSES

The basic processing steps that are used to fabricate various silicon devices, such as diodes, transistors, and integrated circuits (ICs), can be categorized as follows:

1. Diffusion (and ion implantation)
2. Oxidation
3. Photolithography
4. Chemical-vapor deposition (including epitaxy)
5. Metallization

Starting with single-crystal silicon wafers, the processes listed above can be used to produce functioning discrete devices (i.e., individual diodes and transistors) and ICs. These devices or ICs will be in wafer form, with tens, hundreds, or even thousands of devices or ICs on the same silicon wafer. The wafer must be then divided up to obtain the individual *dice* or *chips*.

These chips are then encapsulated or *packaged*, with a wide variety of packages and packaging methods being possible. The purpose of the packaging is twofold: (1) to provide encapsulation of the chip for protection of the chip from environmental effects, and (2) to provide easy access to the various parts of the chip by means of a lead or pin structure such that the device may be conveniently plugged into or attached to the rest of the system.

The basic *wafer fabrication* processes listed above are generally applied a number of times in succession, especially in the case of ICs, where as many as 10 repetitions of the photolithography, oxidation, and diffusion steps may be used.

## 1.2 SILICON WAFER PREPARATION

The starting materials for silicon device processing are single-crystal silicon wafers of suitable conductivity type and doping. The silicon wafer preparation sequence involves the following basic steps:

1. Crystal growth and doping
2. Ingot trimming and grinding
3. Ingot slicing
4. Wafer polishing and etching
5. Wafer cleaning

We will now discuss these steps briefly.

### 1.2.1 Crystal Growth

In the crystal growth process single-crystal silicon ingots of the appropriate doping level and type are produced. The starting material for crystal growth is highly purified polycrystalline silicon. The purity of this material is generally down in the range of more than 99.9999999% or "9-nines" purity. This corresponds to an impurity concentration of less than 1 part per billion atoms ($<1$ ppba), that is, less than one impurity atom for every billion ($10^9$) silicon atoms. The number of silicon atoms per unity volume is $5.0 \times 10^{22}\ \text{cm}^{-3}$, so that an impurity concentration of 1 ppba corresponds to an impurity density of $5 \times 10^{13}\ \text{cm}^{-3}$. Much of this residual impurity density will be acceptor impurities such as boron, and the resistivity corresponding to the impurity density above is approximately 300 $\Omega$-cm. Polycrystalline silicon with impurity concentrations of less than 0.1 ppba is available.

The *Czochralski* crystal growth process is the one most often used for producing single-crystal silicon ingots. The polycrystalline silicon together with an appropriate amount of dopant or doped silicon is put into a quartz crucible, which is then placed inside a crystal growth furnace. The material is then heated to a temperature that is slightly in excess of the silicon melting point of 1420°C. A small single-crystal rod of silicon called a *seed crystal* is then dipped into the silicon melt and slowly withdrawn, as shown in Figure 1.1. The conduction of heat up the seed crystal will produce a reduction in the temperature of the melt in contact with the seed crystal to slightly below the silicon melting point. This silicon will therefore freeze onto the end of the seed crystal, and as the seed crystal is slowly pulled up out of the melt it will pull up with it a solidified mass of silicon that will be a crystallographic continuation of the seed crystal. Both the seed crystal and the crucible are rotated, but in opposite directions during the crystal pulling process in order to produce crystalline ingots of circular cross section.

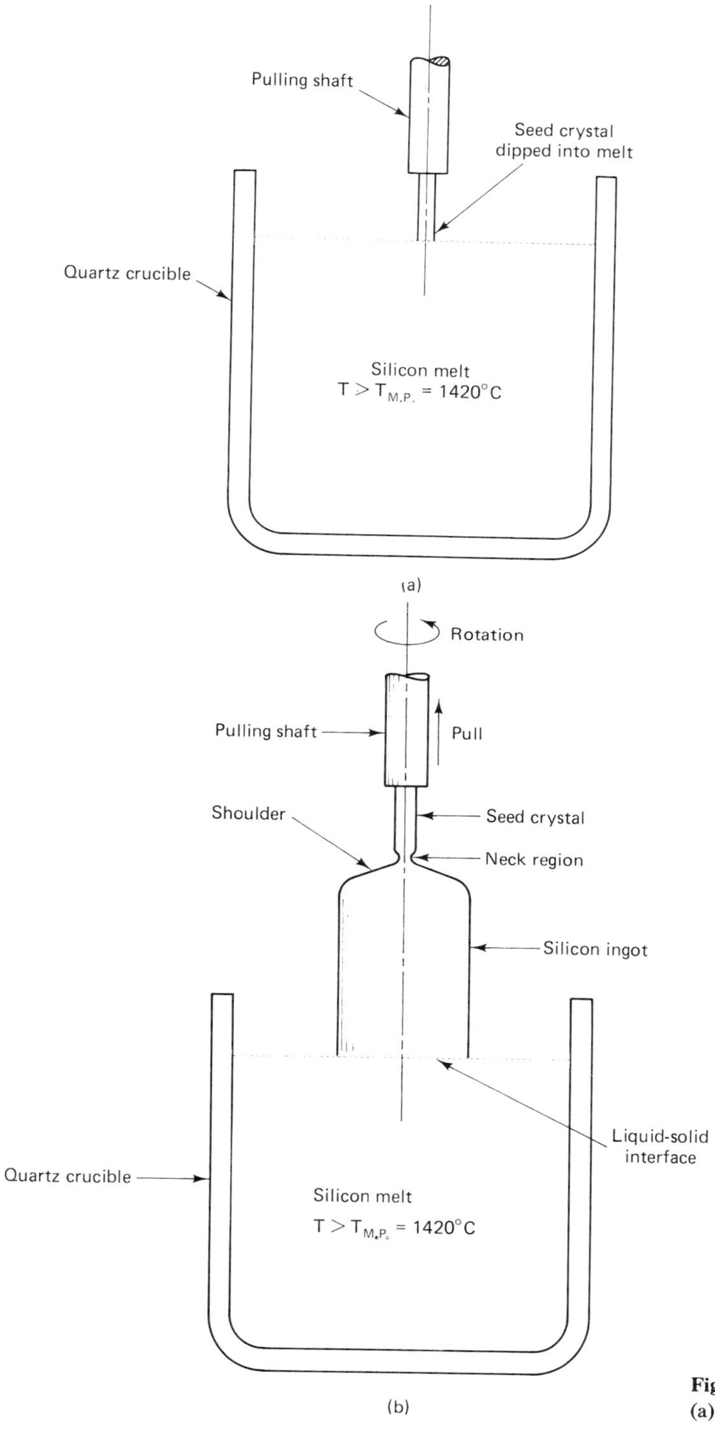


**Figure 1.1** Czochralski crystal growth: (a) seed setting; (b) crystal pulling.

The crystal pulling is done in an inert-gas atmosphere, usually argon or helium, and sometimes a vacuum is used. The diameter of the ingot is controlled by the pulling rate and the melt temperature, with ingot diameters of about 100 to 150 mm (4 to 6 in.) being the most common. The ingot length will generally be of the order of 100 cm, and several hours are required for the "pulling" of a complete ingot.

The *float-zone* process is often used for the production of single-crystal silicon ingots, especially where very high-resistivity material (>100 Ω-cm) is required. In this crystal growth process a rod of polycrystalline silicon about 100 mm in diameter is placed in a quartz tube and a high vacuum is produced. A small single-crystal seed section of silicon is placed in contact with one end of the polycrystalline silicon rod. As shown in Figure 1.2, a water-cooled radio-frequency (RF) induction heating coil induces eddy currents in the silicon and raises the temperature of the region of the rod near the induction heating coil to above the melting point. This produces a molten zone held in place by surface-tension effects. The molten zone is started out at the seed end of the polycrystalline silicon rod, and by moving the induction heating coil the molten zone can be slowly moved through the length of the rod. As the zone moves through the rod the silicon will melt and then resolidify as a crystallographic continuation of the seed crystal.

The float-zone process can also be used for the purification of silicon, this process being known as *zone refining*. Virtually all impurities in silicon tend to stay in the liquid phase rather than be incorporated into the solid phase at a moving liquid–solid interface. If a molten zone is passed through a silicon rod, the impurities will therefore tend to become segregated in the molten zone and thereby be swept to the far end of the rod. If a molten zone is passed along the length of the silicon rod to one end, and then started again at the opposite end and the process repeated a number of times, a very high degree of purification can be achieved.

Float-zone crystals are grown in a vacuum atmosphere and there is no contact of the ingot with any crucible. As a result of this, very high-resistivity crystals with very low content of oxygen and other impurities are possible with the float-zone process. In the Czochralski process the contact of the molten silicon with the quartz crucible will result in contamination of the melt with oxygen and other impurities from the crucible. In Table 1.1 a general comparison of the Czochralski and float-zone silicon processes is presented.

### 1.2.2 Ingot Trimming and Slicing

After the crystal growth process is completed, the extreme top and bottom portions of the ingot are cut off and the ingot surface is ground to produce a constant and exact diameter (usually 75, 100, or 125 mm). A crystallographic orientation flat is also ground along the length of the ingot.

The ingot is then sliced using a large-diameter stainless steel saw blade with industrial diamonds embedded into the inner-diameter cutting edge. This will produce circular slices or wafers that are about 600 to 700 μm (0.023 to 0.028 in.) thick, as shown in Figure 1.3. The orientation flat will serve as a useful reference plane for various device processes to be described later.

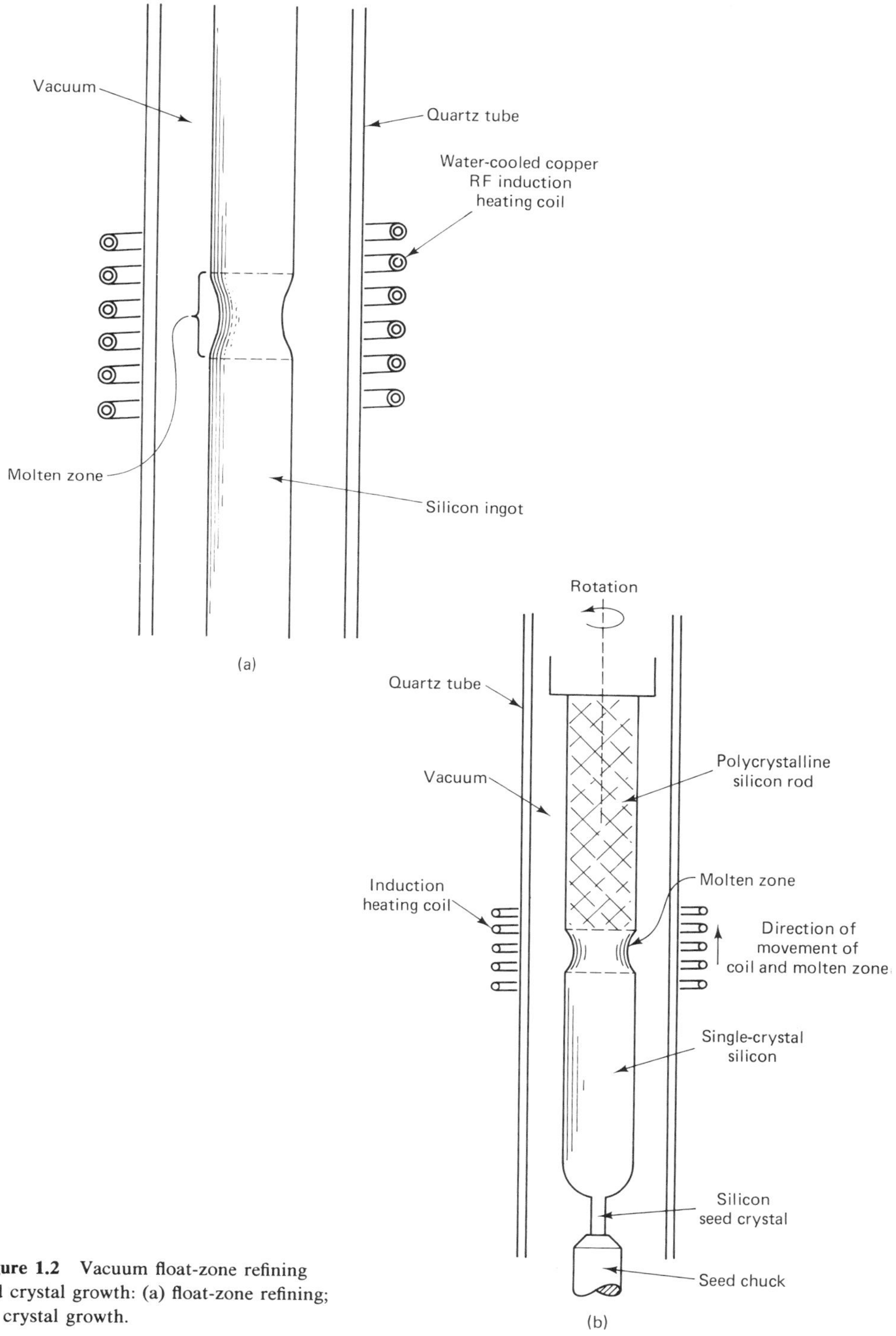


**Figure 1.2** Vacuum float-zone refining and crystal growth: (a) float-zone refining; (b) crystal growth.

**TABLE 1.1 COMPARISON OF CZOCHRALSKI AND FLOAT-ZONE SILICON PROCESSES**

| Parameter | | Czochralski | Float-zone |
|---|---|---|---|
| Resistivity (Ω-cm) | | | |
| N-Type | | As 0.001–25 | 0.001–25 |
| | | P: 0.001–100 | 0.001–500 |
| | | Sb: 0.005–20 | 1.0–50 |
| P-Type | | B: 0.0003–100 | 0.001–10,000 |
| | | Al: 0.060–15 | |
| | | Ga: 0.10–20 | |
| Radial resistivity Gradient (from center to half-radius) (%) | | As: 20 | 15 |
| | | P: 15 | 10 |
| | | Sb: 10 | 15 |
| | | B: 10 | 10 |
| | | Al: 10 | |
| | | Ga: 10 | |
| Lifetime ($\mu$s) | Resistivity (Ω-cm) | $\tau$ ($\mu$s) | $\tau$ ($\mu$s) |
| | 1–5 | 10 | 10 |
| | 5–10 | 25 | 25 |
| | 10–50 | 50 | 50–100 |
| | Above 50 | 100 | 75–200 |
| Dislocation density | 0–1500 $cm^{-2}$ | | 10,000–30,000 $cm^{-2}$ |
| Oxygen content | 5–10 ppm ($2.5 \times 10^{17}$ to $5 \times 10^{17}$ $cm^{-3}$ | | 0.5–1.0 ppm $2.5 \times 10^{16}$ to $5 \times 10^{16}$ $cm^{-3}$ |

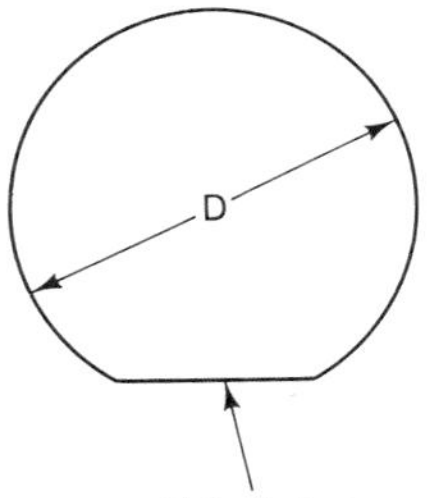


**Figure 1.3** Silicon wafer. $D$ = 100, 125, or 150 mm.

### 1.2.3 Wafer Polishing and Cleaning

The raw cut silicon wafers have very heavily damaged surfaces as a result of the slicing operation. The wafers will now undergo a number of polishing steps for the following purposes:

1. To remove the damaged silicon from the sawn surfaces
2. To produce a highly planar or flat surface that will be required for the photolithographic process, especially when very fine-line geometries are involved

3. To improve the parallelism of the two major surfaces of the wafer, this also being needed for the photolithographic process

In Figure 1.4 some diagrams are shown of the wafer surface contour and thickness following the various polishing steps. Usually, only one side of the wafer is given the final mirror smooth highly polished finish, the other side (i.e., the "back side") being given just a lapping operation to ensure an acceptable degree of flatness and parallelism. The final polishing operation is often a very light chemical etching to remove the last vestiges of mechanical damage arising from the previous polishing operations, and sometimes a combined chemical–mechanical polishing operation is used.

After the wafer polishing operations are completed, the wafers are thoroughly cleaned, rinsed, and dried, and they are now ready to be used for the various device processing steps to be discussed next.

We will now discuss the various device processes separately, and then we will see how they all fit together in the manufacturing of semiconductor devices.

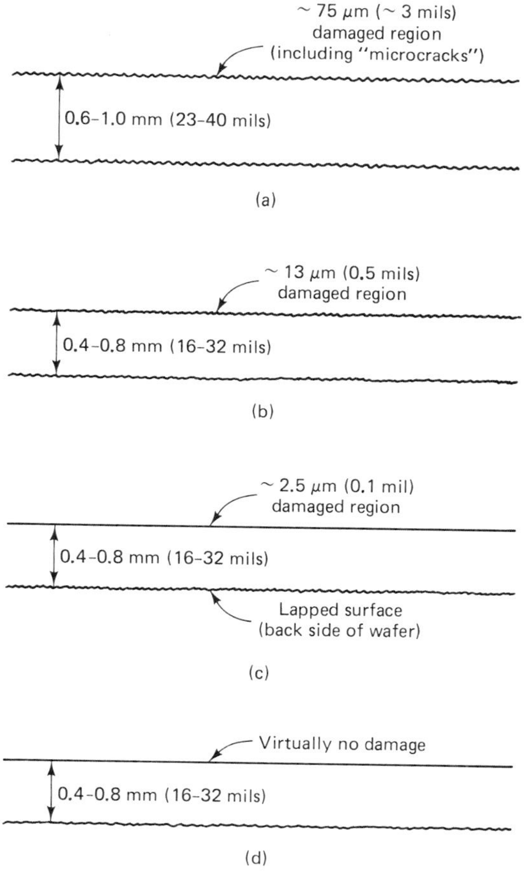


**Figure 1.4** Wafer lapping and polishing: (a) raw cut slice; (b) lapped slice; (c) polished wafer; (d) etched slice.

## 1.3 DIFFUSION

The diffusion process under consideration here is *solid-state impurity diffusion* in which various dopant impurities can be introduced into the surface region of a silicon wafer. For this process to occur at a reasonable rate the temperature of the wafer must be in the range of about 900 to 1200°C. Although these are rather high temperatures, they are still well below the melting point of silicon, which is at 1420°C. The rate at which the various impurities diffuse into silicon will be of the order of 1 μm/h at the temperature range above, and the penetration depths that are involved in most diffusion processes will be of the order of 0.3 to 30 μm. At room temperature the diffusion process will be so extremely slow such that the impurities can be considered to be essentially "frozen" in place.

Dopants of very small atomic or ionic radii, such as lithium ($Li^+$), can fit in the gaps or voids (*interstices*) between silicon atoms, and can therefore diffuse very rapidly. These small-size dopants will be *interstitial diffusants*; the diffusion process is called *interstitial diffusion* and is illustrated in Figure 1.5. Although lithium will act as a donor impurity in silicon, it is not normally used because it will still move around even at temperatures near room temperature, and thus will not be "frozen" in place. This is true of most other interstitial diffusions, so long-term device stability cannot be assured with this type of impurity.

The more useful diffusants in silicon are the larger *substitutional* dopants. These dopant atoms are too big to fit into the interstices, so the only way they can enter the silicon crystal structure is to substitute for a silicon atom. The most commonly used substitutional donor dopants are phosphorus (P), arsenic (As), and antimony (Sb), with phosphorus being used most often. The only acceptor dopant that is used to any extent is boron (B).

For the substitutional diffusion process the presence of vacancies in the crystal structure is of primary importance. A *vacancy* is the absence of an atom from a

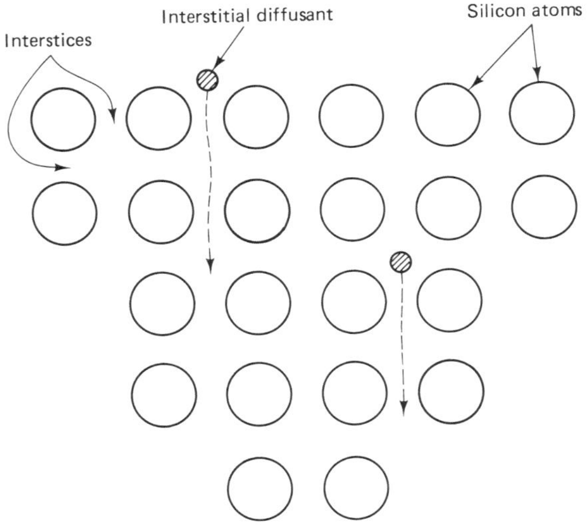


**Figure 1.5** Interstitial diffusion.

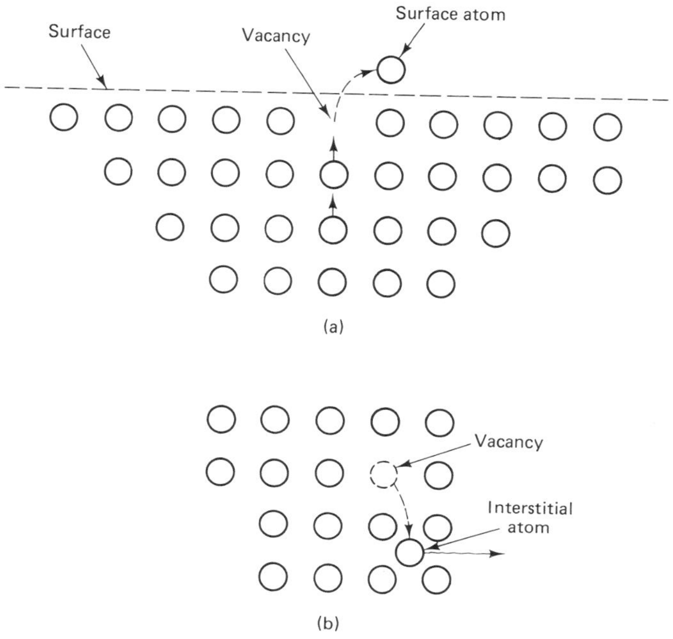


**Figure 1.6** Vacancy generation processes: (a) Schottky defect (surface atom—vacancy pair); (b) Frenkel defect (interstitialcy—vacancy pair).

point in the crystal structure where an atom would ordinarily be present. Although the ideal crystal structure has no vacancies, some vacancies are naturally present in any real crystal. The vacancy density or population can be vastly increased by raising the temperature of the crystal, the vacancy concentration increasing exponentially with temperature. Vacancies can be generated at the surface of the crystal (*Schottky defects*) as shown in Figure 1.6a, or throughout the interior of the crystal (*Frenkel defects*) as shown in Figure 1.6b. The vacancies generated at the surface can diffuse into the interior of the crystal and become uniformly distributed throughout the volume of the crystal.

The presence of vacancies will permit impurity atoms to enter the surface layer of the crystal and then to diffuse slowly into the interior, as shown in Figure 1.7. This impurity diffusion is a very slow step-by-step process since a vacancy must be present in a suitable position adjacent to the impurity atom for it to move down one atomic layer at a time.

### 1.3.1 Diffusion Equations

The diffusion process involves the flow of atoms or other particles under the influence of a concentration gradient, as shown in Figure 1.8. Since there are more particles to the left of the line at $x = x_1$ than to the right of this line, there will be a net flow of particles from left to right. Note that in Figure 1.8 the concentration gradient $dN/dx$ is negative and the particle flow will be in the $+x$ direction. The particle

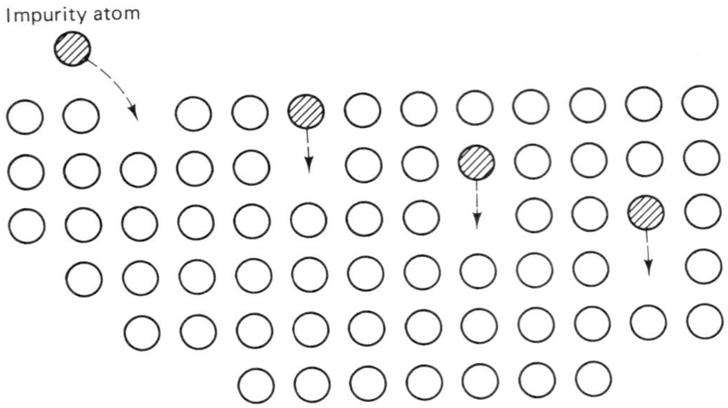


**Figure 1.7** Solid-state impurity diffusion.

flow or flux will be proportional to the concentration gradient given by $F \propto -dN/dx$, where $F$ is the diffusant flux, which is the rate of flow of the particles or atoms per unit area. To make this into an equation we introduce a constant of proportionality to obtain an equation called Fick's first law, given by $F = -D(dN/dx)$. In this equation $D$ is the *diffusion constant* or *diffusion coefficient* and has units of cm²/s. Looking at the overall balance of units, we have

$$F\,(1/\text{cm}^2\text{-s}) = -D\,(\text{cm}^2/\text{s}) \times \frac{dN}{dx}\,(\text{cm}^{-3}/\text{cm}) \tag{1.1}$$

The diffusion constant $D$ will be a function of the type of diffusant, the material into which the diffusion takes place, and will increase exponentially with temperature. An equation for $D$ can be written as $D = D_0 \exp(-qE_A/kT)$, where $q$ is the electronic charge ($1.6 \times 10^{-19}$ C), $k$ is Boltzmann's constant ($1.38 \times 10^{-23}$ J/K), $T$ is the absolute temperature (K), and $D_0$ is a preexponential constant with units of cm²/s. The quantity $E_A$ is an activation energy with units of electron volts (eV) and will be in the range 4 to 5 eV for most substitutional dopant diffusants of interest in silicon, such as P, B, As, and Sb. Figure 1.9 is a graph of the variation of the diffusion constant with temperature for these diffusants. Note the straight-line relation-

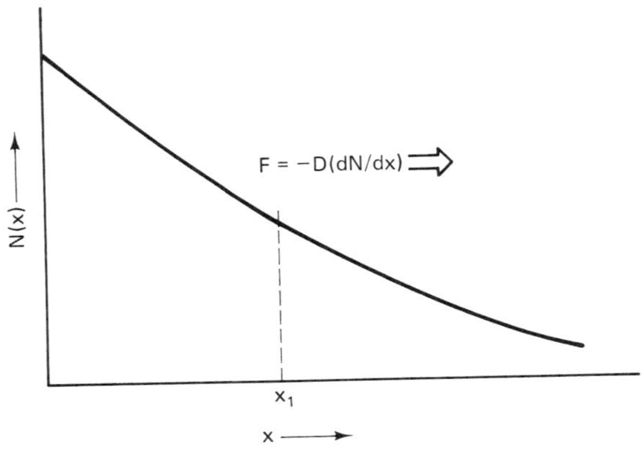


**Figure 1.8** Diffusion: Fick's first law.

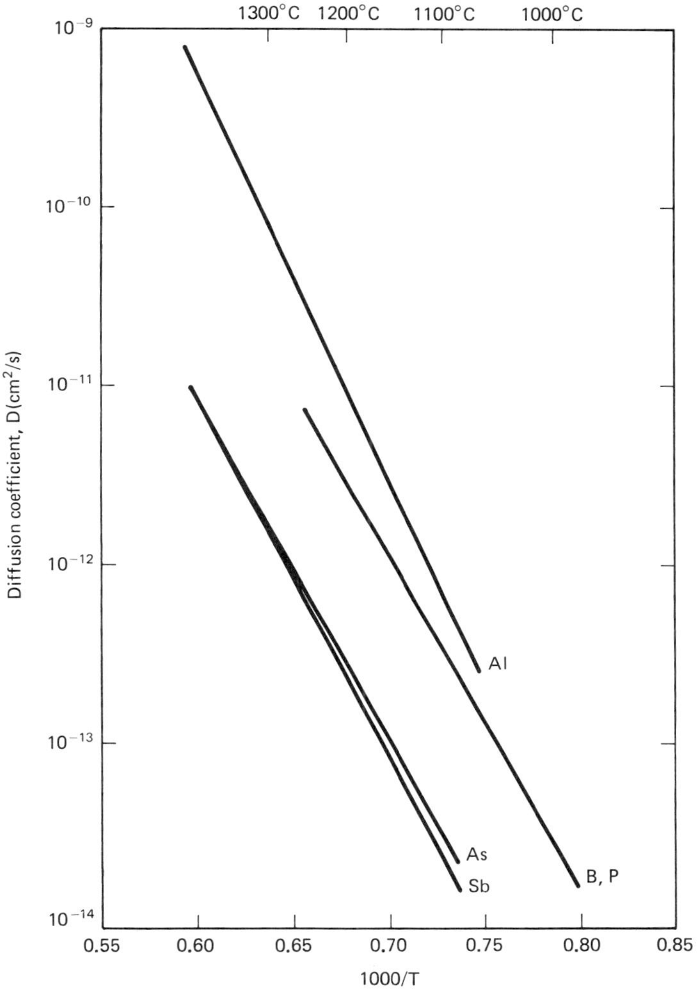


**Figure 1.9** Diffusion coefficients of donor and acceptor impurities in silicon. (Adapted from C. S. Fuller and J. A. Ditzenberger, "Diffusion of Donor and Acceptor Elements in Silicon," *Journal of Applied Physics*, Vol. 27, pp. 544–553, May 1956.)

ship that is obtained by having $D$ on a logarithmic scale and the temperature on a reciprocal temperature, $1/T$, scale.

A second relationship that will be involved in the diffusion process will be the *continuity equation*, given by

$$\frac{\partial N}{\partial t} = \frac{F(x) - F(x + dx)}{dx} = \frac{-\partial F}{\partial x} \tag{1.2}$$

This equation is basically a statement of the conservation of matter and states that the rate of change of atoms or particles $\partial N/\partial t$ in a given volume is equal to the rate of flow of the particles into the volume minus the rate of flow of particles out of the volume.

If we now combine Fick's first law and the continuity equation, we obtain *Fick's second law*, given by

$$\frac{\partial N}{\partial t} = \frac{-\partial F}{\partial x} = \frac{-\partial}{\partial x}\left(-D\frac{\partial N}{\partial t}\right) = D\frac{\partial^2 N}{\partial x^2} \tag{1.3}$$

The solution of this partial differential equation, subject to the appropriate boundary conditions, will give us the distribution or profile of the diffusant atoms $N(x, t)$. The two sets of boundary conditions that are of principal interest for solid-state impurity diffusion are (1) constant surface concentration and (2) limited source conditions.

### 1.3.2 Constant Surface Concentration Diffusion

We will first consider diffusion under constant surface concentration or "infinite source" conditions. For this case the diffusant concentration at the surface ($x = 0$) will be assumed to be constant, so that we will have that $N(0, t) = N_0 = \text{constant}$. The other boundary condition for this case is that as $x \longrightarrow \infty$ the diffusant concentration decays to zero. Subject to these boundary conditions the solution to Fick's second law will be

$$\boxed{N(x, t) = N_0 \operatorname{erfc}\left(\frac{x}{2\sqrt{Dt}}\right)} \tag{1.4}$$

where $N_0$ is the constant surface concentration, $x$ the distance into the material from the surface ($x = 0$), $t$ the diffusion time, and erfc is a mathematical function called the *complementary error function*.

In Figure 1.10 a graph of the complementary error function erfc ($z$) is presented. At $z = 0$ we have that erfc (0) = 1, and as $z \longrightarrow \infty$, erfc ($z$) $\longrightarrow 0$, and we note that erfc ($z$) decreases very rapidly with increasing $z$.

In Figure 1.11 a curve of a erfc type of impurity profile is shown for the example of a P-type boron diffusion into an N-type (phosphorus-doped) substrate. At the position $x = x_J$ the boron diffusant concentration is equal to the phosphorus substrate concentration. In the region of $x < x_J$ the boron (acceptor) concentration is greater than the phosphorus (donor) substrate concentration, so this region is therefore P-type. In the region of $x > x_J$ the donor concentration is greater than the acceptor concentration, so this region is N-type. At $x = x_J$ there is therefore the transition from a net P-type doping to a net N-type doping, so that $x_J$ marks the position of the PN junction.

This is therefore a PN junction produced by the diffusion process. Since the doping level in the P-type diffused layer is relatively heavy, especially near the surface where concentrations of about $10^{20}$ cm$^{-3}$ can be expected, the diffused layer can be

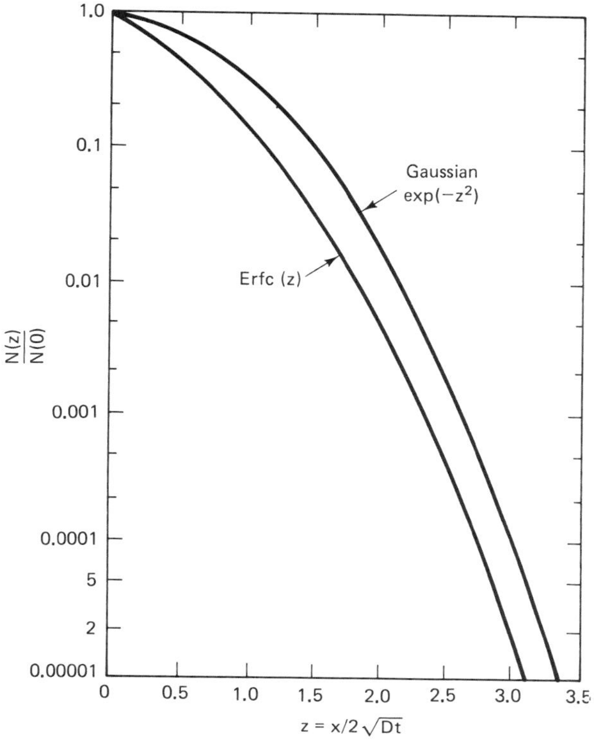


**Figure 1.10** Complementary error function and Gaussian functions.

designated as a $P^+$ region. The surface concentration will usually be fixed by the solid-state solubility of the impurity in silicon. In Figure 1.12 a graph of the solid-state solubilities of various dopants in silicon versus temperature is given. For boron in silicon, the solubility over the range of usual diffusion temperatures (900 to 1200°C) the solubility will be in the range $1 \times 10^{20}$ to $2 \times 10^{20}$ $cm^{-3}$.

In Figure 1.11 a series of impurity profiles are shown for a number of different diffusion times. The PN junction occurs at $x = x_J$, at which point we have that

$$N_A(x_J, t) = N_0 \operatorname{erfc}\left(\frac{x_J}{2\sqrt{Dt}}\right) = N_B \tag{1.5}$$

where $N_B$ represents the background or substrate concentration. From this we see that the junction depth $x_J$ will be proportional to the square root of the diffusion time.

Let us now consider as a representative example a constant surface concentration boron diffusion at 1150°C into a phosphorus-doped N-type silicon substrate. The substrate doping level will be chosen to be $6 \times 10^{15}$ $cm^{-3}$, which corresponds to a resistivity of 0.9 Ω-cm. At 1150°C the solid-state solubility of boron in silicon is approximately $3 \times 10^{20}$ $cm^{-3}$, and the diffusion coefficient of boron in silicon will be $D = 1 \times 10^{-12}$ $cm^2/s$. From the normalized erfc ($z$) graph of Figure 1.10 we see that at the junction we will have that erfc ($z_J$) $= N_B/N_0 = (6 \times 10^{15}\ cm^{-3})/$

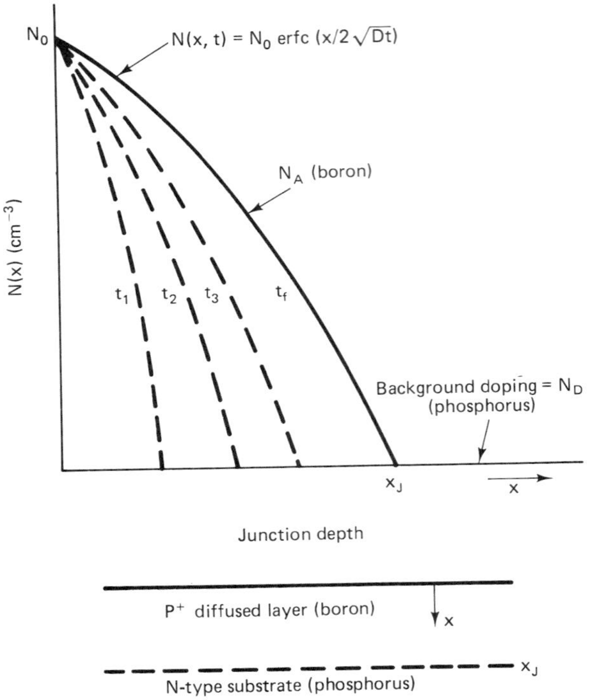


**Figure 1.11** Diffusion profile: constant surface concentration diffusion (deposition diffusion).

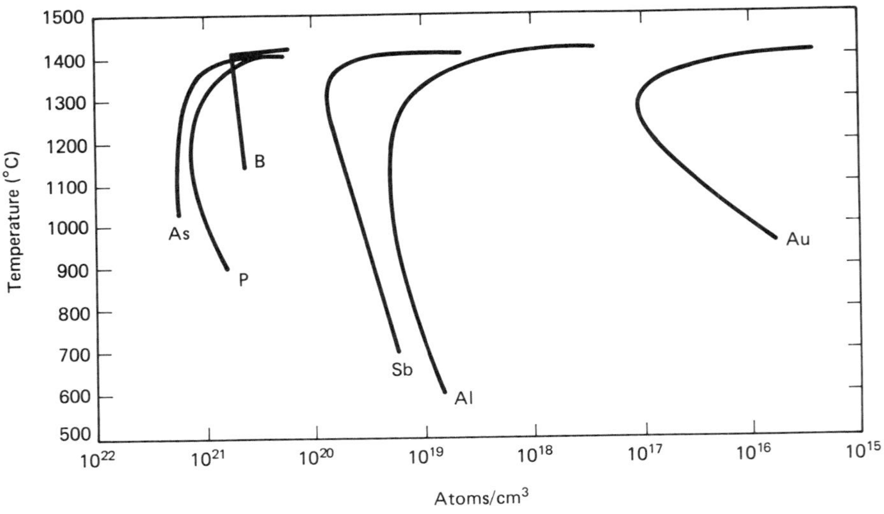


**Figure 1.12** Solid-state solubility of impurities in silicon. (Adapted with permission from F. A. Trumbore, "Solid Solubilities of Impurity Elements in Germanium and Silicon," *Bell System Technical Journal*, Vol. 39, pp. 205–233, January 1960, copyright 1960, AT&T.)

$(3 \times 10^{20}\ \text{cm}^{-3}) = 2 \times 10^{-5}$, so that $z_J = x_J/2\sqrt{Dt} = 3.0$. Solving for $x_J$ this gives $x_J = 3.0 \times 2 \times \sqrt{1 \times 10^{-12}\ \text{cm}^2/\text{s} \times t} = 6 \times 10^{-6}\ \text{cm}\ \sqrt{t/\text{s}}$. Since $1\ \mu\text{m} = 10^{-4}$ cm, this can be reexpressed as $x_J = 0.06\ \mu\text{m}\ \sqrt{t/\text{s}} = 3.6\ \mu\text{m}\ \sqrt{t/\text{h}}$. Therefore, a 1.0-h diffusion at 1150°C will result in a junction depth of 3.6 μm. For a 1.0-μm junction depth the required diffusion time will be 0.53 h = 32 min. At the other extreme, a 10-μm junction depth will require a total diffusion time of 7.72 h, so that we see that deep junctions can require some very long diffusion times.

### 1.3.3 Constant Surface Concentration Diffusion Processes

For the various types of diffusion and oxidation processes a resistance-heated tube furnace is usually used. A tube furnace has a long (about 2 to 3 m) hollow opening into which a quartz tube about 100 to 150 mm (4 to 6 in.) in diameter is placed as shown in Figure 1.13. The temperature within the quartz furnace tube can be controlled very accurately such that a temperature within $\frac{1}{2}$°C of the set-point temperature can be maintained uniformly over a "hot zone" about 1 m in length. The silicon wafers to be processed are stacked up vertically into slots in a quartz carrier or "boat" and inserted into the furnace tube.

A number of gases are metered into the furnace. The principal gas flow in the furnace will be nitrogen ($N_2$), which acts as a relatively inert gas and is used as a carrier gas to be a dilutent for the other more reactive gases. The $N_2$ carrier gas will generally make up some 90 to 99% of the total gas flow. A small amount of oxygen and a very small amount of a compound of the dopant will make up the rest of the gas flow. For boron diffusions such compounds as $B_2H_6$ (diborane) and $BBr_3$ (boron tribromide) are commonly used.

The following reactions will be occurring simultaneously at the surface of the silicon wafers:

$$\text{(1)}\quad Si + O_2 \longrightarrow SiO_2 \text{ (silica glass)} \qquad (1.6)$$

$$\text{(2)}\quad 4BBr_3 + 3O_2 \longrightarrow B_2O_3 \text{ (boron glass)} + 2Br_2, \text{ or} \qquad (1.7)$$

$$2B_2H_6 + 3O_2 \longrightarrow B_2O_3 \text{ (boron glass)} + 6H_2 \qquad (1.8)$$

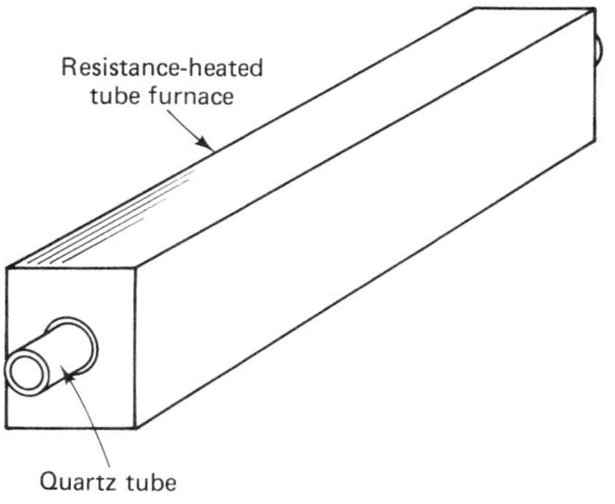


**Figure 1.13** Diffusion furnace.

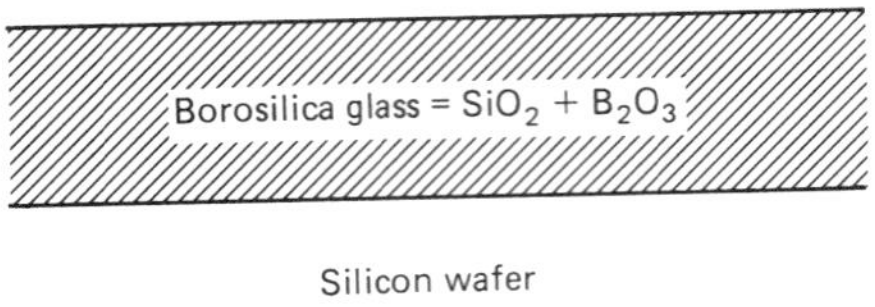


**Figure 1.14** Boron diffusion glass layer on silicon wafer.

This process is the chemical-vapor deposition (CVD) of a glassy layer on the silicon surface which is a mixture of silica glass ($SiO_2$) and boron glass ($B_2O_3$) called *borosilica glass* (BSG). The BSG glassy layer, shown in Figure 1.14, is a viscous liquid at the diffusion temperatures and the boron atoms can move around relatively easily. Furthermore, the boron concentration in the BSG is such that the silicon surface will be saturated with boron at the solid-state solubility limit throughout the time of the diffusion process as long as the BSG remains present.

For phosphorus diffusions such compounds as $PH_3$ (phosphine) and $POCl_3$ (phosphorus oxychloride) can be used. In the case of a diffusion using $POCl_3$ the reactions occurring at the silicon wafer surfaces will be:

$$\text{(1) } Si + O_2 \longrightarrow SiO_2 \text{ (silica glass)} \tag{1.9}$$

$$\text{(2) } 4POCl_3 + 3O_2 \longrightarrow P_2O_5 \text{ (phosphorus glass)} + 6Cl_2 \tag{1.10}$$

This will result in the production of a glassy layer on the silicon wafers that is a mixture of phosphorus glass and silica glass called *phosphorosilica glass* (PSG), which is a viscous liquid at the diffusion temperatures. The mobility of the phosphorus atoms in this glassy layer and the phosphorus concentration is such that the phosphorus concentration at the silicon surface will be maintained at the solid-state solubility limit throughout the time of the diffusion process. Similar processes occur with other dopants, such as the case of arsenic, in which arsenosilica glass is formed on the silicon surface.

The constant surface concentration diffusion results in the deposition of dopant atoms into the surface regions of the silicon wafers, in most cases to depths in the range 0.3 to 3 $\mu$m. This type of diffusion is often called a *deposition diffusion*.

### 1.3.4 Drive-in Diffusion

The deposition diffusion is often followed by a second diffusion process in which the external dopant source (i.e., the diffusion glass) is removed such that *no* additional dopants enter the silicon. During this diffusion process the dopants that are already in the silicon move farther in and are thus redistributed. The junction depth increases, and at the same time the surface concentration decreases. This type of diffusion is called *drive-in*, or *redistribution*, or *limited-source* diffusion.

The impurity profile for the drive-in diffusion can be obtained by solving Fick's second law subject to the boundary condition that the total diffusant density (i.e., diffusant atoms per unit area) remains constant. This condition can be stated mathematically as $Q = \int_0^\infty N(x, t)\, dx =$ constant, where $Q$ is the diffusant density. An alternative way of expressing this boundary condition is to note that under the stated conditions there is no net flow of diffusant atoms into or out of the silicon wafers at the surface ($x = 0$), so that in terms of Fick's first law we have $D(dN/dx)|_{x=0} = 0$. If Fick's second law is solved subject to this boundary condition, the resulting impurity distribution will be given by

$$N(x, t) = \frac{Q}{\sqrt{\pi Dt}} \exp\left(-\frac{x^2}{4Dt}\right) \tag{1.11}$$

This is a Gaussian distribution and a normalized curve of exp $(-z^2)$ is shown in Figure 1.10. Note that the surface diffusant concentration will be $N(0, t) = Q/\sqrt{\pi Dt}$ and thus will not be constant, but rather will decrease with increasing diffusion time. Note also that the slope of the impurity profile, $dN/dx$, at the surface ($x = 0$) will be zero, so that the Gaussian profile will indeed satisfy the boundary conditions for the drive-in diffusion.

The foregoing equation for the Gaussian impurity profile is based on the assumption that at $t = 0$ there is a diffusant density equal to $Q$ right at the silicon surface ($x = 0$). However, if the drive-in diffusion is preceded by a deposition diffusion such that the junction depth resulting from this first diffusion $x_{J_1}$ is considerably smaller than the junction depth $x_{J_2}$ after the completion of the drive-in diffusion such that $x_{J_2}^2 \gg x_{J_1}^2$, the final impurity profile will, to a close approximation, be given by the Gaussian equation above. This two-step diffusion process is illustrated in Figure 1.15.

The diffusant density produced as a result of a deposition diffusion will be given by $Q = 2N_0\sqrt{D_1t_1/\pi}$, where $D_1$ and $t_1$ are the diffusion constant and diffusion time, respectively, for the deposition diffusion. If a short deposition diffusion is followed by a drive-in diffusion, the resulting impurity profile will be approximately a Gaussian profile as given by $N(x, t_2) = Q/\sqrt{\pi D_2t_2}$ exp $(-x^2/4D_2t_2)$, where $D_2$ and $t_2$ refer to the diffusion constant and time of the drive-in diffusion, respectively. The surface concentration at the end of the drive-in diffusion will be

$$N(0, t_2) = \frac{Q}{\sqrt{\pi D_2t_2}} = \frac{2}{\pi}\sqrt{\frac{D_1t_1}{D_2t_2}}\,N_0 \tag{1.12}$$

Thus if the drive-in diffusion is a relatively long-time, high-temperature diffusion compared to the deposition diffusion such that $D_2t_2 \gg D_1t_1$, the surface concentration can be reduced substantially below the value of $N_0$ as imposed by the solid-state solubility limit. In Figure 1.15 some impurity profiles for a number of drive-in times are presented. Note that the total diffusant density $Q$ does not change but is only redistributed or "driven in" such that the junction depth increases and at the same time the surface concentration decreases. This two-step combination of a deposition diffusion followed by a drive-in diffusion is often used to produce the base region of transistors.

In Figure 1.16 a diagram is presented of a typical impurity profile of a double-diffused NPN transistor. Starting with the N-type silicon substrate, a boron deposition diffusion is performed. The boron diffusion glass is then removed and a drive-in diffusion is performed to reduce the boron surface concentration and to increase the junction depth. This produces the P-type base region of the transistor. The silicon wafer is then subjected to a high-concentration phosphorus deposition diffusion that

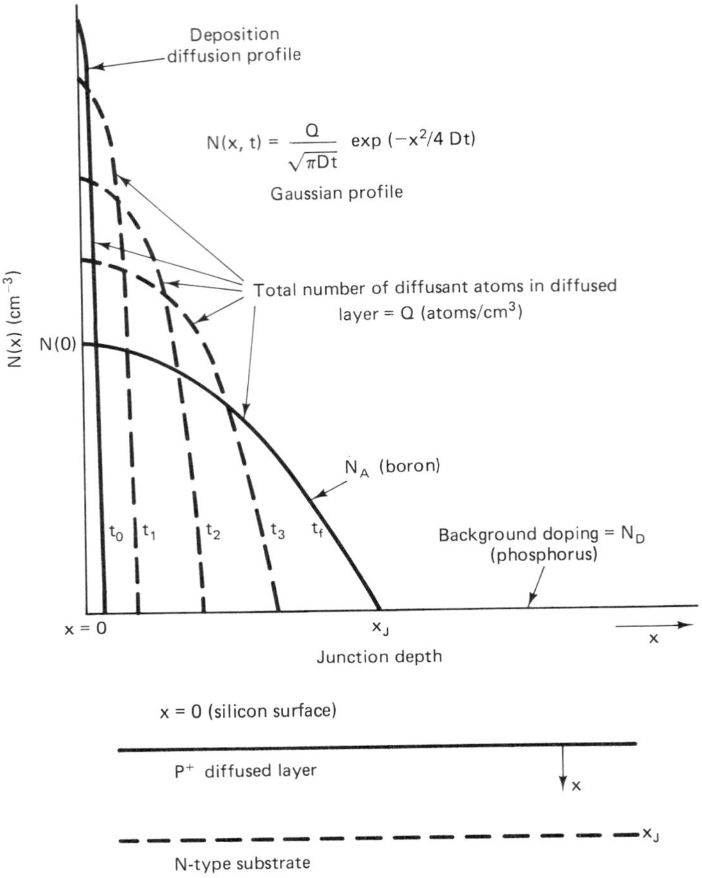


**Figure 1.15** Diffusion profile: drive-in diffusion.

converts part of the P-type diffused layer back to N-type. This diffusion results in a heavily doped $N^+$ region that will be the emitter of the transistor.

After the diffusions have been completed, the collector–base junction depth will be $x_{J(CB)}$ and will generally be about 3 $\mu$m. The emitter–base junction depth $x_{J(EB)}$ will be around 2 to 2.5 $\mu$m. The base width of the transistor will be $W_B = x_{J(CB)} - x_{J(EB)}$ and will generally be in the range 0.3 to 1.0 $\mu$m, although some "supergain" or "super-beta" transistors have base widths as small as in the range 0.1 to 0.2 $\mu$m.

The emitter surface concentration will usually be up in the range of $10^{21}$ cm$^{-3}$, and the base surface concentration will be reduced by the drive-in diffusion down to the range of around $3 \times 10^{18}$ cm$^{-3}$. The doping density at the emitter–base junction will generally fall in the range of $3 \times 10^{17}$ cm$^{-3}$, and the collector doping will typically fall in the range $10^{15}$ to $10^{16}$ cm$^{-3}$, corresponding to resistivities of from 5 $\Omega$-cm down to about 0.5 $\Omega$-cm. The double-diffused transistor structure with a diffused

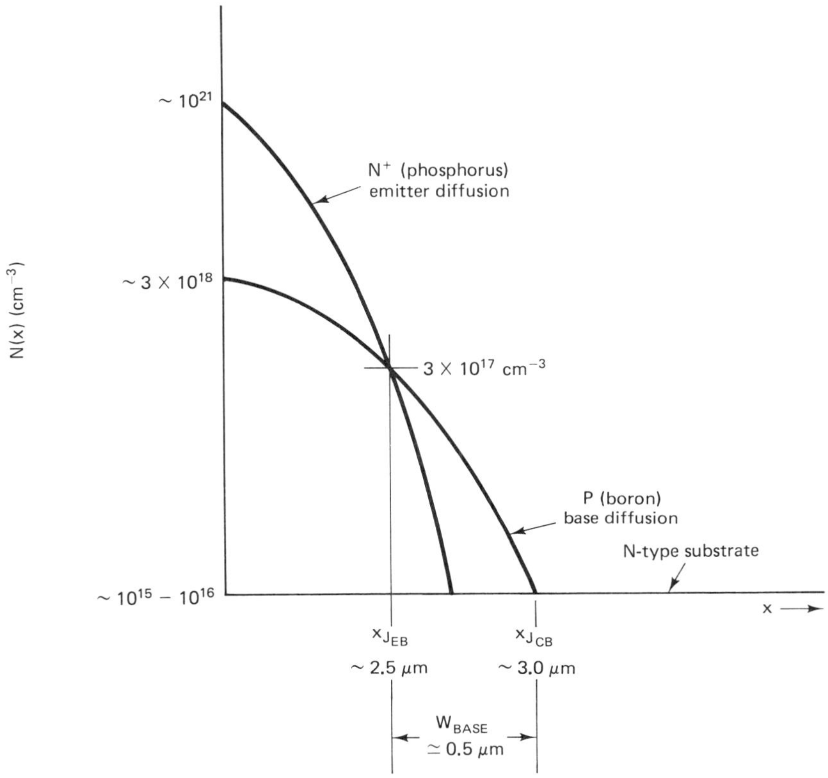


**Figure 1.16** Impurity profile of a double-diffused NPN transistor.

base region followed by a diffused emitter region is the technique used for the manufacture of almost all silicon transistors, including virtually all bipolar IC transistors.

### 1.3.5 Sheet Resistance

The two principal parameters that are used for the evaluation and characterization of diffused layers are the junction depth $x_J$ and the sheet resistance $R_S$. The resistance of a layer of material, as shown in Figure 1.17a, is given by the familiar equation $R = \rho L/A = \rho L/tW$, where ρ is the average resistivity of the layer, $L$ the length, $t$ the layer thickness, and $W$ the width of the layer.

If the shape is such that $L = W$ such that the material is a square, the resistance equation will become

$$\boxed{R = \frac{\rho}{t}} \tag{1.13}$$

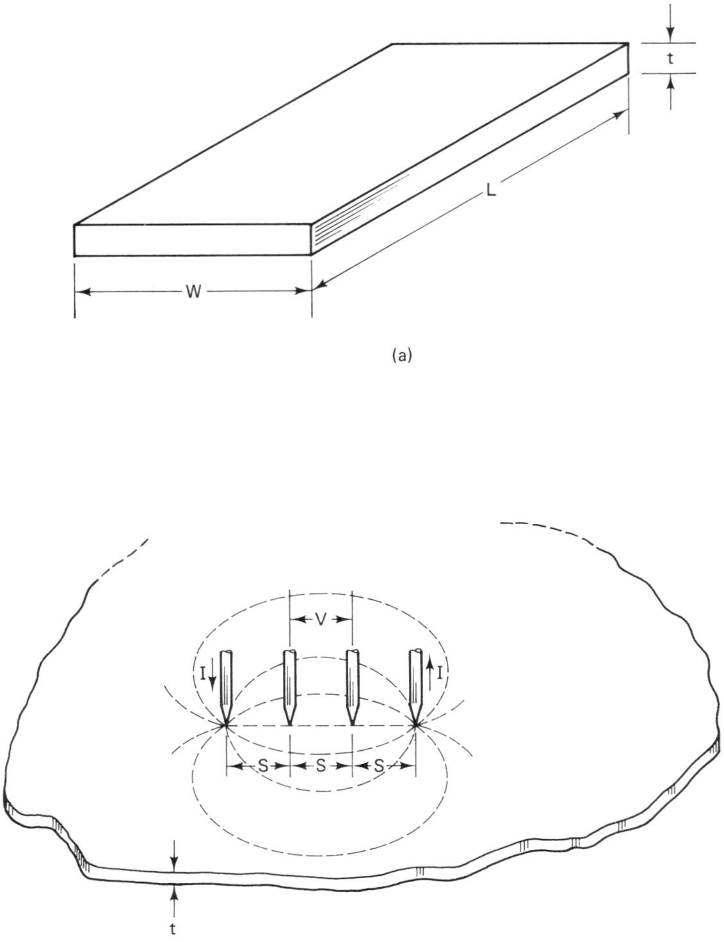


(a)

(b)

**Figure 1.17** (a) Sheet resistance; (b) four-point probe for measurement of sheet resistance.

Notice that the resistance of a "square" is independent of the lateral dimensions, $L$ and $W$. This is called the *sheet resistance*, $R_S$, and is often expressed in such units as Ω/square (or Ω/□). The resistance of a layer of material can thus be expressed in terms of the sheet resistance as

$$\boxed{R = R_S \frac{L}{W}} \tag{1.14}$$

The sheet resistance of thin layers, including diffused layers, can be conveniently measured using a four-point probe apparatus, as shown in Figure 1.17b. If the conditions are satisfied that the layer thickness $t$ is small compared to the probe spacing such that $t \ll s$, and that the edge of the layer be relatively remote from the probe array, the sheet resistance will be given approximately by $R_S = 4.5324(V/I)$. The current $I$ is supplied by the two outer probes, and the voltage $V$ is measured by a high-impedance voltmeter across the two inner probes. As a result of having separate pairs of probes for supplying the current and measuring the voltage drop, the probe contact resistance will not influence the measurement of the sheet resistance. The four-point probe can be used to measure the sheet resistance of various types of diffused layers as well as that of silicon wafers for the measurement of the resistivity.

Sheet resistance values of diffused layers will generally fall in the range from 1 Ω/square up to about 1000 Ω/square. The transistor base diffused layer will have a sheet resistance of typically about 200 Ω/square, and the $N^+$ emitter diffused layer will be down in the range of around 2 Ω/square. In Figure 1.18a and b graphs

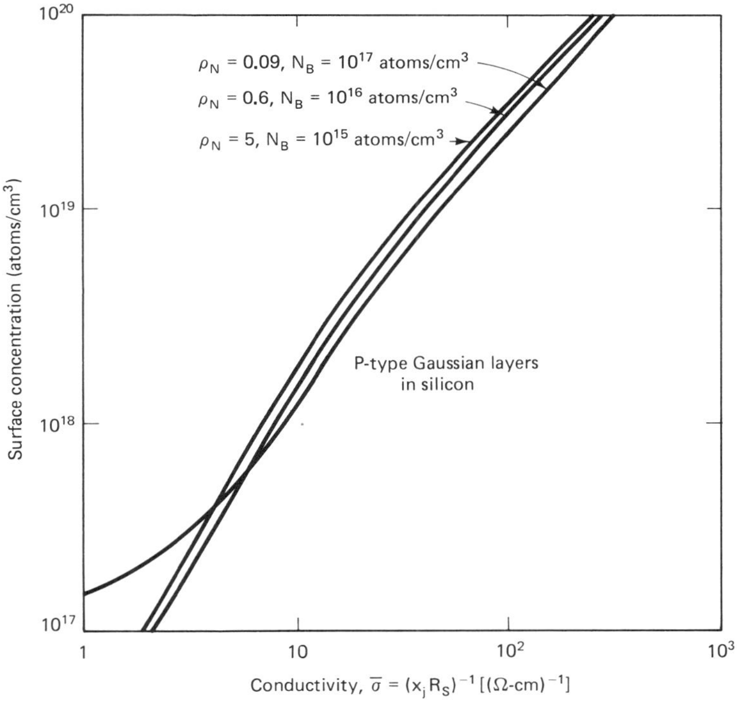


(a)

**Figure 1.18** Average conductivity of diffused layers in silicon: (a) average conductivity versus surface concentration for P-type Gaussian layers; (b) average conductivity versus surface concentration for N-type erfc layers; (c) resistivity versus impurity concentration for silicon. [(a) and (b), Adapted with permission from J. C. Irvin, "Resistivity of Bulk Silicon and of Diffused Layers," *Bell System Technical Journal*, Vol. 41, pp. 387–410, 1962, copyright 1962, AT&T; and S. K. Ghandi, *Theory and Practice of Microelectronics*, Wiley, 1968; (c), Adapted with permission from J. C. Irvin, "Resistivity of Bulk Silicon and of Diffused Layers in Silicon," *Bell System Technical Journal*, Vol. 41, pp. 387–410, 1962, copyright 1962, AT&T.]

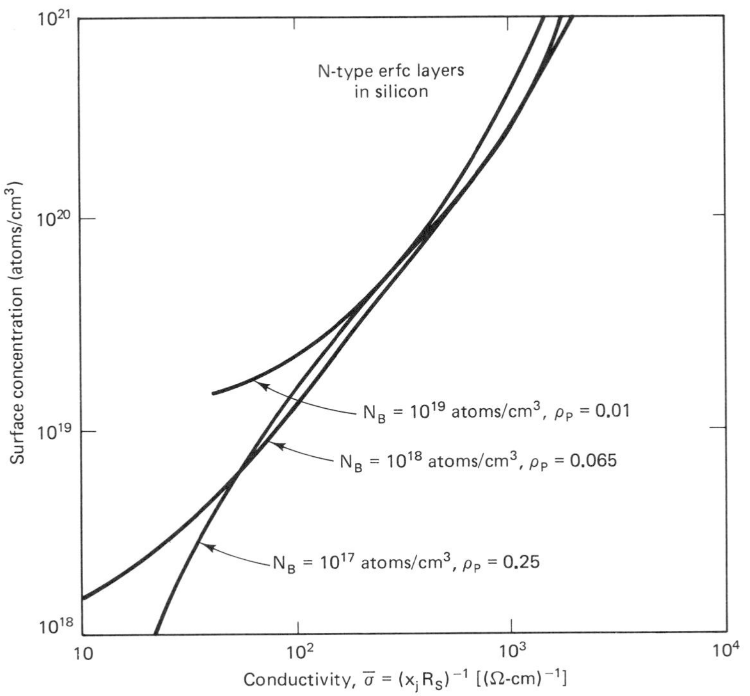
N-type erfc layers in silicon
$N_B = 10^{19}$ atoms/cm$^3$, $\rho_P = 0.01$
$N_B = 10^{18}$ atoms/cm$^3$, $\rho_P = 0.065$
$N_B = 10^{17}$ atoms/cm$^3$, $\rho_P = 0.25$
Surface concentration (atoms/cm$^3$)
Conductivity, $\overline{\sigma} = (x_j R_S)^{-1}$ [($\Omega$-cm)$^{-1}$]

(b)

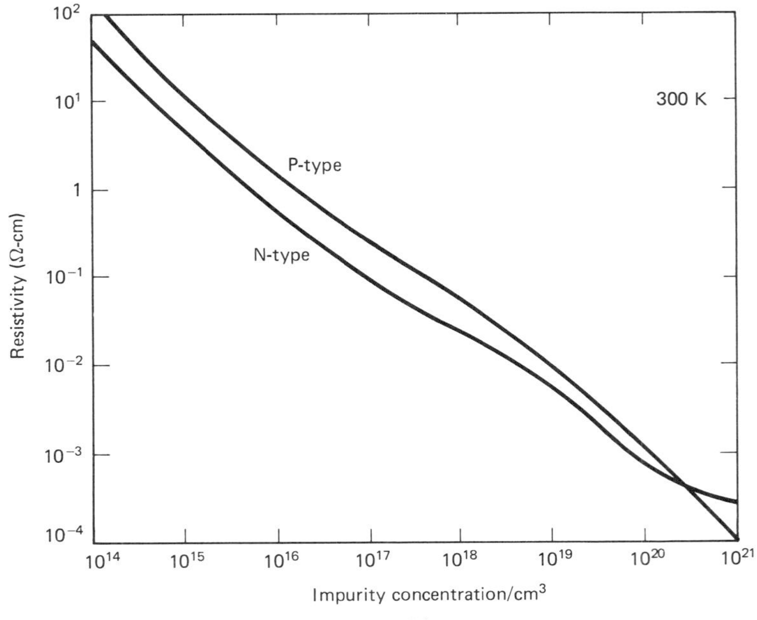
300 K
P-type
N-type
Resistivity ($\Omega$-cm)
Impurity concentration/cm$^3$

(c)

**Figure 1.18** *(cont.)*

are given that can be used to determine the sheet resistance P-type Gaussian and N-type erfc diffused layers in silicon. In Figure 1.18c a graph of resistivity versus impurity concentration for bulk silicon is presented.

### 1.3.6 Ion Implantation

Ion implantation can be used to produce a shallow surface region of dopant atoms deposited into a silicon wafer, and thus can be used as an alternative to a deposition diffusion. In the ion implantation process silicon wafers are placed in a vacuum chamber and are scanned by a beam of high-energy dopant ions, as shown in Figure 1.19. These ions have been accelerated through potential differences of from 20 kV to as much as 250 kV, and as the ions strike the silicon wafers they will penetrate some small distance into the wafer. The depth of penetration of any particular type of ion will increase with increasing accelerating voltage, as shown by Figure 1.20.

The penetration depth will generally be in the range 0.1 to 1.0 $\mu$m. For boron ions ($B^+$) in silicon, for example, the average penetration depth, called the *projected range*, will be only 0.067 $\mu$m for an accelerating voltage of 20 kV. At 100 kV the projected range increases to 0.30 $\mu$m, and at 200 kV it will be 0.52 $\mu$m. Even at a large accelerating potential of 300 kV the projected range is still only 0.70 $\mu$m. For phosphorus ions in silicon the projected ranges are even smaller, being only 0.026 $\mu$m at 20 kV, 0.123 $\mu$m at 100 kV, 0.254 $\mu$m at 200 kV, and 0.386 $\mu$m at 300 kV, as given in Table 1.2.

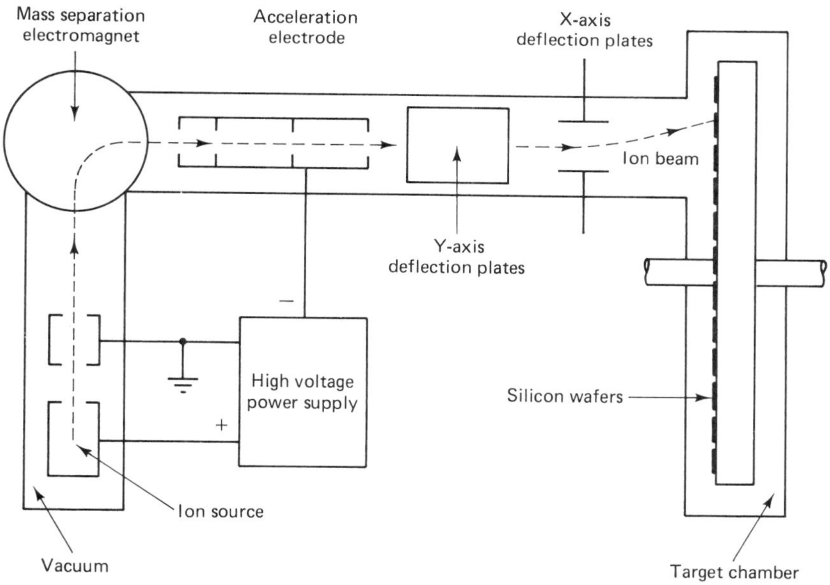


**Figure 1.19** Ion implantation system.

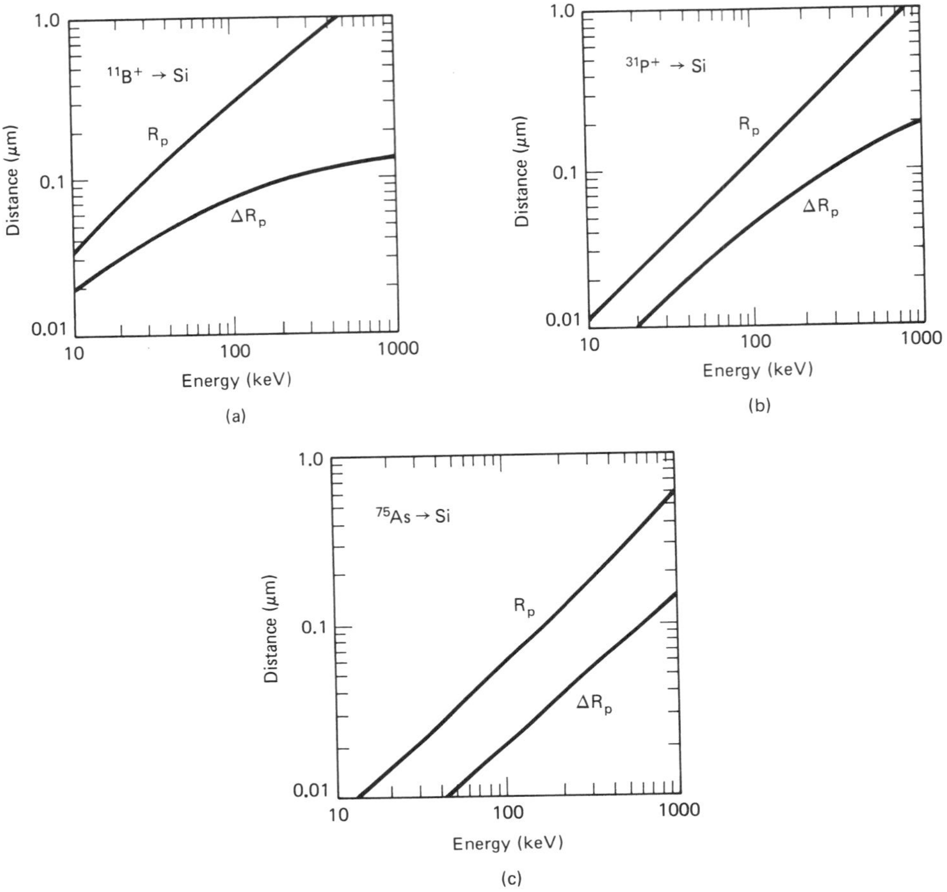


**Figure 1.20** Ion implantation range and straggle of boron, phosphorus, and arsenic in silicon. (Adapted from D. H. Lee and J. W. Mayer, "Ion Implanted Semiconductor Devices," *Proceedings of the IEEE*, pp. 1241–1255, September 1974 © 1974 IEEE.)

The distribution of the implanted ions as a function of distance $x$ from the surface will be a Gaussian distribution, given by

$$N(x) = N_p \exp\left[-\frac{(x - R_p)^2}{2\Delta R_p{}^2}\right] \tag{1.15}$$

where $x$ = distance into substrate from surface
$R_p$ = projected range
$\Delta R_p$ = straggle (standard deviation) of the projected range
$N_p$ = peak concentration of implanted ions

An ion implantation impurity profile is shown in Figure 1.21. The peak implanted ion concentration is related to the *implantation dosage* $Q$ by $N_p = Q/(\sqrt{2\pi}\ \Delta R_p) = 0.4(Q/\Delta R_p)$. The implantation dosage $Q$ is the number of implanted ions per

**TABLE 1.2** PROJECTED RANGES AND STRAGGLE FOR BORON, PHOSPHORUS, AND ARSENIC IN SILICON

| Energy (kV) | Boron | | Phosphorus | | Arsenic | |
|---|---|---|---|---|---|---|
| | $R_p$ ($\mu$m) | $\Delta R_p$ ($\mu$m) | $R_p$ ($\mu$m) | $\Delta R_p$ ($\mu$m) | $R_p$ ($\mu$m) | $\Delta R_p$ ($\mu$m) |
| 10 | 0.0344 | 0.0156 | 0.0144 | 0.0070 | 0.0096 | 0.0037 |
| 20 | 0.0674 | 0.0253 | 0.0260 | 0.0122 | 0.0159 | 0.0060 |
| 30 | 0.0999 | 0.0331 | 0.0375 | 0.0169 | 0.0216 | 0.0081 |
| 40 | 0.1311 | 0.0392 | 0.0490 | 0.0214 | 0.0271 | 0.0101 |
| 50 | 0.1616 | 0.0445 | 0.0610 | 0.0259 | 0.0324 | 0.0120 |
| 60 | 0.1914 | 0.0491 | 0.0732 | 0.0303 | 0.0377 | 0.0139 |
| 70 | 0.2202 | 0.0530 | 0.0855 | 0.0345 | 0.0429 | 0.0158 |
| 80 | 0.2478 | 0.0564 | 0.0980 | 0.0386 | 0.0481 | 0.0176 |
| 90 | 0.2739 | 0.0593 | 0.1106 | 0.0425 | 0.0533 | 0.0194 |
| 100 | 0.2992 | 0.0618 | 0.1233 | 0.0463 | 0.0584 | 0.0211 |
| 110 | 0.3238 | 0.0641 | 0.1362 | 0.0500 | 0.0635 | 0.0229 |
| 120 | 0.3479 | 0.0663 | 0.1491 | 0.0535 | 0.0686 | 0.0246 |
| 130 | 0.3714 | 0.0682 | 0.1621 | 0.0570 | 0.0738 | 0.0263 |
| 140 | 0.3942 | 0.0700 | 0.1752 | 0.0604 | 0.0789 | 0.0279 |
| 150 | 0.4165 | 0.0716 | 0.1883 | 0.0636 | 0.0840 | 0.0296 |
| 160 | 0.4383 | 0.0731 | 0.2014 | 0.0668 | 0.0891 | 0.0312 |
| 170 | 0.4594 | 0.0744 | 0.2146 | 0.0698 | 0.0943 | 0.0329 |
| 180 | 0.4799 | 0.0757 | 0.2277 | 0.0726 | 0.0995 | 0.0345 |
| 190 | 0.500 | 0.0769 | 0.2407 | 0.0753 | 0.1048 | 0.0362 |
| 200 | 0.5198 | 0.0780 | 0.2538 | 0.0780 | 0.1101 | 0.0378 |
| 220 | 0.5582 | 0.0800 | 0.2801 | 0.0831 | 0.1208 | 0.0412 |
| 240 | 0.5954 | 0.0818 | 0.3065 | 0.0880 | 0.1316 | 0.0445 |
| 260 | 0.6314 | 0.0834 | 0.3329 | 0.0927 | 0.1425 | 0.0478 |
| 280 | 0.6663 | 0.0849 | 0.3594 | 0.0972 | 0.1535 | 0.0510 |
| 300 | 0.7001 | 0.0862 | 0.3858 | 0.1015 | 0.1645 | 0.0543 |

unit of surface area as given by such units as ions/cm². The ion density drops off rapidly from the peak value with distance as measured from $R_p$ in either direction as indicated by the following values of the Gaussian function:

| $x$ | $N(x)/N_p$ |
|---|---|
| $R_p$ | 1.0 |
| $R_p \pm 1.2\Delta R_p$ | 0.5 |
| $R_p \pm 2\Delta R_p$ | 0.1 |
| $R_p \pm 3\Delta R_p$ | 0.01 |
| $R_p \pm 3.7\Delta R_p$ | 0.001 |
| $R_p \pm 4.3\Delta R_p$ | 0.0001 |

Note that the Gaussian implanted ion profile will be truncated at $x = 0$.

After the ions have been implanted they are lodged principally in interstitial positions in the silicon crystal structure, and the surface region into which the implantation has taken place will be heavily damaged by the impact of the high-energy ions.

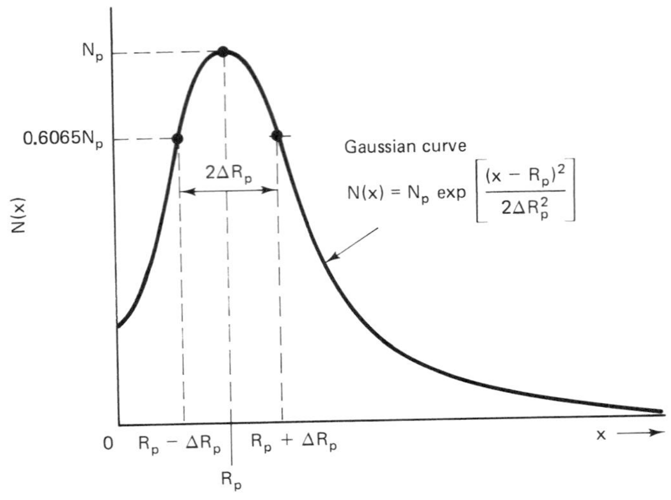


**Figure 1.21** Ion implantation impurity profile.

The disarray of silicon atoms in the surface region is often to the extent that this region is no longer crystalline in structure but rather, amorphous. To restore this surface region back to a well-ordered crystalline state and to allow the implanted ions to go into substitutional sites in the crystal structure, the wafer must be subjected to an *annealing* process. The annealing process usually involves the heating of the wafers to some elevated temperature, often in the range of 1000°C for a suitable length of time such as 30 min.

There are other annealing techniques, such as laser beam and electron-beam annealing, in which only the surface region of the wafer is heated and recrystallized. An ion implantation process is often followed by a conventional-type drive-in diffusion, in which case the annealing process will occur as part of the drive-in diffusion.

Ion implantation is a substantially more expensive process than an ordinary deposition diffusion, both in terms of the cost of the equipment and the throughput. It does, however, offer the advantage of much more precise control over the density of dopants $Q$ deposited into the wafer, and hence the sheet resistance. It also allows for very low values of dosage $Q$ ($<10^{14}$ cm$^{-3}$), so that very large values of sheet resistance ($>1000$ Ω/square) can be obtained. These high sheet resistance values are useful for obtaining large-value resistors ($\gtrsim$ 50 kΩ) for ICs. Very low-dosage, low-energy implantations are also used for the adjustment of the threshold voltage of MOSFETs and other applications.

To determine the ion-implanted pattern on a silicon wafer, a masking layer is needed on the surface of the wafer to stop the ions from penetrating the silicon in the various regions where this is needed. Commonly used materials used for the masking of lower-energy (less than 100 kV) ion implantations are $SiO_2$, $Si_3N_4$, and photoresist. The relative stopping power of these materials with silicon as a reference for both boron and phosphorus implantations are $SiO_2$: 1.25, $Si_3N_4$: 1.62, and photoresist: 0.75. For higher-energy implantations heavy metals such as thin films of gold, platinum, tungsten, or tantalum deposited on top of the $SiO_2$ layer can be used. The relative stopping power for these materials for boron or phosphorus implantations at 1000 kV is in the range 2 to 4.

To consider an example of ion implantation masking, let us take the case of a boron implantation into silicon at an energy of 100 kV. For this implantation the projected range $R_p$ is 0.30 $\mu$m, and the straggle $\Delta R_p$ is 0.062 $\mu$m. The distance into the silicon to the point where the implanted ion concentration is down to less than 0.1% of the peak concentration is $x = R_p + 3.7\Delta R_p + 0.30\ \mu m + 3.7 \times 0.062\ \mu m = 0.53\ \mu m$. To mask against this ion implantation to the extent that the implanted ion concentration in the silicon will not exceed 0.1% of the peak concentration the thickness of the masking layer must be 0.53 $\mu$m/stopping power. If an $SiO_2$ layer is to be used for the patterning of the ion implantation, as is very often the case, the required thickness of the oxide layer will be 0.53 $\mu$m/1.25 = 0.424 $\mu$m = 4240 Å. This thickness of oxide can easily be obtained by the thermal oxidation process.

## 1.4 OXIDATION

If a bare silicon surface is first exposed to air, a very thin oxide layer will almost immediately be formed by the reaction $Si + O_2 \rightarrow SiO_2$. This thin oxide layer is called the *native oxide* and it will be only some 20 to 30 Å thick. For the oxidation process to take place, silicon and oxygen atoms must react to form $SiO_2$. The resultant $SiO_2$ layer that is produced will be a dense continuous protective layer on the silicon surface and will thereby act to inhibit the further growth of the oxide layer, so that the oxidation will be a self-limiting process.

Silicon dioxide ($SiO_2$) layers are used for two principal purposes:

1. To act as a diffusion or implantation mask to produce a desired pattern of the diffused or implanted layers
2. To serve as a protective or *passivating* layer on the silicon surface for the protection of the semiconductor devices

For these two purposes the thickness of the oxide layer that is required will generally be in the range of about 5000 to 10,000 Å. To produce these thick layers of oxide a process called *thermal oxidation* can be used.

The silicon wafers are stacked up in a quartz boat, which is then inserted into a quartz furnace tube inside a tube furnace, very similar to a diffusion furnace. The silicon wafers are raised to a high temperature, generally in the range 950 to 1150°C, and at the same time the wafers are exposed to a gas containing $O_2$ or $H_2O$, or both. The main gas flow in the quartz tube, however, is a $N_2$ carrier gas.

The diffusion rate of $O_2$ and $H_2O$ in $SiO_2$ increases exponentially with temperature, so that at these elevated temperatures the diffusion of these oxidants through the oxide layer will continue such that the required thickness of oxide can be produced

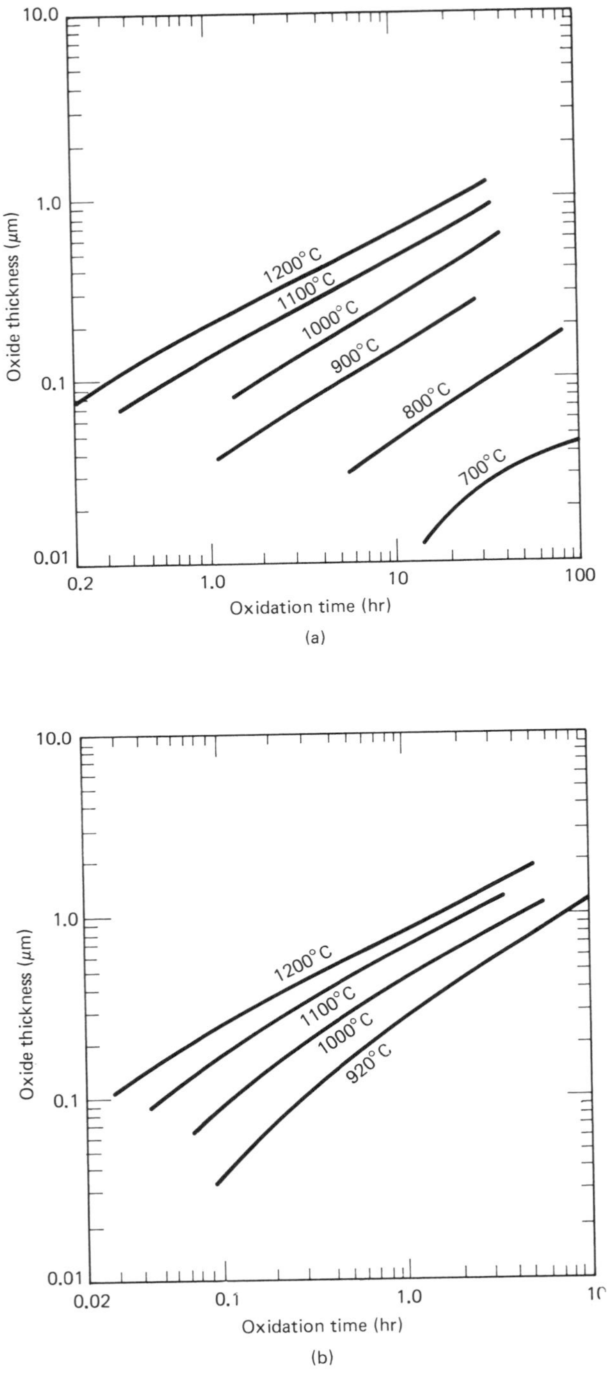


**Figure 1.22** Oxide thickness versus oxidation time: (a) thermal oxidation with dry oxygen; (b) thermal oxidation with wet oxygen (95°C $H_2O$). (From B. E. Deal and A. S. Grove, "General Relationship for the Thermal Oxidation of Silicon," *Journal of Applied Physics*, Vol. 36, no. 12, pp. 3370–78, 1965, and S. K. Ghandi, *Theory and Practice of Microelectronics*, Wiley, 1968.)

within an acceptable period of time, as shown by the graphs of Figure 1.22. The chemical reactions taking place at the silicon surface are:

$$\text{(1) For } O_2\text{: } Si + O_2 \longrightarrow SiO_2 \tag{1.16}$$

$$\text{(2) For } H_2O\text{: } Si + 2H_2O \longrightarrow SiO_2 + 2H_2 \tag{1.17}$$

This oxidation process is called thermal oxidation because of the use of high temperatures to promote the growth of the oxide layer.

The initial growth of the oxide is limited by the rate at which the chemical reaction takes place. After the first 100 to 300 Å of oxide has been produced, the growth rate of the oxide layer will be limited principally by the rate of diffusion of the oxidant ($O_2$ or $H_2O$) through the oxide layer, as illustrated in Figure 1.23. The rate of diffusion of $O_2$ or $H_2O$ through the oxide layer will be inversely proportional to the thickness of the layer, so that we will have that $dx/dt = C/x$, where $x$ is the oxide thickness and $C$ is a constant of proportionality. Rearranging this equation gives $x\,dx = C\,dt$, and then integrating both sides gives $x^2/2 = Ct$. Solving for the oxide thickness $x$ gives $x = \sqrt{2Ct}$. We see that after an initial reaction-rate-

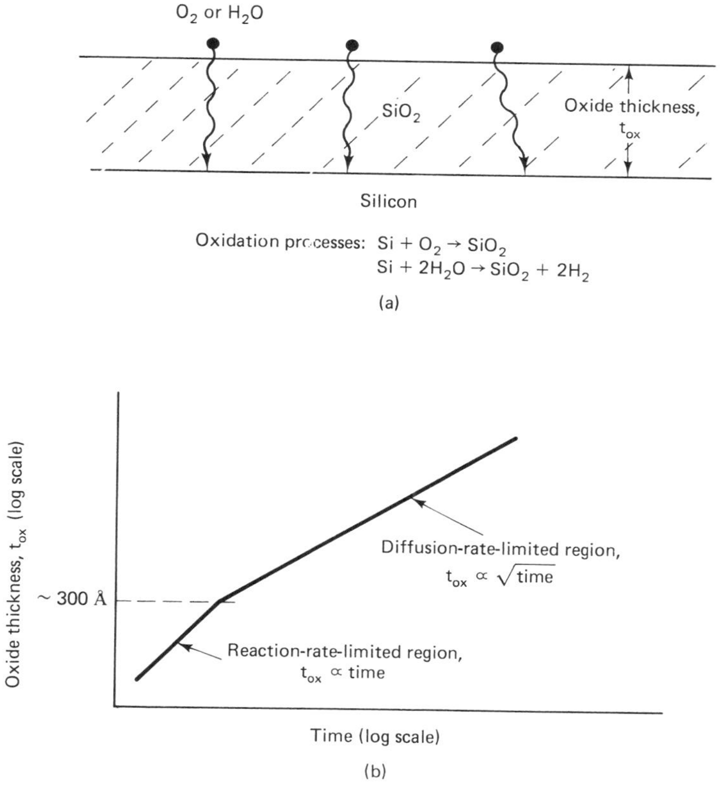


**Figure 1.23** (a) Oxidation of silicon; (b) oxide thickness versus time for thermal oxidation of silicon.

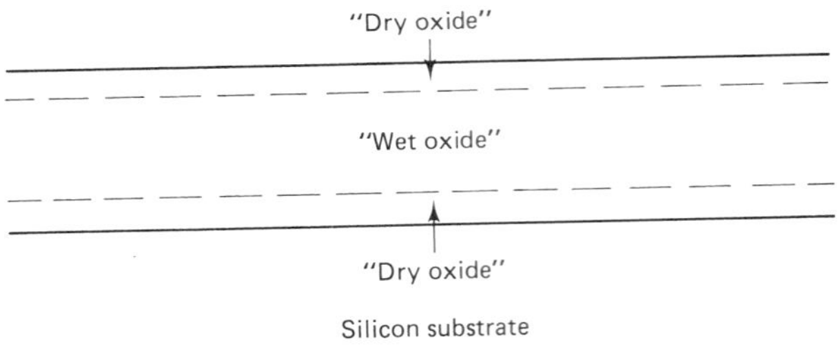


**Figure 1.24** Composite "dry-wet-dry" oxide.

limited linear growth phase, the oxide growth will become diffusion-rate limited, with the oxide thickness increasing as the square root of the growth time.

The oxide growth rate that is obtained when using $H_2O$ as the oxidant will be about four times faster than the rate obtained with $O_2$. This is primarily the result of the $H_2O$ molecule being about one-half the size of the $O_2$ molecule, so that the rate of diffusion of $H_2O$ through the $SiO_2$ layer will be much greater than the $O_2$ diffusion rate. Although the oxide growth rate with $H_2O$ is much faster than with $O_2$, the "dry" ($O_2$) oxide will be a slightly denser oxide with a higher dielectric strength than the "wet" ($H_2O$) oxide. In many cases a "dry–wet–dry" oxidation process is used, starting off the initial oxide growth using $O_2$. This is followed by a $H_2O$ oxide growth phase to produce the bulk of the oxide thickness, and then completed by a final "dry" oxidation. This will produce a composite oxide layer as shown in Figure 1.24 with the denser "dry" oxide regions being adjacent to the silicon surface and serving as a protective cap on top, and the less dense "wet" oxide being sandwiched in between.

In the thermal growth process of an oxide layer some silicon from the substrate is consumed. If the resulting thickness of the $SiO_2$ layer is designated as $t_{ox}$, the thickness of the silicon consumed will be $0.44t_{ox}$, as shown in Figure 1.25.

### 1.4.1 Oxide Masking

An oxide layer can be used to mask an underlying silicon surface against a diffusion or ion implantation process. If an oxide layer is patterned as by the photolithographic process to produce regions where there are openings or "windows" where the oxide has been removed to expose the underlying silicon, these exposed silicon regions will be subjected to the diffusion or implantation of dopants, whereas the unexposed silicon regions will be protected. The pattern of dopant that will be deposited into

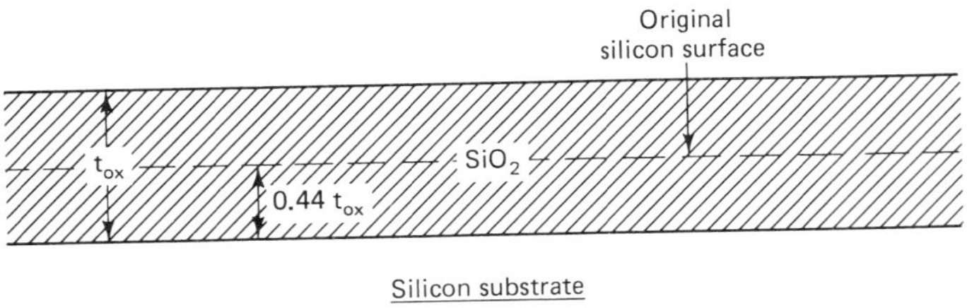


**Figure 1.25** Consumption of silicon during oxide growth.

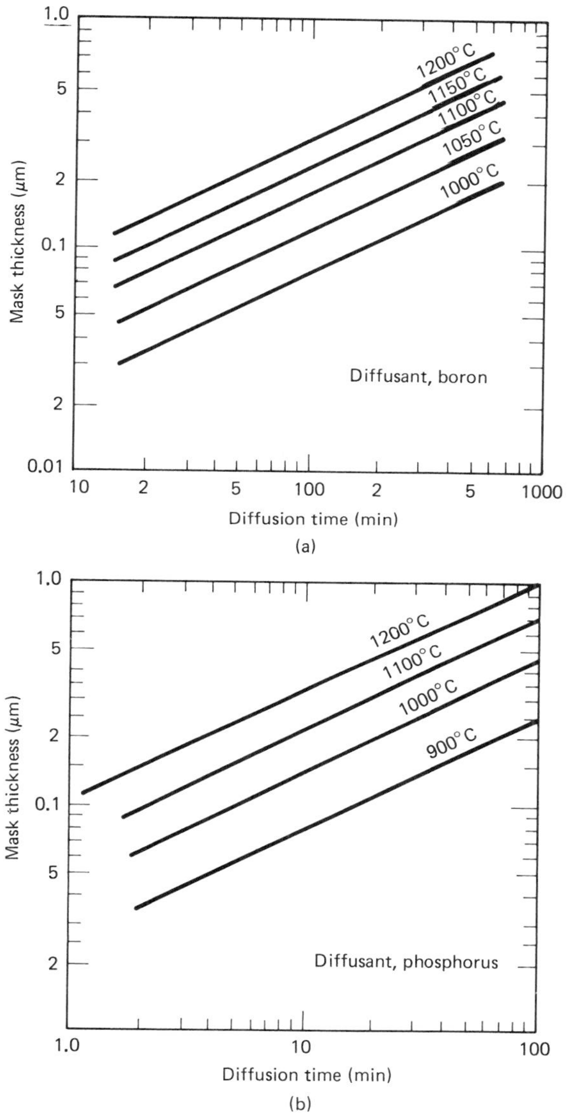


**Figure 1.26** Oxide thickness needed for diffusion masking: (a) mask thickness for boron diffusions; (b) mask thickness for phosphorus diffusions. (From S. K. Ghandi, *Theory and Practice of Microelectronics*, Wiley, 1968.)

the silicon will thus be a replication of the pattern of openings in the oxide layer. The replication is a key factor in the production of microscopic-size solid-state electronic devices.

The thickness of oxide needed for diffusion masking is a function of the type of diffusant and the diffusion time and temperature conditions as shown by Figure 1.26. In particular, we see that an oxide thickness of some 5000 Å will be sufficient to mask against almost all diffusions. This oxide thickness will also be sufficient to block almost all but the highest-energy ion implantations.

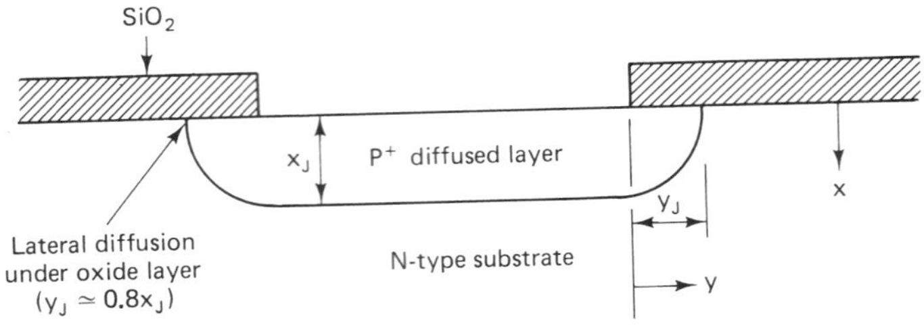


**Figure 1.27** Diffusion masking using an oxide layer.

### 1.4.2 Planar Devices: Oxide Passivation

In Figure 1.27 a cross-sectional view is presented of a PN junction produced by diffusion through an oxide window. Since diffusion is an isotropic process, the diffusion will not only proceed in the downward direction, but also sideways as well. The junction depth in the vertical direction is indicated as $x_J$. The distance from the edge of the oxide window to the junction in the lateral direction underneath the oxide is indicated as $y_J$. The relationship between $y_J$ and $x_J$ is that $y_J \simeq 0.8x_J$, so that the curvature of the junction in the regions underneath the edge of the oxide window is approximately that of a quarter-circle with a radius of curvature that is approximately equal to the junction depth.

We note that the junction intersects the silicon surface well underneath the protective thermally grown oxide layer. This oxide layer will protect the junction against various environmental effects and is therefore called a *passivated junction*. We also note that the locus of intersection of the junction with the silicon surface is entirely within a single geometric plane, and for that reason this type of junction is called a *planar junction*. Note that the junction itself is not flat or plane, but rather there is a curvature of the junction in the regions underneath the edges of the oxide window. This junction curvature will result in an increase in the electric field intensity in these regions which will cause the breakdown voltage of the junction to be lower than that of a corresponding plane junction with the same doping levels.

The production of the pattern of oxide openings or windows involves a number of processing steps that are collectively called the *photolithographic process*. This process will be discussed in a following section.

### 1.4.3 High-Pressure Oxidation

The oxidation rate can be greatly accelerated by carrying out the thermal oxidation process at pressures that are much above atmospheric pressure. The rate of diffusion of the oxidant molecules through an oxide layer is proportional to the ambient pressure. For example, at a pressure of 10 atm the diffusion rate will be increased by a factor of 10 and the corresponding oxidation time can be reduced by approximately the same factor. Alternatively, the oxidation can be done for the same length of time, but the temperature required will be substantially lower. For example, a steam ($H_2O$) oxidation at 1200°C and a 1.0 atm ambient pressure will produce an oxide layer that is 6000 Å (0.6 μm) thick, for an oxidation time of 36 min. This same thickness of oxide can be produced in the same period of time at a temperature of only 920°C if the ambient pressure is increased to 10 atm. As a second example, a 1.0-h steam oxidation at 920°C and at 1.0 atm ambient pressure will result in a 2000-Å-thick oxide. If the ambient pressure is increased to 10 atm, the temperature can be reduced to only 795°C for the same oxide thickness produced in the same period of time.

The possibility of lower-temperature processing is one of the principal advantages of the high-pressure oxidation process. The lower processing temperature reduces the formation of crystalline defects and produces less effect on previous diffusions and other processes. The shorter oxidation time is also advantageous in increasing the system throughput. The major disadvantage of this process is the high initial cost of the system.

## 1.5 PHOTOLITHOGRAPHY

The photolithographic process is the means by which microscopically small circuit and device patterns can be produced on silicon wafers, resulting in as many as 10,000 transistors on a 1 cm × 1 cm chip. With the conventional photolithographic process using ultraviolet light exposure, device dimensions or line widths as small as 2 μm can be obtained. By the use of *electron-beam* or *x-ray* exposure, device details down into the submicron (<1 μm) range can be obtained, with device dimensions as small as 0.2 μm having been achieved.

The photolithographic process encompasses a number of processing steps, which will now be described.

**1. Photoresist application (spinning).** A drop of light-sensitive liquid called *photoresist* is applied to the center of a silicon wafer that is held down by a vacuum chuck. The wafer is then accelerated rapidly to a rotational velocity in the range 3000 to 7000 r/min for some 30 to 60 s. This flings off most of the photoresist, leaving a thin photoresist film of the silicon wafer that is in the range 5000 to 10,000 Å thick, as illustrated in Figure 1.28a. The thickness of the photoresist layer will be approximately inversely proportional to the square root of the rotational velocity.

Sometimes, prior to the application of the photoresist the silicon wafers are given a "bake-out" at a temperature of at least 100°C to drive off moisture from the wafer surfaces so as to obtain better adhesion of the photoresist.

**2. Prebake.** The silicon wafers are now put into an oven at about 80°C for about 30 to 60 min to drive off solvents in the photoresist and to harden it into a semisolid film.

**3. Alignment and exposure.** The photoresist-coated wafer is now placed in an apparatus called a *mask aligner* in very close proximity (about 25 to 125 μm) to a *photomask*. The relative positions of the wafer and the photomask are adjusted such that the photomask is correctly lined up with reference marks or a preexisting pattern on the wafer.

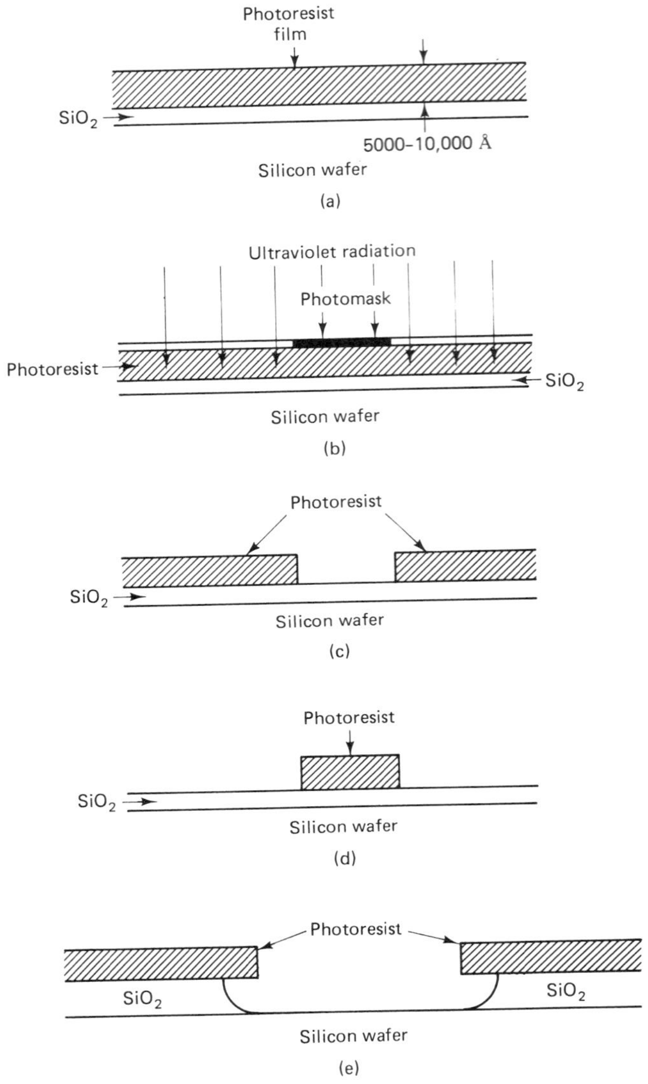


**Figure 1.28** Photolithographic process: (a) photoresist spinning; (b) exposure; (c) development: negative photoresist; (d) development: positive photoresist; (e) oxide etching.

The photomask is a glass plate, typically about 125 mm (5 in.) square and about 2 mm thick. The photomask has a photographic emulsion or thin film metal (generally chromium) pattern on one side. The pattern has just clear or opaque areas, with no shades of gray.

The alignment of the photomask to the wafer is often required to be accurate to within less than 1 $\mu$m, and in some cases to within 0.5 $\mu$m. After proper alignment has been achieved, the wafer is brought into direct contact with the photomask.

A highly collimated ultraviolet light is then turned on and the areas of the silicon wafer that are not covered by the opaque areas of the photomask are exposed to the ultraviolet radiation, as shown in Figure 1.28b. The exposure time will generally be in the range 3 to 30 s and is carefully controlled such that the total ultraviolet radiation dosage in watt-seconds or joules is the required amount.

The photomask is obtained from an original large-size drawing of the desired mask pattern that is then photographically reduced. Generally, two consecutive reductions are used, each being of the order of 10 to 30 times for a total reduction of about 100 to 1000. The second reduction may also include a step-and-repeat exposure process for making a master photomask. This step-and-repeat exposure process reproduces the original mask drawing in reduced form on the photomask in an array containing tens, hundreds, or even thousands of replications of the original drawing. The resulting circuit or device pattern on the wafer will be such that the wafer can be correspondingly divided up into many chips, each one with a complete device or with a complete IC. The master photomask is then used to make many copies that are actually used in the alignment and exposure process.

**4. Development.** There are two basic types of photoresist that are used, negative photoresist and positive photoresist. If a negative photoresist is used, the areas of the photoresist that are exposed to the ultraviolet radiation become polymerized. This polymerization process increases the length of the organic chain molecules that make up the photoresist. This makes the resist tougher and makes it essentially insoluble in the developer solution. The areas of the resist that are unexposed to the ultraviolet radiation, however, remain unpolymerized and are readily dissolved by the developer solution. The resisting photoresist pattern after the development process will therefore be a replication of the photomask pattern, with the clear areas on the photomask corresponding to the areas where the photoresist remains on the wafers, as shown in Figure 1.28c.

With a *positive* photoresist the opposite type of process occurs. Exposure to ultraviolet radiation results in *depolymerization* of the photoresist. This makes these exposed areas of the photoresist readily soluble in the developer solution, whereas the unexposed areas are essentially insoluble. The developer solution will thus remove the exposed or depolymerized regions of the photoresist, whereas the unexposed areas will remain on the wafer, as shown in Figure 1.28d. Thus again there is a replication of the photomask pattern, but this time the clear areas of the photomask produce the areas on the wafer from which the photoresist has been removed.

**5. Postbake.** After development and rinsing the wafers are usually given a postbake in an oven at a temperature of about 150°C for about 30 to 60 min to toughen further the remaining resist on the wafer. This is to make it adhere better to the wafer and to make it more resistant to the hydrofluoric acid (HF) solution used for etching of the $SiO_2$.

**6. Oxide etching.** The silicon wafers are now immersed in or sprayed with a hydrofluoric (HF) acid solution. This solution is usually a diluted solution of typically 10:1 $H_2O$:HF, or more often a 10:1 $NH_4F$ (ammonium fluoride):HF solution. The HF solutions will etch the $SiO_2$ but will not attack the underlying silicon, nor will it attack the photoresist layer to any appreciable extent. The wafers are exposed to the etching solution long enough to remove the $SiO_2$ completely in the areas of the

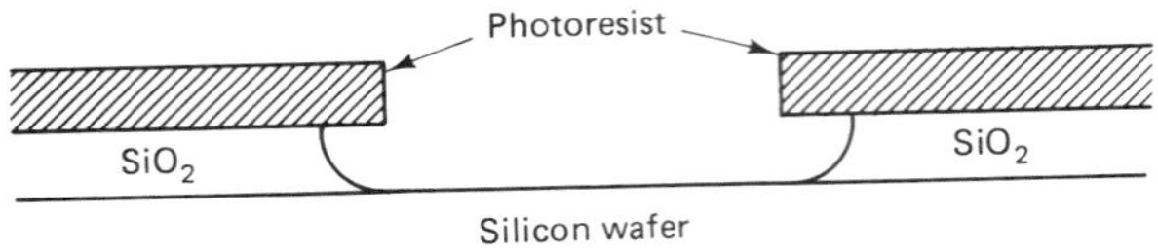


**Figure 1.29** Effect of oxide overetching.

wafer that are not covered by the photoresist, as shown in Figure 1.28e. For the 10:1 buffered HF solution ($NH_4F$:HF) the $SiO_2$ etching rate is about 1000 Å/min at 25°C so that only about 5 min will be required to remove a typical oxide layer of 5000 Å thickness.

The result of the oxide etching process will be a pattern of openings or windows in the $SiO_2$ layer that will replicate the photoresist pattern, and will therefore be a replication of the pattern on the photomask.

The oxide etching time is very carefully controlled so that all of the oxide in the photoresist windows is removed. The etching time, however, should not be overly prolonged beyond this point because it will result in more undercutting underneath the photoresist and thus to a widening of the oxide opening beyond what is desired, as shown in Figure 1.29.

*Plasma Etching.* The oxide etching process just described is a wet etching process and the chemical reagents used are in liquid form. A newer process for oxide etching is a dry etching process called plasma etching. In the plasma etching process the wafers to be etched are placed in a vacuum chamber or enclosure at reduced pressures, often in the range 1 to 10 torr. Suitable reagent gases such as $CF_4$ (Freon 14) or $C_2F_6$ are then metered into the reaction chamber. A radio-frequency (RF) field is then applied which results in the ionization of the gas molecules and the breaking up of some of the gas molecules into highly reactive free radicals. The density of positive and negative ions and electrons will be such that there will be approximately overall charge neutrality, so this is called a gaseous plasma.

The free radicals react with the $SiO_2$ layer to produce $O_2$ and various vapor-phase silicon compounds, which are then vented out of the system by the vacuum pump. By means of this dry etching process the $SiO_2$ in the areas is exposed by openings in the photoresist. The plasma etching will also result in some attack on the photoresist and any exposed silicon, but not at the same rate as the $SiO_2$. In any case the etching process is carefully monitored with some end-point detection method to stop the process shortly after the oxide windows have been cleared. With $CF_4$ the silicon etching rate will actually be larger than the $SiO_2$ rate, but with $C_2F_6$ an $SiO_2$/Si etch ratio of 15:1 is possible.

A major advantage of the dry etching process is that smaller line openings can be achieved ($\lesssim 1$ $\mu$m) than with the wet etching process. With wet etching, surface-tension effects can inhibit the flow of the etchant into very narrow photoresist openings. Such is not the case with dry etching. The wet etching process is isotropic, so that the $SiO_2$ will be attacked equally well in all directions. As a result there will be some undercutting of the photoresist layer, as shown in Figure 1.30. This undercutting will restrict the minimum line width that can reasonably be obtained to a little more than twice the oxide thickness. With an oxide thickness of generally

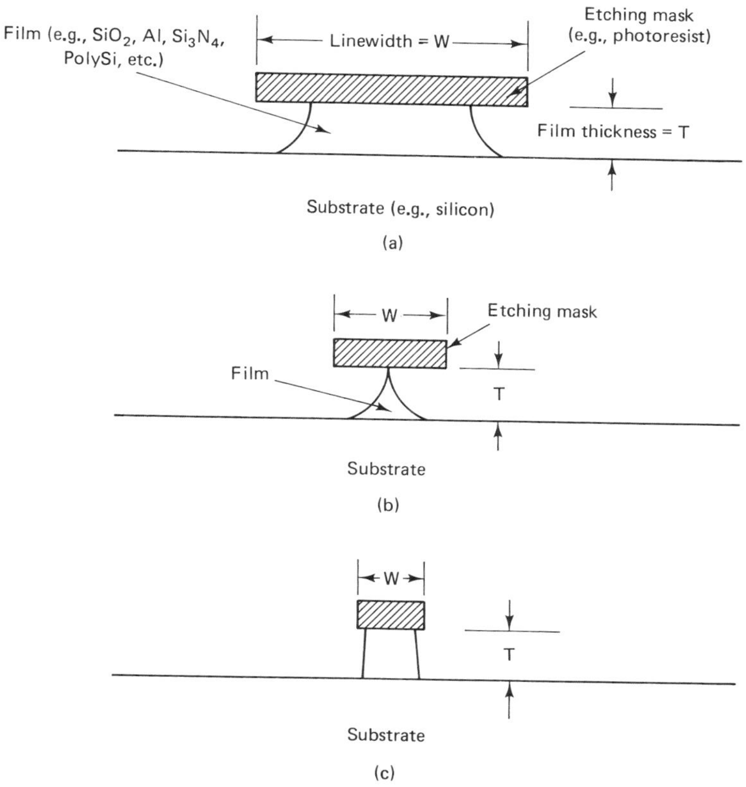


**Figure 1.30** Advantages of etching anisotropy: (a) isotropic etching: line width $\gg$ film thickness, $W \gg T$; (b) isotropic etching: line width $= 2 \times$ film thickness, $W = 2T$; (c) anistropic etching (large etch ratio). In (c), line width can be comparable to or even less than the film thickness.

about 5000 Å being used, this limits the minimum line width to something in excess of 1 $\mu$m.

The dry etching process can be designed to produce some degree of etching anisotropy, as illustrated in Figure 1.30c. With a high degree of etching anisotropy the etching can proceed in the downward direction at a rate substantially faster than sideways etching. This anisotropy in the etching process helps to make possible the attainment of submicron dimensions in ICs.

A dry etching process that offers a relatively high degree of etching anisotropy is *planar plasma etching*. In this process there is an electric field oriented perpendicular to the wafer surface, as shown in Figure 1.31. This electric field accelerates the ions and free radicals of the plasma toward the wafer surface. This directed flow of the ions and free radicals will result in a much greater etching rate in the downward direction than in the sideways direction, so that a large *etch ratio* can be achieved.

Another dry etching process that can produce a large etch ratio is called *ion milling*. The wafers are placed in a vacuum chamber and a beam of inert-gas ions, usually argon ($Ar^+$) is accelerated through a large potential difference of generally

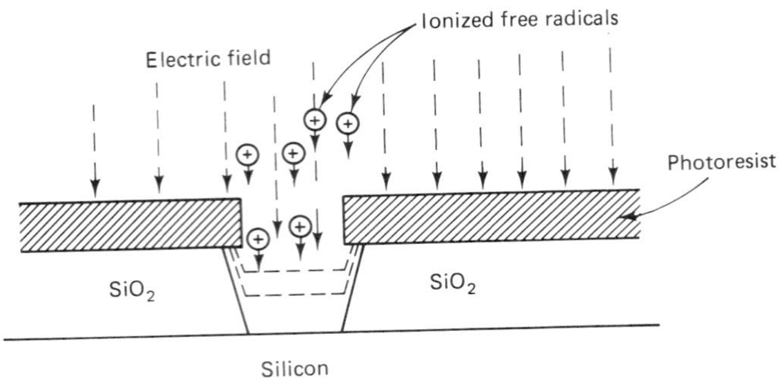


**Figure 1.31** Planar plasma etching.

400 to 1000 V. The beam of $Ar^+$ ions then strikes the wafer surface, and the kinetic energy of the ions is such that some of the atoms at or near the surface are knocked off, this process being called *sputtering*. Sputtering rates of about 100 to 1000 Å/min are commonly obtained.

**7. Photoresist stripping.** The last step in the photolithographic process is the removal of the photoresist, known as photoresist stripping. Positive photoresists can usually be easily removed in organic solvents such as acetone. Negative photoresists are more difficult to remove. A hot sulfuric acid immersion together with mechanical scrubbing is often used with negative photoresists.

Plasma reactors of the type that are used for plasma etching can also be used for photoresist stripping. Oxygen is used as the reagent gas and the organic resist material is oxidized to produce various gases, such as Co, $CO_2$, and $H_2O$, which are then vented out of the system through the vacuum pump.

## 1.6 CHEMICAL-VAPOR DEPOSITION

The chemical-vapor deposition (CVD) process is a means for the deposition of thin films on a substrate. The materials to be deposited enter a reaction chamber in the gaseous or vapor phase and react on or near the surface of the substrates, which are at some elevated temperature. The chemical reaction that occurs produces the atoms or molecules that are deposited on the substrate surface. A number of different materials can be deposited by the CVD process. In the sections to follow some of the techniques that are used to deposit CVD films of importance in semiconductor device fabrication are discussed.

### 1.6.1 Silicon Dioxide Deposition

An important example of the CVD process is the deposition of $SiO_2$ layers on silicon substrates. A number of chemical reactions can be used for this process, and one that is very commonly used is

$$SiH_4 \text{ (silane)} + O_2 \longrightarrow SiO_2 + 2H_2 \tag{1.18}$$

Silane is a gas that is metered into the reaction chamber together with oxygen. The silicon wafers are placed on a heated substrate and raised to a temperature of about 450 to 600°C and the reaction above takes place. This results in a deposited (CVD) oxide layer (called "silox"), compared to the thermally grown oxide layer discussed previously.

The CVD oxide can be deposited in a matter of only a few minutes, but it will not be as dense as the thermally grown oxide. The thermally grown oxide will generally be a higher-quality oxide with higher dielectric strength and serves as a better passivating layer than the CVD oxide. However, the temperature required for the CVD oxide is much lower than that for the thermal oxide, and the time required is much shorter. As a result, the CVD oxide can often be deposited on top of the wafer after previous processing steps such as metallization have been completed without seriously affecting the prior processes. The CVD oxide layer can thus be used for such purposes as "postmetallization passivation" to serve as a protective layer covering the device or IC after all processes, including metallization, have been completed.

### 1.6.2 Silicon Nitride Deposition

Silicon nitride ($Si_3N_4$) thin-film layers can be deposited by the CVD process. These layers can be used for device protection and passivation. In particular, $Si_3N_4$ serves as a very good barrier against the penetration of such contaminants as $Na^+$, $K^+$, and othei ions, against which $SiO_2$ is much less effective. Silicon nitride can also serve as a diffusion or ion implantation mask. An interesting application of silicon nitride is as an oxidation mask. The diffusion rate of the various oxidants, such as $O_2$ and $H_2O$, is very small in silicon nitride, such that a very thin nitride layer (~1000 Å) will be sufficient to prevent the oxidation of the underlying silicon. This is the basis of the *local oxidation of silicon* (LOCOS) process used in the fabrication of some types of high-density ICs. In this process a silicon nitride layer is deposited on a silicon wafer and patterned by a photolithographic process. The wafer is then subjected to a thermal oxidation process, but the thermal oxide is grown only in those areas that are not covered by the nitride film, as shown in Figure 1.32.

The most commonly used process for the deposition of silicon nitride layers uses the CVD reaction given by

$$3SiH_4 \text{ (silane)} + 4NH_3 \text{ (ammonia)} \longrightarrow Si_3N_4 + 12H_2 \qquad (1.19)$$

at temperatures in the range 600 to 800°C.

Silicon nitride films can be etched with a hot (~180°C) phosphoric acid ($H_3PO_4$) solution. Unfortunately, photoresists will not stand up to this etching solution, so a CVD oxide layer is used as the etching mask, as shown in Figure 1.33. The nitride layer is deposited first and then the CVD oxide is deposited on top of the nitride. Then photoresist is spun on the wafer and using the standard photolithographic process, the CVD oxide is patterned. The HF solution that is used for the etching of the CVD oxide will not attack the nitride layer very rapidly, so that openings or windows are etching in only the CVD oxide at this point. The photoresist is now completely removed and the wafers are immersed in the hot phosphoric acid solution,

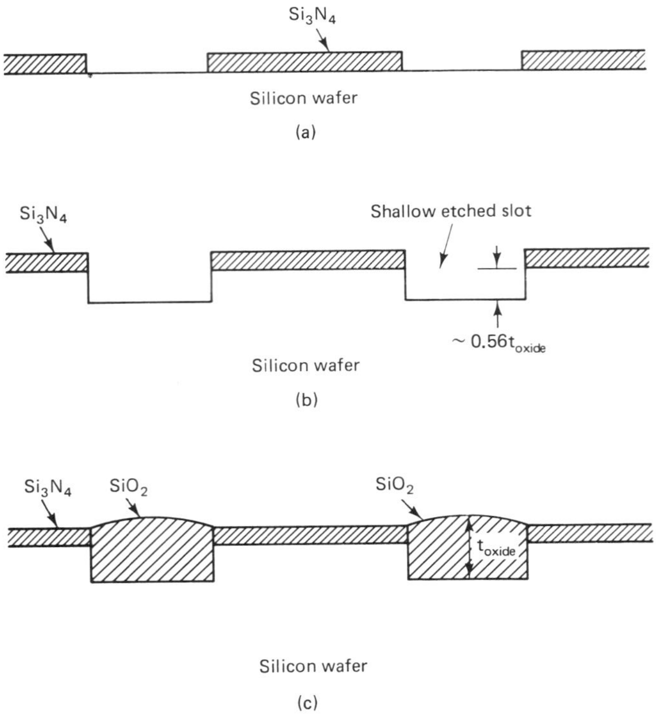


**Figure 1.32** Local oxidation of silicon (LOCOS) process: (a) deposition and patterning of a silicon nitride film; (b) etching slots in silicon using silicon nitride as an etching mask; (c) growth of thermal oxide layer using silicon nitride as an oxidation mask.

which will etch the nitride layer but does not attack the CVD oxide. The photoresist thus serves as an etching mask for the CVD oxide, which in turn serves as an etching mask for the nitride layer.

A more convenient etching process for silicon nitride is the use of plasma etching, in which case photoresist can be used. Silicon nitride can be etched with a $SiF_4/O_2$ gas mixture with an etch ratio of 5:1 compared to silicon and 50:1 with respect to $SiO_2$.

### 1.6.3 Silicon Epitaxial Layers

A special case of chemical-vapor deposition is called *epitaxy* or *epitaxial layer deposition,* in which case the deposited layer is in single-crystal form. This epitaxial process will occur only for certain combinations of substrate and layer materials and under certain deposition conditions.

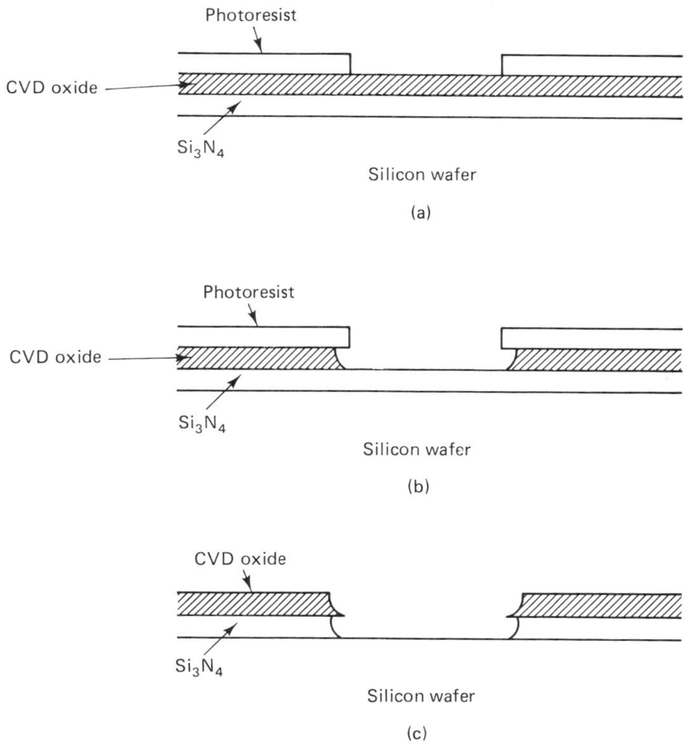


**Figure 1.33** Patterning of silicon nitride layers using CVD oxide as an etching mask: (a) photolithography to produce a patterned photoresist film; (b) etching of CVD oxide using HF solution; (c) removal of photoresist and etching of silicon nitride using hot phosphoric acid.

The most common example of epitaxy is the deposition of a silicon epitaxial layer on a single-crystal silicon substrate. In this case the substrate and layer materials are the same, and this is called *homoepitaxy*. The epitaxial layer becomes a crystallographic continuation of the substrate.

The doping and conductivity type of the epitaxial layer is unrestricted by the substrate, so that such epitaxial layer/substrate combinations as $N/N^+$, N/P, P/N, and $P/P^+$ are possible. With epitaxy a lightly doped epitaxial layer can be deposited on a heavily doped substrate. This is in contrast with the situation with a diffused layer, in which case the diffused layer doping is almost always very much heavier than the substrate doping. Although very thick and very thin epitaxial layers can be deposited, most epitaxial layers are in the thickness range 3 to 30 $\mu m$.

There are a number of different chemical reactions that can be used for the deposition of epitaxial layers, the most important reactions being the following:

| Reaction | Temperature (°C) | Deposition rate ($\mu$m/min) |
|---|---|---|
| 1. $SiCl_4 + 2H_2 \longrightarrow Si + 4HCl$ (silicon tetrachloride) | 1150–1250 | 0.4–1.5 |
| 2. $SiHCl_3 + H_2 \longrightarrow Si + 3HCl$ (trichlorosilance) | 1100–1200 | 0.4–2.0 |
| 3. $SiH_2Cl_2 \longrightarrow Si + 2HCl$ (dichlorosilane) | 1050–1150 | 0.4–3.0 |
| 4. $SiH_4 \longrightarrow Si + 2H_2$ (silane) | 950–1050 | 0.2–0.3 |

The epitaxial layer deposition takes place in a chamber or apparatus called an *epitaxial reactor*. The three basic types of epitaxial reactors are the horizontal reactor, the vertical reactor, and the cylindrical reactor, shown in Figure 1.34. In most cases the means for heating the silicon wafers to the required temperature is radio-frequency (RF) induction heating, although radiant heating using an array of focused high-intensity quartz lamps and resistance heating can also be used. For RF induction heating the silicon wafers rest on a silicon-carbide-coated graphite susceptor. A water-cooled copper induction coil serves as the primary winding of a transformer. The graphite susceptor serves, in effect, as a single-turn secondary winding. The voltage induced in the susceptor produces a circulating eddy current which as a result of heating produced by the $I^2R$ power loss raises the temperature to the required value.

Doped P-type or N-type epitaxial layers can be grown to specified doping levels. A number of gases are metered into the reactor tube, including some very small amounts of doping gases, such as $B_2H_6$ (diborane) for boron doping and $PH_3$ (phosphine) for phosphorus doping of the epitaxial layer. During the epitaxial layer deposition the dopant gas molecules react and become decomposed and the dopant atoms thus produced become incorporated into the epitaxial layer.

Before the start of the epitaxial layer deposition, anhydrous HCl gas is fed into the reactor. This HCl gas reacts with the silicon at the surface of the wafers in reactions that are the reverse of those listed above for epitaxial layer deposition. These reverse reactions result in *vapor-phase etching* of the silicon surface. Vapor-phase etching is done immediately before the deposition process to remove a small amount of silicon and other contaminants from the wafer surfaces to ensure that a clean freshly etched silicon surface will be available for epitaxial layer deposition.

Epitaxial layer deposition takes place at temperatures in the range 950 to 1250°C, and as a result during the deposition as well as during all subsequent high-temperature processing steps there will be diffusion of impurities across the epitaxial layer/substrate interface. This will cause a blurring of the impurity profile in the region of this interface. The most serious problem in this regard will be in the case of a very thin and very lightly doped epitaxial layer that is deposited on a very heavily doped substrate. An example of this would be a 50-Ω-cm ($N_D \simeq 1 \times 10^{14}$ cm$^{-3}$) 5-$\mu$m N-type epitaxial layer that is deposited on a 0.005-Ω-cm ($N_D \simeq 1 \times 10^{19}$ cm$^{-3}$) N$^+$ substrate. The outdiffusion of impurities from the heavily doped substrate into the lightly doped epitaxial layer will obliterate the sharp N/N$^+$ transition that would

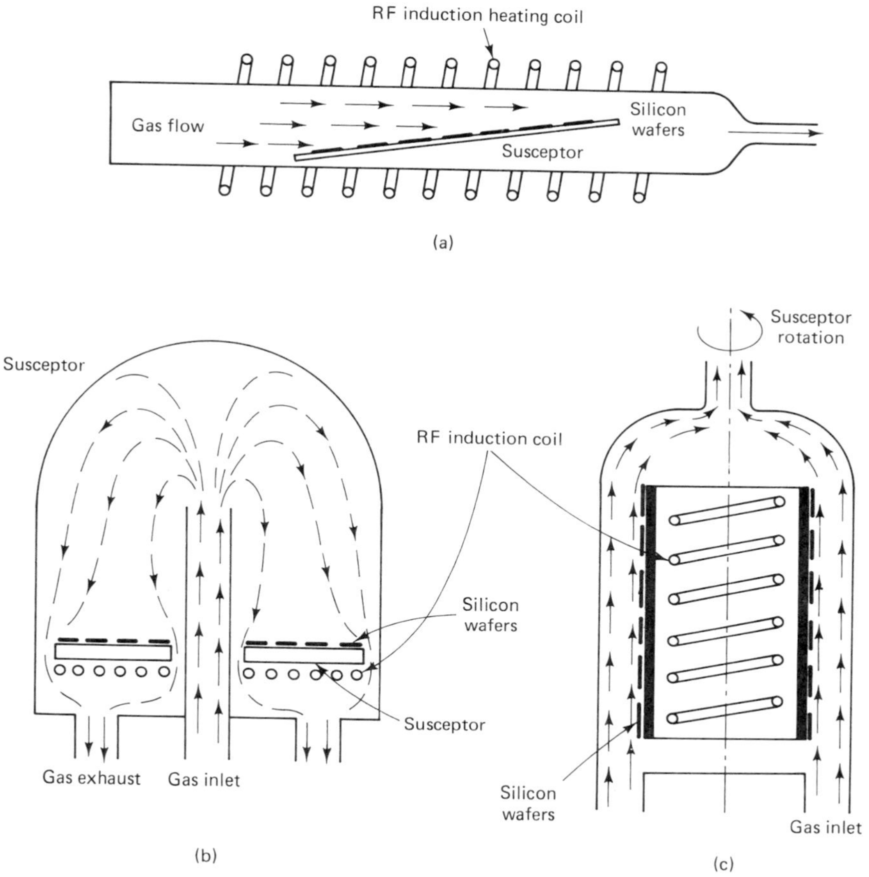


**Figure 1.34** Epitaxial reactors: (a) horizontal reactor; (b) vertical reactor; (c) cylindrical reactor.

otherwise be present at the layer–substrate interface, as shown in Figure 1.35. The influx of donor atoms from the substrate will reduce the effective thickness of the lightly doped epitaxial layer by 1 or 2 $\mu$m. To minimize this problem of outdiffusion from heavily doped $N^+$ substrates, slow donor diffusants such as antimony (Sb) and arsenic (As) are often used for the doping of substrate in preference to phosphorus.

### 1.6.4 Silicon Heteroepitaxy

The crystal structure of sapphire, which is crystalline $Al_2O_3$, is such that under carefully controlled deposition conditions an epitaxial silicon layer can be deposited on a sapphire substrate. This is an example of *heteroepitaxy* since the substrate and epitaxial layer are different materials. For the heteroepitaxial process the crystal structure and atomic spacing of the substrate must be a very close match to that of the layer to be deposited. If this is not the case, the CVD layer that is deposited will be polycrystalline, or in some cases amorphous.

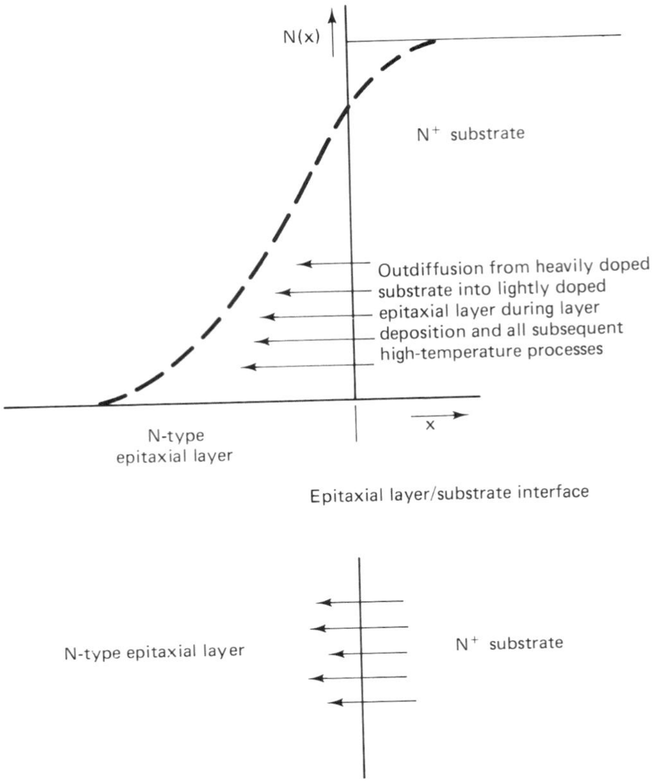


**Figure 1.35** Epitaxial layer/substrate outdiffusion.

The *silicon-on-sapphire* (SOS) process uses a very thin (~1 μm) silicon epitaxial layer on an insulating sapphire substrate. This SOS structure is often used for CMOS ICs (CMOS/SOS). The insulating substrate results in a lower parasitic capacitance, which in turn results in high-speed performance and very low-power-consumption CMOS circuits.

## 1.7 METALLIZATION

The final step in the wafer processing sequence is that of metallization. The purpose of this process is to produce a thin-film metal layer that will serve as the required conductor pattern for the interconnection of the various devices and circuit elements on the chip. The metallization pattern is also used to produce metallized areas called *bonding pads* around the periphery of the chip to provide areas for the bonding of wire leads from the package to the chip. The bonding wires are typically 25-μm (0.001-in.)-diameter gold wires, and the bonding pads are usually made to be around 100 μm × 100 μm (0.004 in. × 0.004 in.) square to accommodate fully the flattened ends of the bonding wires and to allow for some registration errors in the placement of the wires on the pads.

The material used for the metallization of most ICs and discrete diodes and transistors is aluminum (Al). The film thickness is about 1 $\mu$m and conductor widths of about 2 to 25 $\mu$m are commonly used. The use of aluminum offers the following advantages:

1. It is a relatively good conductor.
2. It is easy to deposit thin films of aluminum by vacuum evaporation.
3. Aluminum forms good mechanical bonds with silicon by sintering at about 500°C or by alloying at the eutectic temperature of 577°C.
4. Aluminum forms low-resistance, nonrectifying (i.e., ohmic) contacts with P-type silicon and with heavily doped ($\gtrsim 10^{19}$ cm$^{-3}$) N-type silicon.

For the metallization process the first step is the deposition of a thin film ($\sim$1 $\mu$m) of aluminum on the silicon wafers. This is done by placing the wafers in a vacuum evaporation chamber. The pressure in the chamber is reduced to the range of about $10^{-6}$ to $10^{-7}$ torr (1 atm = 760 torr = 760 mm Hg). The material to be evaporated is placed in a resistance-heated tungsten coil or basket. Electron-beam heating is also very commonly used for vacuum evaporation. The material to be evaporated is placed in a water-cooled crucible. A focused electron beam of very high power density is directed at the surface of the material to be evaporated. This causes a small region of the material to heat up to very high temperatures and to start to vaporize. The molecules that are evaporated from the surface then travel in straight-line paths. The evaporated molecules that hit the substrates will condense there and form a thin-film coating, as shown in Figure 1.36.

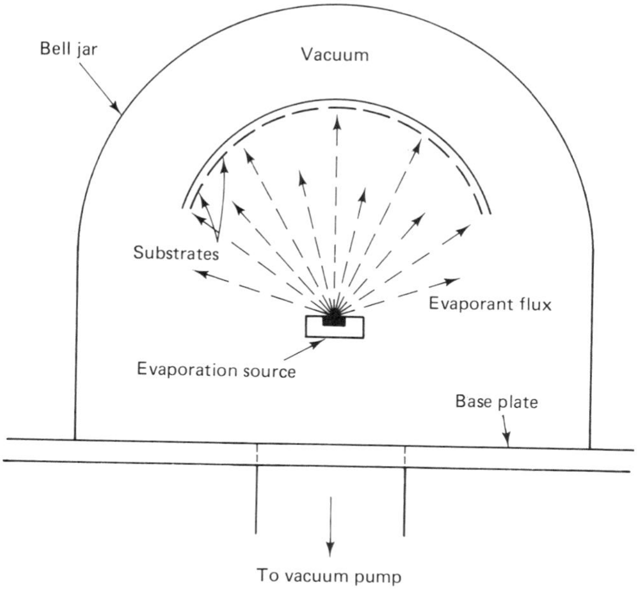


**Figure 1.36** Vacuum evaporation.

A high vacuum is required for the vacuum thin-film deposition process to permit the evaporated molecules to travel to the substrate unimpeded by collisions with gas molecules. The mean free path of gas molecules in a vacuum chamber is related to the pressure by mean free path (mfp) $= 4.5 \times 10^{-3}$ cm/pressure (torr). At a pressure of $10^{-3}$ torr the mfp will be 4.5 cm, increasing to 45 cm at $10^{-4}$ torr, and 450 cm at $10^{-5}$ torr. For a source-to-substrate distance of typically 50 cm a pressure of $10^{-6}$ torr will thus ensure that virtually all of the evaporant flux will reach the substrate unimpeded. The high vacuum is also needed to prevent undesirable chemical reactions between the evaporant molecules and the residual gases in the vacuum chamber.

The thickness of the deposited film can be monitored during the evaporation process by a quartz crystal film thickness monitor. The quartz crystal is exposed to, and samples part of the evaporant flux. As the film thickness builds up on the exposed face of the quartz crystal, it increases the net mass of the crystal. This will result in a decrease in the resonant frequency of the crystal that is proportional to the net mass of the crystal. Consequently, the shift in the oscillator frequency will be directly related to the film thickness that has been deposited on the crystal. When the required film thickness has been reached the evaporant flux can be shut off automatically by means of a shutter that is placed in front of the evaporation source.

After the thin-film metallization has been done the film must be patterned to produce the required interconnection and bonding pad configuration. This is done by a photolithographic process of the same type that is used for producing patterns in $SiO_2$ layers. Aluminum can be etched by a number of acid and base solutions, including HCl, $H_3PO_4$, KOH, and NaOH. The most commonly used aluminum etchant

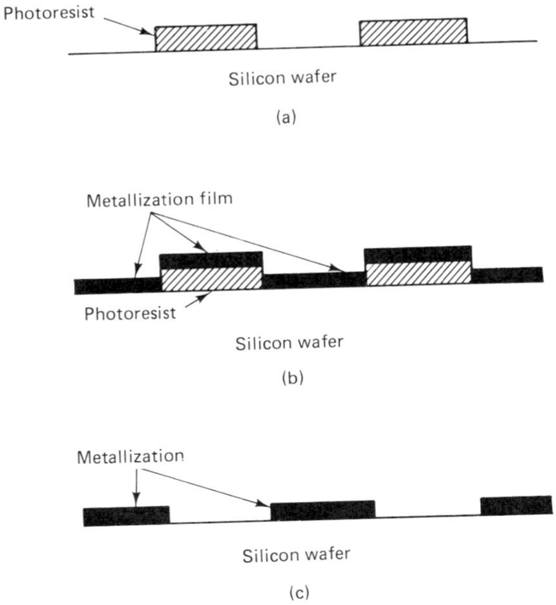


**Figure 1.37** Lift-off process: (a) photolithographic process to pattern photoresist layer; (b) deposition of metallization thin film; (c) photoresist lift-off.

is phosphoric acid ($H_3PO_4$), often with the addition of small amounts of $HNO_3$ (nitric acid) and acetic acid, to result in a moderate etch rate of about 1 μm/min at 50°C. Plasma etching can also be used with aluminum. Using a $CCl_4$/He plasma at a pressure of $3 \times 10^{-4}$ torr, an aluminum etch rate of 0.18 μm/min can be produced. Highly anisotropic aluminum etching with 10:1 (vertical-to-horizontal) etch ratios are possible. As a result, very fine-line (~1 μm) aluminum patterns are possible with aluminum film thicknesses that are substantially greater than the line widths.

An alternative metallization patterning technique is the *lift-off* process. In this process a positive photoresist is spun on the wafer and patterned using the standard photolithographic process. Then the metallization thin film is deposited on *top* of the remaining photoresist. The wafers are then immersed in a suitable solvent such as acetone and at the same time subjected to ultrasonic agitation. This causes swelling and dissolution of the photoresist. As the photoresist comes off it lifts off the metallization on top of it, as shown in Figure 1.37. For this lift-off process to work, the metallization film thickness must generally be somewhat less than the photoresist thickness. This process can, however, produce a very fine-line-width (~1 μm) metallization pattern, even with metallization film thicknesses that are greater than the line width.

With very narrow line widths it is desirable to keep up the film thickness to a value such that the resistance of the interconnections is maintained at an acceptably small value. A large value of interconnection resistance when combined with the parasitic capacitance, as shown in Figure 1.38, will result in an *RC* time constant that will limit the overall speed of the circuit. Indeed, with ICs containing ultrafast small geometry transistors, the principal limitation on the speed of operation can be the interconnection *RC* time constant rather than the devices themselves.

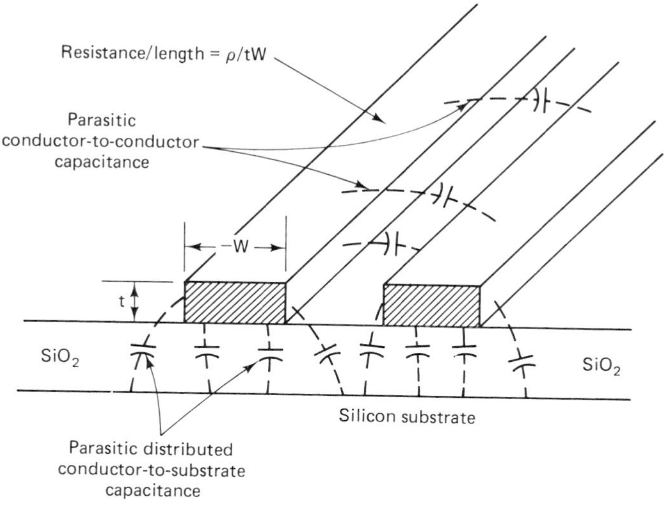


**Figure 1.38** Interconnection *RC* time constant.

## PROBLEMS

### Diffusion Problems

**1.1.** Given: Constant surface concentration diffusion of phosphorus at 1150°C for 60 min into a 1.0-Ω-cm P-type (boron-doped) silicon substrate. The phosphorus surface concentration will be solubility limited. Find:

**(a)** Phosphorus surface concentration, $N_0$. (*Ans.*: ~$1.5 \times 10^{21}$ cm$^{-3}$)

**(b)** Phosphorus diffusion constant, $D$. (*Ans.*: ~$1 \times 10^{-12}$ cm$^2$/s)

**(c)** Substrate doping level, $N_B$. (*Ans.*: ~$1.8 \times 10^{16}$ cm$^{-3}$)

**(d)** Junction depth, $x_J$. (*Ans.*: 3.7 μm)

**(e)** Junction depth if diffusion time is 15 min. (*Ans.*: 1.9 μm)

**(f)** Junction depth for a 2.0-h diffusion. (*Ans.*: 5.2 μm)

**(g)** Time required for $x_J = 10$ μm. (*Ans.*: 7.3 h)

**1.2.** Given: Limited source boron diffusion at 1150°C for 90 min into a 10-Ω-cm N-type silicon substrate. The boron deposited into the silicon surface region has a density of $Q = 1 \times 10^{15}$ cm$^{-2}$. Find:

**(a)** Boron diffusion constant, $D$. (*Ans.*: ~$1 \times 10^{-12}$ cm$^2$/s)

**(b)** Substrate doping. (*Ans.*: ~$4.7 \times 10^{14}$ cm$^{-3}$)

**(c)** Junction depth, $x_J$. (*Ans.*: 4.6 μm)

**(d)** Boron surface concentration after the diffusion. (*Ans.*: $N(O) = 7.7 \times 10^{18}$ cm$^{-3}$)

**(e)** Average conductivity of diffused layer, $\bar{\sigma}$. (*Ans.*: ~24 S/cm)

**(f)** Sheet resistance of diffused layer, $R_S$. (*Ans.*: ~93 Ω/square)

**(g)** End-to-end resistance of the diffused resistor, if the lateral dimensions of this diffused layer are 0.5 mil × 100 mil. (*Ans.*: ~18.5 kΩ)

**1.3.** In order to deposit a boron surface density of $Q = 1 \times 10^{15}$ cm$^{-2}$ into the silicon surface region, a short, low-temperature deposition diffusion step is performed at 950°C.

**(a)** Find the boron diffusion constant for the deposition diffusion. (*Ans.*: ~$7.6 \times 10^{-15}$ cm$^2$/s)

**(b)** Find the surface concentration during deposition diffusion, $N_0$. (*Ans.*: ~$5 \times 10^{20}$ cm$^{-3}$)

**(c)** Find the time required for deposition diffusion. (*Ans.*: 6.9 min)

**(d)** Would it be preferable to have this deposition diffusion done at a somewhat lower temperature such as 900°C? Explain.

**1.4.** The variation of diffusion constant with temperature can be described by the relationship $D = D_0 \exp(-qE_A/kT)$, where $D_0$ is the preexponential constant, $q$ is the electronic charge, $k$ is Boltzmann's constant ($1.384 \times 10^{-23}$ J/K), and $E_A$ is the activation energy for the diffusion process (in electron volts).

**(a)** From the graph of the diffusion constant versus temperature find the activation energy $E_A$ for the diffusion of boron and phosphorus into silicon. (*Ans.*: $E_A$ = 3.61 eV)

**(b)** Show that the fractional change of the diffusion constant with temperature $(1/D)(dD/dT)$ will be given by $(1/D)(dD/dT) = qE_A/kT^2$. Find the value of $(1/D)(dD/dT)$ for the diffusion of boron or phosphorus into silicon for a diffusion temperature of 1100°C. Express the answer in percentage change as well as in fractional change. (*Ans.*: 0.022/°C or 2.2%/°C)

**(c)** Given that the junction depth for either a deposition or drive-in type of diffusion is approximately proportional to $\sqrt{Dt}$, show that the fractional change in the junction depth will be related to the change in the diffusion temperature by $(1/x_J)(dx_J/$

$dT) = \frac{1}{2}(1/D)(dD/dT) = \frac{1}{2}qE_A/kT^2$. Find the fractional and percentage change in the junction depth with temperature for the diffusion of boron or phosphorus into silicon at 1100°C. (*Ans.*: 0.011/°C or 1.1%/°C)

**(d)** Given that the current gain $h_{FE}$ of a transistor is inversely proportional to the base width, show that the fractional change in $h_{FE}$ with diffusion temperature will be given by

$$\frac{1}{h_{FE}}\frac{dh_{FE}}{dT} \simeq \frac{1}{2}\frac{x_J}{W}\frac{qE_A}{kT^2}$$

where $W$ is the base width and $x_J$ can refer to either the emitter–base or the collector–base junction depth of the transistor. (*Hint*: Assume that $W$ is small compared to $x_{J(EB)}$ and $x_{J(CB)}$.)

**(e)** For a transistor with a base width of 0.4 μm and an emitter–base junction depth of 2.4 μm, find the fractional and percentage variation in $h_{FE}$ due to changes in either the emitter or the base diffusion temperature. Use a diffusion temperature of 1100°C. (*Ans.*: 0.066/°C or 6.6%/°C. Note that since a 1°C change in the diffusion temperature is only about a 0.1% change, the value of $h_{FE}$ will change by about 70% for only a 1% change in the diffusion temperature.)

**1.5.** (*Boron ion implantation and drive-in diffusion*) A boron ion implantation followed by a drive-in diffusion at 1150°C produces a diffused layer with a boron surface concentration of $1 \times 10^{18}$ cm$^{-3}$ and a junction depth of 3.0 μm. The substrate phosphorus doping is $1 \times 10^{15}$ cm$^{-3}$. Find:

**(a)** Diffusion time. (*Ans.*: 54.3 min)

**(b)** Ion implantation dosage. (*Ans.*: $1.0 \times 10^{14}$ cm$^{-2}$)

**(c)** Average resistivity of diffused layer. (*Ans.*: ~0.15 Ω-cm)

**(d)** Sheet resistance of the diffused layer. (*Ans.*: ~500 Ω/square)

**(e)** Average hole mobility in the diffused layer. (*Ans.*: ~123 cm²/V-s)

**(f)** Implantation time per wafer if the implantation beam current is 50 μA, the wafer diameter is 100 mm, and the wafers are scanned by the implantation ion beam in a rectangular raster. (*Ans.*: 32 s)

**1.6.** (*Impurity profile evaluation*) A $P^+$ diffused layer is produced by a boron ion implantation followed by a drive-in diffusion at 1150°C to produce a junction depth of 2.5 μm in a $1 \times 10^{15}$ cm$^{-3}$ phosphorus-doped silicon substrate. The impurity profile of this diffused layer is to be evaluated by etching off successive 0.2-μm-thick layers and measuring the sheet resistance after each etching step.

**(a)** Show that the sheet resistance of the removed layer will be given by

$$\frac{1}{\Delta R_S} = \frac{1}{R_{S_b}} - \frac{1}{R_{S_a}}$$

where $\Delta R_S$ is the sheet resistance of the removed layer, $R_{S_b}$ the sheet resistance that is measured *before* the removal of the layer, and $R_{S_a}$ the sheet resistance that is measured *after* the removal of the layer. The initial sheet resistance is 500 Ω/square, and after the removal of the first 0.2-μm layer the sheet resistance has increased to 556 Ω/square.

**(b)** Find the sheet resistance of removed layer. (*Ans.*: 5000 Ω/square)

**(c)** Find the average resistivity of the removed layer. (*Ans.*: 0.1 Ω-cm)

**(d)** Find the surface concentration of the diffused layer, $N$(O). (*Ans.*: $5 \times 10^{17}$ cm$^{-3}$)

**(e)** Find the drive-in diffusion time. (*Ans.*: 42 min)

**(f)** Find the ion implantation dosage. (*Ans.*: $4.4 \times 10^{13}\ cm^{-2}$)

**(g)** By how much will the boron concentration vary over the thickness of the first, second, third, fourth, fifth, and sixth layers that are removed? Does the choice of 0.2 $\mu m$ as the thickness increment appear to be a good one? Explain. (*Ans.*: first: 4.1%; second: 12.8%; third: 22.1%; fourth: 32.3%; fifth: 44.3%; sixth: 55.3%)

**(h)** Find the average boron concentration in the second, fourth, sixth, eighth, and tenth layers that are removed. (*Ans.*: second: $4.6 \times 10^{17}\ cm^{-3}$; fourth: $3.1 \times 10^{17}\ cm^{-3}$; sixth: $1.5 \times 10^{17}\ cm^{-3}$; eighth: $5.3 \times 10^{16}\ cm^{-3}$; tenth: $1.4 \times 10^{16}\ cm^{-3}$)

## Oxidation Problems

**1.7.** Find the time required to grow a 5000-Å $SiO_2$ layer on silicon at a temperature of 1100°C using **(a)** dry oxygen and **(b)** wet oxygen ($H_2O$). (*Ans.*: 10 h, 40 min)

**1.8.** Show that if a silicon wafer is subjected to a number of successive thermal oxidation processes, the resulting final oxide thickness $x_f$ will be given by $x_f^2 = x_1^2 + x_2^2 + x_3^2 + \ldots$, where $x_1, x_2, x_3, \ldots$ are the oxide thicknesses that would result from the application of each oxidation process to an oxidized wafer.

**1.9.** A diffusion window is etched in a field oxide layer of thickness $x_{f1}$ as shown in Figure P1.9a. There is then a regrowth of oxide in the window to a thickness of $x_w$ during which time the field oxide thickness increases to $x_{f2}$, as shown in Figure P1.9b.

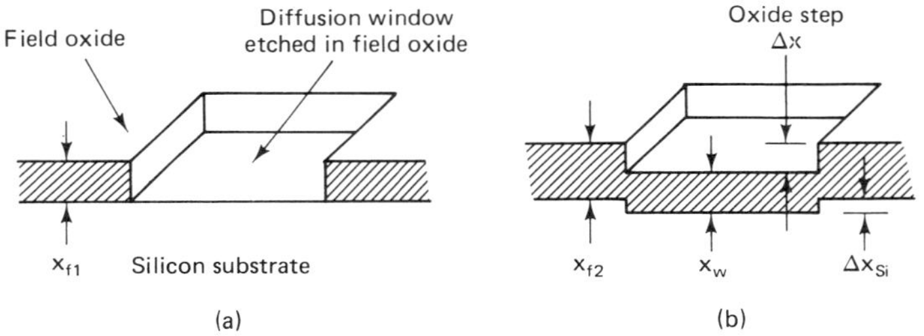


**Figure P1.9**

**(a)** Show that the difference in the oxide thicknesses will be given by

$$x_{f2} - x_w = \sqrt{x_{f1}^2 + x_w^2} - x_w$$

**(b)** Show that the height of the step in the silicon surface will be given by

$$\Delta x_{Si} = 0.44(x_w - x_{f2} + x_{f1})$$

**(c)** Show that the oxide step height will be given by

$$\Delta x = x_{f2} - x_w + \Delta x_{Si}$$

**(d)** If $x_{f1} = 5000$ Å and $x_w = 5000$ Å, find the heights of the silicon step and the oxide step. (*Ans.*: 1289 Å, 3360 Å)

**1.10.** A silicon wafer has a field oxide of 5000 Å thickness. A window is etched in the oxide for a base diffusion and then the oxide is regrown to a thickness of 5000 Å. A

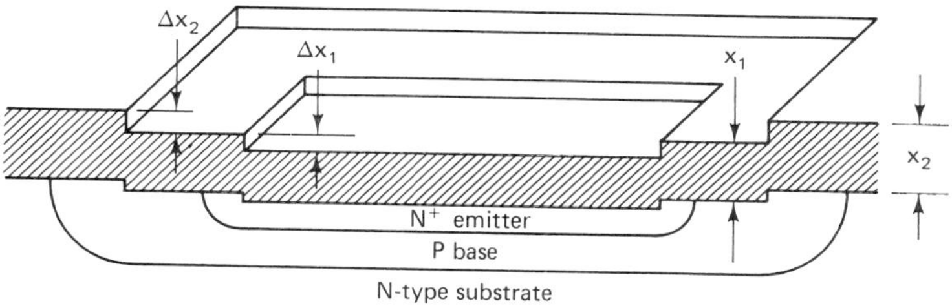


**Figure P1.10**

second window is then etched within the confines of the first window for the emitter diffusion, and then the oxide in this area is regrown to a thickness of 5000 Å, producing the result shown in Figure P1.10. Find $x_2$, $x_1$, $\Delta x_2$, $\Delta x_1$, and the heights of the two silicon steps. (*Ans.*: 8860 Å, 7071 Å, 3090 Å, 3360 Å, 1501 Å, 1289 Å)

**1.11.** An oxidized silicon wafer is subjected to a measurement of the optical reflectance as a function of wavelength.

**(a)** Show that the thickness $t$ of the oxide layer is related to the distance between the adjacent reflectance maxima by

$$\frac{1}{t} = 2n\left(\frac{1}{\lambda_m} - \frac{1}{\lambda_{m+1}}\right)$$

where $n$ is the index of refraction of the oxide layer and $\lambda_m$ and $\lambda_{m+1}$ are the wavelengths of adjacent reflection peaks.

**(b)** Show that the relationship above will also hold true for the reflection minima, where $\lambda_m$ and $\lambda_{m+1}$ are the wavelengths of the adjacent minima.

**(c)** For an $SiO_2$ layer on silicon adjacent reflection peaks are observed at wavelengths of 5000 Å and 6000 Å. If the index of refraction of $SiO_2$ is 1.5, find the thickness of the oxide layer. (*Ans.*: 10,000 Å)

**(d)** For an $SiO_2$ layer adjacent reflection minima are observed at 5454 Å and 6667 Å. Find the thickness of the layer. (*Ans.*: 10,000 Å)

**(e)** An $SiO_2$ layer is 6000 Å thick. Find the wavelengths of the reflection maxima and minima. (*Ans.*: maxima at 18 kÅ, 9 kÅ, 6 kÅ, 4.5 kÅ, 3.6 kÅ, 3.0 kÅ, . . . ; minima at 36 kÅ, 12 kÅ, 7.2 kÅ, 5.14 kÅ, 4.0 kÅ, . . .)

**1.12.** (*Charge carrier mobility*) The mobility of charge carriers (free electrons and holes) in a semiconductor will decrease with increasing doping level due to the scattering of the charge carriers by the ionized dopant impurities. The mobility can be expressed approximately as $1/\mu = 1/\mu_L + 1/\mu_I$, where $\mu$ is the charge carrier mobility, $\mu_L$ is the mobility value that is obtained when there is no ionized impurity scattering so that the only scattering process is lattice scattering, and $\mu_I$ is the mobility value that would be obtained if there were only ionized impurity scattering.

An approximate relationship for $\mu_I$ is given by $\mu_I = KN_I^\alpha$, where $N_I$ is the ionized impurity (dopant) concentration, and $K$ and $\alpha$ are constants. Given the mobility values listed below for silicon at 25°C, find $K$ and $\alpha$ such that the mobility equation will be an exact fit to the mobility versus doping curve at $N_I = 1 \times 10^{16}$ cm$^{-3}$ and at $N_I = 1 \times 10^{18}$ cm$^{-3}$. Then find the percentage error at $N_I = 1 \times 10^{17}$ cm$^{-3}$ and at $N_I = 1 \times 10^{19}$ cm$^{-3}$ for the hole mobility (P-type) and the electron mobility (N-type).

| Dopant density, $N_I$ ($cm^{-3}$) | Hole mobility, $\mu_p$ ($cm^2$/V-s) | Electron mobility, $\mu_n$ ($cm^2$/V-s) |
|---|---|---|
| $1 \times 10^{16}$ | 400 | 1000 |
| $1 \times 10^{17}$ | 250 | 500 |
| $1 \times 10^{18}$ | 100 | 280 |
| $1 \times 10^{19}$ | 28 | 115 |
| Lattice mobility, $\mu_L$ | 500 | 1500 |

(*Ans.*: P-type: $K = 1.164 \times 10^{-13}$, $\alpha = 0.6021$, no error at $1 \times 10^{17}$ $cm^{-3}$, 5.05% error at $1 \times 10^{19}$ $cm^{-3}$; N-type: $K = 1.0 \times 10^{-11}$, $\alpha = 0.470$, +21% error at $1 \times 10^{17}$ $cm^{-3}$, −5.9% error at $1 \times 10^{19}$ $cm^{-3}$)

***1.13.** (*Resistivity versus impurity concentration*) Write a computer program to obtain the resistivity values as a function of doping level for P-type and for N-type silicon at 25°C. Use the mobility results of problem 1.12. Compare the results obtained with the graph of Figure 1.18c.

***1.14.** (*Sheet resistance of diffused layers*) Write a computer program to obtain the sheet resistance of a diffused layer with a Gaussian impurity profile as a function of surface concentration $N(O)$ and junction depth $x_J$. The results of Problems 1.12 and 1.13 can be used in the computer program. Use the computer program to obtain the sheet resistance for the following P-type diffused layers.

**(a)** $N(O) = 1 \times 10^{18}$ $cm^{-3}$, $x_J = 3.0$ $\mu m$, $N_{substrate} = 1 \times 10^{15}$ $cm^{-3}$
**(b)** $N(O) = 1 \times 10^{18}$ $cm^{-3}$, $x_J = 3.0$ $\mu m$, $N_{substrate} = 1 \times 10^{17}$ $cm^{-3}$
**(c)** $N(O) = 1 \times 10^{17}$ $cm^{-3}$, $x_J = 3.0$ $\mu m$, $N_{substrate} = 1 \times 10^{15}$ $cm^{-3}$
**(d)** $N(O) = 1 \times 10^{19}$ $cm^{-3}$, $x_J = 3.0$ $\mu m$, $N_{substrate} = 1 \times 10^{15}$ $cm^{-3}$
**(e)** $N(O) = 1 \times 10^{19}$ $cm^{-3}$, $x_J = 2.0$ $\mu m$, $N_{substrate} = 1 \times 10^{15}$ $cm^{-3}$

Compare the results obtained with those obtained from the graph of Figure 1.18a.

***1.15.** (*Sheet resistance of ion-implanted diffused layers*) Write a computer program to obtain the sheet resistance of diffused layers that are produced by a low-energy ion implantation followed by a drive-in diffusion. The sheet resistance should be obtained as a function of ion implantation dosage, junction depth, and background or substrate doping. Find the sheet resistance for a boron-doped layer with a junction depth of 2.5 $\mu m$, a background N-type doping of $1 \times 10^{15}$ $cm^{-3}$, and the following boron ion implantation dosages ($cm^{-2}$): $1 \times 10^{10}$, $1 \times 10^{11}$, $1 \times 10^{12}$, $1 \times 10^{13}$, $1 \times 10^{14}$, $1 \times 10^{15}$, and $1 \times 10^{16}$. Compare the results obtained with those of Figure 2.54.

## REFERENCES


BAR-LEV, A., *Semiconductors and Electronic Devices*, Prentice-Hall, 1984.

BRODIE, I., *The Physics of Microfabrication*, Plenum Press, 1982.

BURGER, R. M., and R. P. DONOVAN, *Fundamentals of Silicon Integrated Circuit Device Technology*, Vol. 1, Prentice-Hall, 1967.

CAMENZIND, H. R., *Electronic Integrated Systems Design*, Van Nostrand Reinhold, 1972.

COLCLASER, R. A., *Microelectronics: Processing and Device Design*, Wiley, 1980.

CONNELLY, J. A., *Analog Integrated Circuits*, Wiley, 1975.

DeForest, W. S., *Photoresist: Materials and Processes*, McGraw-Hill, 1975.

Doyle, J. M., *Thin Film and Semiconductor Integrated Circuitry*, McGraw-Hill, 1966.

Eimbinder, J., *Application Considerations for Linear Integrated Circuits*, Wiley, 1970.

Einspruch, N. G., *VLSI Electronics: Microstructure Science*, Vols. 1–6, Academic Press, 1982.

Elliott, D. J., *Integrated Circuit Fabrication Technology*, McGraw-Hill, 1982.

Fogiel, M., *Microelectronics: Basic Principles, Circuit Design, Fabrication Technology*, Research & Education Associates, 1972.

Ghandi, S. K. *The Theory and Practice of Microelectronics*, Wiley, 1968.

Glaser, A. B., and G. E. Subak-Sharpe, *Integrated Circuit Engineering*, Addison-Wesley, 1977.

Grebene, A. B., *Analog Integrated Circuit Design*, Van Nostrand Reinhold, 1972.

Grove, A. S., *Physics and Technology of Semiconductor Devices*, Wiley, 1967.

Hamilton, D. J., and W. G. Howard, *Basic Integrated Circuit Engineering*, McGraw-Hill, 1975.

Hnatek, E. R., *A User's Handbook of Integrated Circuits*, Wiley, 1973.

Millman, J., *Microelectronics*, McGraw-Hill, 1979.

Motorola, Inc., *Integrated Circuits: Design Principles and Fabrication*, Vol. 1, McGraw-Hill, 1967.

Runyan, W. R., *Silicon Semiconductor Technology*, McGraw-Hill, 1965.

Streetman, B. G., *Solid State Electronic Devices*, Prentice-Hall, 1980.

Sze, S. M., *VLSI Technology*, McGraw-Hill, 1983.

Till, W. C., and J. T. Luxon, *Integrated Circuits: Materials, Devices, and Fabrication*, Prentice-Hall, 1982.

Veronis, A., *Integrated Circuit Fabrication Technology*, Reston, 1979.

Vossen, J. L., and W. Kern, *Thin Film Processes*, Academic Press, 1978.

2

# INTEGRATED-CIRCUIT DEVICES

## 2.1 JUNCTION CHARACTERISTICS

Before continuing with a discussion of the fabrication sequence for diodes, transistors, and ICs, it is of importance to look at some characteristics of PN junctions, particularly the junction capacitance $C_J$, the breakdown voltage, and the series resistance $R_S$. The results of this discussion will provide insight into the use of epitaxial structures for devices.

### 2.1.1 Junction Capacitance

A reverse-biased PN junction can be considered to be a parallel-plate capacitor with the depletion region being the insulator or dielectric as shown in Figure 2.1. The depletion or space-charge region is the region adjacent to the PN junction that is essentially depleted or devoid of all mobile charge carriers (free electrons and holes), so that it indeed acts like an insulator.

The junction capacitance is given by the parallel-plate capacitance equation as $C_J = \epsilon A/W$, where $\epsilon = \epsilon_r \times \epsilon_0 = 11.8 \times 8.85 \times 10^{-14}$ F/cm $= 1.0443 \times 10^{-12}$ F/cm $=$ permittivity of silicon, $A$ is the junction area, and $W$ is the depletion region width. For the case in which there is uniform doping on both sides of the junction, the depletion region width will be given by

$$\boxed{W = \sqrt{\frac{2\epsilon V_J}{qN}}} \tag{2.1}$$

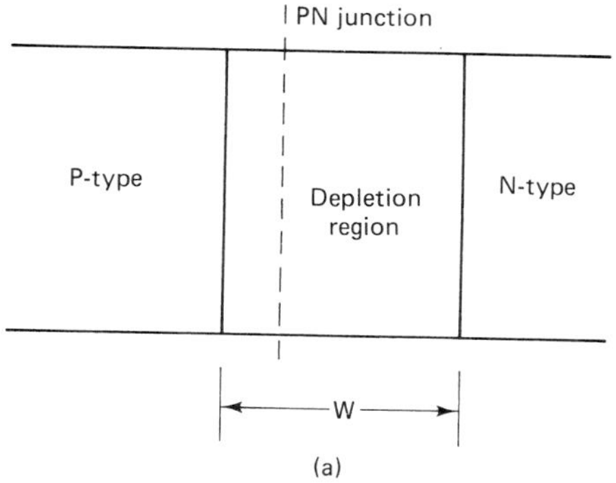


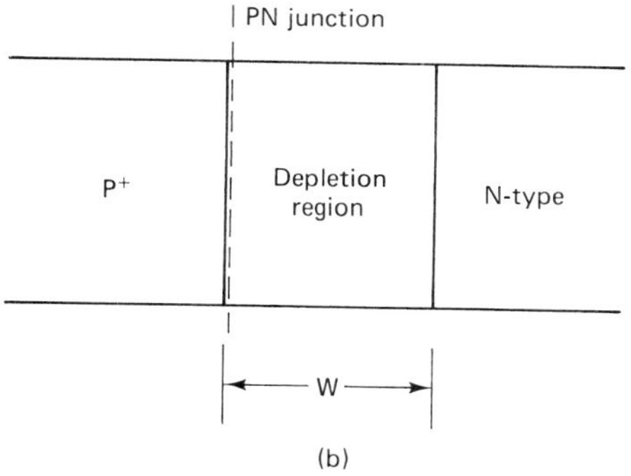


**Figure 2.1** PN junction depletion region width: (a) general case; (b) P+/N "one-sided" junction.

where $V_J$ = junction voltage = $\phi + V_R$ = contact potential (~0.8 V) + applied reverse-bias voltage, and $N = N_A N_D/(N_A + N_D)$. The corresponding equation for junction capacitance will be

$$C_J = A\sqrt{\frac{q\epsilon N}{2V_J}} \tag{2.2}$$

At $N = 1 \times 10^{16}$ cm$^{-3}$ and $V_J = 1.0$ V, these equations give $W = 0.361$ $\mu$m and $C_J = 2.89 \times 10^{-8}$ F/cm$^2$ = 289 pF/mm$^2$. In Figure 2.2 a graph of $C_J$ versus $N$ is given for several different values of $V_J$.

If the junction area for the preceding case is 500 $\mu$m $\times$ 500 $\mu$m (0.020 in. $\times$ 0.020 in.), the junction capacitance will be $C_J$ = 289 pF/mm$^2$ $\times$ (0.5 mm $\times$ 0.5 mm) = 72.3 pF at $V_J = 1.0$ V, which corresponds to a reverse-bias voltage of only about 0.2 V. If $V_J$ increases to 25 V, the capacitance will decrease by a factor of 5 to become 14.5 pF.

For almost all diffused PN junctions the doping on the diffused layer side of the junction will be very much heavier than the doping on the other (substrate) side of the junction, so that most diffused junctions can be considered to be *one-sided* junctions. In Figure 2.1b a one-sided P+N junction is shown. Notice that the

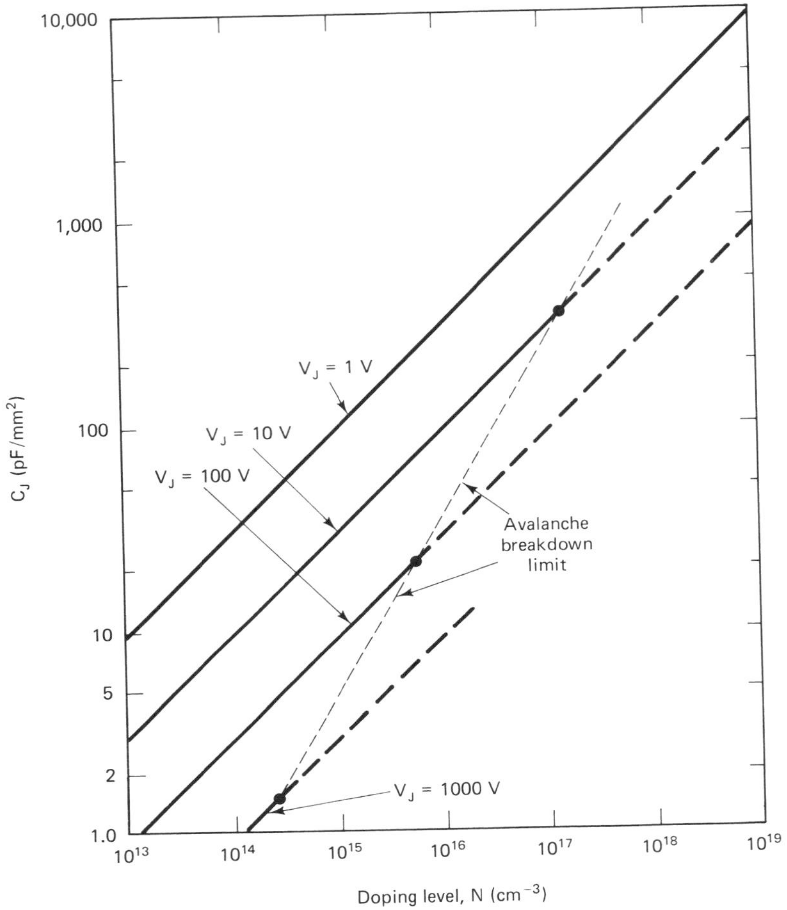


**Figure 2.2** PN junction capacitance.

depletion region is almost entirely on the lightly doped (substrate) side of the junction and extends very little into the diffused layer side. For a $P^+N$ one-sided junction we have that $N = N_A N_D/(N_A + N_D) \simeq N_D$, and for an $N^+P$ one-sided junction we have $N \simeq N_A$.

For many applications it is desirable to have a very small junction capacitance so we see that a lightly doped substrate will be needed. This will be especially true in the case of high-speed and high-frequency device applications.

### 2.1.2 Breakdown Voltage

The breakdown voltage of a PN junction is in general a function of the doping levels on both sides of the junction. For a one-sided junction, the breakdown voltage will be a function principally of the doping level on the more lightly doped side of

**TABLE 2.1 BREAKDOWN VOLTAGE VERSUS DOPING LEVEL**

| Doping level | Resistivity (Ω-cm) | | Breakdown voltage (V) | |
|---|---|---|---|---|
| | N-type | P-type | Equation | Experimental |
| $1 \times 10^{14}$ | 50 | 135 | 1250 | 1250 |
| $3 \times 10^{14}$ | 16 | 45 | 600 | 600 |
| $1 \times 10^{15}$ | 5 | 13.5 | 270 | 300 |
| $3 \times 10^{15}$ | 1.7 | 4.5 | 130 | 150 |
| $1 \times 10^{16}$ | 0.60 | 1.5 | 58 | 65 |
| $3 \times 10^{16}$ | 0.23 | 0.63 | 28 | 30 |
| $1 \times 10^{17}$ | 0.09 | 0.28 | 12.5 | 15 |
| $3 \times 10^{17}$ | 0.05 | 0.14 | 6.0 | 8.0 |
| $1 \times 10^{18}$ | 0.025 | 0.065 | 2.7 | 5.3 |
| $2 \times 10^{18}$ | 0.017 | 0.040 | 1.7 | 5.0 |

the junction. An approximate equation for the breakdown voltage of a one-sided flat (i.e., no junction curvature) junction is: breakdown voltage $= 2.7 \times 10^{12}$ V/$N^{2/3}$, where $N$ is the doping level (cm$^{-3}$). This equation is a good approximation up to doping levels of about $1 \times 10^{17}$ cm$^{-3}$. Table 2.1 demonstrates the results obtained by using the breakdown voltage equation.

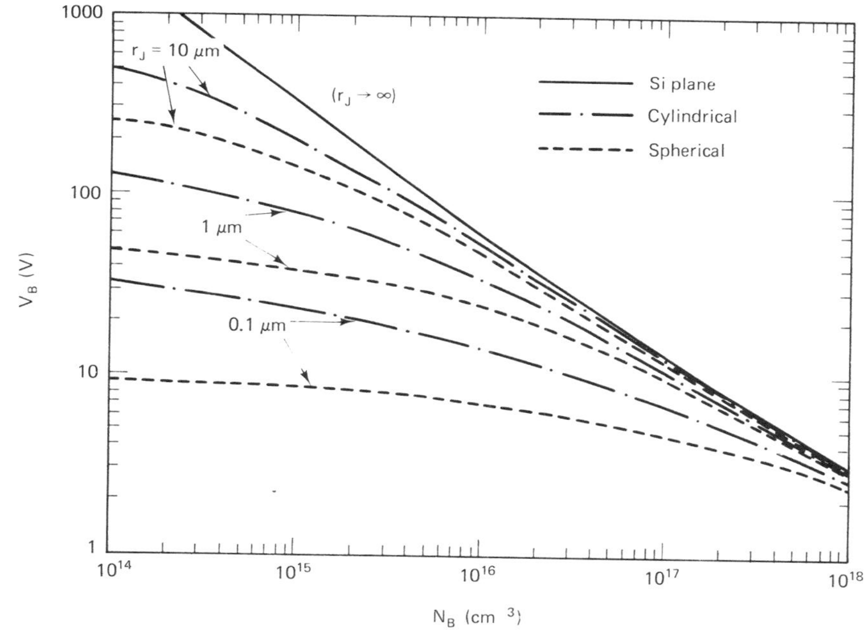


**Figure 2.3** Breakdown voltage as a function of substrate doping and junction curvature for abrupt one-sided junctions in silicon. (Reprinted with permission from S. M. Sze and G. Gibbons, "Effect of Junction Curvature on Breakdown Voltage in Semiconductors," *Solid State Electronics*, Vol 9, p. 831, 1966. Copyright 1966, Pergamon Press, Ltd.)

The breakdown voltage for planar junctions will be somewhat lower than these values, especially at the lighter doping levels, where the breakdown voltage can be very substantially less. This is due to the effect of the junction curvature in the regions underneath the edges of the oxide window, which results in an increase in the electric field intensity. The effect of junction curvature on breakdown voltage is shown in Figure 2.3. Looking at an extreme case, for $N = 1 \times 10^{14}\ \mathrm{cm}^{-3}$ the breakdown voltage for a plane (i.e., flat) junction will be around 1250 V. A junction curvature corresponding to a junction depth of $x_J = 1\ \mu\mathrm{m}$ will drop the breakdown voltage all the way down to about 50 V. Increasing $x_J$ to 3 $\mu$m will increase the breakdown voltage back up to around 100 V, and at $x_J = 10\ \mu\mathrm{m}$ the breakdown voltage will reach around 250 V. At heavier dopings, however, we see that the influence of junction curvature on breakdown voltage becomes less.

From the preceding discussion we see that for a high breakdown voltage a light doping is required. The junction depth should also not be too small, especially for the cases in which very high breakdown voltages ($\gtrsim$100 V) are required.

### 2.1.3 Series Resistance

The bulk series resistance of a PN junction is due to the finite resistivity of the P-type and N-type regions of the junction, outside the depletion region. For a $P^+N$ diffused junction the series resistance due to the $P^+$ diffused layer will be negligible compared to the resistance due to the N-type side of the junction. Thus for a small value of series resistance a low-resisitivity (i.e., heavily doped) substrate is needed.

## 2.2 EPITAXIAL STRUCTURE

From the previous discussion we have the following requirements:

1. For low junction capacitance, $C_J$: low doping (lightly doped substrate)
2. For high breakdown voltage: low doping (lightly doped substrate)
3. For low series resistance, $R_S$: heavy doping (low resistivity substrate)

We see that the series resistance requirement will be incompatible with the capacitance and breakdown voltage requirements.

The epitaxial structure shown by the diode example of Figure 2.4 offers a good way of resolving this incompatibility and simultaneously satisfying the capacitance, breakdown voltage, and series resistance requirements. As long as the depletion region remains entirely within the lightly doped epitaxial layer and does not reach the heavily doped $N^+$ substrate, the capacitance and breakdown voltage will be a function only of the epitaxial layer doping and will be independent of the substrate doping. The series resistance will, however, be determined to a major extent by the $N^+$ substrate doping since the epitaxial layer is very thin ($\sim$10 $\mu$m) compared to the substrate thickness of some 250 to 400 $\mu$m.

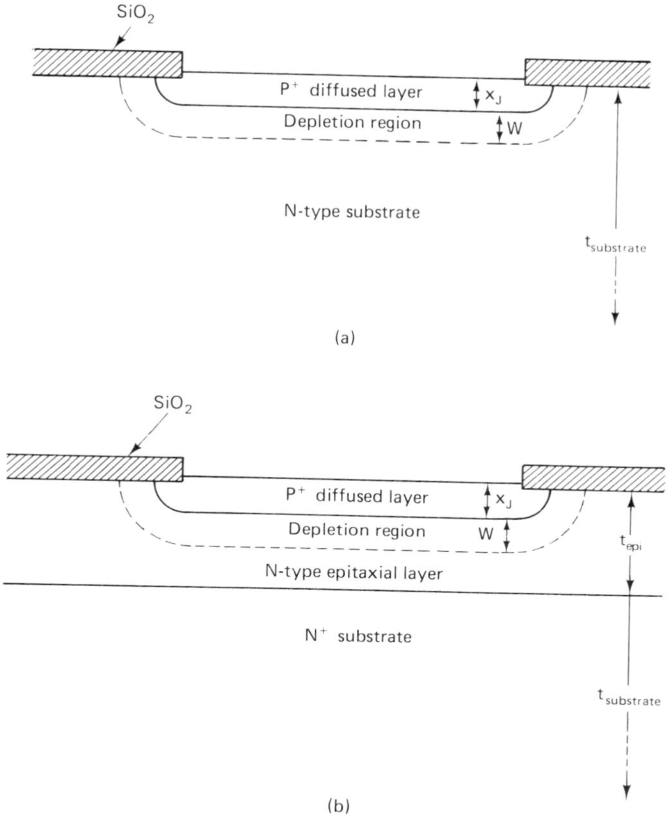


**Figure 2.4** (a) Planar (nonepitaxial) diode; (b) planar epitaxial diode.

The series resistance of the epitaxial diode is given by

$$R_S = R_{\mathrm{epi}} + R_{\mathrm{substrate}} = \frac{\rho_{\mathrm{epi}}(t_{\mathrm{epi}} - x_J - W)}{A} + \frac{\rho_{\mathrm{substrate}} t_{\mathrm{substrate}}}{A} \tag{2.3}$$

Note that $t_{\mathrm{epi}} - x_J - W$ is the thickness of the *undepleted* portion of the epitaxial layer, which is the distance from the edge of the depletion region to the substrate. The heavily doped low-resistivity substrate that constitutes the major part of the overall device thickness can lead to a very substantial reduction in the series resistance.

### 2.2.1 Epitaxial Diode Example: Voltage-Variable Capacitance Diode

As an example of the efficacy of the epitaxial structure in reducing the series resistance, let us consider a planar epitaxial $P^+NN^+$ diode that will be used as a *voltage-variable capacitance* (VVC) or *varactor* (i.e., variable-reactance) diode. A VVC diode makes specific use of the dependence of the diode capacitance on the bias voltage. This device can be used in a variety of applications, including voltage-controlled oscillators

(VCOs) and automatic frequency control (AFC) circuits, in which case the VVC diode is part of an *LC* tuned circuit.

Let us consider a $P^+NN^+$ VVC diode of the following design:

$$\begin{aligned}
&\text{Junction diameter} = 500\ \mu\text{m} = 0.50\ \text{mm}\ (0.020\ \text{in.}) \\
&\text{Junction depth, } x_J = 2.0\ \mu\text{m} \\
&\text{Epitaxial layer thickness, } t_{\text{epi}} = 10\ \mu\text{m} \\
&\text{Epitaxial layer doping, } N_{\text{epi}} = 1 \times 10^{15}\ \text{cm}^{-3}\ (\rho_{\text{epi}} = 5\ \Omega\text{-cm}) \\
&\text{Substrate resistivity, } \rho_{\text{substrate}} = 0.005\ \Omega\text{-cm (Sb-doped)} \\
&\text{Substrate thickness, } t_{\text{substrate}} = 300\ \mu\text{m} = 0.3\ \text{mm}\ (0.012\ \text{in.})
\end{aligned} \tag{2.4}$$

The junction area corresponding to the diameter of 500 $\mu$m will be

$$A = \frac{\pi}{4d^2} = \frac{\pi}{4}(0.5\ \text{mm})^2 = 0.196\ \text{mm}^2 = 0.196 \times 10^{-2}\ \text{cm}^2 \tag{2.6}$$

Under zero-bias conditions we have that $V_J = \phi = 0.8$ V, so that $C_J(0) = 289/\sqrt{10 \times 0.8}$ pF/mm² $\times$ 0.196 mm² = $\underline{20.1\ \text{pF}}$, and the zero-bias depletion layer width will be $W(0) = 0.361 \times \sqrt{10 \times 0.8} = \underline{1.02\ \mu\text{m}}$. The series resistance under these conditions will be given by $R_S(0) = R_{\text{epi}}(0) + R_{\text{sub}} = 5\ \Omega\text{-cm} \times (10^{-2} - 1.02) \times 10^{-4}\ \text{cm}/0.196 \times 10^{-2}\ \text{cm}^2 + 0.005\ \Omega\text{-cm} \times 0.03\ \text{cm}/0.196 \times 10^{-2}\ \text{cm}^2 = 1.78\ \Omega + 0.0765\ \Omega = \underline{1.86\ \Omega}$. If a nonepitaxial structure were used, the series resistance would be $R_S = 5\ \Omega\text{-cm} \times 0.03\ \text{cm}/0.196 \times 10^{-2}\ \text{cm}^2 = 76.5\ \Omega$. The quality factor $Q$ of the diode under zero-bias conditions at a frequency of 50 MHz will be $Q(0) = (1/\omega C_J)/R_S = 1/(2\pi \times 50\ \text{MHz} \times 20.1\ \text{pF} \times 1.86\ \Omega) = \underline{85.5}$. This is to be compared to a value of $Q(0) = 2.0$ for a nonepitaxial structure.

At a reverse-bias voltage of 3.0 V we have that $V_J = 3.8$ V so that the capacitance will now be $C_J(-3\ \text{V}) = 20.1 \times \sqrt{0.8/3.8} = \underline{9.20\ \text{pF}}$ and the depletion layer width will be $W(-3\ \text{V}) = 1.02\ \mu\text{m} \times \sqrt{3.8/0.8} = \underline{2.22\ \mu\text{m}}$. The series resistance will now be given by $R_S = R_{\text{epi}}(-3\text{V}) + R_{\text{sub}} = 5\ \Omega\text{-cm} \times (10^{-2} - 2.22) \times 10^{-4}/0.196 \times 10^{-2} + 0.0765 = 1.47 + 0.0765\ \Omega = \underline{1.551\ \Omega}$. The $Q$-value at 3 V bias will be $Q(-3\ \text{V}) = \underline{233}$, and has increased from the zero-bias value due to the decrease in both the capacitance and the series resistance. If a nonepitaxial structure were to be used, the $Q$-value at this voltage would be only 4.5.

The voltage at which the depletion region will extend all of the way across the epitaxial layer from the junction to the $N^+$ substrate (i.e., "full depletion") will be given by $1.02\ \mu\text{m} \times \sqrt{V_J/0.8\ \text{V}} = t_{\text{epi}} - x_J = 8\ \mu\text{m}$, so that $V_J = 49.2$ V and thus the required reverse-bias voltage will be $V_R = V_J - \phi = \underline{48.4\ \text{V}}$.

When the epitaxial layer is fully depleted between the junction and the $N^+$ substrate the capacitance will level off at a minimum value as given by

$$C_{\text{MIN}} = \frac{\epsilon A}{W_{\text{MAX}}} = \frac{\epsilon A}{t_{\text{epi}} - x_J} = \frac{1.04 \times 10^{-12}\ \text{F/cm} \times 0.196 \times 10^{-2}\ \text{cm}^2}{8 \times 10^{-4}\ \text{cm}} \tag{2.7}$$

$$= \underline{2.55\ \text{pF}}$$

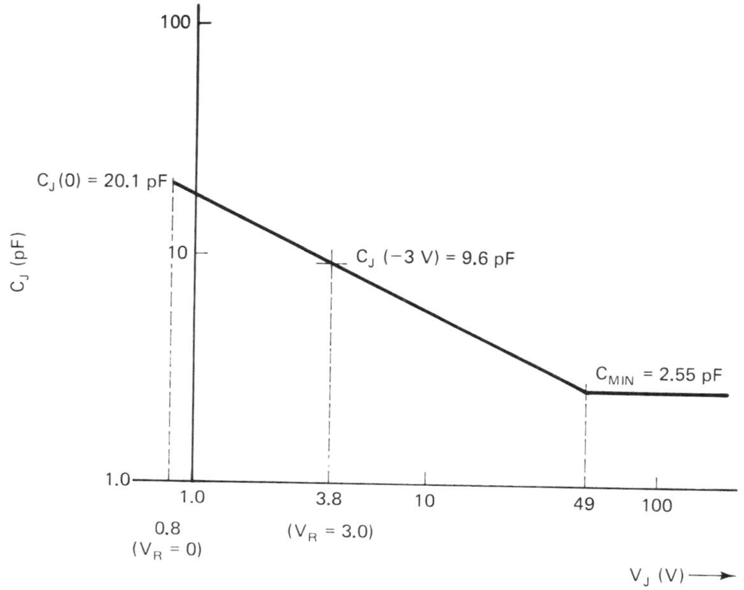


**Figure 2.5** Capacitance versus voltage characteristic for VVC diode example.

The series resistance will now be due to just the substrate resistance since the epitaxial layer is now fully depleted so will now be $R_{S\,(\text{MIN})} = R_{\text{substrate}} = 0.0765\ \Omega$. The $Q$-value at 50 MHz will now be at its maximum value since both the capacitance and resistance have now reached their minimum values and will be $Q_{\text{MAX}} = 16{,}300$. This is to be compared to a $Q$-value of about 136 for the nonepitaxial structure at the same reverse-bias voltage.

From this example we see the clear advantage of the epitaxial structure. In Figure 2.5 a graph of the variation of capacitance with junction voltage is presented for this $P^+NN^+$ epitaxial diode. The capacitance ratio is of interest, and we have that $C_J(0)/C_{\text{MIN}} = \underline{7.87}$ and $C_J(-3\text{V})/C_{\text{MIN}} = \underline{3.61}$, so that quite a large capacitance variation can be obtained.

If this VVC diode is part of an $LC$ tuned circuit as shown in Figure 2.6, with a fixed capacitance of 2.0 pF, the frequency or tuning ratio will be given by

$$\frac{f_{\max}}{f_{\min}} = \frac{\sqrt{1/LC_{\min}}}{\sqrt{1/LC_{\max}}} = \sqrt{\frac{(20.1 + 2)\ \text{pF}}{(2.55 + 2)\ \text{pF}}} = \underline{2.20} \tag{2.8}$$

for the case in which the minimum reverse-bias voltage is 0 V. If the minimum reverse-bias voltage is restricted to −3 V, the tuning ratio will be reduced to $\underline{1.57}$.

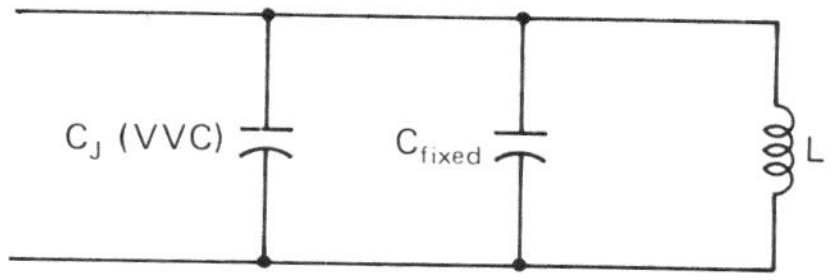


**Figure 2.6** $LC$ tuned circuit with VVC diode.

Thus an appreciable frequency swing will be available using this VVC diode in a tuned circuit.

## 2.3 PLANAR EPITAXIAL DIODE PROCESSING SEQUENCE

Now that we have discussed the basic processing steps for the fabrication of silicon devices, and we have seen the purpose of the epitaxial structure, we will summarize the processing sequence for a number of devices, starting with a planar $P^+N/N^+$ diode.

A processing sequence for a $P^+N/N^+$ epitaxial planar diode of the type shown in Figure 2.4b is summarized below.

1. *Starting material*: $N/N^+$ epitaxial wafers with a 0.005 Ω-cm (Sb-doped) substrate and an epitaxial layer of anywhere from 5 to 25 μm thick and phosphorus doped to resistivities in the range 5 to 50 Ω-cm.
2. *Oxidation*: An oxide layer about 5000 to 8000 Å thick is grown.
3. *First photolithography*: Window openings in the oxide layer for the $P^+$ diffusion are produced.
4. *Boron diffusion*: A $P^+$ diffused layer about 1 to 3 μm thick is produced to be the anode regions of the diodes.
5. *Second photolithography*: Anode contact windows are produced.
6. *Metallization*: Aluminum deposition (~1 μm) produces the anode contacts.
7. *Third photolithography*: Metallization patterned for anode contacts.
8. *Contact sintering or alloying*: This is a heat treatment at about 500 to 600°C for sintering or alloying the metallization film to form a good mechanical bond to the silicon and to produce a low-resistance, nonrectifying ("ohmic") contact.
9. *Back-side metallization*: A thin film of gold is evaporated on to the lapped back sides of the wafers. This is for the eutectic die (chip) bonding of the chips to gold-plated headers or substrates at temperatures in the range 400 to 420°C, the gold/silicon eutectic temperature being 370°C.

## 2.4 PLANAR EPITAXIAL TRANSISTOR

We will now enumerate the processing steps for a representative NPN planar epitaxial transistor of the type shown in Figure 2.7, where a cross-sectional view of the transistor is presented together with a graph showing the impurity profiles.

1. *Starting material*: $N/N^+$ epitaxial wafer with 0.005 Sb-doped substrate and N-type epitaxial layer of about 6 to 12 μm thickness and 0.3 to 3 Ω-cm resistivity.
2. *Oxidation*: An oxide layer of about 5000 to 8000 Å thickness is grown.
3. *First photolithography*: Oxide windows are etched for the base diffusion.
4. *Base diffusion*: A two-step deposition—drive-in boron diffusion is performed

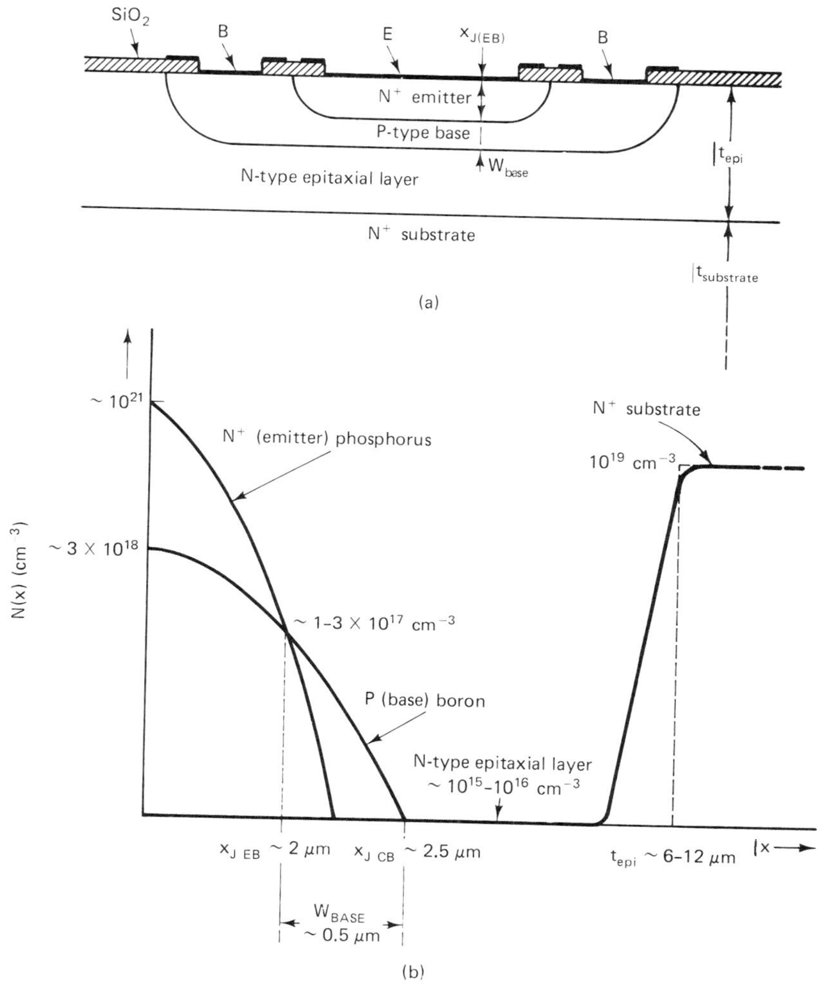


**Figure 2.7** NPN double-diffused planar epitaxial transistor: (a) cross-sectional view; (b) impurity profiles.

to produce a P-type diffused layer with a junction depth of about 2 to 3 μm, a surface concentration of around $3 \times 10^{18}$ cm$^{-3}$, and a sheet resistance of about 200 Ω/square. The drive-in diffusion is performed in an oxidizing ambient ($O_2$) so that oxide is regrown in the windows that were produced in the preceding step.

5. *Second photolithography*: Oxide windows are etched for the emitter diffusion.
6. *Emitter diffusion*: A high surface concentration phosphorus diffusion is performed to produce an $N^+$ diffused layer with a junction depth of about 2 to 2.5 μm, a surface concentration in the range $1 \times 10^{21}$ cm$^{-3}$, and a sheet resistance of about 2 to 2.5 Ω/square.

7. *Third photolithography*: Oxide windows are etched for emitter and base contacts.
8. *Metallization*: An aluminum thin film of about 0.5 to 1 μm thickness is deposited on the front surface of the wafers.
9. *Fourth photolithography*: Windows are etched for emitter and base contact areas.
10. *Contact sintering or alloying*: This is a heat treatment at 500 to 600°C for sintering or alloying metallization.
11. *Back-side metallization*: A gold thin film is deposited on the back side of the wafers.

## 2.5 CHIP SEPARATION AND CHIP BONDING

After the wafer processing sequence for diodes, transistors, or ICs is completed, the wafers must be divided up into individual dice or chips. This can be done by a "scribe-and-break" operation using a diamond-tipped scribe, a high-intensity laser beam (laser scribing), or a high-speed circular saw to produce grooves in the silicon. In the case of the diamond scribe the grooves are very shallow, somewhat deeper with laser scribing, and may extend more than halfway through the wafer with the saw. The wafers will have a pattern of orthogonally oriented "scribing streets" which are kept clear of oxide and metal and are aligned along certain crystallographic directions to promote the easy and smooth cleavage of the wafer.

A popular process for chip separation is to use a wafer saw to cut entirely through the wafer. The wafer is mounted on an adhesive-coated tape prior to the sawing operation so that after sawing the chips will remain in matrix form for convenience in further operations.

The chips or dice are then bonded to either metal headers or ceramic substrates. The metal headers are usually gold-plated Kovar. Kovar is an iron–nickel–cobalt alloy whose thermal expansion coefficient is a close match to that of silicon. The headers are heated to temperatures in the range 400 to 420°C in an inert-gas atmosphere ($N_2$ or a mixture of about 90% $N_2$ and 10% $H_2$). The chips are then bonded to the headers by means of the formation of a gold–silicon alloy that results in a good mechanical bond and a low-resistance electrical contact. This contact will be to the collector of the transistors, the cathodes of the $P^+N/N^+$ diodes, and the substrates of ICs.

### 2.5.1 Lead Bonding and Encapsulation

Small-diameter (~20 to 40 μm or 0.8 to 1.6 mils) gold wires are now bonded from the package leads or terminal posts to the metallized contact areas or bonding pads on the chips. Aluminum bonding wire is also sometimes used, especially for high-current-power devices, where larger-diameter round or flat ribbon leads may be used.

The device is now encapsulated in a metal, ceramic, or plastic package. The plastic package is the lowest in cost, but the metal and ceramic packages offer the advantage of providing a hermetic seal and a higher operating temperature range.

## 2.6 JFET PROCESSING SEQUENCE

An N-channel JFET is shown in Figure 2.8a. The N-type channel of this device is formed by the N-type epitaxial layer region between the $P^+$ diffused layer (gate) and the P-type substrate. The processing sequence for this device follows closely along the lines of that of the double-diffused transistor, and is summarized below.

1. *Starting material*: N/P epitaxial wafers
2. *Oxidation*
3. *First photolithography*: Windows for $P^+$ boron top gate diffusion
4. *Boron diffusion*: $P^+$ top gate diffusion
5. *Second photolithography*: Windows for $N^+$ source and drain diffusion
6. *Phosphorus diffusion*: $N^+$ diffusion to produce the source and drain regions of the JFET
7. *Third photolithography*: Contact windows
8. *Metallization*

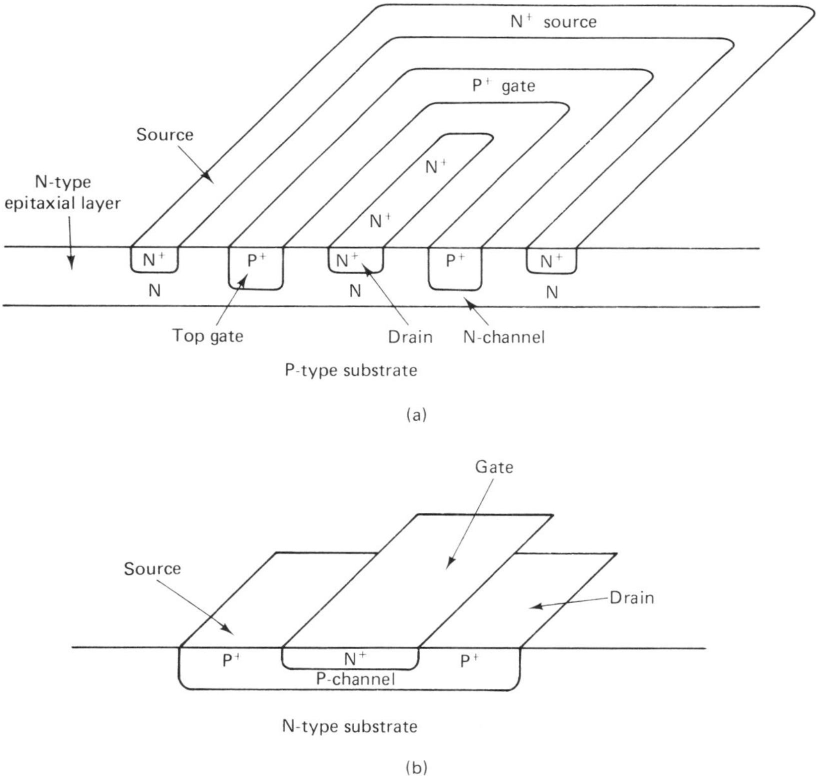


**Figure 2.8** Junction-field-effect-transistor structures: (a) N-channel epitaxial JFET; (b) double-diffused P-channel JFET.

9. *Fourth photolithography*: Metallization patterning for source, drain, and gate contact areas
10. *Contact sintering or alloying*
11. *Back-side metallization*

In Figure 2.8b a P-channel double-diffused JFET is shown. The P-type channel of this device is formed by the P-type diffused region between the $N^+$ diffused layer and the N-type substrate. This device is structurally similar to a double-diffused NPN transistor with the P-type channel corresponding to the P-type base region of the bipolar transistor and the processing sequence is basically the same.

### 2.6.1 MESFET

In Figure 2.9 a diagram of a gallium arsenide (GaAs) MESFET (metal-semiconductor field-effect transistor) is shown. This device operates in essentially the same way as does a junction-gate FET, except that instead of a gate-channel PN junction there is a gate-channel Schottky barrier. The depletion region associated with this Schottky barrier will control the effective height of the conducting channel and can thereby control the drain-to-source current of the device. The width of this depletion region will increase with increasing gate voltage so that we see again that the gate will be the control electrode, and as long as the Schottky barrier is reverse biased, the gate current will be very small.

N-type GaAs offers the advantage of an electron mobility that is much higher than that of silicon (8500 $cm^2/V$-s for GaAs versus 1400 $cm^2/V$-s for lightly doped silicon). This higher electron mobility together with a very short channel length ($\sim$1 $\mu$m) will result in very short channel transit times for electrons and therefore lead to very high-speed devices that can operate well up into the gigahertz range.

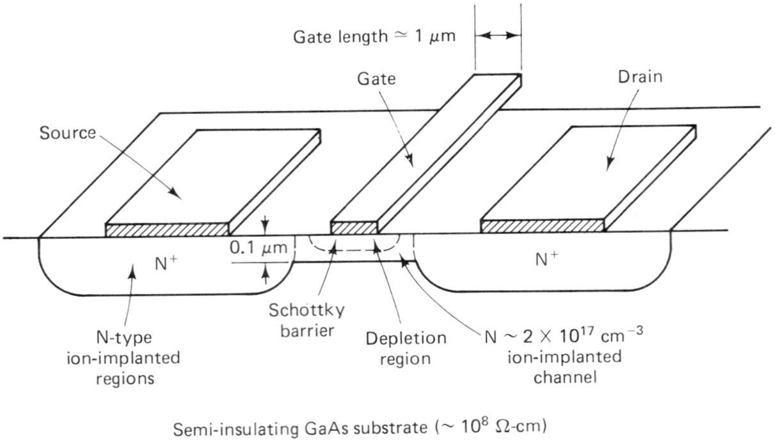


**Figure 2.9** Gallium arsenide MESFET structure.

## 2.7 MOSFET PROCESSING SEQUENCE

A cross-sectional view of a simple N-channel aluminum gate MOSFET is shown in Figure 2.10a. The processing steps for this device are listed below.

1. *Starting material*: P-type silicon, ~10 Ω-cm resistivity.
2. *Oxidation*: Thermally grown oxide of about 10,000 Å thickness.
3. *First photolithography*: Oxide windows for source and drain diffusion.
4. *Phosphorus diffusion*: $N^+$ diffusion to produce the source and drain regions.
5. *Second photolithography*: Oxide removed from channel region between source and drain.

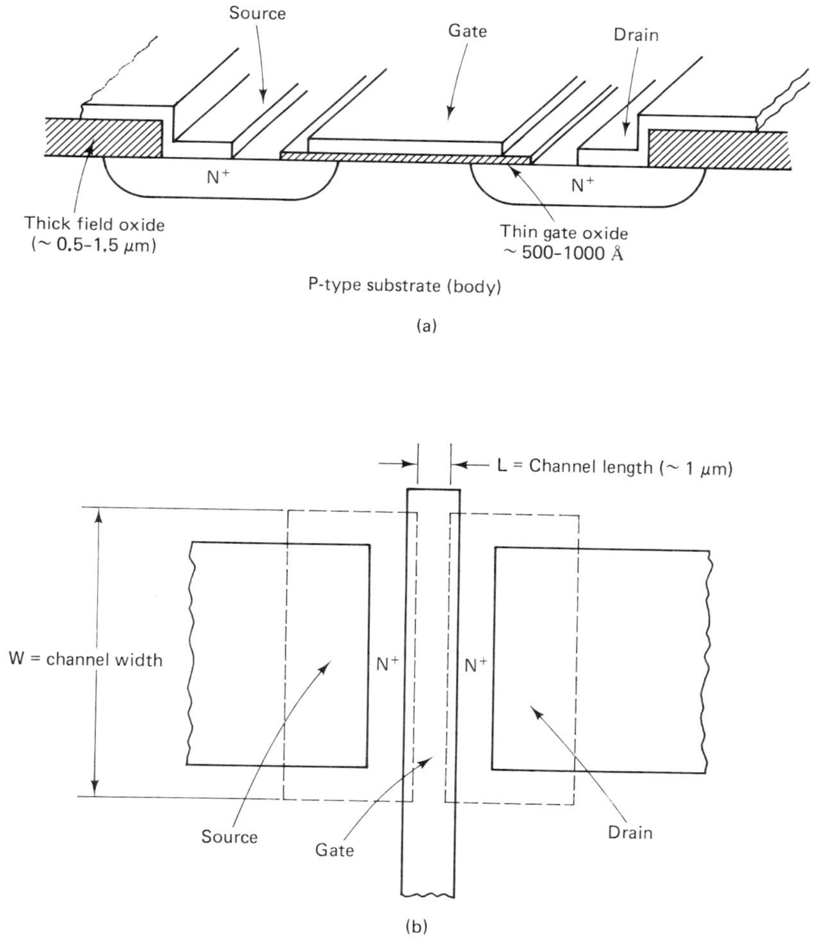


**Figure 2.10** MOSFET: (a) cross-sectional perspective view; (b) top view.

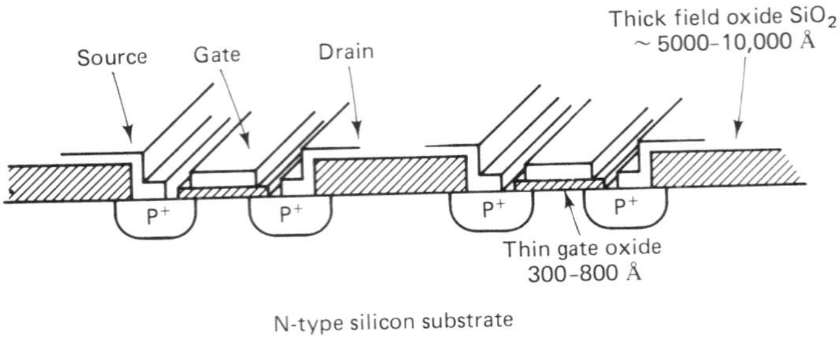


**Figure 2.11** Self-isolating characteristics of MOS transistors.

6. *Oxidation*: Growth of a very thin oxide layer over the channel region. This gate oxide will generally be in the range 300 to 800 Å (Figure 2.11) and is grown under very carefully controlled conditions to minimize contamination of the oxide by such impurities as various alkali ($Na^+$, $K^+$, etc.) ions.
7. *Third photolithography*: Contact windows.
8. *Metallization*: Aluminum thin film.
9. *Fourth photolithography*: Metallization patterning to produce the gate electrode and the source and drain contact areas.
10. *Contact sintering*
11. *Back-side metallization*

### 2.7.1 Self-Aligned Gate MOSFETs: Overlap Capacitance

For a MOSFET to be turned on, a conducting channel in the form of a surface inversion layer must be produced over the entire distance between the source and drain areas. Therefore, the gate electrode must extend all the way between the source and the drain. To allow for possible mask registration errors, the gate is designed to overlap the edges of the source and drain regions by a small amount, often in the range of about 5 μm. This condition will result in a small *overlap capacitance* between the gate and the source ($C_{gs}$) and the source and the drain ($C_{gd}$). These capacitance will generally be of the order of 1 to 3 pF. The gate-to-drain capacitance is of particular interest since it represents a feedback capacitance from output (drain) to input (gate), and its effect on the input capacitance of the MOSFET is increased by the Miller effect.

In Figure 2.12 a MOSFET structure that features a "self-aligned" gate is shown.

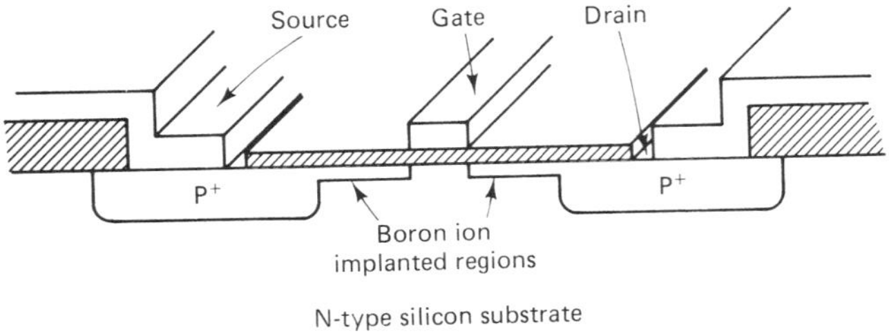


**Figure 2.12** Self-aligned gate MOSFET using ion implantation.

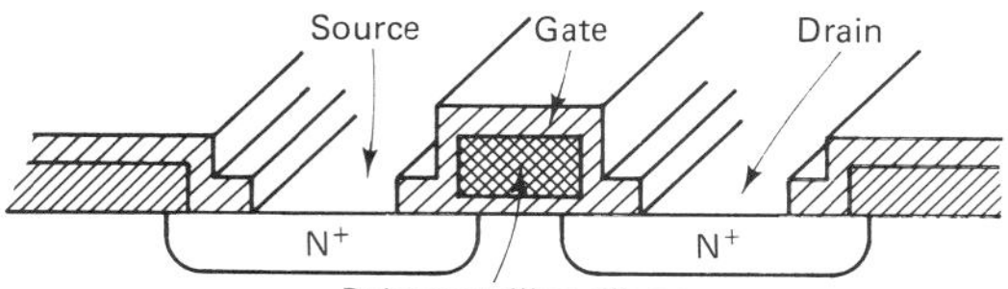


**Figure 2.13** Polysilicon or refractory metal gate MOSFET.

The processing steps for this device are similar to those of a conventional P-channel MOSFET, except that the diffused source and drain regions do not extend all the way to the gate. A boron ion implantation is used to produce P-type extensions of the source and drain regions right up to the edge of the gate. The boron ions, with about 100 kV of energy, are able to penetrate the thin gate oxide, but are blocked by the much thicker gate and by the thick field oxide. The gate electrode thus serves as an implantation mask such that the source and drain regions effectively terminate right under the edges of the gate so that the overlap capacitance is thereby minimized. The annealing of the ion implantation takes place at a relatively low temperature in the range 400 to 500°C such that the lateral diffusion of the implanted boron ions underneath the gate is negligible.

Ion implantation is also useful in MOSFETs for adjusting the threshold voltage $V_T$. For this purpose a very low-dosage ($\sim 10^{11}$ to $10^{12}$ cm$^{-2}$), low-energy ($\sim$30 kV) implantation is to modify the doping of the silicon surface in the channel region. This technique is especially useful for N-channel MOSFETs, where the threshold voltage would otherwise be too low ($\lesssim 1$ V); this is true for many applications, including most digital circuits. A low-dosage boron implantation can be used here to increase the effective doping of the P-type substrate surface region and thus raise the threshold voltage to an acceptable level.

Another self-aligned gate structure is shown in Figure 2.13. This device uses a gate of polycrystalline silicon ("polysilicon") or a refractory metal (Mo, Ta, or W) or metal silicide ($MoSi_2$, $TaSi_2$, $WSi_2$). These gate materials can withstand the high temperatures used in the diffusion process, so that the gate can now serve as a diffusion mask. There will be some lateral diffusion underneath the gate, but the overlap capacitance will still be considerably smaller than in the conventional MOSFET structure.

The polysilicon gate also has the advantage of producing a small favorable shift in the threshold voltages, reducing the P-channel threshold voltage and increasing the threshold voltage of the N-channel devices.

### 2.7.2 Short-Channel MOSFETs

A short channel length is advantageous in a MOSFET since it results in a higher transfer conductance. The higher transfer conductance, in turn, leads to a larger voltage gain and gain–bandwidth product. Also, the drain-to-source current $I_{DS}$ at any given gate voltage will be larger, so that the current-handling capability of the

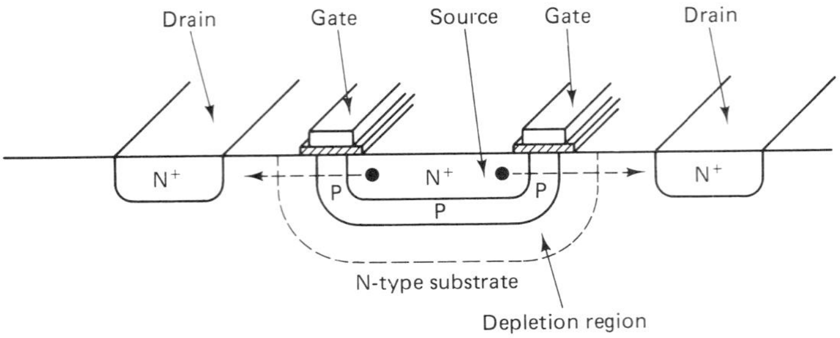


**Figure 2.14** Diffused channel MOS (DMOS) transistor.

device is increased. Indeed, current ratings of up to 10 A are available with some VMOS and vertical DMOS devices.

In Figure 2.14 a diagram of a double-diffused MOSFET (DMOS) is shown. The channel length underneath the gate oxide is the lateral distance between the $N^+P$ junction and the PN substrate junction, and is thus controlled by the junction depths produced by the $N^+$- and P-type diffusions. This is similar to the situation with respect to the base width of a double-diffused bipolar transistor, and the channel length can be made to be as small as 0.5 $\mu$m.

The application of a suitably large positive voltage to the gate (i.e., $V_{GS} > V_t$) will invert the P-type surface region underneath the gate to N-type, and the resulting N-type surface inversion region will serve as a conducting channel for the flow of electrons from source to drain.

The lightly doped N-type substrate and the room available for the expansion of the depletion region between the P-type diffused region and the $N^+$ drain contact region will make possible a relatively high breakdown voltage between drain and source ($BV_{DS}$).

A "vertical" DMOS structure is shown in Figure 2.15. In this case the drain contact region is the $N^+$ substrate. The removal of the $N^+$ drain contact regions from top surface will allow for more parallel-connected channels to be formed so that the transfer conductances and the drain-to-source current capability of the device can be correspondingly increased. With a large high-density array of very many gate electrodes on the top surface, vertical DMOS devices with current ratings of up to 10 A are possible.

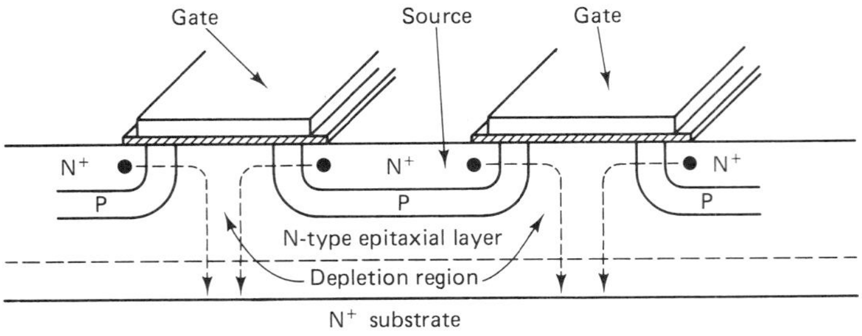


**Figure 2.15** Vertical DMOS (V-DMOS) transistor.

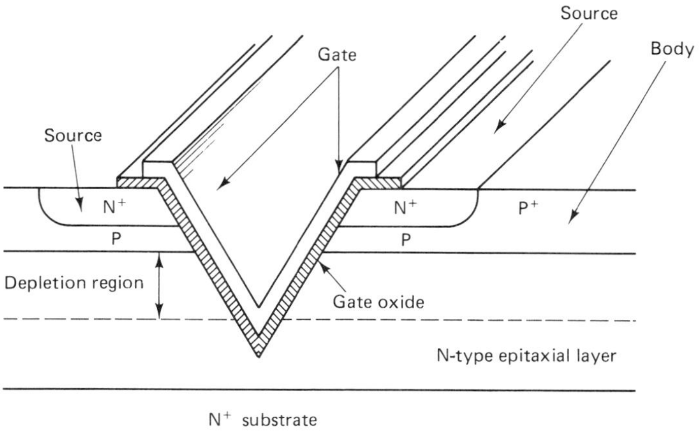


**Figure 2.16** VMOS transistor.

In Figure 2.16 a VMOS transistor is shown. This is again a double-diffused device in which the channel length is set by the difference between the $N^+$- and P-type diffusions. The lightly doped N-type epitaxial layer and the room allowed for the expansion of the depletion region between the $P^+$ diffused layer and the $N^+$ substrate will lead to a high breakdown voltage ($BV_{DS} \gtrsim 50$ V) and a low drain capacitance. At the same time the drain series resistance is kept to a small value as a result of the heavily doped $N^+$ substrate.

The *V*-grooves are produced by an anisotropic or orientation-dependent etching (ODE) process. The etchant that is used, such as KOH at 80 to 100°C, attacks silicon very rapidly in the [100] crystallographic direction, but very slowly in the [111] direction. For (100)-oriented silicon substrates the result will be the production of V-shaped grooves that have (111) sidewalls, as shown in Figure 2.17b. The angle of the (111) groove sidewalls with respect to the (100) silicon surface will be 54.74°. The width of the grooves, *W*, is controlled by the width of the opening in the oxide layer, which is used as an etching mask since $SiO_2$ is attacked only very slowly by the etching solution. In Figure 2.18 the relationships between the principal crystallographic planes are shown. Note that with a (110) silicon substrate, the etching of vertical-walled slots is possible. The orientation-dependent etching process is useful for a number of silicon devices in addition to the VMOS transistors. VMOS transistors with large arrays of very many V-groove gate structures are available with current ratings of up to several amperes.

Vertical DMOS transistors consisting of a very large number of parallel connected cells in a rectangular or hexagonal pattern on a common $N/N^+$ drain region are available with continuous current ratings in excess of 25 A at voltages of up to 500 V, which gives a power handling capability of 12.5 kW. The very short channel length and very large total channel width that can be of the order of 1 million times greater than the channel length can also result in very small values for the drain-to-source resistance $r_{ds(\mathrm{ON})}$, with values as low as 0.12 Ω being achieved.

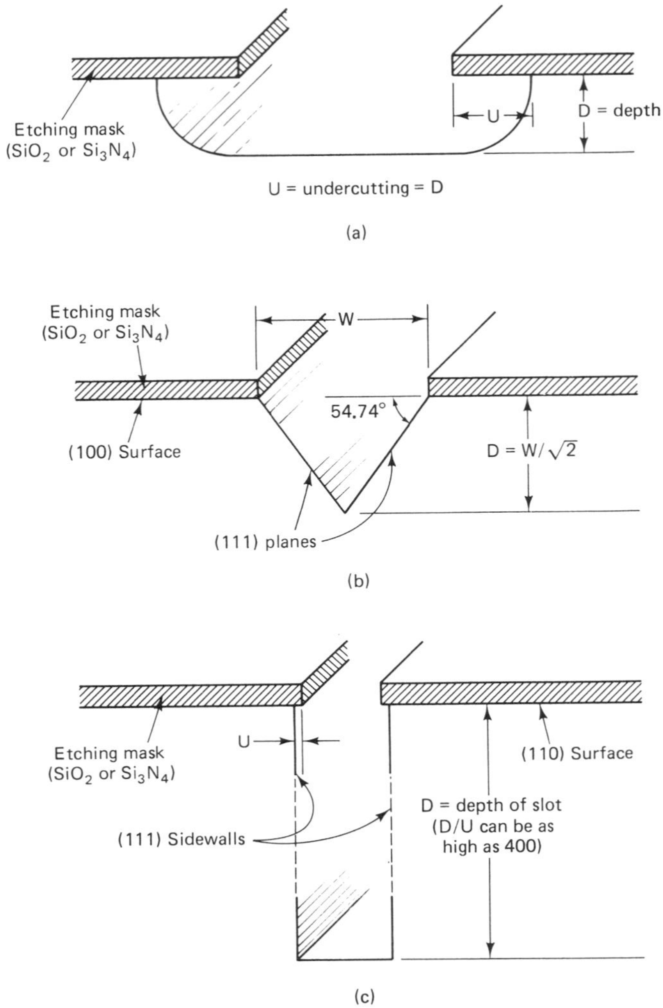


**Figure 2.17** Isotropic and anisotropic etching of silicon: (a) isotropic etching (example: HF:$HNO_3$:acetic acid mixture); (b) anisotropic or orientation-dependent etching on (100) silicon wafer; (c) anisotropic or orientation-dependent etching on (110) silicon wafer.

### 2.7.3 Examples of Ion Implantation for MOS Threshold Voltage Adjustment

It was mentioned previously that ion implantation can be used for the adjustment of the threshold voltage $V_t$ for MOSFETs. As a result of positive-ion contamination in the gate oxide, surface charges at the oxide–silicon interface, and the gate electrode/silicon contact potential, the threshold voltage for NMOS transistors will often be very low ($\lesssim$1 V). In some cases the NMOS threshold voltage can actually be slightly

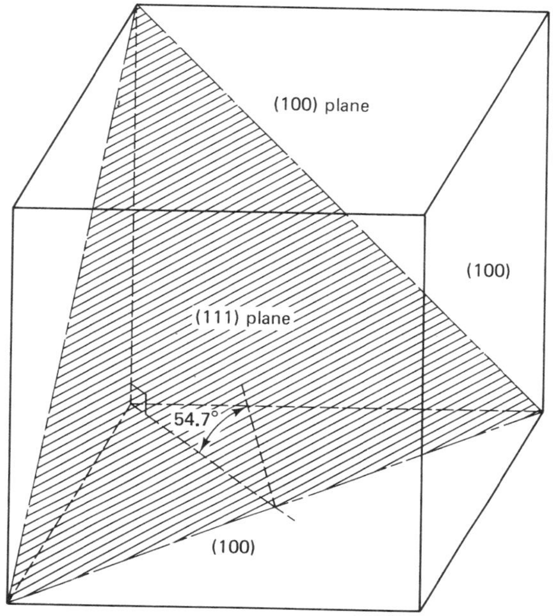


(a)

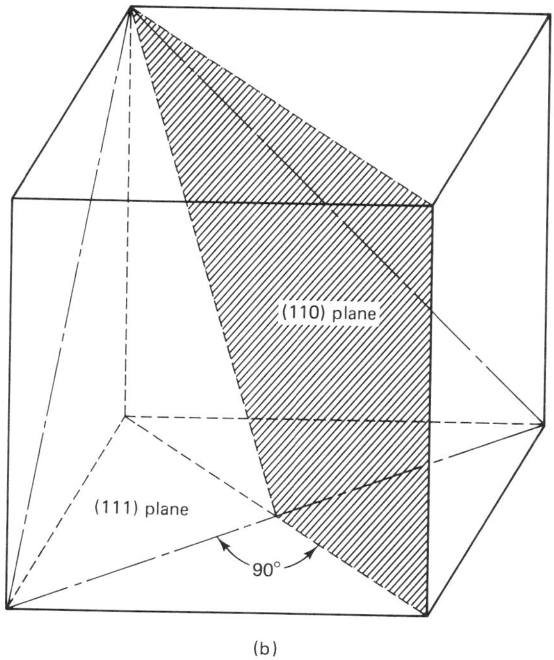


(b)

**Figure 2.18** Principal crystallographic planes: (a) relationship between the (100) and (111) planes; (b) perpendicular set of (110) planes and (111) planes.

negative, so that there will be an N-type inversion layer present even in the absence of any applied gate voltage. At the same time the gate voltage for PMOS devices will be relatively large ($\sim -5$ V) in the negative direction due to these same effects.

A boron ion implantation can be used to shift the threshold voltage in the positive direction for both NMOS and PMOS devices. The boron ions implanted in the silicon will form a shallow surface layer of negatively charged acceptor ions. The charge per unit area due to these implanted boron ions will be $\Delta Q = q \times$ implantation dosage, where $q$ is the electron charge. This surface charge density will result in a shift in the threshold voltage of $\Delta Q = C_{\text{oxide}} \, \Delta V_t$, where $C_{\text{oxide}}$ is the capacitance per unit area of the MOS gate capacitance. This capacitance will be given by $C_{\text{oxide}} = \epsilon_{\text{oxide}}/t_{\text{oxide}}$, where $\epsilon_{\text{oxide}} = \epsilon_r \times \epsilon_0 = 3.8\epsilon_0 = 3.8 \times 8.85 \times 10^{-14}$ F/cm $= 3.363 \times 10^{-13}$ F/cm.

Let us now consider the example of a PMOS transistor with a threshold voltage of $-5.0$ V and a 1000-Å gate oxide. The boron ion implantation dosage needed to shift the threshold voltage down to $-2.5$ V will be given by

$$\Delta Q = q \times \text{dosage} = C_{\text{oxide}} \, \Delta V_t = \frac{3.36 \times 10^{-13} \text{ F/cm}}{0.1 \times 10^{-4} \text{ cm}} \times 2.5 \text{ V} \tag{2.9}$$

$$= 8.408 \times 10^{-8} C$$

so that the required dosage will be $5.25 \times 10^{11}$ boron ions/cm$^2$. If the ion implantation system has a beam current of 1.0 $\mu$A, the time required for the implantation of a 100-mm-diameter wafer will be only 8.4 s.

For a second example, let us consider an NMOS transistor with a threshold voltage of only $+0.5$ V and a 1000-Å gate oxide. The boron ion dosage required to shift the threshold voltage from $+0.5$ V to $+1.5$ V will be $2.1 \times 10^{11}$ boron ions/cm$^2$. The implantation time for a 100-mm wafer and a 1-$\mu$A beam current will be only 3.4 s.

## 2.8 IC DEVICE ISOLATION

In monolithic ICs there are many transistors and other devices sharing the same small single-crystal silicon chip, so there must be some means of providing electrical isolation between devices wherever needed. The most commonly used means of providing this isolation is called *junction isolation*, illustrated in Figure 2.19. Starting with a P-type substrate, an N-type epitaxial layer of about 10 $\mu$m thickness is deposited, as shown in Figure 2.19a. This is followed by a thermal oxidation process and photolithography to produce the line openings in the oxide, as shown in Figure 2.19b. A high-concentration deep boron P$^+$ diffusion is then performed. The boron diffusion proceeds long enough to penetrate all the way through the epitaxial layer, down into the P-type substrate, as shown in Figure 2.19c and d.

The N-type epitaxial layer has now been subdivided up into separate N-type regions. Between every N-type region and every other N-type region there are two PN junctions. As long as all of these PN junctions are reverse biased there will be electrical isolation between the various N-type regions and between the N-type regions

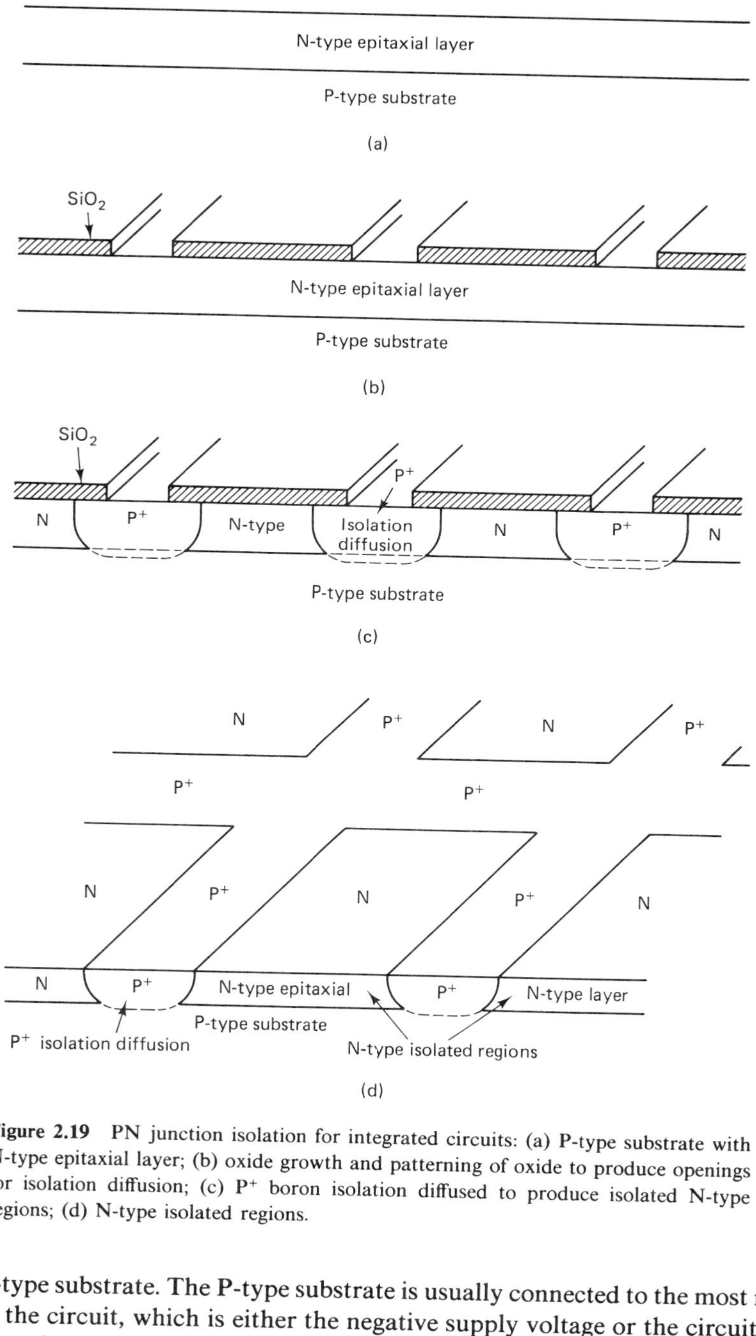


**Figure 2.19** PN junction isolation for integrated circuits: (a) P-type substrate with N-type epitaxial layer; (b) oxide growth and patterning of oxide to produce openings for isolation diffusion; (c) $P^+$ boron isolation diffused to produce isolated N-type regions; (d) N-type isolated regions.

and the P-type substrate. The P-type substrate is usually connected to the most negative voltage in the circuit, which is either the negative supply voltage or the circuit ground in the case of single-supply operation. This substrate connection will automatically ensure that all of the N-region to P-substrate junctions will be reverse biased and there will therefore be electrical isolation between the N-type regions.

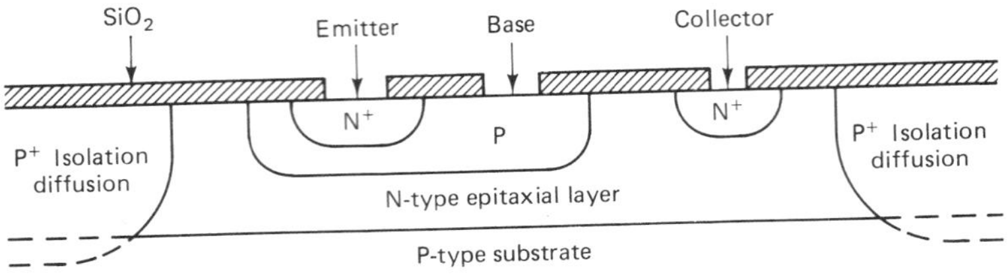


**Figure 2.20** IC NPN transistor: cross-sectional view.

This electrical isolation is not perfect, however, and there will be a very small amount of leakage current between the N-type isolated regions and the P-type substrate that will be of the order of 1 nA/mm² at 25°C. Most IC transistors will have areas in the range 0.001 to 0.01 mm², so that the corresponding leakage current will be in the range of only 1 to 10 pA and so will be negligible for most applications.

The principal isolation problem will be the presence of the capacitance of the N-type isolated region/P-type substrate junction. This capacitance will be a function of the doping levels and the bias voltage, but will generally be in the range of 300 pF/mm² (0.19 pF/mil²) at zero bias, decreasing to 120 pF/mm² (0.08 pF/mil²) at 5 V reverse bias, and about 95 pF/mm² (0.06 pF/mil²) at 10 V bias. For a typical bipolar transistor 100 $\mu$m $\times$ 100 $\mu$m = 0.01 mm² in area, the parasitic N-type region (collector) to substrate capacitance $C_{CS}$ will therefore be about 1 pF at bias voltages in the range 5 to 15 V.

A cross-sectional view of an IC NPN transistor is shown in Figure 2.20. Note that the contacts to all three regions—emitter, base, and collector—of the transistor are on the front or top surface, unlike the case of an individual or discrete transistor, in which the collector contact is on the back side of the chip. As a result of the topside collector contact there will be a large collector series resistance $r_{cc'}$ present due to the high sheet resistance of the N-type collector region—about 1000 to 10,000 Ω/square. This resistance can, however, be reduced very substantially by the use of an *$N^+$ buried layer* or *subcollector diffusion*, as shown in Figure 2.21. Since the collector–base junction is still bounded by the more lightly doped N-type epitaxial region, the collector–base junction capacitance $C_{CB}$ can be kept acceptably low, and the collector–base breakdown voltage can be maintained at a high value (~50 V). The basic processing sequence for monolithic ICs using this type of NPN transistor is given in Table 2.2.

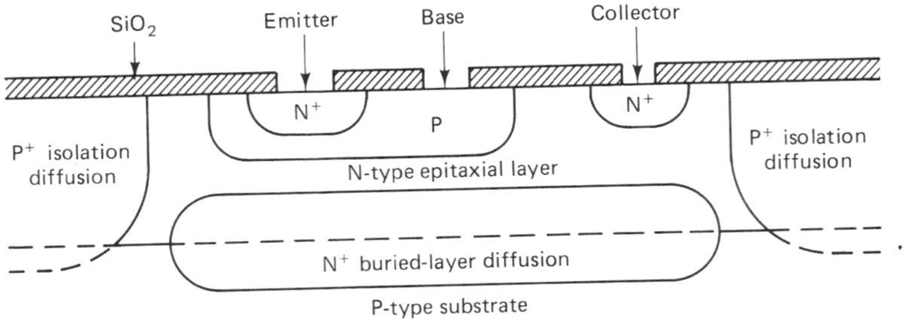


**Figure 2.21** IC NPN transistor with buried layer.

**TABLE 2.2** SILICON MONOLITHIC IC PROCESSING SEQUENCE

1. *Starting material*: semiconductor grade ($\lesssim$1 ppba) polycrystalline silicon
2. *Crystal growth*: Czochralski (or float-zone) process to produce P-type single-crystal silicon ingots about 100 mm in diameter and boron doped to a resistivity in the range 1 to 10 Ω-cm
3. *Mechanical preparation of silicon wafers*: ingot slicing, wafer lapping, and polishing to produce wafers about 0.4 mm (16 mils) thick
4. *Oxidation*: thermally grown $SiO_2$ layer about 5000 Å thick
5. *First photolithography*: windows etched in oxide for buried-layer diffusion
6. *Buried-layer diffusion*: diffusion of high-concentration Sb-doped $N^+$ layer about 3 μm deep in oxidizing ambient followed by oxide removal
7. *Epitaxial layer deposition*: deposition of phosphorus-doped N-type epitaxial layer of 0.1 to 1 Ω-cm resistivity and 5 to 15 μm thick
8. *Oxidation*: growth of thermal oxide about 5000 to 10,000 Å thick
9. *Second photolithography*: oxide windows etched for the $P^+$ isolation diffusion
10. *Isolation diffusion*: $P^+$ boron diffusion all of the way through the epitaxial layer down to the P-type substrate with the boron drive-in performed in an oxidizing ambient
11. *Third photolithography*: oxide window opening for base region (and resistor) diffusion
12. *Base and diffused resistor diffusion*: a two-step $P^+$ boron deposition–drive-in diffusion to produce a junction depth of about 2 to 3 μm and a sheet resistance of 200 Ω/square, with the drive-in done in an oxidizing ambient
13. *Fourth photolithography*: windows etched in oxide for emitter and collector contact $N^+$ diffusion
14. *Emitter and collector contact diffusion*: a phosphorus $N^+$ diffusion to a junction depth of about 2 μm and a sheet resistance of about 2.2 Ω/square, with the drive-in done in an oxidizing ambient
15. *Fifth photolithography*: contact windows etched in oxide
16. *Contact and interconnection metallization*: an aluminum thin film of about 0.5 to 1.0 μm deposited on the front surface of the IC wafers
17. *Sixth photolithography*: etching of metallization pattern
18. *Contact sintering or alloying*: heat treatment at 500 to 600°C to produce good mechanical bond and low-resistance electrical contact
19. *Back-side metallization*: gold thin film deposited on wafer backs
20. *Wafer probing*: testing of ICs with an automatic testing system that inks bad ICs for later identification
21. *Die separation*: wafer sawed or broken up into individual IC chips or dice
22. *Die mounting*: IC chips bonded to metallic headers, lead frames, or ceramic substrates using a eutectic alloy, solder preform, or epoxy
23. *Lead bonding*: small-diameter wire leads, usually ~1-mil gold wire, bonded from the package leads to the bonding pads on the IC chip
24. *Encapsulation*: IC sealed in a metal can (TO-5), ceramic, or plastic (DIP or dual-in-line) package
25. *Testing*: complete testing of ICs to determine if they meet the device specifications; for high-reliability and military-specification devices an extensive testing program, including operation under full power condition at various temperature extremes for an extended period of time ("burn-in") and temperature cycling, may be carried out

The purpose of the $N^+$ collector contact regions that are formed at the same time as the $N^+$ emitter regions is to ensure a low-resistance, nonrectifying (ohmic) contact between the aluminum metallization and the collector region. In the absence of these $N^+$ regions the alloying of the aluminum contacts will result in a shallow region of silicon underneath the aluminum contact being saturated with aluminum to a concentration of up to around $10^{19}$ cm$^{-3}$. Since aluminum acts as an acceptor dopant in silicon, this region will be converted to P-type if the donor doping in

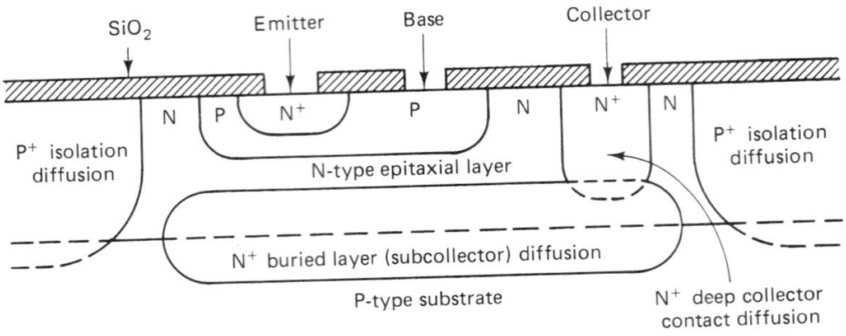


**Figure 2.22** IC NPN transistor with buried layer and deep collector contact diffusion.

this region is less than the aluminum concentration. A parasitic PN junction can thus be formed and will appear in series with the collector of the transistor.

To prevent the formation of this P-type region, the N-type donor doping in the collector region underneath the collector contact metallization should be well in excess of $10^{19}$ cm$^{-3}$. This can easily be done by producing an N$^+$ diffused layer in the collector contact regions by opening oxide windows at the same time that the emitter diffusion windows are opened. The resulting emitter diffusion that takes place will result in a surface concentration in the $10^{21}$ cm$^{-3}$ range of phosphorus so that the contact regions will not be converted to P-type.

For an even lower collector series resistance $r_{cc'}$ the transistor structure of Figure 2.22 can be used, in which there is a deep collector contact diffusion that extends all of the way through the epitaxial layer to intersect the N$^+$ buried layer. This will produce a major reduction in $r_{cc'}$, but at the expense of requiring an additional photolithography and N$^+$ diffusion step.

## 2.9 TRANSISTOR LAYOUT AND AREA REQUIREMENTS

In Figure 2.23 a cross-sectional view and a top view of a single-base-stripe IC NPN transistor is shown. From the top view we see that the total chip area taken up by the transistor is very much greater than the "active area" of the transistor, which is the area of the N$^+$ emitter region, for it is in this area that the electron flow from emitter to collector takes place across the thin base region. In particular, the oxide opening for the P$^+$ isolation diffusion and the lateral spread of the diffusion underneath the edges of the oxide opening will take up an appreciable part of the total transistor area. If the epitaxial layer thickness is designated as $t_{epi}$, this lateral diffusion will be approximately equal to $t_{epi}$ if the P$^+$ isolation diffusion is to be deep enough to intersect the P-type substrate. To make absolutely certain that the P$^+$ isolation diffusion will indeed pass all the way down through the N-type epitaxial layer down to the P-type substrate in spite of variations in epitaxial layer thickness and diffusion conditions, the isolation diffusion is extended for a somewhat longer period of time. The resulting lateral diffusion under the oxide mask will be $nt_{epi}$ where $n > 1$ and typically around 1.5.

A minimum clearance requirement is needed between the various diffused re-

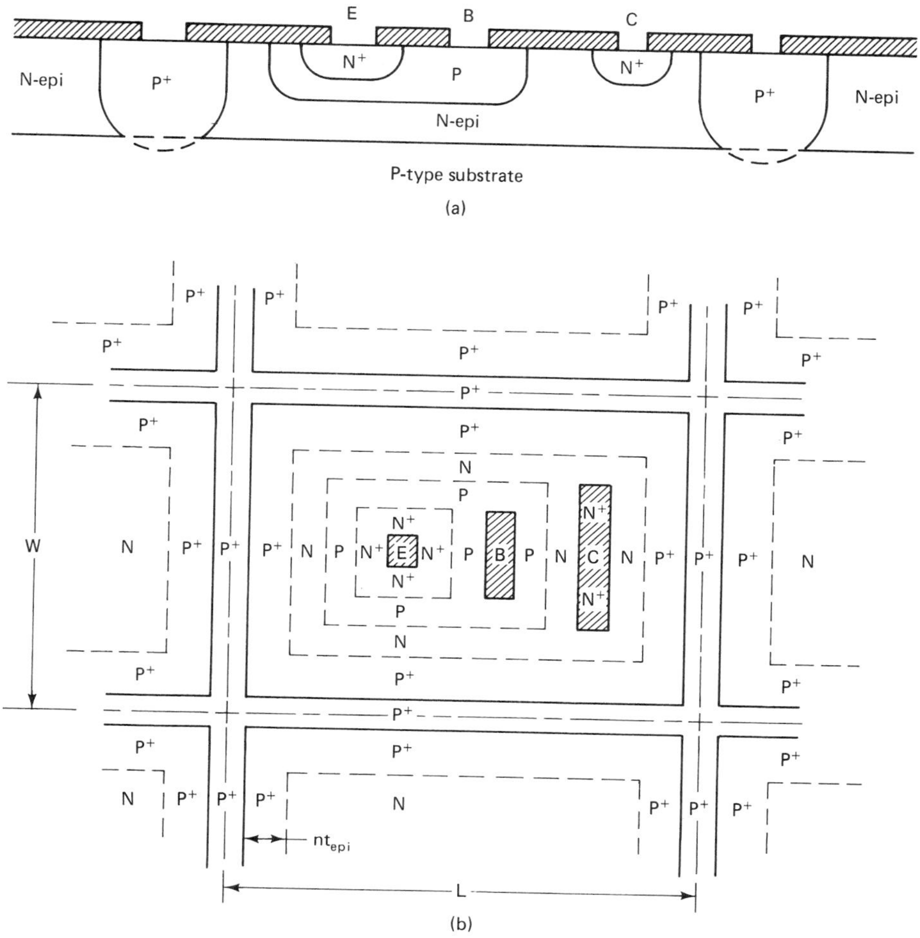


**Figure 2.23** Single-base-stripe IC NPN transistor: (a) cross-sectional view; (b) top view.

gions, contact openings, and metallized contacts to allow for mask registration errors and the minimum line resolution that is available with the photolithographic process. If this minimum clearance or "design rule" is specified as $d$ for all of the above, and if $t_{epi} = d$ and $n = 1.5$, for the transistor of Figure 2.23 we will have the overall dimensions $L = 15d$ and $W = 11d$. The transistor area will be $A = LW = 165d^2$. Of this total transistor area, the active area or $N^+$ emitter region is only $9d^2$, which represents only 5.5% of the total transistor area.

If a 10-$\mu$m design rule is used, the transistor area will be $A = LW = 150$ $\mu$m $\times$ 110 $\mu$m = 0.0165 mm² = 26 mil², corresponding to a device density of 60 transistors/mm². If the photolithography resolution and mask alignment is such that a 5-$\mu$m design rule can be used, the transistor area can be reduced to about 0.004 mm² = 6.4 mil², for a density of about 250 transistors/mm². Thus a small

IC chip measuring only 1 mm × 1 mm (40 mils × 40 mils) can accommodate a large number of transistors.

The base spreading resistance $r_{bb'}$ can be reduced by about a factor of 4 by using a double-base-stripe geometry in which there is a base contact strip on both sides of the emitter. This will, however, be at the expense of increasing dimension $L$ by $2d$, so that the transistor area will now be $A = 17d \times 11d = 187d^2$. For a 10-$\mu$m design rule, this gives an area of 0.019 mm² = 29 mil², for a device density of 53 transistors/mm².

The lowest base and collector series resistance will be obtained by using a ring-type contact geometry for both the base and collector. This type of contact geometry will increase $L$ by $4d$ and $W$ by $8d$, so that $A = LW = 19d \times 19d = 361d^2$. For a 10-$\mu$m design rule, this gives an area of 0.036 mm³ = 56 mil², for a device density of 28 transistors/mm².

## 2.10 PNP TRANSISTORS

The favored type of bipolar transistor for ICs is the NPN transistor, for two basic reasons.

1. The electron mobility in silicon is about 2.5 times higher than the hole mobility, so that the base transit time will generally be shorter in NPN transistors than in PNP transistors. This will lead to a somewhat higher current gain and to improved high-frequency performance for the NPN transistors.

2. The solid-state solubility of the donor dopants, phosphorus and arsenic, is substantially larger than that of the acceptor dopant, boron. It is desirable to make the emitter region much more heavily doped than the base region, so that when the emitter–base junction is forward biased most of the current flow across the junction will be due to charge carriers emitted by the emitter into the base rather than the flow of charge carrier from base to emitter. It is only the charge carriers that are emitted by the emitter into the base that can contribute to the collector current. The opposite flow of carriers from base to emitter only adds to the base current. As a result of the higher solubility of the donor dopants compared to boron, a more efficient emitter–base structure can be obtained in the case of an NPN transistor than in the PNP case, this leading to a higher current gain. As a result the structure of most bipolar ICs is based on the use of NPN transistors.

For many applications, however, PNP transistors are needed at various places in the circuit. In Figure 2.24a an IC PNP transistor that is compatible with the NPN transistor fabrication sequence is shown. No extra processing steps are needed to produce this PNP transistor than those needed for the NPN transistor. The transistor of Figure 2.24a is called a *vertical* or *substrate PNP* transistor. The collector of the transistor is the P-type substrate and the current flow of holes emitted by the $P^+$ emitter is in the downward or vertical direction through the N-type epitaxial base region to the P-type collector.

The base width of the vertical PNP transistor is the epitaxial layer thickness between the $P^+$ diffused layer and the P-type substrate and will generally be about 5 $\mu$m. This is to be compared to the base width of the NPN transistor of only

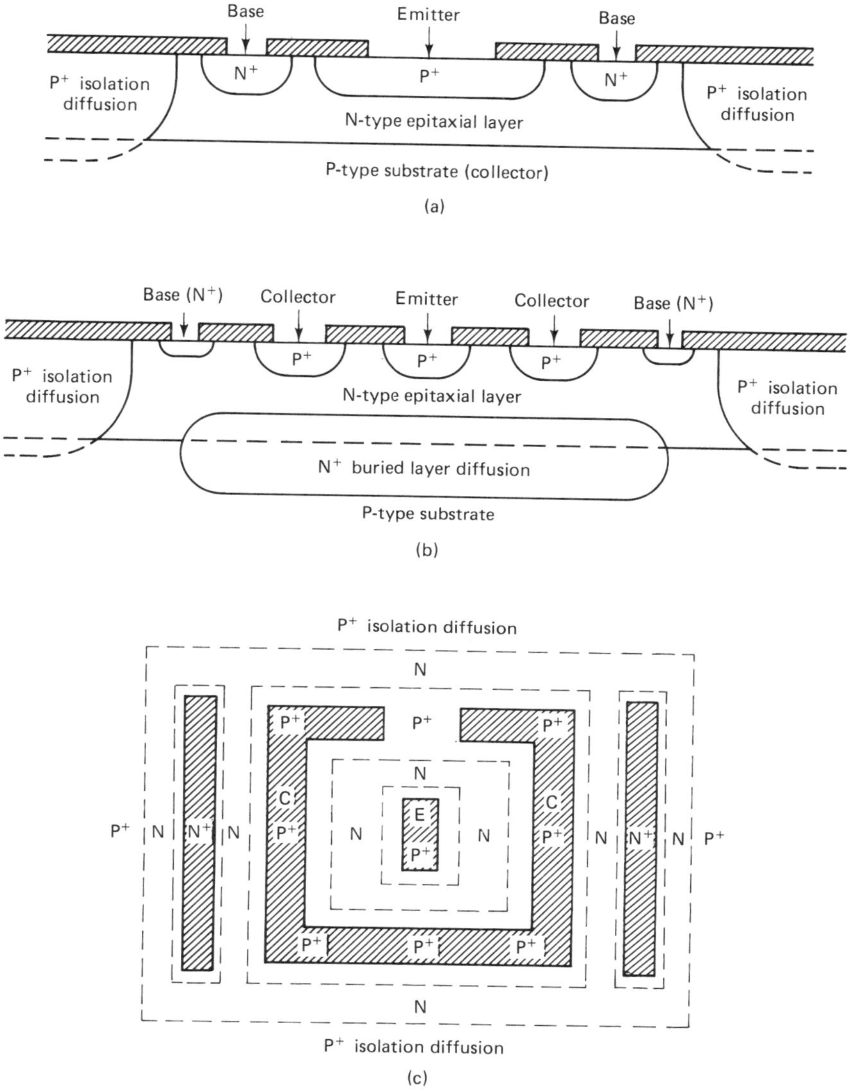


**Figure 2.24** IC PNP transistors: (a) vertical (substrate) PNP transistor; (b) lateral PNP transistor (cross-sectional view); (c) lateral PNP transistor (top view).

about 0.5 μm. The large base width of the PNP transistor will result in a long emitter-to-collector transit time for the holes traveling across the base region, which in turn will produce a low value for the current gain, $\beta$ (~5 to 30), and a poor high-frequency response with $f_T$ values in the range 10 to 30 MHz, compared to about 500 MHz for the NPN transistor.

The collector of this transistor is the P-type substrate, which will be connected to the negative supply voltage or to ground, so that in either case the substrate will be a-c ground. This condition will restrict the operation of this PNP transistor to only the common-collector (or emitter-follower) configuration. In spite of this rather severe limitation, this transistor structure does find application in some ICs.

In Figure 2.24b another PNP transistor structure is shown. This transistor is called a *lateral PNP* transistor and it is also compatible with the NPN transistor fabrication sequence. This PNP transistor is called a lateral PNP transistor because the direction of flow of the holes from the $P^+$ emitter to the $P^+$ collector region is parallel to the surface. The base width of this device is the spacing between the edges of the $P^+$ emitter and collector regions. Because of the limitations of the photolithographic process, this base width will be limited to a minimum of about 3 to 5 $\mu$m. This relatively large base width will again result in a long emitter-to-collector hole transit time across the base region. In addition, some of the holes emitted by the $P^+$ emitter will leave from the bottom area of the emitter and be lost by recombination with electrons in the N-type base region and will therefore not contribute to the collector current. Furthermore, many of the holes traveling across the base region near the surface will be lost to surface recombination and not reach the collector. All of these facts will lead to a low current gain ($\beta \sim 5$ to 20) as compared to $\beta$ values in the range 50 to 200 for NPN transistors. The large base width will also be responsible for poor high-frequency response, with $f_T$ values in the range 1 to 10 MHz. The lateral PNP, however, unlike the vertical PNP transistor, can be used in any circuit configuration.

In Table 2.3 a summary of the characteristics of IC NPN and PNP transistors is presented. It is possible to produce IC PNP transistors with characteristics that are comparable to those of NPN transistors. These higher-performance PNP transistors will, however, require extra processing steps beyond those required for NPN transistors, and so add to cost of the IC.

**TABLE 2.3** TYPICAL IC BIPOLAR TRANSISTOR PARAMETER VALUES

| Transistor type | Current gain, $\beta$ | $f_T$ (MHz) | $BV_{EBO}$ (V) | $BV_{CBO}$ (V) |
|---|---|---|---|---|
| NPN | ~50–200 | ~500 | ~6–8 | ~50 |
| Vertical PNP | ~5–30 | ~10–30 | ~50 | ~50 |
| Lateral PNP | ~5–20 | ~1–10 | ~50 | ~50 |

## 2.11 IC JUNCTION-FIELD-EFFECT TRANSISTORS

In Figure 2.25 some IC JFET structures are shown. The N-channel JFET structure of Figure 2.25a is compatible with the NPN transistor fabrication sequence. Another view of this N-channel JFET is shown in Figure 2.26a, where it is to be noted that the top $P^+$ gate region extends beyond the N-type epitaxial layer region to make contact with the P-type substrate bottom gate. The N-type channel is thus completely encircled by the gate structure and the application of a suitably large negative voltage

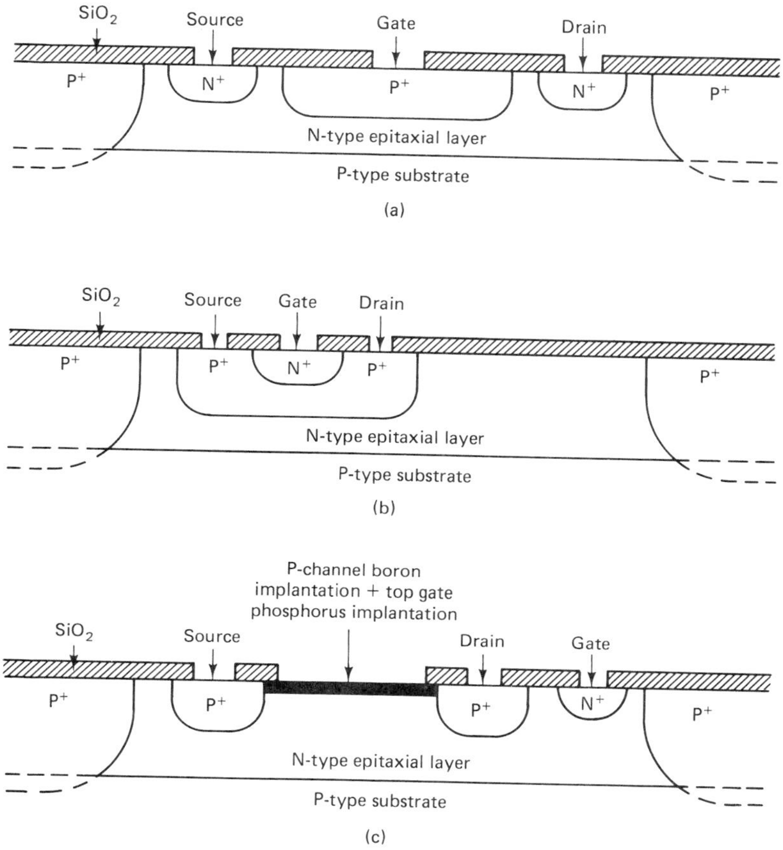


**Figure 2.25** Integrated-circuit JFET structures: (a) N-channel JFET; (b) P-channel JFET; (c) ion implanted P-channel JFET.

to the gate can pinch the channel off and reduce the drain-to-source current to essentially zero. If the $P^+$ top gate did not extend out to overlap the P-type substrate, the N-type channel would not be completely encircled by the gate structure and it would not be possible to cut off the drain-to-source current. The major disadvantage, however, of the JFET structure of Figure 2.26a is that the gate is connected to the P-type substrate, which is at a-c ground potential. This can severely restrict the use of this JFET to only the common-gate configuration.

Another N-channel JFET is shown in Figure 2.26b. The fabrication of this JFET is similar to the one just considered, the principal difference being in the top surface geometry. In this JFET the $P^+$ top gate is in the form of an annular ring that completely encloses the drain region of the JFET. The only current path source to drain will be underneath the $P^+$ top gate. Therefore, the application of a suitably large negative voltage to the top gate can pinch off the channel and reduce the drain-to-source current to essentially zero.

A P-channel JFET is shown in Figure 2.25b. The $N^+$ gate region of this device

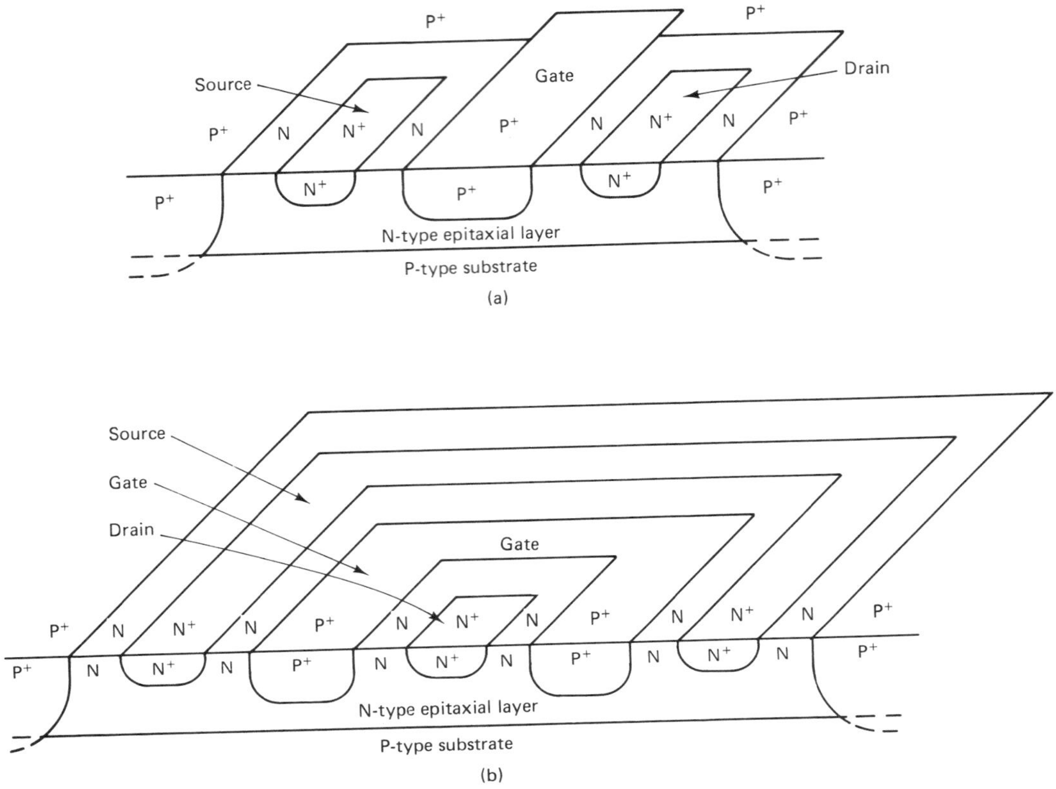


**Figure 2.26** IC N-channel JFET geometries: (a) common top and bottom gate structure; (b) annular ring gate structure.

extends out beyond the $P^+$ source/drain/channel region to overlap the N-type epitaxial layer so that the gate completely encircles the channel. The same processing sequence can be used for this JFET as for the NPN transistor, but if this is done, the gate-to-channel breakdown voltage (corresponding to $BV_{EBO}$) will be down in the range 6 to 8 V and the full pinch-off of the channel may not be possible. As a result, a specially tailored low-concentration boron diffusion will be required to produce a higher gate-channel breakdown voltage and lower channel doping, so that the channel can be pinched off at a voltage that is conveniently less than the breakdown voltage. This, however, involves some extra processing steps and adds to the cost of the IC.

In Figure 2.25c a P-channel JFET that features a boron-ion-implanted channel region is shown. Since the ion implantation dosage can be very precisely controlled, the JFET parameters, such as the pinch-off voltage $V_P$ and the value of $I_{DSS}$ ($I_{DS}$ at $V_{GS} = 0$), can be closely set to the values desired. This JFET uses the same processing steps as the NPN transistor, with the addition of a photolithography, boron ion implantation, and annealing step.

## 2.12 IC MOSFETS

In Figure 2.27 some P-channel MOSFETs sharing a common N-type silicon substrate are shown to illustrate the self-isolating characteristics of these devices. The oxide underneath the gate electrode will be only about 500 to 1000 Å thick, compared to the field oxide, which will be about 1.0 to 1.5 $\mu$m thick. The application to the MOSFET gate of a negative voltage that is larger than the threshold voltage $V_T$ will produce a P-type surface inversion layer underneath the gate oxide. This P-type inversion layer will serve as a conducting channel between the $P^+$ source and drain regions. The voltage required to invert the N-type silicon under the thick field oxide will be very much greater than the gate threshold voltage $V_t$, which is generally in the range $-2$ to $-10$ V. This is a direct result of the greater thickness of the field oxide compared to the very thin gate oxide. The voltage required to invert the N-type silicon underneath the field oxide will be in excess of maximum negative voltage in the circuit, so that the metallization on top of the field oxide will not be able to produce an inversion layer. There will thus not be any P-type conducting channels formed between adjacent MOSFETs. Therefore, the MOSFETs are self-isolating and no special isolation diffusions or isolated regions are necessary. As a result of this and the simple geometry of the MOSFETs, the area required for a MOSFET on the IC chip will be much smaller than that for a bipolar transistor.

In Figure 2.28 a cross-sectional view and a top view of a minimum-area P-channel MOSFET (PMOS) is shown. If a design rule dimension of $d$ is assumed for all clearances and spacings, the overall PMOS transistor dimensions will be given approximately by $L = 7.5d$ and $W = 3d$ for an area of $A = LW = 22.5d^2$ for each transistor. This is to be compared to the area of $165d^2$ for the minimum size (single base strip–single collector stripe) bipolar transistor, so we see that a much higher transistor density will be available with MOSFETs.

For a 10-$\mu$m design rule, the PMOS transistor area will be $A = 0.0023$ mm$^2$ = 3.5 mil$^2$, for a density of 444 transistors/mm$^2$. If a 5-$\mu$m design rule can be used, the transistor area can be reduced to about 0.0006 mm$^2$ = 0.9 mil$^2$, for a device density of 1800 transistors/mm$^2$. A 1 cm × cm IC chip therefore can contain as many as 180,000 transistors.

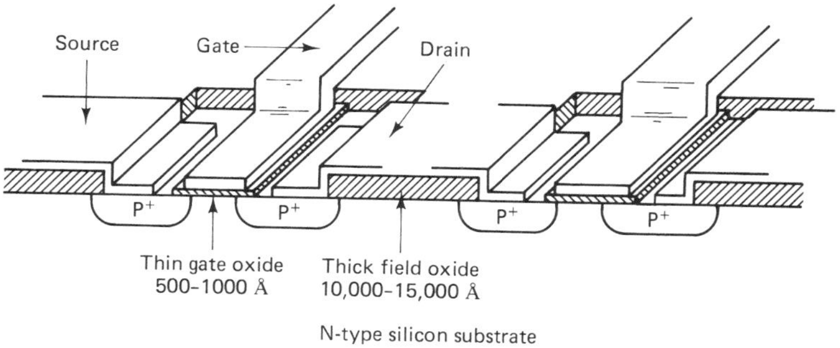


**Figure 2.27** Self-isolating characteristic of MOSFETs.

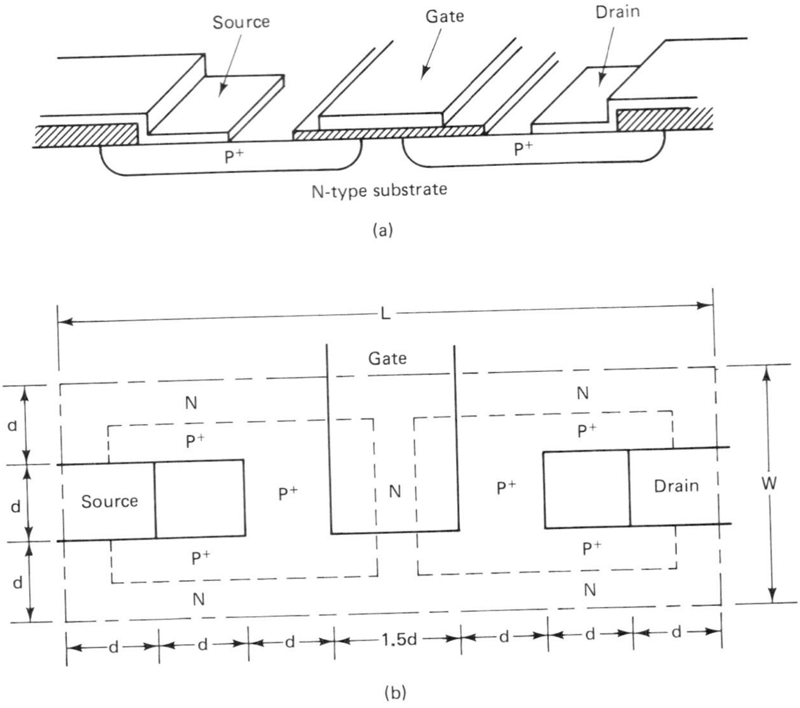


**Figure 2.28** (a) P-channel IC MOSFET; (b) top surface layout of P-channel MOSFET.

With PMOS devices there will usually be no problem with device isolation, but there can be a problem with NMOS transistors, especially on the more lightly doped substrates. Positive ions (such as $Na^+$ and $K^+$) trapped in the oxide can act to reduce the threshold voltage such that conducting channels may be formed between the various NMOS transistors. To prevent this, $P^+$ guard rings or "channel stoppers" may be used around the NMOS transistors. Another technique is to use a boron ion implantation to increase the net acceptor doping level in the surface region of the P-type substrate between the NMOS transistors so that the surface is prevented from becoming inverted. Some guard ring structures for IC NMOS transistors are shown in Figure 2.29.

### 2.12.1 Depletion Mode MOSFETs-FET Transfer Characteristics

The type of MOSFET discussed up to this point is the enhancement mode MOSFET in which the conducting channel between source and drain is produced by the action of the electric field that results from the application of a gate voltage. If the gate voltage is greater than the threshold voltage $V_t$ then a surface inversion layer will be formed at the silicon surface adjacent to the silicon-gate oxide interface. This surface inversion layer will be of opposite conductivity type to that of the silicon

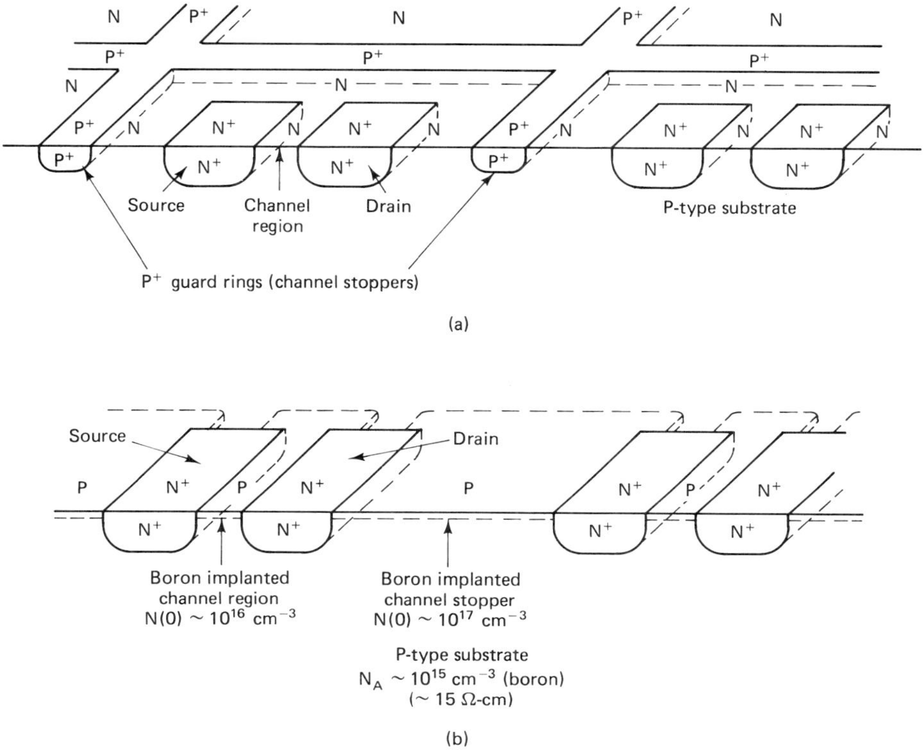


**Figure 2.29** Guard ring structures for NMOS transistors: (a) NMOS structure using diffused $P^+$ channel stoppers; (b) NMOS structure using a boron-ion-implanted channel stopper.

substrate (or body) and constitutes a conducting channel between the source and drain regions.

In contrast to the enhancement mode MOSFET, a *depletion mode* MOSFET has a "built-in" channel between source and drain. This built-in channel is not a field-induced channel, but rather is formed by producing a shallow, lightly doped region at the surface, connecting the source and drain and is of the same doping type as the source and drain regions. This channel can conveniently be formed by means of a shallow, low dosage ion implantation. The depletion mode MOSFET (or D-MOSFET) combines some of the features of JFETS and enhancement mode MOSFETs. Taking an N-channel D-MOSFET as an example, the application of a gate-to-source voltage of positive polarity will draw additional electrons into the already existing N-type channel. This will increase the conductance of the channel and thereby increase the drain-to-source current. Application of a negative voltage will produce the opposite effect. Electrons will be repelled from the channel and the channel will become thinner. This will decrease the conductance of the channel and decrease the drain current. The gate voltage at which the channel becomes fully depleted of electrons is called the *pinch-off voltage* ($V_P$ or $V_{GS(OFF)}$), just as in the

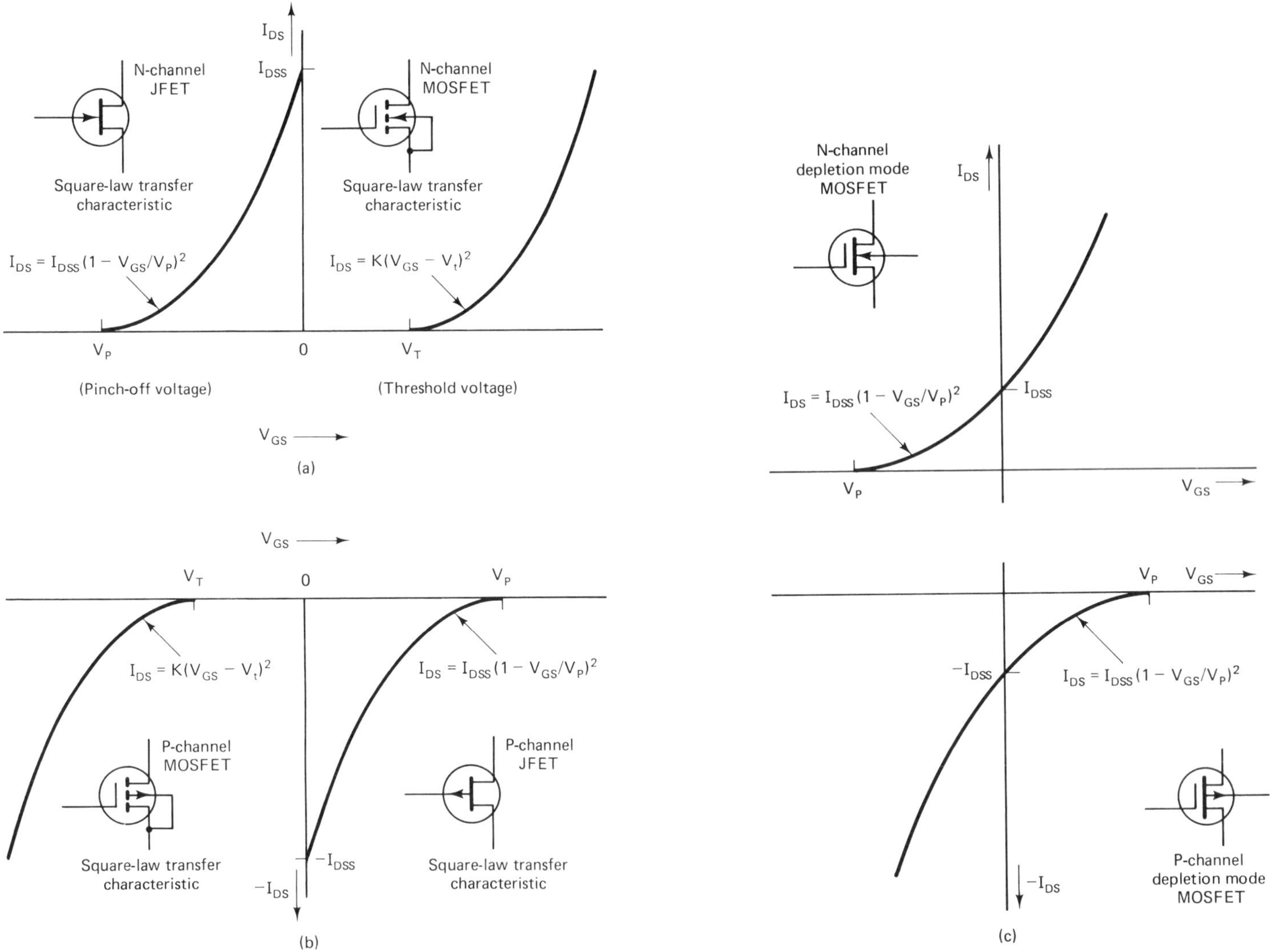


**Figure 2.30** Field-effect-transistor transfer characteristics: (a) N-channel FETs; (b) P-channel FETs; (c) depletion mode MOSFETs.

case of the JFETs. Depletion mode MOSFETs can be used as active loads and as current regulator diodes in various MOSFET circuits.

In Figure 2.30 graphs are presented of the $I_{DS}$ versus $V_{GS}$ transfer characteristics of the various types of FETs. Note that the D-MOSFETs have a transfer characteristic very similar to that of JFETs, except that both polarities of $V_{GS}$ are permitted and as a result currents that are in excess of $I_{DSS}$ can be obtained. Note also the symbols used for the enhancement mode MOSFETs and the depletion mode MOSFETs. The solid line between the source and drain regions in the depletion mode MOSFET is symbolic of the preexisting channel, whereas the dashed line for the enhancement mode devices indicates a channel that is produced only by means of the application of a gate voltage in excess of the threshold voltage.

From Figure 2.30 we note that both the JFETs and the MOSFETs exhibit a "square-law" transfer characteristic. An exception of this general rule, however, is that of the short-channel ($L \lesssim 1$ $\mu$m) MOSFETs for which an approximately linear transfer characteristic is obtained.

## 2.13 COMPLEMENTARY-SYMMETRY MOSFETS

In Figure 2.31 a complementary-symmetry pair of MOSFETs (CMOS) is shown. The CMOS pair consists of an NMOS transistor and a PMOS transistor, both sharing the same N-type silicon substrate, together with other CMOS transistor pairs.

The NMOS transistor is made by first producing a deep P-type boron-doped diffused layer with a low surface concentration called the "P-well." The source and drain regions are then formed by a $N^+$ phosphorus diffusion. The PMOS transistor is formed by the conventional processing sequence.

In Figure 2.32 a CMOS structure is shown in which there are $P^+$ and $N^+$ guard rings surrounding the NMOS and PMOS devices, respectively. This is to prevent a "latch-up" condition that could otherwise occur as a result of a four-layer $N^+$-P-N-$P^+$ switching action.

In Figure 2.33 a diagram of the basic CMOS transistor circuit is given together with the individual $I_{DS}$ versus $V_{DS}$ output characteristics for the NMOS and PMOS devices. In Figure 2.33c the characteristics of the NMOS and PMOS devices are superimposed with a family of curves for various values of the input voltage. In Figure 2.33d the corresponding CMOS transfer characteristic is given.

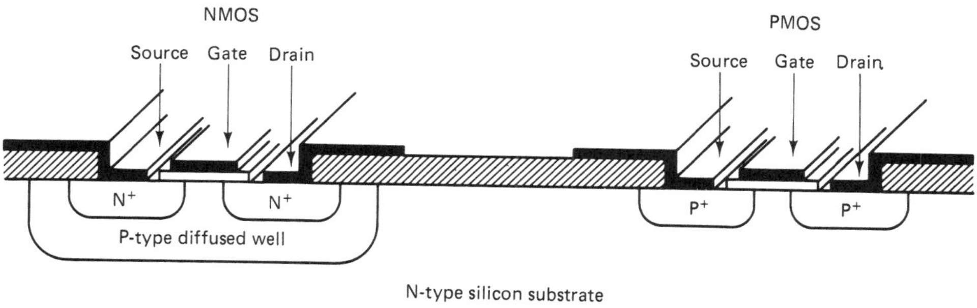


**Figure 2.31** Complementary-symmetry MOSFETs (CMOS).

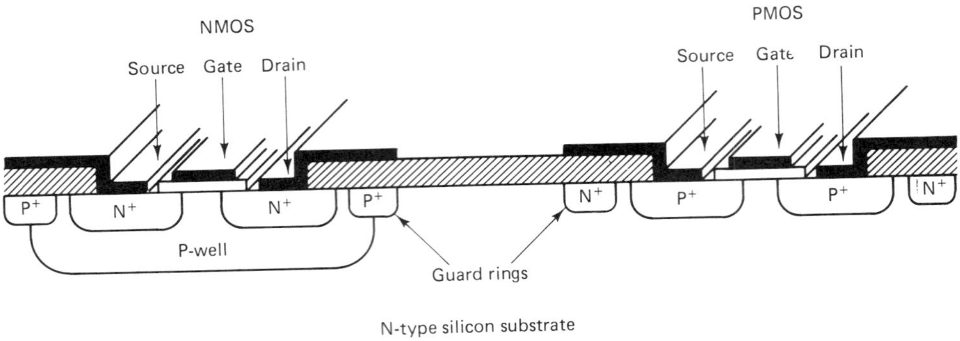


**Figure 2.32** CMOS with guard rings.

When the input voltage $V_{IN}$ is "low" such that the NMOS transistor is off, the PMOS device will be on and the output voltage will be pulled up to the "high" state near the positive supply voltage $V_{DD}$. When $V_{IN}$ goes "high" the NMOS is now turned on and the PMOS goes off, so that the output voltage will now drop down to the low state near ground potential.

In either of the foregoing conditions one of the transistors of the CMOS pair will be off, so that the current through the CMOS pair will be negligible. If the load driven by this CMOS circuit is CMOS devices or MOSFETs, the only significant power consumption will occur during the switching transitions. This very low power dissipation of CMOS devices is a principal advantage of this transistor configuration, especially for large digital systems. Another favorable feature of the CMOS circuit is the large logic voltage swing, with the high-state output voltage being very close to $+V_{DD}$, and the low-state output voltage dropping to very close to either ground potential or the negative supply voltage $-V_{SS}$.

In Figure 2.34 a basic CMOS inverter is shown and the corresponding transfer characteristic is given in Figure 2.35. In Figure 2.36 a CMOS NAND gate is shown, and a CMOS NOR gate is presented in Figure 2.37.

## 2.13.1 CMOS Silicon-on-Sapphire (CMOS/SOS)

In Figure 2.38 a heteroepitaxial CMOS structure is shown. A thin (~1 $\mu$m) single-crystal silicon epitaxial layer is deposited on a highly polished single-crystal sapphire substrate. The silicon film is doped $N^+$ and $P^+$ by phosphorus and boron ion implantations, followed by an annealing or drive-in diffusion step, as required. The thin silicon film is then etched into many separate NMOS and PMOS devices and interconnected by the metallization pattern.

This CMOS/SOS structure is an example of a dielectrically isolated IC, and the use of the insulating substrate results in a greatly reduced parasitic capacitance. This leads to very high-speed device performance and very low power dissipation. These features are especially useful for very high-speed, high-density ICs.

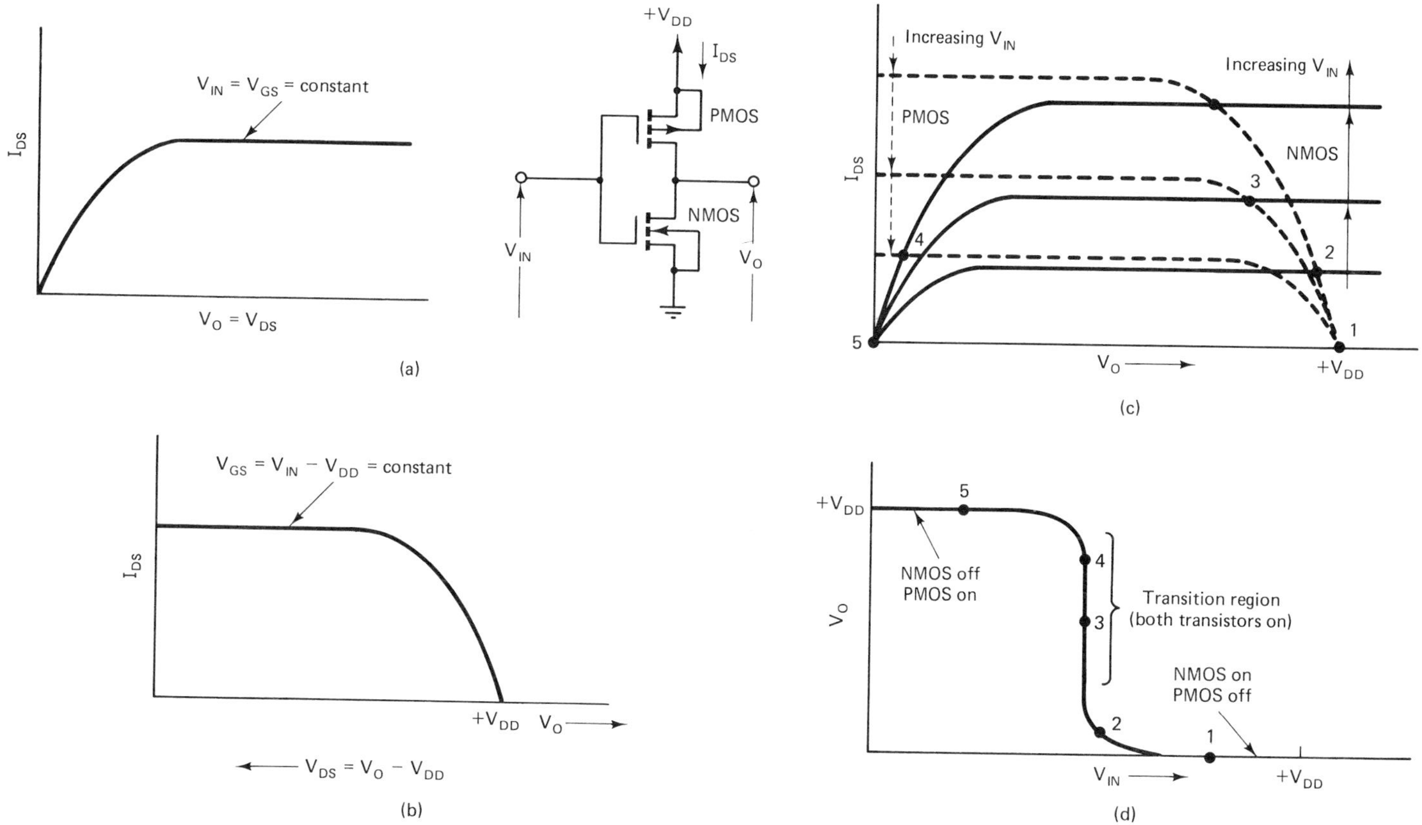


**Figure 2.33** CMOS $I_D$ versus $V_{DS}$; characteristics: (a) NMOS $I_{DS}$ versus $V_{DS}$; (b) PMOS $I_{DS}$ versus $V_{DS}$; (c) composite NMOS and PMOS characteristics; (d) CMOS transfer characteristics.

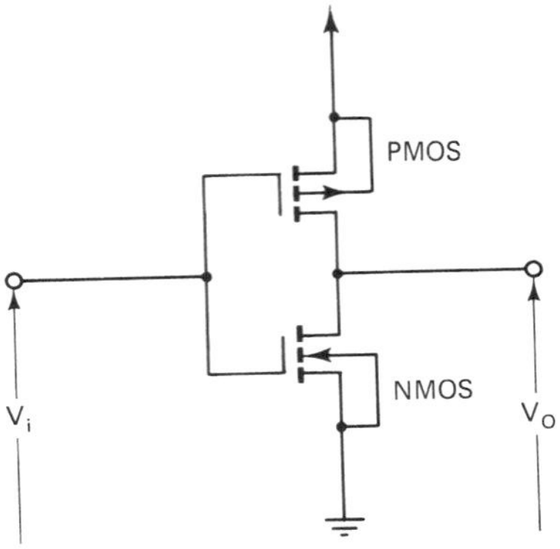


**Figure 2.34** CMOS (complementary MOS) inverter.

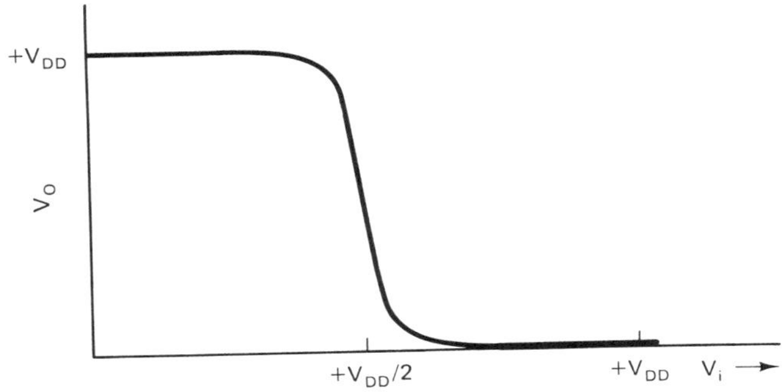


**Figure 2.35** CMOS transfer curve.

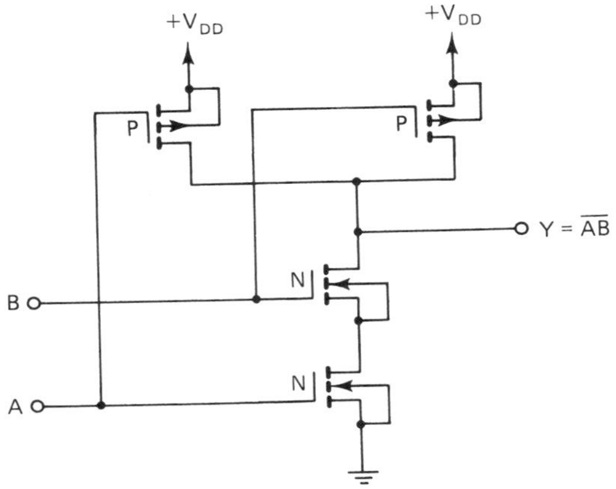


**Figure 2.36** CMOS NAND gate.

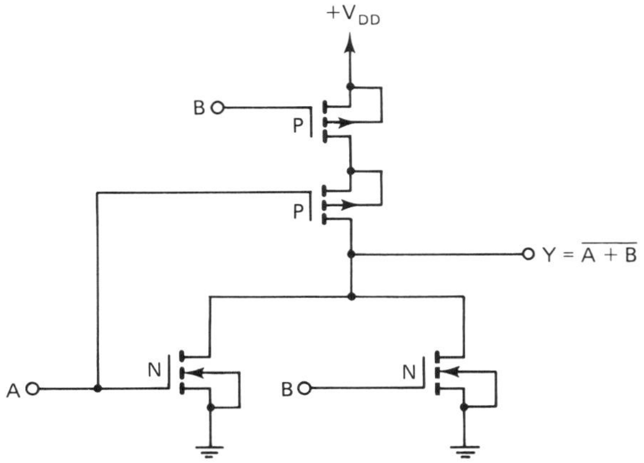


**Figure 2.37** CMOS NOR gate.

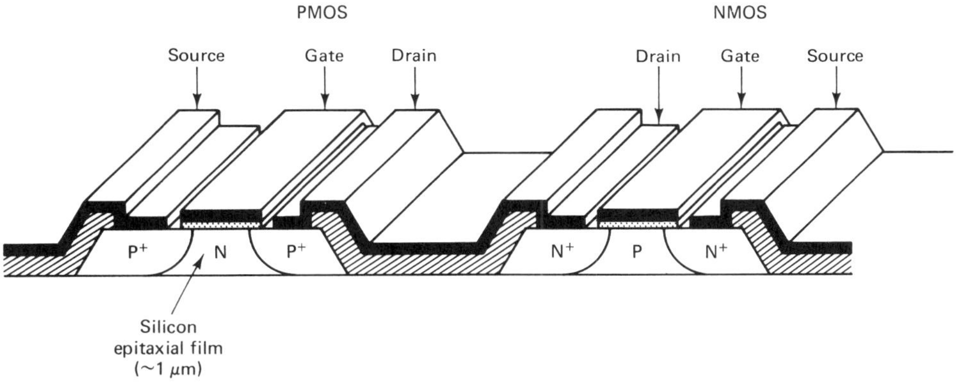


**Figure 2.38** CMOS silicon-on-sapphire (CMOS/SOS).

## 2.14 IC DIODES

There are a number of IC diode structures that are compatible with the NPN transistors; that is, no additional processing steps are required beyond those for the NPN transistor. Six basic diode configurations are shown in Figure 2.39, together with some of the basic characteristics as series resistance, forward voltage drop, breakdown voltage, and storage time. The base spreading resistance $r_{bb'}$ will generally be about 100 Ω and the collector resistance $r_{cc'}$ will be about 30 Ω. The emitter–base breakdown voltage $BV_{EBO}$ will be in the range 6 to 8 V, and the collector–base breakdown voltage $BV_{CBO}$ will usually be about 50 V.

From Figure 2.39 we see that the $V_{CB} = 0$ diode configuration 4 will offer the lowest series resistance and forward voltage drop, and also the lowest storage time. The small value of minority carrier storage time will mean a fast switching transition from forward-bias conditions to reverse bias conditions. The breakdown voltage is, however, limited to $BV_{EBO}$ or only 6 to 8 V. Thus for all applications in

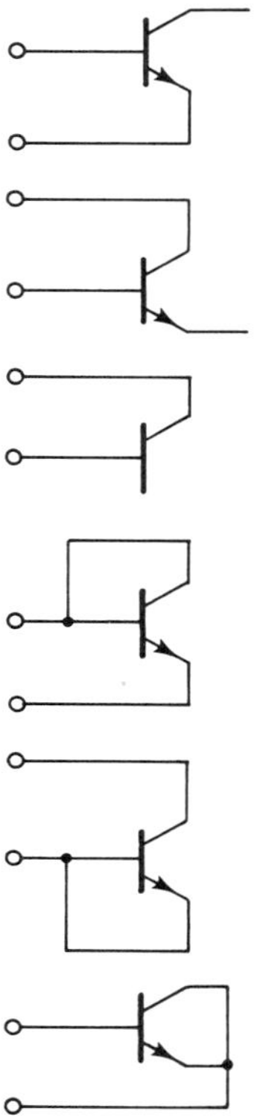

| | Diode configuration | Series resistance | $V_F$ at 10 mA (mV) | Breakdown voltage | Storage time (ns) |
|---|---|---|---|---|---|
| 1. | $I_C = 0$ | $r_{bb'}$ | ~ 960 | $BV_{EBO}$ | ~ 70 |
| 2. | $I_E = 0$ | $r_{bb'} + r_{cc'}$ | ~ 950 | $BV_{CBO}$ | ~ 130 |
| 3. | $I_E = 0$ (no emitter $N^+$ diffusion) | $r_{bb'} + r_{cc'}$ | ~ 950 | $BV_{CBO}$ | ~ 80 |
| 4. | $V_{CB} = 0$ | $r_{bb'}/\beta$ | ~ 850 | $BV_{EBO}$ | ~ 6 |
| 5. | $V_{EB} = 0$ | $r_{bb'}/\beta_R + r_{cc'}$ | ~ 940 | $BV_{CBO}$ | ~ 90 |
| 6. | $V_{CE} = 0$ | $r_{bb'}$ | ~ 920 | $BV_{EBO}$ | ~ 150 |

**Figure 2.39** Diodes for integrated circuits.

which the diode will not be subjected to reverse-bias voltages in excess of about 6 V, this will be the preferred diode configuration.

In this diode circuit ($V_{CB} = 0$) most of the diode forward current $I_F$ will be the collector current, the base current $I_B$ being smaller than the collector current by a factor of $\beta$, the transistor current gain. As a result, the diode forward voltage drop due to the base spreading resistance $r_{bb'}$ will be given by $I_B r_{bb'} = (I_C/\beta) r_{bb'} =$

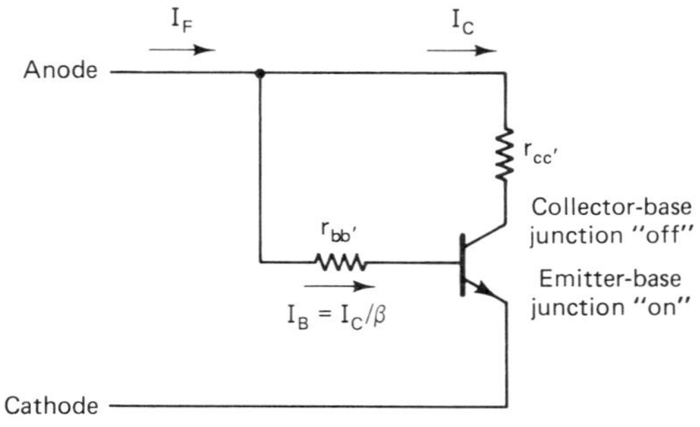


**Figure 2.40** IC diode ($V_{CB} = 0$) series resistance.

$I_F(r_{bb'}/\beta)$. Since the collector current will remain relatively independent of the collector voltage as long as the transistor operates in the active mode of operation, the voltage drop due to the collector series resistance $I_C r_{cc'}$ will not affect the diode voltage drop. Therefore, the effective series resistance of the diode will be approximately $r_{bb'}/\beta$, which usually will be only a few ohms. The $I_B$ and $I_C$ components of the total diode forward-bias current flow $I_F$ are illustrated in Figure 2.40, in which the resistances $r_{bb'}$ and $r_{cc'}$ are shown as elements external to the transistor, although they are really inside the transistor structure.

### 2.14.1 Storage Time and Schottky-Barrier Diodes

The switching speed or *reverse recovery time* of diodes is limited by the storage time of the minority carriers. When a PN junction is forward biased there will be electrons injected from the N-type side of the junction into the P-type side, where they become minority carriers, and similarly there will be holes emitted by the P-type side into the N-type side, where they also become minority carriers. As the injected minority carriers travel deeper into each side, the population of these injected minority carriers will steadily drop off with increasing distance as a result of recombination with the majority carriers. The average lifetime of these injected minority carriers will range from much less than 1 $\mu$s for heavily doped ($N^+$ or $P^+$) silicon to about 10 $\mu$s for light to moderately doped silicon.

If the bias voltage across the PN junction is now suddenly switched from forward bias to reverse bias, many of the injected minority carriers can turn around and start flowing back to the opposite side of the junction, from which they first were injected. Thus a large reverse current can flow through the diode for a period of time called the *storage time*, which will often constitute the major portion of the reverse recovery time of the diode. The excess minority carrier density that is responsible for this large reverse current flow will dissipate after a short period of time due to the withdrawal of carriers back into their region of origin and due also to the recombination of the minority carriers with the majority carriers.

The short storage time of the $V_{CB} = 0$ diode configuration is due to the fact that when the emitter–base junction is forward biased most of the current flow across the emitter–base junction will be the result of electrons emitted from the $N^+$ emitter into the P-type base region, rather than vice versa. The electrons emitted into the base will then very rapidly travel across the very short base region and be collected by the collector. The transit time of the electrons across the base region will be very short ($\lesssim$100 ps), so that there will not be many electrons at any one time in transit across the base region. It is because of this very small density of minority carriers (electrons) in the base region that this diode configuration will have a very short storage time ($\sim$6 ns). This is to be compared to the storage time of some 70 ns for the $I_C = 0$ diode configuration. In this case there will be no net flow of electrons out of the collector and there will be consequently a buildup of minority carrier (electrons) density in the base region. In Figure 2.41 a diagram is presented of the minority carrier distribution in the transistor operating in the active mode, as is the case for the $V_{CB} = 0$ diode configuration.

A Schottky-barrier diode is shown in Figure 2.42a. In this type of diode when

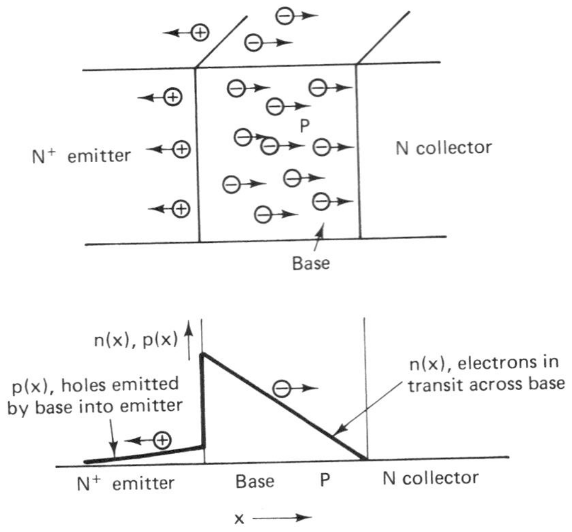


**Figure 2.41** Minority carrier storage for transistor in active mode of operation.

the metal-semiconductor contact (Schottky-barrier) is forward biased there will be only the injection of electrons from the N-type semiconductor into the metal. For every electron injected from the semiconductor into the metal there will be a corresponding electron flowing out of the metal, around the external circuit, and back into the semiconductor. There will therefore be no minority carrier injection or storage, and the storage time for the Schottky-barrier diodes will be essentially zero. These diodes will thus exhibit a very fast switching action.

In addition to the fast switching characteristics of Schottky-barrier diodes, the forward voltage drop will be about one-half of that of the PN junction diodes, typically being about 250 to 350 mV for currents in the range 1 to 10 mA as shown in Figure 2.42b.

In the Schottky-barrier diode of Figure 2.43 there is a large localized intensification of the electric field strength at the edges of the oxide window due to the sharp curvature of the metal contact in this region. This can lead to a decreased breakdown voltage. The Schottky-barrier diode of Figure 2.44 has a $P^+$ guard ring around the edge of the oxide window. The purpose of this guard ring is to spread out and decrease the electric field intensity in the region of the oxide window edge, and thereby increase the breakdown voltage and reduce the junction leakage current. An IC Schottky-barrier diode is shown in Figure 2.45.

Schottky-barrier diodes are often placed across the collector–base junction of transistors as a clamping diode to increase the transistor switching speed in going from the saturation region to the active region. Since the voltage required to turn on a Schottky-barrier diode is only around 0.25 V compared to 0.5 V for the collector–base PN junction, the collector–base junction will be prevented from turning on when the transistor–Schottky-barrier clamping diode combination goes into saturation. As a result there will be no injection of electrons (for the NPN case) from the collector back into the base, and therefore the base will not be inundated with a great increase in the minority carrier (electrons) density as would otherwise be the

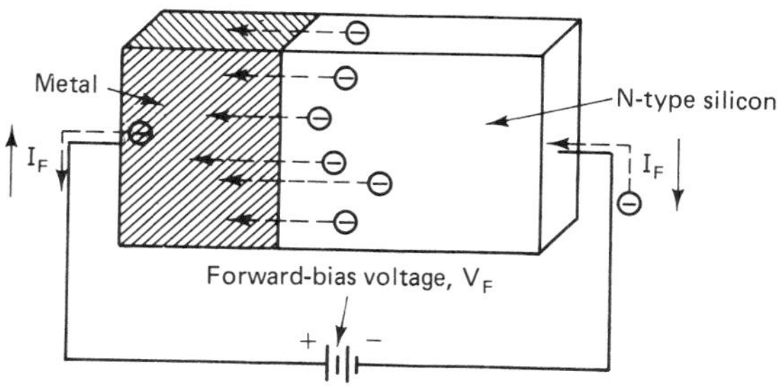


(a)

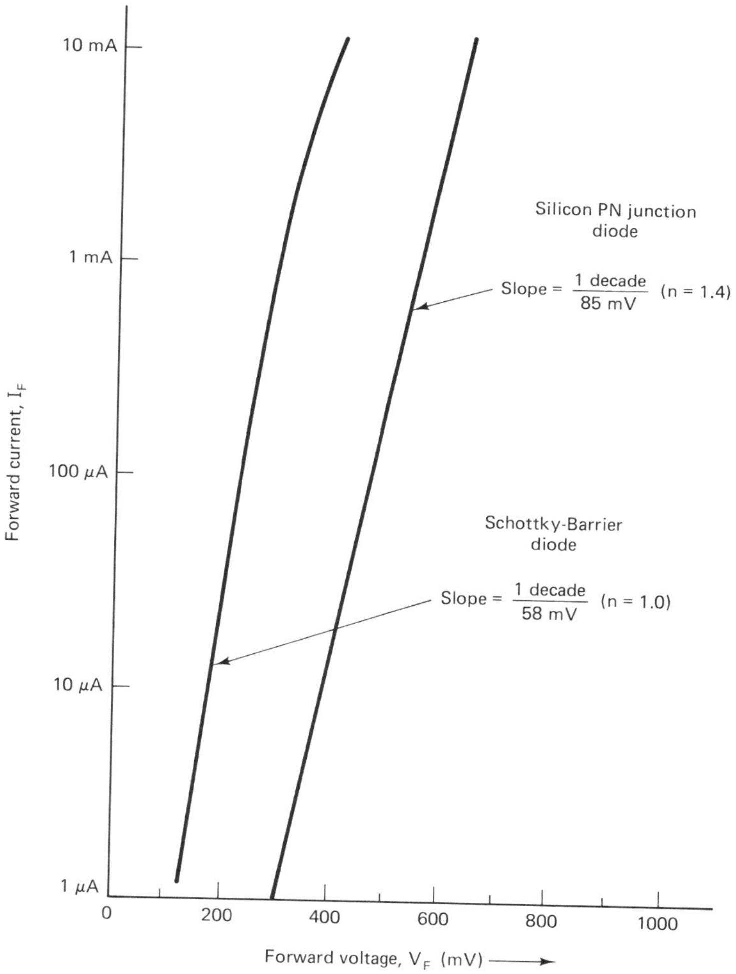


(b)

**Figure 2.42** (a) Forward-biased Schottky-barrier diode; (b) comparison of Schottky-barrier and PN-junction silicon diode characteristics.

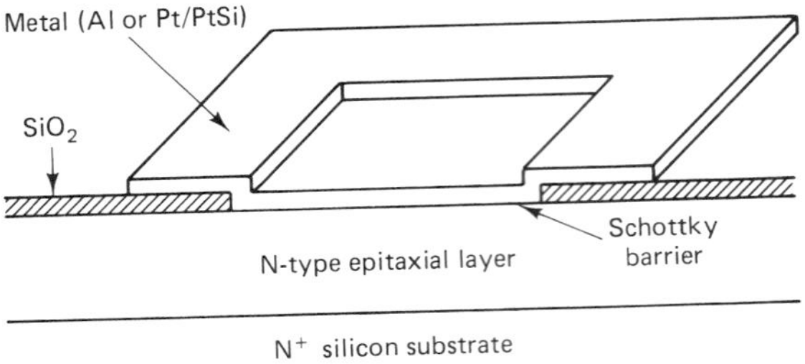


**Figure 2.43** Schottky barrier diode.

case. Consequently, the transistor storage time will be kept small and it will be possible to bring the transistor rapidly out of the saturation region.

The Schottky-barrier diode is usually incorporated as an integral part of the transistor structure, as is illustrated in Figure 2.46. Schottky diode clamped transistors are used in many switching-type circuits, including voltage comparators and TTL logic gates such as the 54LS, 74LS, 54S, and 74S series of TTL gates.

### 2.14.2 Zener Diodes

There are three PN junctions associated with the IC NPN transistor structure: the emitter–base, collector–base, and collector–substrate junctions. The collector–base and collector–substrate junctions will have breakdown voltages of around 50 V. Since this will be in excess of the voltages available in the circuit, these two PN junctions cannot be used as a zener or voltage regulator diode in ICs. The emitter–base junction, however, will have a breakdown voltage in the range 6 to 8 V and it therefore can be used as a zener diode, as shown by Figure 2.47.

The *temperature coefficient of the zener voltage* or $TCV_Z$, given by $TCV_Z = dV_Z/dT$, will be approximately +3 mV/°C for the emitter–base junction. The temperature coefficient of the forward voltage drop across the emitter–base junction, $TCV_{BE} = dV_{BE}/dT$, will be negative and approximately −2.2 mV/°C. The negative temperature coefficient of the forward-biased emitter–base junction can therefore be used to compensate partially for the positive temperature coefficient of the zener diode by the use of a series combination of a zener diode and a forward-biased diode, as shown in Figure 2.48a. In Figure 2.48b the IC structure of this temperature-compensated zener diode is shown. Note that the two diodes share a common N-type collector region and a common P-type base region.

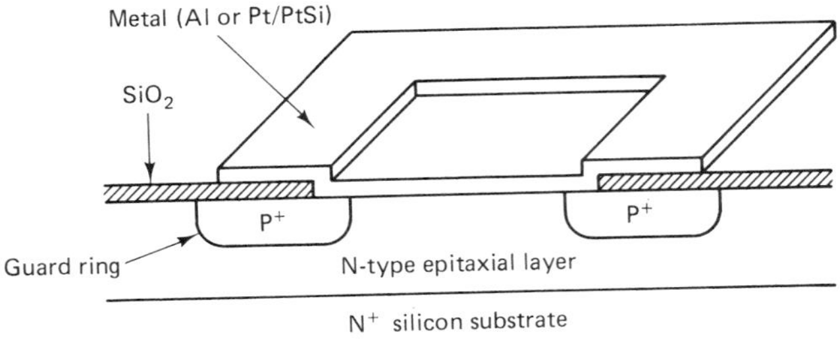


**Figure 2.44** Schottky barrier diode with guard ring.

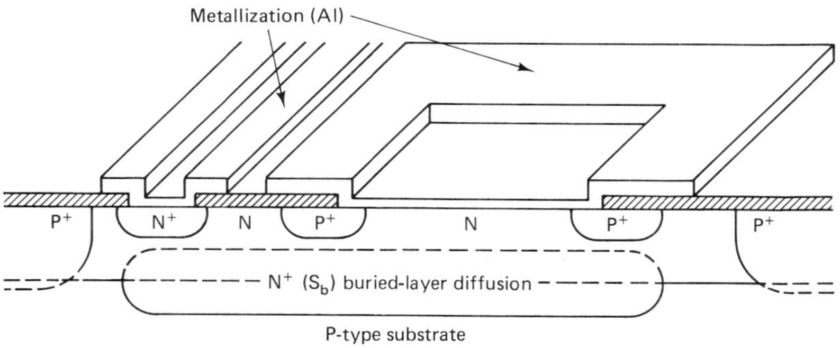


**Figure 2.45** Integrated-circuit Schottky-barrier diode.

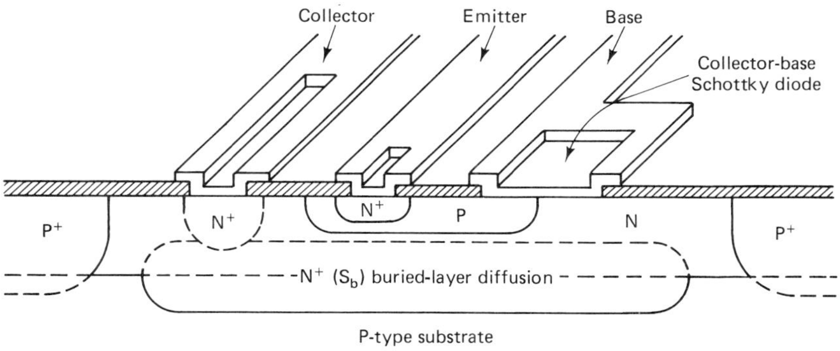


**Figure 2.46** Schottky diode clamped IC transistor.

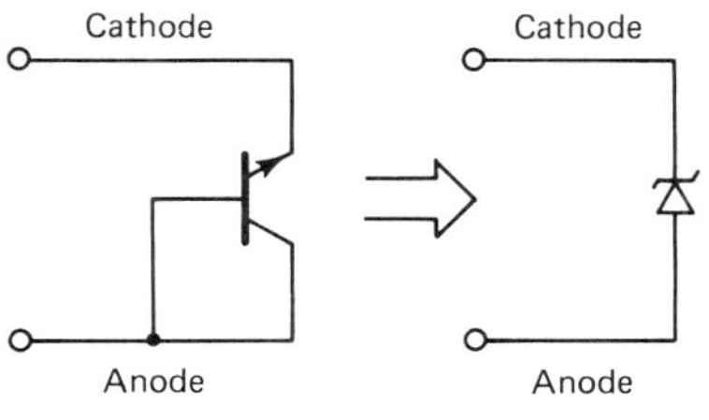


**Figure 2.47** IC zener diode.

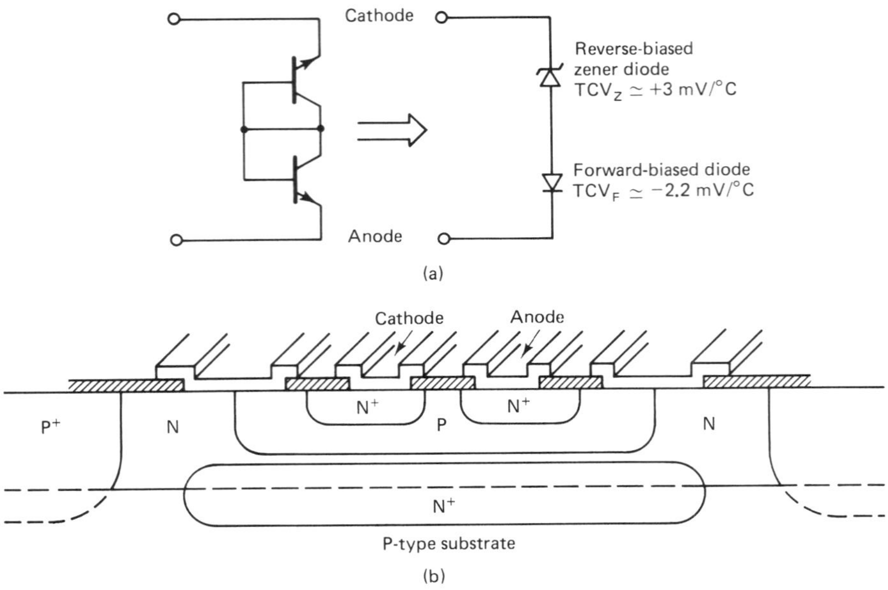


**Figure 2.48** IC temperature-compensated zener diode: (a) temperature-compensated zener diode circuit; (b) IC structure for temperature-compensated zener diode.

## 2.15 IC RESISTORS

In Figure 2.49 a number of different IC resistor configurations are shown together with some typical sheet resistance values. The most widely used resistor type is the diffused resistor, which uses the sheet resistance of the transistor base diffused layer, as shown in Figure 2.49a. The sheet resistance will typically be about 200 Ω/square and the *temperature coefficient of resistance TCR*, given by $TCR = (1/R)\ dR/dT$, will be about +2000 ppm/°C = 2 × $10^{-3}$/°C = 0.2%/°C. In Figures 2.50 and 2.51 the top surface layout of a diffused resistor is shown.

The N-type epitaxial layer isolated region is normally tied to the positive supply voltage and the P-type substrate is connected to the negative supply or ground, so that not only will the N-layer/P-substrate junction be reverse biased, but also the P-resistor/N-layer junction. These two reverse-biased junctions eliminate the possibility of the parasitic PNP (base–collector–substrate) transistor being turned on, and also minimizes the parasitic junction capacitance of the collector–base and the collector–substrate junctions. In addition, many resistors can share the same common N-type isolated region as shown in Figure 2.50b.

The resistance value of a resistor can be expressed in terms of the sheet resistance $R_S$ as $R = R_S(L/W)$, where $L$ is the total length of the resistor and $W$ is the width. The $L/W$ ratio is, in effect, the number of squares in series. There will be statistical deviations in the resistor value from the mean or design center value due in part to variations in the sheet resistance from wafer to wafer, and even over the surface of any given wafer. For resistors of widths less than about 25 μm, however,

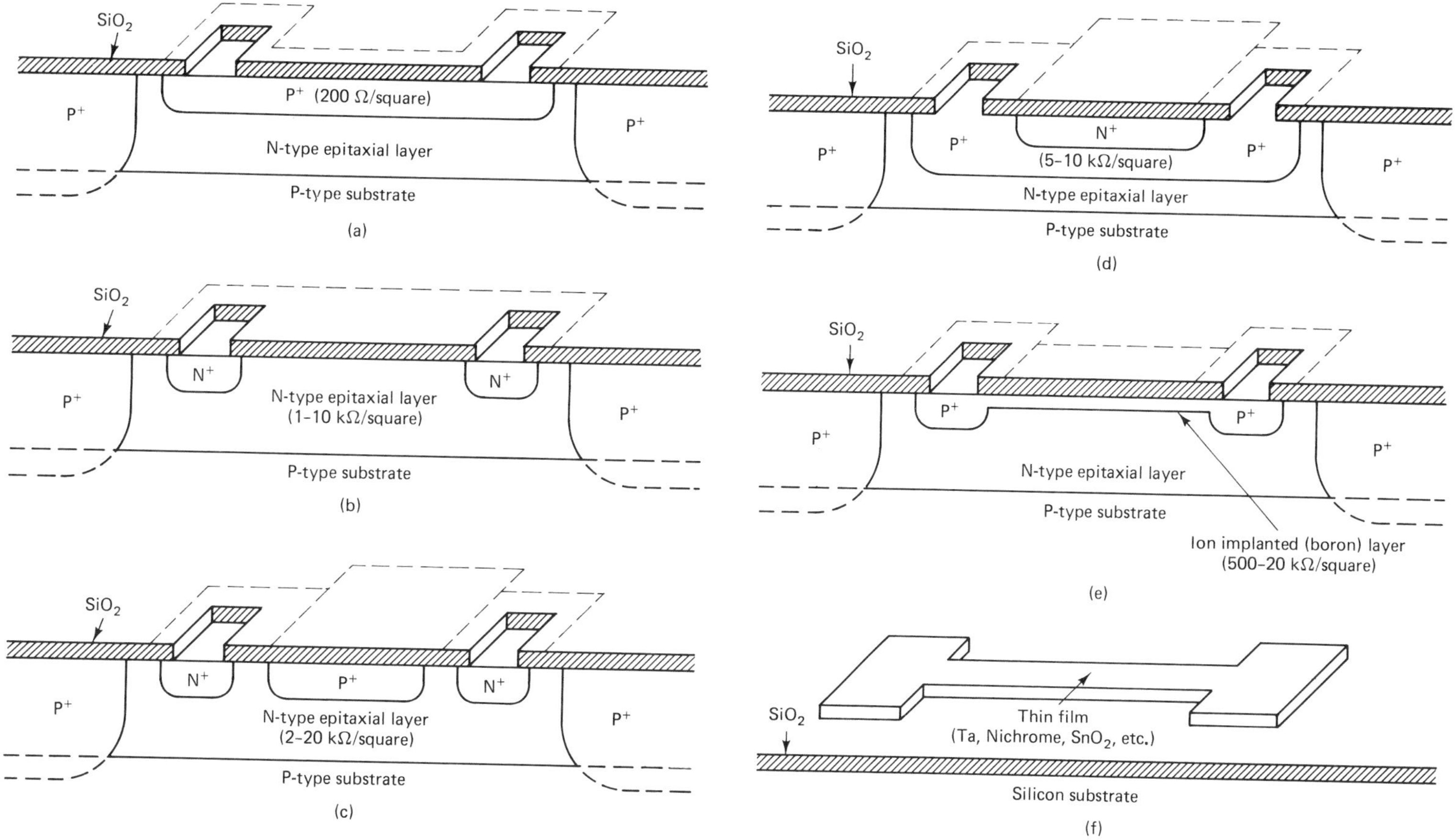


**Figure 2.49** Integrated-circuit resistors: (a) diffused resistor (base diffusion); (b) epitaxial layer resistor; (c) "pinched" epitaxial layer resistor; (d) "pinched" base diffusion resistor; (e) ion-implanted resistor; (f) thin-film resistor.

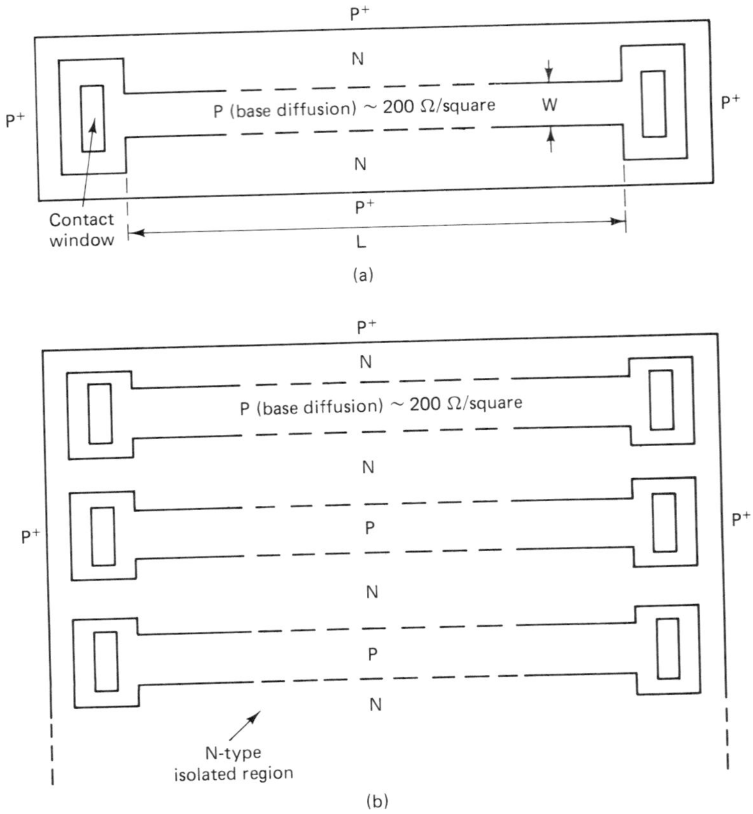


**Figure 2.50** IC diffused resistor top surface geometry: (a) small-value resistor; (b) several resistors sharing the same N-type isolated region.

the principal cause of resistor value deviations will be statistical fluctuations in the line width $W$. These width fluctuations are the result of variations in the amount of oxide undercutting, the extent of the lateral diffusion underneath the oxide, and other variations in the photolithographic process. These statistical deviations will generally be in the range $\pm 1$ or 2 $\mu$m. Thus, if a resistor line width of only 5 $\mu$m is chosen, these variations in the line width can result in a resistor tolerance or standard deviation of as much as $\pm 30\%$. An increase in the line width will result in a reduced standard deviation, but for widths beyond about 25 $\mu$m the standard deviation will level out at about $\pm 5\%$ due to the fluctuations in the sheet resistance.

The greater resistor line widths, however, will be at the price of a larger chip area that is occupied by the resistor. The length of the resistor will be given by $L = (R/R_S)W$, so the chip area occupied by the resistor will be proportional to $W^2$. In Table 2.4 some of the important characteristics of the various IC resistors are listed.

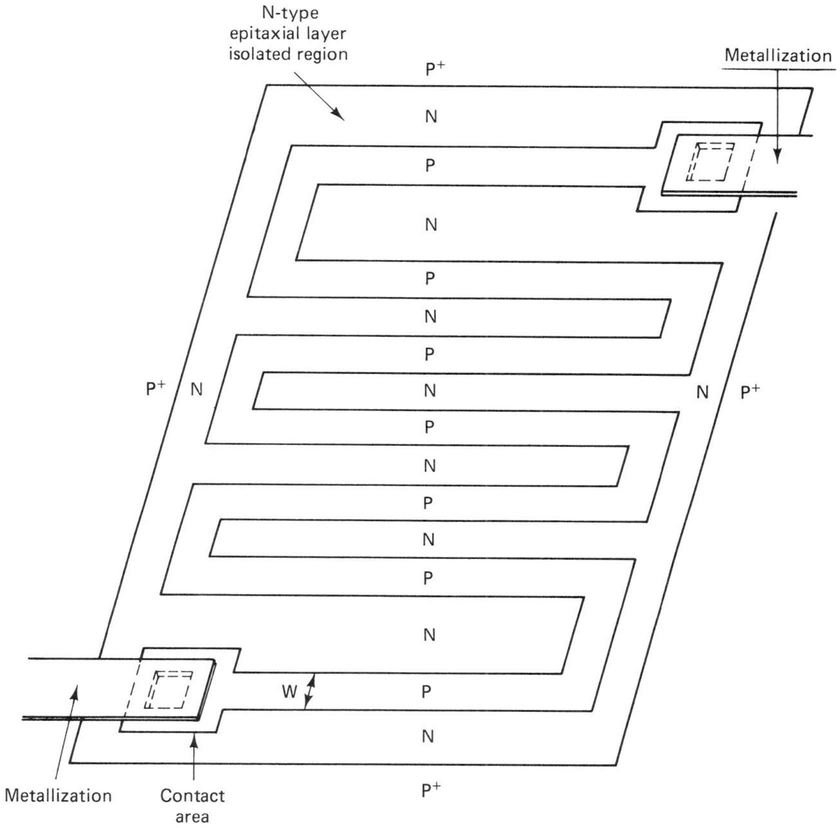


**Figure 2.51** Large-value diffused resistor layout.

**TABLE 2.4** IC RESISTOR CHARACTERISTICS

| Resistor type | Sheet resistance (Ω/square) | Temperature coefficient, TCR (ppm/C) | Tolerance Absolute value (%) | Ratio (%) (1:1) |
|---|---|---|---|---|
| 1. Diffused (base) | 100–300 | +1000 to +3000 | ±10 | ±1 |
| 2. Diffused (emitter) | 2–3 | +100 | ±10 | ±1 |
| 3. Epitaxial layer | 1–10k | +3500 to +5000 | ±30 | ±5 |
| 4. "Pinched" expitaxial | 2–20k | +4000 | ±50 | ±10 |
| 5. "Pinched" base diffusion | 5–10k | +3000 to +5000 | ±40 | ±6 |
| 6. Ion implanted | 500Ω–20k | +200 to +1000 | ±6 | ±2 |
| 7. Thin film | | | | |
| Tantalum (Ta) | 200Ω–5k | 100 | ±5 | ±1 |
| Nichrome (Ni-Cr) | 40–400 | 100 | ±5 | ±1 |
| Tin Oxide ($SnO_2$) | 80Ω–4k | 0 to −1500 | ±8 | ±2 |

The *ratio tolerance* of an IC resistor is the statistical deviation in the ratio of two similar-type IC resistors on the same chip from the mean or design center value. Since the resistors on the same chip that are thus being compared are within a very small distance from each other and have undergone identical processing, it is to be expected that the resistor characteristics will indeed closely match each other. Thus the ratio tolerance will usually be very much smaller than the absolute value tolerance. Considering a pair of IC P-type diffused resistors of equal nominal or design center value, the 1:1 ratio tolerance for wide-line-width ($\geq 25$ $\mu m$) resistors will typically be down in the range $\pm 0.5\%$. If the line width is decreased to 7 $\mu m$, the ratio tolerance will increase to about $\pm 2\%$. Also, as the resistance ratio increases, the ratio tolerance will increase, from 0.5% for a 1:1 ratio to 1.5% for a 5:1 ratio for wide-line-width resistors. For very low-value resistors ($\lesssim 10\ \Omega$) and for diffused cross-overs as shown in Figure 2.57, the $N^+$ emitter diffusion with a sheet resistance of about 2.2 $\Omega$/square can be used.

The epitaxial layer resistor of Figure 2.49b offers a high sheet resistance in the range 1000 to 10,000 $\Omega$/square, but it has a high temperature coefficient and exhibits a large standard deviation. This large standard deviation or tolerance range is the result of variations in the epitaxial layer thickness and doping level. An even higher sheet resistance is available with the "pinched" epitaxial layer resistor of Figure 2.49c. In this case the epitaxial layer resistor cross-sectional area is reduced by the intrusion of the P-type (base) diffused layer. Unfortunately, the resistance standard deviation will be even greater in this case as a result of the greater effect of variations in the epitaxial layer thickness and the additional effect of the variations in the junction depth of the P-type diffused layer.

The "pinched" P-type base diffused layer of Figure 2.49d offers the highest sheet resistance of all of the diffused resistor structures, but at the expense of a very large standard deviation in the resistance value due to the combined effects of variations in the $N^+$ emitter and P-type base diffusion depths.

The ion-implanted resistor of Figure 2.49e offers the combination of a high sheet resistance and a relatively low standard deviation of the resistance value. This is the result of the very close control over the implanted ion dosage that is possible. This is at the expense, however, of an extra photolithographic and ion implantation processing step.

The thin-film resistors of the type shown in Figure 2.49f are deposited on *top* of the $SiO_2$ layer. Relatively high sheet resistances are available with a variety of thin-film materials, together with a low standard deviation of the resistance value. One particularly attractive feature of these nonsilicon resistors are the low temperature coefficients, only about 100 ppm/°C or 0.01%/°C. This is at the expense again of extra processing steps. As a result, these thin-film resistors are not used much on silicon monolithic IC chips, but are used extensively for thin-film hybrid integrated circuits.

An interesting feature of thin-film resistors when used with either silicon monolithic ICs or with hybrid ICs is their ability to be trimmed or adjusted in value. The resistance value can be increased by cutting a slot part way across the width of the resistor, as shown in Figure 2.52, usually by means of a high-energy laser beam. The resistance value can thus be increased to some desired value, usually by

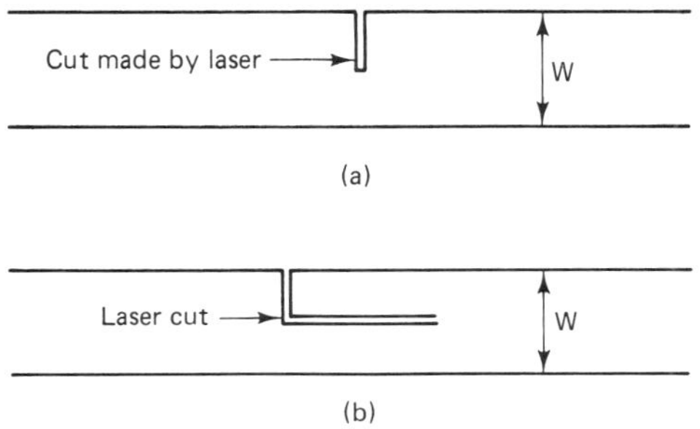


**Figure 2.52** Laser trimming of thin-film resistors: (a) straight cut for small resistance increase; (b) right-angle cut for larger resistance increase.

means of a closed-loop laser trimming system in which the resistance value is monitored during the trimming operation. In many IC circuits a pair of resistors are involved, and the value of one of the resistors is increased to balance the circuit. This technique is often used for offset voltage nulling in high-precision operational amplifiers.

### 2.15.1 Resistor Area

Large-value IC resistors will generally take the form of a meandering pattern, as shown in Figure 2.51. The total resistor length $L$ will be given by $L = (R/R_S)W$, and each corner counts as 0.65 square. If the spacing between the P-type diffused regions is chosen to be equal to the resistor line width $W$, the area required by a large value ($\gtrsim 10$ k$\Omega$) resistor will be given approximately by $A = L \times 2W = 2(R/R_S)W^2$. The resistor line width will generally be in the range of 7 to 25 $\mu$m depending on the resistance tolerance and power density requirements.

For example, a 50-k$\Omega$ diffused resistor with $R_S = 200$ $\Omega$/square and a line width of $W = 15$ $\mu$m (0.6 mil) will require an area on the IC chip of about $A = 2(50\ \text{k}\Omega/200\ \Omega) \times (15\ \mu\text{m})^2 = 0.1125\ \text{mm}^2 = 174\ \text{mil}^2$. This is to be compared to the area required by a single-base-stripe NPN transistor with a 10-$\mu$m design rule, which is 0.0165 mm$^2$, so the resistor occupies an area equivalent to that required by seven bipolar transistors. A MOSFET with a 10-$\mu$m design rule will need an area of only 0.00225 mm$^2$, so that the 50-k$\Omega$ resistor occupies an area that can accommodate about 50 MOSFETs. This area requirement for resistors is one of the reasons for the vast preponderance of transistors over resistors in the design of both analog and digital ICs.

### 2.15.2 Parasitic Capacitance

For the various IC resistors there will be a significant parasitic capacitance distributed along the length of the resistor that will have an important effect on the circuit behavior of the resistor. For the P-type diffused resistor (base region diffusion) and for the P-type ion-implanted resistor there will be a distributed parasitic capacitance along the length of the resistor resulting from the capacitance of the PN junction formed between the P-type layer and the N-type epitaxial layer. In Figure 2.53 an equivalent circuit for the resulting distributed $RC$ structure is shown. The distributed $RC$ transmission line of Figure 2.53a can be approximated by the lumped circuit of

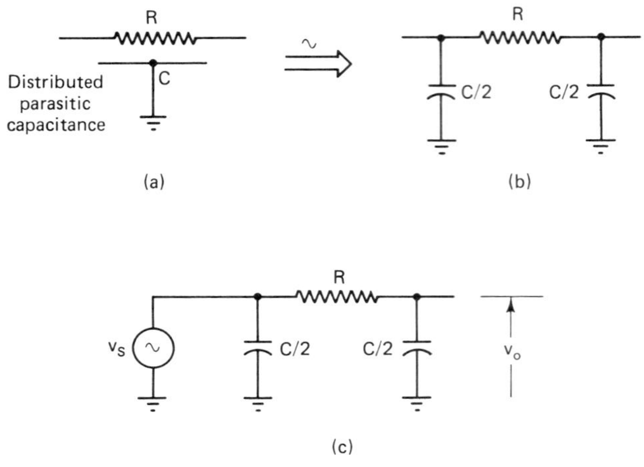


**Figure 2.53** IC resistors: distributed *RC* network: (a) distributed *RC* network; (b) approximate equivalent circuit; (c) approximate equivalent circuit used for evaluation of transfer function.

Figure 2.53b, in which the total parasitic capacitance is divided into two equal parts and concentrated at the two extreme ends of the resistor.

If the approximate equivalent circuit is connected as shown in Figure 2.53c, we see that it will take the form of a simple *RC* low-pass network and the time constant $\tau$ will be $\tau = RC/2$, where $C$ is the total parasitic capacitance. Since $R = R_S(L/W)$ and $C = (C/A)LW$, where $C/A$ is the capacitance per unit area, we have that

$$\tau = \frac{RC}{2} = \frac{R_S(C/A)L^2}{2} = \frac{R_S(C/A)(R/R_S)^2W^2}{2} = \frac{(R^2/R_S)(C/A)W^2}{2} \tag{2.10}$$

The 10 to 90% rise time will be $2.2\tau$ and the 3-dB bandwidth or breakpoint frequency will be $f_1 = 1/(2\pi\tau)$.

**TABLE 2.5** RESISTOR FREQUENCY RESPONSE CHARACTERISTICS

$R_S = 200\ \Omega$/square
$W = 15\ \mu$m (0.6 mil)
$C/A = 100$ pF/mm$^2$

| Resistance value (kΩ) | Time constant | Rise time | Bandwidth |
|---|---|---|---|
| 1 | 113 ps | 248 ps | 1.41 GHz |
| 10 | 11.3 ns | 24.8 ns | 14.1 MHz |
| 20 | 45 ns | 99 ns | 3.5 MHz |
| 30 | 101 ns | 223 ns | 1.6 MHz |
| 50 | 281 ns | 619 ns | 566 kHz |
| 70 | 551 ns | 1.21 μs | 289 kHz |
| 100 | 1.13 μs | 2.48 μs | 141 kHz |
| 150 | 2.53 μs | 5.57 μs | 62 kHz |
| 200 | 4.50 μs | 9.90 μs | 35 kHz |
| 300 | 10.13 μs | 22.3 μs | 16 kHz |

To consider an example, let us use as representative values $R_S = 200\ \Omega$/square, $C/A = 100$ pF/mm$^2$ (at about 10 V bias), and a resistor width of 15 $\mu$m (0.6 mil). In Table 2.5 some values of $\tau$, rise time, and 3-dB bandwidth are listed for various values of total resistance. We see that the parasitic capacitance can be an important factor in limiting the usefulness of large-value resistors. This is an additional reason for the emphasis on using mostly transistors and very few resistors in the design of integrated circuits. In many ICs very few resistors are used, and in some MOSFET ICs no resistors at all are used. We also take note of the important role of the resistor width $W$ and the trade-offs involved between resistor area and high-frequency performance, on the one hand, and the absolute value and ratio tolerances, on the other.

### 2.15.3 Ion Implantation Dosage for Resistors and Base Region

It is of interest to do some calculations of the ion implantation dosage required for various P-type resistors and for the base region of an NPN transistor. The *sheet conductance* $G_S$ of a P-type ion implanted layer of thickness $x_J$ will be given by

$$G_S = \frac{1}{R_S} = \int_0^{x_J} \sigma(x)\, dx = \int_0^{x_J} \mu_p q p(x)\, dx \simeq \int_0^{x_J} \mu_p q N_A(x)\, dx \tag{2.11}$$

where $N_A(x)$ is the acceptor (boron) density in the P-type implanted layer, $p(x)$ is the corresponding hole density, $\mu_p$ is the hole mobility, and $e$ is the electronic charge. The hole mobility $\mu_p$ will be a function of the doping level, decreasing at high doping concentrations due to the scattering of holes by the ionized dopant impurities. The hole mobility will start off at around 450 cm$^2$/V-s for light to moderately doped silicon ($N_A \lesssim 10^{16}$ cm$^{-3}$), decreasing to around 250 cm$^2$/V-s at a doping concentration of $1 \times 10^{17}$ cm$^{-3}$, and about 100 cm$^2$/V-s at doping levels of $1 \times 10^{18}$ cm$^{-3}$.

If we can use an appropriately weighted average value for the hole mobility $\mu_p$, the equation for the sheet conductance can be written as

$$G_S = \overline{\mu}_p q \int_0^{x_J} N_A(x)\, dx = \overline{\mu}_p q \phi \tag{2.12}$$

where $\phi$ is the ion implantation dosage. For the sheet resistance we therefore have that $R_S = 1/G_S = 1/(\overline{\mu}_p q \phi)$.

Let us now find the dosage required for some high-value resistors. Since the doping levels involved will be in the moderate range we will use an average mobility value of $\overline{\mu}_p = 200$ cm$^2$/V-s for this calculation. For an ion-implanted P-type resistor with a sheet resistance of 10 k$\Omega$/square we will have that

$$10\ \text{k}\Omega = \frac{1}{200\ \text{cm}^2/\text{V-s} \times 1.6 \times 10^{-19}\ \text{C} \times \phi} \tag{2.13}$$

so that the required dosage will be $\phi = 3 \times 10^{12}$ cm$^{-2}$. For a sheet resistance of 1 k$\Omega$/square the required dosage will be somewhat more than $\phi = 3 \times 10^{13}$ cm$^{-2}$, due to the decrease in the mobility at the higher doping levels.

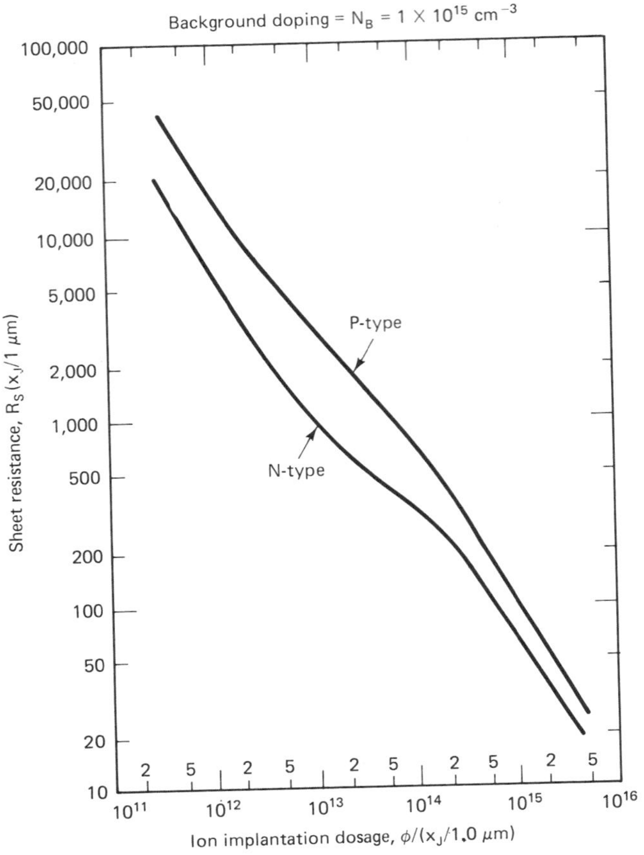


**Figure 2.54** Sheet resistance versus ion implantation dosage in silicon.

Let us now consider the case of the base region of an NPN transistor that is produced by a low-energy boron ion implantation followed by a drive-in diffusion for a resulting sheet resistance of 200 Ω/square. As a result of the higher doping the average mobility in the layer will now be substantially lower than in the previous cases, so a choice of $\mu_p = 100\ \text{cm}^2/\text{V-s}$ will be appropriate. From this we obtain an implantation dosage of $\phi = 4 \times 10^{14}\ \text{cm}^{-2}$.

In Figure 2.54 curves are given of the sheet resistance versus implantation dosage for P-type and N-type layers that are produced by an ion implantation followed by a drive-in diffusion/annealing step. The drive-in diffusion results in a junction depth $x_J$ in silicon with a background doping of $1 \times 10^{15}\ \text{cm}^{-3}$. A background doping that is not $1 \times 10^{15}\ \text{cm}^{-3}$ will generally result in only a small change in the sheet resistance. The effect of a different background doping can be approximately taken into account by changing the ion implantation dosage that is required for a given sheet resistance by $\Delta\phi = \Delta N_B\ x_J$, where $\Delta N_B - (1 \times 10^{15}\ \text{cm}^{-3})$ and $N_B$ is the background doping.

We will now compare the approximate values of the implantation dosage obtained in the preceding examples with the more exact values obtained from Figure 2.54, based on a junction depth of 2.0 $\mu$m. For $R_S = 10$ k$\Omega$/square we obtain $\phi = 1.6 \times 10^{12}$ cm$^{-2}$, for $R_S = 1.0$ k$\Omega$/square the dosage is $\phi = 4.5 \times 10^{13}$ cm$^{-2}$, and for $R_S = 200$ $\Omega$/square the required dosage is $\phi = 4 \times 10^{14}$ cm$^{-2}$.

## 2.16 IC CAPACITORS

There are a number of capacitor configurations available for ICs as shown in Figure 2.55. In Table 2.6 the approximate values for the capacitance per unit area and breakdown voltage for these capacitors are listed. The emitter–base junction capacitor, $C_{BE}$, offers the highest capacitance density, about 1000 pF/mm$^2$, but the breakdown voltage is only about 7 V, which severely limits the use of this capacitor configuration.

The *metal-oxide-silicon* (MOS) capacitor offers the next highest capacitance density, about 350 pF/mm$^2$ for a 1000 Å = 0.1 $\mu$m oxide thickness. The breakdown electric field strength of $SiO_2$ is about 600 V/$\mu$m, so a 1000-Å oxide layer will have a breakdown voltage of about 60 V. As a result of the low resistance of the $N^+$ diffused layer (the emitter region of transistors), the parasitic series resistance of this capacitor will be very low compared to that of the other capacitor configurations. For a typical 50-pF MOS capacitor with $L = W$ and a sheet resistance of the $N^+$ diffused layer of 2 $\Omega$/square, the effective series resistance of the capacitor will be about 1 $\Omega$, and at a frequency of 1 MHz the $Q$-factor will be about 3000.

The area occupied by this 50-pF MOS capacitor will be 0.15 mm$^2$, which will be equivalent to the area needed by about 10 bipolar transistors and about 100 MOSFETs. We therefore see that large capacitances, especially above about 100 pF, will be undesirable components for ICs, due to the large chip area required.

Associated with the various IC capacitors there will not only be a parasitic series resistance, but also a parasitic capacitance, as shown in Figure 2.56. When the capacitors are used as bypass capacitors with one end tied to ground the parasitic capacitance will pose no problem since they will either be shorted out or will just add to the total bypass capacitance. For other circuit applications, however, the para-

**TABLE 2.6** IC CAPACITORS

| Capacitor type | Capacitance per unit area (5V reverse bias) pF/mm$^2$ | pF/mil$^2$ | Breakdown voltage (V) |
|---|---|---|---|
| Collector–base junction, $C_{CB}$ | 125 | 0.08 | 50 |
| Emitter–base junction, $C_{BE}$ | 1000 | 0.65 | 7 |
| Collector–substrate junction, $C_{CS}$ | | | |
| With $N^+$ buried layer | 90 | 0.06 | 50 |
| No $N^+$ buried layer | 60 | 0.04 | 50 |
| MOS capacitor (0.1 $\mu$m oxide thickness) | 350 | 0.22 | 60 |

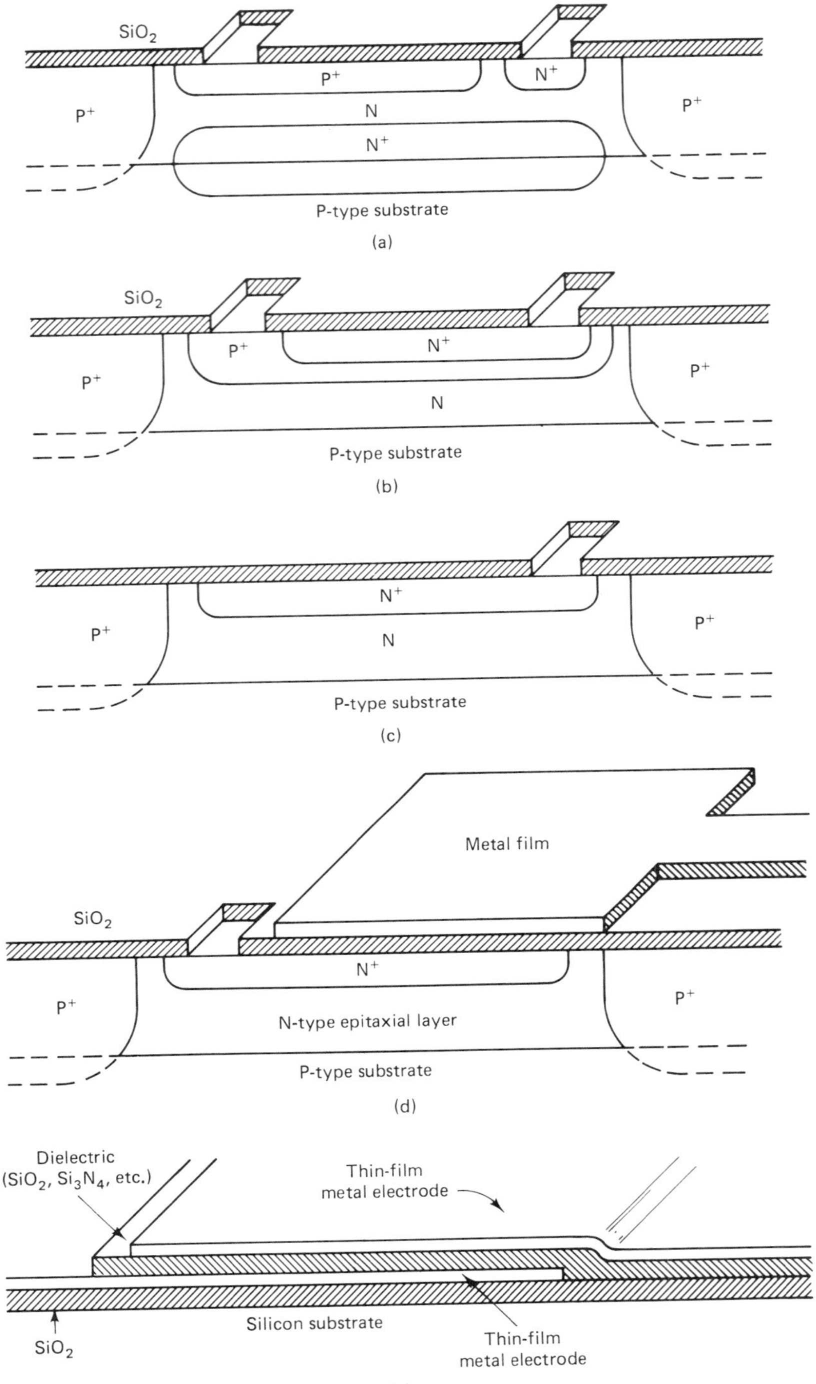


**Figure 2.55** Integrated-circuit capacitors: (a) collector–base junction capacitor; (b) emitter–base junction capacitor; (c) collector–substrate capacitance; (d) metal-oxide-silicon (MOS) capacitor; (e) thin-film capacitor.

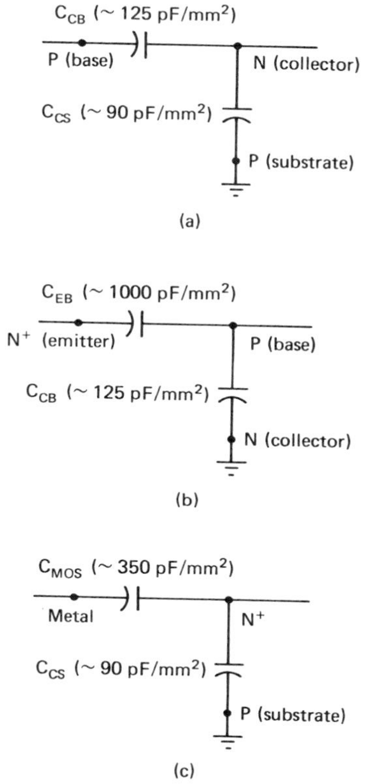


**Figure 2.56** IC capacitor parasitic capacitance: (a) collector–base junction capacitance; (b) emitter–base junction capacitance; (c) MOS capacitor.

sitic capacitance can result in a substantial signal loss. For example, in the case of the collector–base junction capacitance $C_{CB}$ of Figure 2.56a the parasitic collector–substrate junction capacitance is almost equal to the collector–base capacitance, so that the voltage transfer ratio of this capacitor configuration will be only a little better than one-half. For the MOS capacitor of Figure 2.56c the ratio of the MOS capacitance to the collector–substrate capacitance is about 350 pF/90 pF $\simeq$ 4, so that a voltage transfer ratio of about 0.8 will be obtained.

A further advantage of the MOS capacitor is that its capacitance is substantially independent of the voltage across the capacitor. This is unlike the situation with the junction capacitors ($C_{CB}$, $C_{BE}$, and $C_{CS}$), for which the capacitance will be a function of the bias voltage. Furthermore, the MOS capacitor is a *nonpolarized* capacitor which can operate with either polarity of applied voltage, whereas the junction capacitors are polarized in that the PN junction must not be allowed to become forward biased. The MOS capacitor thus offers a number of advantages over the IC junction capacitors. It does, however, require some extra processing steps. These include a photolithographic step to produce an open area in the thick passivating oxide and an extra oxidation process to grow the thin oxide layer, about 1000 Å thick, that serves as the dielectric of the MOS capacitor. Nevertheless, the MOS capacitors are generally the most widely used type of capacitor for monolithic ICs.

## 2.17 IC INDUCTORS

Inductors pose a special problem for ICs. IC devices are essentially two-dimensional, in that the depth dimension is usually very small ($\sim$1 to 10$\mu$m) compared to the lateral dimensions. IC inductors can be made in the form of a flat metallic thin-film spiral, but the inductance will be only a few nanohenries. The combination of the very low inductance value and the appreciable series resistance of the thin-film metallization spiral will result in $Q$-factors that are very low, so that this type of inductor will be of very limited usefulness. For any reasonable inductance value a three-dimensional coil structure is needed to obtain the large number of turns needed to produce a large magnetic flux density and a very large number of flux linkages.

In most cases the necessity for inductances is eliminated by the design of the IC system. In many cases feedback circuits using $RC$ networks can take the place of $LC$ tuned circuits or produce a net input admittance that looks inductive. For other applications, such as RF and IF circuits, where inductors are an absolute necessity, inductors external to the IC package will have to be used. An exception to this will be in the case of thin-film hybrid microwave integrated circuits (MICs), where thin-film inductor spirals may be of use.

## 2.18 IC CROSSOVERS

In the conventional IC all of the metallization for device interconnection is in the same geometric plane. In almost all ICs there will be various places in the circuit where there must be crossovers, although the number of crossovers can be minimized by the careful design layout of the IC.

Although all of the metallization is in the same plane, a diffused region in the silicon can be used as a connecting link in one of the metallization runs that would otherwise intersect. Such a crossover is shown in Figure 2.57, in which an $N^+$ layer is used to tunnel under the conductor to form a connecting link for another conductor. This $N^+$ diffused layer can be produced at the same time as the emitter diffusion and thus does not require any additional processing steps. It does, however, occupy a substantial amount of chip area and adds a significant amount of parasitic series resistance and shunt capacitance to the circuit. For an $N^+$ crossover with a sheet resistance of 2 $\Omega$/square, the added series resistance will be about 10 $\Omega$.

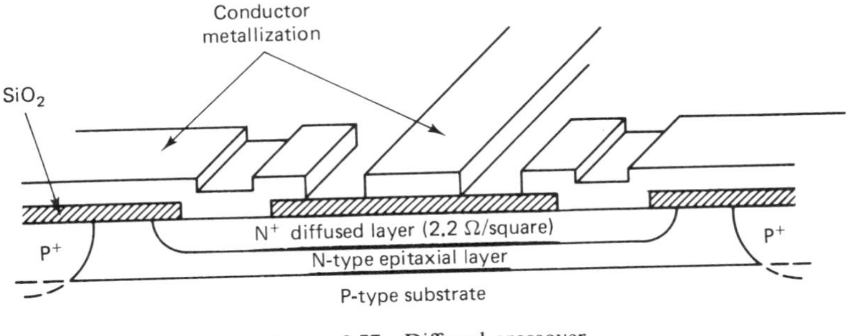


**Figure 2.57** Diffused crossover.

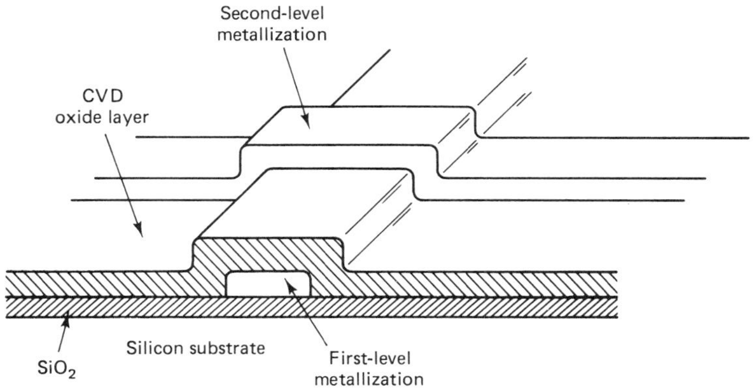


**Figure 2.58** Crossover using two-level metallization.

A diffused crossover that takes up no additional chip area and adds no extra series resistance or shunt capacitance can be obtained by using a P-type resistor that is already part of the circuit. In a high-density IC there will be the need for a great many crossovers. The use of the $N^+$ type of "crossunder" will be very undesirable because of the area requirements and the added series resistance, and there may not be a sufficient number of suitably located P-type resistors in the circuit to be used for crossovers. For this case a multiple-level metallization structure may be the best solution.

A two-level metallization IC is shown in Figure 2.58. The first metallization level is produced by conventional means. Then a low-temperature (~400°C) CVD $SiO_2$ layer is deposited and contact windows or "vias" to the first metallization level are produced by a photolithographic process. The second metallization layer is then deposited and patterned. Some ICs will have as many as three levels of metallization.

## 2.19 DIELECTRIC ISOLATION PROCESSES

In the dielectrically isolated IC the various devices on the chip are electrically isolated from the substrate and from each other by an insulating or dielectric layer. This is to be compared to the junction isolated IC, in which the isolation is by means of reverse-biased PN junctions.

A method for producing a dielectrically isolated IC is illustrated in Figure 2.59. Starting with an N-type substrate, an $N^+$ diffusion is performed. This is followed by the growth of an $SiO_2$ layer, which is then patterned to form a grid of intersecting line openings in the oxide. A cross-sectional view is presented in Figure 2.59a. The wafer is then subjected to an orientation-dependent etching (ODE) process using the patterned oxide layer as the etching mask. This results in the V-shaped grooves as shown in Figure 2.59b, in which the (111) plane sidewalls are at an angle of 54.74° with respect to the (100) top surface of the silicon wafer. The depth of the V-groove will thus be related to the width of the oxide opening by $D = W/\sqrt{2}$.

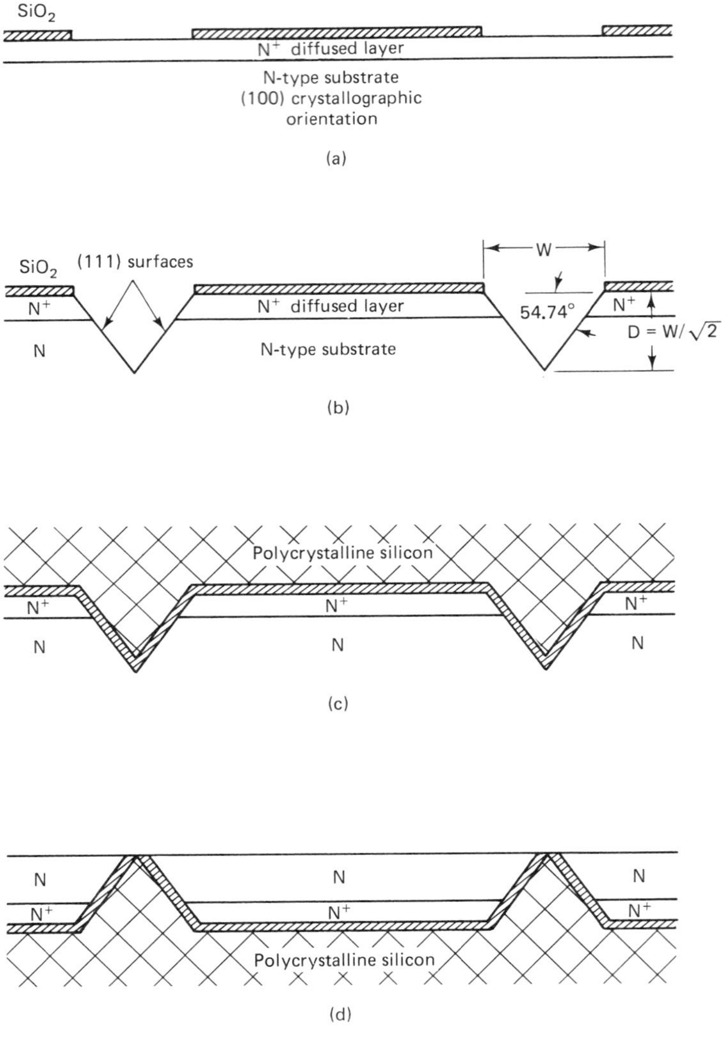


**Figure 2.59** Dielectric isolation for integrated circuits: (a) $N^+/N$ wafers with openings in the oxide layer; (b) orientation-dependent etching to produce V-shaped slots; (c) thermal oxidation followed by deposition (CVD) of polycrystalline silicon layer; (d) lapping of N-type silicon substrate from back side to expose vertices of the V-grooves.

The wafer now undergoes a thermal oxidation process to cover the sidewalls of the V-groove with an oxide layer. This is followed by the deposition of a very thick layer of polycrystalline silicon, as shown in Figure 2.59c. This CVD silicon layer will not be single-crystal because it is not deposited directly on the silicon substrate, but on the $SiO_2$ film, which has an amorphous structure, so that a polycrystalline silicon layer results.

The next step is the most critical and difficult one. The silicon wafers are mounted on a lapping plate with the polycrystalline side of the wafers down, and the N-type silicon substrate is then carefully lapped down to the level at which the vertices of the V-grooves become exposed, producing the result shown in Figure 2.59d. This results in an array of N-type single-crystal silicon regions that are isolated from the polycrystalline silicon substrate and from each other by the thermally grown oxide layer. The polycrystalline silicon now becomes the substrate and serves to provide the mechanical support for the IC. Although the polycrystalline silicon serves no electrical function, it is a very suitable material for this application because its thermal expansion coefficient is a very good match to that of single-crystal silicon and it can withstand the high processing temperatures.

Note that it is very important that the lapping operation remove the N-type silicon all of the way down to the vertices of the V-grooves. If it falls short of this, the N-type regions will be connected and device isolation will not be achieved. If, on the other hand, the lapping operation proceeds too far, the N-type regions will be too thin, or possibly even removed completely. This will require very close control of the lapping process over the entire surface of the wafer, since the wafer diameter will be about 100 mm and the V-groove depth will be only about 10 $\mu$m.

The $N^+$ diffused layer that started off on the top surface of the N-type silicon wafer now ends up at the bottom of the N-type isolated regions. The $N^+$ layer serves as a buried layer to reduce the collector series resistance of the NPN transistors. The rest of the processing sequence for the dielectrically isolated ICs follows along the same line as for the conventional junction isolated IC.

The processing of dielectrically isolated ICs is much more expensive than for junction isolated ICs. The dielectrically isolated ICs are, however, very useful for such applications as high-voltage and radiation-resistant ICs. The dielectric strength of $SiO_2$ is about 600 V/$\mu$m, so that a 5000-Å layer of oxide will result in an N-type isolated region-to-substrate breakdown voltage in the region of 300 V. In comparison, the N-region-to-substrate breakdown voltage of the junction isolated IC is only about 50 V. As a result, with dielectrically isolated ICs, high-voltage transistors and diodes are now possible.

A burst of high-energy ionizing radiation of x-rays or gamma rays can produce very large numbers of excess free electrons and holes in silicon as a result of the photogeneration process. X-rays and gamma rays consist of high-energy photons, above 100 eV for x-rays and in excess of 100 keV for gamma rays. To remove an electron from a covalent bond in silicon and thereby produce a free electron and a hole requires an energy $E_G$ equal to 1.1 eV. Therefore, each x-ray or gamma-ray photon can result in the photogeneration of a large number of free electrons and holes. These excess photogenerated electrons and holes will result in large increases in the leakage current of the PN junctions in the IC. This can produce severe problems in junction isolated ICs, where a burst of ionizing radiation can result in large current transients or spikes. The dielectrically isolated IC is more resistant to the effects of ionizing radiation because of the presence of the insulating oxide layer between the N-type isolated regions and the polycrystalline silicon substrate.

Another type of dielectrically isolated IC is the *silicon-on-sapphire* (SOS) structure, discussed earlier in conjunction with the CMOS/SOS device structure. The crystallographic mismatch that exists at and near the silicon/sapphire interface will result in severe stresses in the silicon and crystallographic defects in the thin (~1-μm) silicon layer. This will produce a very low minority carrier lifetime in the silicon film, especially in the region close to the sapphire substrate. This low minority carrier lifetime will limit the usefulness of the SOS structure for bipolar devices, but this structure will still be very useful for MOS devices, as in the case of the CMOS/SOS configuration. In this case a principal feature provided by the insulating substrate is a large reduction in the parasitic capacitance, thus increasing the high-speed performance of the circuit.

### 2.19.1 Silicon-on-Insulator Technology

A more recently developed technique for dielectrically isolated devices is the *silicon-on-insulator* (SOI) process, in which a thin layer of single-crystal silicon can be produced on top of a thermal $SiO_2$ layer on a silicon wafer. The oxide layer is photolithographically patterned to produce islands or strips of oxide. A thin CVD layer of silicon is then deposited on the wafer. In the regions where the deposited silicon layer overlays the oxide, it will be polycrystalline, but it will be single-crystal in the regions where it is in direct contact with the silicon substrate. The silicon layer is then directionally recrystallized using a scanned laser, electron beam, or resistance-heated strip heater. The silicon film that is in direct contact with the substrate recrystallizes, with the silicon substrate serving as the nucleation center. As the heated zone is scanned across the wafer the crystal growth propagates from these nucleation regions to the regions of the silicon film on top of the oxide islands or strips. The result will be a complete single-crystal layer of silicon.

A related SOI process is the *epitaxial lateral overgrowth* (ELO) process. The starting material is again a thermally oxidized silicon wafer in which the oxide layer has been photolithographically patterned into islands or strips of oxide. A repeated sequence of carefully controlled CVD silicon deposition and vapor-phase etching cycles is then carried out to produce a single-crystal silicon film on the silicon substrate in the exposed regions between the oxide islands. the vapor-phase etching process preferentially removes any polycrystalline silicon that is deposited on top of the $SiO_2$. As successive cycles of the CVD deposition and vapor-phase etching processes continue, the single-crystal silicon that is formed in the oxide windows starts to extend over the adjoining oxide regions, and at the same time any polycrystalline silicon that is deposited on top of the oxide is removed by the vapor-phase etching process. The end result will be the formation of a complete single-crystal layer of silicon. This heteroepitaxial silicon layer can have mobility and minority carrier lifetime values that are comparable to those obtained in homoepitaxial layers and are suitable for the fabrication of good-quality field-effect and bipolar transistors.

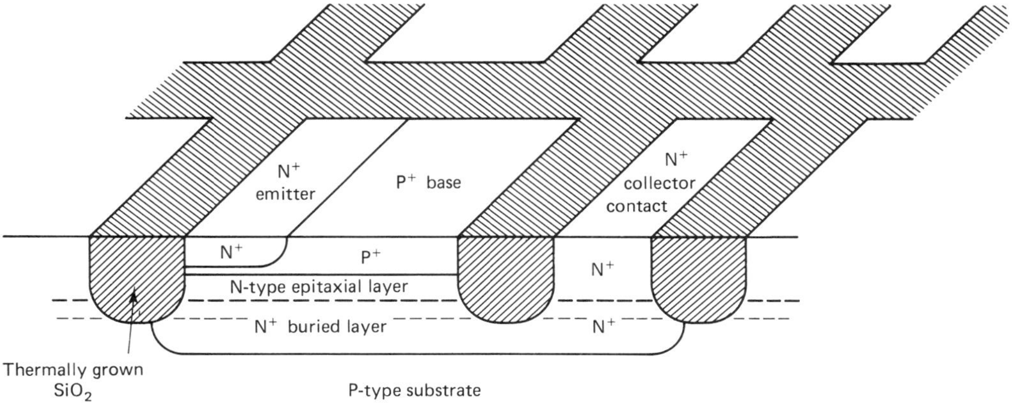


**Figure 2.60** Isoplanar integrated-circuit process using the local oxidation of silicon technique (LOCOS).

## 2.20 ISOPLANAR AND OTHER IC STRUCTURES

An isoplanar IC structure is shown in Figure 2.60. It is made by using the local oxidation of silicon (LOCOS) process, which was discussed earlier. This process produces oxidized isolation moats between the N-type silicon islands. This is not a dielectrically isolated IC, however, since there is a PN junction between each N-type island and the common P-type substrate. The oxide isolation moat can, however, result in a reduced collector-to-substrate capacitance and an increased breakdown volatge due to the absence of the heavily doped $P^+$ regions. A main advantage of the isoplanar process is in the increased device density since the $N^+$ emitter and $P^+$ base regions can be put directly up against the oxide isolation regions as shown in Figure 2.60.

### 2.20.1 Double-Diffused Isolation

In the conventional junction isolated IC the $P^+$ isolation diffusion must extend downward all the way through the N-type epitaxial layer to the P-type substrate. As a result of this deep diffusion there will be a large lateral diffusion underneath the oxide so that the $P^+$ isolation diffusion will occupy a rather large strip and therefore take up a large amount of chip area. For a 10-$\mu$m epitaxial layer thickness and a 10-$\mu$m-wide oxide opening, the total width of the $P^+$ region will be some 30 to 50 $\mu$m.

To reduce the chip area taken up by the isolation diffusion, the structure shown in Figure 2.61 can be used. Prior to the deposition of the N-type epitaxial layer there will be the usual $N^+$ (Sb) buried layer diffusion, and in this case also a $P^+$ (B) diffusion performed into the P-type substrate. During the subsequent epitaxial layer deposition and high-temperature diffusion and oxidation processes there will be outdiffusion from the buried $N^+$ and $P^+$ regions into the epitaxial layer. As a result of the $P^+$ outdiffusion, the $P^+$ isolation diffusion that comes in from the top

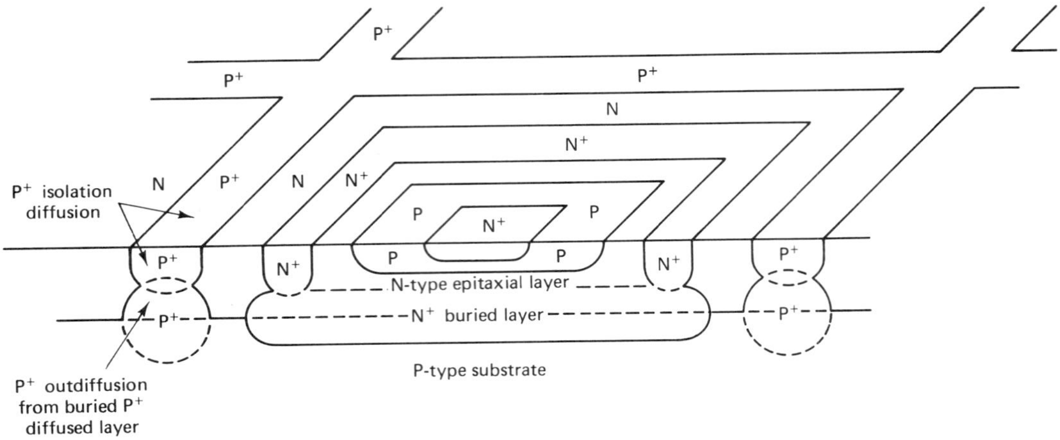


**Figure 2.61** Double-diffused isolation.

surface will have to proceed only less than halfway through the epitaxial layer before it merges with the P$^+$ outdiffusion from the substrate and thereby completes the isolation of the N-type regions. Although this process can substantially reduce the area occupied by the P$^+$ isolation regions, it does require some extra processing steps and also requires a critical pattern alignment such that the P$^+$ outdiffusion from the substrate will be lined up correctly with the P$^+$ isolation diffusion from the top surface.

## 2.20.2 Collector-Diffused Isolation

A very effective way of increasing the device density in ICs is to use the collector-diffused isolation process in which the P$^+$ isolation is eliminated altogether, as shown in Figure 2.62. The P-type substrate has an N$^+$ buried layer diffusion as in the conventional case, but now a very thin P-type epitaxial layer only about 2 $\mu$m in thickness

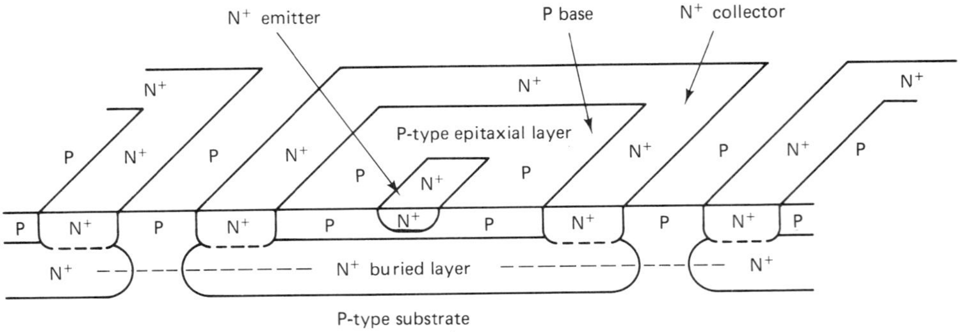


**Figure 2.62** Collector-diffused isolation.

is used. An $N^+$ diffusion is performed to penetrate all the way through the P-type epitaxial layer to merge with the $N^+$ outdiffusion from the substrate. This $N^+$ diffusion from the top surface serves the dual function of providing device isolation and being a deep collector contact diffusion to minimize the collector series resistance. The devices are isolated by means of the reverse-biased PN junctions formed between the $N^+$ diffused layer and the P-type epitaxial layer.

The resulting NPN transistors will have an epitaxial base region rather than a diffused base. The base width will be determined by a combination of the epitaxial layer thickness, the $N^+$ outdiffusion from the substrate, and the $N^+$ emitter diffusion from the top surface. As a result, there will generally be a wider spread in the base widths for this type of transistor than in the case of the double-diffused devices. This will lead to a larger spread in the current gains and poorer transistor matching. In addition, the heavily doped $N^+$ collector region that is directly adjacent to the P-type base region will lead to a lower collector-to-emitter breakdown voltage.

## 2.21 BONDING PADS AND CHIP EDGE CLEARANCE

Around the periphery of any IC chip there will be an array of bonding pads which are metallized regions to which the small-diameter wires from the package are bonded. These wires are typically 1-mil (25-$\mu$m)-diameter gold wires and are bonded to the bonding pads by a process known as *thermocompression bonding*, which uses a combination of heat (~250°C) and pressure to obtain a good mechanical bond and a very low-resistance electrical contact. In the thermocompression and similar types of bonding process the end of the wire becomes expanded to about two-to-three times its original wire diameter on the bonding pad, so that the bonding pad must be made large enough to accommodate this as well as to allow for alignment errors during the lead bonding operation. As a result, the bonding pads will generally be in the range 100 to 150 $\mu$m (4 to 6 mils) square, as shown in Figure 2.63.

For the wafer sawing or scribing operation it is generally desirable to keep the areas where the sawing or scribing will take place clear of oxide or metal. To do this the IC photomask set is designed to provide for a region about 75 to 100 $\mu$m (3 to 4 mils) in width between the adjacent ICs, and the bare silicon is kept exposed. This area is often referred to as a "scribing street."

There will be a substantial amount of mechanical damage occurring to the edge of the chips during the wafer sawing or scribing operation, including the generation of microcracks that will extend a short distance into the chip from the edge. To keep the active circuitry of the IC well away from this region, as well as to accommodate the bonding pads, scribing streets, and the required clearances, there will be a strip about 250 $\mu$m (10 mils) wide around the periphery of the chip that will not have any active circuitry on it as shown in Figure 2.63. For small ICs in which the chip size is only 1 mm $\times$ 1 mm (40 mils $\times$ 40 mils) this can amount to about one-half of the total chip area.

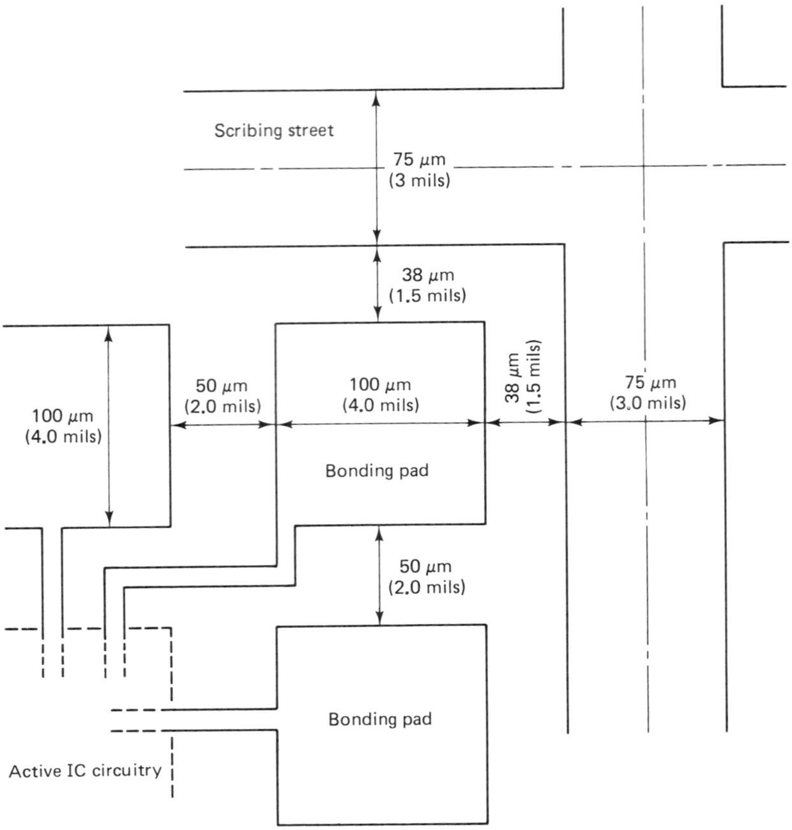


**Figure 2.63** Bonding pad and scribing street layout for ICs.

## 2.22 IC CHIP SIZE AND CIRCUIT COMPLEXITY

The first transistor was developed in 1948 and was a germanium-alloy junction transistor. Silicon devices were developed in the mid-to-late 1950s and integrated circuits were first made in the early 1960s. Since that time the size and complexity of ICs has increased very rapidly, as shown by the following brief chronology.

| | |
|---|---|
| Invention of the transistor (germanium) | 1948 |
| Development of silicon transistors | 1955–1959 |
| Planar processing | 1959 |
| First ICs, small-scale integration (SSI) 3 to 30 gates/chip | ~1960 |
| Medium-scale integration (MSI) 30 to 300 gates/chip | ~1965–1970 |
| Large-scale integration (LSI) 300 to 3000 gates/chip | ~1970–1975 |
| Very large-scale integration (VLSI) more than 3000 gates/chip | ~1975 |

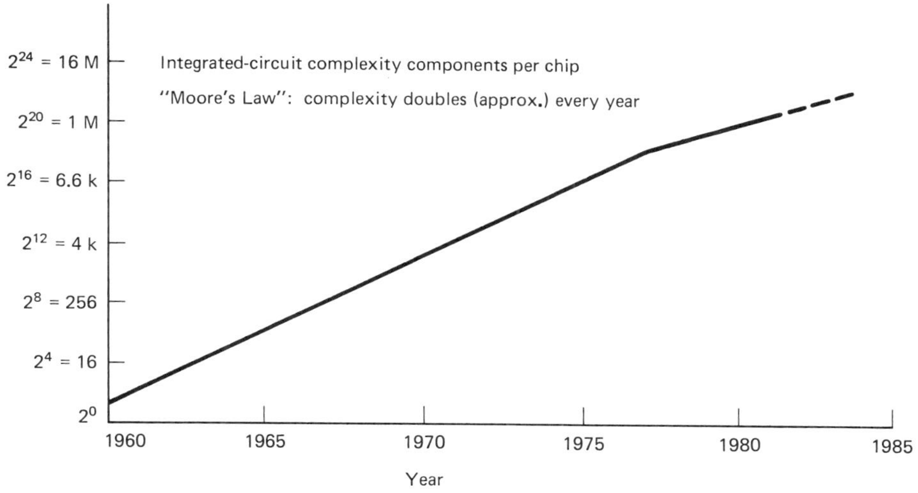


**Figure 2.64** Increases in the number of devices on an IC chip. (Adapted from G. E. Moore, "Progress in Digital Integrated Electronics," *IEEE, Intl. Electron Devices Meeting Tech. Digest*, pp. 11–13, Washington, D.C., 1975, © 1975 IEEE.)

| | |
|---|---|
| First commercial VLSI chip: 64K RAM | Late 1970s |
| 256K RAMs | Early 1980s |
| 512K RAMs, 1M ROMs, Very high speed GaAs ICs, "Three-dimensional" (multilayer) ICs, silicon-on-insulator technology | Middle 1980s |

The evolution of ICs over the years has been in terms of some very large increases in the device density together with some increases in the chip area. Figure 2.64 shows the dramatic and approximately exponential increase with time of the number of devices on an IC chip. The graph essentially follows "Moore's law," which states that there has been an approximate doubling of the number of devices per chip every year. Some of this rapid increase is due to an increase in the chip size, but most of it is due to improvements in the resolution of the photolithographic process with device dimensions and line widths shrinking from the range 10 to 15 $\mu$m in the 1960s down to range 1 to 2$\mu$m in the late 1970s and early 1980s. Advances in x-ray and electron-beam lithography can result in continued increases in IC complexity, although probably not at the same rate as in the 1960s and 1970s.

### 2.22.1 X-Ray and Electron-Beam Lithography

With the conventional ultraviolet (UV) photolithography process in which the UV wavelengths used are in the range 0.3 to 0.4 $\mu$m the minimum device dimensions or line widths are limited by diffraction effects to around five wavelengths or about 2 $\mu$m. This is what puts an upper limit on the IC device density using UV photolithography.

In the mid-to-late 1970s x-ray and electron-beam lithography techniques came into use to produce device dimensions down into the submicron ($<1$ $\mu$m) range. This is due to the much shorter wavelengths involved. In the case of x-rays, wavelengths of the order of 10 to 100 Å (0.001 to 0.01 $\mu$m) are available, and the wavelengths for electron-beam exposure are even less. With these techniques MOSFETs with gate lengths as small as 0.25 $\mu$m have been made.

The cost of the x-ray or electron-beam exposure equipment is very high and the exposure times are very much longer than with UV photolithography, so that this becomes a much more expensive process, and is used only when the very small device dimensions ($\lesssim 1$ $\mu$m) are needed.

The use of shorter-wavelength UV radiation, called "deep UV," can result in device dimensions down to about 1.25 $\mu$m. The deep UV sources that are used for photoresist exposure have wavelengths down in the range 2000 to 3000 Å.

In the case of x-ray lithography the mask often used consists of a very thin transparent plastic (Mylar) membrane only about 3 $\mu$m (0.1 mil) in thickness. The Mylar membrane has a thin film of gold, about 0.5 $\mu$m thick, deposited on it and then patterned to produce the desired x-ray exposure mask. Gold is used because it is a good absorber of x-rays.

For electron-beam lithography, the electron beam that strikes the photoresist produces the same type of polymerization or depolymerization effects that UV or x-ray photon irradiation produces. For electron-beam exposure an electron beam that is focused down to a spot size of about 0.1 to 0.2 $\mu$m is scanned across the wafer surface. The beam can be scanned across the wafer in a raster pattern and is turned on only when passing over the areas to be exposed. This is similar to the raster type of pattern that is used in the production of a TV picture on a CRT, except that there is no gray scale, the electron beam being turned completely on or completely off. An alternative technique is to use a vector scan in which the electron beam is kept turned on all of the time, but is directed only to those areas on the wafer that are to be exposed.

With electron-beam lithography an exposure mask is not required and the desired circuit patterns can be "written" directly on the wafer by a computer-controlled electron-beam exposure system. This is without the need for any set of masks or even any artwork of the circuit configuration. The equipment cost is, however, very high and the exposure times are generally very long, up to as much as several hours for one wafer. The electron-beam lithography process is very useful for producing the photomasks that can be used in the x-ray and UV photolithography processes.

### 2.22.2 IC Chip Sizes

In Figure 2.65 representative small (SSI), medium (MSI), and large (LSI or VLSI) IC chips are shown. The chip areas range from 1 $mm^2$ (1600 $mil^2$) for the SSI chip to 1 $cm^2$ (160,000 $mil^2$) for the LSI chip. In each case, room must be allowed on the chip for the bonding pads, scribe streets, and other clearances. In addition, the active devices on the chip must be set back some safe distance from the edge of the chip due to the mechanical damage that occurs in this region during the chip separation process.

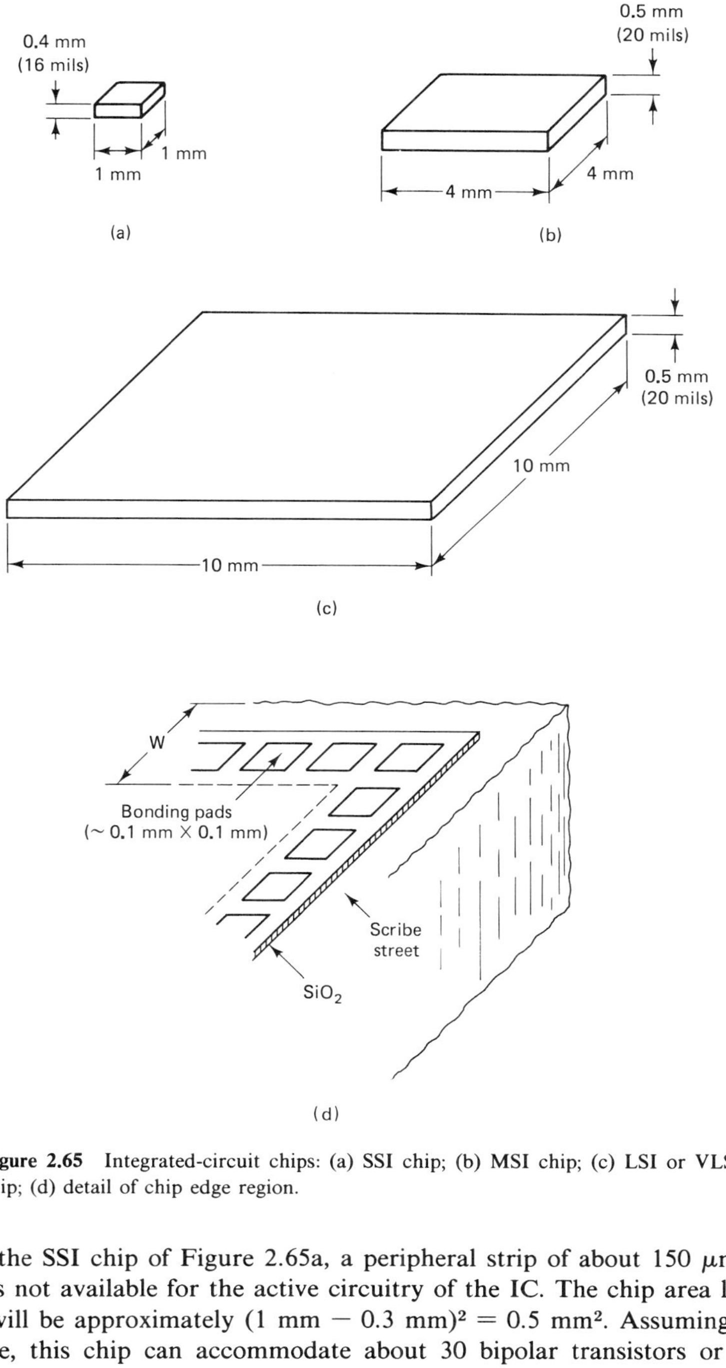


**Figure 2.65** Integrated-circuit chips: (a) SSI chip; (b) MSI chip; (c) LSI or VLSI chip; (d) detail of chip edge region.

For the SSI chip of Figure 2.65a, a peripheral strip of about 150 $\mu$m (6 mils) in width is not available for the active circuitry of the IC. The chip area left of this circuitry will be approximately $(1 \text{ mm} - 0.3 \text{ mm})^2 = 0.5 \text{ mm}^2$. Assuming a 10-$\mu$m design rule, this chip can accommodate about 30 bipolar transistors or over 200 MOSFETs. Approximately 7000 of these small IC chips can be obtained from a 100-mm wafer.

For the large 1-cm² LSI or VLSI chip, about 70 ICs can be obtained from a 100-mm wafer. If a 200-$\mu$m (8-mil) peripheral strip is used for the bonding pads, scribing streets, and other clearances, the remaining chip area for the active circuitry will be approximately $(10 \text{ mm} - 0.4 \text{ mm})^2 = 92 \text{ mm}^2$. This area will accommodate as many as 6000 bipolar transistors or 40,000 MOSFETs, assuming a 10-$\mu$m design rule. With a 4-$\mu$m design rule the number of MOSFETs can reach 250,000.

## 2.23 THERMAL DESIGN CONSIDERATIONS

The dissipation of power in semiconductor devices will result in a temperature rise of the device above the ambient or surrounding temperature. An excessive device temperature can result in the device operating beyond the specified temperature range so that the device will not operate within the manufacturer's specifications, and in particular there will be an excessive junction leakage current. Another more serious consequence of an excessive temperature rise can be the permanent irreversible damage of the device.

The *net* electrical power delivered to the device must under equilibrium conditions be equal to the heat power dissipated by the device, so that we have $P_d = P_{in}$, where $P_{in}$ is the *net* electrical power delivered to the device and $P_d$ is the heat power that is dissipated by the device, as shown in Figure 2.66. The heat will flow out of the device to its surroundings by means of the heat conduction and convection processes. Normally, heat radiation will not be a significant factor, due to the relatively low temperature rises involved.

The temperature rise will therefore be directly proportional to the heat flow so that $\Delta T = \Theta P_d$, where $\Theta$ is the *thermal impedance* and $\Delta T = T_J - T_A$ is the temperature rise of the device junction above the ambient temperature $T_A$. The critical point is at the PN junctions of the device, hence the reference that is made to the

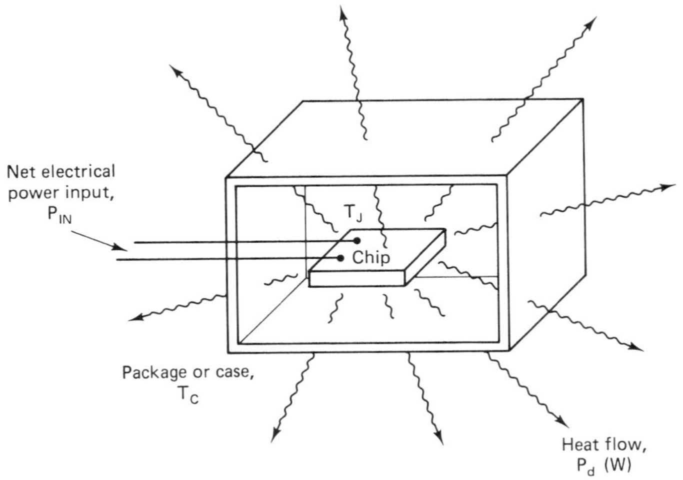


**Figure 2.66** Semiconductor device heat flow and temperature rise.

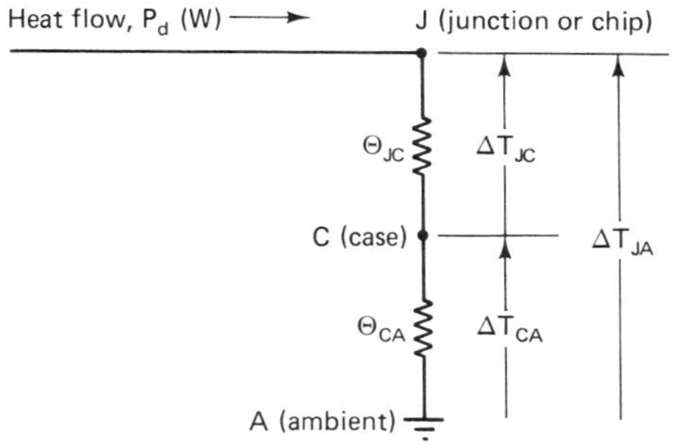


**Figure 2.67** Thermal equivalent circuit.

junction temperature $T_J$. However, as a result of the small size of the silicon chip and the high thermal conductivity of silicon, the average temperature of the chip will be generally within a few degrees of the junction temperature.

In Figure 2.67 an electrical analog thermal equivalent circuit is shown. The heat flow $P_d$ through the thermal impedances from the chip (or junction) to the case (or package) $\Theta_{JC}$, and from the case to the ambient $\Theta_{CA}$ will produce temperature drops given by $\Delta T_{JC} = T_J - T_C = \Theta_{JC}P_d$ and $\Delta T_{CA} = T_C - T_A = \Theta_{CA}P_d$. The total temperature drop from junction to ambient will be

$$\Delta T_{JA} = T_J - T_A = \Delta T_{JC} + \Delta T_{CA} = (\Theta_{JC} + \Theta_{CA})P_d = \Theta_{JA}P_d \tag{2.14}$$

The thermal impedances will have units of °C/W, although the thermal impedance is sometimes given in units of °C/mW.

In Figure 2.68 a variety of IC packages are shown together with values of the thermal impedance. The most popular package type is the *dual-in-line package* or DIP. For the lowest-cost plastic DIP packages the thermal impedance ranges from 200°C/W for the 8-pin DIP (or "mini-DIP") to 160°C/W for the 14-pin DIP and 150°C/W for the 16-pin DIP. The ceramic DIP, which is more expensive than the plastic DIP, has a substantially lower thermal impedance, being 100°C/W for the 14-pin DIP and 90°C/W for the 16-pin DIP. The Flat-Pack has as a principal feature a low profile, being a maximum of only 0.07 in. (1.8 mm) high. However, together with the small package volume there is a relatively high thermal impedance, being about 300°C/W for both the 14-pin and 16-pin packages. The metal can packages, including the 8-pin TO-99 and the 10-pin TO-100 packages, have a thermal impedance of about 160°C/W. These impedance values are the junction-to-ambient thermal impedance $\Theta_{JA}$ under "free air" conditions; that is, the device is cooled by natural convection and there is no heat sink attached to the device. The junction-to-case thermal impedance for most of the IC packages will be in the range 20 to 40°C/W.

Let us consider now as a representative example a 14-pin plastic DIP package that has a thermal impedance of 160°C/W. The usual upper temperature limit for plastic package devices is 150°C, due in part to the softening of the plastic package material. If the ambient temperature is 25°C, the maximum allowable power dissipation rating $P_{d(\text{MAX})}$ will be obtained from $\Delta T_{\text{MAX}} = T_{J(\text{MAX})} - T_A = \Theta P_{d(\text{MAX})}$ so that

$$P_{d(\text{MAX})} = (T_{J(\text{MAX})} - T_A)/\Theta \tag{2.15}$$

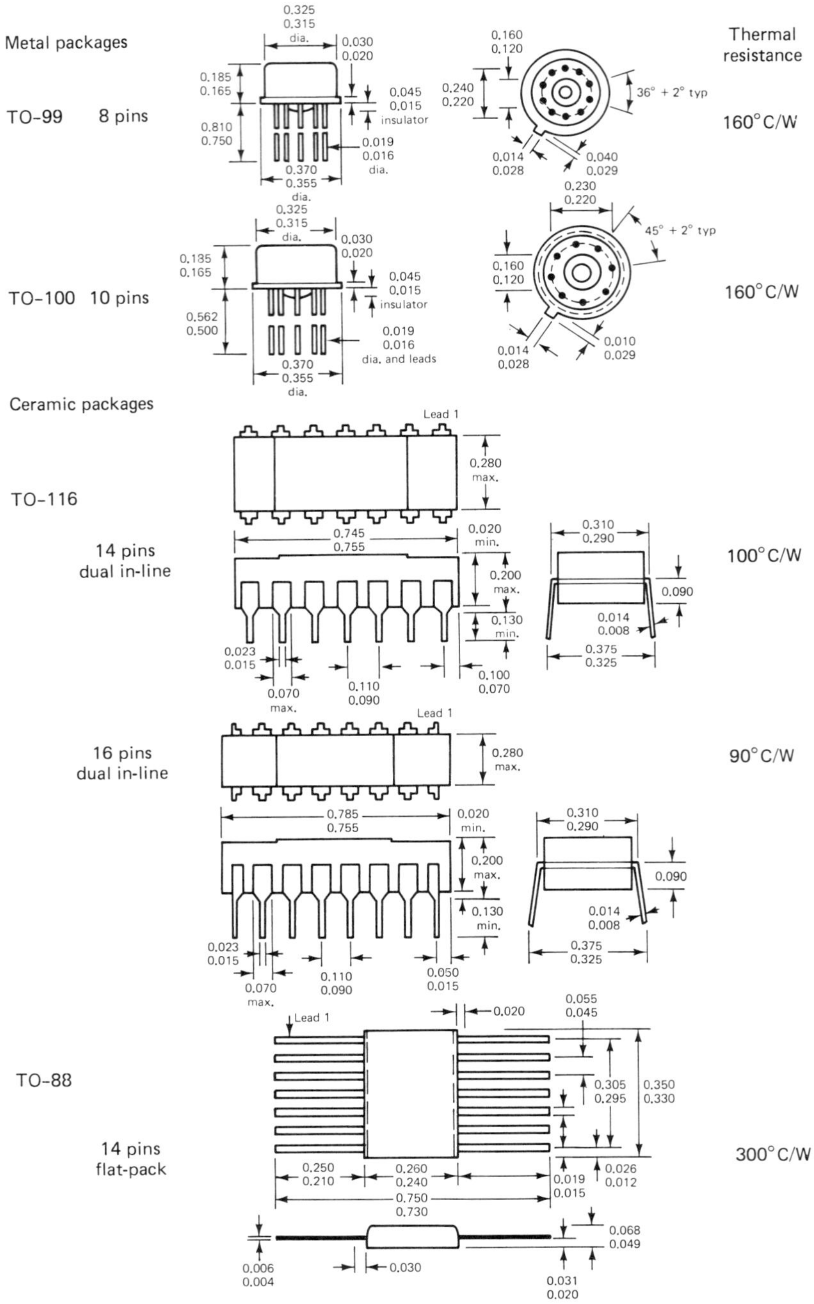


Most often used integrated circuit package configurations.

**Figure 2.68** Integrated-circuit packages. (From Hans. R. Camenzind, *Electronic Integrated Systems Design*, Van Nostrand Reinhold, 1972.)

16 pins
flat-pack

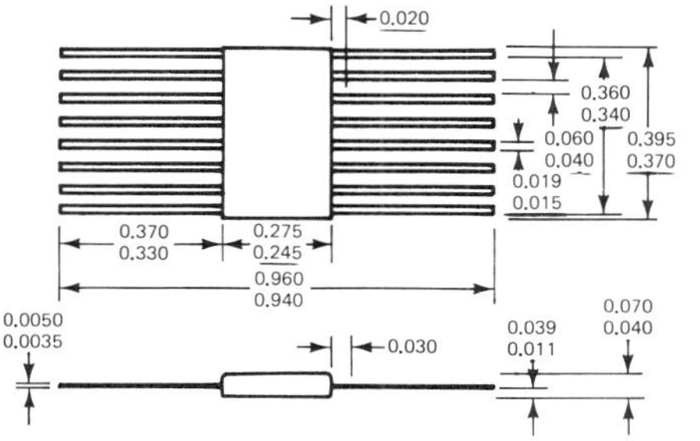


300°C/W

Plastic packages

8-pins
dual in-line

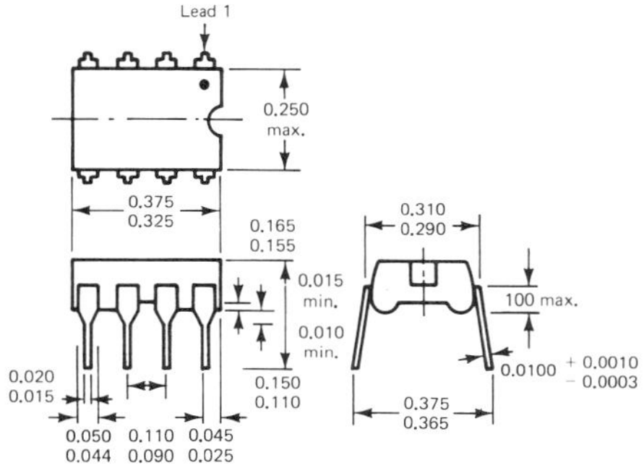


200°C/W

TO-116

14 pins
dual in-line

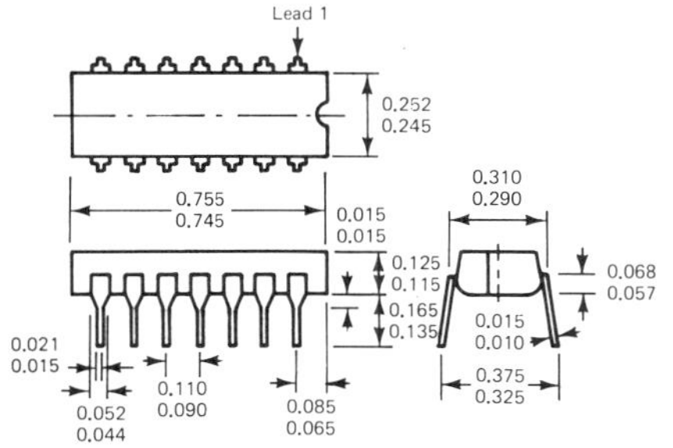


160°C/W

16 pins
dual in-line

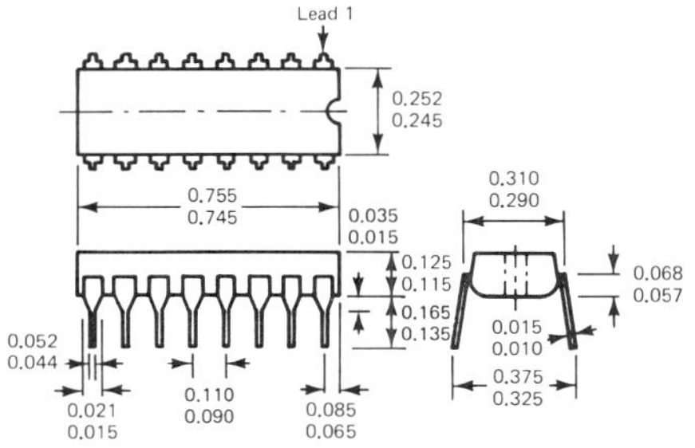


150°C/W

**Figure 2.68** *(cont.)*

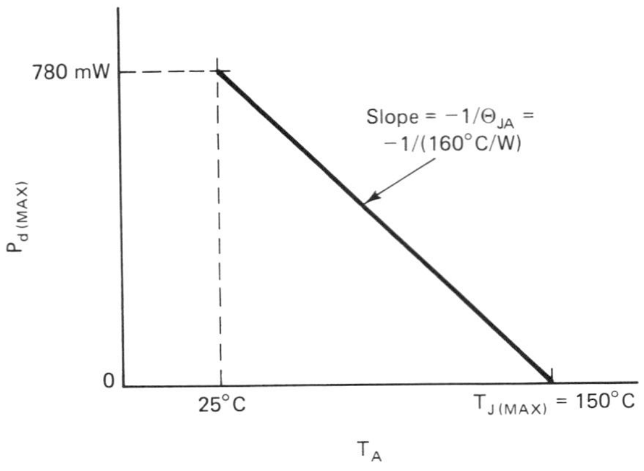


**Figure 2.69** Maximum power dissipation derating characteristic.

and thus

$$P_{d\,(\text{MAX})} = \frac{(150\text{–}25)°\text{C}}{160°\text{C/W}} = \underline{0.78\text{ W}} = \underline{780\text{ mW}} \tag{2.16}$$

If the ambient temperature is higher than 25°C, the maximum power dissipation rating will be reduced, and as the ambient temperature $T_A$ approaches the maximum allowable junction temperature $T_{J\,(\text{MAX})}$ the value of $P_{d\,(\text{MAX})}$ will approach zero. In Figure 2.69 a *derating curve* is shown for this example, showing the decrease in the maximum power dissipation rating of the device as the ambient temperature increases. The slope of the derating line will be equal to the negative of the reciprocal of the thermal impedance, $-1/\Theta_{JA}$.

For devices in the metal or ceramic packages the usual upper temperature limit for the junction temperature, $T_{J\,(\text{MAX})}$, will be about 175°C. Thus for the metal TO-99 (8-pin) and TO-100 (10-pin) packages with a thermal impedance of 160°C/W and an ambient temperature of 25°C, the maximum power dissipation rating will be

$$P_{d\,(\text{MAX})} = \frac{(175\text{–}25)°\text{C}}{160°\text{C/W}} = \underline{0.94\text{ W}} = \underline{940\text{ mW}} \tag{2.17}$$

A further advantage of the metal package is the possibility of obtaining a hermetic gastight seal so that better long-term device protection can be obtained. The metal package is, however, much more costly than the plastic package. Indeed, a major part of the cost of many semiconductor devices will be the packaging cost.

### 2.23.1 Heat Sinks

The maximum power dissipation rating of semiconductor devices can be increased substantially by the use of a *heat sink*. A heat sink is a metal part, usually some type of finned structure, that can be clipped, clamped, or bonded to the semiconductor device package, and aids the convective heat transfer process. There are a wide variety of heat sinks available for various device package configurations.

The use of a heat sink can reduce the case-to-ambient thermal impedance $\Theta_{CA}$ by a very large factor so that the maximum power dissipation rating of the device can be greatly increased, often by a factor as large as 3 or 4. With a heat sink present the case-to-ambient thermal impedance is reduced from the "free-air" value to a value determined by the heat sink so that $\Theta_{CA}$ is now equal to $\Theta_{HS}$, which is the heat sink thermal impedance. Depending on the design of the heat sink, $\Theta_{HS}$ values ranging from 100°C/W for very small clip-on heat sinks down to around 10°C/W for larger area and more massive heat sinks are available. Simple clip-on heat sinks for DIPs are available that can provide thermal impedances down in the range 20 to 30°C/W. In addition to the clip-on heat sinks there are heat sinks that can be bonded to the semiconductor device package by a thermally conductive epoxy adhesive. The heat sink adhesives will have thermal impedances in the range 70°C-cm/W. For an area of 1 cm², the thermal impedance of an adhesive film of thickness $t$(mm) will be $\Theta = 7°\text{C/W} \times t(\text{mm})$. Thus an adhesive film of $t = 0.1$ mm $=$ 0.004 in. in thickness and 1 cm² in area will have a thermal impedance of only 0.7°C/W.

For a representative example of the advantage to be gained by the use of a heat sink, let us consider a device with $\Theta_{JA} = 150°\text{C}$ (free air), $\Theta_{JC} = 40°\text{C/W}$, and $T_{J(\text{MAX})} = 175°\text{C}$. For an ambient temperature of 25°C the free-air maximum power dissipation rating will be $P_{d(\text{MAX})} = (175–25)°\text{C}/(150°\text{C/W}) = \underline{1.0\text{ W}}$. If a heat sink is now used that has a heat sink-to-ambient thermal impedance of $\Theta_{HS}$ of 20°C/W, the case-to-ambient thermal impedance will be reduced from 110°C/W to the 20°C/W impedance of the heat sink. The total thermal impedance will now be $\Theta_{JA} = \Theta_{JC} + \Theta_{HS} = (40 + 20)°\text{C/W} = 60°\text{C/W}$, and the corresponding value of $P_{d(\text{MAX})}$ will be increased to $P_{d(\text{MAX})} = 150°\text{C}/(60°\text{C/W}) = \underline{2.5\text{ W}}$. In the limiting case of an ideal "infinite heat sink" that has a zero thermal impedance, the case-to-ambient thermal impedance is reduced to zero, so that $\Theta_{JA} = \Theta_{JC} = 40°\text{C/W}$. The maximum power dissipation rating reaches a value of $P_{d(\text{MAX})} = 150°\text{C}/(40°\text{C/W}) = \underline{3.75\text{ W}}$.

## PROBLEMS

### Voltage-Variable Capacitance Diode Problems: P⁺NN⁺ Epitaxial Diode (Problems 2.1–2.14)

**2.1** Given that the VHF TV bands are 54 to 88 MHz (channels 2 to 6) and 174 to 216 MHz (channels 7 to 13) and each channel is 6.0 MHz wide, can one VVC diode be used for tuning over the entire VHF TV band? Assume that the minimum allowable reverse-bias voltage is 3.0 V and the maximum voltage available in the circuit is 30 V. Explain your answer.

**2.2** Can one VVC diode be used for tuning over the TV band above (channels 2 to 13) if the inductor is switched, with one value inductance $L_1$ being used for the VHF "low band" (channels 2 to 6) and a smaller inductance $L_2$ being used for the VHF "high band" (channels 7 to 13)? Explain.

**2.3.** If the switched inductance technique described above is used, and there is a fixed capacitance of 4.0 pF in the circuit, find the required minimum and maximum diode capacitance values required. [*Ans.*:

$$C_{J\,(\mathrm{MIN})} = 34.62\ \mathrm{pF}\ \text{(full depletion of epitaxial layer)}$$
$$C_{J\,(\mathrm{MAX})} = C_J(-3\ \mathrm{V}) = 98.56\ \mathrm{pF}$$

**2.4.** If the epitaxial layer doping is $N_{\mathrm{epi}} = 1 \times 10^{15}\ \mathrm{cm}^{-3}$ (5 Ω-cm), find the junction area required. (*Ans.*: $A = 2.10\ \mathrm{mm}^2$)

**2.5.** If the junction depth is $x_J = 2.0\ \mu\mathrm{m}$, find the required thickness of the epitaxial layer. (*Ans.*: $t_{\mathrm{epi}} = 8.33\ \mu\mathrm{m}$)

**2.6.** Find the maximum value of the series resistance, $R_S(3\ \mathrm{V})$. [*Ans.*: $R_S(3\ \mathrm{V}) = 0.105\ \Omega$]

**2.7.** Find the minimum value of $Q$ (i.e., at 3 V and 174 MHz). (*Ans.*: $Q = 88.4$)

**2.8.** What value of $Q$ will be needed (at 174 MHz) for a 6-MHz bandwidth? Will the VVC $Q$-value be satisfactory for this? (*Ans.*: 29, yes)

**2.9.** If the junction area is increased to twice the minimum value, by what factor will these parameters change?

(a) $C_{J\,(\mathrm{MAX})}$. (*Ans.*: 2:1)
(b) $C_{J\,(\mathrm{MIN})}$. (*Ans.*: 2:1)
(c) Diode capacitance ratio. (*Ans.*: unaffected by an area change)
(d) Series resistance, $R_S$. (*Ans.*: 1:2)
(e) $Q$. (*Ans.*: unaffected by an area change)
(f) $C_{\mathrm{TOTAL\ MAX}}/C_{\mathrm{TOTAL\ MIN}}$. (*Ans.*: 2.746:2.656)
(g) Tuning ratio, $f_{\mathrm{MAX}}/f_{\mathrm{MIN}}$. (*Ans.*: 1.657:1.630)

**2.10.** If a nonepitaxial 5-Ω-cm N-type wafer were used instead of the N/N$^+$ wafer, what will the series resistance be at 3 V bias? What will the $Q$ be at 3 V and 174 MHz? Will this $Q$-value be satisfactory? [*Ans.*: $R_S = 7.1\ \Omega$, $Q(-3\ \mathrm{V},\ 174\ \mathrm{MHz}) = 1.30$, no]

**2.11.** Will the use of the complementary N$^+$PP$^+$ epitaxial diode structure offer any advantages over the P$^+$NN$^+$ structure? Explain. How will the $Q$-values compare for the two structures, assuming equal doping levels and dimensions? (*Ans.*: no, the $Q$ will be lower by a factor of about 2.5)

**2.12.** The UHF TV band extends from 470 to 890 MHz (channels 14 to 83). If the minimum allowable reverse-bias voltage is 3.0 V and the maximum voltage available is 30 V, can one VVC diode be used for tuning over the UHF TV band?

**2.13.** To what value will the minimum diode bias voltage be reduced to allow for tuning over the entire UHF TV band? (*Ans.*: $V_{R\,(\mathrm{MIN})} = 1.6\ \mathrm{V}$)

**2.14.** The junction capacitance of a VVC diode will exhibit a temperature dependence. The rate of change of capacitance with temperature can be designated as $TCC_J = (1/C_J)(dC_J/dT)$. A representative value for $TCC_J$ is 400 ppm/°C = 0.04%/°C°.

(a) What will the corresponding temperature coefficient of the frequency of the tuned circuit be? [*Ans.*: $TCF = (1/f)(df/dT) = 200$ ppm/°C = 0.02%/°C]
(b) Will this frequency shift cause any serious problems for a UHF tuner, considering channel 83 (890 MHz) as the "worst-case" condition? (*Ans.*: $\Delta f = 0.18$ MHz/°C, or only 3%/°C of the 6.0-MHz channel width, so a temperature change limited to a few °C should not cause any serious problem, although a larger temperature change can be a problem)
(c) Repeat part (b) for the case of an FM tuner (88 to 108 MHz) with a 200-kHz bandwidth

per FM channel. (*Ans*.: $\Delta f = 22$ kHz/°C or about 11%/°C of the channel bandwidth, so a temperature change of even a few degrees can produce a substantial mistuning if not compensated for)

**2.15.** (*PIN switching diode*) Given: $P^+NN^+$ (PIN) planar epitaxial VHF switching diode as shown in Figure P2.15. The package capacitance is 0.05 pF.

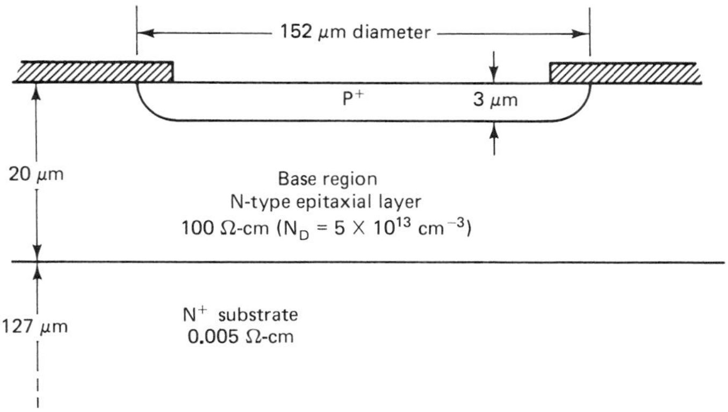


**Figure P2.15**

**(a)** Find the minimum capacitance, $C_{\text{min}}$. (*Ans*.: 0.162 pF)

**(b)** Find the value of the base (epitaxial layer) region component of the series resistance, $R_o$, for very small forward currents. (*Ans*.: ~936 Ω)

**(c)** Find the substrate component of the series resistance. (*Ans*.: 0.35 Ω)

**(d)** As a result of the injection of holes from the $P^+$ region into the very lightly doped N-type epitaxial layer under forward-bias conditions, there will be additional electrons drawn into the epitaxial layer from the $N^+$ substrate. These electrons will be drawn in to neutralize the charge density produced by the holes in transit across the epitaxial layer or base region. This flow of electrons and holes into the base region will result in an increased conductivity and therefore a decreased series resistance for the base region. This effect is called *conductivity modulation*. As a result of this conductivity modulation the base-region series resistance will be given by

$$R_{\text{base}} = \frac{V_T}{I(1+b)} \ln\left[1 + \frac{IR_o(1+b)}{V_T}\right]$$

where $b = \mu_u/\mu_p$ is the mobility ratio (2.5 for silicon) and $R_o$ is the unmodulated value of the base resistance ($R_{\text{base}}$ at $I$ = O). Using $b = 2.5$ and $V_T = 26$ mV (27°C), find the base-region resistance at the following forward current values: 100 nA, 1.0 μA, 10 μA, 100 μA, 1 mA, 10 mA, 30 mA, 50 mA, 100 mA, and 200 mA. (*Ans*.: 926, 878, 604, 194, 36, 5.3, 2.04, 1.30, 0.70, 0.376 Ω)

**(e)** The total dynamic forward resistance of this diode will be given by

$$r_d = \frac{nV_T}{I} + R_{\text{series}} = \frac{nV_T}{I} + R_{\text{base}} + R_{\text{substrate}}$$

Find the diode dynamic resistance for the current values given in part (d). Use $n$ = 1.6. (*Ans*.: 417kΩ, 42.5kΩ, 4.76kΩ, 610Ω, 78Ω, 9.75Ω, 3.72Ω, 2.42Ω, 1.41Ω, 0.88Ω)

**(f)** What reverse-bias voltage is required for the full depletion of the base region such that the junction capacitance is at its minimum value? Assume 0.8 V for the contact potential. (*Ans.*: 10.3 V)

**(g)** If the complementary structure (i.e., $N^+ PP^+$) were used for this switching diode, but with the same resistivity values and dimensions as before, how would the equation for $R_{\text{base}}$ be changed? Calculate $R_{\text{base}}$ and $r_d$ for the $N^+PP^+$ diode at a forward current of 100 mA. Find the value of reverse-bias voltage required for the full depletion of the base region. (*Ans.*: $R_{\text{base}} = 1.58\ \Omega$, $r_d = 2.35\ \Omega$, $V_R = 31.7$ V)

**(h)** The $P^+NN^+$ diode is to be used as a series switch in a 50-Ω system (50-Ω source, 50-Ω load) at 100 MHz. Find the *isolation* (in dB) with a 15-V reverse bias, and the *insertion loss* (in dB) when there is a forward current of 50 mA through the diode. (*Ans.*: 39.85 dB, 0.208 dB)

**(i)** Repeat part (h) for the case of the PIN diode being used as a *shunt* switch. (*Ans.*: 21.1 dB, ~O dB)

**(j)** Repeat part (h) for the case of a *series-shunt* switch in which two PIN diodes are used, one in the series position and one in the shunt position. (*Ans.*: 66.6 dB, 0.21 dB)

**2.16.** (*Short-channel MOSFET*) In a short-channel MOSFET the drain-to-source electric field strength in the channel region will be such that the charge carriers will travel at drift velocities close to the saturation velocity $v_{\text{sat}}$. The total mobile charge in the channel will be $Q_{\text{channel}} = C_{\text{ox}}(V_{GS} - V_{\text{threshold}})$, where $C_{\text{ox}}$ is the gate-to-channel capacitance. The drain-to-source current $I_{DS}$ will be given by $I_{DS} = Q_{\text{channel}}/t_{\text{transit}}$, where $t_{\text{transit}}$ is the source-to-drain transit time and will be given approximately by $t_{\text{transit}} = L/v_{\text{sat}}$, with $L$ being the channel length.

**(a)** Show that $I_{DS}$ will be given by

$$I_{DS} = g_{fs}(V_{GS} - V_{\text{threshold}})$$

where $g_{fs} = (\epsilon_{\text{ox}}/t_{\text{ox}})Wv_{\text{sat}}$ and $W$ is the channel width.

**(b)** A short-channel MOSFET has an oxide thickness of $t_{\text{ox}} = 1000$ Å. Find $g_{fs}$ per centimeter of channel width. Use $\epsilon_{r\,(\text{oxide})} = 3.8$ and $v_{\text{sat}} = 1 \times 10^6$ cm/s for electrons in the N-type channel. (*Ans.*: 33.6 S/cm)

**(c)** A VMOS short-channel transistor is on a 1 mm × 1 mm chip. The V-grooves have a center-to-center spacing of 40 μm. Using the same parameters as given above find the $g_{fs}$ of the transistor. (*Ans.*: 168 S)

**(d)** For the above VMOS transistor the threshold voltage is +2.0 V. Find $I_{DS}$ at $V_{GS} = +12$ V. (*Ans.*: $I_{DS} = 1.68$ A)

**(e)** For the MOSFET described above the junction depth of the $N^+$ source region is 2.0 μm. Find the gate-to-source capacitance $C_{gs}$. (*Ans.*: 41 pF)

**2.17.** (*Ion-implanted P-channel JFET*) The P-type channel of a JFET is formed by a boron ion implantation into an N-type epitaxial layer as shown in Figure 2.25c.

**(a)** Show that the pinch-off voltage $V_P$ of the JFET will be given approximately by $V_P = q\phi^2/2\epsilon N_D - V_{CP}$, where $\phi$ is the ion implantation dosage, $N_D$ is the epitaxial layer doping, and $V_{CP}$ is the contact potential.

**(b)** Find the ion implantation dosage required to produce a JFET with a pinch-off voltage of +4.0 V if the epitaxial layer doping is $N_D = 1 \times 10^{16}$ cm$^{-3}$ (0.6 Ω-cm) and $V_{CP} = 0.8$ V. (*Ans.*: $\phi = 7.91 \times 10^{11}$ cm$^{-2}$)

**(c)** If the average hole mobility in the implanted layer is 200 cm²/V-s, find the sheet resistance of the layer. Assume bias conditions such that the channel resistance will be at its minimum value (i.e., essentially zero bias or a slight forward bias at the gate-to-channel junction). (*Ans.*: 39,454 Ω/square)

(d) If the JFET channel length is 8 μm and the channel width is 160 μm, find $r_{ds(ON)}$, $I_{DSS}$, and $g_{fso}$. The following approximate relationship can be used for $I_{DSS}$: $I_{DSS} = -V_P/[3r_{ds(ON)}]$. (*Ans.*: 1973 Ω, −0.676 mA, 0.338 mS)

(e) Repeat the preceding problems for an epitaxial layer doping of $N_D = 1 \times 10^{17}$ cm$^{-3}$ (0.09 Ω-cm). (*Ans.*: $\phi = 2.502 \times 10^{12}$ cm$^{-2}$, $R_S$ = 12,476 Ω/square, $r_{ds(ON)}$ = 624 Ω, $I_{DSS}$ = −2.14 mA, $g_{fso}$ = 1.069 mS)

(f) Design a JFET with an ion-implanted P-type channel to meet the following specifications: $V_P$ = +2.2 V and $I_{DSS}$ = −100 μA. The epitaxial layer doping is $1 \times 10^{16}$ cm$^{-3}$ (0.6 Ω-cm) and the channel length is 10 μm. Specify the ion implantation dosage $\phi$ and the channel width $W$. Assume an average hole mobility of 200 cm$^2$/V-s in the P-type channel and a contact potential of 0.8 V. (*Ans.*: $\phi = 6.25 \times 10^{11}$ cm$^{-2}$, $W$ = 50 μm)

**2.18.** (*Ion-implanted depletion-mode MOSFET*) The device structure of Figure 2.25c is modified into a depletion-mode MOSFET by the formation of a thin oxide layer over the P-type implanted channel and the deposition of a gate electrode on top of the oxide layer. The P-type channel is formed by a low-energy, low-dosage boron ion implantation followed by a low-temperature (~700°C) annealing step to activate the implanted ions by causing them to go into substitutional sites in the silicon crystal lattice.

(a) Show that the pinch-off voltage will be given approximately by $V_P = (q\phi t_{ox}/\epsilon_{ox}) - V_{CP}$, where $\phi$ is the ion implantation dosage and $V_{CP}$ is the gate-to-silicon contact potential (generally in the range 0 to 0.5 V).

(b) Find the ion implantation dosage $\phi$ required for a pinch-off voltage of +4.0 V if the oxide thickness $t_{ox}$ is 800 Å. Assume that $V_{CP} \simeq 0$, as for the case of a silicon gate device. (*Ans.*: $1.05 \times 10^{12}$ cm$^{-2}$)

(c) Find the minimum value of the sheet resistance of the P-type channel region (i.e., $V_{GS} = 0$). Assume that the average hole mobility in the channel is 200 cm$^2$/V-s. (*Ans.*: 29.74 kΩ/square)

(d) If the channel length is $L$ = 8.0 μm and the width is $W$ = 40 μm, find $r_{ds(ON)}$, $I_{DSS}$, and $g_{fso}$. (*Ans.*: 5.55 kΩ, 224 μA, 112 μS)

(e) Show that $I_{DSS}$ is given approximately by $I_{DSS} \simeq (\epsilon_{ox}\bar{\mu}_p W/t_{ox}L)V_P^2$ and $g_{fso}$ will be given approximately by $g_{fso} \simeq \frac{2}{3}(\epsilon_{ox}\bar{\mu}_p W/t_{ox}L)V_P$.

(f) Compare this ion-implanted depletion-mode MOSFET with the ion-implanted JFET of Problem 2.17. What are the relative advantages and disadvantages of the two devices?

**2.19.** (*Ion-implanted IC resistor*) A boron ion implantation is followed by a drive-in diffusion at 1150°C to produce a P-type layer with a sheet resistance of 5000 Ω/square and a junction depth of 2.0 μm. The substrate is doped with $1 \times 10^{15}$ cm$^{-3}$ phosphorus. Find:

(a) Average resistivity of the layer. (*Ans.*: 1.0 Ω-cm)

(b) Boron surface concentration. (*Ans.*: $4 \times 10^{16}$ cm$^{-3}$)

(c) Drive-in diffusion time. (*Ans.*: 45.2 min)

(d) Ion implantation dosage. (*Ans.*: $3.7 \times 10^{12}$ cm$^{-2}$)

(e) Average hole mobility in the layer. (*Ans.*: 338 cm$^2$/V-s)

(f) Ion implantation time per 100-mm-diameter wafer if the beam current is 5 μA with a rectangular raster scan. (*Ans.*: 11.8 s)

(g) Chip area required for a 1.0-MΩ resistor if the resistor line width is to be 15 μm and the line-to-line spacing is to be 15 μm. (*Ans.*: ~0.1 mm$^2$)

(h) Capacitance per unit area and the total parasitic capacitance for the 1.0-MΩ resistor at a reverse-bias voltage of 10 V. (*Ans.*: 27.8 pF/mm$^2$, ~1.25 pF)

(i) Cutoff frequency (3-dB bandwidth) for the resistor. (*Ans.*: ~254 kHz)

**2.20.** (*Gallium arsenide MESFET*) A GaAs MESFET similar to that of Figure 2.9 has a channel region formed by a low-energy, low-dosage ion implantation followed by a short drive-in diffusion/annealing step.

**(a)** If the Gaussian impurity profile of the channel can be approximated by a rectangular-shaped profile, show that the pinch-off voltage will be given by $V_P = q\phi t_c/\epsilon - V_{cp}$, where $\phi$ is the implantation dosage, $t_c$ is the channel thickness, and $V_{cp}$ is the metal-semiconductor contact potential (typically ~0.8 V for metal-GaAs Schottky barriers).

**(b)** If the implanted channel has a surface concentration of $N(0) = 2 \times 10^{17}$ cm$^{-3}$ and at a distance of 0.25 $\mu$m from the surface the concentration is down to $2 \times 10^{15}$ cm$^{-3}$, find:

**(1)** The ion implantation dosage. (*Ans.*: $2.065 \times 10^{12}$ cm$^{-2}$)

**(2)** The pinch-off voltage using the approximate relationship of part (a) with $V_{cp} = 0.8$ V and $t_c = 0.25$ $\mu$m ($\epsilon_r = 10.9$ for GaAs). (*Ans.*: $-7.77$ V)

**(3)** $r_{ds(ON)}$, $I_{DSS}$, and $g_{fso}$ for a channel length of $L = 2.0$ $\mu$m and a width of $W = 10$ $\mu$m. Use an average electron mobility of 5000 cm$^2$/V-s for the channel. (*Ans.*: 121 $\Omega$, 21 mA, 5.5 mS)

**(4)** The minimum source-to-drain transit time and the corresponding frequency. The electron saturation velocity in GaAs is ~$10^7$ cm/s. (*Ans.*: 20 ps, 8.0 GHz)

***(c)** Using the equation for the Gaussian impurity profile and Posson's equation, write a computer program to obtain the pinch-off voltage. Find the pinch-off voltage and compare the value obtained with the approximate result obtained in part (b,2).

## REFERENCES


Bar-Lev, A., *Semiconductors and Electronic Devices*, Prentice-Hall, 1984.

Burger, R. M., and R. P. Donovan, *Fundamentals of Silicon Integrated Circuit Device Technology*, Vol. 1, Prentice-Hall, 1967.

Camenzind, H. R., *Electronic Integrated Systems Design*, Van Nostrand Reinhold, 1972.

Colclaser, R. A., *Microelectronics: Processing and Device Design*, Wiley, 1980.

Connelly, J. A., *Analog Integrated Circuits*, Wiley, 1975.

Eimbinder, J., *Application Considerations for Linear Integrated Circuits*, Wiley, 1970.

Einspruch, N. G., *VLSI Electronics: Microstructure Science*, Academic Press, 1982.

Fogiel, M., *Microelectronics: Basic Principles, Circuit Design, Fabrication Technology*, Research & Education Association, 1972.

Ghandi, S. K., *The Theory and Practice of Microelectronics*, Wiley, 1968.

Glaser, A. B., and G. E. Subak-Sharpe, *Integrated Circuit Engineering*, Addison-Wesley, 1977.

Gray, P. R., and R. G. Meyer, *Analysis and Design of Analog Integrated Circuits*, Wiley, 1984.

Grebene, A. B., *Analog Integrated Circuit Design*, Van Nostrand Reinhold, 1972.

Grove, A. S., *Physics and Technology of Semiconductor Devices*, Wiley, 1967.

Hamilton, D. J., and W. G. Howard, *Basic Integrated Circuit Engineering*, McGraw-Hill, 1975.

Hnatek, E. R., *A User's Handbook of Integrated Circuits*, Wiley, 1973.

JASTRZEBSKI, L., A. C. IPRI, and J. F. CORBOY, "Device Characterization on Monocrystalline Silicon Grown over $SiO_2$ by the ELO (Epitaxial Lateral Overgrowth) Process," *IEEE Electron Device Letters*, Vol. EDL-4, No. 2, February 1983.

MILLMAN, J., *Microelectronics*, McGraw-Hill, 1979.

MORTENSON, K. E., *Variable Capacitance Diodes*, Artech House, 1974.

MOTOROLA, INC., *Analysis and Design of Integrated Circuits*, McGraw-Hill, 1967.

MULLER, R. C., *Device Electronics for Integrated Circuits*, Wiley, 1977.

OXNER, E. S., *Power FETs and Their Applications*, Prentice-Hall, 1982.

RICHMAN, P., *MOS Field-Effect Transistors and Integrated Circuits*, Wiley, 1973.

SEVIN, L. J., *Field-Effect Transistors*, McGraw-Hill, 1965.

SHAH, R. R., S. A. EVANS, D. L. CROSTHWAIT, and R. L. YEAKLEY, "Laser Recrystallized Polysilicon for High Performance $I^2L$," *IEEE Trans. Electron Devices*, Vol. ED-28, No. 12, December 1981.

STREETMAN, B. G., *Solid State Electronic Devices*, Prentice-Hall, 1980.

SZE, S. M., *Physics of Semiconductor Devices*, Wiley, 1969.

TILL, W. C., and J. T. LUXON, *Integrated Circuits: Materials, Devices, and Fabrication*, Prentice-Hall, 1982.

VERONIS, A., *Integrated Circuit Fabrication Technology*, Reston, 1979.

WALLMARK, J. T., and H. JOHNSON, *Field-Effect Transistors: Physics, Technology, and Applications*, Prentice-Hall, 1966.

WOLF, H. F., *Semiconductors*, Wiley, 1971.

# 3

# CONSTANT-CURRENT SOURCES, VOLTAGE SOURCES, AND VOLTAGE REFERENCES

## 3.1 CONSTANT-CURRENT SOURCES

The ideal constant-current source is an electric circuit element that provides a current to a load that is independent of the voltage across the load, or equivalently the load impedance. Note that the adjective "constant" in the term "constant-current source" refers to the fact that the current will be independent of the load conditions. Therefore, under this definition a constant-current source can produce a current that is time varying, that is, an a-c current. The constant-current source can also be a controlled source such that the strength of the current source is a function of some other voltage or current in the system, but not of the voltage across the load supplied by the constant-current source under consideration.

In electronic circuits, especially for integrated circuits, there are many applications for constant-current sources, and in particular d-c constant-current sources. Although it is never possible to have an ideal constant-current source in a real electronic circuit, there are ways to produce circuits that provide a very close approximation to the ideal constant-current source.

For integrated-circuit applications these constant-current sources generally make use of the fact that for a transistor in the active mode of operation (i.e., emitter–base junction "on," collector–base junction "off"), the collector current will be relatively independent of the collector voltage. For a transistor in the active mode or region of operation, the voltage across the transistor from collector to emitter, $V_{CE}$, should be greater than approximately 0.2 V but less than the collector-to-emitter breakdown voltage, $BV_{CEO}$, which is typically at least 50 V for IC transistors. Over this range of voltages the collector current, $I_C$, will be relatively independent of the collector-to-emitter voltage, $V_{CE}$, of the transistor. FETs operating in the saturated mode in which the drain current is relatively independent of the drain-to-source voltage can also be used for current sources. Both JFET and MOSFET current sources will be presented in this chapter.

As a preface to looking at some constant-current sources, let us first consider the basic circuit shown in Figure 3.1. We will first consider the case of the two transistors being identical in every respect. Noting that the bases of the two transistors are tied together and that the two emitters go to the same point, we have that $V_{B1} = V_{B2}$, and $V_{E1} = V_{E2}$, so that $V_{BE1} = V_{BE2}$. As a result the two transistors will have *exactly* the same base-to-emitter voltages. Transistor $Q_1$ is a *diode-connected* transistor with the collector shorted to the base so that $V_{CB} = 0$. The base–emitter junction of $Q_1$ will be biased "on" as a result of the current $I_1$ that is forced through this transistor. As a result of the fact that $V_{CB} = 0$, the collector–base junction will be "off." Therefore, transistor $Q_1$ will be operating in the active region.

Transistor $Q_2$ will also be in the active region as long as the voltage across $Q_2$, $V_{CE2}$, is greater than 0.2 V but less than the breakdown voltage, $BV_{CEO}$. Since we have two identical transistors, both in the active region with identical base-to-emitter voltages, the collector currents of the two transistors will be approximately equal, so that $I_{C2} = I_{C1}$. Since $I_1 = I_{C1} + I_{B1} + I_{B2} = I_C + 2I_C/\beta = I_C\ (1 + 2/\beta)$, as a result we have that $I_{C2} = I_{C1} = I_1/(1 + 2/\beta)$. Since the current gain $\beta$ (or $h_{fe}$) values of IC transistors are typically much greater than unity, we can say that $I_{C2} = I_{C1} \simeq I_1$. For a typical current gain value of 100, the effect of the base current will be to produce only a 2% difference between $I_C$ and $I_1$. Even a current gain as low as 50 will result in only a 4% difference between $I_C$ and $I_1$. Therefore, for most applications we can safely neglect the effects of the base current and say that $I_{C_2} = I_{C_1} \simeq I_1$.

The circuit that we have been considering so far can be called a *current mirror*, since the current that flows through the left-hand side of the circuit produces essentially a mirror image on the right-hand side. This circuit will be the basis of most of the current source circuits that we will be looking at, and it will also be the basis of most of the differential amplifier active load circuits to be considered later.

The foregoing analysis of the current mirror transistor pair was made on the basis of the two transistors being identical in every respect. We will now consider what happens in an actual situation in which this is not the case, for even in the case of two integrated-circuit transistors of identical design right next to each other

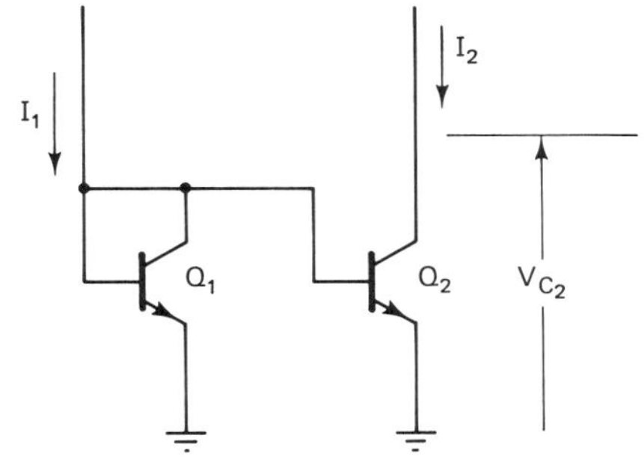


**Figure 3.1** Current mirror circuit.

on the same IC chip, there will still be small differences in the electrical characteristics of the two transistors.

The most important difference that will exist between two nominally identical transistors will be the base width, $W$. Differences in the base widths of two otherwise identical transistors will result in differences in the current gain values and will produce an offset voltage, $V_{OS}$.

For the current mirror circuit the differences in the current gain values will not be important since the base current is small anyway. The offset voltage, however, can be of importance. The concept of the offset voltage of a pair of transistors will also be important in the discussion of the differential amplifier. For two transistors that are characterized in the active region by the relationships

$$I_{C_1} = I_{TO_1} \exp\left(\frac{V_{BE_1}}{V_T}\right) \qquad \text{for } Q_1 \tag{3.1}$$

$$I_{C_2} = I_{TO_2} \exp\left(\frac{V_{BE_2}}{V_T}\right) \qquad \text{for } Q_2 \tag{3.2}$$

the offset voltage, $V_{OS}$, is given by the equation $\exp(V_{OS}/V_T) = I_{TO_1}/I_{TO_2}$. Thus, for a current mirror pair (i.e., $V_{BE_1} = V_{BE_2}$), the two collector currents will not be exactly equal, but will differ by the ratio $I_{C_1}/I_{C_2} = I_{TO_1}/I_{TO_2} = \exp(V_{OS}/V_T)$. For nominally identical IC transistors offset voltages of about $\pm 1$ mVolt are typical. This will correspond to a current ratio of $I_{C_1}/I_{C_2} = \exp(V_{OS}/V_T) = \exp(\pm 1 \text{ mV}/25 \text{ mV}) = 1 \pm \frac{1}{25} = 1 \pm 0.04$, or a $\pm 4\%$ difference in the two collector currents of the pair of transistors.

Let us now consider the simple constant-current source circuit shown in Figure 3.2. Current $I_1$ will be given by

$$I_1 = \frac{V^+ - V_{BE}}{R_1} \tag{3.3}$$

where $V^+$ is the positive d-c supply voltage for the circuit. Since $I_2 = I_{C_2} = I_{C_1} \simeq I_1$, we see that the design of this circuit is relatively simple. If, for example, we want this current source to supply a current of $I_2 = 1.0$ mA, we need $I_1 = 1.0$ mA. If $V^+ = 15$ V, we have that $R_1 = (V^+ - V_{BE})/I_1 = (15 - 0.7)$ V/1.0 mA $= 14.3$ kΩ.

It is important to note that $I_2$ will stay approximately constant at 1.0 mA

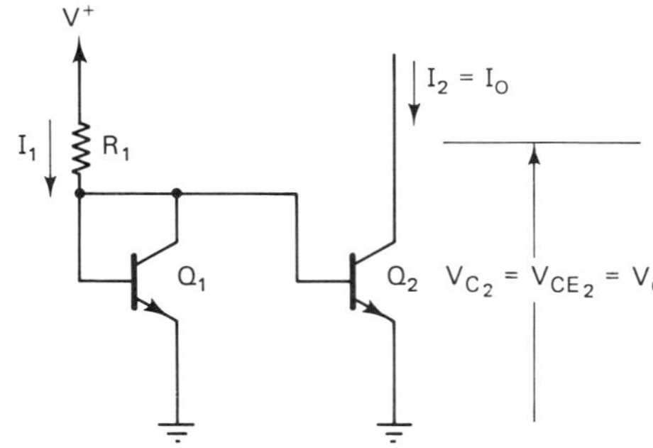


**Figure 3.2** Basic current source circuit.

only over the range of $V_{C_2}$ values such that transistor $Q_2$ remains in the active region. If $BV_{CEO} = 50$ V, the range of $V_{C_2}$ voltages for constant-current source operation will be from +0.2 to +50 V. Outside of this range of voltages the circuit will no longer act, even approximately, as a constant-current source.

The voltage range over which a circuit acts as an approximate constant-current source is called the *voltage compliance range*. For the circuit under consideration the voltage compliance range is thus from +0.2 to +50 V.

Even within the voltage compliance range, however, this circuit is only a good approximation to an ideal current source. In Figure 3.3 the output characteristics of this current source circuit are compared to those of an ideal constant-current source. Note that within the voltage compliance range the output current of this current source does increase slightly with increasing voltage across the current source, $V_O = V_{C_2}$. The $I_o$ versus $V_o$ characteristic curve of the current source circuit has a relatively constant slope over most of the voltage compliance range. This slope is given mathematically by *slope* $= dI_O/dV_O = g_o =$ *dynamic output conductance of the current source*. The reciprocal of $g_o$ is $r_o$ as given by $r_o = 1/g_o$ and is the *dynamic output resistance* of the current source. We note that the ideal current source has a dynamic output conductance, $g_o$, of zero (i.e., $g_o = 0$), and correspondingly an infinite dynamic output resistance.

In Figure 3.4 a circuit representation of the real constant-current source is shown and contrasted with the ideal current source for which $g_o = 0$. Note that this representation holds only within the voltage compliance range of the current source.

We will now proceed to determine some typical values for the output conductance, $g_o$. For $g_o$ we have

$$g_o = \frac{dI_o}{dV_o} = \frac{dI_{C2}}{dV_{C2}} = \frac{dI_{C2}}{dV_{CE_2}} = g_{ce2} = \frac{I_C}{V_A} \tag{3.4}$$

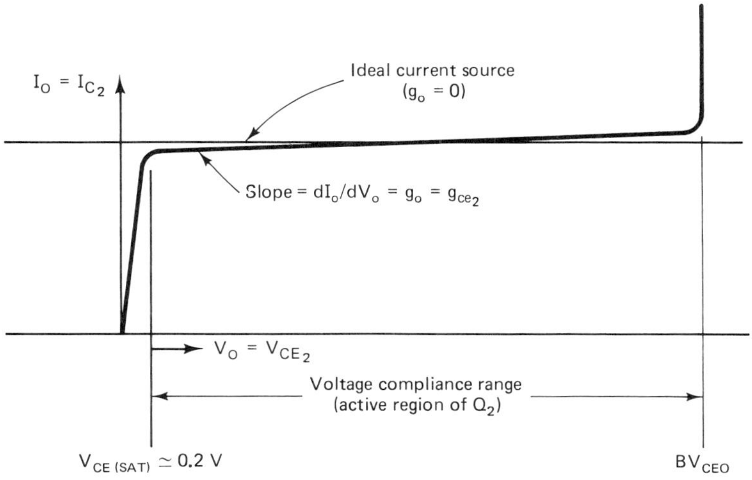


**Figure 3.3** Current source output characteristics.

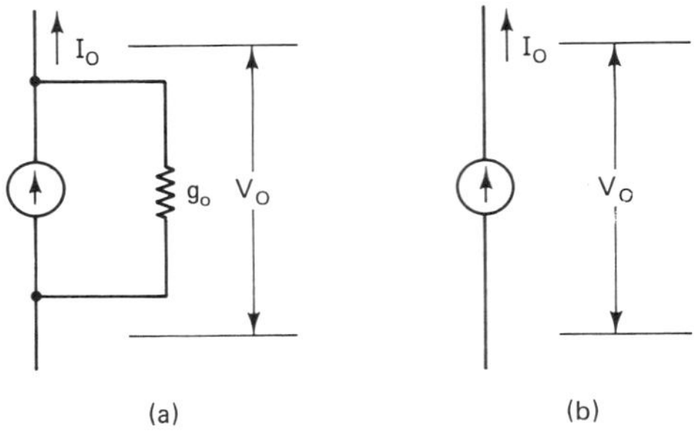


**Figure 3.4** Circuit representations of current sources: (a) real current source; (b) ideal current source ($g_o = 0$).

where $g_{ce_2}$ is the dynamic collector-to-emitter conductance of $Q_2$, and $V_A$ is the *Early voltage* or *base-width modulation coefficient* of $Q_2$ as discussed in Appendix B. Typical values of $V_A$ for IC transistors are in the range 100 to 300 V, with 250 V being selected here as a representative value. Since $I_{C_2} = 1.0$ mA we have that $g_o = g_{ce_2} = I_C/V_A = 1.0$ mA/250 V $= 1000\ \mu$A/250 V $= 4\ \mu$A/V $= 4\ \mu$S, and correspondingly $r_o = 250$ V/1.0 mA $= 250$ kΩ. Therefore, we see that $I_O = I_{C_2}$ will change by 4 $\mu$A for every 1-V change in $V_O$. On a percentage-wise basis the percentage change in $I_O$ per 1-V change in $V_O$ will be given by $(1/I_C)(dI_C/dV_C) \times 100\% = 4\ \mu\text{A/V}/1000\ \mu\text{A} \times 100\% = \underline{0.4\ \%/\text{V}}$. Therefore, we see that the output current of this current source will increase by 0.4% for every 1-V increase in the voltage across the current source.

On a more general level we note that since $g_o = I_O/V_A$, the *fractional* change in the output current per 1-V change in the output voltage will be given by

$$\frac{1}{I_O}\frac{dI_O}{dV_O} = \frac{1}{I_O}g_o = \frac{1}{I_O}\frac{I_O}{V_A} = \frac{1}{V_A} \tag{3.5}$$

Correspondingly, the percentage change in $I_O$ per 1-V change in $V_O$ will be given by

$$\frac{1}{I_O}\frac{dI_O}{dV_O} \times 100\% = \frac{100\%}{V_A} \tag{3.6}$$

Note that this is *independent* of the output current level.

To summarize the characteristics of the basic current source circuit, it has a basic current output of 1.0 mA with a voltage compliance range of +0.2 to +50 V. The dynamic output conductance is 4 $\mu$S, or on a percentage basis, 0.4%/V. Note that the direction of current flow (at the top, or ungrounded terminal) is *into* the current source circuit. As a result of this direction of current flow, this type of constant-current circuit is often referred to as a current *sink*, as contrasted to a current *source* in which the current flows *out* of the top (i.e., ungrounded) terminal.

To obtain a current "source" equivalent to this current "sink" the same basic circuit configuration can be used with the supply voltage $V^+$ being replaced by a negative d-c voltage $V^-$, and the NPN transistors being replaced by PNP transistors, giving the result shown in Figure 3.5. This current source will have a voltage compliance range of from −0.2 to −50 V.

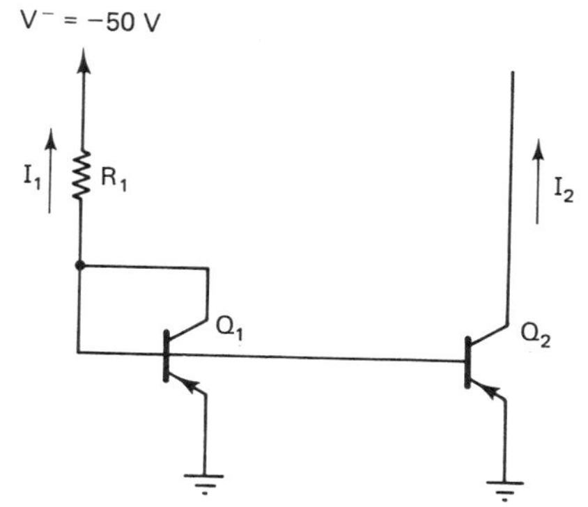


**Figure 3.5** Current source circuit.

### 3.1.1 Current Source with Bipolar Voltage Compliance Range

For many IC applications it is desirable to have a current source that has a "bipolar" compliance range, that is, a compliance range that extends over both positive and negative voltages. The circuit of Figure 3.6 is a simple current sink circuit that has a bipolar voltage compliance range. The basic principles of operation of this circuit are the same as the circuit previously considered. Since the emitter of $Q_2$ now goes $V^-$ rather than ground the voltage compliance range of this current sink is shifted in the negative direction by $V^-$ volts.

To consider an example, if $V^+ = 15$ V and $V^- = -15$ V, the voltage compliance range will be from $-14.8$ to $+35$ V. Thus the voltage compliance range extends through ground potential to negative voltage values.

For $R_1$ we now have the relationship that $I_1 = (V^+ - V^- - V_{BE})/R_1$. Thus, for $I_1 = 1.0$ mA, the corresponding value that is needed for $R_1$ will be $R_1 = (30 - 0.7)$ V/1.0 mA $= 29.3$ kΩ. If, on the other hand, the top end of $R_1$ is returned directly to ground rather than to $V^+$, the required value of $R_1$ can be reduced to 14.3 kΩ, as before.

The dynamic output conductance of this circuit will be the same as the circuit

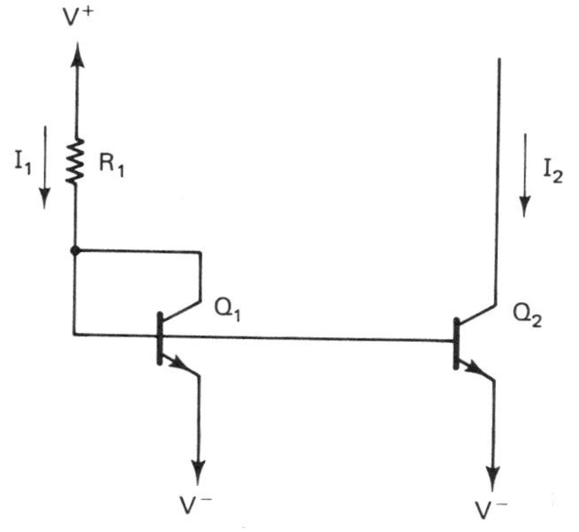


**Figure 3.6** Current source with bipolar voltage compliance range.

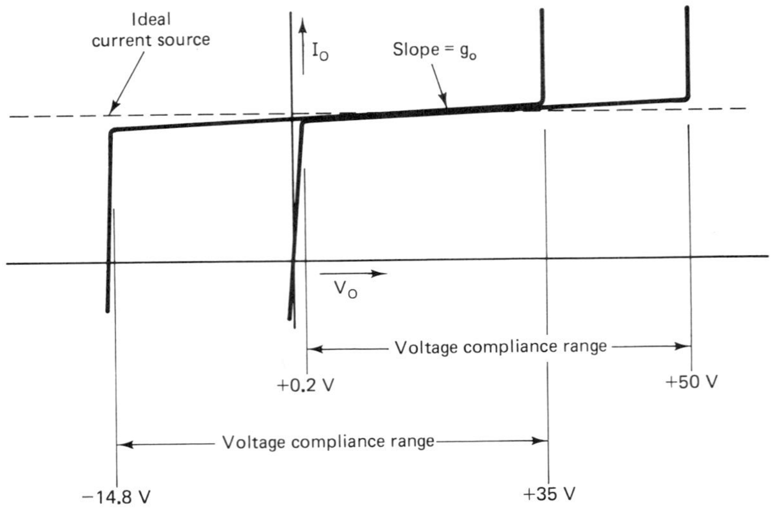


**Figure 3.7** Comparison of voltage compliance ranges.

previously considered, namely 4 $\mu$S (or 0.4%/V). In Figure 3.7 the output characteristics of this current source are compared to those of the first current source, and to that of an ideal current source.

### 3.1.2 Current Source for Low Current Levels

In the current source circuits just considered a resistance $R_1$ of 14.3 k$\Omega$ was required for a current level of 1.0 mA. For smaller current levels the value of $R_1$ will have to be increased proportionally. For many IC applications currents in the microampere range, or less are required. For a current of 1.0 $\mu$A a resistance value of $R_1 =$ 14.3 M$\Omega$ would be needed. If $R_1$ were to be an ordinary (i.e., discrete) resistor this would pose no major problem since a 14-M$\Omega$ resistor costs about the same, and is about the same size, volume, and weight as a 14-k$\Omega$ resistor. Integrated-circuit resistors, however, will take up an area on the silicon chip that is roughly proportional to the value of the resistor. Therefore, the prorated "cost" of the resistor in terms of the amount of "real estate" on the chip that it occupies will increase with increasing resistance. Generally, anything over 50 k$\Omega$ is to be avoided for IC applications, if at all possible.

As a result, for current levels below about 1.0 mA, a modification of the current source previously analyzed will now be considered. This new current source is shown in Figure 3.8, and the modification consists of the insertion of a resistor $R_2$ in series with the emitter of $Q_2$. With this change we will no longer have that $V_{BE_2} = V_{BE_1}$ due to the voltage drop across $R_2$. Indeed, we now have that

$$V_{BE_2} = V_{BE_1} - I_2 R_2 \tag{3.7}$$

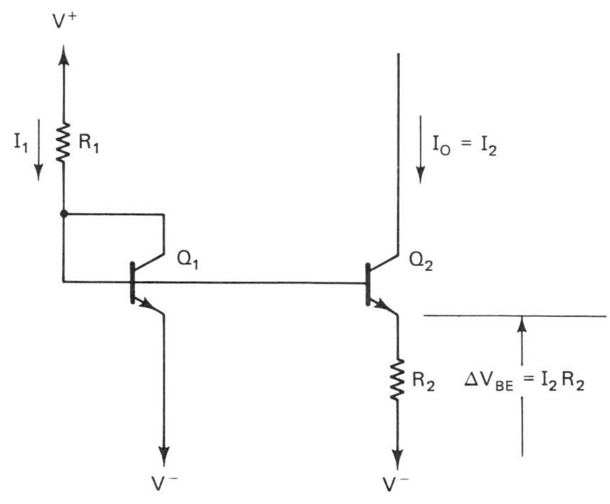


**Figure 3.8** Current source for low current levels.

To analyze this new situation we must go back to the basic exponential relationship for a transistor in the active region between the collector current and the base-to-emitter voltages as given by

$$I_C = I_{TO} \exp\left(\frac{V_{BE}}{V_T}\right) \tag{3.8}$$

For two identical transistors the ratio of the collector currents will be given by

$$\frac{I_{C_1}}{I_{C_2}} = \frac{\exp\,(V_{BE_1}/V_T}{\exp\,(V_{BE_2}/V_T} = \exp\left(\frac{\Delta V_{BE}}{V_T}\right) \tag{3.9}$$

where $\Delta V_{BE} = V_{BE_1} - V_{BE_2}$. For the circuit under consideration here we have that $\Delta V_{BE} = I_2R_2$, so that $I_1/I_2 = I_{C_1}/I_{C_2} = \exp\,(I_2R_2/V_T)$. We now see that as a result of the introduction of $R_2$ into the circuit, $I_2$ no longer will be equal to $I_1$, and indeed we can make $I_2$ much smaller than $I_1$.

To consider a specific example, let us require that $I_1 = 1.0$ mA, as before, but now let us design for an $I_O = I_2 = 10\ \mu\text{A} = 0.01$ mA. Since the current ratio is $I_1/I_2 = 1.0\ \text{mA}/0.01\ \text{mA} = 100 = \exp\,(I_2R_2/V_T)$, we have that $I_2R_2/V_T = \ln 100 = 4.6$, so that $R_2 = 4.6 \times 25\ \text{mV}/10\ \mu\text{A} = 11.5\ \text{k}\Omega$. Since $R_1$ is 14.3 kΩ, as before, for a total resistance $R_1 + R_2 = 26$ kΩ, a current level for $I_o = 10\ \mu$A can be obtained, whereas if $R_2$ was not present, $R_1$ would have to be increased to 14.3 V/10 μA = 1.43 MΩ. Thus with this circuit current levels down in the microampere range with acceptable total circuit resistance values (less than 50 kΩ) are possible.

We will now consider the effect that this circuit modification has on other current source parameters, such as voltage compliance range and the output conductance. As a result of the voltage drop across $R_2$ the total voltage compliance span will be shifted by $I_2R_2$. This generally will represent only a very small shift in the compliance range. For example, in the case just considered, $I_2R_2 = 4.6 \times 25$ mV = 115 mV. Thus for $V^+ = 15$ V and $V^- = -15$ V, the compliance range will be shifted from −14.8 to +35 V to −14.7 to +34.9 V, an almost negligible change.

The addition of $R_2$ to the circuit will, on the other hand, produce a very major change in the output conductance, and a beneficial one at that. For the case of the circuit with $R_2$ we must consider the full equation for the transistor output conductance as derived in Appendix B and given by

$$g_o = g_c = \frac{I_C}{V_A} \frac{1 + (I_C/V_T)(Z_E + Z_B)/\beta}{1 + (I_C/V_T)[Z_E + (Z_E + Z_B)/\beta]} \tag{3.10}$$

For the circuit under consideration $Z_E = R_2$, and $Z_B$ is the dynamic impedance that is seen looking out from the base of $Q_2$. This dynamic impedance will consist of the parallel combination of $R_1$ and the dynamic resistance of the diode-connected transistor $Q_1$. The dynamic resistance of the diode-connected transistor will be given by $r_d = V_T/I_1 = 25\ \text{mV}/1.0\ \text{mA} = 25\ \Omega$. Therefore, $Z_B = R_1 \parallel r_d = 14.3\ \text{k}\Omega \parallel 25\ \Omega \simeq 25\ \Omega$. Since $I_C = I_{C_2} = I_O = 10\ \mu\text{A}$, and again using $V_A = 250$ V, we obtain

$$g_o = g_c = \frac{10\ \mu\text{A}}{250\ \text{V}} \frac{1 + (10\ \mu\text{A}/25\ \text{mV})(11.5\ \text{k}\Omega)/\beta}{1 + (10\ \mu\text{A}/25\ \text{mV})(11.5\ \text{k}\Omega + 11.5\ \text{k}\Omega/\beta)} \tag{3.11}$$

Since $25\ \text{mV}/10\ \mu\text{A} = 2.5\ \text{k}\Omega$, for $\beta \gtrsim 100$, $(11.5\ \text{k}\Omega/2.5\ \text{k}\Omega)/\beta \ll 1$, so that

$$g_o = g_c \simeq 0.04\ \mu\text{S} \frac{1}{1 + (11.5/2.5)} = \frac{40\ \text{nS}}{5.6} = 7.14\ \text{nS} \tag{3.12}$$

Thus $g_o = 7.14$, nS $= 7.14$ nA/V, and $r_o = 1/g_o = 140\ \text{M}\Omega$. On a percentage basis this becomes

$$\frac{1}{I_O}\frac{dI_O}{dV_O} = \frac{7.14\ \text{nA/V}}{10\ \mu\text{A}} \times 100\% = \underline{0.0714\%/\text{V}} \tag{3.13}$$

Thus we see that the output current will increase by only 0.07% for every volt increase in the voltage across the current source. For a 10-V change in $V_o$ the change in $I_o$ will still be less than 1%, so that this current source represents a very close approximation indeed to an ideal current source. In comparison with the circuit with $R_2 = 0$ we note that for that case the normalized change in the output current was 0.4%/V, so that this circuit with $R_2$ represents an improvement by a factor of 5.6.

### 3.1.3 Dependence of Current Source Current on Supply Voltage

It is desirable that the strength of the current source be as independent of the d-c supply voltage as possible for most applications. We will now consider the dependence of the strength of the current source, $I_O$, on the supply voltage using the circuit just considered for the "current source for low current levels" as presented in Figure 3.8. Using the exponential relationship between the collector currents and the base-to-emitter voltages, we have previously obtained the relationship

$$\frac{I_1}{I_2} = \exp\left(\frac{\Delta V_{BE}}{V_T}\right) = \exp\left(\frac{I_2 R_2}{V_T}\right) \tag{3.14}$$

so that $I_1 = I_2 \exp(I_2R_2/V_T)$. Taking the derivative $dI_1/dI_2$ gives

$$\begin{aligned}\frac{dI_1}{dI_2} &= \exp\left(\frac{I_2R_2}{V_T}\right) + \frac{R_2}{V_T} I_2 \exp\left(\frac{I_2R_2}{V_T}\right)\\ &= \frac{I_1}{I_2} + \frac{I_1}{I_2}\ln\frac{I_1}{I_2}\\ &= \frac{I_1}{I_2}\left(1 + \ln\frac{I_1}{I_2}\right)\end{aligned} \tag{3.15}$$

The reciprocal quantity becomes

$$\frac{dI_2}{dI_1} = \frac{I_2}{I_1}\frac{1}{1 + \ln(I_1/I_2)} \tag{3.16}$$

In terms of the fractional change in the currents, $dI_2/I_2$ and $dI_1/I_1$ we now have

$$\frac{dI_2}{I_2} = \frac{dI_1}{I_1}\frac{1}{1 + \ln(I_1/I_2)} \tag{3.17}$$

For example, if $I_1 = 1.0$ mA and $I_2 = 10\ \mu$A $= 0.01$ mA as considered before, ln $(I_1/I_2) = \ln 100 = 4.6$, so that

$$\frac{dI_2}{I_2} = \frac{dI_1}{I_1}\frac{1}{5.6} \tag{3.18}$$

and thus we see that the fractional or percentage change in the current source output current $I_O = I_2$ will be 5.6 times *smaller* than the change in $I_1$.

In contrast with this, for the case in which $R_2 = 0$ we will have that $I_2 = I_1$, so that $\ln(I_1/I_2) = 0$, and the fractional change in $I_2$ will thus be equal to the change in $I_1$.

The current $I_1$ is related to the supply voltage by

$$I_1 = \frac{V^+ - V^- - V_{BE}}{R_1} \simeq \frac{V^+ - V^-}{R_1} = \frac{V_{\text{supply}}}{R_1} \tag{3.19}$$

From this we can easily see that

$$\frac{dI_1}{I_1} \simeq \frac{dV_{\text{supply}}}{V_{\text{supply}}} \tag{3.20}$$

that is, the fractional change in $I_1$ will be equal to the fractional change in the supply voltage. Therefore, for the fractional change in $I_2$ we will have

$$\frac{dI_2}{I_2} \simeq \frac{dV_{\text{supply}}}{V_{\text{supply}}}\frac{1}{1 + \ln(I_1/I_2)} \tag{3.21}$$

Thus for a 100:1 current ratio, the fractional (or percentage) change in $I_2$ will be approximately 5.6 times *smaller* than the fractional (or percentage) change in the total supply voltage.

As an example, if $V^+ = +15$ V and $V^- = -15$ V, we have that $V_{\text{supply}} = 30$

V. For every 1-V change in the total supply voltage the percentage change in the current source current, $I_O = I_2$ will be

$$\frac{dI_o}{I_o} \times 100\% = \frac{1\text{ V}}{30\text{ V}} \times \frac{1}{5.6} \times 100\% = \underline{0.6\%/\text{V}} \tag{3.22}$$

Thus the current source current will change by approximately 0.6% for every 1-V change in the total supply voltage. This dependence of the strength of the current source on the supply voltage will be important when such things as the dependence of amplifier gain on the supply voltage and the power supply rejection ratio (**PSRR**) are considered.

### 3.1.4 Temperature Coefficient of the Current Source

We will now consider the effect of temperature on the current source. We will start with the basic interrelationship between the two transistor currents given by $I_2 = I_1 \exp(-I_2R_2/V_T)$. Noting that $I_1$ itself will have a temperature dependence and that $V_T = kT/q$, taking the derivative of $I_2$ with respect to temperature gives

$$\frac{dI_2}{dT} = \frac{dI_1}{dT}\frac{I_2}{I_1} + I_2\left[\frac{-I_2R_2}{V_T}\frac{1}{I_2}\frac{dI_2}{dT} + \frac{I_2R_2}{TV_T} - \frac{I_2}{V_T}\frac{dR_2}{dT}\right] \tag{3.23}$$

so that dividing through by $I_2$ and collecting terms in $I_2$ on the left-hand side gives

$$\frac{1}{I_2}\frac{dI_2}{dT}\left(1 + \ln\frac{I_1}{I_2}\right) = \frac{1}{I_1}\frac{dI_1}{dT} + \frac{1}{T}\ln\frac{I_1}{I_2} - \ln\frac{I_1}{I_2}\frac{1}{R_2}\frac{dR_2}{dT} \tag{3.24}$$

Since $I_1 = (V_{\text{supply}} - V_{BE})/R_1$, we have that

$$TCI_1 = \frac{1}{I_1}\frac{dI_1}{dT} = -\frac{1}{R_1}\frac{dR_1}{dT} - \frac{dV_{BE}/dT}{V_{\text{supply}} - V_{BE}} \simeq -TCR_1 - \frac{dV_{BE}/dT}{V_{\text{supply}}} \tag{3.25}$$

where $TCR_1 = (1/R_1)(dR_1/dT)$ = temperature coefficient of $R_1$. Substituting the expression for the temperature coefficient of $I_1$, $TCI_1$, into the expression for the temperature coefficient of $I_2$ gives

$$\begin{aligned} TCI_2 &= \frac{1}{I_2}\frac{dI_2}{dT} \\ &\simeq \frac{-TCR_1 - [(dV_{BE}/dT)/V_{\text{supply}}] + (1/T)\ln(I_1/I_2) - \ln(I_1/I_2)TCR_2}{1 + \ln(I_1/I_2)} \end{aligned} \tag{3.26}$$

We will now consider a typical example. For IC resistors the temperature coefficient is typically in the range +2000 ppm/°C = $+2 \times 10^{-3}$/°C. For $dV_{BE}/dT$ a value of −2.2 mV/°C is appropriate. We will again choose $I_1/I_2 = 100$ so that $\ln(I_1/I_2) = 4.6$, and $V_{\text{supply}} = 30$ V. Substituting values into the equation for $TCI_2$ gives

$$\begin{aligned} TCI_2 &= \frac{-2 \times 10^{-3} + (2.2\text{ mV}/30\text{ V}) + (1/300)(4.6) - (4.6)(2 \times 10^{-3})}{1 + 4.6} \\ &= 0.751 \times 10^{-3}/°\text{C} = 751\text{ ppm}/°\text{C} = \underline{0.0751\%/°\text{C}} \end{aligned} \tag{3.27}$$

Note that the most important terms in the expression for $TCI_2$ are the $TCR_1$, $TCR_2$, and the $(1/T)$ ln $(I_1/I_2)$ terms. A calculation of $TCI_2$ based on just these three terms will give a result of 0.074%/°C.

As a result of the calculation of the temperature coefficient of the current source output current we see that for the preceding example the current will increase by only 0.075%/°C. This variation of the current source current with temperature will contribute to the variation of the amplifier gain, and to a small extent to the variation of offset voltage with temperature.

### 3.1.5 Wilson Current Mirror

We will now consider a compound type of current mirror known as the Wilson current mirror as shown in Figure 3.9. It will be demonstrated that this type of current mirror can offer some significant advantages over the simple type of current mirror considered earlier.

**Base current cancellation.** For this analysis we will consider all transistors to be essentially identical. Since $Q_1$ and $Q_2$ have identical base-to-emitter voltages, then $I_{C_1} = I_{C_2}$. Noting that the base currents will be very small compared to the collector currents we can say that $I_3 \simeq I_2 \simeq I_{C_2}$, so that all of the base currents will be approximately equal to each other. We can therefore write the following nodal equations:

$$(1)\ I_1 = I_{C_1} + I_B \tag{3.28}$$

$$(2)\ I_2 = I_{C_2} + 2I_B \tag{3.29}$$

$$(3)\ I_2 = I_3 + I_B \tag{3.30}$$

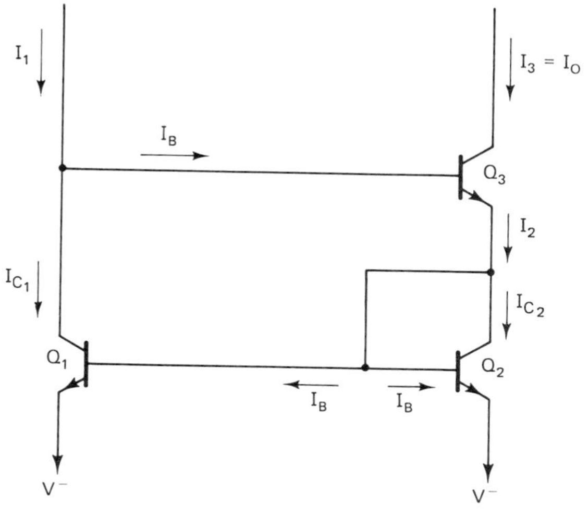


**Figure 3.9** Wilson current mirror.

and the relationship given above that

$$(4)\ I_{C_1} = I_{C_2} \tag{3.31}$$

Combining equations (2) and (3) gives

$$I_3 = I_2 - I_B = I_{C_2} + 2I_B - I_B = I_{C_2} + I_B = I_{C_1} + I_B \tag{3.32}$$

Identifying this last result with equation (1) gives $I_3 = I_1$. We note that there has been a cancellation of the effect of the finite base currents. In actuality, the base current cancellation will not be exact due to mismatches in the transistors, but the resulting difference between $I_3$ and $I_1$ will be extremely small.

**Voltage compliance range.** For this circuit to operate properly all three transistors must be operating in the active region. Since the voltage drop across $Q_2$ will be $V_{BE}$ or about 0.6 V, and the voltage required to keep $Q_3$ out of saturation will be about +0.2 V, the total voltage required across $Q_2$ and $Q_3$ will be approximately 0.8 V. For example, if $V^- = -15$ V, the lower limit of the voltage compliance range will be −14.2 V.

**Dynamic output conductance of Wilson current mirror.** To determine the dynamic output conductance of this circuit we will represent the dynamic collector-to-emitter conductance, $g_{ce}$, of $Q_3$ as a conductance external to the transistor as shown in Figure 3.10. We will assume a change in the output voltage of $\Delta V_o$ and determine the corresponding change in output current, $\Delta I_O$, that results. Then taking the ratio of $\Delta I_O$ to $\Delta V_O$ will give us the output conductance as $g_o = \Delta I_O / \Delta V_O$.

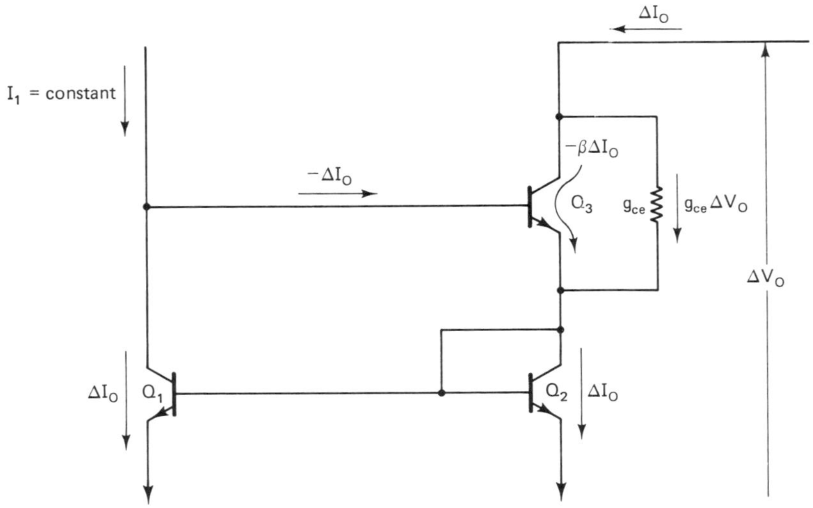


**Figure 3.10** Dynamic output conductance of Wilson current mirror.

The change in the output current $\Delta I_O$ in passing through $Q_2$ will produce an equal change in the current through $Q_1$. If we assume that the supply current $I_1$ remains constant, the change in the base current of $Q_3$ will be $-\Delta I_O$. This change in the base current of $Q_3$ will produce a change in the collector current of $-\beta \Delta I_O$.

The change in the output voltage, $\Delta V_O$, will produce a change in the current through $g_{ce}$ of $g_{ce} \Delta V_O$. If we now sum the currents at the collector or $Q_3$ we obtain

$$\Delta I_O = -\beta \Delta I_O + g_{ce} \Delta V_O \tag{3.33}$$

After collecting terms in $I_O$ on the left-hand side we obtain

$$\Delta I_O(1 + \beta) = g_{ce} \Delta V_O \tag{3.34}$$

so that we now have

$$\text{dynamic output conductance} = g_o = \frac{\Delta I_O}{\Delta V_O} = \frac{g_{ce}}{1+\beta} = \frac{I_O/V_A}{1+\beta} \tag{3.35}$$

If, for example, $I_O = I_{C3} = 10\ \mu\text{A}$ and $V_A = 250$ V (the Early voltage), and $\beta = 100$, we obtain for $g_o$

$$g_o = \frac{10\ \mu\text{A}/250\ \text{V}}{101} = 0.4\ \text{nA/V} = 0.4\ \text{nS} \tag{3.36}$$

This is an extremely small output conductance. On a normalized or percentage basis we obtain

$$\frac{1}{I_O}\frac{dI_O}{dV_O} = \frac{1}{V_A(1+\beta)} = 4 \times 10^{-5}/\text{V} = \underline{4 \times 10^{-3}\%/\text{V}} \tag{3.37}$$

Thus the output current changes by only 0.004% per 1-V change in the output voltage.

### 3.1.6 Compound Current Sink (Current-Sink-Biased Current Sink)

In Figure 3.11 another type of compound current source is illustrated. This circuit can be considered to be a "current-sink-biased current sink" in which $Q_5$ and $Q_2$ (together with $R_1$ and $R_2$) operate as a conventional current source circuit, and in turn the collector current of $Q_2$ acts to bias $Q_1$, which acts as a current source. Transistors $Q_3$ and $Q_4$ are operated as diodes and the total voltage drop across these two diode-connected transistors, $2V_{BE} = 1.3$ V, is used to provide for the base-to-emitter voltage drop needed to keep $Q_1$ in the active region and the collector-to-base voltage required to keep $Q_2$ also in the active region. If $Q_3$ and $Q_4$ were not present (i.e., replaced by a short-circuit) it would not be possible for $Q_1$ and $Q_2$ to be simultaneously in the active region of operation.

To consider a design example, let us choose $I_o = 10\ \mu\text{A}$ and $I_1 = 1.0$ mA

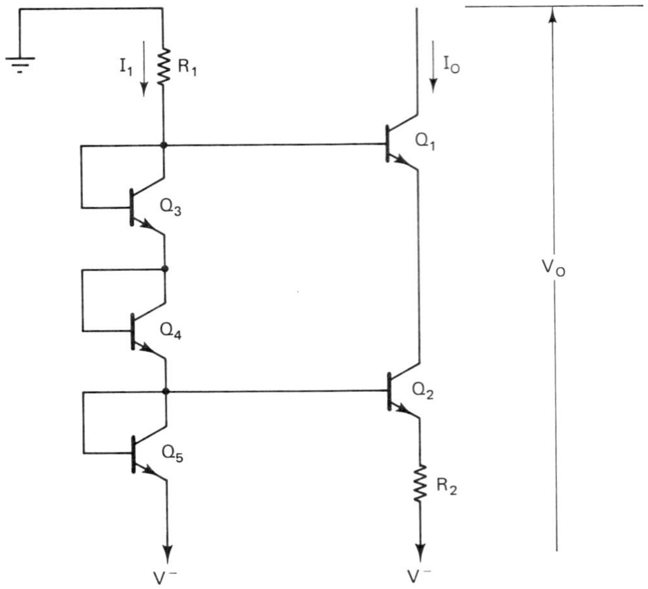


**Figure 3.11** Compound current sink (current sink biased current sink).

with $V^- = -15$ V. Since $I_1 = (V^- - 3V_{BE})/R_1$ and assuming a $V_{BE}$ of 0.7 V, we obtain for $R_1$,

$$R_1 = \frac{(15 - 2.1)\text{ V}}{1.0\text{ mA}} = 12.9\text{ k}\Omega$$

Since the current ratio, $I_1/I_2 = 100$, the voltage drop across $R_2$ will be given by

$$I_2R_2 = \Delta V_{BE} = V_T \ln 100 = 25\text{ mV} \times 4.6 = 115\text{ mV}$$

Since $I_2 = I_O = 10\ \mu$A, we have that $R_2 = 115$ mV/10 $\mu$A $= 11.5$ k$\Omega$. Thus we see that the design of this circuit is very similar to that of the simple current source for low current levels considered earlier.

**Voltage compliance range.** For this current sink to operate within its voltage compliance range, all the transistors in the circuit must be in the active mode of operation. The lower limit of the compliance range will be given by

$$\begin{aligned} V_{\text{lower limit}} &= V^- + V_{BE_5} + V_{BE_4} + V_{BE_3} + V_{CB(\text{SAT})1} \\ &= -15 + 3(0.7) - 0.5 = -13.4\text{ V} \end{aligned} \tag{3.38}$$

so that the compliance range extends to 1.6 V above the negative supply voltage. The upper limit of the voltage compliance range will be determined by the collector-to-base breakdown voltage of $Q_1$ as given by

$$V_{\text{upper limit}} = BV_{CBO_1} + 3(V_{BE}) + V^- \tag{3.39}$$

For a 50-V breakdown voltage this gives

$$V_{\text{upper limit}} = 50 + 2.1 - 15 = +37.1\text{ V} \tag{3.40}$$

**Dynamic output conductance of current-sink-biased current sink.** The principal feature of the current-sink-biased current sink compared to the simpler current source circuits is its very low output conductance. This is due to the very high impedance in series with the emitter of $Q_1$, which is the output impedance of $Q_2$.

If we look at the equation for the output or collector conductance of a transistor as given by

$$g_c = \frac{I_C}{V_A}\frac{1 + (I_C/V_T)(Z_E + Z_B)/\beta}{1 + (I_C/V_T)[Z_E + (Z_E + Z_B)/\beta]} \tag{3.41}$$

we see that in the limiting case as $Z_E$ gets very large, $g_c$ will approach $g_c = (I_C/V_A)/\beta$. Since that is approximately the case here, we see that this circuit will be characterized by a very small dynamic output conductance.

If we use $V_A = 250$ V as a representative value, and $\beta = 100$, we obtain

$$g_c \simeq \frac{I_C/V_A}{\beta} = \frac{10\ \mu\text{A}/250\ \text{V}}{100} = 0.4\ \text{nA/V} = \underline{0.4\ \text{nS}} \tag{3.42}$$

The corresponding value for $r_o$ is $r_o = 1/g_o = 1/(0.4\ \text{nS}) = \underline{2.5\ \text{G}\Omega}$. On a percentage basis, the percent change in $I_O$ per 1-V change in $V_O$ will be given by

$$\frac{1}{I_O}\frac{dI_O}{dV_O} \times 100\% = \underline{0.004\%/\text{V}} \tag{3.43}$$

This circuit is, therefore, within its compliance range, a very good approximation to the ideal current source.

**Dynamic output admittance.** The very low value of the dynamic output conductance of this current sink will be evident only at low frequencies. As we go to higher frequencies we must consider the parasitic capacitances in the circuit, and in doing so the dynamic output *admittance* of this current sink will be given by $y_o = g_o + j\omega C_{\text{total}}$. If this IC transistor has a collector–base junction capacitance of 1.0 pF and a collector-to-substrate capacitance of $C_{CS} = 2.0$ pF, the total capacitance seen looking into this current sink (i.e., looking into the collector of $Q_1$) will be $C_{\text{total}} = 3.0$ pF. The output admittance will therefore be given by

$$y_o = g_o + j\omega C_{\text{total}} = 0.4\ \text{nS} + j\omega 3.0\ \text{pF} \tag{3.44}$$

The breakpoint frequency for the admittance will be given by

$$f_{\text{breakpoint}} = \frac{g_o}{2\pi C_{\text{total}}} = 21\ \text{Hz} \tag{3.45}$$

Therefore, only for frequencies below about 10 Hz will the output admittance appear as approximately $g_o = 0.4$ nS. Above about 50 Hz, the dominant term in the output admittance will be the susceptive term $j\omega C_{\text{total}}$. For example, at $f = 1.0$ kHz the output admittance will be $y_O = j19$ nS, and correspondingly the dynamic output impedance will be $-j53$ M$\Omega$. At $f = 1.0$ MHz, the corresponding values will be $y_o = +j19\ \mu$S and $z_o = -j53$ k$\Omega$.

### 3.1.7 Multiple Current Sinks

In Figure 3.12 a multiple, or ganged, group of current sinks is illustrated. Transistor $Q_2$ and resistor $R_2$ serve as the reference for the current sink transistors $Q_3$ through $Q_6$. We note that the bases of transistors $Q_2$ through $Q_6$ are tied together so that all of these transistors will share a common base voltage. Furthermore, the bottom end of resistors $R_2$ through $R_6$ all go to the same negative supply voltage terminal.

Now, if the emitter areas of transistors $Q_2$ through $Q_6$ are scaled so that the *current density* is the same in all of these transistors, then the base-to-emitter voltage drops, $V_{BE}$, of all these transistors will be the same. As a result, the voltage across resistors $R_2$ through $R_6$ will be the same. Since the collector currents will be approximately equal to the emitter currents, and the voltage drop across all of the resistors is the same, we can say that $I_3R_3 = I_4R_4 = I_5R_5 = I_6R_6 = I_1R_2$. Therefore, the currents will be inversely proportional to the emitter resistances.

Let us now consider a design example. Let us specify the following current levels: $I_1 = 1.0$ mA, $I_3 = 1.0$ mA, $I_4 = 2.0$ mA, $I_5 = 4.0$ mA, and $I_6 = 8.0$ mA. We will also specify that the voltage drop across all of the resistors, except $R_1$, be 4.0 V. Therefore, for the resistors we have the following values:

(1) $R_2 = R_3 = 4.0\ \text{V}/1.0\ \text{mA} = 4.0\ \text{k}\Omega$

(2) $R_4 = 4.0\ \text{V}/2.0\ \text{mA} = 2.0\ \text{k}\Omega$

(3) $R_5 = 4.0\ \text{V}/4.0\ \text{mA} = 1.0\ \text{k}\Omega$

(4) $R_6 = 4.0\ \text{V}/8.0\ \text{mA} = 500\ \Omega$

If $V^+ = +10$ V and $V^- = -10$ V, the total voltage drop across $R_1 + R_2$ will be $20 - 1.4 = 18.6$ V. Since $R_2 = 4.0\ \text{k}\Omega$, this gives $R_1 = 14.6\ \text{k}\Omega$.

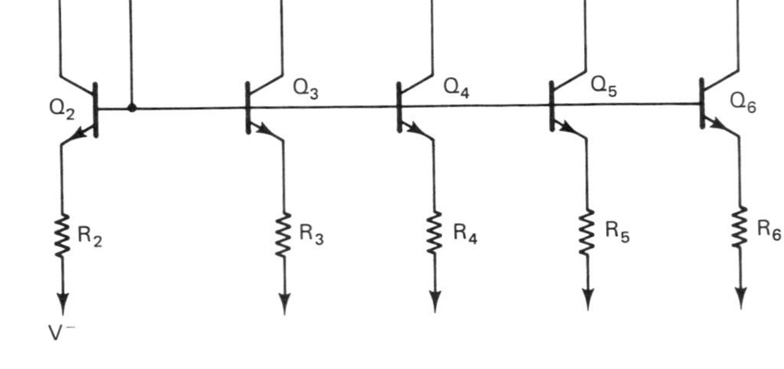

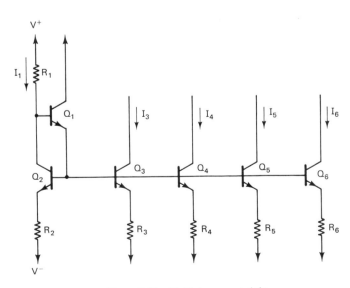

**Figure 3.12** Multiple current sinks.

Assuming a representative current gain value of 100, the total base current of transistors $Q_2$ through $Q_6$ will be $I_{B(\text{total})} = 16\text{ mA}/100 = 0.16\text{ mA}$. If transistor $Q_1$ were not present, the current through $R_1$, $I_1$, would be 0.16 mA larger than the current through $Q_2$. To reduce the difference between these two currents, transistor $Q_1$ is used to reduce the total base current by a factor of its current gain. If the current gain of $Q_1$ is 100, the total base current of 0.16 mA is reduced to 0.16 mA/100 = 1.6 $\mu$A. Therefore, the difference between the current of $Q_2$ and $I_1$ is now only 1.6 $\mu$A or 0.16%.

To reduce the amount of area taken up on the IC chip by the transistors, it may be desirable not to scale the transistor emitter areas. This would be especially of interest when very large current ratios are needed. If the transistor areas are not scaled to produce equal current densities, the base-to-emitter voltage drops of the transistors will not be equal, and as a result the resistor values must be adjusted to account for this, especially if very accurate current ratios are desired.

To consider an example in which the areas are not scaled, let us return to the previous example. The current through $Q_4$ is twice that through $Q_3$ and $Q_2$, so that the $V_{BE}$ of $Q_4$ will be greater than that of $Q_3$ and $Q_2$ by $V_T \ln (I_4/I_3) = 25$ mV ln 2 = 17 mV. Therefore, the voltage drop across $R_4$ will be 17 mV less than that across $R_3$ and $R_2$, so that $R_4$ must be reduced by 17 mV/2 mA = 8.7 Ω from 2000 to 1991 Ω. In a similar fashion the $V_{BE}$ of $Q_5$ will be $V_T \ln 4 = 35$ mV less than that of $Q_3$ and $Q_2$, so that $R_5$ must be reduced by 35 mV/4 mA = 8.7 Ω, from 1000 to 991 Ω. Finally, the $V_{BE}$ of $Q_6$ will be $V_T \ln 8 = 52$ mV less than that of $Q_3$ and $Q_2$, so that $R_6$ should be reduced by 52 mV/8 mA = 6.5 Ω from 500 to 493.5 Ω. These small adjustments in the resistance values will serve to compensate for the differences in the $V_{BE}$ values when the transistor areas are not scaled. Since the differences in the $V_{BE}$ values will vary with temperature, the current ratios will have a slight temperature dependence.

### 3.1.8 Current Source Independent of Supply Voltage

For some applications it is desirable to have a current source whose strength is almost completely independent of the supply voltage. An example of such a circuit is shown in Figure 3.13. For this analysis we will assume all transistors of the same type to be identical. We will also assume that $R_1 = R_3$, so that $I_1 = I_3$.

We note that $I_3$ and $I_2$ are related by $I_2/I_3 = \exp (I_3R_3/V_T)$. We also note that the ratio of $I_5$ to $I_6$ is determined by the ratio of $R_6$ to $R_5$ as given by $I_5/I_6 = R_6/R_5$. Since the base currents will be small compared to the collector and emitter currents, we will have that $I_2 = I_5$ and $I_3 = I_6$, so that we can write

$$\frac{I_2}{I_3} = \exp\left(\frac{I_3R_3}{V_T}\right) = \frac{I_5}{I_6} = \frac{R_6}{R_5} \tag{3.46}$$

Solving this last equation for $I_3$ gives

$$I_1 = I_3 = \frac{V_T}{R_3} \ln \frac{R_6}{R_5} \tag{3.47}$$

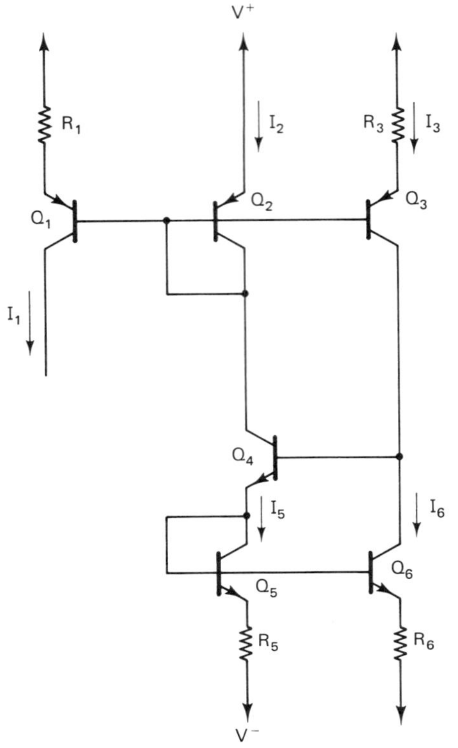


**Figure 3.13** Current source that is independent of supply voltage.

Note that the output current of the current source will be independent of the supply voltage. In order to have the best accuracy for the $I_5/I_6$ current ratio, the emitter areas of $Q_5$ and $Q_6$ should be scaled such that the two transistors will have equal current densities and therefore equal $V_{BE}$ drops.

To consider an example, let us choose a current ratio of $I_5/I_6 = 2$, so that $R_6/R_5 = 2$. For greatest accuracy $Q_5$ should have twice the emitter active area as $Q_6$. For this case we have that the current source output current will be given by

$$I_{\text{output}} = I_1 = \frac{V_T}{R_3} \ln 2 \tag{3.48}$$

From an output current of 1.0 $\mu$A, the required value of $R_3 = R_1$ will be

$$R_1 = R_3 = \frac{25 \text{ mV}}{1 \ \mu\text{A}} \ln 2 = 17.3 \text{ k}\Omega$$

Note that a very small current has been obtained using only a very moderate resistance value. Note also that if the active areas of $Q_5$ and $Q_6$ are properly scaled, it is possible to eliminate $R_5$ and $R_6$ from the circuit entirely. The current ratio $I_5/I_6$ will then be determined entirely the transistor active area ratio as given by $I_5/I_6 = A_5/A_6$. In this case the current source output current will be given by

$$I_{output} = I_1 = I_3 = \frac{V_T}{R_1} \ln \frac{A_5}{A_6} \tag{3.49}$$

Although the output current has been shown to be independent of the supply voltage, there is a minimum supply voltage level needed for the circuit to operate properly. For the case in which resistors $R_5$ and $R_6$ are not present, this minimum supply voltage will be $2V_{BE} + V_{CE(SAT)}$ or about 1.6 V.

The voltage compliance range on the output will range from an upper limit of $V^+ - I_1R_1 - V_{CE(SAT)_1} = V^+ - 0.2$ V, to a lower limit of $V^+ - BV_{CEO}$. For $V^+ = 10$ V and $BV_{CEO} = 50$ V, this will be a voltage compliance range of from +9.8 to −40 V.

This current source is characterized by a constant-current output over a very wide range of supply voltages, an ability to operate with total supply voltages as low as 1.6 V, and microampere current levels with only moderate resistance values. These characteristics make this type of current source especially suitable for micropower operational amplifier applications.

### 3.1.9 JFET Current Regulator Diodes

A diode-connected junction field-effect transistor (JFET) as shown in Figure 3.14 can be used as a current regulator diode or current source. The JFET is connected as a diode with the gate shorted to the source so that $V_{GS} = 0$. In Figure 3.15 the $V$–$I$ characteristic of this diode is shown, and it is basically the $I_{DS}$ versus $V_{DS}$ drain characteristic of the JFET with $V_{GS} = 0$. Note that when the drain-to-source voltage $V_{DS}$ goes above the "pinch-off" voltage $V_P$, the drain-to-source current $I_{DS}$ saturates at a current level of $I_{DSS}$. This is not a complete saturation, for as $V_{DS}$ continues to increase, the drain current will increase slightly. However, from the point where $V_{DS} = V_P$ up to where $V_{DS} = BV_{DS}$, which is the drain-to-source breakdown voltage, the drain current stays approximately constant, so that this region represents the voltage compliance range of this device. The dynamic conductance in this region is the drain-to-source conductance of the JFET $g_{ds}$ and is the slope of the curve as given by $g_{ds} = dI_{DS}/dV_{DS}$. For an ideal current regulator diode or current source, this slope would be zero.

The current regulator diode can be considered the circuit *dual* of the voltage regulator or zener diode. In Figure 3.16 the $V$–$I$ characteristics of a current regulator

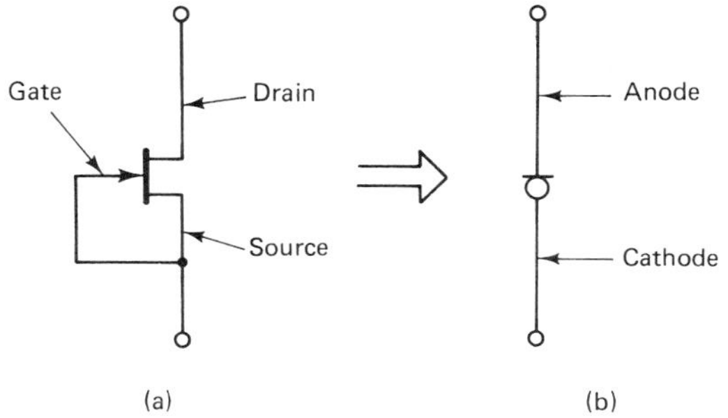


**Figure 3.14** Diode-connected J-FET as a current regulator diode: (a) diode-connected J-FET ($V_{GS} = 0$); (b) current regulator diode symbol.

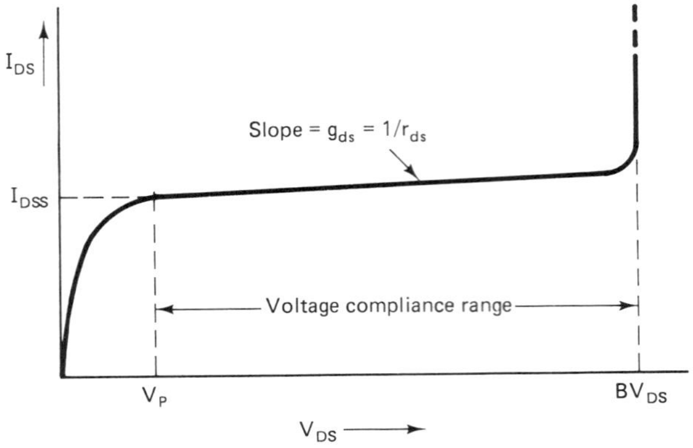


**Figure 3.15** Current regulator diode *V–I* characteristic.

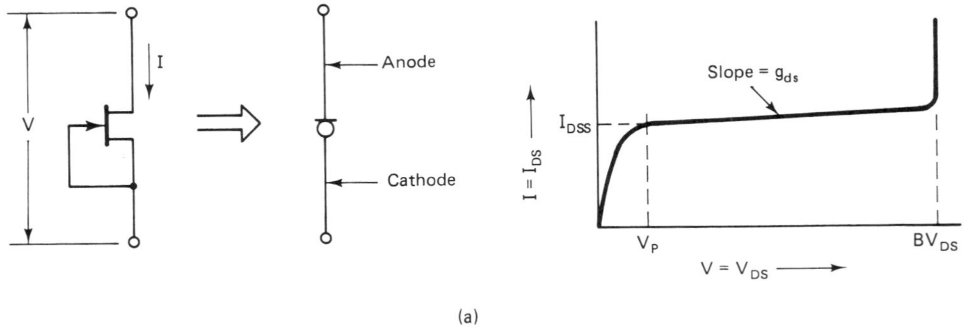


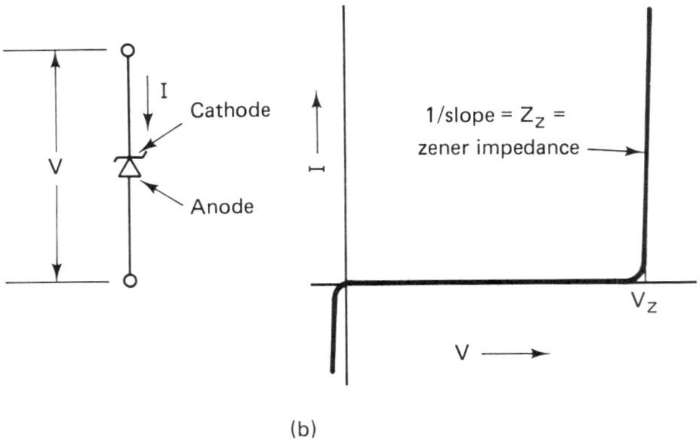


**Figure 3.16** Current regulator diode and voltage regulator diode comparison: (a) current regulator diode; (b) voltage regulator diode (zener diode).

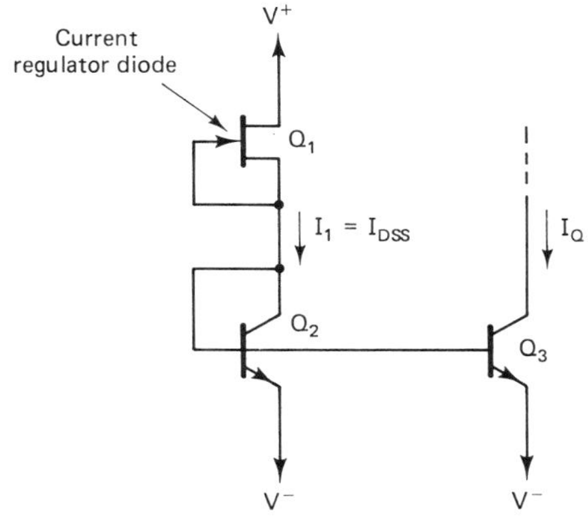


**Figure 3.17** Current sink using a current regulator diode.

diode and a voltage regulator diode are compared. Whereas the voltage regulator diode when operated in the breakdown region acts to keep the *voltage* drop across it at a constant value, the current regulator diode when operated within its voltage compliance range acts to keep the current through it at a constant value.

In Figure 3.17 an example is presented of the use of a current regulator diode. The current regulator diode $Q_1$ regulates the current through $Q_2$ at a value of $I_{DSS}$. Since $Q_2$ and $Q_3$ are a current mirror, the collector current of $Q_3$ which is the current sink current $I_Q$ will also be equal to $I_{DSS}$. As a consequence of the current through $Q_1$ being relatively independent of the voltage across $Q_1$, and therefore of the supply voltage, the current $I_Q$ will also be relatively independent of the supply voltage.

To consider an example, if $I_{DSS} = 1.0$ mA and $r_{ds} = 50$ kΩ, the rate of change of $I_Q$ with supply voltage will be given by

$$\begin{aligned}\frac{dI_Q}{dV_{\text{supply}}} &= \frac{dI_Q}{dI_{DS}} \times \frac{dI_{DS}}{dV_{DS}} \times \frac{dV_{DS}}{dV_{\text{supply}}} \\ &= \quad 1 \quad \times \quad g_{ds} \quad \times \quad 1 \\ &= 1/50\ \text{k}\Omega = \underline{20\ \mu\text{S}} = \underline{20\ \mu\text{A/V}}\end{aligned} \tag{3.50}$$

On a normalized basis this will be

$$\frac{1}{I_Q}\frac{dI_Q}{dV_{\text{supply}}} = (1/1000\ \mu\text{A}) \times 20\ \mu\text{A/V} = \underline{0.02/\text{V}} \text{ or } \underline{2\%/\text{V}} \tag{3.51}$$

so that $I_Q$ will increase by only 2% for every 1-V increase in the total supply voltage.

### 3.1.10 MOSFET Current Sources

A MOSFET can also be used as a constant current source. In Figure 3.18 the output or drain characteristic of a MOSFET is shown for the case of a fixed value of gate-to-source voltage $V_{GS}$ that is greater than the threshold voltage $V_t$. As the drain-to-source voltage $V_{DS}$ increases the drain current $I_{DS}$ increases. However, as $V_{DS}$ increases the voltage across the gate oxide at the drain end of the channel decreases.

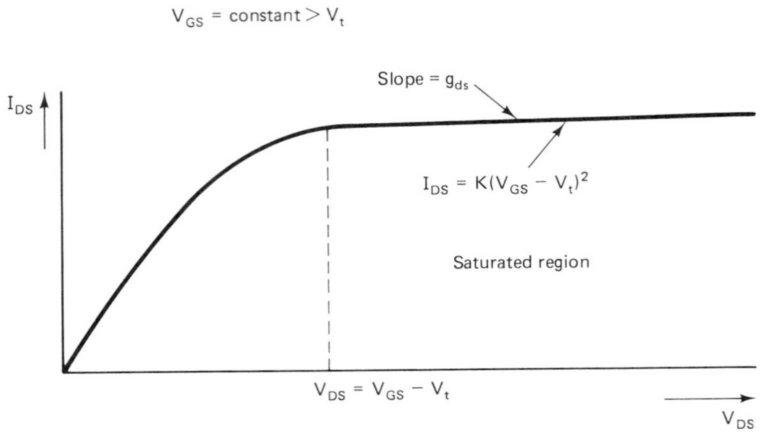


**Figure 3.18** MOSFET output characteristic.

This will result in a decrease in the mobile charge carrier population of the surface inversion layer at the drain end of the channel. This will lead to a corresponding decrease in the channel conductance so that as $V_{DS}$ increases the rate of increase of the drain current, $dI_{DS}/dV_{DS}$, which corresponds to the slope of the curve, will decrease.

When $V_{DS}$ reaches a value given by $V_{DS} = V_{GS} - V_t$, the voltage across the gate oxide at the drain end of the channel will be $V_{GD} = V_{GS} - V_{DS} = V_t$. Further increases in $V_{DS}$ will cause $V_{GD}$ to drop below the threshold voltage $V_t$ and the channel will become "pinched-off" for a very short distance at the drain end. This will result in the drain current leveling off or saturating at a value given by $I_{DS} = K\,(V_{GS} - V_t)^2$. In this equation $K = (\mu\ C_{ox}/2)(W/L)$ where $\mu$ is the carrier mobility in the surface inversion layer channel, $C_{ox}$ is the capacitance per unit area of the gate oxide MOS capacitor, and $W/L$ is the channel width to length ratio.

In the "saturated" region the drain current increases only very slowly with increasing drain voltage. The slope of the $I_{DS}$ versus $V_{DS}$ curve in the saturated region is the dynamic drain-to-source conductance $g_{ds}$. This dynamic conductance will be directly proportional to the drain current and can be expressed approximately as $g_{ds} = I_{DS}/V_A$ where $V_A$ is a transistor parameter having units of volts and is analogous to the Early voltage of a bipolar transistor. The reciprocal quantity $1/V_A$ is the channel length modulation coefficient and is similarly closely related to the base width modulation coefficient of a bipolar transistor (see Appendix B). The $V_A$ parameter for MOSFETs has about the same range of values as found for the Early voltage of bipolar transistors, and will generally be in the range of 30 to 150 V.

There are many MOSFET current source configurations that are similar to the bipolar transistor current source circuits considered previously. A simple example is the circuit of Figure 3.19 which uses a MOSFET current mirror. For current $I_1$ we have $I_1 = (V^+ - V^- - V_{GS})/R_1$. If transistors $Q_1$ and $Q_2$ are a matched pair then $I_2 = I_1$, subject to the requirement that transistor $Q_2$ be operating in the saturated

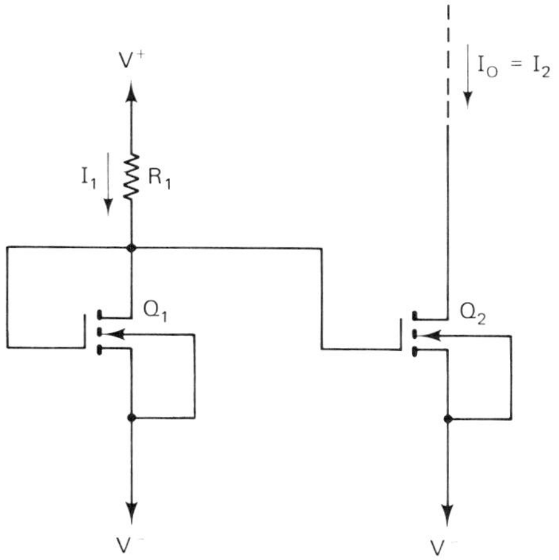


**Figure 3.19** MOSFET current source.

region. To satisfy this requirement we must have the $V_{DS2} \geq V_{GS} - V_t$, and this will also determine the lower limit of the voltage compliance range. For example, if $V_t = +2$ V, $V_{GS} = +4$ V, and $V^- = -10$ V, then for the operation of $Q_2$ in the saturated region $V_{DS2} \geq +2$ V, and the voltage compliance range will extend down to $-8$ V.

The fractional change in the output current $I_O$ per voltage change in the output voltage $V_O$ is given by $(1/I_O)dI_O/dV_O = (1/I_O)g_o$. Since $g_o = g_{ds} = I_{DS2}/V_A = I_O/V_A$, this becomes $(1/I_O)dI_O/dV_O = 1/V_A$. If $V_A = 50$ V this gives $(1/I_O)dI_O/dV_O = 1/50\text{ V} = 0.02/\text{V} = 2\%/\text{V}$.

If transistors $Q_2$ and $Q_1$ are identical in every respect except that the channel widths are in the ratio of $W_2/W_1$ then the current ratio will be given by $I_2/I_1 = W_2/W_1$. This is similar in effect to a bipolar transistor current mirror circuit of scaling the transistor active areas, $A_2$ and $A_1$, to produce a current ratio given by $I_2/I_1 = A_2/A_1$.

Another example of a MOSFET current source is the compound current source of Figure 3.20. This is essentially the MOSFET counterpart of the bipolar transistor circuit of Figure 3.11. The principal advantage of this MOSFET current source over the simpler circuit just considered is a substantially lower dynamic output conductance and therefore much better current regulation. This, however, will be at the expense of a somewhat reduced voltage compliance range.

The dynamic output conductance $g_o$ will be given by $g_o = g_{ds1}/(1 + g_{fs1}Z_{S1})$ where $Z_{S1} = 1/g_{ds2}$ (see Appendix B). Since $g_{ds1} = g_{ds2} = I_O/V_A$ and $g_{fs1} = 2K(V_{GS} - V_t) = 2I_O/(V_{GS} - V_t)$ we have that $g_{fs1}Z_{S1} = g_{fs1}/g_{ds2} = 2I_O/(V_{GS} - V_t) \times V_A/I_O = 2V_A/(V_{GS} - V_t)$. For $g_o$ we now have $g_o = dI_O/dV_O = (I_O/V_A)/[1 + 2V_A/(V_{GS} - V_t)]$. The fractional change in $I_O$ per voltage change in $V_O$ or current regulation will be given by

$$\textit{current regulation} = \frac{(1/I_O)dI_O}{dV_O} = \frac{g_o}{I_O} = \frac{1/V_A}{1 + 2V_A/(V_{GS} - V_t)} \tag{3.52}$$

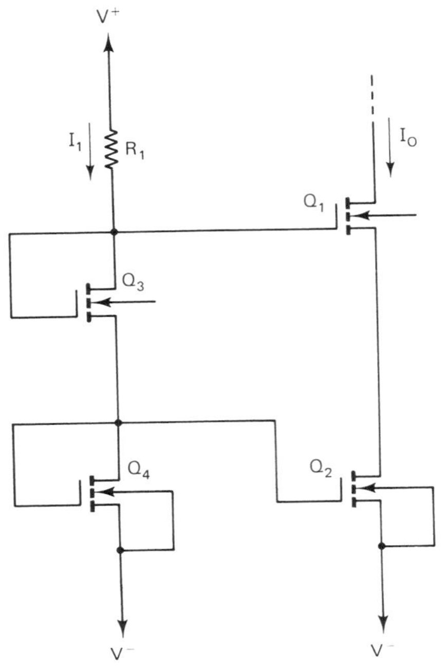


**Figure 3.20** MOSFET compound current source.

As a representative example let us take $V_{GS} - V_t = 2$ V and again assume $V_A = 50$ V. The current regulation will be given by

$$\textit{current regulation} = \frac{(1/I_O)dI_O}{dV_O} = \frac{1/50\text{ V}}{1 + 2 \times 50\text{ V}/2\text{ V}} = \frac{1/50\text{ V}}{1 + 50} \tag{3.53}$$

$$= \frac{1}{2{,}500\text{ V}} = \underline{0.04\ \%/\text{V}}$$

This represents a substantial improvement over the current regulation value of 2%/V obtained for the first MOSFET current source circuit that was considered. At an output current level of $I_O = 100\ \mu$A the corresponding value of output conductance will be $g_o = 100\ \mu\text{A}/2{,}500\text{ V} = \underline{40\text{ nS}}$.

We will now determine the lower limit of the voltage compliance range. We will again use the values of $V_{GS} - V_t = +2$ V and $V_t = +2$ V and we will assume $V^- = -10$ V. We now obtain

$$V_{D4} = V_{G2} = V^- + V_{GS2} = -10 + 4\text{ V} = -6\text{ V} \tag{3.54}$$

and

$$V_{G1} = V_{D3} = V_{G3} = V_{D4} + V_{GS3} = -6 + 4\text{ V} = -2\text{ V} \tag{3.55}$$

For $Q_1$ to be in the saturated region we must have that $V_{GD1} = V_{G1} - V_{D1} \le V_t = +2$ V. Since $V_{G1} = -2$ V, then $V_O = V_{D1} \ge -4$ V, so that the voltage compliance range will extend down to −4 V.

### 3.1.11 MOSFET Current Source Relatively Independent of Supply Voltage

In Figure 3.21 a MOSFET current source that produces an output current that is relatively independent of the supply voltage is shown. The output current $I_O$ will be given by $I_O = I_{DS2} = V_{GS1}/R_2$ and $I_1$ will be given by $I_1 = (V^+ - V^- - V_{GS2} - V_{GS1})/R_1$. For the usual case in which $V^+ - V^- \gg V_{GS}$, this can be simplified to $I_1 \simeq (V^+ - V^-)/R_1$.

The change in $I_O$ with respect to changes in the supply voltage ($V^+$ or $V^-$) will be

$$\frac{dI_O}{dV^+} = \frac{dI_O}{dV_{GS1}} \times \frac{dV_{GS1}}{dI_1} \times \frac{dI_1}{dV^+} \tag{3.56}$$

$$\simeq \frac{1}{R_2} \times \frac{1}{g_{fs1}} \times \frac{1}{R_1} \tag{3.57}$$

For $g_{fs1}$ we have

$$g_{fs1} = 2K(V_{GS1} - V_t) = \frac{2I_1}{V_{GS1} - V_t} \simeq \frac{2(V^+ - V^-)}{(V_{GS1} - V_t)R_1} \tag{3.58}$$

so that $dI_O/dV^+$ can now be expressed as

$$\frac{dI_O}{dV^+} \simeq \frac{1}{R_2} \times \frac{(V_{GS} - V_t)R_1}{2(V^+ - V^-)} \times \frac{1}{R_1} = \frac{V_{GS} - V_t}{2R_2(V^+ - V^-)} \tag{3.59}$$

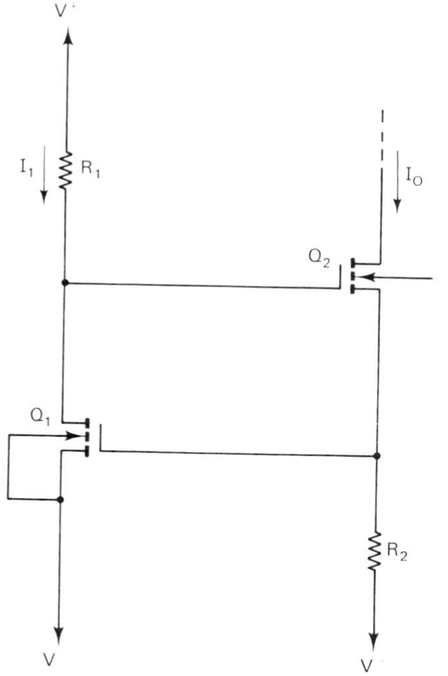


**Figure 3.21** MOSFET current source that is relatively independent of the supply voltage.

The fractional change in $I_O$ can now be related to the fractional change in the supply voltage by

$$\frac{dI_O}{I_O} \simeq \left[\frac{V_{GS} - V_t}{2R_2}\right]\left[\frac{dV^+}{V^+ - V^-}\right]\frac{1}{I_O} = \left[\frac{V_{GS1} - V_t}{2V_{GS1}}\right]\frac{dV^+}{V^+ - V^-} \tag{3.60}$$

For example, let us take $V_t = 1$ V and $V_{GS} = 2$ V, for which we obtain $dI_O/I_O \simeq (1/4)\, dV^+/(V^+ - V^-)$. If $V^+ = +10$ V and $V^- = -10$ V then a 1 V change in $V^+$ (or a $-1$ V change in $V^-$) will result in a change in $I_O$ of only 1.25%.

**Depletion mode MOSFETs as current sources.** Depletion mode MOSFETs can be used as current regulator diodes in a manner similar to the JFET current regulator diodes. In Figure 3.22a an example of a depletion MOSFET current sink is shown. The output conductance $g_o$ of this current sink will be equal to the dynamic drain-source conductance $g_{ds}$ of the transistor. For a representative example we will use $I_{DSS} = 100\ \mu$A, $V_P = -4.0$ V, $V_A = 80$ V, $BV_{DS} = 40$ V, and $V^- = -10$ V. The output current $I_O$ will be $I_O = I_{DSS} = 100\ \mu$A. The dynamic output conductance will be given by

$$g_o = \frac{dI_O}{dV_O} = \frac{dI_{DS}}{dV_{DS}} = g_{ds} = \frac{I_O}{V_A} = \frac{100\ \mu\text{A}}{80\ \text{V}} = \underline{1.25\ \mu\text{S}} \tag{3.61}$$

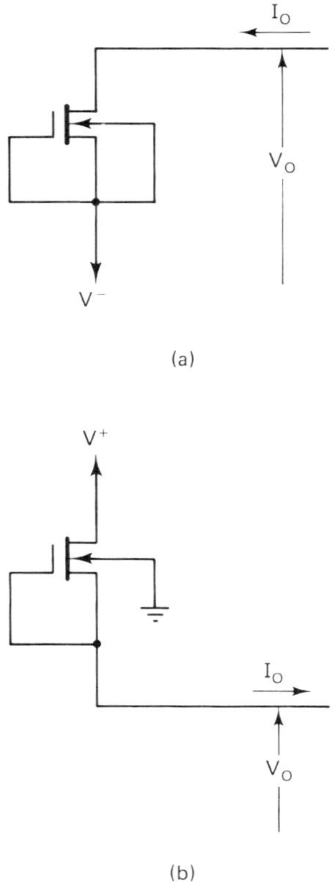


**Figure 3.22** Depletion mode MOSFETs used as current regulator diodes: (a) depletion mode MOSFET current sink; (b) depletion mode MOSFET current source.

The current regulation will be

$$current\ regulation = \frac{1}{I_O}\frac{dI_O}{dV_O} \times 100\% = \frac{100\%}{V_A} = \underline{1.25\ \%/\text{V}} \tag{3.62}$$

The voltage compliance range will extend from −6 to +40 V.

In Figure 3.22b the same transistor is shown used as a current source. In many IC situations the body of the N-channel MOSFETs will be the common P-type substrate of the IC and at a-c ground potential, and it is shown as such in Figure 3.16b. The resulting body-to-source voltage $V_{BS}$ of the transistor will have a major effect on the dynamic output conductance. In Appendix B there is a discussion of the *substrate bias effect* or *body effect* and an equation for the output conductance for this case is given as $g_o = g_{ds} + g_{fs}\,K_{BS}$. If we assume the same parameters as in the foregoing discussion and use a representative value for the body effect coefficient of $K_{BS} = 0.08$ we obtain

$$g_o = \frac{100\ \mu\text{A}}{80\ \text{V}} + \frac{2 \times 100\ \mu\text{A}}{4\ \text{V}} \times 0.08 = 1.25\ \mu\text{S} + 4.0\ \mu\text{S} = \underline{5.25\ \mu\text{S}}. \tag{3.63}$$

The corresponding current regulation will be 5.25 %/V. We see that this type of current source has a relatively high output conductance and a poor current regulation as a result of the body effect.

**Current sources used as active loads.** A very important application of current sources for both analog and digital ICs is as an active load for various amplifier and digital logic circuits. Examples of the most commonly used active load configurations are shown in Figure 3.23. The current mirror active load circuit that is widely used for both bipolar transistor and FET differential amplifiers will be discussed in Chapter 4.

Equations for the small signal a-c voltage gain for each of the active load circuits of Figure 3.23 will now be given. The voltage gain expression will be valid for the case in which the driving transistor is operating in the active region. The external load that is driven by each circuit is represented by a conductance of value $g_L$. The dynamic forward transfer conductance of the driving transistors is represented by $g_{m1}$ for the bipolar transistor case and $g_{fs1}$ for the MOSFET circuits. The output conductances of the driving and load transistors are given as $g_{o1}$ and $g_{o2}$, respectively. The parameters $V_{A1}$ and $V_{A2}$ represent the Early voltage, which is the base width modulation coefficient for bipolar transistors and the channel length modulation coefficient for MOSFETS (see Appendix B), for the driving and load transistors respectively.

**1. NPN driver—PNP current source active load.**

$$A_V = -\frac{g_m}{g_{o1} + g_{o2} + g_L} = -\frac{(I_C/V_T)}{(I_C/V_{A1}) + (I_C/V_{A2}) + g_L} \tag{3.64}$$

$$\text{For } g_L \simeq 0\text{: } A_V \simeq -\frac{1/V_T}{(1/V_{A1}) + (1/V_{A2})} \tag{3.65}$$

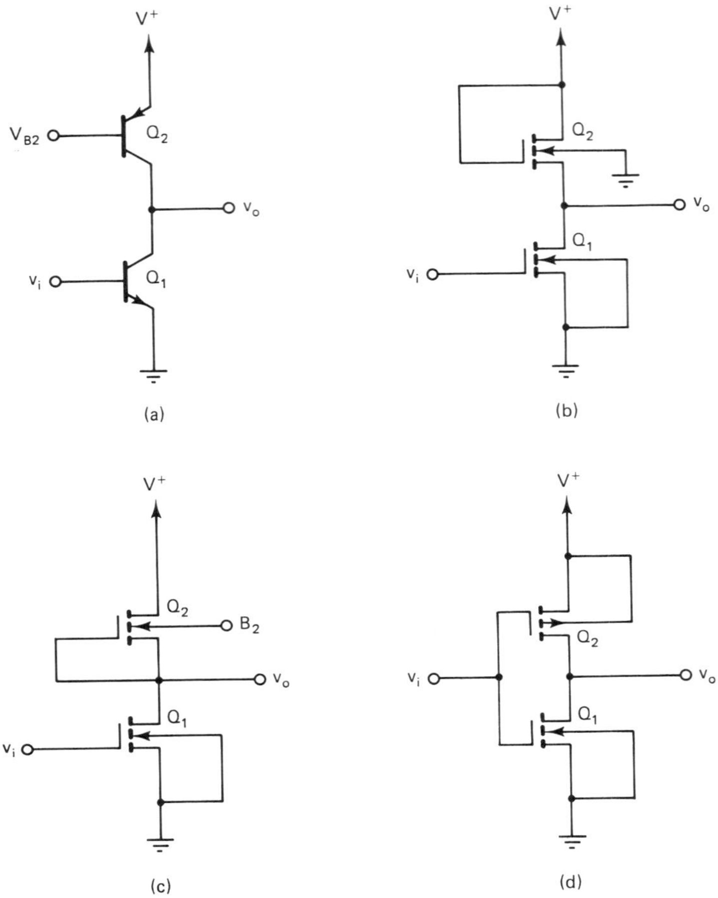


**Figure 3.23** Active load circuits: (a) NPN driver–PNP current source active load; (b) NMOS driver–NMOS load; (c) NMOS driver-depletion mode NMOS current source active load; (d) Complementary-symmetry (CMOS).

## 2. NMOS driver—NMOS load.

$$A_V = -\frac{g_{fs\,1}}{g_{o\,1} + g_{fs\,2} + g_L} \simeq -\frac{g_{fs\,1}}{g_{fs\,2}} \tag{3.66}$$

If the driver and load transistors are identical, except for different channel width $W$ to length $L$ ratios, this can be written as

$$A_V \simeq -\frac{g_{fs\,1}}{g_{fs\,2}} = -\frac{2\sqrt{K_1 I_{DS}}}{2\sqrt{K_2 I_{DS}}} = -\sqrt{\frac{K_1}{K_2}} = -\sqrt{\frac{(W/L)_1}{(W/L)_2}} \tag{3.67}$$

### 3. NMOS driver—depletion NMOS active load

$$A_V = -\frac{g_{fs\,1}}{g_{o\,1} + g_{o\,2} + g_L} = -\frac{2I_{DS}/(V_{GS\,1} - V_t)}{(I_{DS}/V_{A\,1}) + (I_{DS}/V_{A\,2}) + g_L} \quad \underline{(\text{If } V_{B\,2} = V_{S\,2})} \tag{3.68}$$

If the body of the load transistor is the IC substrate which will be at a-c ground potential, then as a result of the body effect $g_{o\,2} = g_{ds\,2} + g_{fs\,2}K_{BS}$ the voltage gain expression becomes

$$A_V = -\frac{g_{fs\,1}}{g_{o\,1} + g_{o\,2} + g_L} = -\frac{2I_{DSS}/(V_{GS\,1} - V_t)}{\dfrac{I_{DSS}}{V_{A\,1}} + \dfrac{I_{DSS}}{V_{A\,2}} + \dfrac{2I_{DSS}K_{BS}}{-V_P} + g_L} \tag{3.69}$$

where $I_{DSS}$ is the saturated drain current ($V_{GS} = 0$) of $Q_2$
$V_P$ is the pinch-off voltage of $Q_2$
and $V_t$ is the threshold voltage of $Q_1$.

**Complementary symmetry MOS (CMOS).** For the CMOS circuit transistors $Q_1$ (NMOS) and $Q_2$ (PMOS), both simultaneously perform the dual functions of driving transistor and active load. The voltage gain is given by

$$A_V = -\frac{g_{fs\,1} + g_{fs\,2}}{g_{o\,1} + g_{o\,2} + g_L} \tag{3.70}$$

If the two transistors have threshold voltages of equal magnitude and if the K-values are the same, then at the midpoint of the active (or high-to-low transition) region we have that

$$g_{fs\,1} = g_{fs\,2} = \frac{2I_{DS}}{V_{GS} - V_t} = \frac{2I_{DS}}{\frac{1}{2}V^+ - V_t} \tag{3.71}$$

If, in addition, we have that $g_L \simeq 0$, the expression for the voltage gain can be written as

$$A_V \simeq -\frac{2g_{fs}}{g_{o\,1} + g_{o\,2}} = -\frac{\dfrac{4I_{DS}}{\frac{1}{2}V^+ - V_t}}{\dfrac{I_{DS}}{V_{A\,1}} + \dfrac{I_{DS}}{V_{A\,2}}} = -\frac{\dfrac{4}{\frac{1}{2}V^+ - V_t}}{\dfrac{1}{V_{A\,1}} + \dfrac{1}{V_{A\,2}}} \tag{3.72}$$

If the two transistors have equal channel length modulation coefficients such that $V_{A\,1} = V_{A\,2}$, then the voltage gain expression for the CMOS circuit reduces to simply

$$A_V \simeq -\frac{\dfrac{4}{\frac{1}{2}V^+ - V_t}}{\dfrac{2}{V_A}} = -\frac{2\,V_A}{\frac{1}{2}V^+ - V_t} \tag{3.73}$$

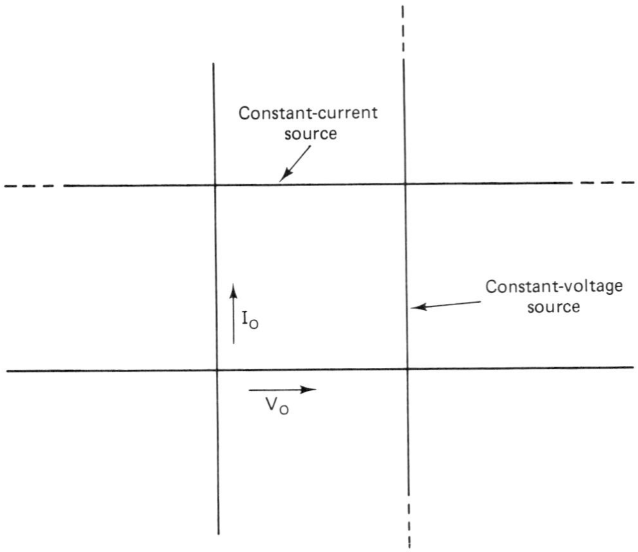


**Figure 3.24** Characteristics of ideal constant current source and constant voltage source.

## 3.2 VOLTAGE SOURCES

A voltage source is an electric circuit element that produces an output voltage, $V_o$, that is independent of the load driven by the voltage source, or equivalently, of the output current. The voltage source is the circuit dual of the constant-current source. In Figure 3.24 the characteristics of the ideal voltage source and the ideal current are compared.

As is the case with the constant-current source, although it is not possible to produce an exact or ideal voltage source, it is possible to have electronic circuits which closely approximate the behavior of the ideal voltage source. There are two principal electronic techniques that can be used, individually, or in combination, to produce a voltage source. One technique is to use the impedance transforming properties of the transistor, which in turn is related to the current gain of the transistor. The other technique is the use of an amplifier with negative feedback.

### 3.2.1 Impedance Transformation

To investigate the impedance-transforming characteristics of the transistor, let us turn our attention to the circuit of Figure 3.25. The base of the transistor is driven by a source represented by a voltage source $V_S$ and a series resistance $R_S$. We will now investigate what happens to $V_O$ as $I_O$ increases. Let us assume an increase in $I_O$ of amount $dI_O$. This will produce an increase in the base current of amount $dI_B = dI_E/(\beta + 1) = dI_O/(\beta + 1)$. This increase in the base current will increase the voltage drop across the source resistance by an amount $dI_B R_S = dI_O R_S/(\beta + 1)$. There will also be an increase in the base-to-emitter voltage drop of the transistor as given by $dV_{BE} = (dV_{BE}/dI_E)\, dI_E$. The quantity $dV_{BE}/dI_E$ is the dynamic emitter-

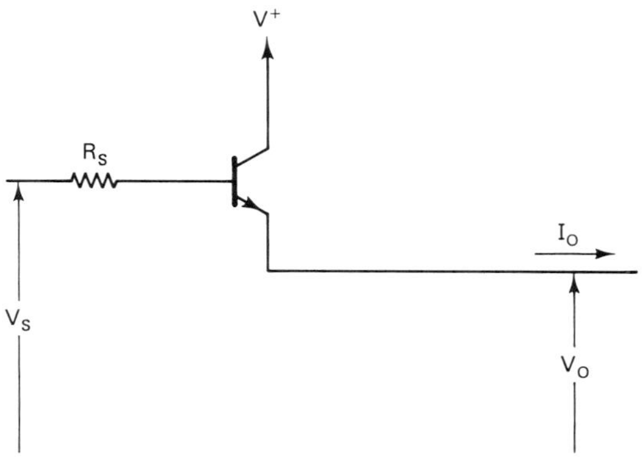


**Figure 3.25** Voltage source using transistor impedance transformation.

to-base resistance of the transistor. Its value can easily be determined from the following relationships. For the transistor in the active mode we have that $I_E \simeq I_C = I_{TO} \exp(V_{BE}/V_T)$, so that $dI_E/dV_{BE} \simeq dI_C/dV_{BE} = I_E/V_T$, and therefore $dV_{BE}/dI_E = r_{eb} = V_T/I_E$.

The total change in the output voltage will therefore be given by

$$dV_O = -\,dI_B R_S - dV_{BE} = -\left[\frac{dI_O R_S}{\beta + 1} + dI_O r_{eb}\right] \tag{3.74}$$

The effective output resistance of this circuit, as seen looking into the emitter of the transistor, will be given by

$$\text{output resistance} = r_o = \frac{-dV_O}{dI_O} = \frac{R_S}{\beta + 1} + r_{eb} \tag{3.75}$$

We see that as a result of the current gain of the transistor, the effective value of $R_S$ as seen looking into the emitter of the transistor is $R_S/(\beta + 1)$. Since the current gain, $\beta$, of the transistor will generally be large, typically of the order of 100, this will represent a very large impedance transformation.

To consider an example, let us assume that $R_S = 1000\ \Omega$ and $I_O = 5$ mA with $V_O = 10$ V and a current gain of 100 for the transistor. The effective output resistance will be given by

$$r_o = \frac{1000}{\beta + 1} + \frac{25\text{ mV}}{5\text{ mA}} = 10\ \Omega + 5\ \Omega = 15\ \Omega \tag{3.76}$$

The change in $V_O$ will therefore be a drop of 15 mV per 1-mA increase in $I_O$. On a percentage basis this will be (15 mV/mA)/10 V × 100% = 0.15%/mA, so that the output voltage decreases by 0.15% for every 1-mA increase in the output current.

The relationships above will hold true only for small changes in the output current. We will now investigate the changes in $V_O$ that result from large changes in the output current. Starting with the exponential relationship between current and base-to-emitter voltage for the transistor, we have that $I_E \simeq I_C = I_{TO} \exp(V_{BE}/V_T)$, so that $V_{BE} = V_T \ln(I_E/I_{TO})$. The change in $V_{BE}$ resulting from the output, or emitter current changing from $I_{E_1}$ to $I_{E_2}$, will be given by $\Delta V_{BE} = V_T \ln(I_{E_2}/$

$I_{E_1}) = V_T \ln (I_{O_2}/I_{O_1})$. Therefore, the total change in the output voltage that results from the output current increasing from a value of $I_{O_1}$ to $I_{O_2}$ will be given by

$$\Delta V_O = -\left[\frac{(I_{O_2} - I_{O_1})R_S}{\beta + 1} + V_T \ln \frac{I_{O_2}}{I_{O_1}}\right] \tag{3.77}$$

Thus, if in the example above $I_O$ changes from $I_{O_1} = 1.0$ mA up to a full-load value of $I_{O_2} = 5.0$ mA, the change in the output voltage will be given by

$$\Delta V_O = -\left[\frac{4.0 \text{ mA} \times 1000\ \Omega}{101} + 25 \text{ mV} (\ln 5)\right] \tag{3.78}$$

$$= -(40 \text{ mV} + 40 \text{ mV}) = -80 \text{ mV}$$

which represents a voltage decrease of 0.8%.

The *load regulation* of a voltage source or a voltage regulator is the change or decrease in the output voltage as the output current goes from some specified "no load" current value to a full-load current. The load regulation will be directly related to the output impedance of the voltage source or voltage regulator. In Figure 3.26 the output characteristics of the ideal voltage source are compared to those of a real voltage source. In Figure 3.27 the corresponding equivalent circuits are shown.

### 3.2.2 Use of Negative Feedback to Reduce the Output Impedance

In Figure 3.28 a simple feedback circuit is presented to illustrate how an amplifier with negative feedback can be used to produce a very low value of output impedance and thus ensure a very good load regulation for the voltage source. In this circuit $A$ is the gain (open-loop) of the amplifier and $R_S$ represents the open-loop output

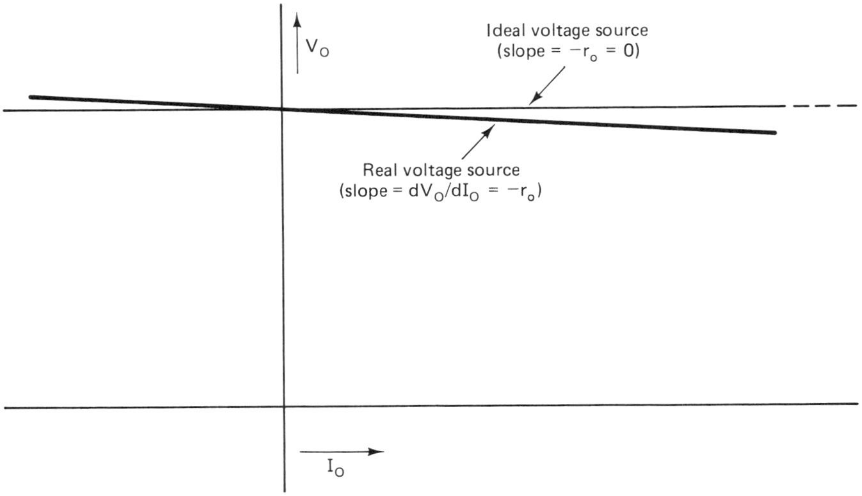


**Figure 3.26** Voltage source graphical output characteristics.

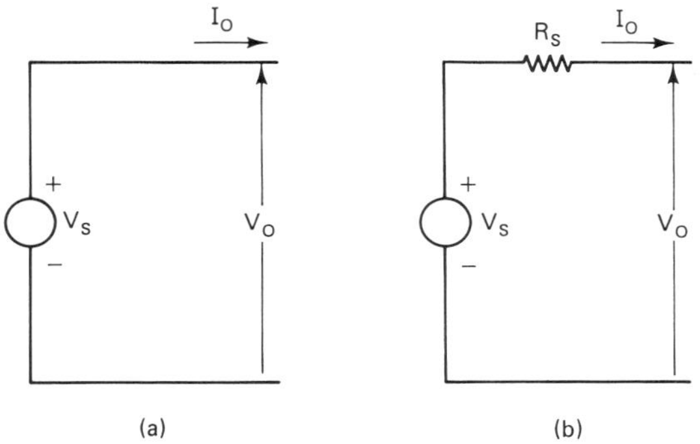


**Figure 3.27** Voltage source equivalent circuits: (a) ideal voltage source; (b) real voltage source.

impedance of the amplifier. The resistance $R_S$ is actually internal to the amplifier, but for purposes of analysis is represented as a resistor external to the amplifier.

The output voltage of this circuit, $V_O$, will be given by $V_O = (V_S - V_O)A - I_O R_S$, so that collecting terms in $V_O$ on the left-hand side gives

$$V_O(1 + A) = V_S A - I_O R_S$$

and thus

$$\begin{aligned} V_O &= V_S \frac{A}{1+A} - I_O \frac{R_S}{1+A} \\ &= V_S \frac{A}{1+A} - I_O R'_S \end{aligned} \tag{3.79}$$

where $R'_S = R_S/(1 + A)$ is the *closed-loop* output resistance of this circuit. For the usual case of a very large open-loop gain such that $A \gg 1$, we will have that $V_O \simeq V_S - R'_S I_O$, where $R'_S \simeq R_S/A$. For this case we see that the closed-loop output resistance will be very much smaller than the open-loop value.

Very often the output stage of the amplifier is an emitter-follower impedance-transforming stage of the type just studied. In addition, an emitter-follower stage is often added to the amplifier output to increase the current output range. An example of such a circuit is shown in Figure 3.29. The open-loop output resistance has been shown previously to be given by

$$R_o = \frac{R_S}{\beta + 1} + r_{eb} = \frac{R_S}{\beta + 1} + \frac{V_T}{I_O} \tag{3.80}$$

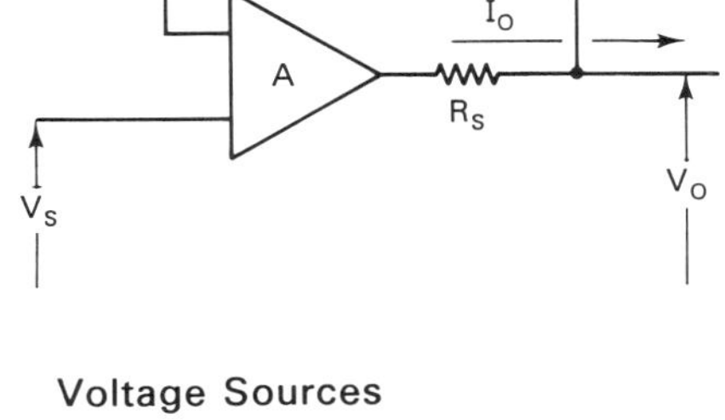


**Figure 3.28** Use of amplifier with negative feedback to reduce output resistance.

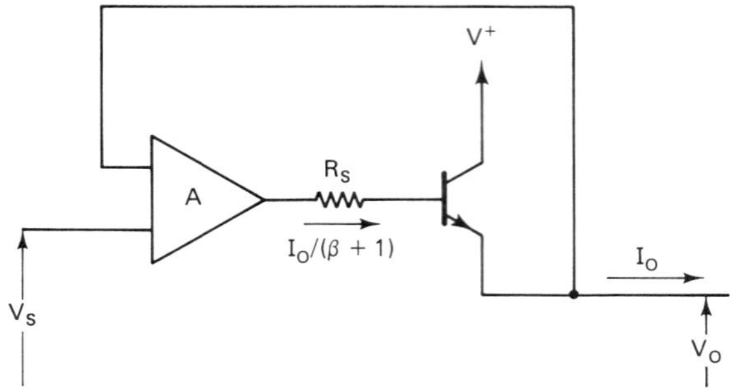


**Figure 3.29** Feedback amplifier with emitter-follower output circuit.

Under the feedback conditions of Figure 3.24 the corresponding closed-loop output resistance will be given by $R'_O = R_O/(A + 1)$. For $R_S = 1000\ \Omega$ and $I_O = 1.0$ mA, for example, we will have that $R_O = (1000/101) + (25\ \text{mV}/1.0\ \text{mA}) = 35\ \Omega$. If the amplifier open-loop gain at low frequencies is 10,000, the *closed-loop* output resistance will be

$$R'_O = \frac{R_O}{A+1} = \frac{35\ \Omega}{10\ \text{k}\Omega} = 3.5\ \text{m}\Omega = 3.5\ \mu\text{V/mA} \tag{3.81}$$

We see that that output resistance of this circuit will have the very low value of 3.5 m$\Omega$ or 3.5 $\mu$V/mA. This means that the decrease in the output voltage for an increase in the output current of 1.0 mA will be only 3.5 $\mu$V. This represents a very small change indeed, and thus it can be said that this circuit represents a very close approximation to an ideal voltage source.

The value of the dynamic output resistance obtained above will be the dynamic output impedance of the voltage source circuit only at low frequencies. At higher frequencies the decreasing gain and phase angle of the voltage gain of the amplifier will produce corresponding changes in both the magnitude and the angle of the output impedance. For most amplifiers, and especially for operational amplifiers, the voltage gain can be expressed in the form $A(f) = A(0)/[1 + j(f/f_1)]$, where $A(0)$ is the zero-frequency open-loop gain and $f_1$ is the breakpoint frequency. For frequencies that are more than one-half of a decade above this breakpoint frequency, the open-loop voltage gain can be expressed approximately as $A(f) \simeq A(0)f_1/jf = f_u/jf$, where $f_u$ is the unity-gain frequency as given by $f_u = A(0)f_1$. The closed-loop output impedance will as a result be given by

$$Z'_S = \frac{R_O}{A+1} \simeq \frac{R_O}{f_u/jf} = jf\frac{R_o}{f_u}$$

We thus see that in the higher-frequency region the output impedance will begin to look reactive and the closed-loop output impedance given by

$$Z'_S \simeq jf\frac{R_O}{f_u} = j\omega\frac{R_O}{2\pi f_u} = j\omega L'_S \tag{3.82}$$

where $L'_S = R_O/2\pi f_u = R_o/\omega_u$ is the effective output inductance of the circuit. To continue with the previous example, and assuming an amplifier unity gain frequency of 1.0 MHz, we will have that $Z'_S \simeq jf(35\ \Omega/1\ \text{MHz})$. The breakpoint frequency $f_1$ will be given by $f_1 = f_u/A(0) = 1.0\ \text{MHz}/10\ \text{k}\Omega = 100\ \text{Hz}$. Thus at a frequency of 1.0 kHz the dynamic output impedance will be $j35\ \text{m}\Omega = j35\ \mu\text{V/mA}$ as compared to the low-frequency value of 3.5 mΩ. At 10 kHz, the output impedance will have risen to $+j0.35\ \Omega$, and at 100 kHz it will be approximately $+j3.5\ \Omega$. The equivalent circuit of this voltage source can at these frequencies be represented as a d-c voltage source, $V_S$, in series with a resistance of 3.5 mΩ and an inductance of 5.6 μH.

### 3.2.3 Power Supply Rejection (Line Regulation)

We have seen that one very desirable feature of a voltage source is that it have a very low dynamic output impedance so that the output voltage changes very little with changes in the output current. Another desirable feature for voltage sources or voltage regulators is that the output voltage be as independent as possible of the supply voltage. A simple example of a circuit in which this is accomplished is shown in Figure 3.30. In this circuit the voltage regulator diode or "zener diode" is biased by a current source. The characteristics of the voltage regulator diode are such that when it is biased in the breakdown region beyond the knee of the curve, the voltage drop across the diode will be relatively independent of the current through the diode. The voltage drop across the diode will not, however, be completely independent of the current. The change in the voltage drop across the diode divided by the change in the current is called the *zener impedance*, given by $Z_Z = dV_Z/dI_Z$, where $V_Z$ is the voltage across the diode and $I_Z$ is the current through the diode. Typical values of the zener impedance $Z_Z$ range from a few ohms to a few tens of ohms.

In the circuit of Figure 3.30 a change in the supply voltage $dV_{\text{supply}}$ will result in a small change in the current through the current source of amount $dI_O = g_o\, dV_{\text{supply}}$, where $g_o$ is the dynamic output conductance of the current source. This will result in a change in the current through the voltage regulator diode of amount $dI_Z = dI_O$, which in turn will change the voltage drop across the voltage regulator diode by an amount $dV_Z = Z_Z\, dI_Z = Z_Z\, dI_O = g_o Z_Z\, dV_{\text{supply}}$. This ratio of the

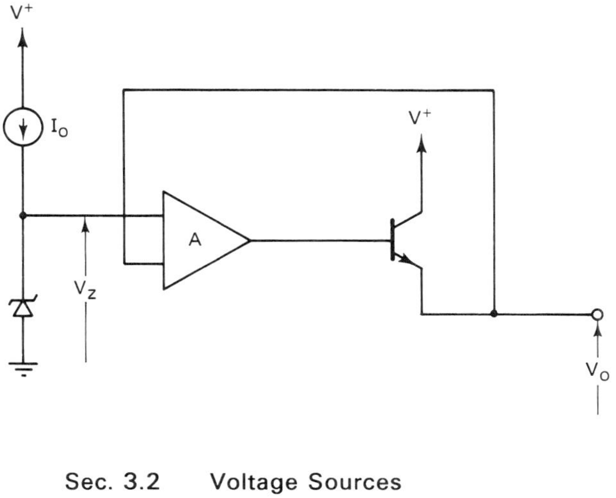


**Figure 3.30** Voltage source with current source biasing for supply voltage rejection.

change in the voltage across the voltage regulator diode, and therefore of the output voltage, $V_O$, to the change in the supply voltage will be given by

$$\frac{dV_O}{dV_{\text{supply}}} = \frac{dV_Z}{dV_{\text{supply}}} = g_o Z_Z \tag{3.83}$$

To consider a representative example, let us choose $Z_Z = 10\ \Omega$ and $g_o = 100$ nS, so that $dV_O/dV_{\text{supply}} = 100\ \text{nS} \times 10\ \Omega = 1 \times 10^{-6}$. Thus a change in the supply voltage of 1.0 V will result in a change in the output voltage of only 1 $\mu$V.

### 3.2.4 Voltage Source Using $V_{BE}$ as the Reference Voltage

In Figure 3.31 a voltage source that uses the base-to-emitter voltage drop as the reference voltage is shown. If we assume the base current to be small, the output voltage will be related to the $V_{BE}$ of the transistor by the simple resistive voltage divider of $R_1$ and $R_2$ as given by the relationship

$$\frac{V_{BE}}{V_O} = \frac{R_2}{R_1 + R_2}$$

so that

$$V_O = \frac{V_{BE}(R_1 + R_2)}{R_2} = V_{BE}\left(1 + \frac{R_1}{R_2}\right) \tag{3.84}$$

The dynamic output resistance can be obtained by assuming a change in the output voltage $V_O$ of amount $dV_O$ and finding the change in $I_O$ that results. Taking the ratio of $dV_O$ to $dI_O$ will give $dV_O/dI_O = r_O =$ dynamic output resistance. For purposes of simplicity we will again assume the base current to be small.

A change in the output voltage of amount $dV_O$ will change the current through $R_1 + R_2$ by an amount $dV_O/(R_1 + R_2)$. This current change will in turn produce a change in the $V_{BE}$ of the transistor by an amount given by $dV_{BE} = [dV_O/(R_1 +$

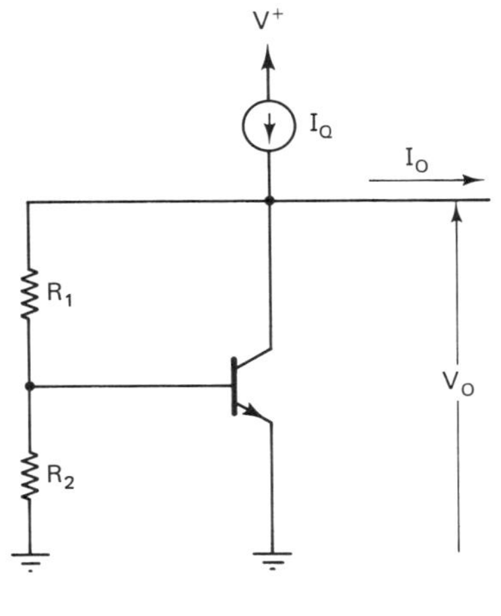


**Figure 3.31** Voltage source using $V_{BE}$ as the reference voltage.

$R_2)]R_2$, which will produce a change in the current flowing through the transistor by an amount $dI_C = g_m\, dV_{BE} = [g_m R_2/(R_1 + R_2)]\, dV_O$. The total change in the output current will be the sum of the change of current through $R_1 + R_2$ and the change in the transistor current, so that we will have

$$dI_O = dV_O\left(\frac{1}{R_1 + R_2} + \frac{g_m R_2}{R_1 + R_2}\right) = dV_O \frac{1 + g_m R_2}{R_1 + R_2} \tag{3.85}$$

The dynamic output resistance will therefore be given by

$$r_o = \frac{dV_O}{dI_O} = \frac{R_1 + R_2}{1 + g_m R_2} \tag{3.86}$$

Since

$$\frac{V_O}{V_{BE}} = \frac{R_1 + R_2}{R_2} \tag{3.87}$$

we can rewrite $r_o$ in the form

$$r_o = \frac{V_O}{V_{BE}} \frac{R_2}{1 + g_m R_2} \tag{3.88}$$

If $g_m R_2 \gg 1$, as is usually the case, then we can write $r_o$ as

$$r_o \simeq \frac{V_O}{V_{BE}} \frac{1}{g_m} = \frac{V_O}{V_{BE}} \frac{V_T}{I_C} \tag{3.89}$$

If, for example, $I_C = 1.0$ mA and $V_O = 1.0$ V, and $V_{BE} = 650$ mV at this collector current, we will have that

$$r_o \simeq \frac{1000 \text{ mV}}{650 \text{ mV}}\left(\frac{25 \text{ mV}}{1.0 \text{ mA}}\right) = \underline{38.5\ \Omega} \tag{3.90}$$

To minimize the effect of the base current we should make the current through $R_1$ and $R_2$ at least 10 times, and preferably 20 times as large as the maximum expected base current. For a current gain of $\beta = 50$ (min.), this means that the current through $R_1$ and $R_2$ should be about 20(1.0 mA/50) = 0.4 mA. Since $V_O = 1.0$ V we have that $R_1 + R_2 = 1.0$ V/0.4 mA = 2.5 kΩ. For $R_2$ we have that $R_2 = 650$ mV/0.4 mA = 1.625 kΩ, and thus $R_1 = 2.5$ kΩ − 1.625 kΩ = 875 Ω. The current source $I_Q$ should be designed to supply the required amount of current through $R_1 + R_2$ plus the transistor collector and the maximum required load current. If the maximum required load current is, for example, 2.0 mA, the current source should be 3.4 mA.

The output voltage of this voltage source will be relatively independent of the supply voltage by virtue of the small dynamic conductance that can be exhibited by the current source. A change $dV_{\text{supply}}$ in the supply voltage will change the current source current by $dI_Q = g_o\, dV_{\text{supply}}$, where $g_o$ is the dynamic output conductance of the current source. This current change will have the same effect on the output voltage, $V_O$, as will an equal change in $I_O$, so that we will have that $dV_O = r_o$

$dI_Q = r_o g_o\, dV_{\text{supply}}$, so that we end up with $dV_O/dV_{\text{supply}} = g_o r_o$. If $g_o$, for example, is 1.0 $\mu$S, we will have

$$\frac{dV_O}{dV_{\text{supply}}} = 10^{-6}\text{ mS} \times 39\ \Omega = 39 \times 10^{-6} = 39\ \mu\text{V/V}$$

Thus a change in the supply voltage of 1 V will result in a change in the output voltage of only 39 $\mu$V.

## 3.3 VOLTAGE REFERENCES

A voltage reference is an electric circuit designed to produce an output voltage that is independent of temperature. In practice, it is never possible to achieve this complete independence of temperature, especially over an extended temperature range.

The variation of the output voltage of the voltage reference circuit with temperature is expressed in terms of the temperature coefficient or "temco," given by

$$TC_{V(\text{REF})} = \frac{dV_{\text{REF}}}{dT} = \text{temperature coefficient of the reference voltage}$$

The most important characteristic of a voltage reference is the temperature coefficient of the output voltage, $TC_{V(\text{REF})}$. In most cases it is also desirable that the reference voltage be as independent of the supply voltage as possible, that is, that there be a good power supply rejection. Another desirable feature is that the output voltage be as independent of the loading or output current as possible; that is, the circuit should have a low output impedance. Indeed, in many cases a *voltage reference* circuit is used to bias a *voltage source* circuit, the combination for many applications being called a *voltage regulator*. A voltage regulator therefore combines the characteristics of a low temperature coefficient, a low output impedance (i.e., good load regulation), and good power supply rejection characteristics (i.e., good line regulation).

Since all electronic components exhibit temperature coefficients, the basic technique employed in voltage reference circuits is to design the circuit such that there are canceling effects so as to produce, at least nominally, a zero temperature coefficient at a given temperature. An example of a circuit in which such a cancellation technique is used is shown in Figure 3.32. In this circuit transistors $Q_1$ through $Q_3$ are operated as diode-connected transistors. The current source, $I_Q$, drives a current through $Q_1$ in the reverse-bias direction such that $Q_1$ is operated in the reverse-bias breakdown region as a voltage regulator or zener diode. The voltage drop produced across $Q_1$ is the emitter–base junction breakdown voltage, so that we have $V_Z = BV_{EBO}$. This voltage is typically around 6 or 7 V.

The voltage at the emitter of $Q_2$ will be $V_{E_2} = V_Z - V_{BE_4} - V_{BE_2}$. The voltage at the collector of $Q_3$ will be $V_{C_3} = V_{BE_3}$. The output voltage of this circuit will be therefore given by

$$V_{\text{REF}} = V_O = V_{E_2}\frac{R_2}{R_1 + R_2} + V_{C_3}\frac{R_1}{R_1 + R_2} \tag{3.91}$$

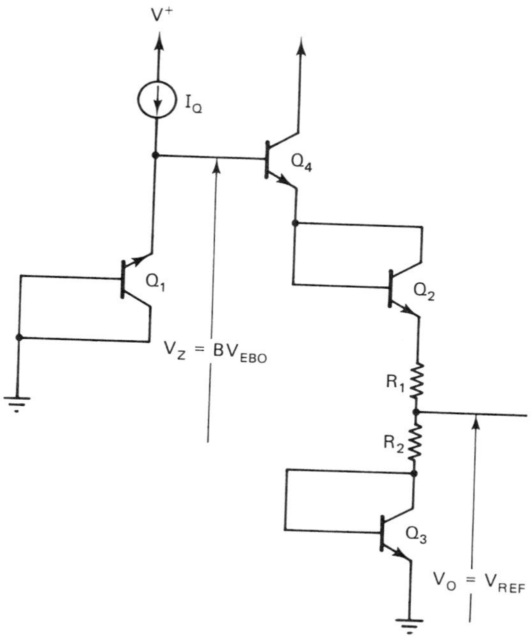


**Figure 3.32** Temperature-compensated voltage reference.

this result being obtained by the simple application of the voltage-division rule and the superposition theorem.

Since the currents through $Q_4$, $Q_2$, and $Q_3$ will be essentially the same, and these transistors are IC transistors of similar construction, the $V_{BE}$ voltage drop, and more important, the temperature coefficient of the $V_{BE}$ drop, $dV_{BE}/dT$, will be approximately the same for these three transistors. We therefore can write $V_O$ as

$$V_{\mathbf{REF}} = V_O = \frac{(V_Z - 2V_{BE})R_2 + V_{BE}R_1}{R_1 + R_2} = \frac{V_Z R_2 - V_{BE}(2R_2 - R_1)}{R_1 + R_2} \tag{3.92}$$

The temperature coefficient of the reference voltage will therefore be

$$TC_{V(\mathbf{REF})} = \frac{dV_{\mathbf{REF}}}{dT} = \frac{R_2(dV_Z/dT) - (dV_{BE}/dT)(2R_2 - R_1)}{R_1 + R_2} \tag{3.93}$$

We must certainly note here that $R_1$ and $R_2$ will be temperature dependent. Nevertheless, $R_1$ and $R_2$ are IC resistors of similar construction, so that they will have identical temperature coefficients, so that for any given temperature change, both resistors will increase by the same fractional or percentage amount. Therefore, the ratio of $R_1$ to $R_2$ will not change with temperature, so that in looking at $TC_{V(\mathbf{REF})}$ we do not have to consider the temperature variation of $R_1$ and $R_2$.

Looking at the equation for $TC_{V(\mathbf{REF})}$ we now note the possibility of producing a zero temperature coefficient by suitable choice of the resistance ratio such that the numerator becomes equal to zero. For this to be the case we require that

$$R_2 \frac{dV_Z}{dT} = (2R_2 - R_1)\frac{dV_{BE}}{dT} \tag{3.94}$$

so that

$$\frac{2R_2 - R_1}{R_2} = \frac{dV_Z/dT}{dV_{BE}/dT} \tag{3.95}$$

Since $dV_{BE}/dT$ will be negative, this can be expressed in the most convenient form as

$$\frac{R_1}{R_2} - 2 = \frac{dV_Z/dT}{-dV_{BE}/dT}$$

or as

$$\boxed{\frac{R_1}{R_2} = 2 + \frac{dV_Z/dT}{-dV_{BE}/dT}} \tag{3.96}$$

To consider a representative example, we will use the following quantities

$$\begin{aligned} V_Z &= BV_{EBO} = 6.3 \text{ V} \qquad & \frac{dV_Z}{dT} &= +3.0 \text{ mV/°C} \\ V_{BE} &= 0.70 \text{ V} & \frac{dV_{BE}}{dT} &= -2.3 \text{ mV/°C} \end{aligned} \tag{3.97}$$

For a zero temperature coefficient for $V_{\mathrm{REF}}$, the required resistor ratio will be given by

$$\frac{R_1}{R_2} = 2 + \frac{dV_Z/dT}{-dV_{BE}/dT} = 2 + \frac{+3 \text{ mV/°C}}{+2.3 \text{ mV/°C}} = 3.30 \tag{3.98}$$

The equation for $V_{\mathrm{REF}}$ given earlier can be rewritten in terms of the $R_1/R_2$ ratio by dividing through by $R_2$, giving

$$V_{\mathrm{REF}} = V_Z \frac{1}{1 + R_1/R_2} - V_{BE} \frac{2 - R_1/R_2}{1 + R_1/R_2} \tag{3.99}$$

Insertion of the appropriate quantities gives for $V_{\mathrm{REF}}$ the value of 1.68 V. Note that the value of $V_{\mathrm{REF}}$ cannot be selected at will, for its value is a consequence of the requirement that $TC_{V(\mathrm{REF})} = 0$.

Using this circuit for a reference voltage source, it is possible to produce a $TC_{V(\mathrm{REF})}$ that is *nominally* zero. That is, the temperature coefficient will be zero if all of the circuit parameters have values that correspond exactly to the values used in the design. If any of the circuit parameters do not in actuality correspond to the value used in the design, the temperature coefficient will not be zero, although it may still be very small.

In order to examine the possible effects of deviations of circuit parameters from their design center values of $TC_{V(\mathrm{REF})}$, let us take the equation for $TC_{V(\mathrm{REF})}$ given earlier, and by dividing through by $R_2$ rewrite it as

$$TC_{V(\mathrm{REF})} = \frac{(dV_Z/dT) - (dV_{BE}/dT)(2 - R_1/R_2)}{1 + R_1/R_2} \tag{3.100}$$

If all quantities had their design center values, the numerator would cancel out and $TC_{V(\mathrm{REF})}$ will be equal to zero. Now let us see what the resulting temperature coefficient will be if $TC_{V_Z} = dV_Z/dT$ differs from its design center value by ±5%. If this is the case, then for $TC_{V(\mathrm{REF})}$ we will have

$$TC_{V(\mathrm{REF})} = \frac{\pm 0.05(dV_Z/dT)}{1+3.3} = \pm 0.0116 \times 3\ \mathrm{mV/^\circ C} \tag{3.101}$$
$$= \pm 0.035\ \mathrm{mV/^\circ C} = \pm 35\ \mu\mathrm{V/^\circ C}$$

Thus, instead of being zero, the actual value of $TC_{V(\mathrm{REF})}$ may be in the range 35 $\mu$V/°C, or on a fractional basis 0.0021%/°C. Although the temperature coefficient of $V_{\mathrm{REF}}$ has not been reduced to zero, it still will be satisfactorily small for most applications.

To consider now a second example, let us see what temperature coefficient for $V_{\mathrm{REF}}$ will result from a deviation of ±0.1 mV/°C in the value of $dV_{BE}/dT$ for which a design center value of −2.3 mV/°C was chosen. After substitution into the equation for $TC_{V(\mathrm{REF})}$, we now have

$$\begin{aligned} TC_{V(\mathrm{REF})} &= \frac{\pm 0.1\ \mathrm{mV/^\circ C}(2 - R_1/R_2)}{1 + R_1/R_2} \\ &= \frac{\pm 0.1\ \mathrm{mV/^\circ C}(2 - 3.3)}{4.3} = \pm 0.030\ \mathrm{mV/^\circ C} \\ &= \pm 30\ \mu\mathrm{V/^\circ C} \end{aligned} \tag{3.102}$$

and on a percentage basis, ±0.0018%/°C. This is, again, a small residual temperature coefficient and should prove satisfactory for most applications.

Finally, let us investigate the effect of a small deviation of the $R_1/R_2$ resistance ratio from the design center value. For integrated-circuit resistors, the ratio tolerance of similarly constructed resistors on the same IC chip is generally quite small, generally less than 5%. For this example, let us look at the effect of a ±2% difference between the actual resistance ratio and the design center value.

For this case, by suitable substitution into the equation for $TC_{V(\mathrm{REF})}$, we obtain

$$TC_{V(\mathrm{REF})} = \frac{(-dV_{BE}/dT)(\pm 0.02)}{1 + R_1/R_2} = \frac{\pm 0.02 \times 2.3\ \mathrm{mV/^\circ C}}{4.3} \tag{3.103}$$
$$= \pm 0.011\ \mathrm{mV/^\circ C} = \pm 11\ \mu\mathrm{V/^\circ C}$$

and on a percentage basis, ±0.00064%/°C. This again is a very small residual temperature coefficient, so that this circuit should prove to perform adequately for most applications.

### 3.3.1 Band-Gap Voltage Reference

In Figure 3.33 a very interesting and useful voltage reference circuit is presented. To proceed forthwith with the analysis of this circuit we have that $V_{\mathrm{REF}} = V_{BE_3} + I_2R_2$. In this analysis we will assume all transistors to be identical, so that for the

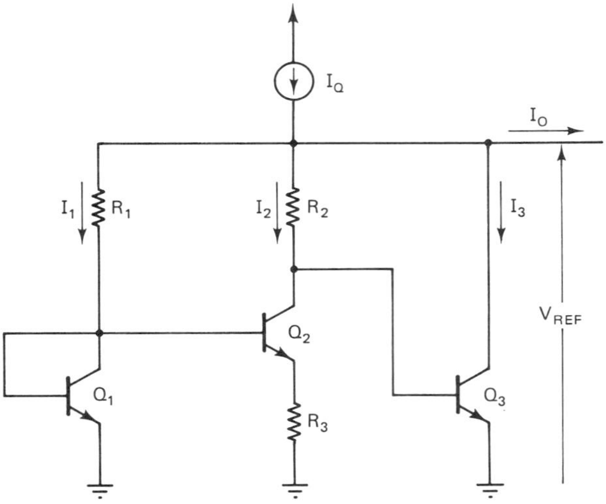


**Figure 3.33** Band-gap voltage reference circuit.

relationship between $I_1$ and $I_2$ we will have that $I_1 = I_2 \exp(I_2R_3/V_T)$. In this analysis we will assume the base currents to be small enough to be neglected. From the equation just given we can solve for $I_2R_3$ as $I_2R_3 = V_T \ln(I_1/I_2)$, so that

$$I_2R_2 = \frac{R_2}{R_3} I_2R_3 = \frac{R_2}{R_3} V_T \ln \frac{I_1}{I_2} \tag{3.104}$$

Substituting this into the expression for $V_{\text{REF}}$ gives

$$V_{\text{REF}} = V_{BE_3} + \frac{R_2}{R_3} V_T \ln \frac{I_1}{I_2} \tag{3.105}$$

We now will take note of the fact that $V_{BE}$ will have a negative temperature coefficient (i.e., decreases with increasing temperature), whereas the last term in the equation will have a positive temperature coefficient since $V_T = kT/q$. Therefore, by suitable choice of the resistance ratio and the current ratio it should be possible to produce a cancellation of the two temperature coefficients, and thereby end up with a zero net temperature coefficient.

To investigate this further, let us now obtain an expression for the temperature coefficient of $V_{\text{REF}}$ as

$$TC_{V(\text{REF})} = \frac{dV_{\text{REF}}}{dT} = \frac{dV_{BE}}{dT} + \frac{R_2}{R_3}\frac{k}{q} \ln \frac{I_1}{I_2} \tag{3.106}$$

Note again that since $R_1$ and $R_2$ are two IC resistors of similar construction in close thermal contact on the same IC chip, the fractional change in both resistors will be the same, so that the ratio of the two resistances will be independent of temperature. To proceed further we will now need to make a slight digression in order to obtain an expression for $TC_{V(BE)} = dV_{BE}/dT$.

**Temperature coefficient of $V_{BE}$.** For a transistor in the active region we have the now familiar exponential relationship between the collector current and the base-to-emitter voltage as given by

$$I_C = I_{TO} \exp\left(\frac{V_{BE}}{V_T}\right) \tag{3.107}$$

where $V_T$ is the thermal voltage $(kT/q)$ and the preexponential constant, $I_{TO}$, is a strong function of temperature as given by $I_{TO} = CT^3 \exp(-qE_{GO}/kT)$. The quantity $E_{GO}$ is the value of the energy band gap at absolute zero (0 K) as obtained by a linear extrapolation from room temperature (300 K) to absolute zero. The value of $E_{GO}$ is 1.205 V = 1205 mV.

In order to obtain the temperature coefficient of $V_{BE}$, we first obtain an expression for $V_{BE}$ as given by $V_{BE} = V_T \ln(I_C/I_{TO})$, and then take the derivative of $V_{BE}$ with respect to temperature for constant $I_C$. Doing this gives

$$\begin{aligned}\frac{dV_{BE}}{dT} &= \frac{k}{q}\ln\frac{I_C}{I_{TO}} - V_T\frac{d(\ln I_{TO})}{dT}\\ &= \frac{V_{BE}}{T} - V_T\frac{d(\ln I_{TO})}{dT}\end{aligned} \tag{3.108}$$

Since $\ln I_{TO} = \ln C + 3\ln T - qE_{GO}/kT$, we have that $d(\ln I_{TO})/dT = (3/T) + qE_{GO}/kT^2$, so that we now have

$$\begin{aligned}\frac{dV_{BE}}{dT} &= \frac{V_{BE}}{T} - V_T\left(\frac{3}{T} + \frac{qE_{GO}}{kT^2}\right),\\ &= \frac{V_{BE}}{T} - 3\left(\frac{k}{q} + \frac{E_{GO}}{T}\right)\\ &= -\left(\frac{E_{GO} - V_{BE}}{T} + 3\frac{k}{q}\right)\end{aligned} \tag{3.109}$$

Since

$$\begin{aligned}3\frac{k}{q} &= 3(1.38 \times 10^{-23}\ \text{J/K}/1.602 \times 10^{-19}\ \text{C})\\ &= 2.6 \times 10^{-4}\ \text{V/K} = 0.26\ \text{mV/°C}\end{aligned} \tag{3.110}$$

and

$$E_{GO} = 1205\ \text{mV}$$

we have

$$\boxed{TC_{V(BE)} = \frac{dV_{BE}}{dT} = -\left(\frac{1205\ \text{mV} - V_{BE}}{T} + 0.26\ \text{mV/°C}\right)} \tag{3.111}$$

For a representative value of $V_{BE} = 650$ mV, we obtain a temperature coefficient of $TC_{V(BE)} = -2.1$ mV/°C.

**Conditions for a zero temperature coefficient.** Now that we have obtained an analytical expression for $TC_{V(BE)}$ we can insert it into the expression for $TC_{V(\mathrm{REF})}$ to obtain

$$TC_{V(\mathrm{REF})} = \frac{dV_{\mathrm{REF}}}{dT} = \frac{dV_{BE}}{dT} + \frac{R_2}{R_3}\frac{k}{q}\ln\frac{I_1}{I_2} \tag{3.112}$$

For $TC_{V(\mathrm{REF})} = 0$ we require that $(R_2/R_3)(k/q)\ln(I_1/I_2) = -dV_{BE}/dT$, so that the corresponding value of $V_{\mathrm{REF}}$ will be given by

$$\begin{aligned} V_{\mathrm{REF}} &= V_{BE} + \frac{R_2}{R_3}V_T\ln\frac{I_1}{I_2} = V_{BE} - T\frac{dV_{BE}}{dT} \\ &= V_{BE} + (E_{GO} - V_{BE}) + 3\frac{k}{q}T \\ &= E_{GO} + 3V_T = 1205\text{ mV} + 78\text{ mV} \qquad \text{at } 300°\text{C} \\ &= \underline{1283\text{ mV}} = \underline{1.283\text{ V}} \end{aligned} \tag{3.113}$$

Thus this voltage reference circuit when operating under the condition that $TC_{V(\mathrm{REF})} = 0$ will have an output voltage of 1.28 V. Notice that the reference voltage is determined principally by the energy band gap value ($E_{GO} = 1205$ mV for silicon).

In order to produce the zero-temperature condition, the basic circuit requirement is given above as

$$\frac{R_2}{R_3}\frac{k}{q}\ln\frac{I_1}{I_2} = -\frac{dV_{BE}}{dT} = \frac{E_{GO} - V_{BE}}{T} + 3\frac{k}{q} \tag{3.114}$$

If we now multiply both sides of this equation by $T$, we obtain

$$\frac{R_2}{R_3}V_T\ln\frac{I_1}{I_2} = E_{GO} - V_{BE} + 3V_T = 1283\text{ mV} - V_{BE} \tag{3.115}$$

Dividing through by $V_T = 25.9$ mV at 300 K gives

$$\frac{R_2}{R_3}\ln\frac{I_1}{I_2} = \frac{1283\text{ mV} - V_{BE}}{25.9\text{ mV}} \tag{3.116}$$

For a representative value of $V_{BE} = 650$ mV, this will yield the condition $(R_2/R_3)\ln(I_1/I_2) = 24.5$.

**Design example.** We will now consider a design example of the "band-gap" voltage reference. For this example we will make the following arbitrary, but reasonable choices: $I_1 = I_3 = 1.0$ mA, $I_1/I_2 = 5$. Since $I_2R_3 = V_T\ln(I_1/I_2) = 41.6$ mV, we have that

$$R_3 = 41.6\text{ mV}/0.2\text{ mA} = \underline{208\ \Omega}$$

For the resistance ratio we have that $R_2/R_3 = 24.5/\ln(I_1/I_2) = 15.2$, so that $R_2 = 15.2 \times 208\ \Omega = \underline{3166\ \Omega}$. For $R_1$ we have that

$$R_1 = \frac{V_{\text{REF}} - V_{BE}}{I_1} = \frac{1283\text{ mV} - 650\text{ mV}}{1.0\text{ mA}} = \underline{633\ \Omega} \tag{3.117}$$

All three resistance values have now been determined, and we note that they are all of reasonable and acceptable magnitude for high-accuracy IC resistors in not being excessively large or excessively small. The current values obtained are also within a reasonable range. The strength of the current source should be such that $I_Q = I_1 + I_2 + I_3 + I_{O(\text{MAX})} = 2.2\text{ mA} + I_{O(\text{MAX})}$, where $I_{O(\text{MAX})}$ is the maximum output current that this voltage reference current will have to supply. Normally, for purposes of minimizing the effects of loading on the reference voltage, this output current will be very small.

**Effect of Loading on $V_{\text{REF}}$.** To determine the effect of changes on the output current of the band-gap voltage reference circuit on $V_{\text{REF}}$ we note that changes in $V_{\text{REF}}$ will result in changes in $I_1$, $I_2$, and $I_3$. However, due to the stabilizing effects of the resistors, the change in the current through $Q_3$ will be the predominant effect. Or to put it another way, changes in $I_O$ will be matched by almost equal but opposite changes in $I_3$. Corresponding to a change in $I_3$ there will be a change in the base-to-emitter voltage of $Q_3$, as given by the relationship $dI_3/dV_{BE\,3} = g_{m\,3} = I_3/V_T$. As a result we can write

$$\frac{dV_{\text{REF}}}{dI_o} = \frac{-dV_{\text{REF}}}{dI_3} = \frac{-dV_{BE\,3}}{dI_3} = \frac{1}{g_{m\,3}} \tag{3.118}$$

so therefore we have that

$$r_o = \frac{-dV_{\text{REF}}}{dI_O} = \frac{1}{g_{m\,3}} = \frac{V_T}{I_3} \tag{3.119}$$

Thus if $I_3 = 1.0$ mA as in the example above, then $r_o = 25\ \Omega$. If, for example, we wanted to limit the variations in $V_{\text{REF}}$ due to variations in the output current to no more than 1.0 mV, we will have a corresponding limitation on variations in $I_O$ as given by

$$\frac{\Delta V_{\text{REF}}}{\Delta I_O} = 25\ \Omega$$

so that

$$\Delta I_{O(\text{MAX})} = \frac{1\text{ mV}}{25\ \Omega} = 40\ \mu\text{A} \tag{3.120}$$

As a result we see that in order to obtain the full benefits available from this voltage reference we should generally insert a high-input-impedance buffer circuit between this voltage reference and the load that is to be driven.

We can also note, at this point, the desirability of using a current source to provide the overall biasing of this circuit in order to prevent variations in the supply voltage, $V^+$, from having an undue effect on $V_{\text{REF}}$. To illustrate the benefit afforded by the use of a current source with a low dynamic output conductance, let us consider the variation in the reference voltage with respect to the supply voltage as given by

$$\frac{dV_{\text{REF}}}{dV^+} = \frac{dV_{\text{REF}}}{dI_3} \frac{dI_3}{dI_Q} \frac{dI_Q}{dV^+} \tag{3.121}$$

wherein again we make use of the fact that changes in $I_Q$ will be taken up principally by changes in the current through $Q_3$, so that as a result, $dI_3/dI_Q \simeq 1$. Since $dV_{\text{REF}}/dI_3 = 1/g_{m3} = V_T/I_3$, and $dI_Q/dV^+ = g_o$, we now have that

$$\frac{dV_{\text{REF}}}{dV^+} \simeq \frac{1}{g_{m3}} g_o = \frac{V_T}{I_3} g_o \tag{3.122}$$

If, for example, $I_3 = 1.0$ mA as before and $g_o = 100$ nS $= 1 \times 10^{-7}$ S, we have

$$\begin{aligned} \frac{dV_{\text{REF}}}{dV^+} &= 25\ \Omega \times 1 \times 10^{-7}\ \text{S} = 2.5 \times 10^{-6} \\ &= 2.5\ \mu\text{V/V} \end{aligned} \tag{3.123}$$

so that the change in $V_{\text{REF}}$ will be only 2.5 $\mu$V for every 1-V change in the supply voltage. This represents a very good degree of power supply rejection, indeed, so that this circuit is well insulated from the two external effects of temperature variation and supply voltage variation.

**Effect of error in parameter values on temperature coefficient.** In this circuit, as was the case for the voltage reference circuit considered earlier, we must take note of the fact that although the nominal temperature coefficient will be zero, the actual temperature coefficient will not be zero, due to the inevitable deviation in the component values from their exact design center values. To illustrate this point again, let us consider the effect of a deviation in the $R_2/R_3$ resistance ratio from the design center value. For this example we will choose a representative deviation of $\pm 2\%$.

The equation for $TC_{V(\text{REF})}$ has been given as

$$TC_{V(\text{REF})} = \frac{dV_{BE}}{dT} + \frac{R_2}{R_3} \frac{k}{q} \ln \frac{I_1}{I_2} \tag{3.124}$$

When $R_2/R_3$ is at the design center value we will assume that $TC_{V(\text{REF})} = 0$, so that the two terms on the right-hand side are equal in magnitude but opposite in algebraic sign. Therefore, a $\pm 2\%$ deviation in $R_2/R_3$ from the design center value will produce correspondingly a $\pm 2\%$ divergence in the magnitude of the entire quantity of which $R_2/R_3$ is a factor. Since at the design center point this quantity is equal in magnitude to $dV_{BE}/dT$, the $\pm 2\%$ variation in $R_2/R_3$ will result in a variation equal to $\pm 2\%$ of $dV_{BE}/dT$ in $TC_{V(\text{REF})}$. Since the design center value of $TC_{V(\text{REF})}$ is zero, the $\pm 2\%$ divergence of the resistance ratio will result in a residual temperature coefficient for $V_{\text{REF}}$ given by

$$TC_{V(\text{REF})} = \pm 0.02 \frac{dV_{BE}}{dT} = \pm 0.02 \times 2.1\ \text{mV/°C} = 42\ \mu\text{V/°C} \tag{3.125}$$

and on a percentage basis, $\pm 0.0033\%/°\text{C}$.

If a very small temperature coefficient is desired, a technique such as laser trimming can be utilized. If such a technique as laser trimming is to be employed with a band-gap regulator of the type under consideration here, resistors $R_2$ and $R3$ would be in the form of thin-film resistors deposited by a vacuum deposition technique on the IC chip. A high-energy laser beam can then be used to cut a notch or slot in the thin-film resistor and thereby increase the resistance value as a consequence of the reduced effective width of the resistor. The laser is generally part of a feedback loop that senses the temperature coefficient of the reference voltage and controls the depth of the laser cut. If the temperature coefficient is too high in the positive direction, $R_3$ can be trimmed to increase its value and thereby bring the temperature coefficient down toward zero. If, on the other hand, the temperature coefficient is negative, $R_2$ would be trimmed to bring the temperature coefficient up toward zero. By this means very small residual temperature coefficients can be achieved, perhaps in the range of a few microvolts per degree.

**Band-Gap voltage reference with larger reference voltage.** The reference voltage produced by the band-gap voltage reference under consideration here is $V_{\text{REF}} = E_{GO} + 3V_T = 1283$ mV. A larger reference voltage can be obtained by adding some diodes, in the form of diode-connected transistors, to the circuit at the lower end of $R_2$ in Figure 3.28. If $n$ diodes are added to the circuit, the equation for $V_{\text{REF}}$ becomes

$$V_{\text{REF}} = (n+1)V_{BE} + \frac{R_2}{R_3} V_T \ln \frac{I_1}{I_2} \tag{3.126}$$

and correspondingly

$$TC_{V(\text{REF})} = \frac{dV_{\text{REF}}}{dT} = (n+1)\frac{dV_{BE}}{dT} + \frac{R_2}{R_3}\frac{k}{q} \ln \frac{I_1}{I_2} \tag{3.127}$$

For zero temperature coefficient we thus have the requirement that

$$\frac{R_2}{R_3}\frac{k}{q} \ln \frac{I_1}{I_2} = -(n+1)\frac{dV_{BE}}{dT} \tag{3.128}$$

When this condition is satisfied, the reference voltage will be given by

$$V_{\text{REF}} = (n+1)V_{BE} - (n+1)T\frac{dV_{BE}}{dT} = (n+1)\left(V_{BE} - \frac{T\,dV_{BE}}{dT}\right) \tag{3.129}$$

For the previous analysis we have seen that $V_{BE} - T(dV_{BE}/dT) = E_{GO} + 3V_T$, so that we now have that $V_{\text{REF}} = (n+1)(E_{GO} + 3V_T) = (n+1)(1.283\text{ V})$. Thus if one additional diode (diode-connected transistor) is used, $V_{\text{REF}} = 2 \times 1.283\text{ V} = 2.566\text{ V}$, and for two additional diodes ($n = 2$) we have that $V_{\text{REF}} = 3 \times 1.283\text{ V} = 3.849\text{ V}$. Since

$$\frac{dV_{BE}}{dT} = -\left[\frac{E_{GO} + 3(k/q) - V_{BE}}{T}\right] \tag{3.130}$$

the design condition for the resistance ratio will be

$$\frac{R_2}{R_3}\ln\frac{I_1}{I_2}=(n+1)\frac{1283\text{ mV}-V_{BE}}{25.9\text{ mV}} \tag{3.131}$$

For a current ratio of 5 and for $n = 2$ the required resistance ratio will be

$$\frac{R_2}{R_3}=(3)\frac{1283\text{ mV}-650\text{ mV}}{25.9\ln 5}=45.6 \tag{3.132}$$

The large value of resistance ratio required in this case is not very desirable from the standpoint of IC circuit design and the resistance ratio tolerance will probably be somewhat larger than in the previous example. As a result the residual temperature coefficient on a percentage basis will likely be larger than in the case in which no additional diodes are used.

### 3.3.2 Voltage Reference with Feedback Amplifier

In Figure 3.34 the combination of a voltage reference and a feedback amplifier is shown. This circuit can provide for the step-up of the reference voltage to a value needed for the particular application, and for the isolation of the voltage reference from the load, so that changes in the output current will not change the reference voltage, and produce only very minimal changes in the output voltage due to the very small value of the closed-loop output impedance of the feedback amplifier.

In this circuit a fraction of the output voltage given by the voltage division ratio $R_5/(R_4 + R_5)$ is fed back to the inverting input terminal of the amplifier. It is compared in the amplifier to the reference voltage which is applied to the noninverting input terminal. The amplifier by means of the feedback loop acts to maintain the voltage at the output such that $V_O R_5/(R_4 + R_5) = V_{\text{REF}}$. As a result, we have that

$$V_O = V_{\text{REF}}\left(1+\frac{R_4}{R_5}\right) \tag{3.133}$$

As a result, $V_{\text{REF}}$ can be stepped up to any reasonable voltage to meet the circuit requirements. Note that whatever residual temperature coefficient there is for $V_{\text{REF}}$ will be also stepped up, and by exactly the same factor as $V_{\text{REF}}$ itself, so that we will have that $TC_{V_O} = dV_O/dT = (dV_{\text{REF}}/dT)(1 + R_4/R_5)$. Nevertheless, on a

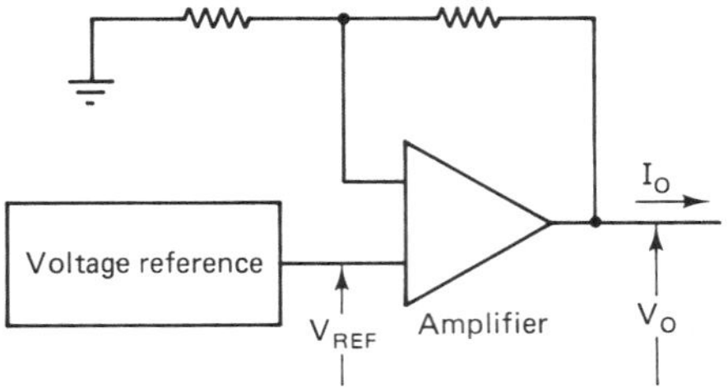


**Figure 3.34** Voltage reference with feedback amplifier for load isolation and voltage step-up.

percentage or fractional basis the normalized temperature coefficient of $V_O$ will be the same as that of $V_{\text{REF}}$ as given by

$$\frac{1}{V_O}\frac{dV_O}{dT} = \frac{1}{V_{\text{REF}}}\frac{dV_{\text{REF}}}{dT} \tag{3.134}$$

This last result is based on the assumption that the resistance ratio $R_4/R_5$ does not change with temperature. To ensure compliance with this, these two resistors should be resistors with closely matched temperature coefficients and should be placed in the circuit such that they will be at the same temperature, so that as a result the resistance variations of the two resistors will track each other.

### 3.3.3 Voltage Reference Diode

We will consider a two-terminal device based on the band-gap voltage reference circuit that produces a constant, temperature-compensated voltage drop across its terminals. This device thus is a diode characterized by a constant voltage drop and a very low temperature coefficient, and we will call it a *reference diode*.

The circuit of the reference diode is shown in Figure 3.35 and is a slight simplification of the circuit of the LM113 reference diode (National Semiconductor). In this circuit transistors $Q_1$ through $Q_3$ and resistors $R_1$ through $R_3$ perform the same respective functions as in the basic band-gap voltage reference circuit of Figure 3.28. Therefore, the diode voltage drop $V_D$ will be equal to 1.283 V.

The basic function of the remainder of the circuit is to provide for a low output

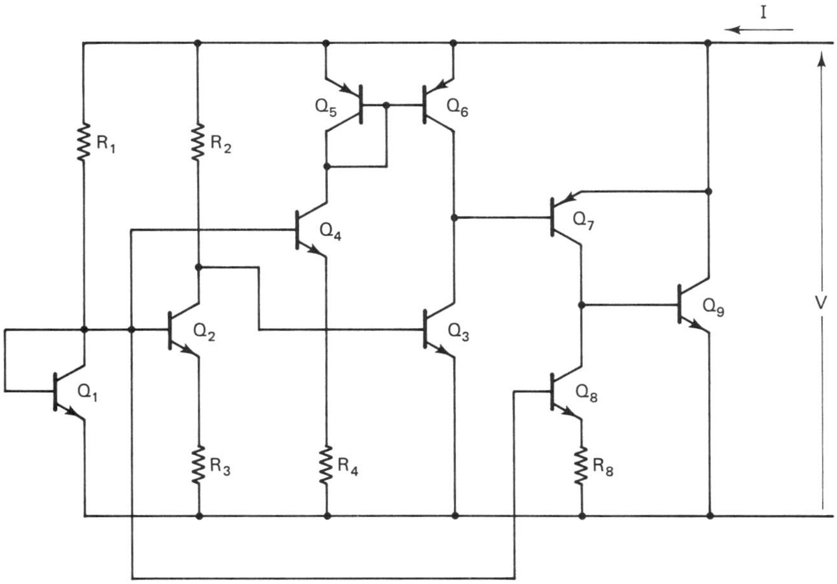


**Figure 3.35** Voltage reference diode (LM113 National Semiconductor).

impedance and to allow for a relatively large diode current. Transistor $Q_4$ in conjunction with resistor $R_4$ acts as a current source and biases the current mirror comprised of $Q_5$ and $Q_6$. Transistor $Q_6$ acts as a current source and is the active load for $Q_3$. Transistor $Q_3$ is in the common-emitter configuration and drives transistor $Q_7$, which is also in the common-emitter configuration. Transistor $Q_8$ in conjunction with resistor $R_8$ is a current sink and provides an active load for $Q_7$. Transistor $Q_9$ in the common-emitter configuration is driven by $Q_7$, and provides most of the current passing through the diode.

The basic operation of this diode comprised of transistors $Q_4$ through $Q_9$ and the associated resistors can be understood as follows. An increase in $V_D$ will produce a corresponding increase in the base-to-emitter voltage of $Q_3$, since the voltage drop across $R_2$ will remain relatively constant. The increase in the current through $Q_3$ will be given by $\Delta I_3 = g_{m\,3}\,\Delta V_{BE\,3} = g_{m\,3}\,\Delta V_D$. Since $Q_6$ is a current source, current $I_6$ will not change, and thus the change in the current through $Q_3$ will produce an equal change in the base current of $Q_7$ as given by $\Delta I_{B\,7} = \Delta I_3$. The change in the base current of $Q_7$ will produce a change in the collector current of this transistor by an amount $\Delta I_7 = \beta_7\,\Delta I_{B\,7} = \beta_7\,\Delta I_3$. Since $Q_8$ is a constant-current sink, the change in the collector current of $Q_7$ will result in an equal change in the base current of $Q_9$, so that $\Delta I_{B\,9} = \beta_7\,\Delta I_3$. This base current change will produce a change in the collector current of amount $\Delta I_9 = \beta_9\beta_7\,\Delta I_3$. Since $\Delta I_3$ has been related to $\Delta V_D$ by $\Delta I_3 = g_{m\,3}\,\Delta V_D$, we now have that $\Delta I_9 = \beta_9\beta_7 g_{m\,3}\,\Delta V_D$. As a result of the current gains of $Q_7$ and $Q_9$, the current through $Q_9$ will be much larger than that through $Q_7$ or $Q_3$, so that the change in the diode current will be due mostly to the change in the current through $Q_9$, so that we will have $\Delta I_D \simeq \Delta I_9 = \beta_9\beta_7 g_{m\,3}\,\Delta V_D$. As a result, the output resistance for this diode (i.e., the diode dynamic resistance) can be written as

$$r_d = \frac{\Delta V_D}{\Delta I_D} = \frac{1}{\beta_9\beta_7 g_{m\,3}} \tag{3.135}$$

Since $g_{m\,3} = I_3/V_T$, this can be rewritten as

$$\text{diode dynamic resistance} = r_d = \frac{V_T/I_3}{\beta_9\beta_7} \tag{3.136}$$

As a result of the current-gain product in the denominator of this expression, very low values of diode dynamic resistance can be obtained, as will be seen in the example to follow.

**Voltage reference diode design example.** For an example of the design of the voltage reference diode we will require that the diode dynamic resistance be no greater than 0.1 Ω. We will assume that all the transistors will have current gains of $\beta = 50$ minimum. We therefore have that

$$r_d = 0.1\ \Omega = \frac{25\ \text{mV}/I_3}{50 \times 50} \tag{3.137}$$

so that we have for $I_3$ the value of

$$I_3 = \frac{25 \text{ mV}}{0.1\ \Omega \times 50 \times 50} = 100\ \mu\text{A} \tag{3.138}$$

In accordance with this value of $I_3$, a reasonable choice for $I_1$ is 200 $\mu$A, so that assuming a $V_{BE}$ of 600 mV, we will have

$$R_1 = \frac{(1.283 - 0.600)\text{ V}}{0.2 \text{ mA}} = \underline{3.415 \text{ k}\Omega} \tag{3.139}$$

We will select a current ratio of $I_1/I_2 = 5$, so that for $R_3$ we have the relationship

$$\frac{I_2 R_3}{V_T} = \ln \frac{I_1}{I_2} = \ln 5$$

so that

$$R_3 = \frac{25 \text{ mV}}{40\ \mu\text{A}} \ln 5 = \underline{1.006 \text{ k}\Omega} \tag{3.140}$$

We can now easily solve for $R_2$ as

$$R_2 = \frac{V_D - V_{BE}}{I_2} = \frac{(1.283 - 0.600)\text{ V}}{40\ \mu\text{A}} = \underline{17.1 \text{ k}\Omega} \tag{3.141}$$

Since $I_3 = I_6 = I_5 = I_4 = 100\ \mu$A, we have for $R_4$ the relationship

$$\frac{I_4 R_4}{V_T} = \ln \frac{I_1}{I_4} = \ln 2 \tag{3.142}$$

so that

$$R_4 = \frac{25 \text{ mV} \ln 2}{100\ \mu\text{A}} = \underline{173\ \Omega} \tag{3.143}$$

Based on the choices for the other currents, an acceptable choice for $I_8$ is 100 $\mu$A, so that $R_8 = R_4 = \underline{173\ \Omega}$.

If we add up all the bias currents, we have a total quiescent current of $I_Q = I_1 + I_2 + I_5 + I_6 + I_8 = (200 + 40 + 100 + 100 + 100)\ \mu\text{A} = 540\ \mu\text{A}$. Therefore, it should be expected that for this diode to be operating in its constant-voltage temperature-compensated regime, a current, $I_D$, of at least 540 $\mu$A should pass through the diode. If the current level is below this, the circuit will not function properly, giving rise to a larger dynamic resistance and a significant temperature coefficient.

### 3.3.4 Thermally Stabilized Voltage Reference

A thermally stabilized voltage reference is a voltage reference device whose temperature is maintained or stabilized at a constant value. As a result, the voltage reference output voltage will be almost completely independent of the ambient temperature, with temperature coefficients of less than 1 ppm/°C being obtainable.

With any voltage reference circuit there will inevitably be some deviation of the device and component parameters away from the design center values such that there will be a small net temperature coefficient. One approach to reducing the temperature coefficient of a voltage reference is to thermally isolate it from the ambient temperature changes by maintaining the device temperature at some constant value. As the ambient temperature changes, the device temperature remains almost constant, and accordingly the reference voltage change will be extremely small. As a result, the temperature coefficient of the device, which is defined as the rate of change of reference voltage with respect to the ambient temperature, will correspondingly be extremely small.

To look at it mathematically, we have that the temperature coefficient of the reference voltage will be given by

$$TC_{V(\mathrm{REF})} = \frac{dV_{\mathrm{REF}}}{dT_{\mathrm{ambient}}} = \frac{dV_{\mathrm{REF}}}{dT_{\mathrm{chip}}} \frac{dT_{\mathrm{chip}}}{dT_{\mathrm{ambient}}} \tag{3.144}$$

As a result of the thermal stabilization of the chip temperature, the factor $dT_{\mathrm{chip}}/dT_{\mathrm{ambient}}$ will be very much smaller than unity. Therefore, the temperature coefficient of the reference voltage (with respect to the ambient temperature) will be much smaller than what would be obtained without stabilization of the chip temperature.

In all thermally stabilized voltage references, the chip temperature is maintained above the ambient temperature by an electrothermal feedback loop. The chip temperature is typically maintained at about 90 to 100°C by a feedback loop that controls the amount of electrical power dissipated on the chip and senses the resulting temperature rise. The temperature stabilization circuit and the voltage reference circuit is on the same small silicon chip, so that the two are in very good thermal contact, especially when the high thermal conductivity of the silicon and the small dimensions of the chip are considered.

**Thermally stabilized voltage reference example.** We will now consider a specific example of a thermally stabilized voltage reference. This voltage reference is the LM199/299/399 series (National Semiconductor). In Figure 3.36a the thermal stabilization circuit is shown, and in Figure 3.36b the voltage reference circuit is given, with both circuits being located on the same chip to form a monolithic IC.

We will first consider the thermal stabilization circuit. In this circuit $Q_4$ acts as a thermal shutdown transistor that controls the power dissipation of the chip by turning the Darlington configuration of $Q_1$ and $Q_2$ off whenever the temperature rises above a certain value. Whenever the temperature is substantially below the stabilization temperature, $Q_1$ and $Q_2$ will be in full conduction, limited only by the current-limit transistor $Q_3$. As the temperature of the chip increases, transistor $Q_4$ starts to conduct and diverts base drive away from the base of $Q_1$. This causes a reduction in the current through $Q_1$ and thus a reduction in the power dissipation and of the rate of rise of chip temperature. As a result of this feedback action, the chip temperature will asymptotically approach the desired stabilization temperature.

The temperature-sensing circuitry for the thermostatic action of this circuit consists of transistors $Q_4$, $Q_6$, and $Q_7$, and zener diode $D_1$, which is a diode-connected transistor using the emitter–base breakdown voltage of 6.3 V. Assuming a $V_{BE}$ of 0.65 V for $Q_6$ and $Q_7$, the net voltage across $R_1$ and $R_2$ will be $V_Z - 2V_{BE} =$

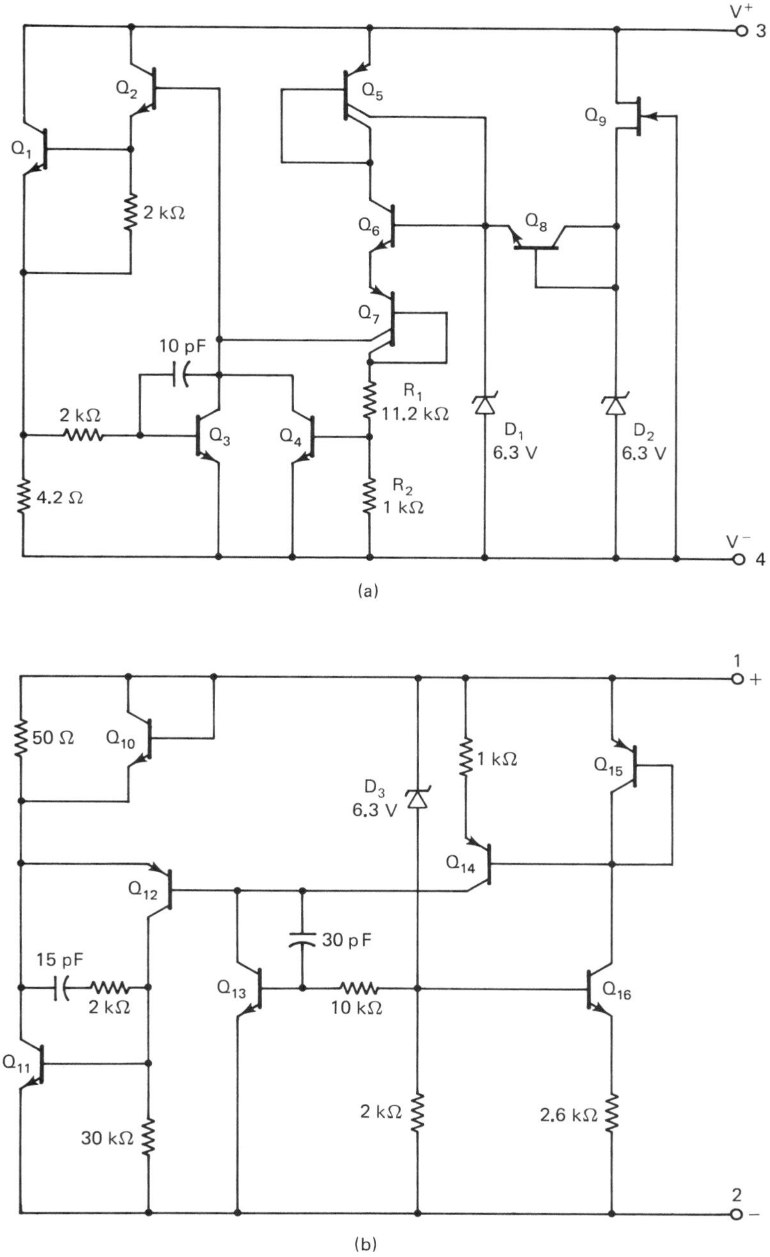


**Figure 3.36** Thermally stabilized voltage reference: (a) temperature stabilization circuit; (b) voltage reference circuit (National Semiconductor).

$6.3 - 2(0.65) = 5.0$ V. The current through $R_1$ and $R_2$ resulting from this voltage will be 5.0 V/12.2 kΩ = 0.41 mA. This current will produce a voltage drop of 410 mV across $R_2$ at 25°C. Therefore, we see that at room temperature (25°C) the base-to-emitter voltage of $Q_4$ will be insufficient to turn $Q_4$ on.

Transistor $Q_7$ is a multiple-collector lateral PNP transistor. The construction of $Q_7$ is such that the ratio of the collector currents is 0.3, so that when 410 μA flows out of the second collector, 120 μA will be the current supplied by the first collector. This current will become the base drive of $Q_2$. This base drive is sufficient to produce a relatively large collector current for $Q_1$. Since this current can be quite large, as much as about 600 mA or even possibly 1 A, current limiting is incorporated into the design of the circuit. The current limiting is produced by the action of $Q_3$ in conjunction with the 4.2-Ω current-limit resistor. When the voltage drop across the current-limit resistor (4.2 Ω) approaches 600 mV, $Q_3$ is biased into full conduction and diverts an increasingly large proportion of the 120 μA collector current supplied by $Q_7$ away from the base of $Q_2$. This current-limiting action limits the current through $Q_1$ to about 140 mA.

If the circuit starts initially at 25°C, the initial current through $Q_1$ will be 140 mA. With a terminal voltage of 15 V the power dissipation will be 2.1 W. The thermal impedance of this IC with a polysulfone thermal shield is about 220°C/W, so we see that if this current were to continue for any length of time there would be an excessive temperature rise.

As the temperature increases, the voltage across $R_2$ increases due to the positive temperature coefficient of the zener diode, about 3 mV/°C, and the negative temperature coefficient of the base-to-emitter voltage drops of $Q_6$ and $Q_7$, about −2.3 mV/°C each. After considering the voltage divider action of $R_1$ and $R_2$ we see that the voltage across $R_2$ will increase at a rate of (3 mV/°C + 4.6 mV/°C)(1 kΩ/12.2kΩ) = 0.62 mV/°C. At the same time that this is happening, the required base-to-emitter voltage to turn $Q_4$ on is decreasing at a rate of about 2.3 mV/°C. If we recall that at 25°C the voltage across $R_2$ was 410 mV and note that at this temperature the voltage necessary to turn $Q_4$ on is about 600 mV, we see that the temperature increase required to turn $Q_4$ on will be given by

$$\Delta T = \frac{600 - 410}{(0.62 + 2.3)\ \text{mV/°C}} = \frac{190\ \text{mV}}{2.92\ \text{mV/°C}} = 65\text{°C} \qquad (3.145)$$

so that we should expect that $Q_4$ will be turned on at about 90°C.

Actually, what happens is that as the temperature increases, the current through $Q_4$ starts to increase. This diverts more and more base drive away from $Q_2$, reducing the current through $Q_1$ and thus lessening the chip power dissipation. The rate of increase of chip temperature correspondingly decreases and the chip temperature asymptotically approaches an equilibrium value of about 90°C. Thus the thermal stabilization circuit involves an electrothermal feedback loop that senses the chip temperature and controls the chip power dissipation to set the chip temperature to the desired value.

Transistors $Q_8$ and $Q_9$ together with diode $D_2$ is a startup circuit to provide the initial bias to the temperature-stabilization circuit. Once diode $D_1$ has been biased to breakdown ($V_Z$), the voltage across $Q_8$ will be zero and $Q_8$ will turn off, which will disconnect the startup circuit from the rest of the thermal stabilization circuit. After this happens the current through $D_1$ will be provided by the multiple-collector PNP transistor $Q_5$. The current through $Q_5$ is controlled by $D_1$, $Q_6$, $Q_7$, $R_1$, and $R_2$ and thus will be almost completely independent of the supply voltage. We see, therefore, that this circuit will be "self-biasing," so that changes in the supply voltage will have virtually no effect on the control of the chip temperature and thus on the reference voltage.

We have seen that the initial turn-on current of the thermal stabilization circuit will be 140 mA at 25°C. As the temperature approaches the stabilization temperature of 90°C, this current will decrease. Since the temperature rise from 25 to 90°C is 65°C, and the thermal impedance is 220°C/W, the steady-state power dissipation of the chip when thermally stabilized will be 65°C/(200 W/°C) = 300 mW. If a supply voltage of 15 V for the stabilization circuit of 15 V is used, the corresponding current will be about 20 mA.

As a result of the small thermal mass and of the large thermal impedance of the LM199 series devices, the warm-up time is quite short. The warm-up time for the regulator output voltage to reach within 0.05% of its operating voltage that is nominally 6.95 V is only 3 s starting from 25°C and with a supply voltage of 30 V.

**Voltage reference circuit.** The voltage reference circuit is shown in Figure 3.36b and uses a temperature-compensated zener diode as the basic voltage reference. The zener diode is $D_3$, and its positive temperature coefficient is compensated by the negative temperature coefficient of the base-to-emitter drop of $Q_{13}$. The reference voltage consisting of the voltage drop of the zener diode of 6.3 V and the base-to-emitter drop of $Q_{13}$ of 0.65 V appears across the external terminals of the device as a total voltage of 6.95 V typical (6.8 minimum, 7.1 maximum).

Most of the device current, however, does not go through $D_3$ and $Q_{13}$ but rather through $Q_{11}$. Any change in the output voltage will produce a corresponding change in the $V_{BE}$ of $Q_{13}$. Transistor $Q_{14}$ acts as a current source active load for $Q_{13}$, and thus changes in the collector current of $Q_{13}$ will produce an equal change in the base current of $Q_{12}$. This base current change is multiplied by the current gain of $Q_{12}$ and drives the base of $Q_{11}$, where it is again multiplied by the current gain. Thus any changes in the output voltage result principally in changes in the current through $Q_{11}$. As a result of the high gain of $Q_{13}$ and the current gains of $Q_{12}$ and $Q_{11}$, the rate of change of output current with respect to output voltage, $dI_O/dV_O$, which is the dynamic output conductance, will be relatively large. The large value of the dynamic output conductance, $g_o$, will thus mean that the device will have a small dynamic output resistance, $r_o$. For the LM199 series of devices, this is specified as 0.5 Ω typical, 1.0 Ω maximum, at a current level of 1 mA.

The temperature coefficient of the terminal voltage is specified as 0.3 ppm/°C typical, 1.0 ppm/°C maximum. This corresponds to 2 μV/°C typical, 7 μV/°C maximum. This compares very favorably with the temperature coefficients of the nonthermally stabilized voltage references, which generally are in the range 10 to 100 μV/°C.

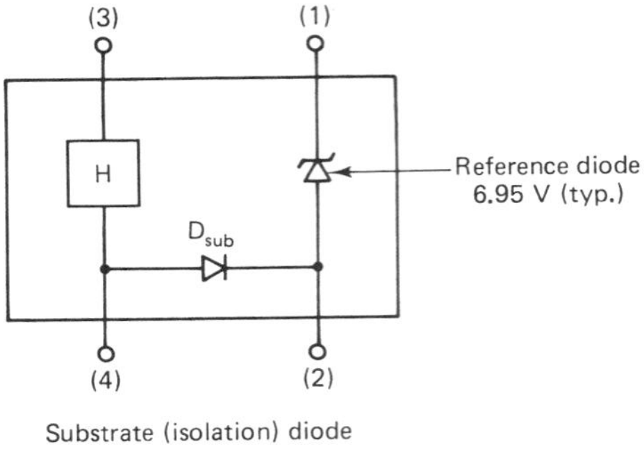


**Figure 3.37** Functional diagram of thermally stabilized voltage reference.

**Applications of the thermally stabilized voltage reference.** In Figure 3.37 a functional block diagram of the thermally stabilized voltage reference is shown. Diode $D_{sub}$ is a consequence of the fact that the thermal stabilization circuit and the voltage reference circuit share the same IC chip. For proper operation of the device, this diode should never be forward biased in order to keep the two circuits electrically isolated. In addition, the breakdown voltage of this diode is 40 V (min.), so that no more than 40 V reverse bias should be applied across this diode.

In Figure 3.38 a buffered voltage reference is shown using the thermally stabilized voltage reference and an operational amplifier to provide isolation of the voltage reference device from the load. The operational amplifier circuit also provides for step-up of the reference voltage to the desired value, as given by $V_O = V_{REF}(1 + R_2/R_1)$. To preserve the low-temperature-coefficient characteristics of the voltage reference, the operational amplifier should be chosen to have a very low offset-voltage temperature coefficient, $TC_{V(OS)}$. This should preferably be down in the range of about 1 $\mu$V/°C or less.

Since the output voltage will be a function of the $R_2/R_1$ resistor ratio, great care should be taken in the selection and placement of these two resistors. The resistors should preferably have low temperature coefficients of resistance ($TCR$), and even

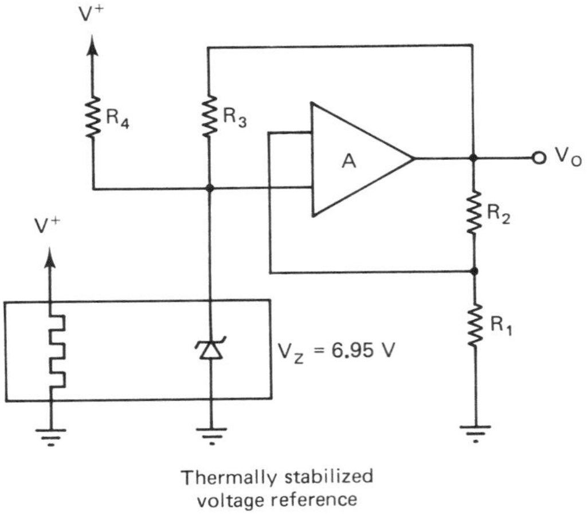


**Figure 3.38** Buffered voltage reference using thermally stabilized voltage reference.

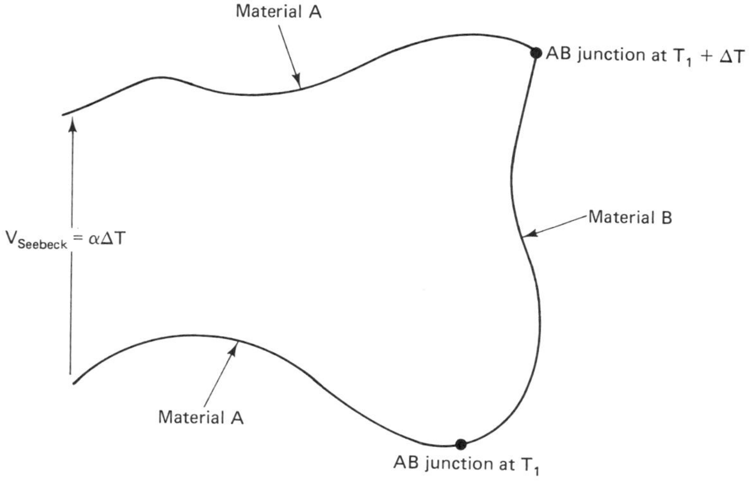


**Figure 3.39** Seebeck effect: a thermocouple.

more important, very well matched temperature coefficients, so that the resistor values will track each other with temperature changes such that the resistance ratio remains constant. The placement of these resistors in the circuit with respect to considerations of heat flow and temperature differentials is also very important. In general, the placement of the resistors should be such that the temperature difference between the two resistors is minimized.

Another consideration that can be of importance when very low temperature coefficients are of interest is that of thermocouple effects, that is, the thermoelectric voltage that is generated as the result of electrical contact between two dissimilar materials (i.e., the Seebeck effect). The Seebeck coefficient is the thermoelectric voltage produced per degree Celsius temperature difference for a junction between two given materials, as illustrated in Figure 3.39. In this figure, $\alpha$ is the Seebeck coefficient ($\mu$V/°C), so the thermoelectric voltage that results from a temperature difference of $\Delta T$ between the two junctions between metals A and B will be $V_{\text{Seebeck}} = \alpha \Delta T$.

The Seebeck coefficient between the Kovar pins of the IC and the copper leads or the printed-circuit-board copper metallization pattern is about 30 $\mu$V/°C. Therefore, as shown in Figure 3.40, a temperature differential of as little as 0.1°C between the pin connections of the voltage reference can lead to a thermoelectric voltage with a temperature coefficient of about 3 $\mu$V/°C, which can represent a serious degradation in the voltage reference performance. Therefore, very careful layout of the circuit is important, with great attention paid to minimizing thermal gradients and temperature differentials.

To consider a simple design example for the buffered voltage reference of Figure 3.33, we will assume a supply voltage of +15 V, and that the desired scaled-up reference voltage output of the circuit is 10 V. The specified operating range of the LM199 series is 0.5 to 10 mA. Here we will choose a current of 1.0 mA to flow

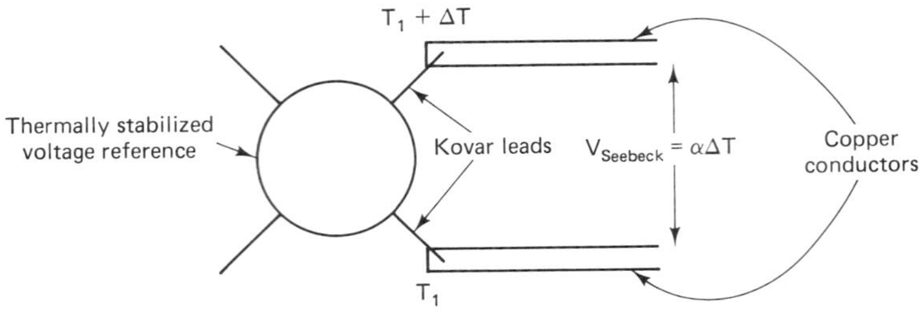


**Figure 3.40** Generation of Seebeck voltage due to temperature differential at pins of thermally stabilized reference.

through the voltage reference circuit so that $R_3$ becomes $R_3 = (V_0 - V_Z)/1.0$ mA $= (10 - 6.95)/1.0$ mA $= 3$ kΩ.

Resistor $R_4$ is used for the initial startup of the circuit. It is used to supply a small initial bias current to the voltage reference circuit to produce a sufficiently large voltage drop across the voltage reference such that the ouput voltage, $V_O$, will be large enough such that sufficient current will flow through $R_3$ to bring the reference circuit up to its full reference voltage. Resistor $R_4$ should be small enough such that the initial current flow will be large enough to initiate operation of this circuit, but on the other hand it should not be too small such that variations in the supply voltage $V^+$ will result in significant variations in the current through the reference circuit. Since a current of 10 to 100 μA should be large enough to initiate operation of the circuit, a suitable value for $R_4$ will be in the range of about 80 to 800 kΩ.

For the design of a resistive voltage divider the basic requirement is that $1 + R_2/R_1 = V_O/V_Z$. This therefore fixes the ratio of the two resistors. For the absolute values of $R_1$ and $R_2$ an important consideration is the effect of the operational amplifier input bias current and the temperature coefficient thereof. The values of $R_1$ and $R_2$ should be chosen to be small enough such that the maximum expected bias current will not produce a significant change in the output voltage due to the flow of the bias current through $R_1$ and $R_2$. Perhaps even more important, the values of the two resistors should be small enough such that the temperature coefficient of the bias current will not produce an unacceptable degradation of the overall temperature coefficient of the circuit.

A reasonable choice of operational amplifier for this circuit would be the LM108A or 208A, or equivalent. This amplifier is characterized by a very low offset voltage [0.3 mV (typ.), 0.5 mV (max.)] and even more important, a very low offset-voltage temperature coefficient of 1.0 μV/°C (typ.) and 5 μV/°C (max). For a 10-V output voltage, this temperature coefficient will correspond to 0.14 ppm /°C (typ.) and 0.72 ppm/°C (max.), so that this operational amplifier will not produce any significant increase in the overall temperature coefficient.

The input bias current of this operational amplifier is 0.8 nA (typ.) and 2.0 nA (max.). The temperature coefficient of the bias current is about 6 pA/°C (typ.). The contribution to the temperature coefficient of the output voltage due to the bias current temperature coefficient will be given by $dV_O/dT = R_2(dI_B/dT)$. We see therefore that as long as $R_2$ is not excessively large (i.e., less than 1 MΩ), this contribution to the overall temperature coefficient will be small.

A reasonable choice for $R_1 + R_2$ will be 50 kΩ, as this will not draw off an excessive current from the amplifier output. Since $1 + R_2/R_1 = (R_1 + R_2)/R_1 =$ 10 V/6.95 V, we have that $R_1 = 50$ kΩ × 0.695 = 34.75 kΩ. Therefore, $R_2 = 50$ kΩ − 34.75 kΩ = 15.25 kΩ.

Some other thermally stabilized voltage references are the LM199A/299A/399A series with a temperature coefficient of only 0.2 ppm/°C (typ.), 0.5 ppm/°C (max.), and the LM3999, with a temperature coefficient of 2 ppm/°C (typ.), 5 ppm/°C (max.), These voltage references have internal circuitry that is identical to that of the LM199 and have nominal reference voltages of 6.95 V.

It is of interest now to compare these thermally stabilized voltage references with the LM129/329 series of voltage references. These devices have a voltage reference circuit that is identical to that of the above-mentioned thermally stabilized references and differ from them only in that there is no thermal stabilization circuit.

As a result of the chip temperature not being stabilized, the temperature coefficient of the reference voltage will be substantially greater than that of the thermally stabilized voltage references. The temperature coefficient for the best of the LM129/329 series is that of the LM129A and 329A, which is 6 ppm/°C (typ.) and 10 ppm/°C (max.). For the LM329D the temperature coefficient is 50 ppm/°C (typ.) and 100 ppm/°C (max.). The nominal reference voltage is 6.95 V for all of these devices.

We see therefore that the temperature coefficient is much greater in the absence of thermal stabilization of the chip temperature. Indeed, the temperature coefficient is greater by a factor in the range of about 30 to 100. Since $dV_{REF}/dT = (dV_{REF}/dT_{chip})(dT_{chip}/dT_{ambient})$, we can conclude that the thermal stabilization circuit has a temperature stabilization coefficient in the range 0.01 to 0.03. This means that a 1°C change in the ambient temperature will result in a chip temperature change of only 0.01 to 0.03°C.

## PROBLEMS

For all problems use the following parameters unless otherwise indicated or implied.

1. Assume all transistors of the same type to be identical.
2. For NPN transistors: $\beta = 100$ (typ.), 50 (min.). For PNP transistors: $\beta = 50$ (typ.), 25 (min.).
3. Early voltage $V_A$: 200 V for both NPN and PNP transistors.
4. Base-to-emitter voltage temperature coefficient, $TC_{V(BE)} = dV_{BE}/dT = -2.2$ mV/°C.
5. Collector–base breakdown voltage: 50 V (min.).
6. $V_{BE} = 0.7$ V.
7. $V_T$ (thermal voltage) = 25 mV.
8. Resistor temperature coefficient, $TCR = (1/R)\ (dR/dT) = +2000$ ppm/°C.

9. Assume all resistors to have the same temperature coefficient and to be at the same temperature.
10. Supply voltages: $V^+ = +10$ V, $V^- = -10$ V.

## Current Source Problems

**3.1.** (*Diode-biased current sink*) Refer to Figure P3.1.

**(a)** Show that $I_2 = I_1/(1 + 2/\beta)$.

**(b)** For $I_1 = 1.0$ mA, find $R_1$. (*Ans.*: 19.3 kΩ)

**(c)** Find $I_2$ (typ., max., and min.). [*Ans.*: 0.96 (min.), 0.98 (typ.), 1.00 (max.) mA]

**(d)** Show that the temperature coefficient of $I_2$, $TC_{I_2} = (1/I_2)(dI_2/dT)$ will be given by $TC_{I_2} \simeq -\ TC_{R_1} - (1/V_{\text{supply}})(dV_{BE}/dT)$, where $V_{\text{supply}} = V^+ - V^-$.

**(e)** Find $TC_{I_2}$. (*Ans.*: −1890 ppm/°C or −0.189%/°C)

**(f)** Find the dynamic output resistance $r_o$ and the dynamic output conductance $g_o$ of the current sink. (*Ans.*: $r_o$ = 200 kΩ, $g_o$ = 5 μS)

**(g)** Find the percent change in the output current per 1-V change in the output voltage. (*Ans.*: 0.5%/V)

**(h)** Find the voltage compliance range. (*Ans.*: −9.8 to +40 V)

**(i)** If the offset voltage between $Q_1$ and $Q_2$ is ±1.0 mV (max.), find the resulting difference produced between $I_1$ and $I_2$. (*Ans.*: 0.041 mA or 4.1%)

**3.2.** (*Wilson current mirror current sink*) Refer to Figure P3.2.

**(a)** Show that $I_2 = I_1[1 + (1/\beta_1) + (1/\beta_3) - (2/\beta_2)]$, neglecting the second-order terms in $1/\beta$ (i.e., the $1/\beta^2$ terms).

**(b)** If $\beta = 100$ and the maximum $\beta$ mismatch between any two transistors is ±5%, find the maximum mismatch between $I_2$ and $I_1$. (*Ans.*: ±0.105%)

**(c)** Find $R_1$ for $I_1 = 1.0$ mA. (*Ans.*: 18.6 kΩ)

**(d)** Find the voltage compliance range. (*Ans.*: −9.1 to +41 V)

**(e)** Show that $TC_{I_2} = -TC_{R_1} - (2/V_{\text{supply}})(dV_{BE}/dT)$.

**(f)** Find $TC_{I_2}$. (*Ans.*: −1780 ppm/°C or −0.18%/°C)

**(g)** Find $r_o$ and $g_o$. (*Ans.*: 20 MΩ, 50 nS)

**(h)** Find the percent change in $I_2$ per 1-V change in $V_{C_2}$. (*Ans.*: 0.005%/V)

**3.3.** (*Current sink for low current levels*) Refer to Figure P3.3.

**(a)** Show that $I_2 = (V_T/R_2)\ln(I_1/I_2)$.

**(b)** Show that $dI_2/I_2 = (dI_1/I_1)/[1 + \ln(I_1/I_2)]$.

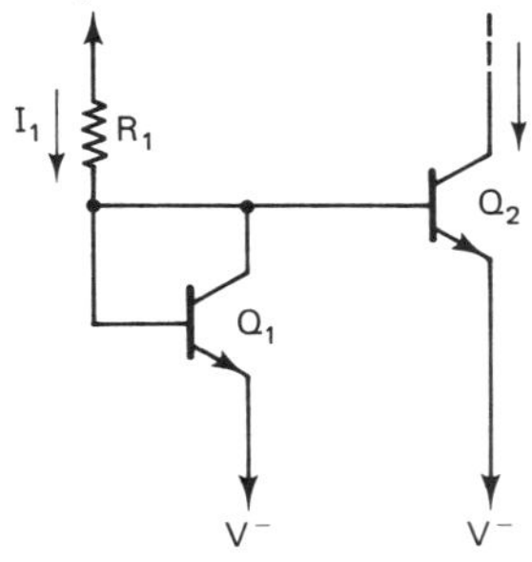


**Figure P3.1**

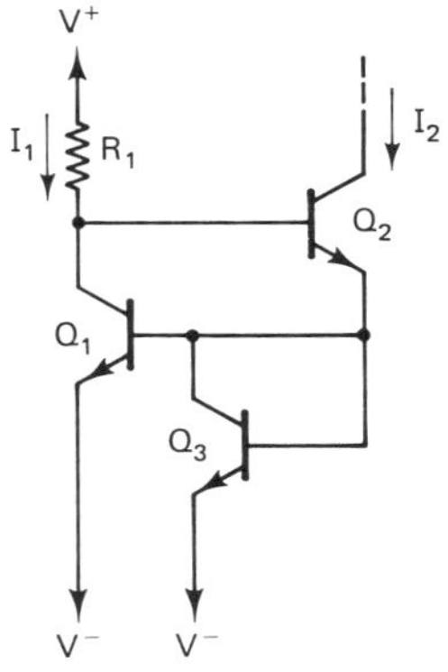


**Figure P3.2**

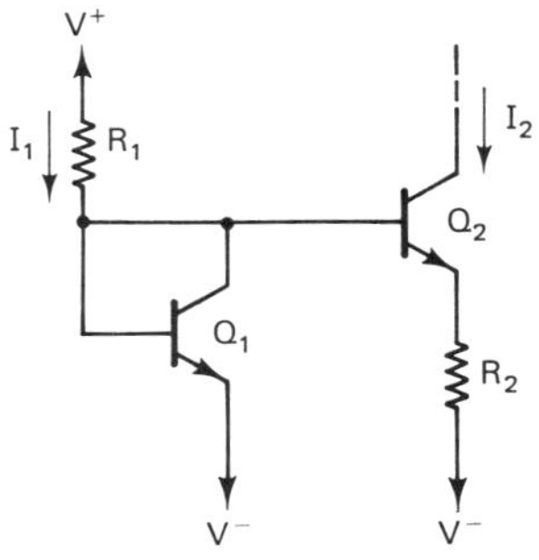


**Figure P3.3**

(c) Show that $dI_1/I_1 = dV_{supply}/V_{supply}$ and that as a result, $dI_2/I_2 = (dV_{supply}/V_{supply})/[1 + \ln(I_1/I_2)]$.

(d) Find $R_1$ for $I_1 = 1.0$ mA. (*Ans*.: 19.3 kΩ)

(e) Find $R_2$ for $I_2 = 10\ \mu$A. (*Ans*.: 11.5 kΩ)

(f) Find the percent change in $I_2$ per 1-V change in the supply voltage. (*Ans*.: 0.89%/V)

(g) Find the voltage compliance range. (*Ans*.: −9.7 to +41 V)

(h) Find $r_o$ and $g_o$ and the percent change in $I_2$ per 1-V change in $V_{C2}$. (*Ans*.: 108 MΩ, 9.26 nS, 0.0926%/V)

(i) Show that $TC_{I_2}$ is given approximately by

$$TC_{I_2} \simeq \frac{(1/T) - (TC_{R_2}) - \dfrac{(TC_{R_1}) + (1/V_{supply})\, dV_{BE}/dT}{\ln(I_1/I_2)}}{1 + 1/\ln(I_1/I_2)}$$

(j) Find $TC_{I_2}$. (*Ans*.: +758 ppm/°C or +0.0758%/°C)

**3.4.** (*Multiple current sinks*) Refer to Figure P3.4. *Given*: $R_B = 4$ kΩ, $I_A = 1.0$ mA, $I_1 = 1.0$ mA, $I_2 = 2.0$ mA, $I_3 = 10$ mA, and $I_4 = 0.1$ mA.

(a) Find $R_A$, $R_1$, $R_2$, $R_3$, and $R_4$, assuming that the transistor emitter areas are scaled such that the *current densities are equal in* $Q_A$, $Q_1$, $Q_2$, $Q_3$, and $Q_4$. (*Ans*.: 14.6 kΩ, 4 kΩ, 2 kΩ, 400, 40 kΩ)

(b) If all of the transistor emitter–base junciton areas are *equal*, find $R_1$ through $R_4$. (*Ans*.: 4 kΩ, 1.991 kΩ, 394, 40.575 kΩ)

(c) Find the voltage compliance range. (*Ans*.: −5.8 to +44 V)

(d) Find the maximum difference that there will be between $I_A$ and $I_1$ due to the effect of the base currents. (*Ans*.: 0.564%)

(e) Show that the $g_o$ for all of the current sinks will be given approximately by $g_o = I_C/(1.24 \times 10^4\ \text{V})$.

(f) Find the percent change in the output current of each of the current sinks per 1-V change in output voltage. (*Ans*.: 0.00807%/V)

(g) What is the function of $Q_B$?

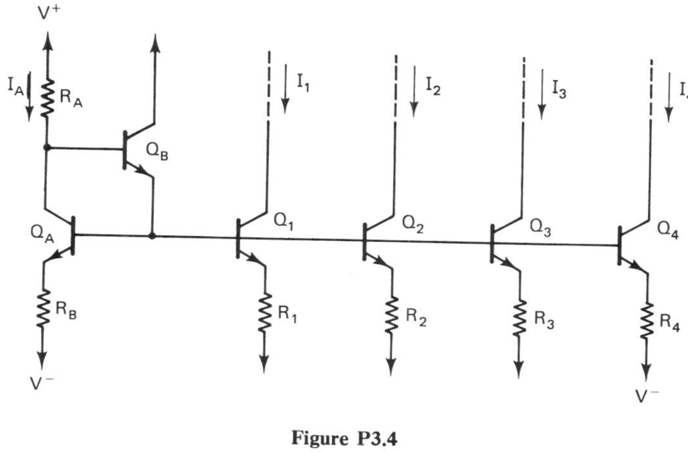


Figure P3.4

**3.5.** (*Current source using PNP transistor*)   Refer to Figure P3.5.
   **(a)** Show that $I_2 \simeq \alpha_{pnp}(I_o - I_1)$ and that $I_o = (1/R_1)(V^+ - V_{EB} - V_R)$.
   **(b)** For $V_R = +5.0$ V, $V_1 = 0$ V, $I_1 = 1.0$ mA, and $\alpha_{pnp} = 0.98$, find $R_1$ for $I_2 = 1.0$ mA.   (*Ans.:* 2.13 kΩ)
   **(c)** Find the voltage compliance range.   (*Ans.:* −45 to +5.5 V)
   **(d) (1)** Find the transfer conductance $dI_2/dV_R$.   (*Ans.:* $-\alpha/R_1 = -460$ μS)
       **(2)** Find the transfer coefficient $dI_2/dI_1$.   (*Ans.:* $-\alpha_{pnp} = -0.98$)

**3.6.** (*Current source using PNP transistor*)   Refer to Figure P3.6.
   **(a)** Show that $I_2 \simeq \alpha_{pnp}(V^+ - V_1)/R_2$.
   **(b)** Given that $V_1 = +5.0$ V, find $R_1$ for $I_2 = 1.0$ mA.   (*Ans.:* 4.9 kΩ)
   **(c)** Find the minimum allowable value for $I_1$ if $\alpha_{pnp} = 0.95$ (min.).   (*Ans.:* 50 μA)
   **(d)** If the $I_1$ current sink is replaced by a resistor $R_1$, find $R_1$ for $I_1 = 1.0$ mA.   (*Ans.:* $R_1 = 14.3$ kΩ)
   **(e)** Find the voltage compliance range.   (*Ans.:* −45 to +4.8 V)
   **(f)** Find the transfer conductance $dI_2/dV_1$.   (*Ans.:* −0.20 mS)

**3.7.** (*Current source using PNP transistor*)   Refer to Figure P3.7.
   **(a)** Show that $I_2 = I_o - I_1$ and therefore is independent of the current gain of the PNP transistor.
   **(b)** Show that $I_o = (V^+ - V_1)/R_1$.
   **(c)** If $V_1 = +5.0$ V, $I_1 = 1.0$ mA, and $I_2 = 1.0$ mA, find $R_1$.   (*Ans.:* 2.5 kΩ)
   **(d)** Find the transfer conductance $dI_2/dV_1$.   (*Ans.:* $g = -1/R_1 = -0.40$ mS)
   **(e)** Minimum allowable value for $I_1$ if $\alpha_{pnp} = 0.95$ (min.).   (*Ans.:* 50 μA)
   **(f)** Find the voltage compliance range.   (*Ans.:* −45 to +4.8 V)
   **(g)** Find the dynamic output resistance, conductance, and the percentage change in $I_2$ per 1-V change in $V_{C_2}$.   (*Ans.:* 10.1 MΩ, 99 nS, 0.01%/V or 100 ppm/V)
   **(h)** Find the temperature coefficient of $I_2$ if $TC_{I_1} = -500$ ppm/°C.   (*Ans.:* −1500 ppm/°C)

**3.8.** (*Current sink biased current sink*)   Refer to Figure P3.8.
   **(a)** Find $R_1$ for $I_1 = 1.0$ mA.   (*Ans.:* 17.9 kΩ)
   **(b)** Find $R_4$ for $I_5 = 10$ μA.   (*Ans.:* 11.5 kΩ)
   **(c)** Find the voltage compliance range.   (*Ans.:* −8.4 to +42 V)
   **(d)** Find $r_o$ and $g_o$.   (*Ans.:* 2.02 GΩ, 0.495 nS)
   **(e)** Find the percent change in $I_5$ per 1-V change in $V_{C_5}$.   (*Ans.:* 0.005%/V or 49.5 ppm/V)

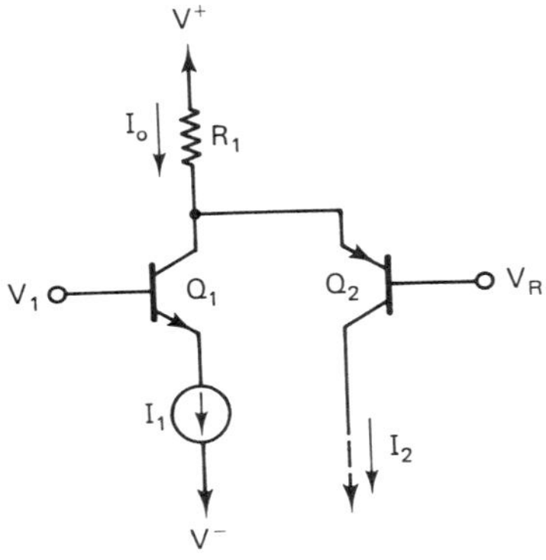

**Figure P3.5**

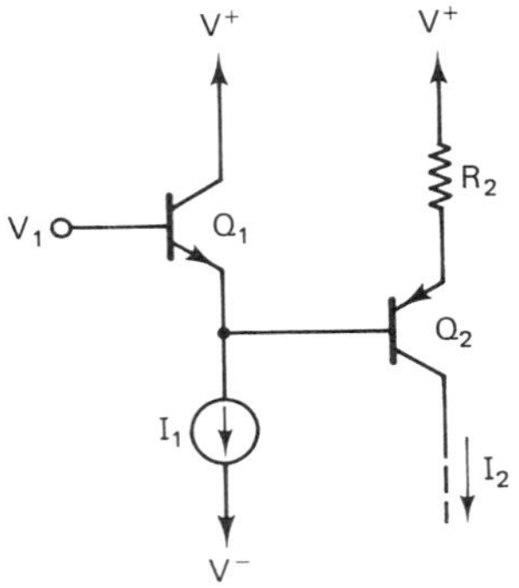

**Figure P3.6**

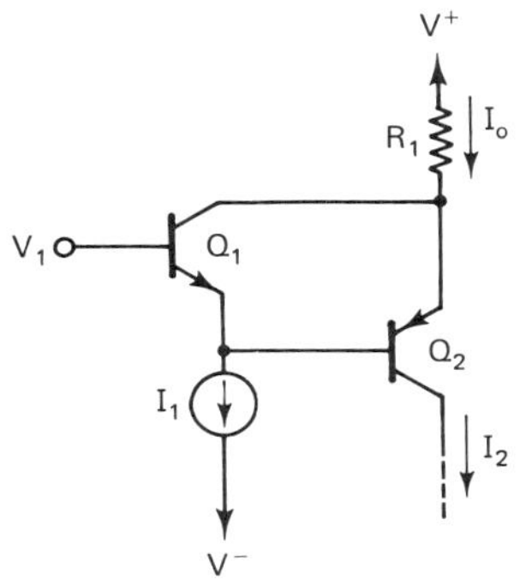


Figure P3.7

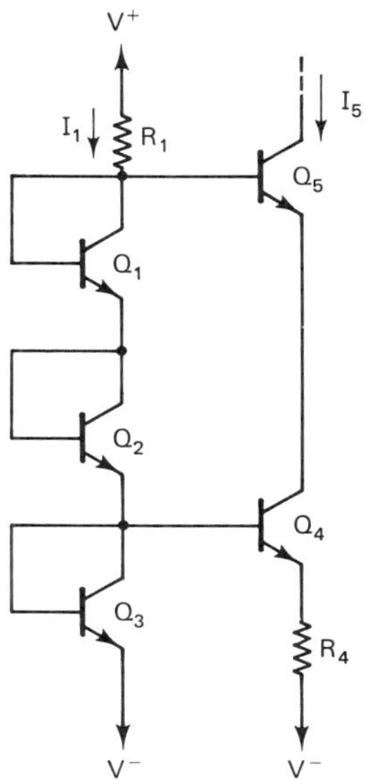


Figure P3.8

**(f)** If $C_{cb} + 2.0$ pF and $C_{cs} = 2.0$ pF, find the dynamic output impedance and admittance of this current sink at a frequency of 1.0 MHz. (*Ans.*: $y_o = +j25.1\ \mu\text{S}$, $z_O = -j39.8\ \text{k}\Omega$)

**3.9.** (*Diode-compensated current source*) Refer to Figure P3.9.

**(a)** Show that for a very large open-loop gain $A_{OL}$ that $I_N = (V_{\text{REF}} - V^-)/R_N$, where $N = 1, 2, 3, \ldots$, and that the base-to-emitter voltage temperature dependence is compensated for by the diode-connected transistor in the feedback loop of the operational amplifier. Assume that transistor active (emitter) areas are scaled for equal current densities in all transistors.

**(b)** Show that the exact expression for $I_N$ will be given by

$$I_N = \frac{1}{R_N}\left(\frac{V_{\text{REF}}}{1 + 1/A_{OL}} - V^- - V_{BE}\frac{1}{A_{OL} + 1}\right)$$

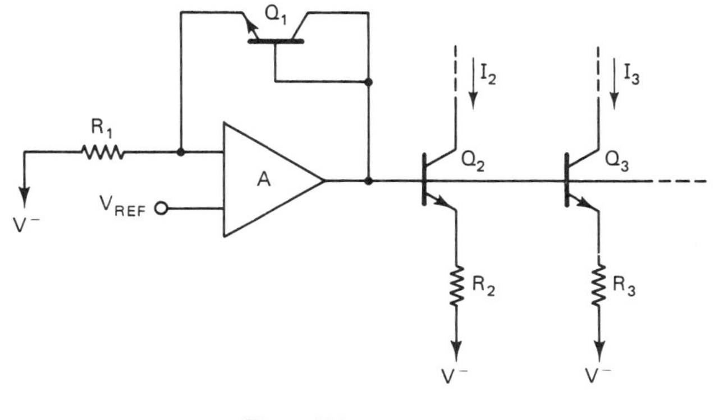


Figure P3.9

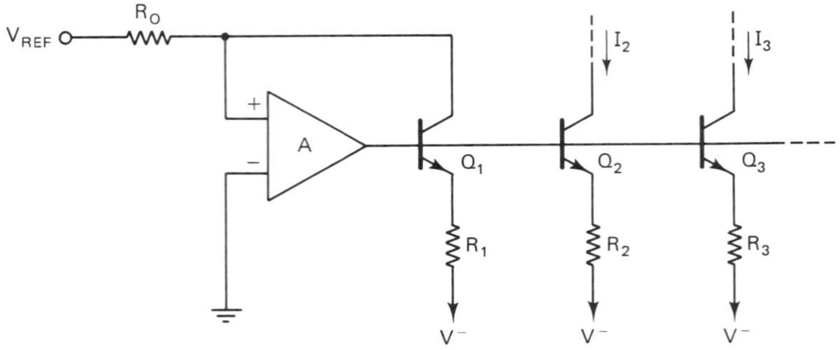


**Figure P3.10**

**(c)** Find the voltage compliance range for $V_{\mathbf{REF}} = -5.0$ V. (*Ans*.: −4.8 to +45 V)
**(d)** Find the minimum allowable value for $V_{\mathbf{REF}}$. (*Ans*.: −10 V)
**(e)** If the resistors $R_2$, $R_3$, . . . are temperature-compensated resistors with a *TCR* of ±50 ppm/°C (max.) and $V_{\mathbf{REF}}$ has a temperature coefficient of ±25 ppm/°C (max.), find the temperature coefficient of $I_N$. [*Ans*.: $TC_{I_N} = \pm 75$ ppm/°C (max.)]

**3.10.** (*Feedback-compensated current source*) Refer to Figure P3.10.
**(a)** Show that if $A_{OL}$ is very large, $I_N = (R_1/R_N)(V_{\mathbf{REF}}/R_O)$, assuming that the transistor active areas are scaled such that the current densities are equal in all transistors.
**(b)** Show that the more exact expression for $I_N$ will be

$$I_N = \frac{R_1}{R_N}\left[\frac{V_{\mathbf{REF}}}{R_O + R_1/A_{OL}} - \frac{V_{BE} + V^-}{A_{OL}\,(R_O + R_1/A_{OL})}\right]$$

**3.11.** (*Multiple-collector lateral PNP current source*) Refer to Figure P3.11.
**(a)** If all collector active "lengths" are equal, show that

$$I_B = I_C = I_D = \frac{I_A}{1 + (4/\beta_{\mathbf{pnp}})}$$

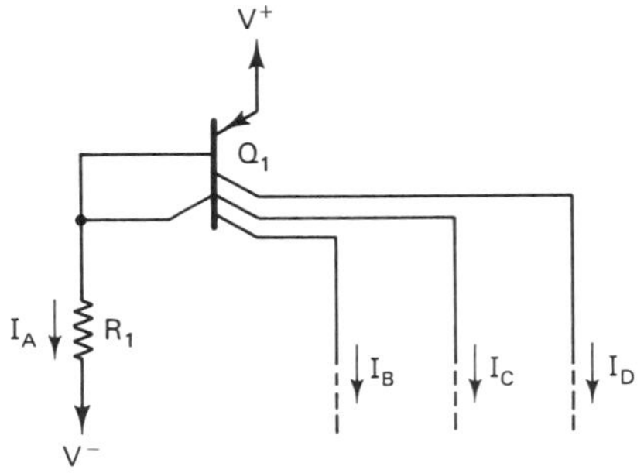


**Figure P3.11**

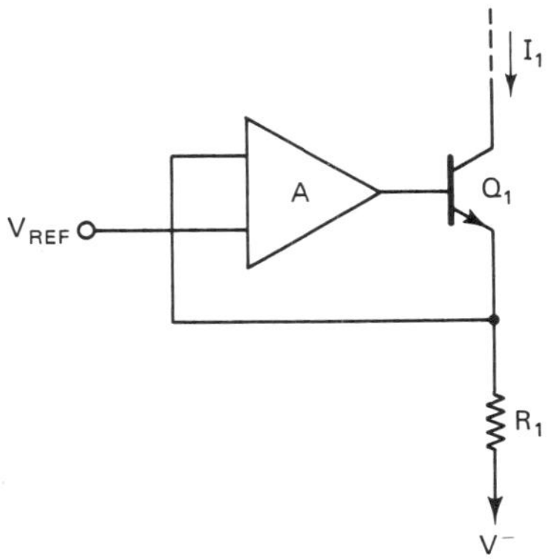


**Figure P3.12**

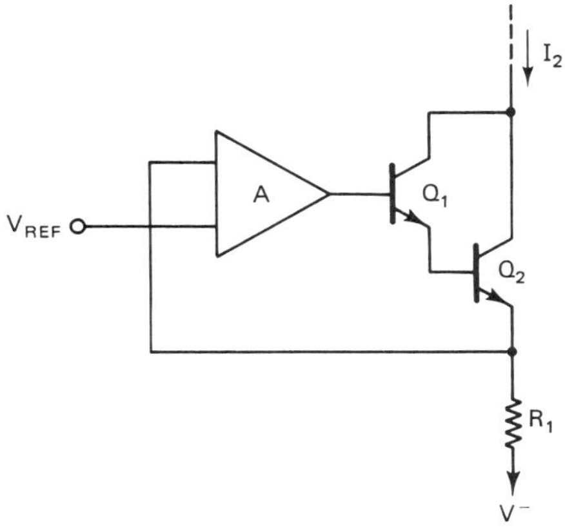


**Figure P3.13**

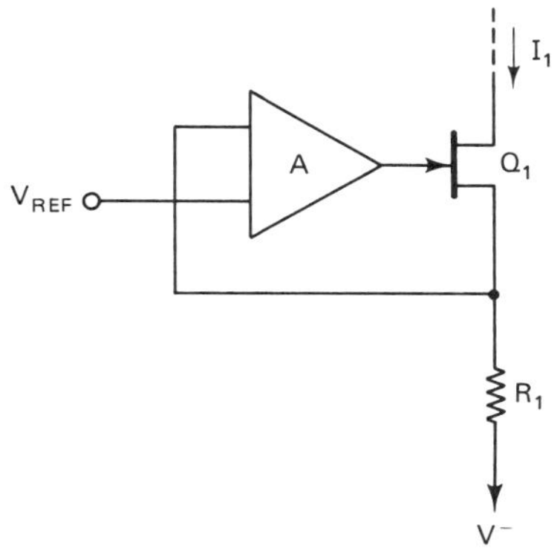


**Figure P3.14**

**(b)** If $\beta_{\text{pnp}} = 25$ (min.), find the maximum difference between $I_A$ and $I_B$. (*Ans.*: 13.8%)

**(c)** Find $R_1$ for $I_A = 1.0$ mA. (*Ans.*: 19.3 kΩ)

**(d)** Find the voltage compliance range. (*Ans.*: −41 to +9.8 V)

**3.12.** (*Operational-amplifier feedback current sink*) Refer to Figure P3.12.

**(a)** Show that for large $A_{OL}$, $I_1 = (V_{\text{REF}} - V^-)/R_1$.

**(b)** Show that a more exact expression for $I_1$ is given by

$$I_1 = \frac{1}{R_1}\left(\frac{V_{\text{REF}} - V^-}{1 + 1/A_{OL}} - \frac{V_{BE} + V^-}{A_{OL} + 1}\right)$$

**(c)** If $A_{OL} = 100$ dB (min.) and $\beta = 25$ (min.), 50 (typ.), find the maximum difference or error between $I_1$ and $(V_{\text{REF}} - V^-)/R_1$. [*Ans.*: 2% (typ.), 4% (max.)]

**(d)** Find the voltage compliance range for $V_{\text{REF}} = 5.0$ V. (*Ans.*: −4.8 to +45 V)

**3.13** (*Darlington feedback current source*) Refer to Figure P3.13.

**(a)** If $A_{OL} = 100$ kΩ (min.), find the maximum difference or error between $I_2$ and $(V_{\text{REF}} - V^-)/R_1$ for $V_{\text{REF}} = 5.0$ V. [*Ans.*: Difference due to finite $A_{OL}$ is $0.04 \times 10^{-3}\%$ (max.) and difference due to $\beta$ is 0.04% (max.), 0.01% (typ.), total difference will be 0.04% (max.)]

**(b)** Find the voltage compliance range for $V_{\text{REF}} = -1.0$ V. (*Ans.*: −4.1 to +45 V)

**3.14.** (*JFET feedback current source*) Refer to Figure P3.14.

**(a)** Show that

$$I_1 = \frac{1}{R_1}\left(\frac{V_{\text{REF}} - V^-}{1 + 1/A_{OL}} - \frac{V^- + V_{GS}}{A_{OL} + 1}\right)$$

**(b)** For $V_{\text{REF}} = +5.0$ V and $V_{GS} = -2.0$ V, find the maximum difference or error between $I_1$ and $(V_{\text{REF}} - V^-)/R_1$. Use $A_{OL} = 100$ dB (min.). [*Ans.*: $2 \times 10^{-4}\%$ (max.)]

**(c)** If $I_{DSS} = 10$ mA and $V_P = -6.0$ V, find $V_{GS}$ when $I_1 = 2.0$ mA and find the maximum value that $I_1$ can have. (*Ans.*: −3.32 V, 10 mA)

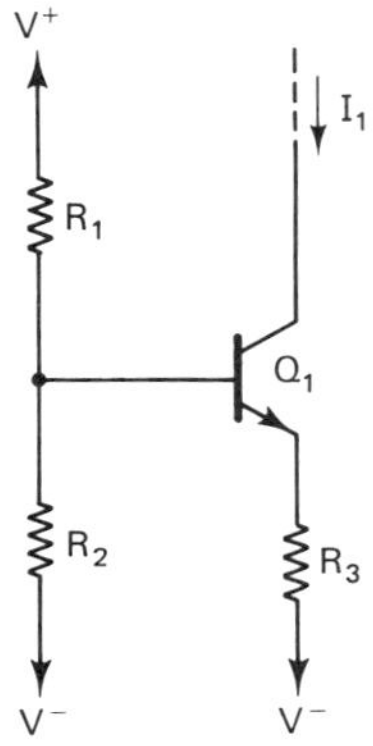


**Figure P3.15**

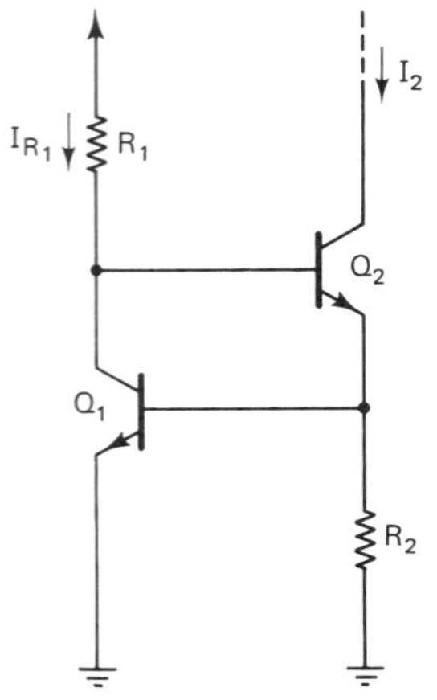


**Figure P3.16**

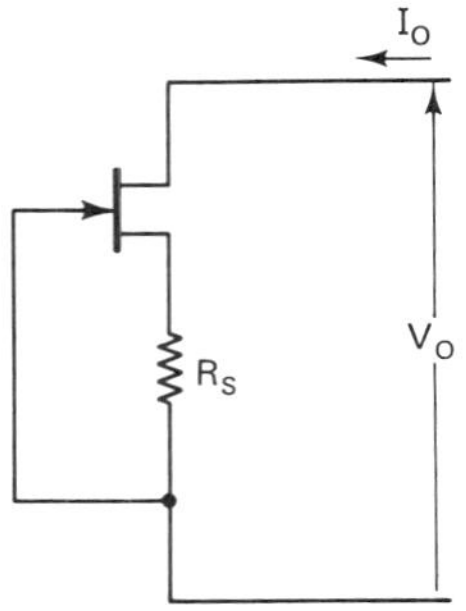


**Figure P3.17**

**3.15.** (*Resistor-biased temperature-compensated current sink*) Refer to Figure P3.15.

**(a)** Show that

$$I_1 = \frac{1}{R_3}\left(\frac{V_{\text{supply}} R_2}{R_1 + R_2} - V_{BE}\right)$$

where $V_{\text{supply}} = V^+ - V^-$.

**(b)** Assuming that all resistors have identical *TCR* values and that they track in temperature, show that the temperature coefficient for $I_1$ will be given by

$$TC_{I_1} = \frac{1}{I_1}\frac{dI_1}{dT} = -TCR - \frac{1}{V_{R_3}}\frac{dV_{BE}}{dT}$$

**(c)** Find $V_{R_3}$ for a zero temperature coefficient for $I_1$. (*Ans.*: 1.1 V)

**(d)** For $I_1 = 1.0$ mA and $I_{R_1} = I_{R_2} = 50 I_{B(\text{MAX})}$, find $R_1$, $R_2$, and $R_3$. (*Ans.*: 18.2 kΩ, 1.8 kΩ, 1.1 kΩ)

**3.16.** (*Current source that is relatively independent of supply voltage*) Refer to Figure P3.16.

**(a)** Show that $I_2$ will be given by

$$I_2 = \frac{(V_{BE_1}/R_2) + (I_{R_1}/\beta_1)}{1 + (1/\beta_2) + (1/\beta_1\beta_2)} \simeq \frac{V_{BE_1}}{R_2}$$

**(b)** If $V_{BE} = 660$ mV at 1.0 mA for $Q_1$ and $Q_2$, find $R_1$ and $R_2$ for $I_1 = 250\ \mu$A, $I_2 = 100\ \mu$A, and $V^+ = 10$ V. (*Ans.*: 35.1 kΩ, 6.25 kΩ)

**(c)** Given that $\beta = 50$ (min.) for $Q_1$ and $Q_2$, as $\beta$ varies from its minimum value to infinity, what will be the resulting variation in $I_2$? (*Ans.*: $\Delta I_2 = -2.9\ \mu$A or $-2.9\%$)

**(d)** Find the change in $I_2$ per 1-V change in the supply voltage $V^+$. Express the result in both absolute and percentage terms. (*Ans.*: 0.432 μA/V or 0.432%/V)

**(e)** If $TCR = +2000$ ppm/°C and $TCV_{BE} = -2.2$ mV/°C, find $TC_{I_2}$. (*Ans.*: −5720 ppm/°C or −0.572%/°C)

**3.17.** (*JFET current regulator diode*) Refer to Figure P3.17. Given: JFET with $I_{DDS} = 1.0$ mA and $V_P = -4.0$ V. The dynamic drain-to-source resistance $r_{ds}$ is 25 kΩ at $I_{DS} = 1.0$ mA. Find: $R_S$, $r_{ds}$, the dynamic output resistance $r_o$, and the percent change in $I_O$ per 1-V change in $V_O$ for the following values of $I_O$:

**(a)** 1.0 mA. (*Ans.*: $R_S = 0$, $r_{ds} = 25$ kΩ, $r_o = 25$ kΩ, 4.0%/V)

**(b)** 0.50 mA. (*Ans.*: 2.34 kΩ, 50 kΩ, 91.4 kΩ, 2.19%/V)

**(c)** 0.25 mA. (*Ans.*: 8.0 kΩ, 100 kΩ, 300 kΩ, 1.33%/V)

**(d)** 0.10 mA. (*Ans.*: 27.4 kΩ, 250 kΩ, 1.33 MΩ, 0.751%/V)

**(e)** 0.03 mA. (*Ans.*: 110 kΩ, 833 kΩ, 8.79 mΩ, 0.379%/V)

***3.18.** Given: Figure P3.3 with $I_1 = 1.0$ mA. Use a computer program to determine $I_2$ for the following values of $R_2$: 300 Ω, 1000 Ω, 3000 Ω, 10 kΩ, and 30 kΩ.

## Voltage Source Problems

**3.19.** Refer to Figure P3.19.

**(a)** Find $R_1$ and $R_2$ for a load voltage $V_L = +5.0$ V and $I_{R_1} \simeq I_{R_2} = 1.0$ mA. (*Ans.*: $R_1 = 4.3$ kΩ, $R_2 = 5.0$ kΩ)

**(b)** Find the change in the load voltage, $\Delta V_L$, for the output current going from a no-load value of 1.0 mA to a full-load value of 10 mA. [*Ans.*: *load regulation* = $\Delta V_L = -0.265$ V (typ.), $-0.473$ V (max.)]

**(c)** Find the *dynamic output resistance* for a load current of 10 mA. [*Ans.*: $r_o$ = 25.6 Ω (typ.), 48.7 Ω (max.)]

**(d)** Find the *dynamic output resistance* at a load current of 1.0 mA. [*Ans.*: $r_o$ = 48.1 Ω (typ.), 71.2 Ω (max.)]

**(e)** Find the no-load to full-load *load regulation* if $Q_1$ is replaced by a Darlington configuration. [*Ans.*: $\Delta V_L = -0.117$ V (typ.), $-0.123$ V (max.)]

**(f)** If the Darlington configuration of part (e) is as shown in Figure P3.19b with $Q_{1A}$ biased by $R_3$ at a quiescent level of 1.0 mA, find $R_3$ and the *load regulation*. [*Ans.*: $R_3 = 700$ Ω, load regulation = $\Delta V_L$ (with $I_{NL} = 2$ mA and $I_{FL} = 11$ mA): $-0.062$ V (typ.), $-0.070$ V (max.)]

**(g)** Find the *dynamic output resistance* $r_o$ for the circuit of part (f) at a load current of 10 mA. [*Ans.*: $r_o = 3.26$ Ω (typ.), 4.20 Ω (max.)].

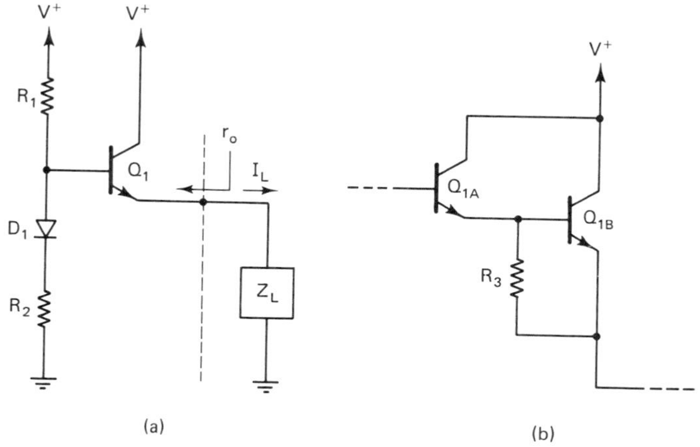


**Figure P3.19**

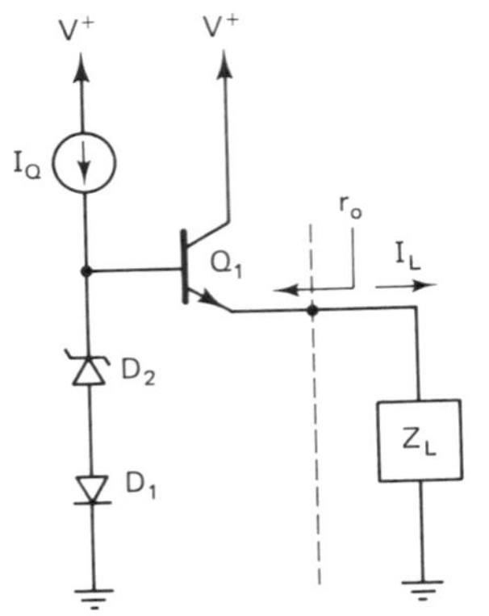


Figure P3.20

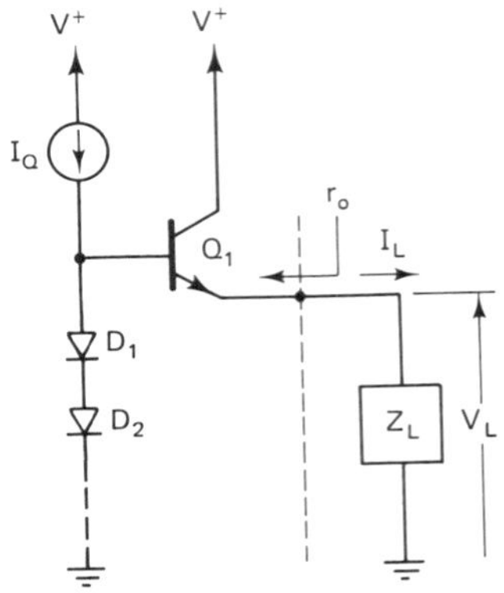


Figure P3.21

**3.20.** Refer to Figure P3.20.

**(a)** If $V_Z$ (breakdown voltage) of $D_2$ is 6.2 V, find $V_L$. (*Ans.*: 6.2 V)

**(b)** If $I_Q = 1.0$ mA and $Z_z = 10\ \Omega$, find the *load regulation* going from $I_{L\,(\mathrm{NL})} = 1.0$ mA to $I_{L\,(\mathrm{FL})} = 10$ mA. [*Ans.*: $\Delta V_L = -0.061$ V (typ.), $-0.064$ V (max.)]

**(c)** Find the dynamic output resistance for $I_L = 10$ mA. [*Ans.*: $r_o = 2.85\ \Omega$ (typ.), $3.20\ \Omega$ (max.)]

**(d)** If $Q_1$ is replaced by the Darlington configuration of Figure P3.19b with the quiescent current of $Q_{1A}$ being set by $R_3$ at 10 mA, find $R_3$ and the *dynamic output resistance* at $I_L = 100$ mA. [*Ans.*: $r_o = 0.2785\ \Omega$ (typ.), $0.314\ \Omega$ (max.), $R_3 = 70\ \Omega$]

**(e)** Find the *load regulation* for the case of part (d) with $I_{L\,(\mathrm{NL})} = 20$ mA and $I_{L\,(\mathrm{FL})} = 110$ mA. [*Ans.*: $\Delta V_L = -0.0578$ V (typ.), $-0.0581$ V (max.)]

**3.21.** Refer to Figure P3.21.

**(a)** If there are six diodes (i.e., $D_1$ to $D_6$) driven by $I_Q$, find $V_L$. (*Ans.*: $V_L = 3.5$ V)

**(b)** If $I_Q = 2.5$ mA, find the *load regulation* for $L_{L\,(\mathrm{NL})} = 1.0$ mA and $I_{L\,(\mathrm{FL})} = 10$ mA. [*Ans.*: ($\Delta V_L = -6.30$ mV (typ.), $-68.4$ mV (max.)]

**(c)** Find the *dynamic output resistance* for $I_L = 10$ mA. [*Ans.*: $r_o = 3.1\ \Omega$ (typ.), $3.7\ \Omega$ (max.)]

**3.22.** Refer to Figure P3.22. Given: $I_Q = 12$ mA and $I_{R1} \simeq I_{R2} = 1.0$ mA.

**(a)** Find $R_1$ and $R_2$ for $V_O = 2.0$ V. (*Ans.*: $R_1 = 1300\ \Omega$, $R_2 = 700\ \Omega$)

**(b)** Find the *current compliance range*. (*Ans.*: $I_L = 0$ to almost $+11$ mA)

**(c)** Find the *load regulation* for $I_{\mathrm{NL}} = 0$ and $I_{\mathrm{FL}} = 10$ mA. (*Ans.*: $\Delta V_O = -171$ mV or $-8.6\%$)

**(d)** Show that the dynamic output conductance, $g_o = dV_O/dI_O$, will be given by $g_o = 1/(R_1 + R_2) + R_2/(R_1 + R_2)g_m$.

**(e)** Find $g_o$ and $r_o$ for $I_O = 10$ mA. (*Ans.*: $g_o = 14.5$ mS, $r_o = 69.0\ \Omega$)

**(f)** Find $g_o$ and $r_o$ for $I_O = 0$. (*Ans.*: $g_o = 154.5$ mS, $r_o = 6.5\ \Omega$)

**3.23.** Refer to Figure P3.23. Given: $I_Q = 12$ mA, $I_{R1} = I_{R2} = 1.0$ mA, $V_Z = 6.3$ V, and $V_Z = 10\ \Omega$.

**(a)** Find $R_1$ and $R_2$ for $V_O = 10$ V. (*Ans.*: $R_1 = 3$ k$\Omega$, $R_2 = 7$ k$\Omega$)

**(b)** Find the *current compliance range*. (*Ans.*: $I_O = 0$ to almost 11 mA)

**(c)** Find the *load regulation* for $I_{\mathrm{NL}} = 0$ and $I_{\mathrm{FL}} = 10$ mA. (*Ans.*: $\Delta V_O = -0.228$ V or $-2.3\%$)

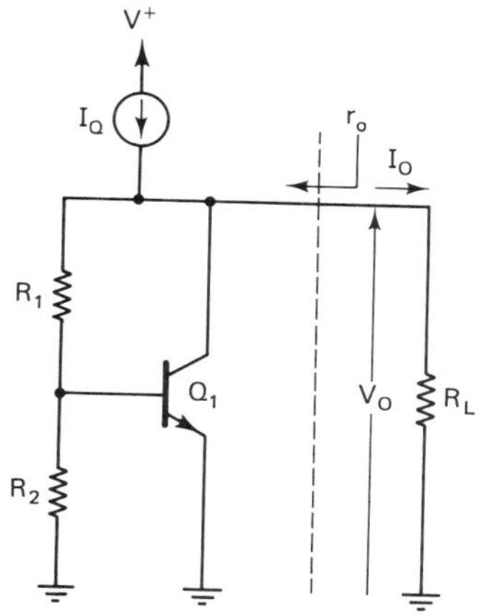


Figure P3.22

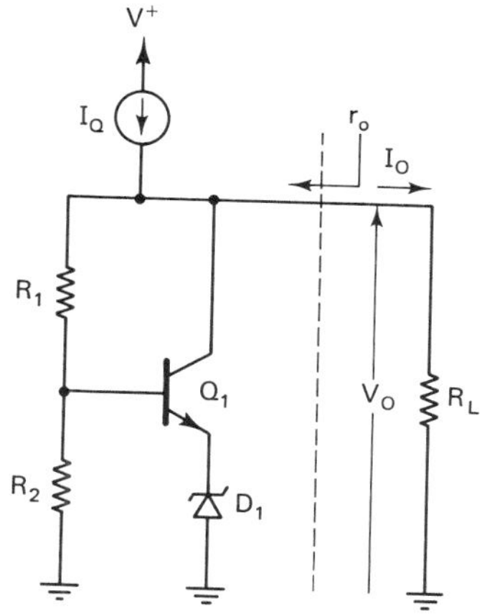


Figure P3.23

**(d)** Show that the dynamic output conductance will be given by

$$g_o = \frac{1}{R_1 + R_2} + \frac{R_2}{R_1 + R_2}\frac{g_m}{1 + g_m Z_Z}$$

**(e)** Find $g_o$ and $r_o$ at $I_O = I_{FL} = 10$ mA. (*Ans.*: $g_o = 20.1$ mS, $r_o = 49.75$ Ω)

**(f)** Find $g_o$ and $r_o$ at $I_O = I_{NL} = 0$. (*Ans.*: $g_o = 57.1$ mS, $r_o = 17.5$ Ω)

**3.24.** Refer to Figure P3.24. Given: $I_{DSS}(Q_4) = 6.0$ mA, $I_{R_1} \simeq I_{R_2} = 1.0$ mA, $I_2 = 5.0$ mA (quiescent current of $Q_2$).

**(a)** Find $R_1$ and $R_2$ for $V_O = 6.0$ V. (*Ans.*: $R_1 = 1.9$ kΩ, $R_2 = 700$ Ω)

**(b)** Find $R_3$. (*Ans.*: $R_3 = 140$ Ω)

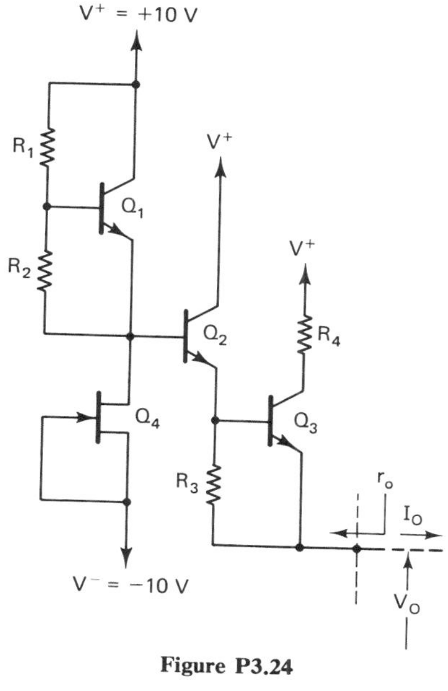


Figure P3.24

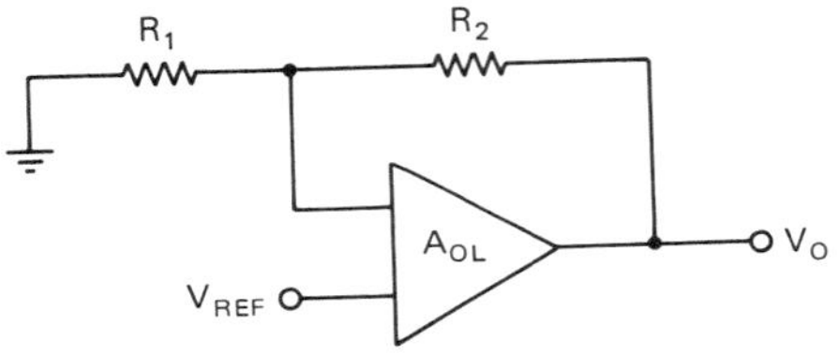


Figure P3.25

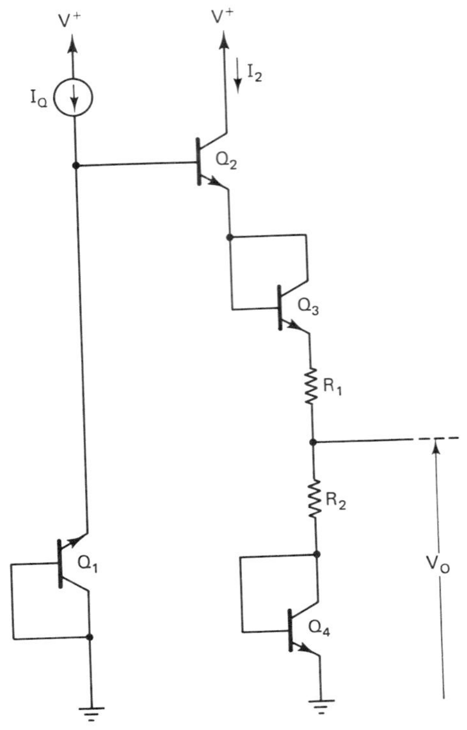


Figure P3.26

(c) Find the *load regulation* for $I_{O(NL)} = 7.0$ mA and $I_{O(FL)} = 105$ mA. (*Ans.*: $\Delta V_o = -98$ mV or $-1.63\%$)

(d) Find the *dynamic output resistance* at $I_O = 50$ mA. [*Ans.*: $r_o = 0.611\ \Omega$ (max.), $0.553\ \Omega$ (typ.)]

**3.25.** Refer to Figure P3.25. Given: $V_{REF} = +5.00$ V, $I_{R_2} = 1.0$ mA.

(a) Find $R_1$ and $R_2$ for $V_O = +10.0$ V. (*Ans.*: $R_1 = 5$ k$\Omega$, $R_2 = 5$ k$\Omega$)

(b) If $A_{OL}(O) = 100$ dB (min.) and the open-loop output resistance is 100 $\Omega$, find the dynamic output resistance, $r_o$. [*Ans.*: $r_o = 2.0$ m$\Omega$ (max.)]

(c) Find the *load regulation* for $I_{O(NL)} = 1.0$ mA and $I_{O(FL)} = 20$ mA. [*Ans.*: $\Delta V_O = -38\ \mu$V (max.) $= -3.8$ ppm $= -0.00038\%$]

(d) If $TC_{V(REF)} = 10$ ppm/°C, find $TC_{V_O}$. (*Ans.*: 100 $\mu$V/°C or 10 ppm/°C)

## Temperature-Compensated Voltage Reference Problems

**3.26.** Refer to Figure P3.26.

(a) Show that $V_O = V_Z[R_2/(R_1 + R_2)] - V_{BE}[(2R_2 - R_1)/(R_1 + R_2)]$.

(b) Show that for $TC_{V_O} = dV_O/dT = 0$ the requirement for the $R_1/R_2$ resistor ratio is $R_1/R_2 = 2 + [(dV_Z/dT)/(-dV_{BE}/dT)]$.

(c) Given that $V_Z = 6.3$ V, $dV_Z/dT = +3.0$ mV/°C, $dV_{BE}/dT = -2.2$ mV/°C, and

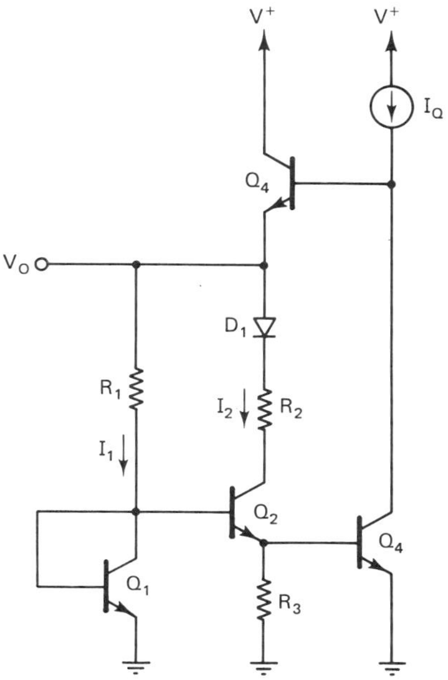


Figure P3.27

$I_2 = 1.0$ mA, find $R_1$, $R_2$, and $V_O$ for $TC_{V_O} = 0$. (*Ans.*: $R_1 = 3.2375$ kΩ, $R_2 =$ 962.5 Ω, $V_O = 1.66$ V)

**(d)** If the temperature coefficient of the zener voltage $dV_Z/dT$ or of the base-to-emitter forward voltage drop $dV_{BE}/dT$ is off from the design center value by ±10%, find the resulting value of $TC_{V_O}$. (*Ans.*: $TC_{V_O} = \pm 69$ μV/°C or 41 ppm/°C)

**(e)** If the $R_1/R_2$ resistance ratio is off from the design center value by ±1%, find the resulting value of $TC_{V_O}$. (*Hint*: Let $R_2$ be 1% greater than the value given above for the $TC_{V_O} = 0$ condition and solve for $TC_{V_O}$. Check you answer by now letting $R_1$ be 1% greater than the value given above for the $TC_{V_O} = 0$ condition and solve for $TC_{V_O}$). (*Ans.*: $TC_{V_O} = \pm 17$ μV/°C or ±10.1 ppm/°C)

**3.27.** Refer to Figure P3.27. Given: $I_1 = 1.0$ mA, $I_2 = 0.10$ mA, and $V_{BE} = 0.7$ V.

**(a)** Find $R_1$, $R_2$, $R_3$, and $V_O$ for the zero temperature coefficient of $V_O$. (*Ans.*: $R_1 =$ 1.866 kΩ, $R_2 = 11.66$ kΩ, $R_3 = 576$ Ω, $V_O = 2.566$ V)

**(b)** If the *ratio tolerance* of the IC resistors is ±2%, what will $TC_{V_O}$ be (assuming that otherwise $TC_{V_O}$ would be zero)? (*Ans.*: $TC_{V_O} = 80$ μV/°C or 31 ppm/°C or 0.0031%/°C)

**(c)** If the *absolute value tolerance* of the IC resistors is ±5%, find $TC_{V_O}$ (assuming that otherwise $TC_{V_O}$ would be zero). [*Hint*: Note that $TC_{V_O} = dV_O/dT = (n + 1)(dV_{BE}/dT) + (R_2/R_3)(V_T/T)\ln(I_1/I_2)$ and that $I_1/I_2 = \exp(I_2R_3/V_T)$, so that $V_T \ln(I_1/I_2) = I_2R_3$ and $TC_{V_O} = (n + 1)(dV_{BE}/dT) + (R_2R_3)I_2R_3/T$.] (*Ans.*: $TC_{V_O} = 201$ μV/°C or 78 ppm/°C or 0.0078%/°C)

**(d)** If the deviation of $V_{BE}$ from the design center value is ±10 mV, find $TC_{V_O}$ (assuming that otherwise $TC_{V_O}$ would be zero). (*Ans.*: $TC_{V_O} = 67$ μV/°C or 26 ppm/°C or 0.0026%/°C)

## REFERENCES


BOYLESTAD, R., and L. NASHELSKY, *Electronic Devices and Circuit Theory*, Prentice-Hall, 1982.

GLASER, A. B., and G. E. SUBAK-SHARPE, *Integrated Circuit Engineering: Design Fabrication and Applications*, Addison-Wesley, 1977.

GRAY, P. R., and R. G. MEYER, *Analysis and Design of Analog Integrated Circuits*, Second Edition, Wiley, 1984.

GREBENE, A. B., *Analog Integrated Circuit Design*, Van Nostrand Reinhold, 1972.

HAMILTON, D. J., and W. G. HOWARD, *Basic Integrated Circuit Engineering*, McGraw-Hill, 1975.

MANASSE, F. K., *Semiconductor Electronics Design*, Prentice-Hall, 1977.

# 4

# DIFFERENTIAL AMPLIFIERS

The differential amplifier is a very important transistor amplifier stage configuration and is widely used in various types of analog ICs, such as operational amplifiers, voltage comparators, voltage regulators, video amplifiers, and balanced modulators and demodulators. Differential amplifiers are also the basis of the emitter-coupled digital logic (ECL) gates. The differential amplifier will be the first or input stage of operational amplifiers and other ICs and will therefore determine many of the important performance characteristics of the IC, such as the offset voltage $V_{OS}$, the input bias current $I_B$, the input offset current $I_{OS}$, the input impedance, and the common-mode rejection ratio (CMRR).

In this chapter the basic operation of differential amplifiers is investigated. Differential amplifiers using bipolar transistors are studied first, followed by an examination of differential amplifiers using JFETs and MOSFETs. This will be followed by a discussion of active load circuits for differential amplifiers and the presentation of representative "current mirror" active load circuits. For some applications the use of the Darlington compound transistor configuration for differential amplifiers is beneficial, and this will be the subject of a brief discussion in the latter part of this chapter.

## 4.1 ANALYSIS OF THE BIPOLAR TRANSISTOR DIFFERENTIAL AMPLIFIER

For the analysis of the differential amplifier we will consider the basic differential-amplifier circuit shown in Figure 4.1. In Figure 4.2 the differential amplifier with load resistances is shown. For this analysis it will be assumed that both transistors

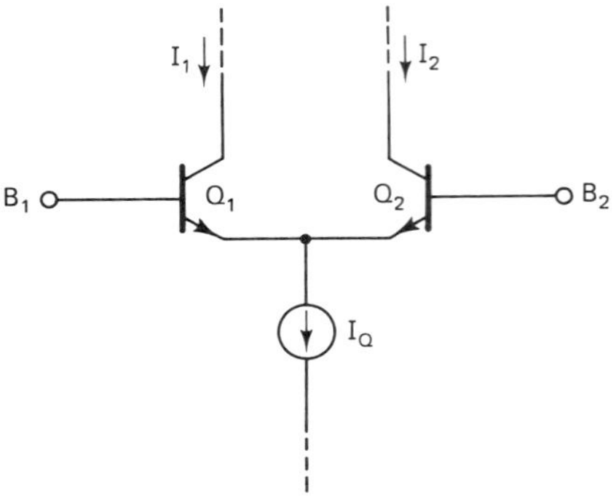


**Figure 4.1** Basic differential amplifier.

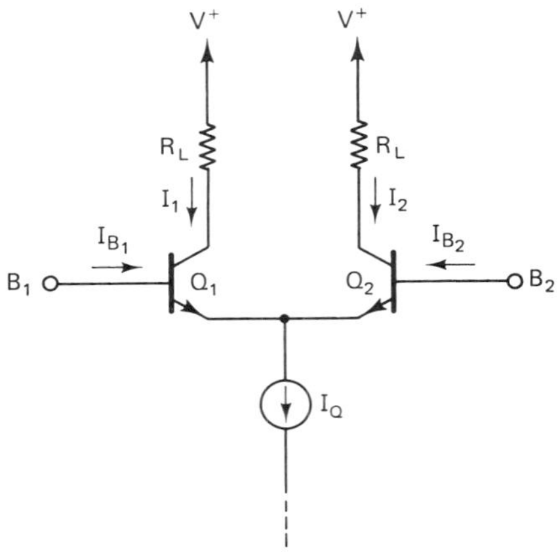


**Figure 4.2** Differential amplifier with load resistances.

of the differential-amplifier pair, $Q_1$ and $Q_2$, are operated in the active mode and that the base currents will be small compared to the collector currents. The fundamental relationship to be used for this analysis will be the exponential relationship between the collector current and the base-to-emitter voltage of a transistor. For the collector current of $Q_1$ we can write

$$I_1 = I_{C_1} = I_{TO_1} \exp\left(\frac{V_{BE_1}}{V_T}\right) \tag{4.1}$$

where the preexponential constant in this relationship for $Q_1$ is $I_{TO_1}$, and $V_{BE_1}$ is the base-to-emitter voltage of $Q_1$ so that $V_{BE_1} = V_{B_1} - V_{E_1}$. In a similar fashion we can write for $Q_2$ the equation

$$I_2 = I_{C_2} = I_{TO_2} \exp\left(\frac{V_{BE_2}}{V_T}\right)$$

If transistors $Q_1$ and $Q_2$ were exactly identical and operated with identical collector voltages, $I_{TO_1}$ would be equal to $I_{TO_2}$. However, even two transistors right next to each other on an IC chip will not be exactly identical. For purposes of convenience let us define the *offset voltage*, $V_{OS}$, by the following relationship:

$$\frac{I_{TO_2}}{I_{TO_1}} = \exp\left(\frac{V_{OS}}{V_T}\right)$$

so that

$$V_{OS} = V_T \ln \frac{I_{TO_2}}{I_{TO_1}} \tag{4.2}$$

Note that if the two transistors are exactly identical, $V_{OS} = 0$. Using the definition of the offset voltage we can rewrite the equation for $I_2$ as

$$I_2 = I_{TO_2} \exp\left(\frac{V_{BE_2}}{V_T}\right) = I_{TO_1} \exp\left(\frac{V_{BE_2} + V_{OS}}{V_T}\right)$$

Since $I_1 + I_2 = I_Q$, we have

$$I_Q = I_{TO_1} \exp\left(\frac{V_{BE_1}}{V_T}\right) + \exp\left(\frac{V_{BE_2} + V_{OS}}{V_T}\right) \tag{4.3}$$

so that

$$I_{TO_1} = \frac{I_Q}{\exp(V_{BE_1}/V_T) + \exp[(V_{BE_2} + V_{OS})/V_T]} \tag{4.4}$$

If we now substitute this expression for $I_{TO_1}$ into the expression for $I_1$, we obtain

$$I_1 = \frac{I_Q \exp(V_{BE_1}/V_T)}{\exp(V_{BE_1}/V_T) + \exp[(V_{BE_2} + V_{OS})/V_T]} \tag{4.5}$$

Dividing numerator and denominator by $\exp(V_{BE_1}/V_T)$ yields

$$I_1 = \frac{I_Q}{1 + \exp[(V_{BE_2} - V_{BE_1} + V_{OS})/V_T]} \tag{4.6}$$

If we now solve for $I_2$, we obtain

$$I_2 = I_{TO_2} \exp\left(\frac{V_{BE_2}}{V_T}\right) = I_{TO_1} \exp\left(\frac{V_{OS}}{V_T}\right) \exp\left(\frac{V_{BE_2}}{V_T}\right)$$

$$= \frac{I_Q \exp(V_{BE_2} + V_{OS}/V_T)}{\exp(V_{BE_1}/V_T) + \exp[(V_{BE_2} + V_{OS})/V_T]} \tag{4.7}$$

so that

$$I_2 = \frac{I_Q}{1 + \exp[(V_{BE_1} - V_{BE_2} - V_{OS})/V_T]} \tag{4.8}$$

Since $V_{BE_1} = V_{B_1} - V_{E_1}$ and $V_{BE_2} = V_{B_2} - V_{E_2}$, and noting that $V_{E_1} = V_{E_2}$, we have that $V_{BE_1} - V_{BE_2} = V_{B_1} - V_{B_2}$.

Let us define the *differential input voltage* or *difference-mode input voltage* as $V_i = V_{B_1} - V_{B_2}$. In terms of the differential input voltage, $V_i$, we can express $I_1$ and $I_2$ as

$$\boxed{I_1 = \frac{I_O}{1 + \exp[(-V_i + V_{OS})/V_T]} = \frac{I_O}{1 + \exp[-(V_i - V_{OS})/V_T]}} \tag{4.9}$$

and

$$\boxed{I_2 = \frac{I_O}{1 + \exp[(V_i - V_{OS})/V_T]}} \tag{4.10}$$

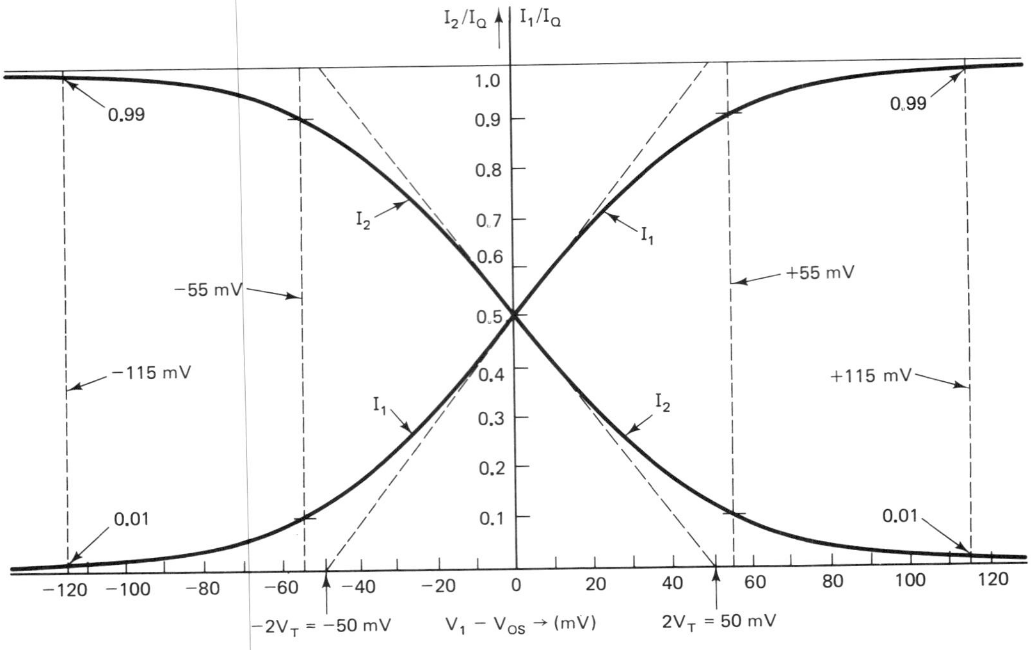


**Figure 4.3** Differential-amplifier transfer characteristics.

In Figure 4.3 graphs of $I_1$ and $I_2$ versus $V_i - V_{OS}$ are presented. Note that when $V_i = V_{OS}$, $I_1$ and $I_2$ will be equal to each other and equal to $I_Q/2$. The value of $V_{OS}$ for IC transistors is normally of the order of 1 or 2 mV. Thus when $V_i = V_{OS}$, the differential amplifier is balanced in that the current of the current source (or "sink") $I_Q$ splits equally between the two transistors of the differential-amplifier pair.

Let us now determine the value of the differential input voltage that will cause 90% of the total current to flow through $Q_1$ and the remaining 10% through $Q_2$. For this condition we have that $I_1 = 0.9I_Q$ and $I_2 = 0.1I_Q$ giving

$$0.1I_Q = \frac{I_Q}{1 + \exp\left[(V_i - V_{OS})/V_T\right]}$$

so that

$$\exp\left(\frac{V_i - V_{OS}}{V_T}\right) = 9 \tag{4.11}$$

and therefore,

$$V_i - V_{OS} = V_T \ln 9 = 25 \text{ mV} \times 2.1972 \simeq \underline{55 \text{ mV}} \tag{4.12}$$

For the opposite condition, that is, $I_1 = 0.1I_Q$ and $I_2 = 0.9I_Q$, we would have $V_i - V_{OS} = \underline{55 \text{ mV}}$. The total voltage change $\Delta V_1$ required to shift the differential-amplifier current distribution from $I_1 = 0.9I_Q$ and $I_2 = 0.1I_Q$ to the opposite case of $I_1 = 0.1I_Q$ and $I_2 = 0.9I_Q$ is called the *transition voltage* and will therefore have a value of about $2 \times 55$ mV = 110 mV. We see that it does not require a very large voltage to produce a major shift in the current distribution of a differential amplifier. Referring to the equations for $I_1$ and $I_2$ and looking at the graphical presentation of the transfer characteristics of the differential amplifier, we see that as $V_1$ increases in either direction, more and more current flows through one transistor and less and less through the other. At no point, however, does all of the current flow through one transistor, with the other one being cut off.

To consider another example for $I_1 = 0.99I_Q$ and $I_2 = 0.01I_Q$, we have

$$0.01 = \frac{1}{1 + \exp[(V_i - V_{OS})/V_T]} \tag{4.13}$$

so that $V_i - V_{OS} = 115$ mV. For $I_1 = 0.01I_Q$ and $I_2 = 0.99I_Q$ we would similarly have $V_i - V_{OS} = -115$ mV. For an even more extreme case, if $I_1 = 0.999I_Q$ and $I_2 = 0.001I_Q$, the required differential input voltage will be $V_i - V_{OS} = 173$ mV.

If the biasing source for the differential amplifier that supplies the bias current $I_Q$ is an ideal constant-current source, then $I_Q$ will be independent of the voltage across the current source and therefore will be independent of the input voltages $V_{B1}$ and $V_{B2}$. Looking at the equations for $I_1$ and $I_2$, we see that if indeed $I_Q$ is a constant, then $I_1$ and $I_2$ will be a function only of the differential input voltage, $V_i = V_{B1} - V_{B2}$, and will not respond to any common-mode component of the input voltage. Thus the amplifier is a true *differential*, or *difference amplifier*, responding only to the difference in the voltages applied to the two input terminals, $B_1$ and $B_2$, and not responding at all to any voltage that is common to the two inputs. The *differential*, or *difference-mode input voltage* has already been defined as $V_i = V_{B1} - V_{B2}$. Let us define the *common-mode input voltage* as the average of the two input voltages so that the *common-mode input voltage* = $V_{CM} = (V_{B1} + V_{B2})/2$. If $V_{B1} = -V_{B2}$, the common-mode component of the input voltage will be zero and we have a pure difference-mode input voltage. If, on the other hand, $V_{B1} = V_{B2}$, the difference-mode component will be zero and we now have a pure common-mode input voltage.

### 4.1.1 Transfer Conductances

Looking at the equations for $I_1$ and $I_2$ and the graph of $I_1$ and $I_2$ versus $V_i$, we see that the differential amplifier is truly a nonlinear device in terms of the relationship between the output currents, $I_1$ and $I_2$, and the input voltage, $V_i$. However, over a limited region of the transfer characteristic curves for $I_1$ versus $V_i$ or $I_2$ versus $V_i$ we see that the relationship between the currents and the input voltage can be considered to be approximately linear. Looking at the transfer characteristics graph, we see that we have a characteristic curve that is approximately linear over a range of input voltages of from about $V_i - V_{os} = -30$ mV to $V_i - V_{os} = +30$ mV, or a total span of about 60 mV. Thus for small-signal a-c conditions the device can be considered to act as an approximately linear device as far as the relationships between the a-c input voltage and the a-c output currents are concerned.

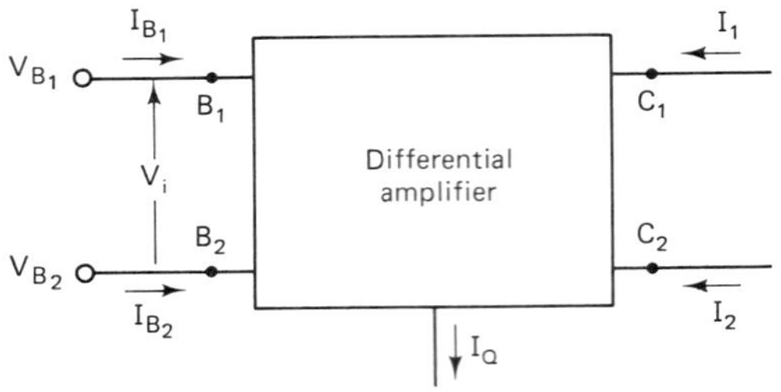


**Figure 4.4** Differential-amplifier currents and voltages.

In this section, expressions for the differential-amplifier transfer conductances will be obtained. A *transfer conductance* is the rate of change of a current at one port of a multiport device with respect to the voltage applied at another port. If we refer to Figure 4.4, which shows the differential amplifier in block diagram form, we can define several transfer conductances as

$$g_{f\,11} = \frac{dI_1}{dV_{B\,1}} \qquad g_{f\,21} = \frac{dI_2}{dV_{B\,1}}$$
$$g_{f\,12} = \frac{dI_1}{dV_{B\,2}} \qquad g_{f\,22} = \frac{dI_2}{dV_{B\,2}} \tag{4.14}$$

We shall see in the forthcoming analysis that all of these transfer conductances for the differential amplifier will be equal in magnitude, so that $|g_{f\,11}| = |g_{f\,22}| = |g_{f\,12}| = |g_{f\,21}| = g_f$ and will therefore differ only in terms of the algebraic sign.

The expression for $I_1$ has been given as

$$I_1 = \frac{I_Q}{1 + \exp\,[-(V_i - V_{os})/V_T]} \tag{4.15}$$

where $V_i = V_{B\,1} - V_{B\,2}$, so that

$$g_{f\,11} = \frac{dI_1}{dV_{B\,1}} = \frac{dI_1}{dV_i}\frac{dV_i}{dV_{B\,1}} = \frac{dI_1}{dV_i} \tag{4.16}$$

Performing the indicated operation yields

$$g_{f\,11} = \frac{dI_1}{dV_i} = \frac{I_1(1/V_T)\exp\,[-(V_i - V_{os})/V_T]}{1 + \exp\,[-(V_i - V_{os})/V_T]} = \frac{I_1/V_T}{1 + \exp\,[(V_i - V_{os})/V_T]} \tag{4.17}$$

Noting that

$$I_2 = \frac{I_Q}{1 + \exp\,[(V_i - V_{os})/V_T]}$$

We can express $g_{f\,11}$ as

$$g_{f\,11} = \frac{I_1 I_2}{I_Q V_T} \tag{4.18}$$

Since $g_{f_{22}}$ can be obtained by simply interchanging subscripts, we have

$$g_{f_{22}} = \frac{I_2 I_1}{I_Q V_T} = g_{f_{11}}$$
(4.19)

Since $I_1$ and $I_2$ will have the same algebraic sign as $I_Q$, $g_{f_{11}} = g_{f_{22}}$ will be algebraically positive, so that $g_{f_{11}} = g_{f_{22}} = g_f$. For $g_{f_{12}}$ we have

$$g_{f_{12}} = \frac{dI_1}{dV_{B_2}} = \frac{dI_1}{dV_i} \frac{dV_i}{dV_{B_2}}$$
(4.20)

Since $V_i = V_{B_1} - V_{B_2}$,

$$\frac{dV_i}{dV_{B_2}} = -1$$

so that

$$g_{f_{12}} = \frac{-dI_1}{dV_i} = \frac{-dI_1}{dV_{B_1}} = -g_{f_{11}} = \frac{-I_1 I_2}{I_Q V_T}$$
(4.21)

Again upon interchanging of subscripts we obtain $g_{f_{21}}$ as

$$g_{f_{21}} = \frac{-I_2 I_1}{I_Q V_T} = g_{f_{12}} = -g_{f_{11}} = -g_{f_{22}}$$
(4.22)

We can therefore summarize the expressions for the transfer conductances of the differential amplifier very simply as

$$\boxed{g_f = g_{f_{11}} = g_{f_{22}} = -g_{f_{12}} = -g_{f_{21}} = \frac{I_1 I_2}{I_Q V_T}}$$
(4.23)

Since $I_1 + I_2 = I_Q$ we can also write $g_f$ as

$$g_f = \frac{I_1(I_Q - I_2)}{I_Q V_T} = \frac{I_2(I_Q - I_1)}{I_Q V_T}$$
(4.24)

The value of $g_f$ will be a maximum for $I_1 = I_2 = I_Q/2$ and be equal to

$$\boxed{g_{f(\text{max})} = \frac{I_Q}{4 V_T} = \frac{I_Q}{100 \text{ mV}}}$$
(4.25)

In Figure 4.5 a graph of the transfer conductance as a function of the quiescent current through either $Q_1$ or $Q_2$ is presented. Note that there is a broad maximum where $I_1 = I_2 = I_Q/2$, but when either $I_1$ or $I_2$ becomes small compared to $I_Q$, the transfer conductance can become very small.

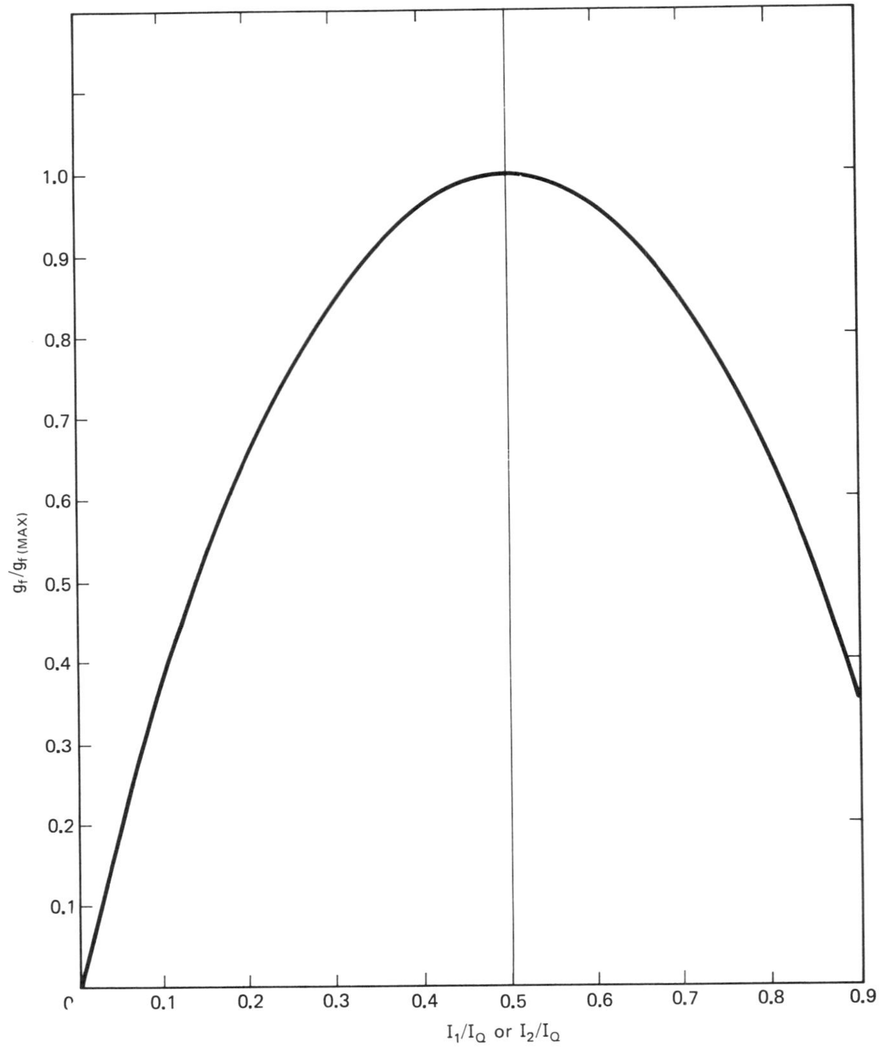


**Figure 4.5** Variation of transfer conductance with quiescent current levels.

The transfer conductance can also be expressed in terms of the quiescent voltage between the two input terminals, $V_i = V_{B1} - V_{B2}$. Since

$$I_1 = \frac{I_Q}{1 + \exp[-(V_i - V_{os})/V_T]} \quad \text{and} \quad I_2 = \frac{I_Q}{1 + \exp[(V_i - V_{os})/V_T]} \tag{4.26}$$

we have that

$$g_f = \frac{I_1 I_2}{I_Q V_T} = \frac{I_Q}{V_T} \frac{1}{[1 + \exp(-x)][1 + \exp(+x)]} \quad \text{where } x = \frac{V_i - V_{os}}{V_T} \tag{4.27}$$

Since

$$[1 + \exp(-x)][1 + \exp(x)] = 2 + \exp(x) + \exp(-x)$$

$$= 2\left[1 + \frac{\exp(x) + \exp(-x)}{2}\right] = 2(1 + \cosh x) \tag{4.28}$$

we have

$$\boxed{g_f = \frac{I_Q/V_T}{2[1 + \cosh(V_i - V_{os})/V_T]}} \tag{4.29}$$

In Figure 4.6 a graph is presented of the variation of the transfer conductance $g_f$ as a function of a d-c bias voltage $V_i$ applied between the differential amplifier input terminals. Note that $g_f$ is a maximum when the bias voltage cancels out the offset voltage (i.e., $V_i - V_{OS} = 0$) so as to balance the circuit and produce the condition that $I_1 = I_2 = I_Q/2$. The transfer conductance decreases rapidly for values of $V_i - V_{OS}$ greater than 10 mV in magnitude. This variation in the transfer conductance will produce a corresponding variation in the voltage gain of the differential amplifier circuit and will be useful for *automatic gain control* (AGC) or *automatic volume control* (AVC) applications.

### 4.1.2 Output Voltage and Voltage Gain

In Figure 4.4 a block diagram of a differential amplifier is shown. In Figure 4.7 a simple a-c small-signal equivalent circuit for the circuit of Figure 4.4 is presented. Load resistances of value $R_L$ are placed in series with the collectors of $Q_1$ and $Q_2$ to produce a-c output voltages from the a-c component of the collector currents of $Q_1$ and $Q_2$. The a-c output voltage at the collector of $Q_1$ will be $v_{o1} = -g_f R_L v_i$, so that the corresponding a-c voltage gain will be

$$\boxed{A_{V1} = \frac{v_{o1}}{v_i} = -g_f R_L} \tag{4.30}$$

Similarly, the a-c voltage at the collector of $Q_2$ will be $v_{o2} = +g_f R_L v_i$ and the voltage gain with respect to this output voltage is

$$\boxed{A_{V2} = \frac{v_{o2}}{v_i} = +g_f R_L} \tag{4.31}$$

Note that these two output voltages, and therefore the corresponding voltage gains, are equal in magnitude but opposite in algebraic sign. These two output voltages are called *single-ended* or *unbalanced outputs* since they are taken with one side of the output voltage at ground potential.

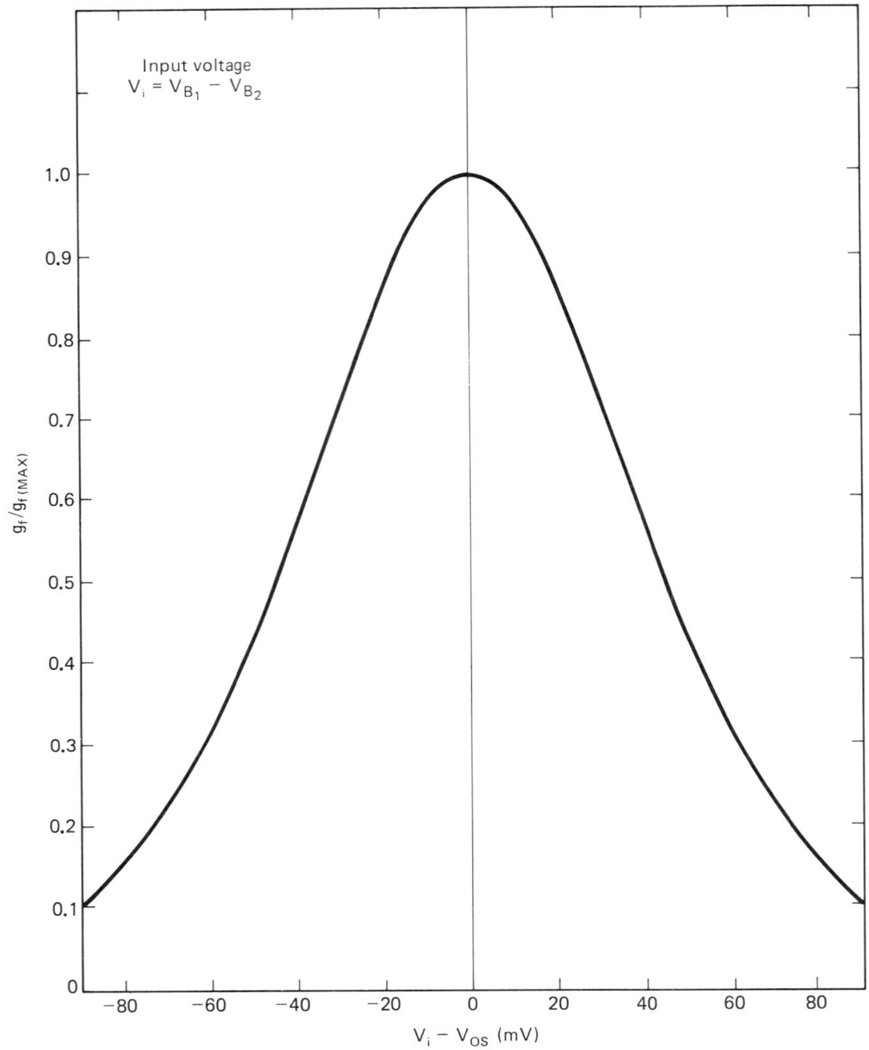


**Figure 4.6** Variation of differential-amplifier transfer conductance with quiescent voltage.

In Figure 4.8 the case of a *double-ended* or *balanced output* is presented wherein the output voltage is taken between the two collectors instead of between either collector and ground. The output voltage for this case is $v_o = v_{o1} - v_{o2} = -2g_f R_L v_i$ and the corresponding voltage gain is

$$\boxed{A_V = \frac{v_o}{v_i} = 2g_f R_L} \tag{4.32}$$

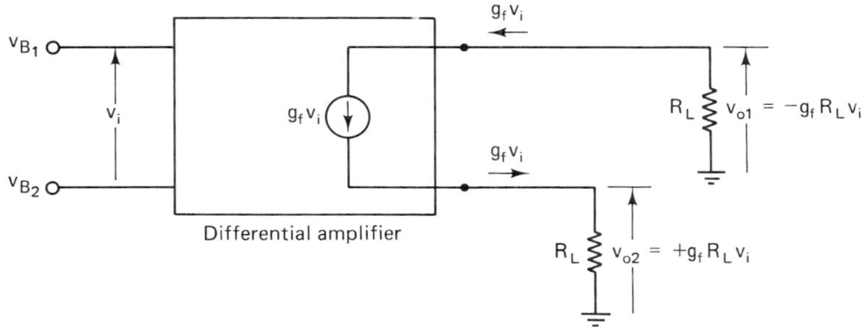


**Figure 4.7** Differential amplifier with single-ended outputs.

Note that the output voltage and the voltage gain for the double-ended case is exactly twice that obtained for the single-ended situation.

If this differential amplifier is to be used to drive a balanced load, such as another differential amplifier with input resistance $R_i$, the net load resistance driven by the differential amplifier between the two collectors will be given by $R_{L\,(\mathrm{NET})} = 2R_L \| R_i$ and the voltage gain will be

$$A_V = -g_f R_{L\,(\mathrm{NET})} = \frac{-g_f}{G_{L\,(\mathrm{NET})}} = \frac{-g_f}{G_L/2 + G_i} \tag{4.33}$$

In the more general case in which complex impedances are involved, we will have

$$A_V = -g_f Z_{L\,(\mathrm{NET})} = \frac{-g_f}{Y_{L\,(\mathrm{NET})}} = \frac{-g_f}{Y_L/2 + Y_i} \tag{4.34}$$

### 4.1.3 Common-Mode Transfer Conductance

From the previous discussion of the differential-amplifier transfer characteristics it appears that the ideal differential amplifier will be sensitive only to the difference-mode component of the input signal, and will be completely unresponsive to the

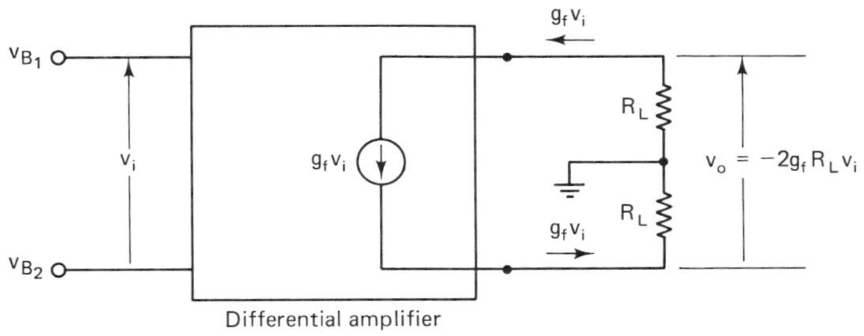


**Figure 4.8** Differential amplifier with double-ended (balanced) output.

common-mode signal. That this will be the case can be verified by investigating the differential-amplifier response to a common-mode signal in which the same voltage is applied to both bases.

If the same voltage is applied to the bases of $Q_1$ and $Q_2$ of Figure 4.1, the voltage at the emitters will follow the voltage at the bases and will therefore appear across the current source $I_Q$. If the $I_Q$ current source is indeed an ideal current source, then this voltage change across the current source will *not* result in any change of current. Since the total differential-amplifier current will remain constant at $I_Q$ and the division of current between $Q_1$ and $Q_2$ is not affected by the common-mode input voltage, we see that the collector currents $I_1$ and $I_2$ will thus be unaffected by the common-mode input signal. The *common-mode transfer conductance*, which is defined as the rate of change of collector current with respect to a common-mode input voltage, will therefore be zero.

We will now consider the case of a nonideal current source in which a change in the voltage will produce a change in the current. A real current source can be represented as the parallel combination of an ideal current source and a small dynamic conductance $g_o$. This conductance is the *dynamic output conductance* of the current source, and the change in the current of the current source $\Delta I_Q$ will be equal to the change in the voltage across the current source times the output conductance $g_o$.

The application of a common-mode input voltage $\Delta v_{CM}$ to the differential amplifier will produce a change in the voltage across the current source of approximately $\Delta v_{CM}$ since the voltage at the two emitters will closely follow the common-mode base voltage. The change in the current of the current-source will therefore be $\Delta I_Q = \Delta v_{CM} \times g_o$. If the differential amplifier is balanced, this current change will be split equally between $Q_1$ and $Q_2$, and thus the change in the collector currents will be $\Delta I_1 = \Delta I_2 = \Delta I_Q/2 = \Delta v_{CM}(g_o/2)$. The *common-mode dynamic transfer conductance* $g_{f(CM)}$ will therefore be given by

$$\boxed{g_{f(CM)} = \frac{\Delta I_1}{\Delta v_{CM}} = \frac{\Delta I_2}{\Delta v_{CM}} = \frac{g_o}{2}} \tag{4.35}$$

For the more general situation in which the differential amplifier is not balanced such that the two collector currents are not equal, we have that

$$I_1 = \frac{I_Q}{1 + \exp\left[-(V_i - V_{OS})/V_T\right]} \quad \text{and} \quad I_2 = \frac{I_Q}{1 + \exp\left(V_i - V_{OS}\right)/V_T} \tag{4.36}$$

We can therefore express the common-mode transfer conductance as

$$\boxed{g_{f1(CM)} = \frac{dI_1}{dV_{CM}} = \frac{dI_1}{dI_Q}\frac{dI_Q}{dV_{CM}} = \frac{1}{1 + \exp\left[-(V_i - V_{OS})/V_T\right]}\,g_o = \frac{I_1}{I_Q}\,g_o} \tag{4.37}$$

and similarly,

$$g_{f2(CM)} = \frac{dI_2}{dV_{CM}} = \frac{I_2}{I_Q} g_o \qquad (4.38)$$

The common-mode transfer conductance will be very much smaller than the difference-mode transfer conductance, usually by a factor of $10^4$ or $10^5$. It is this large ratio of the difference-mode transfer conductance $g_f$ to the common-mode transfer conductance $g_{f(CM)}$ that is principally responsible for the very large common-mode rejection ratio (CMRR) values of operational amplifiers.

### 4.1.4 Input Conductance

The discussion in the previous sections has dealt with the differential-amplifier dynamic transfer conductances, which involve the rate of change of the *output* currents $I_1$ and $I_2$ with respect to the input voltages $V_{B1}$ and $V_{B2}$. We will now consider the dynamic input conductance, which is the rate of change of the *input* currents $I_{B1}$ and $I_{B2}$ to the input voltages. If we refer back to Figure 4.2, we note that the input or base currents of the differential amplifier are related to the output or collector currents by the a-c current gain of the transistor $\beta$ or $h_{fe}$ as given by $dI_1/dI_{B1} = \beta$ and $dI_2/dI_{B2} = \beta$. This relationship will be used in the derivation of the expressions for the input conductance.

The *difference-mode input conductance* $g_i$ will be defined as the rate of change of the input current with respect to the difference-mode input voltage and is given by $g_i = dI_{B1}/dV_i = -dI_{B2}/dV_i = dI_{B1}/dI_1 \times dI_1/dV_i$. Since $dI_{B1}/dI_1 = 1/\beta$ and $dI_1/dV_i$ is the transfer conductance $g_f$, the equation for $g_i$ can be written very simply as

$$g_i = \frac{g_f}{\beta} \qquad (4.39)$$

If we note that $g_f = I_1I_2/I_QV_T$, we see that $g_i$ can be written as $g_i = I_1I_2/\beta I_QV_T$. The *difference-mode input resistance* $r_i$ is the reciprocal of the input conductance, so that $r_i = 1/g_i = \beta I_QV_T/I_1I_2$. Since $I_Q = I_1 + I_2$, this can be rewritten as $r_i = \beta V_T(I_1 + I_2)/I_1I_2 = \beta V_T(1/I_1 + 1/I_2) = V_T(\beta/I_1 + \beta/I_2)$. Since $I_1/I_{B1} \simeq I_2/I_{B2} \simeq \beta$, we can now express $r_i$ very simply as

$$r_i = \frac{V_T}{I_{B1}} + \frac{V_T}{I_{B2}} \qquad (4.40)$$

For the usual case of $I_1 \simeq I_2$ so that $I_{B1} \simeq I_{B2}$, this will become

$$r_i = \frac{2V_T}{I_B} \qquad (4.41)$$

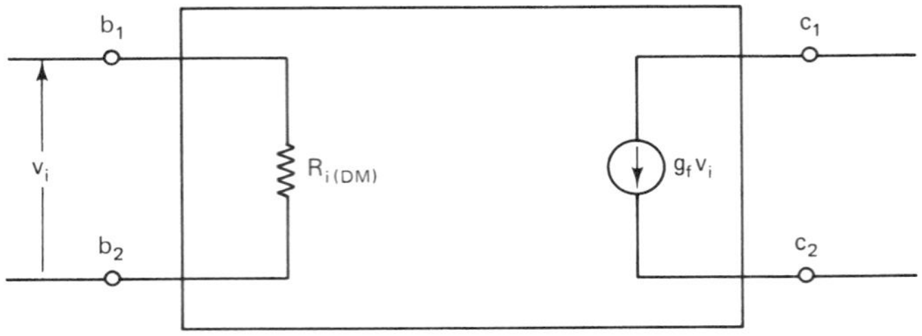


**Figure 4.9** Differential-amplifier a-c small-signal equivalent circuit, including the difference-mode input resistance.

where $I_B$ ($\simeq I_{B1} \simeq I_{B2}$) is the input bias current of the differential amplifier. We see, therefore, that there is a very simple and direct relationship between the input bias (or base) current of the differential amplifier and the dynamic input resistance. For example, if the input bias current is 50 nA, the dynamic difference-mode input resistance will be given by $r_i = 2V_T/I_B = 50\text{ mV}/50\text{ nA} = 1\text{ M}\Omega$. In Figure 4.9 an a-c small-signal equivalent circuit for the differential amplifier is shown in which $r_i$ is included.

The *common-mode input conductance* $g_{i(CM)}$ will be defined as the rate of change of the input current with respect to the common-mode input voltage. Thus we have that

$$g_{i1(CM)} = \frac{dI_{B1}}{dV_{CM}} = \frac{dI_{B1}}{dI_1}\frac{dI_1}{dV_{CM}} = \frac{g_{f1(CM)}}{\beta} \tag{4.42}$$

and similarly,

$$g_{i2(CM)} = \frac{dI_{B2}}{dV_{CM}} = \frac{dI_{B1}}{dI_2}\frac{dI_2}{dV_{CM}} = \frac{g_{f2(CM)}}{\beta} \tag{4.43}$$

For the usual case in which $I_1 \simeq I_2 \simeq I_Q/2$, we will have that $g_{f1(CM)} = g_{f2(CM)} = g_o/2\beta$ and thus

$$\boxed{g_{i1(CM)} = g_{i2(CM)} = g_{i(CM)} = \frac{g_o}{2\beta}} \tag{4.44}$$

This is the dynamic conductance seen looking into each base terminal.

The *common-mode input resistance* $r_{i(CM)}$ is the reciprocal of the input conductance so that

$$\boxed{r_{i(CM)} = \frac{1}{g_{i(CM)}} = \frac{2\beta}{g_o} = 2\beta r_o} \tag{4.45}$$

where $g_o$ is the dynamic output conductance of the $I_Q$ current source and $r_o = 1/g_o$ is the dynamic output resistance of the current source. In Figure 4.10 the equiva-

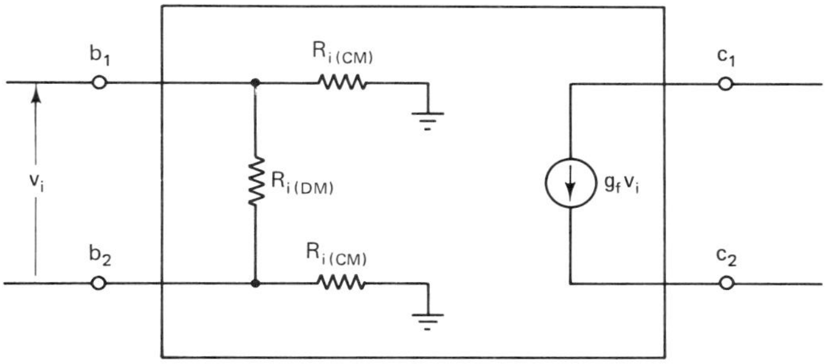


**Figure 4.10** Differential-amplifier equivalent circuit, including difference-mode and common-mode input resistances.

lent circuit for the differential amplifier is again presented, this time including both the difference-mode and common-mode input resistances. The common-mode input resistance will be a very large quantity, often in the gigaohm ($10^9\ \Omega$) range.

### 4.1.5 Balance of the Bipolar Differential Amplifier: Input Offset Voltage

Looking at the equation for the output currents, $I_1$ and $I_2$, of the differential amplifier, we note that when $V_i = 0$ the two currents will not be equal. To make the two currents equal, a small voltage equal to the offset voltage $V_{OS}$ must be applied between the two input terminals such that $V_i = V_{OS}$.

The offset voltage has been defined by the equation

$$\exp\left(\frac{V_{OS}}{V_T}\right) = \frac{I_{TO\,2}}{I_{TO\,1}} \qquad \text{so that} \qquad V_{OS} = V_T \ln \frac{I_{TO\,2}}{I_{TO\,1}} \tag{4.46}$$

The principal factor in $I_{TO}$ that will be responsible for the variation of this quantity from one transistor to another is the effective base width, $W_B'$. Since $I_{TO}$ is proportional to $1/W_B'$, we can express $V_{OS}$ in terms of the base widths of transistors $Q_1$ and $Q_2$ as $V_{OS} = V_T \ln (W_{B\,1}'/W_{B\,2}')$. The base widths, although not equal, will normally be relatively close in value, generally within 10% of each other. Therefore, if we let $W_{B\,1}' = W + \Delta W$ and $W_{B\,2}' = W$, we will have that

$$V_{OS} = V_T \ln \frac{W + \Delta W}{W} = V_T \ln \left(1 + \frac{\Delta W}{W}\right) \simeq V_T \frac{\Delta W}{W} \tag{4.47}$$

if $\Delta W/W << 1$, as will usually be the case.

The current gain of a transistor $\beta$ is inversely proportional to the base width $W_B'$. The primary cause of the unit-to-unit difference in the current gain will be the difference in the base widths. Therefore, we can write $\beta_2/\beta_1 \simeq W_{B\,1}'/W_{B\,2}'$ and thus the fraction difference in the current gains $\Delta\beta/\beta$ will be equal to the fractional difference in the base widths as given by $\Delta\beta/\beta \simeq \Delta W/W$. As a result, we see that there is some relationship between the offset voltage and the $\beta$ mismatch.

To consider a representative example, let us assume that there is a 10% difference in the base widths of a pair of IC transistors. The corresponding offset voltage will be $V_{OS} \simeq V_T \, \Delta W/W = 25 \text{ mV} \times 0.1 = 2.5 \text{ mV}$. In addition to this offset voltage, the difference in the current gains of these two transistors will be expected to be about 10%.

**Temperature coefficient of the offset voltage.** The temperature coefficient of the offset voltage is defined as $TCV_{OS} = dV_{OS}/dT$. We have already seen that $V_{OS}$ can be expressed in terms of the ratio of the effective base widths as $V_{OS} = V_T \ln (W'_{B1}/W'_{B2})$. Since the ratio of the base widths will be relatively independent of temperature, and $V_T = kT/q$ so that $dV_T/dT = k/q = V_T/T$, we will have that

$$\boxed{TCV_{OS} = \frac{dV_{OS}}{dT} = \frac{V_T}{T} \ln \frac{W'_{B1}}{W'_{B2}} \simeq \frac{V_{OS}}{T}} \tag{4.48}$$

For example, if $V_{OS} = 1.5 \text{ mV} = 1500 \ \mu\text{V}$, the temperature coefficient will be given by

$$TCV_{OS} = \frac{V_{OS}}{T} = \frac{1500 \ \mu\text{V}}{300 \text{ K}} = 5 \ \mu\text{V/K} = \underline{5 \ \mu\text{V/°C}} \tag{4.49}$$

In percentage terms we can say that the percent change in the offset voltage per degree temperature change will be given by

$$\frac{1}{V_{OS}} \frac{dV_{OS}}{dT} \times 100\% = \frac{1}{V_{OS}} \frac{V_{OS}}{T} \times 100\% = \frac{100\%}{300\text{K}}$$
$$= \underline{0.33\%/°\text{C}} \tag{4.50}$$

Note that this last expression for the percent change in the offset voltage with temperature is independent of the magnitude of the offset voltage. Thus for a 10°C temperature rise the offset voltage will change by about 3.3%, and the offset voltage will change by about 10% for a 30°C temperature rise. This small but significant change in the offset voltage with temperature can be important, especially in applications requiring offset voltage compensation or nulling.

### 4.1.6 Effect of Collector Voltage on Differential-Amplifier Balance

As a result of the base-width modulation effect (also known as the Early effect), a difference in the collector voltages of the two transistors that comprise the differential-amplifier pair can produce differences in the collector currents. Alternatively, the effect of the difference in the collector voltages can be expressed in terms of an input offset voltage.

The effect of the collector-to-base voltage, $V_{CB}$, of a transistor on the effective (i.e., electrical) base width of the transistor can be expressed in terms of the *base-width modulation factor* as $-(1/W)(dW/dV_{CB}) = 1/V_A$, where $V_A$ is a *constant* having units of volts (see Appendix B). We note that $1/V_A$ is the fractional change in the base width (i.e., $dW/W$) per volt change in the collector-to-base voltage ($V_{CB}$).

As a result of a difference in the collector voltages of two transistors of amount $\Delta V_{CB} = V_{CB1} - V_{CB2}$, there will be a difference in the base widths of the two transistors that will be given approximately by $-(\Delta W/W) = \Delta V_{CB}/V_A$. When the offset voltage of a pair of transistors is considered, it can be expressed in terms of the effective base widths of the two transistors as

$$V_{OS} = V_T \ln \frac{W_2}{W_1} = V_T \ln \frac{W_1 + \Delta W}{W_1} = V_T \ln \left(1 + \frac{\Delta W}{W_1}\right) \simeq V_T \frac{\Delta W}{W} \quad \text{for } \frac{\Delta W}{W} \ll 1 \tag{4.51}$$

The contribution of the difference in the collector-to-base voltages to the offset voltage will therefore be given by

$$\Delta V_{OS} = V_T \left(\frac{-\Delta V_{CB}}{V_A}\right) = \frac{V_T(V_{CB2} - V_{CB1})}{V_A} \tag{4.52}$$

For an example, let us consider a pair of transistors for which the base width modulation voltage, $V_A$, has a value of 250 V, which is a typical value for IC NPN transistors. The change in the offset voltage, $\Delta V_{OS}$, produced by a difference in the collector-to-base voltages of the two transistors, will be given by

$$\Delta V_{OS} = V_T \left(\frac{-\Delta V_{CB}}{V_A}\right) = \frac{25 \text{ mV}}{250 \text{ V}}(-\Delta V_{CB}) = (0.10 \text{ mV/V})(-\Delta V_{CB})$$

Therefore, the offset voltage will change by 0.10 mV for every 1-V change in $\Delta V_{CB}$. A difference of 10 V in the $V_{CB}$ values for the two transistors will therefore be responsible for a shift of 1.0 mV in the offset voltage.

## 4.2 FIELD-EFFECT TRANSISTOR DIFFERENTIAL AMPLIFIERS

Differential amplifiers using field-effect transistors operate in a fashion that is basically similar to bipolar transistor differential amplifiers. The FET differential amplifiers offer the advantages of a very high input impedance ($\sim 10^9$ to $10^{12}$ $\Omega$) and a very low input bias current ($\sim 10^{-9}$ to $10^{-12}$ A). A disadvantage of the FET differential amplifiers is the lower transfer conductance and therefore the lower voltage gain that is available. Another disadvantage is the somewhat higher offset voltage of a FET transistor pair compared to that of a pair of bipolar transistors.

We will first consider the JFET differential amplifier shown in Figure 4.11. For simplicity we will assume that both transistors have identical characteristics. For each transistor operating in the active region we will have the transfer relationship given by

$$I_{DS} = I_{DSS}\left(1 - \frac{V_{GS}}{V_P}\right)^2$$

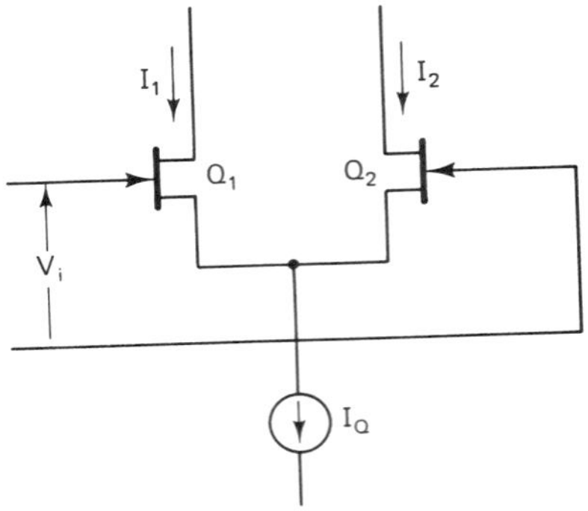


**Figure 4.11** Junction field-effect transistor differential amplifier.

so that

$$1 - \frac{V_{GS}}{V_P} = \sqrt{\frac{I_{DS}}{I_{DSS}}} \tag{4.53}$$

and thus

$$V_{GS} = V_P\left(1 - \sqrt{\frac{I_{DS}}{I_{DSS}}}\right) \tag{4.54}$$

Under quiescent conditions the two JFET currents will be equal and be given by $I_1 = I_{DS\,1} = I_Q/2$ and $I_2 = I_{DS\,2} = I_Q/2$. In response to a difference mode input voltage $V_i$, the two drain currents will change from the quiescent value by an amount $\Delta I$, given by $I_1 = I_Q/2 + \Delta I$ and $I_2 = I_Q/2 - \Delta I$. The corresponding gate-to-source voltages of $Q_1$ and $Q_2$ will be given by

$$V_{GS\,1} = V_P\left(1 - \sqrt{\frac{I_Q/2 + \Delta I}{I_{DSS}}}\right) \quad \text{and} \quad V_{GS\,2} = V_P\left(1 - \sqrt{\frac{I_Q/2 - \Delta I}{I_{DSS}}}\right) \tag{4.55}$$

The relationship between $\Delta I$ and the difference-mode input voltage $V_i$ will be given by

$$\begin{aligned} V_i = V_{G\,1} - V_{G\,2} = V_{GS\,1} - V_{GS\,2} &= V_P\left(\sqrt{\frac{I_Q/2 - \Delta I}{I_{DSS}}} - \sqrt{\frac{I_Q/2 + \Delta I}{I_{DSS}}}\right) \\ &= V_P\sqrt{\frac{I_Q}{I_{DSS}}}\left(\sqrt{\frac{1}{2} - \frac{\Delta I}{I_Q}} - \sqrt{\frac{1}{2} + \frac{\Delta I}{I_Q}}\right) \end{aligned} \tag{4.56}$$

where $I_Q$ must be limited to a value less than $I_{DSS}$. In Figure 4.13 a normalized graph of the transfer characteristic for the JFET differential amplifier is presented.

For small values of $V_i$ and $\Delta I/I_Q$, a linear approximate relationship between $V_i$ and $\Delta I$ can be obtained by using the approximation that $\sqrt{1+x} \simeq 1 + (x/2)$ for $x \ll 1$. Applying this to the relationship between $V_i$ and $\Delta I$ gives

$$\begin{aligned} V_i &\simeq V_P\sqrt{\frac{I_Q}{I_{DSS}}}\frac{1}{\sqrt{2}}\left[\left(1 - \frac{\Delta I}{I_Q}\right) - \left(1 + \frac{\Delta I}{I_Q}\right)\right] \\ &= \frac{-V_P(2\Delta I)}{\sqrt{2I_QI_{DSS}}} \end{aligned} \tag{4.57}$$

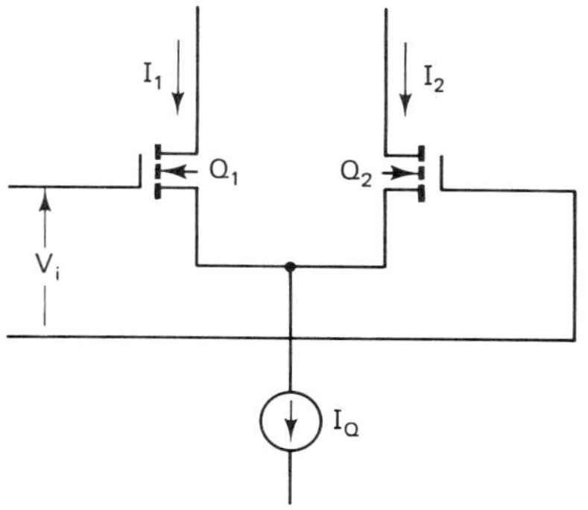


Figure 4.12 MOSFET differential amplifier.

The dynamic transfer conductance of the differential amplifier, $g_f$, will therefore be given by

$$g_f = \frac{\Delta I}{V_i} = \frac{\sqrt{2 I_Q I_{DSS}}}{-2 V_P} = \frac{\sqrt{I_{DSS} I_Q / 2}}{-V_P} \tag{4.58}$$

### 4.2.1 MOSFET Differential Amplifier

A MOSFET differential amplifier is shown in Figure 4.12. We will again assume that the two transistors will have identical characteristics. In the active region of operation the MOSFET transfer characteristic is given by $I_{DS} = K(V_{GS} - V_t)^2$,

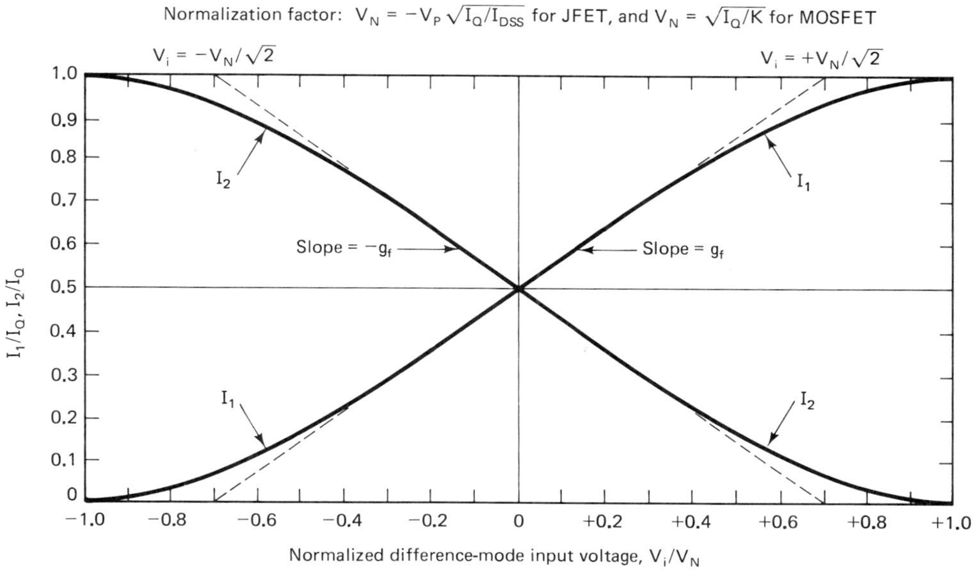


Figure 4.13 Field-effect transistor differential amplifier transfer characteristic.

where $V_t$ is the threshold voltage for channel formation (not the thermal voltage). Solving this equation for $V_{GS}$ gives $V_{GS} = V_t + \sqrt{I_{DS}/K}$.

In response to a difference-mode input voltage $V_i$, the two drain currents will become $I_1 = I_{DS_1} = I_Q/2 + \Delta I$ and $I_2 = I_{DS_2} = I_Q/2 - \Delta I$, so that

$$V_{GS_1} = V_t + \sqrt{\frac{I_Q/2 + \Delta I}{K}} \qquad \text{and} \qquad V_{GS_2} = V_t + \sqrt{\frac{I_Q/2 - \Delta I}{K}} \tag{4.59}$$

The relationship between $\Delta I$ and the difference-mode input voltage $V_i$ can thus be written as

$$\begin{aligned} V_i = V_{G_1} - V_{G_2} = V_{GS_1} - V_{GS_2} &= \sqrt{\frac{I_Q/2 + \Delta I}{K}} - \sqrt{\frac{I_Q/2 - \Delta I}{K}} \\ &= \sqrt{\frac{I_Q}{K}}\left(\sqrt{\frac{1}{2} + \frac{\Delta I}{I_Q}} - \sqrt{\frac{1}{2} - \frac{\Delta I}{I_Q}}\right) \end{aligned} \tag{4.60}$$

We note that this relationship has the same mathematical form as that obtained for the JFET differential-amplifier transfer characteristic. The normalized transfer characteristic shown in Figure 4.13 is also applicable to the MOSFET differential amplifier.

We can again obtain the differential-amplifier transfer conductance $g_f$ by using the approximation that $\sqrt{1 + x} \simeq 1 + (x/2)$ for $x \ll 1$, giving

$$\begin{aligned} V_i &\simeq \sqrt{\frac{I_Q}{2K}}\left[\left(1 + \frac{\Delta I}{I_Q}\right) - \left(1 - \frac{\Delta I}{I_Q}\right)\right] \\ &= \sqrt{\frac{I_Q}{2K}}\,\frac{2\Delta I}{I_Q} = \sqrt{\frac{2}{I_Q K}}\,\Delta I \end{aligned} \tag{4.61}$$

The transfer conductance will therefore be given by

$$\boxed{g_f = \frac{\Delta I}{V_i} = \sqrt{\frac{KI_Q}{2}} = \frac{I_Q/2}{V_{GS} - V_t}} \tag{4.62}$$

If we look at the transfer characteristic graph of the FET differential amplifier, we note that over a large part of the normalized input voltage range an approximately linear transfer relationship is obtained. This approximately linear range extends from about $V_i = -0.5\,V_N$ to $V_i = +0.5\,V_N$, where $V_N$ is the normalization factor as given by $V_N = -V_P\sqrt{I_Q/I_{DSS}}$ for a JFET differential amplifier, and $V_N = \sqrt{I_Q/K} = (V_{GS} - V_t)\sqrt{2}$ for a MOSFET differential amplifier.

The common-mode transfer conductance for FET differential amplifiers will be controlled by the output admittance of the $I_Q$ current source, as is the case for the bipolar transistor differential amplifiers. The common-mode transfer conductance will be given by the same equation as for the bipolar case, that is, $g_{f(CM)} = g_o/2$, where $g_o$ is the output conductance of the current source that is used for the biasing of the differential amplifier.

### 4.2.2 FET Offset Voltage and Temperature Coefficient

Let us consider the case of a pair of IC JFETs, $Q_1$ and $Q_2$, both biased at $V_{GS} \cong 0$. The two drain currents will be designated as $I_{DS1} = I_{DSS}$ and $I_{DS2} = I_{DSS} + \Delta I_{DSS}$, respectively, where $\Delta I_{DSS}$ represents the small difference between the two zero bias drain currents. To equalize the two drain currents, a small gate voltage differential or offset voltage, $\Delta V_{GS} = V_{OS}$, must be applied between the two gates. The offset voltage required to cancel the difference between the two drain currents $\Delta I_{DSS}$ will be given by

$$\Delta I_{DSS} = g_{fso} \Delta V_{GS} = g_{fso} V_{OS} \tag{4.63}$$

where $g_{fso} = 2I_{DSS}/(-V_P)$. Solving for the offset voltage gives

$$V_{OS} = -\frac{\Delta I_{DSS}}{2\, I_{DSS}} V_P \tag{4.64}$$

If the IC processing variation is such that the saturated drain currents of a pair of JFETs differ by one percent, and the pinch-off voltage is $-4.0$ V, the corresponding offset voltage will be $V_{OS} = 0.01 \times 4\ V/2 = 20$ mV. For the bipolar transistor case the offset voltage is given by $V_{OS} = V_T(\Delta W_B/W_B)$ so that a processing variation that results in a one percent difference in the base widths of the two transistors will produce an offset voltage of only 0.25 mV. We can therefore see that the offset voltage for JFETs will generally be considerably larger than for bipolar transistors. An analysis of the offset voltage for MOSFETs will lead to results similar to those obtained for JFETs.

The temperature coefficient of the offset voltage of a JFET pair, $TCV_{OS}$, will be principally due to the temperature coefficient of the pinch-off voltage $V_P$, which will be mostly the result of the temperature dependence of the contact potential $\phi$. For $TCV_{OS}$ we will have

$$TCV_{OS} = \frac{dV_{OS}}{dT} = -\left(\frac{\Delta I_{DSS}}{2I_{DSS}}\right)\left(\frac{dV_P}{dT}\right) = V_{OS}\frac{1}{V_P}\frac{dV_P}{dT} = \frac{V_{OS}}{V_P}\frac{d\phi}{dT} \tag{4.65}$$

The temperature coefficient of the contact potential, $d\phi/dT$, will generally be about 1 mV/°C so that

$$TCV_{OS} \simeq \left(\frac{V_{OS}}{V_P}\right) 1\ \text{mV/°C} \tag{4.66}$$

If, for example, $V_{OS} = 20$ mV and $V_P = 4.0$ V, then $TCV_{OS} \simeq (20\ \text{mV}/4\text{V}) \times 1$ mV/°C = 5 $\mu$V/°C.

For a bipolar transistor pair the temperature coefficient of the offset voltage has been given as $TCV_{OS} = V_{OS}/T$, so that if $V_{OS} = 1$ mV the corresponding temperature coefficient of the offset voltage will be approximately 3 $\mu$V/°C. From this we see that although the offset voltage for JFETs is often considerably larger than that for bipolar transistors, the temperature coefficient of the offset voltage for the two types of transistors can be quite comparable.

For a MOSFET transistor pair, an analysis similar to the foregoing one for the JFET pair can be carried out and similar results will be obtained. An additional complication in the case of MOSFETs, however, is the drift of the threshold voltage due to ionic migration in the gate oxide. This can lead to both short term as well as long term drifts in the offset voltage.

## 4.3 ACTIVE LOADS

The application of a difference-mode input voltage $v_i$ to a differential amplifier will produce a-c collector currents of $i_1 = -i_2 = g_f v_i$. In order to transform these a-c currents into an output voltage, a "load" must be used. The load can take the form of a *passive load,* which uses a pair of load resistors $R_L$ connected between the collectors of the differential amplifier and the d-c supply as shown in Figure 4.2, or an *active load*, which uses transistors for the current-to-voltage transformation.

The simplest type of load is the passive load using load resistors $R_L$. With this type of load the a-c output voltages at the two collectors will be given by $v_{o_1} = -i_1 R_L = -g_f R_L v_i$ and $v_{o_2} = -i_2 R_L = g_f R_L v_i$, and the a-c voltage gain for a single-ended output will be $A_V = g_f R_L$. Thus for a large voltage gain a large value of $R_L$ will be required. This can be seen more clearly by introducing the expression for $g_f$ as $g_f = I_Q/4V_T$ so that $A_V = g_f R_L = I_Q R_L/4V_T = I_Q R_L/0.1$ V. Since very small values of $I_Q$ are often used in differential amplifiers, such as down in the range of a few microamperes, we see that very large values of $R_L$ will indeed be required ($\sim$1 M$\Omega$) for substantial voltage gains. However, the use of these large values of load resistance carries with it some major disadvantages, especially for ICs.

1. For ICs the area taken up by large resistors will be roughly proportional to the resistance value, so that very large resistors will take up an excessive amount of room on the IC chip.
2. Large IC resistors will have large parasitic capacitances associated with them. The combination of a large resistance and a large parasitic capacitance value will result in a very large $RC$ time constant, which in turn will severely limit the frequency response of the amplifier.
3. For the differential amplifier to operate properly, the transistors must remain in the active region and not go into saturation. This limits the maximum input voltage that can be applied to the bases of $Q_1$ and $Q_2$, for the base-to-collector junction must not be allowed to become forward biased by more than 0.5 V. A large value of load resistance will produce a large d-c voltage drop $(I_Q/2)R_L$, so that the voltage at the collectors will be $V_C = V^+ - (I_Q/2)R_L$ and will thus be substantially less than the supply voltage $V^+$. This will correspondingly reduce the input voltage range of the differential amplifier.

It is for these reasons that most IC differential amplifiers use *active loads*, which involve the use of transistors rather than resistors. A simple active-load configuration for a differential amplifier is the *current mirror active load* shown in Figure 4.14.

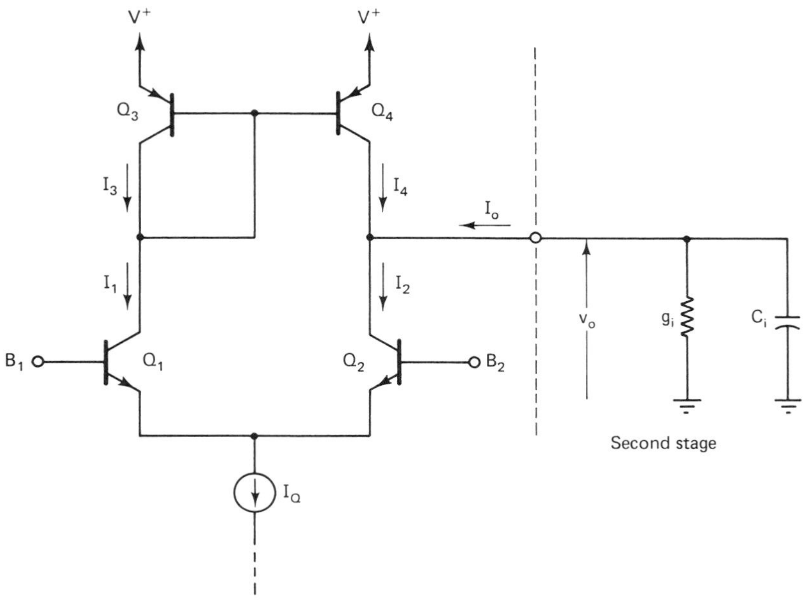


**Figure 4.14** Differential amplifier with current mirror active load.

The active load is composed of transistors $Q_3$ and $Q_4$, with $Q_3$ being a "diode-connected" transistor. This combination of $Q_3$ and $Q_4$ is called a *current mirror*. We will assume $Q_3$ and $Q_4$ to be essentially identical transistors and we note that the base-to-emitter voltages of these two transistors are the same. Since we have two identical transistors with identical base-to-emitter voltages, the collector currents of these two transistors will be equal, so that $I_3 = I_4$. Therefore, whatever current flows through $Q_3$ will be "mirrored" by the same amount of current through $Q_4$.

We will first analyze this active-load circuit under quiescent conditions (i.e., no signal applied) and we will assume that the differential amplifier is balanced such that $I_1 = I_2$. Since $I_1 = I_2$ and $I_3 = I_4$, we will have from Kirchhoff's current law that $I_O = I_{B_3} + I_{B_4} = (I_3 + I_4)/\beta_{\text{PNP}} = (I_1 + I_2)/\beta_{\text{PNP}} = I_Q/\beta_{\text{PNP}}$.

If an a-c input voltage $v_i$ is now applied to the differential amplifier, the various a-c currents will be given by $i_4 = i_3 = i_1 = g_f v_i$ and $i_2 = -g_f v_i$. Since $i_o = i_2 - i_4$, we will have $i_o = -g_f v_i - g_f v_i = -2g_f v_i$. If the net resistance driven by this differential amplifier is $R_L$, the output voltage that will be produced will be $v_o = -i_o R_L = -2g_f R_L v_i$, so that the a-c voltage gain of this differential-amplifier stage will be $A_V = v_o/v_i = 2g_f R_L$. Note that even though the output of this stage is connected to only one side of the differential amplifier, both sides of the differential amplifier will contribute to the output current and voltage by means of the action of the current mirror active load.

For a more detailed analysis of the voltage gain of the differential amplifier with a current mirror active load, we will make use of the a-c small-signal equivalent-circuit diagram of part of the circuit, as shown in Figure 4.15. In this diagram $g_{o2}$

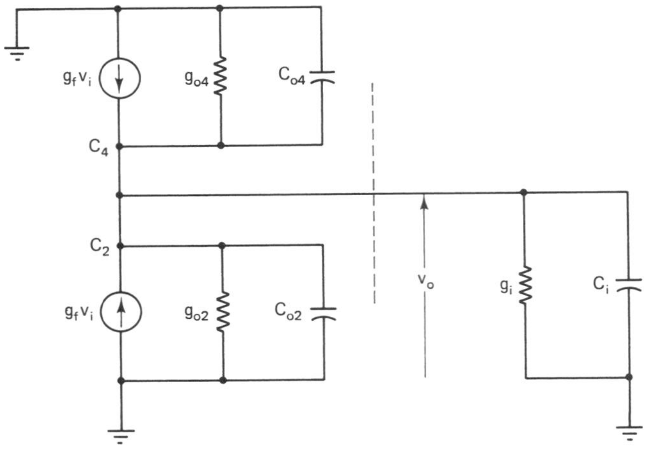


**Figure 4.15** A-c small-signal equivalent circuit at the $C_2$–$C_4$ node.

and $g_{o4}$ represent the dynamic output or collector conductances of $Q_2$ and $Q_4$, respectively, and $C_{o2}$ and $C_{o4}$ represent the output capacitances. The conductance $g_i$ and the capacitance $C_i$ represent the input conductance and capacitance of the following stage. The $g_f v_i$ current source associated with the $C_4$ node represents the a-c current produced in $Q_4$, which is a reflection of the a-c current through $Q_3$ and therefore of the a-c current through $Q_1$.

Writing a node voltage equation (Kirchhoff's current law) for the $C_2$–$C_4$ node gives

$$2g_f v_i = v_o[(g_{o2} + g_{o4} + g_i) + j\omega(C_{o2} + C_{o4} + C_i)]$$

so that

$$A_{V1} = \frac{v_o}{v_i} = \frac{2g_f}{(g_{o2} + g_{o4} + g_i) + j\omega(C_{o2} + C_{o4} + C_i)} = \frac{2g_f}{g_{\text{total}} + j\omega C_{\text{total}}} \tag{4.67}$$

where $g_{\text{total}} = g_{o2} + g_{o4} + g_i$ and $C_{\text{total}} = C_{o2} + C_{o4} + C_i$. The low-frequency gain will be given by

$$A_{V1}(0) = \frac{2g_f}{g_{\text{total}}} \tag{4.68}$$

The general expression for the gain of the differential-amplifier stage can be expressed as

$$\begin{aligned} A_{V1}(f) &= \frac{2g_f}{g_{\text{total}}[1 + j\omega(C_{\text{total}}/g_{\text{total}})]} = \frac{A_{V1}(0)}{1 + j\omega\tau} \\ &= \frac{A_{V1}(0)}{1 + j(\omega/\omega_1)} = \frac{A_{V1}(0)}{1 + j(f/f_1)} \end{aligned} \tag{4.69}$$

where $\tau = C_{\text{total}}/g_{\text{total}}$ and $f_1 = 1/(2\pi\tau)$. We see that the gain will roll off with increasing frequency, with a breakpoint at $f_1$.

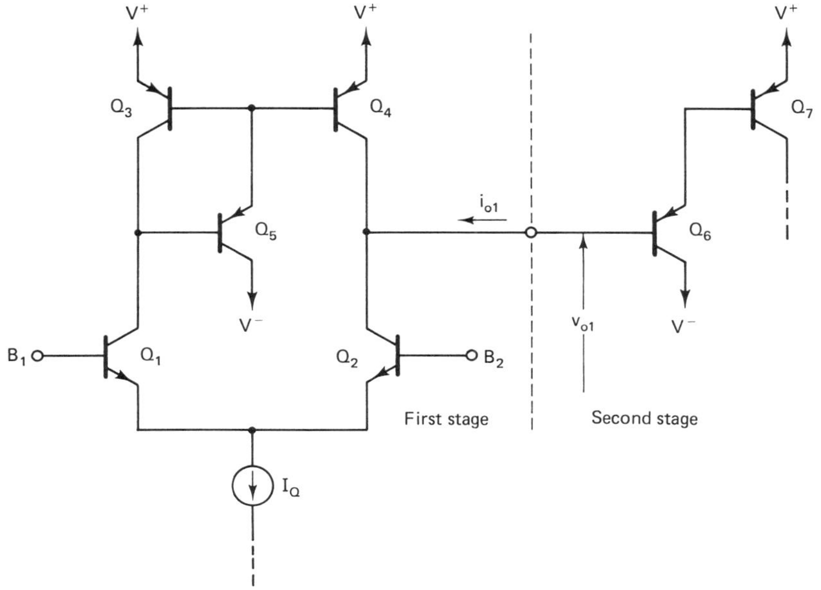


**Figure 4.16** Differential amplifier with current mirror active load.

Let us now consider the circuit of Figure 4.16, in which part of the second gain stage comprised of transistors $Q_6$ and $Q_7$ is shown. We will again assume that all transistors of the *same type* are identical. Under quiescent conditions we see that since $I_1 = I_2$ and $I_3 = I_4$, then $I_{B_5} = I_{B_6}$ and therefore $I_{E_5} = I_{E_6}$ since the transistor current gains will be equal. Since $I_{E_5} = I_{B_3} + I_{B_4}$ and $I_{E_6} = I_{B_7}$, we have that $I_{B_7} = I_{B_3} + I_{B_4}$ and thus $I_{C_7} = I_3 + I_4$, again because the transistor current gains are equal. Since $I_3 + I_4 = I_1 + I_2$, we therefore have that $I_{C_7} = I_1 + I_2 = I_Q$. Thus the current source $I_Q$ not only sets the quiescent current of the differential amplifier, but also sets the quiescent current of the $Q_6$–$Q_7$ second stage.

The a-c low-frequency voltage gain of the first (differential amplifier) stage will be given by $A_{V_1}(0) = 2g_f/g_{\text{total}}$, where $g_{\text{total}} = g_{o_2} + g_{o_4} + g_{i_6}$. For the transistor output conductances we have the following equations:

$$g_{o_2} = \frac{I_2}{2V_{A\,(\text{NPN})}} = \frac{I_Q}{2V_{A\,(\text{NPN})}} \quad \text{and} \quad g_{o_4} = \frac{I_4}{V_{A\,(\text{PNP})}} = \frac{I_Q}{2V_{A\,(\text{PNP})}} \tag{4.70}$$

where $V_A$ is the transistor *Early voltage* or *base-width modulation coefficient* (see Appendix B).

The input conductance of the second stage $g_{i_6}$ as seen looking into the base of $Q_6$ will be $g_{i_6} = I_{B_6}/2nV_T$, where $n$ is a dimensionless factor typically of about 1.5. Since $I_{B_6} = I_{C_6}/\beta_6 = I_{C_7}/\beta_6\beta_7 = I_Q/\beta_6\beta_7$, this equation can be rewritten as $g_{i_6} = I_Q/(\beta_6\beta_7 2nV_T)$. The factor of 2 multiplying $V_T$ comes from the fact that the second stage is a compound transistor (Darlington) configuration with the input voltage being applied across two base-to-emitter junctions in series.

We will now consider a specific example and we will choose the following representative parameter values: $V_{A\,(\mathrm{NPN})} = V_{A\,(\mathrm{PNP})} = 200$ V (min.) and $\beta = 50$ (min.) for all transistors, and we will choose $I_Q = 20\ \mu$A as a reasonable value. With these values we will have

$$
\begin{aligned}
g_f &= \frac{I_Q}{4V_T} = \frac{20\ \mu\text{A}}{100\ \text{mV}} = 200\ \mu\text{S} \\
g_{o\,2} &= \frac{I_Q}{2V_{A\,(\text{NPN})}} = \frac{20\ \mu\text{A}}{400\ \text{V}} = 0.05\ \mu\text{A/V} = 0.05\ \mu\text{S} \\
&= 50\ \text{nS (max.)} \\
g_{o\,4} &= \frac{I_Q}{2V_{A\,(\text{PNP})}} = \frac{20\ \mu\text{A}}{400\ \text{V}} = 0.05\ \mu\text{A/V} = 0.05\ \mu\text{S} \\
&= 50\ \text{nS (max.)} \\
g_{i\,6} &= \frac{I_Q}{\beta_6\beta_7 2nV_T} = \frac{20\ \mu\text{A}}{(50 \times 50 \times 2 \times 1.5 \times 25\ \text{mV})} \\
&= 107\ \text{nS (max.)}
\end{aligned}
\tag{4.71}
$$

Therefore, the total conductance $g_{\text{total}}$ will be $g_{\text{total}} = g_{o\,2} + g_{o\,4} + g_{i\,6} = 50$ nS + 50 nS + 107 nS = 207 nS (max.). The low-frequency voltage gain of the first (differential amplifier) stage will therefore be

$$A_{V\,1}(0) = \frac{2g_f}{g_{\text{total}}} = 2 \times 200\ \mu\text{S}/207\ \mu\text{S (max.)} = \underline{1{,}932\ \text{(min.)}} \tag{4.72}$$

so we see that with an active load a very large voltage gain can indeed be obtained from just a single amplifier stage.

From this analysis we see that the dynamic conductance seen looking into the active load is only 50 nS (max.), which corresponds to a dynamic resistance of $r_{o\,4} = 20$ MΩ (min.), which permits the first stage to have a very large voltage gain. In spite of this very large dynamic resistance value, the d-c voltage drop across the active load will be only $2V_{BE} \simeq 1.2$ V. If a load resistance $R_L$ of this large value of 20 MΩ were actually used in this circuit, the d-c voltage drop that would result will be 10 μA × 20 MΩ = 200 V! Furthermore, this large resistor would take up an enormous amount of area on the IC chip, and the associated parasitic capacitance would be extremely large. The combination of the large resistance and the very large parasitic capacitance will lead to a very poor frequency response.

We note that the voltage drop across the active-load transistors $Q_3$ and $Q_4$ will be $2V_{BE} \simeq 1.2$ V, so that the voltage at the collectors of $Q_1$ and $Q_2$ will be $V^+ - 2V_{BE} \simeq V^+ - 1.2$ V. The voltage drop across the base–emitter junction of a transistor varies logarithmically with current level, with a 10:1 current change resulting in an increase of only 60 mV in the $V_{BE}$ drop. As a result, the voltage drop across the active load under actual operating conditions will remain fairly constant at about 1.2 V.

The base voltage of $Q_1$ and $Q_2$ can go up to 0.5 V above the collector voltage before these transistors go into saturation. Therefore, the input voltage range of the differential amplifier will extend up to $V^+ - 1.2 + 0.5 = V^+ - 0.7$ V and thus up to about 0.7 V away from the positive supply voltage.

The active load uses only two or three transistors and so takes up very little area on the IC chip. The parasitic capacitance of the active load, which is the output or collector capacitance of $Q_4$, will be only about 3 to 10 pF and will thus be relatively small. The active load can permit the amplifier stage to have a gain well in excess of 1000, but at the same time results in a d-c voltage drop of no more than about 1.2 V. The active load therefore suffers none of the disadvantages of the passive load. In addition, it is of interest to note that all of the conductances associated with the first stage will be proportional to $I_Q$, so that the voltage gain of this stage will be independent of the quiescent current. As a result, very small values of $I_Q$ can be used ($\lesssim 20\ \mu$A) while still maintaining a large voltage gain. The small value of $I_Q$ is desirable, for it will result in a small value for $I_{\text{BIAS}}$ and a large value for the input resistance. For the example just considered, the input bias (or base) current will be $I_B = 10\ \mu\text{A}/50\ (\text{min.}) = 200$ nA (max.) and the difference-mode input resistance will be $r_i = 2V_T/I_B = 50\ \text{mV}/0.2\ \mu\text{A} = 250\ \text{M}\Omega$ (min.). A very small value of $I_Q$ will, however, be undesirable because it will result in a poorer frequency domain and time domain response for the amplifier. In many cases where a small value for $I_{\text{BIAS}}$ is needed the best approach is to use a JFET or MOSFET differential amplifier that is operated at a relatively high value of $I_Q$.

Finally, it should be noted that with this active load the voltage level at the collectors of the two transistors of the differential amplifier will remain approximately constant and equal. This will minimize the contribution to the offset voltage that results from large differences in the collector voltages of a pair of transistors.

### 4.3.1 MOSFET Differential Amplifier with Current Mirror Active Load

Let us now consider the MOSFET differential amplifier with a MOSFET current mirror active load as shown in Figure 4.17. The low-frequency voltage gain of this stage will be given by

$$A_{V1}(0) = \frac{v_{o1}}{v_i} = \frac{2g_f}{g_{\text{total}}} = \frac{2g_f}{(g_{o2} + g_{o4})} \tag{4.73}$$

where $v_i = v_{g1} - v_{g2}$ is the difference-mode input voltage; $g_f$ is the differential amplifier transfer conductance as given by $g_f = \sqrt{KI_Q/2} = (I_Q/2)/(V_{GS} - V_t)$; and $g_{o2}$ and $g_{o4}$ are the dynamic output conductances of $Q_2$ and $Q_4$, respectively. These output conductances will be given by

$$g_{o2} = g_{ds2} = \frac{I_{DS2}}{V_{A\,(\text{NMOS})}} = \frac{(I_Q/2)}{V_{A\,(\text{NMOS})}}$$

and (4.74)

$$g_{o4} = g_{ds4} = \frac{I_{DS4}}{V_{A\,(\text{PMOS})}} = \frac{(I_Q/2)}{V_{A\,(\text{PMOS})}}$$

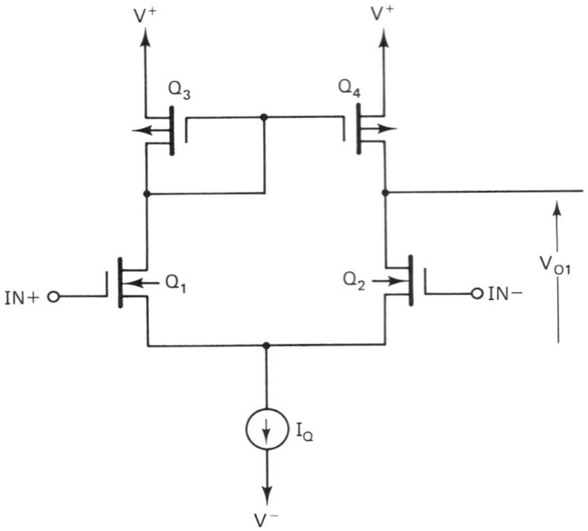


**Figure 4.17** MOSFET differential amplifier with current mirror active load.

The transistor parameter $1/V_A$ is the channel length modulation coefficient as described in Appendix B. The factor of 2 in the equation for $A_{V1}(O)$ comes from the current doubling effect of the current mirror, just as in the case of the bipolar counterpart of this circuit.

After substitution of the above relationships for $g_f$, $g_{o2}$, and $g_{o4}$, the equation for $A_{V1}$ becomes

$$A_{V1} = \frac{\dfrac{I_Q}{V_{GS} - V_t}}{\dfrac{I_Q}{2}\left[\dfrac{1}{V_{A\,(\mathrm{NMOS})}} + \dfrac{1}{V_{A\,(\mathrm{PMOS})}}\right]}$$

$$= \frac{2\left[\dfrac{1}{V_{GS} - V_t}\right]}{\dfrac{1}{V_{A\,(\mathrm{NMOS})}} + \dfrac{1}{V_{A\,(\mathrm{PMOS})}}} \tag{4.75}$$

As an example, if $V_{GS} - V_t = 1$ V and $V_{A\,(\mathrm{NMOS})} = V_{A\,(\mathrm{PMOS})} = 60$ V, we obtain a low-frequency voltage gain of $A_{V1}(0) = 60$. This is to be compared to voltage gains into the range of 1000 that can be obtained for the bipolar transistor differential amplifiers.

We note that the voltage gain is directly proportional to $1/(V_{GS} - V_t)$. Since $I_{DS} = K(V_{GS} - V_t)^2$ the voltage gain will be proportional to $1/\sqrt{I_{DS}} = \sqrt{2/I_Q}$, so that by operating the amplifier at a lower quiescent current level the low-frequency voltage gain can be increased. This increase in the gain will, however, be at the expense of a reduced bandwidth. If we consider the effect of a capacitative load on the performance of this circuit, the equation for $A_{V1}$ will become $A_{V1} = 2g_f/(g_{\mathrm{total}} + j\omega C_L)$ where $C_L$ is the net capacitance at the $D_2$–$D_4$ node and includes not only the load capacitance, but also the output capacitances of $Q_2$ and $Q_4$. The equation for $A_{V_1}$ can now be rewritten as

$$A_{V_1} = \frac{2g_f}{g_{\text{total}}\,[1 + j(\omega C_L/g_{\text{total}})]} = \frac{A_{V_1}(0)}{1 + j(\omega/\omega_1)} = \frac{A_{V_1}(0)}{1 + j(f/f_1)} \tag{4.76}$$

where

$$\omega_1 = g_{\text{total}}/C_L = \left[\frac{I_Q}{2C_L}\right]\left[\frac{1}{V_{A\,(\text{NMOS})}} + \frac{1}{V_{A\,(\text{PMOS})}}\right]$$

and

$$f_1 = \omega_1/(2\pi) = 3 \text{ dB bandwidth} \tag{4.77}$$

From this we see that the bandwidth will be directly proportional to the quiescent current $I_Q$.

As an example let us again take $V_{A\,(\text{NMOS})} = V_{A\,(\text{PMOS})}$, and we will use a value of $I_Q = 60\ \mu\text{A}$ and $C_L = 5$ pF. The 3 dB bandwidth of this differential-amplifier stage will be $f_1 = \underline{32\text{ kHz}}$.

For frequencies well above $f_1$ the voltage gain will become approximately $A_{V_1} = 2g_f/(j\omega C_L)$. The unity gain frequency $f_u$ will therefore be given by $f_u = 2g_f/(2\pi C_L) = \dfrac{I_Q/(V_{GS} - V_t)}{2\pi C_L}$. If we again use $I_Q = 60\ \mu\text{A}$, $C_L = 5$ pF, and $V_{GS} - V_t = 1$ V, the unity gain frequency will be $f_u = 1.91$ MHz. Note that this same result can be obtained from $f_u = A_{V_1}(0)\, f_1 = 60 \times 32\text{ kHz} = 1.91$ MHz.

## 4.4 DIFFERENTIAL AMPLIFIERS USING COMPOUND TRANSISTORS

For many applications the use of a compound-transistor configuration is found to be advantageous for differential amplifiers. In particular, the use of the Darlington configuration differential amplifier, as shown in Figure 4.18, offers the possibility of a much higher input impedance and a much lower input bias current than would

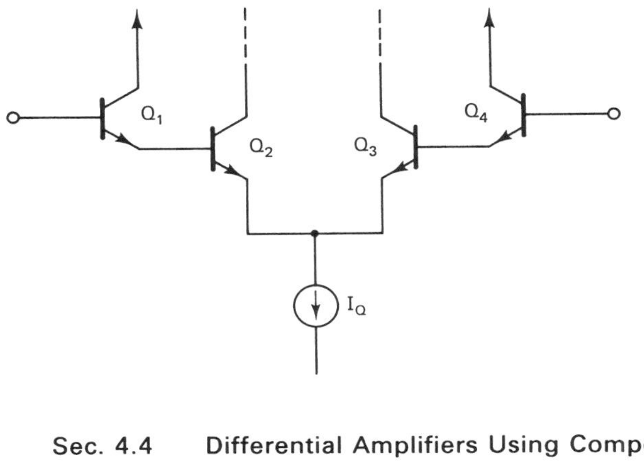


**Figure 4.18** Darlington differential amplifier.

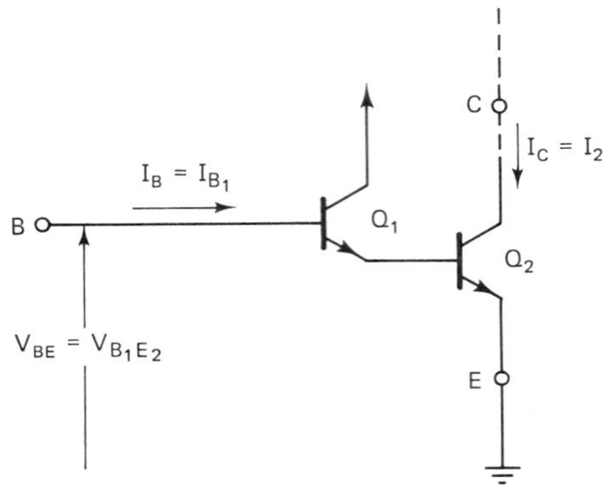


**Figure 4.19** Darlington compound transistor configuration.

otherwise be the case. One drawback of the Darlington differential amplifier is the somewhat higher offset voltage, $V_{OS}$, since four transistors are now involved in the differential amplifier. On a statistical basis, the offset voltage will be expected to be, on the average, about $\sqrt{2}$ times larger than for the equivalent, two-transistor, differential amplifier.

It can be shown that the Darlington circuit, as shown in Figure 4.19, will have an exponential relationship between the output current, $I_2$, and the input voltage, $V_{BE} = V_{B_1E_2}$, as given by $I_2 \propto \exp(V_{BE}/2V_T)$, since the voltage $V_{BE} = V_{B_1E_2} = V_{B_1E_1} + V_{B_2E_2}$ appears across two base-to-emitter PN junctions in series. Note that the exponential relationship just given will be valid only if both transistors that comprise the Darlington circuit are operated in the active mode (i.e., EB junction "on" and CB junction "off").

If we now compare this exponential characteristic given for the Darlington circuit with that obtained for a single transistor, we see that the relationship for the Darlington case differs only by having the factor of $2V_T$ in the exponent rather than $V_T$ as in the single-transistor case. As a result, if we examine the equations developed previously for the differential amplifier, we see that all of them can be applied to the case of the Darlington differential amplifier by simply replacing $V_T$ by $2V_T$ in all of the equations.

The most important difference in the two types of differential amplifiers has to do not with the change from $V_T$ to $2V_T$, but rather with the extraordinarily large current gain of the Darlington configuration. For the Darlington circuit shown in Figure 4.18, the overall current gain, $\beta = I_2/I_B$, will be equal to the product of the current gains of the two transistors that make up the Darlington circuit, as given by

$$\text{current gain} = \beta = \frac{I_2}{I_{B1}} = \frac{I_2}{I_{B2}}\frac{I_{B2}}{I_{E1}}\frac{I_{E1}}{I_{B1}} = \beta_2 \times 1 \times (\beta_1 + 1) \simeq \beta_2\beta_1 \qquad (4.78)$$

Since the current-gain values of the individual transistors will generally be of the order of 100, the overall current gain, $\beta$, of the Darlington circuit will be of the order of 10,000.

As a result of the very high current gain values available with the Darlington circuit, the input bias current, $I_{bias} = I_B$, will be much lower than that of the corresponding two-transistor differential amplifier. Since the input resistance of the differential amplifier is inversely proportional to the bias (base) current, it will be correspondingly much higher than for the two-transistor differential-amplifier case.

## 4.5 DIFFERENTIAL AMPLIFIER WITH INPUT VOLTAGE RANGE THAT INCLUDES GROUND POTENTIAL

One interesting possibility that is made available with the use of the Darlington differential-amplifier configuration is shown in Figure 4.20. This differential amplifier can be operated with the input base terminals ($B_1$ and $B_4$) at, or even slightly below, ground potential (down to about −0.5 V). As a result, this differential amplifier can be operated with just a single supply voltage.

In Figure 4.20 some of the circuit voltages have been indicated. For transistors $Q_1$ and $Q_4$ to be in the active mode, we see that the necessary condition is that $V_{B_1}$ and $V_{B_4}$ both be (algebraically) above −0.5 V, so that the collector–base junctions of these two transistors remain turned "off." If $V_{B_1} = V_{B_4} \geqslant -0.5$ V, we will have that $V_{B_2} = V_{B_3} \geqslant -0.5 + 0.6 = +0.1$ V. Since $V_{C_2} = V_{C_3} = +0.6$ V, the collector-to-base voltages of $Q_2$ and $Q_3$ will be $V_{BC_2} = V_{BC_3} \geqslant -0.5$ V, so that the collector–base junctions of $Q_2$ and $Q_3$ will also be "off." Thus as long as the two input terminals, $B_1$ and $B_4$, are no more than 0.5 V below ground potential, all of the transistors in the differential-amplifier circuit will be in the active mode.

We have seen that this differential amplifier can be operated with the input voltage down to as much as 0.5 V below ground. Because of this, for a-c signal

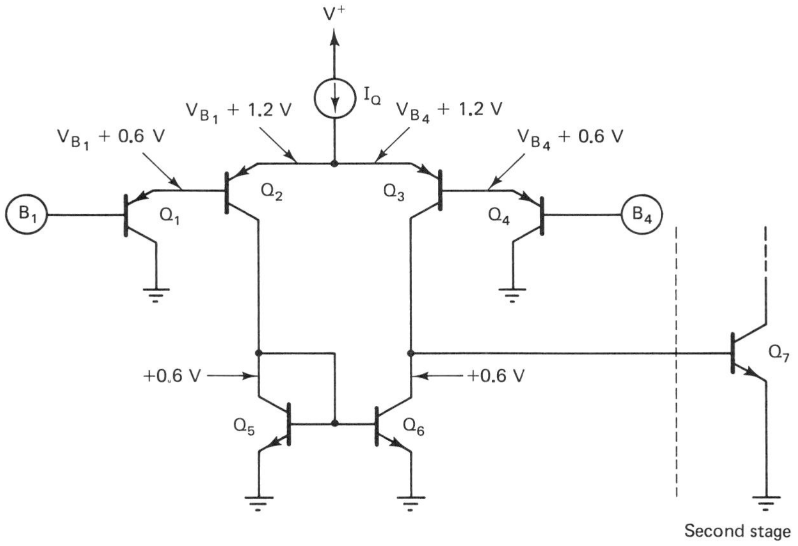


**Figure 4.20** Differential amplifier with input voltage range that includes ground potential.

amplitudes not exceeding 0.5 V, operation of this circuit with a single supply voltage is possible.

## 4.6 AXIS OF SYMMETRY

The differential amplifier can be analyzed in many cases by "splitting" the circuit down the axis of symmetry into two equivalent amplifier circuits. Let us consider the circuit of Figure 4.21a. If the quiescent currents $I_1$ and $I_2$ are equal, the application of two equal-in-magnitude but opposite-in-algebraic-sign, small-signal ($\lesssim$25 mV) a-c voltages to the two bases ($+v_i/2$ to $B_1$ and $-v_i/2$ to $B_2$) will result in a-c currents that are equal in magnitude but opposite in algebraic sign for $Q_1$ and $Q_2$. The net a-c current flow through the current source output resistance, $r_o$, will be *zero*, so that as a result the a-c voltage drop across the current source will be zero. The emitters of $Q_1$ and $Q_2$ will therefore be at *a-c ground potential*.

The circuit of Figure 4.21a can now be redrawn in the form of Figure 4.21b. Notice that one-half of the total base-to-base input voltage of $v_i$ appears across each transistor. If $v_i$ is the total (base-to-base) input voltage, the two a-c collector currents are

$$i_1 = g_m \frac{v_i}{2} = \frac{(I_Q/2)(v_i/2)}{V_T} = \frac{I_Q}{4V_T} v_i = g_f v_i$$

and (4.79)

$$i_2 = g_m \frac{-v_i}{2} = \frac{(I_Q/2)(-v_i/2)}{V_T} = -\frac{I_Q}{4V_T} v_i = -g_f v_i$$

where $g_f$ is the dynamic transfer conductance of the differential amplifier as obtained by the general analysis presented earlier. Note that these results are entirely consistent with that obtained in that general analysis.

The case just considered is for a small-signal a-c difference-mode input signal. Let us now consider what happens for the case of a common-mode input voltage, $v_{CM}$, applied to the two bases. For this case the a-c currents on both sides of the differential amplifier are exactly the same in terms of both magnitude and algebraic sign, so that the a-c current through current source output resistance will be twice that of either transistor. The a-c voltage across the current source will therefore be $2i_C r_O$. The differential amplifier can now be split down the axis of symmetry into two equal halves, as shown in Figure 4.21c. Note that the a-c voltage across $r_O$ in the equivalent circuit of Figure 4.20c is the same as of the original differential amplifier.

The a-c collector current, $i_c$, of either half-circuit will be related to the common-mode input voltage by

$$i_c = \frac{g_m v_{CM}}{1 + g_m 2r_O} \simeq \frac{v_{CM}}{2r_O} = \frac{v_{CM} g_O}{2} \qquad \text{since } g_m(2r_O) \gg 1 \tag{4.80}$$

If we now inspect the corresponding results obtained earlier in the general differential amplifier analysis we see that the same equation is obtained.

Chang, Yong

Last First initial

- Patron's name (print in caps!)



- Date due ______________

LOAN CONDITIONS - OhioLINK

- Loan is 21 days from date of pickup.
- One renewal for 21 days (you must bring the book to the Circulation Desk on or before the due date).
- Late return fee is $.50 per day ($2 per day if lending library needs the book back).
- Replacement charge is $75 plus a $25 processing charge.
- If you incur charges but don't pay, you'll lose your borrowing privileges.

L.S./M.K./F.W. 5-96

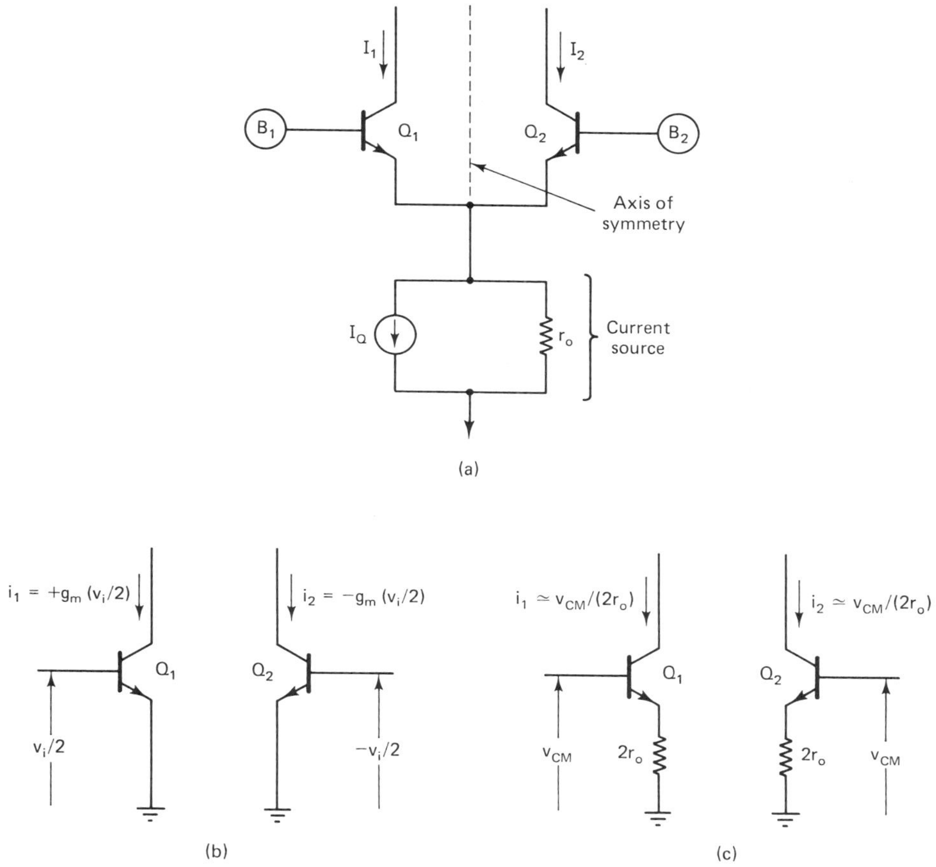


**Figure 4.21** Analysis of differential amplifier using the axis of symmetry: (a) differential amplifier; (b) A-c small-signal equivalent circuit for pure difference-mode input signal; (c) small-signal equivalent circuit for common-mode input signal.

The general case of a combination of small-signal input voltages, $v_1$ and $v_2$, which are neither purely difference-mode nor common-mode signals can be handled by resolving the input voltage into difference-mode and common-mode components using the following relationships:

$$v_{DM} = v_1 - v_2 \qquad \text{and} \qquad v_{CM} = \frac{v_1 + v_2}{2} \tag{4.81}$$

Note that the technique just described of splitting the differential amplifier down the axis of symmetry can be used *only* when the quiescent currents ($I_1$ and $I_2$) on the two sides of the amplifier are equal, and the difference-mode component of the input voltage is a small signal (i.e., amplitude limited to about 25 mV). If these two conditions are not satisfied, the more general analysis must be used.

The foregoing discussion has been in terms of a bipolar transistor type of differential amplifier. The same axis of symmetry technique can, however, be applied equally well to both the JFET and MOSFET differential amplifiers. To use this technique for the FET differential amplifiers, the difference-mode input voltage amplitude should not exceed about $0.5 V_N$, where $V_N = V_P\sqrt{I_Q/I_{DSS}}$ for a JFET differential amplifier, and $V_N = \sqrt{I_Q/K}$ for a MOSFET differential amplifier.

From the foregoing discussion it is now easy to see the relationship between the dynamic forward transfer conductance $g_f$ of a differential amplifier and the dynamic transfer conductance, $g_m$ or $g_{fs}$, of the transistors that comprise the differential amplifier. For the case of the balanced differential amplifier that can be split into two equal halves down the axis of symmetry, the situation is the same as two single-ended transistors, each with an input voltage of one-half of the difference-mode input signal, $v_i/2$. The resulting collector current is given by $i_c = g_m\ v_i/2 = (g_m/2)v_i$ for the bipolar transistor case. Similarly, for an FET differential amplifier the drain current will be given by $i_{ds} = g_{fs}\ v_i/2 = (g_{fs}/2)v_i$. Since the transfer conductance of the differential amplifier is the ratio of the output current to the difference-mode input voltage, we see that the differential amplifier transfer conductance is equal to one-half of the transfer conductance of the individual transistors that comprise the differential amplifier.

## PROBLEMS

**4.1.** (*Differential amplifier*) Given: NPN differential amplifier with current sink biasing and passive load (Figure P4.1); $V_{\text{SUPPLY}} = \pm 15$ V; $I_Q = 200\ \mu$A; $I_{Q_3} = 1.0$ mA; $h_{FE} = 200$ (min.); $V_{BE} = 650$ mV at 1.0 mA; $BV_{EBO} = 7.0$ V (min.); $V_{C_1(Q)} = V_{C_2(Q)} + 9.0$ V; assume all transistors identical, unless indicated otherwise; output conductance of current sink ($Q_4$) = 307 nS; $V_A = 250$ V.

- **(a)** Find $R_1$, $R_2$, and $R_C$. (*Ans.*: 14.35 kΩ, 201 Ω, 60 kΩ)
- **(b)** Find the difference-mode dynamic transfer conductance, $g_f$. (*Ans.*: $g_f = 2.0$ mS)
- **(c)** Find the difference-mode voltage gains: $v_{c_1}/v_{b_1}$, $v_{c_2}/v_{b_1}$, and $(v_{c_1} - v_{c_2})/v_{b_1}$, where $v_{b_1}$ is the small-signal input voltage. (*Ans.*: −120 V, +120 V, −240 V)
- **(d)** Find the input bias current, $I_{\text{BIAS}}$. [*Ans.*: $I_{\text{BIAS}} = 500$ nA (max.)]
- **(e)** Find the difference-mode dynamic input resistance, $R_i$. [*Ans.*: 100 kΩ (min.)]
- **(f)** Find the common-mode input resistance, $R_{i(CM)}$. [*Ans.*: 1.3 GΩ (min.)]
- **(g)** Find the common-mode dynamic transfer conductance, $g_{f(CM)}$. (*Ans.*: 153 nS)
- **(h)** Find the common-mode voltage gain and CMRR. (*Ans.*: $A_{V(CM)} = -0.0092$; CMRR = 82.3 dB)
- **(i)** Find the input voltage range. (*Ans.*: (1) Either $V_{B_1}$ or $V_{B_2}$, or both, must be greater than −14.2 V; (2) if $V_{B_1} = V_{B_2}$, then $V_B$ must be less than +9.5 V, and if $V_{B_1} \neq V_{B_2}$, then both must be less than +3.5 V; (3) the maximum voltage between $B_1$ and $B_2$ must not exceed ±7.7 V)
- **(j)** If $V_{OS} = 0$, $V_{B_1} = +10$ mV, and $V_{B_2} = +30$ mV, find $I_1$ and $I_2$. (*Ans.*: 62 μA, 138 μA)
- **(k)** Find $g_f$ and the difference-mode voltage gain (single-ended output) for the conditions of part (j). (*Ans.*: 1.71 mS, 103)
- **(l)** If $V_{OS} = 2.0$ mV, find $V_{C_1(Q)}$ and $V_{C_2(Q)}$ ($V_i = V_{B_1} - V_{B_2} = 0$). (*Ans.*: +9.24 V, +8.76 V)

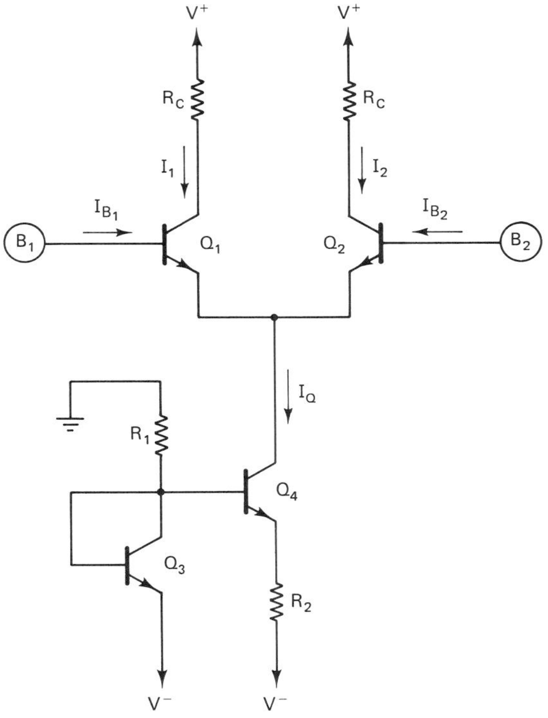


**Figure P4.1**

**(m)** If $V_{OS} = 2.0$ mV, find $TC_{V(OS)}$. *(Ans.*: 6.7 μV/°C)

**(n)** If this differential amplifier is to drive another differential amplifier that has a difference-mode input resistance of 100 kΩ, find the voltage gain (balanced output), $A_V = (v_{c_1} - v_{c_2})/v_{b_1}$. *(Ans.*: −109 V)

**(o)** Find $V_{C_1}$ and $V_{C_2}$ for the conditions of $I_{B_1} = 0$ and $V_{B_2} = 0$. *(Ans.*: +15 V, +3.0 V)

**(p)** If $V_{B_1} = 0$ and $V_{B_2} = -50$ mV, find $I_1$ and $I_2$. *(Ans.*: 176 μA, 24 μA)

**(q)** Find the output conductance of the current sink ($Q_4$). *(Ans.*: 307 nS)

**4.2.** (*JFET differential amplifier*) Given: a JFET differential amplifier as shown in Figure P4.2 with $I_Q = 0.5$ mA. The differential amplifier transistors, $Q_1$ and $Q_2$, and the current source active load transistors, $Q_3$ and $Q_4$, have the following parameters: $I_{DSS} = 1.0$ mA, $V_P = -4.0$ V, and $r_{ds} = 200$ kΩ.

**(a)** Find the dynamic transfer conductance of the differential amplifier, $g_f$. *(Ans.*: $g_f =$ 0.125 mS)

**(b)** Find the a-c small-signal voltage gain for a balanced output, $v_o/v_i$. *(Ans.*: $A_V =$ 25)

**(c)** What difference mode input voltage is required to shift 90% of the total differential-amplifier current $I_Q$ to one side of the differential amplifier? *(Ans.*: ±1.79 V)

**(d)** Repeat part (c) for 99% of the total current. *(Ans.*: ±1.89 V)

**4.3.** (*MOSFET differential amplifier*) Given: A differential amplifier uses MOSFETs that have a threshold voltage of +3.0 V and $I_{DS} = 1.0$ mA at $V_{GS} = +7.0$ V. The total differential-amplifier current is $I_Q = 1.0$ mA.

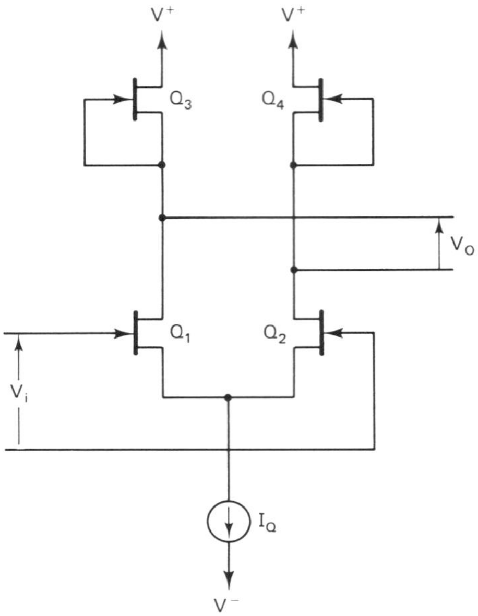


**Figure P4.2**

**(a)** Find the dynamic transfer conductance of the differential amplifier, $g_f$. (*Ans.*: 0.177 mS)

**(b)** Find the input voltage required to shift 90% of the current to one side of the differential amplifier. (*Ans.*: ±2.52 V)

**(c)** Find the input voltage required to shift 99% of the current to one side. (*Ans.*: ±3.58 V)

**4.4.** (*Differential amplifier with current mirror active load*) Given: Differential amplifier with current mirror active load circuit of Figure 4.16.

**(a)** Show that a general expression for the a-c small-signal voltage gain, $A_v = v_{o_1}/v_i$, will be given by

$$A_v = \frac{\beta_6\beta_7}{(1/n) + \beta_6\beta_7 V_T[(1/V_{A_N}) + (1/V_{A_P})]}$$

**(b)** If $\beta_6 = \beta_7 = 70$ (min.), $n = 1.5$, and $V_{A_N} = V_{A_P} = 150$ V (min.), find $A_V$. [*Ans.*: 2130 (min.)]

**(c)** Repeat part (b) for $\beta_6 = \beta_7 = 50$ (min.). [*Ans.*: 1667 (min.)]

**(d)** Repeat part (b) for $\beta_6 = \beta_7 = 30$ (min.). [*Ans.*: 931 (min.)]

**(e)** If $I_Q = 40\ \mu$A and $V^+ = 12$ V, find $I_{BIAS}$ for $\beta_1 = \beta_2 = 50$ (min.). [*Ans.*: 400 nA (max.)]

**(f)** Find the difference-mode input resistance for the above conditions. [*Ans.*: 125 kΩ (min.)]

**(g)** Find the upper limit of the input voltage range for this differential amplifier. (*Ans.*: +11.1 V)

**(h)** For parts (b) to (d), find the maximum voltage gain that is obtained as the transistor current gain $\beta$ goes to infinity. (*Ans.*: 3000)

**(i)** If transistors $Q_1$ and $Q_2$ are each replaced by a Darlington transistor pair, find the voltage gain if $\beta = 50$ (min.) for all transistors, and find the voltage gain that is approached as $\beta$ goes to infinity. [*Ans.*: 1071 (min.), 1500 (max.)]

**(j)** Find $I_{BIAS}$ and the difference-mode input resistance for the case of part (i) with $I_Q = 40\ \mu A$. [*Ans.*: 8 nA (max.), 12.5 MΩ (min.)]

**4.5.** (*MOSFET differential amplifier with current mirror active load*) Given: MOSFET differential amplifier circuit of Figure 4.17. For the NMOS transistors $Q_1$ and $Q_2$ the threshold voltage $V_t$ is +1.0 V and at $I_{DS} = 1.0$ mA the gate-to-source voltage $V_{GS}$ is +3.0 V. The net capacitance at the $Q_2$–$Q_4$ node, including the load capacitance is 5 pF. Assume $V_{A\,(NMOS)} = V_{A\,(PMOS)} = 50$ V.

Find the low-frequency voltage gain $A_{V1}(O)$, the 3 dB bandwidth, and the unity gain frequency $f_u$ for the following values of quiescent current $I_Q$:

**(a)** $I_Q = 2.0$ mA (*Ans.*: 25, 1.27 MHz, 31.8 MHz)

**(b)** $I_Q = 200\ \mu A$ (*Ans.*: 79, 127 kHz, 10.0 MHz)

**(c)** $I_Q = 60\ \mu A$ (*Ans.*: 144, 38 kHz, 5.5 MHz)

**(d)** $I_Q = 20\ \mu A$ (*Ans.*: 250, 12.7 kHz, 3.18 MHz)

**4.6.** (*Exponential current-voltage relationship for the Darlington transistor configuration*) Show that the relationship between the output collector current $I_C$ and the input base-to-emitter voltage $V_{BE}$ for a Darlington transistor configuration as shown in Figure 4.19 is given by

$$I_C = I_O \exp(V_{BE}/2V_T)$$

where $I_O$ is a pre-exponential constant having units of current and $V_T = kT/q$ is the thermal voltage.

## REFERENCES


BOLYSTAD, R., and L. NASHELSKY, *Electronic Devices and Circuit Theory*, Prentice-Hall, 1982.

FITCHEN, F. C., *Electronic Integrated Circuits and Systems*, Van Nostrand Reinhold, 1970.

GIACOLETTO, L. J., *Differential Amplifiers*, Wiley, 1970.

GLASER, A. B., and G. E. SUBAK-SHARPE, *Integrated Circuit Engineering*: Addison-Wesley, 1977.

GRAEME, J. G., G. E. TOBEY, and L. P. HUELSMAN, *Operational Amplifiers—Design and Applications*, McGraw-Hill, 1971.

GRAY, P. R., and R. G. MEYER, *Analysis and Design of Analog Integrated Circuits*, Second Edition, Wiley, 1984.

GRINICH, V. H., and H. G. JACKSON, *Introduction to Integrated Circuits*, McGraw-Hill, 1975.

HAMILTON, D. J., and W. G. HOWARD, *Basic Integrated Circuit Engineering*, McGraw-Hill, 1975.

LENK, J. D., *Manual for M.O.S. Users*, Reston, 1975.

MIDDLEBROOK, R. D., *Differential Amplifiers*, Wiley, 1963.

MILLMAN, J., *Microelectronics*, McGraw-Hill, 1979.

MOTOROLA, INC., *Analysis and Design of Integrated Circuits*, McGraw-Hill, 1967.

ROBERGE, J. K., *Operational Amplifiers*, Wiley, 1975.

WAIT, J. V., *Introduction to Operational Amplifiers: Theory and Applications*, McGraw-Hill, 1975.

5

# OPERATIONAL-AMPLIFIER CHARACTERISTICS AND APPLICATIONS

## 5.1 INTEGRATED CIRCUITS

An integrated circuit (IC) is an electronic device in which there is more than one circuit component in the same package. Most ICs contain many transistors together with diodes, resistors, and capacitors. Integrated circuits may contain tens, hundreds, or even many thousands of transistors. A *monolithic IC* is one in which all of the components are contained on a single-crystal chip of silicon. This chip typically measures from 1 mm $\times$ 1 mm $\times$ 0.25 mm thick for the smaller ICs to 5 mm $\times$ 5 mm $\times$ 0.3 mm thick for the larger size ICs.

In the case of a *hybrid IC* there will be more than one chip in the package. The chips may be monolithic ICs, discrete transistors, or diode chips, or some combination thereof. There can also be chip capacitors. The chips are usually mounted on an insulating ceramic substrate, usually alumina ($Al_2O_3$) and are interconnected by a thin-film or thick-film conductor pattern that has been deposited on the ceramic substrate. There can also be thin-film or thick-film resistor patterns deposited on the ceramic substrate. Thin-film patterns are deposited by vacuum evaporation techniques and are usually about 1 $\mu$m is thickness, whereas thick-film patterns are applied by a silk screening method and are usually 10 to 30 $\mu$m in thickness.

Integrated circuits can also be classified as to function, the two principal categories being *digital* and *analog* (or *linear*) ICs. A digital IC is one in which all of the transistors operate in a switching mode being either off (i.e., cutoff mode) or on (saturation mode) to represent the high and low (or 1 and 0) digital logic levels. The transistors during the switching transient pass very rapidly through the active region. Virtually all digital ICs are of the monolithic type and contain just transistors and resistors, with both bipolar and MOSFET transistors being used. Some MOSFET digital ICs contain as many as several hundred thousand transistors, all on one silicon chip.

*Analog* or *linear* ICs operate on signal voltages and currents in analog form and the transistors operate mostly in the active (or linear) mode of operation. There are many different types of analog ICs, such as operational amplifiers, audio power amplifiers, voltage regulators, voltage references, video amplifiers, radio-frequency amplifiers, voltage comparators, modulators and demodulators, logarithmic converters, multipliers, function generators, voltage-controlled oscillators, phase-locked loops, digital-to-analog and analog-to-digital converters, and other devices. The majority of analog ICs are of the monolithic construction, although there are many hybrid ICs of importance. In this chapter operational amplifiers are studied, and in later chapters other types of analog ICs are investigated.

## 5.2 INTRODUCTION TO OPERATIONAL AMPLIFIERS

An operational amplifier (op amp) is an integrated circuit that produces an output voltage $v_o$ that is an amplified replica of the difference between two input voltages, $v_1$ and $v_2$. An op amp can be ideally characterized by the input–output transfer function given by $v_o = A_{OL}(v_1 - v_2)$, where $A_{OL}$ is the *open-loop gain* of the op amp. Most op amps are of the monolithic type, although there are literally hundreds of different types of op amps and op amps are available from dozens of different manufacturers.

The term "operational amplifier" comes from one of the earliest uses of these devices, dating back to the early and middle 1960s in analog computers. The op amps were used in conjunction with other circuit components, principally resistors and capacitors, to perform various mathematical operations, such as addition, subtraction, multiplication, integration, and differentiation—hence the name "operational amplifier." The applications of op amps over the years since then has vastly expanded from that, however, and op amps are used to perform a multitude of tasks, as will be seen later.

A basic circuit representation of an op amp is shown in Figure 5.1a. The triangular-shaped "box" represents the op amp, which itself is a multistage amplifier containing generally 10 to 100 transistors. The output voltage $v_o$ is related to the two input voltages by the basic relationship $v_o = A_{OL}(v_1 - v_2)$. Note that the op amp is in

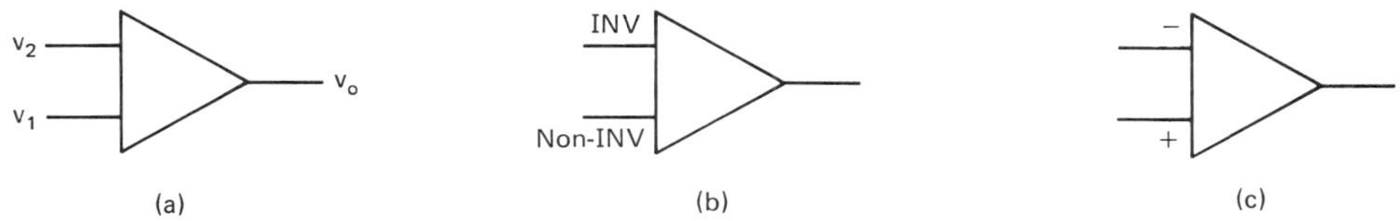


**Figure 5.1** Operational-amplifier symbol: (a) basic operational-amplifier symbol; (b) symbol with input polarities indicated explicitly; (c) symbol with input polarities indicated explicitly.

theory sensitive only to the difference between the two input voltages, which is called the *difference-mode* input voltage, $v_i = v_1 - v_2$. The op amp is ideally completely insensitive to any voltage component that is common to the inputs, which is called the *common-mode* input voltage, and is given by $v_{i(CM)} = (v_1 + v_2)/2$.

The open-loop gain of the op amp is a positive dimensionless constant, and is usually very large, often in the range $10^5$ to $10^6$ at low frequencies (0 to 30 Hz). If voltage $v_1$ is applied alone and $v_2 = 0$, we have that $v_o = A_{OL}v_1$, so that $v_o$ will be an amplified *noninverted* replica of the input voltage $v_1$. If, on the other hand, $v_2$ is now applied alone with $v_1 = 0$, then $v_o = -A_{OL}v_2$ and the output voltage $v_o$ will now be an amplified and *inverted* replica of the input voltage. Hence we can call the lower ($v_1$) input terminal the *noninverting* or *positive-gain input terminal*, and the upper input terminal ($v_2$) can be called the *inverting* or *negative-gain terminal*. This is the usual convention and will be assumed to be the case unless otherwise indicated. The two input terminals may be designated explicitly as shown in Figure 5.1b and c. If there is no explicit indication with respect to the inverting and noninverting input terminals, the usual convention will be assumed to hold.

The op amp is a multistage electronic amplifier containing many transistors. As is the case with any type of electronic amplifier or system, there must be a d-c power supply voltage applied to bias the various transistors properly. This power supply connection is shown explicitly in Figure 5.2a. Usually, however, the power supply connections are omitted from the diagram in order to simplify the diagram, although these connections are always assumed to be present. Most op amps are operated with a *dual* or *split* power supply, as shown in Figure 5.2a. The two supply voltages, $V^+$ and $V^-$, are usually of the same magnitude (but of opposite algebraic sign). The supply voltages can usually range from as little as $\pm 5$ V up to a maximum of $\pm 18$ V, with $\pm 10$ to $\pm 15$ V being the most commonly used range of values for the supply voltages. Some op amps can operate with supply voltages as small as $\pm 3$ V, and other "high-voltage" op amps can operate with supply voltages substantially greater than $\pm 18$ V.

The use of a dual or split supply offers the advantage of allowing the input and output voltages to swing both above and below ground potential for a *bipolar* output voltage swing. Some op amps can be operated with a single supply, as shown in Figure 5.2b. This has the advantage of a simpler power supply configuration, but the output voltage can now swing only in the positive direction and cannot go below ground potential (0 V), so that only a *unipolar* output voltage swing can be obtained. Furthermore, the allowable input voltage range is usually limited such that the input voltage cannot be allowed to drop more than about 0.5 V below ground for proper operation of the circuit.

In any case, the output voltage of an op amp cannot swing beyond either supply voltage. Usually, the maximum attainable output voltage swing will be about 1 V less than the supply voltage in either direction, although under heavily loaded conditions the output voltage swing may be reduced even further. In the case of single supply operation the upper limit remains as stated above, but the lower limit of the output voltage will be ground potential (0 V).

The ideal input–output transfer characteristic of an op amp is shown in Figure 5.3. Note the linear or amplifying region wherein $v_o = A_{OL}v_i$, which is bounded

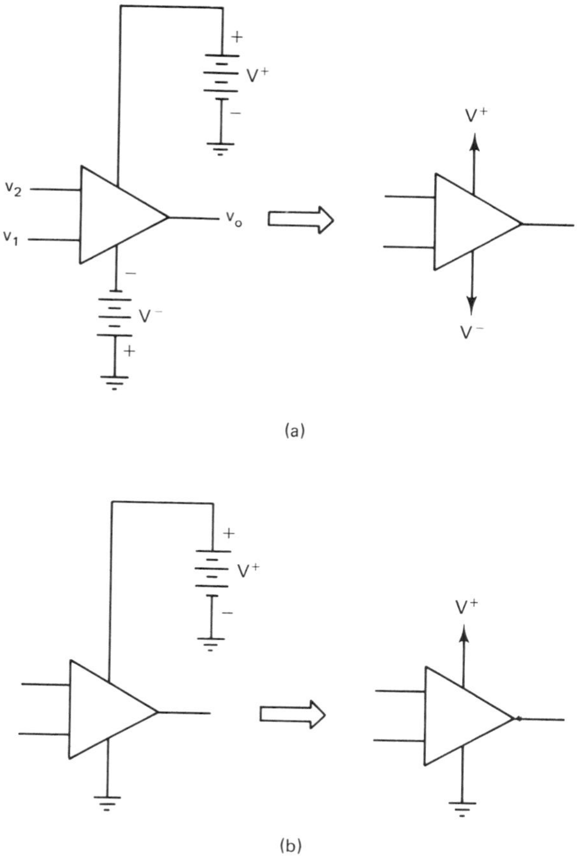


**Figure 5.2** (a) Split-supply operation of operational amplifier; (b) single-supply operation.

on either extreme by the "saturation" regions, in which the output voltage is limited by the supply voltage and is no longer responsive to changes in the input voltage. Since the value of $A_{OL}$ is so very large, especially at the lower frequencies, where values of $10^5$ to $10^6$ are possible, the width of the linear or amplifying region will be very small and is given by $\Delta v_i \simeq (V^+ - V^- - 2\text{ V})/A_{OL}$. Thus typically we will have a supply voltage of around $\pm 10$ V, so that $\Delta v_i \simeq 20\text{ V}/A_{OL} \simeq 20$ to 200 $\mu$V. Therefore, if the output voltage is to be an amplified replica of the input voltage, the amplitude of the input voltage must be kept very small, generally less than 1 mV. If this is not the case, the op amp will enter the saturated region during part of the output-voltage swing and the output-voltage waveform will be "clipped," usually resulting in severe signal distortion. It is because of this limitation as well as a number of other factors to be discussed later that the op amp is usually operated in a *feedback* or *closed-loop* configuration in which a fraction of the output voltage is fed back to the inverting input terminal as shown in Figure 5.4. This produces a condition of *negative feedback* and offers a number of very substantial advantages.

Under these conditions the op amp is operating as a closed-loop system and the fraction of the output voltage that is fed back to the inverting input terminal

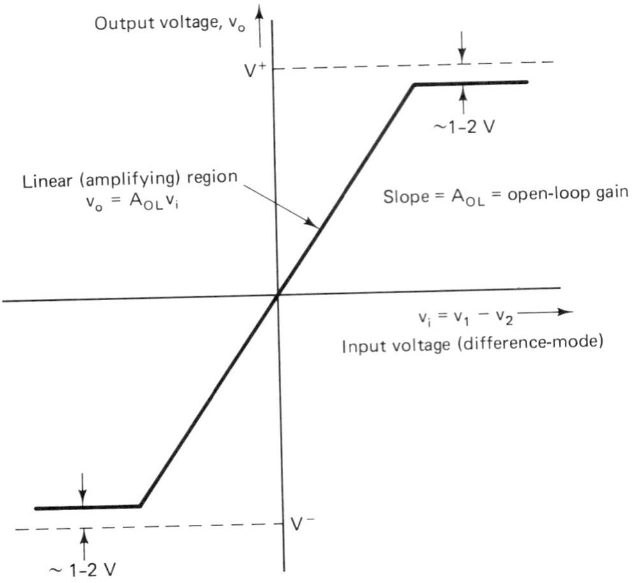


**Figure 5.3** Operational-amplifier input–output transfer characteristic.

is called the *feedback factor F* and for the circuit of Figure 5.4 is given by $F = v_{fb}/v_o = Z_1/(Z_1 + Z_2)$ as obtained from the simple voltage-division relationship. Note that although the basic relationship that $v_o = A_{OL}v_i$ is still valid, the difference-mode input voltage $v_i$ is *no longer* given by $v_i = v_1 - v_2$, but is now given by $v_i = v_1 - v_2' - v_{fb} = v_1 - v_2' - Fv_o$, so that $v_o = A_{OL}v_i = A_{OL}(v_1 - v_2' - Fv_o)$ and thus $v_o(1 + FA_{OL}) = A_{OL}(v_1 - v_2')$. Solving for $v_o$ gives

$$v_o = \frac{A_{OL}}{1 + FA_{OL}}(v_1 - v_2') = A_{CL}(v_1 - v_2') \tag{5.1}$$

where $A_{CL}$ is the *closed-loop gain*. Note that since the signal is no longer applied directly to the inverting input terminal, but rather by means of the $Z_1$–$Z_2$ voltage divider, the voltage $v_2'$ is related to the signal voltage $v_2$ by $v_2' = v_2[Z_2/(Z_1 + Z_2)]$, as again obtained from the simple voltage-division relationship.

For this case of negative feedback we see that the closed-loop gain $A_{CL} = A_{OL}/(1 + FA_{OL})$ will be smaller than the open-loop gain, $A_{OL}$. The quantity $FA_{OL}$ is called the *loop gain*. For the case of a large loop gain such that $FA_{OL} \gg 1$, we will have that $A_{CL} \simeq A_{OL}/FA_{OL} = 1/F$, so that the closed-loop gain will be relatively

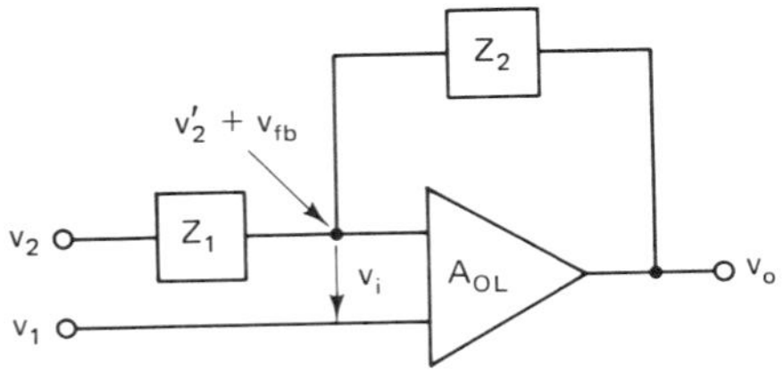


**Figure 5.4** Closed loop (negative feedback) operational-amplifier system.

independent of the open-loop gain $A_{OL}$ and dependent principally on the parameters of the feedback network. For the circuit of Figure 5.4 we have that $F = Z_1/(Z_1 + Z_2)$, so that for large loop gains we will have that $A_{CL} = 1/F = (Z_1 + Z_2)/Z_1 = 1 + Z_2/Z_1$ and thus $A_{CL}$ is controlled principally by the ratio of the two impedances, $Z_1$ and $Z_2$. The gain that the signal applied to the noninverting input terminal, $v_1$, obtains will therefore be $1 + (Z_2/Z_1)$. The net gain with respect to the other input signal $v_2$ will, however, be negative and also will be modified by the $Z_1$–$Z_2$ voltage divider to become

$$\frac{Z_2}{Z_1 + Z_2}\left[-\left(1 + \frac{Z_2}{Z_1}\right)\right] = -\frac{Z_2}{Z_1}$$

In the sections to follow we consider the analysis of closed-loop op-amp systems using the *node voltage circuit equations*, which represent basically applications of Kirchhoff's current law. This will first be applied for the case in which the open-loop gain is assumed to be very large ($A_{OL} \to \infty$), which will be called the *infinite open-loop gain case*. Although this case will never be obtained in practice, the results so obtained will be applicable as an excellent approximation for most actual situations. Then this node voltage method will be applied for the case in which the open-loop gain is assumed to be finite. After this the effects of some of the nonideal op-amp characteristics on the circuit performance of op amps are considered.

## 5.3 ANALYSIS OF CLOSED-LOOP OPERATIONAL-AMPLIFIER CIRCUITS USING THE NODE VOLTAGE EQUATIONS

For the op-amp analysis to follow we will use the node voltage equations, which are basically statements of Kirchhoff's current law. We will assume that the op amp is ideal in that there will be no currents flowing into or out of the two input terminals. We will use the circuit of Figure 5.5 and for the first analysis we will let the open-loop gain of the op amp go to infinity (i.e., the "infinite-gain approximation"), and then an analysis will be provided for the case of finite open-loop gain.

**Ideal operational amplifier with infinite gain.** If we let the open-loop voltage gain of the op amp go to infinity, then for a finite output voltage $v_o$ the input voltage $v_i$ will go to zero as given by $v_i = v_o/A_{OL} \to 0$ as $A_{OL} \to \infty$. Therefore, for the node $x$ and node $y$ voltage we have that $v_x = v_y = v_1$. Writing nodal

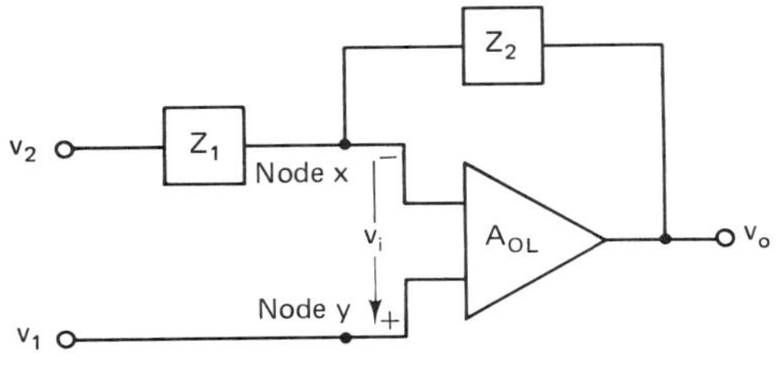


**Figure 5.5** Circuit for node voltage analysis of operational-amplifier circuit.

equations for the currents through the admittances $Y_1$ and $Y_2$ gives $(v_2 - v_1)Y_1 = (v_1 - v_o)Y_2$, so that $v_oY_2 = -v_2Y_1 + v_1(Y_1 + Y_2)$. Solving for $v_o$ gives

$$\boxed{v_o = v_1\left(1 + \frac{Y_1}{Y_2}\right) - v_2\frac{Y_1}{Y_2}} \tag{5.2}$$

and in terms of impedances this will be

$$\boxed{v_o = v_1\left(1 + \frac{Z_2}{Z_1}\right) - v_2\frac{Z_2}{Z_1}} \tag{5.3}$$

**Ideal operational amplifier but with finite open-loop gain.** For the case of a finite value of $A_{OL}$ we have that $v_i = v_o/A_{OL}$, so that $v_x = v_1 - v_o/A_{OL}$. Writing the node voltage equations for this case gives $(v_2 - v_1 + v_o/A_{OL})Y_1 = (v_1 - v_o/A_{OL} - v_o)Y_2$, so that upon collecting terms in $v_o$ we obtain $v_o(Y_2 + Y_2/A_{OL} + Y_1/A_{OL}) = -v_2Y_1 + v_1(Y_2 + Y_1)$. Solving this for the output voltage $v_o$ gives

$$\boxed{\begin{aligned} v_o &= \frac{v_1(Y_2 + Y_1) - v_2Y_1}{Y_2 + (1/A_{OL})(Y_2 + Y_1)} = \frac{v_1(1 + Y_1/Y_2) - v_2(Y_1/Y_2)}{1 + (1/A_{OL})(1 + Y_1/Y_2)} \\ &= \frac{v_1(1 + Z_2/Z_1) - v_2(Z_2/Z_1)}{1 + (1/A_{OL})(1 + Z_2/Z_1)} \end{aligned}} \tag{5.4}$$

Let us consider the simple op-amp system of Figure 5.6. For the output voltage $v_o$ we have that $v_o = A_{CL}v_1 = (1 + R_2/R_1)v_1$. In Figure 5.7 both the open-loop and closed-loop input–output transfer characteristics are shown. Note that since $A_{CL}$ can be made to be much smaller than $A_{OL}$, the input voltage range for linear operation in the closed-loop case can be made to be very much larger than that for the open-loop case.

The open-loop gain of op amps usually exhibit a very large unit-to-unit variation among devices of the same type, with the manufacturers specifying only minimum and sometimes maximum values for $A_{OL}$. The $A_{OL}$ unit-to-unit range can often extend over a span of 3:1 or even 10:1. The open-loop gain also exhibits a strong frequency dependence and can vary from values in the $10^6$ range at low frequencies (0 to 10 Hz) all the way down to values below unity at frequencies in excess of a few MHz.

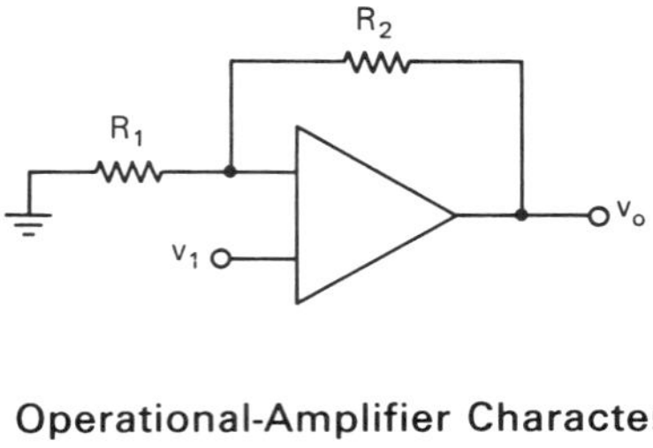


**Figure 5.6** Noninverting operational amplifier circuit.

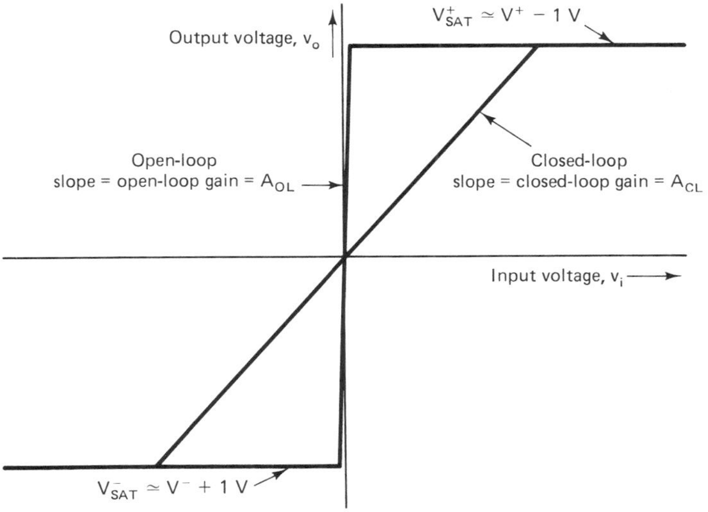


**Figure 5.7** Input–output transfer characteristics.

Furthermore, the open-loop gain is sensitive to supply voltage variations and temperature effects.

In the negative-feedback closed-loop situation we see that the closed-loop gain will become relatively independent of $A_{OL}$ and dependent principally on the parameters of the feedback loop. In the specific case under consideration we have that $A_{CL} = 1 + R_2/R_1$. Since the resistance ratio can be very accurately set, and will be relatively independent of temperature, voltage level, and frequency, we see that the use of negative feedback can produce a closed-loop gain that is both *accurately set* and *stable*.

In the case of Figure 5.6 we have a noninverting amplifier with a gain of $A_{CL} = 1 + R_2/R_1$. An interesting special case of this is the circuit of Figure 5.8, for which the gain is +1. This op-amp circuit is called a *voltage follower* since the output voltage accurately follows the voltage. It might at first be thought that this circuit would be of little or no practical value since the gain is unity. It turns out, however, that this is a very useful and very widely used circuit. This is a result of the very large input impedance $Z_i$ and very small output impedance $Z_o$ of this op-amp configuration. This will allow relatively low-impedance loads to be driven by relatively high-impedance sources without any substantial signal loss, as illustrated by the example shown in Figure 5.9. In the situation of Figure 5.9a, where a 100-kΩ source is driving a load impedance of only 100 Ω, the ratio of output voltage to signal voltage will be given by $v_o/v_s = 100\ \Omega/(100\ \Omega + 100\ \text{k}\Omega) \simeq 0.001$, so that we see that a very

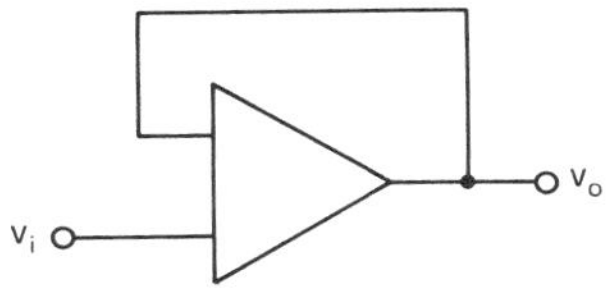


**Figure 5.8** Voltage follower.

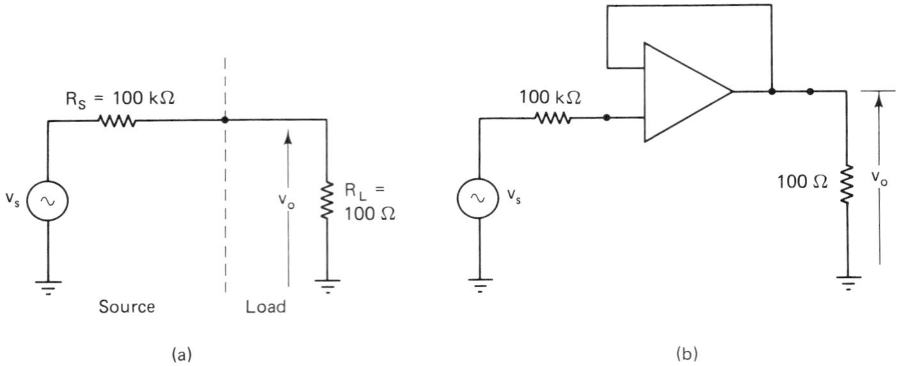


**Figure 5.9** Application of voltage follower.

severe attenuation of the signal will occur. In Figure 5.9b a voltage follower is interposed between source and load. As long as $Z_i \gg 100\ \text{k}\Omega$ and $Z_o \ll 100\ \Omega$, which are conditions that are readily obtainable in practice, we will have that $v_o \simeq v_s$ and there will be very little signal attenuation.

In Figure 5.10a another simple op-amp circuit is shown. For this circuit $v_o = A_{CL}v_2 = -(R_2/R_1)v_2$, so that this is an *inverting amplifier* with a voltage gain of $-R_2/R_1$. Again we note that the gain is controlled by the resistance ratio and is relatively independent of the op-amp open-loop gain. We also note that the gain magnitudes for the two cases just considered are not equal, the gain for the noninverting case being $1 + R_2/R_1$ and the gain for the inverting case being $-R_2/R_1$.

Let us now consider the circuit of Figure 5.10b. For this circuit the gain accorded voltage $v_2$ will be $-R_2/R_1$, as before. We note, however, that the signal voltage $v_1$ is attenuated by the $R_1$–$R_2$ voltage divider so that the voltage that appears at the noninverting input terminal of the op amp will be $v_1R_2/(R_1 + R_2)$. The output voltage resulting from $v_1$ will therefore be given by $v_o = v_1R_2/(R_1 + R_2)(1 + R_2/R_1) = v_1(R_2/R_1)$. The net output voltage resulting from the simultaneous application of both $v_1$ and $v_2$ will produce an output voltage given by

$$v_o = \frac{R_2}{R_1}v_1 - \frac{R_2}{R_1}v_2 = \frac{R_2}{R_1}(v_1 - v_2) \tag{5.5}$$

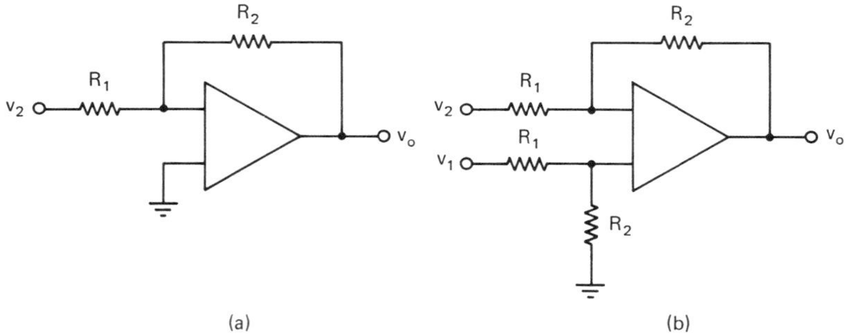


**Figure 5.10** (a) Inverting amplifier; (b) difference amplifier.

so that we see that the voltage gain accorded both $v_1$ and $v_2$ will now be equal in magnitude (but still opposite in algebraic sign). This amplifier configuration can be described as a *difference amplifier* since the output voltage is a function only of the *difference-mode* component of the input signals, and is insensitive to any *common-mode* signal component.

### 5.3.1 Superposition Theorem and Virtual Ground

The superposition theorem and the concept of virtual ground can both be very useful for the analysis of operational-amplifier circuits. Let us consider as an example the circuit of Figure 5.11. For simplicity, we will assume the case of an ideal, infinite-gain operational amplifier.

Making use of the superposition theorem, we will determine the output voltage resulting from each of the input voltages acting alone, the others being set equal to zero, and then obtain the algebraic summation of these separate contributions to the net output voltage. With $V_1$ acting alone, we have the situation shown in Figure 5.11b. Since $V_X - V_Y = V_O/A_{OL}$ goes to zero as $A_{OL}$ goes to infinity, and $V_Y = V_4 = 0$, then $V_X = 0$ and the inverting input terminal will be at ground potential. Since the inverting input terminal will be at ground potential, but there is no direct connection between this terminal and the circuit ground, this will be called a *virtual ground*. We have that $I_1 = (V_1 - V_X)/R_1 = V_1/R_1$ since $V_X = 0$.

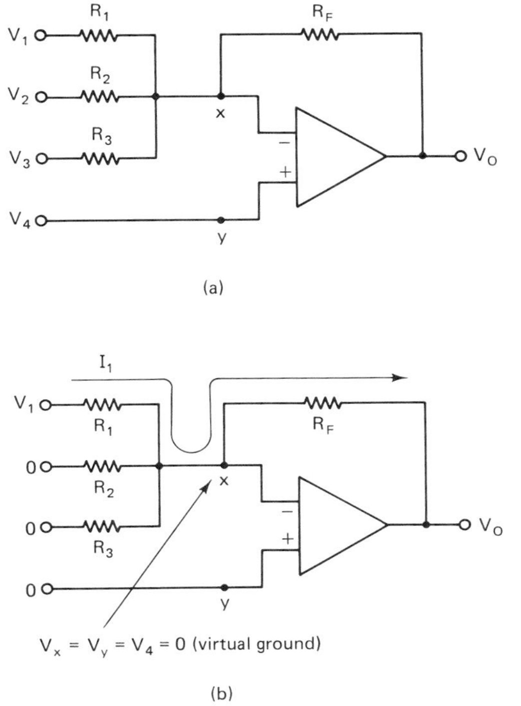


**Figure 5.11** Operational-amplifier circuit analysis example: (a) operational-amplifier circuit with four input voltages; (b) circuit with $V_1$ active and $V_2 = V_3 = V_4 = 0$.

For the case of an ideal operational amplifier, there will be no current flowing into or out of the input terminals, so that the current through $R_F$ will be equal to $I_1$. We therefore will have that $V_O = V_X - I_1R_F = -I_1R_F$, since again $V_X = 0$. Therefore, the output voltage due to $V_1$ acting alone will be $V_O = -I_1R_F = -V_1(R_F/R_1)$. Similarly, with $V_2$ acting alone we have that $V_O = -V_2(R_F/R_2)$, and for $V_3$ we have $V_O = -V_3(R_F/R_3)$.

With $V_4$ active and $V_1 = V_2 = V_3 = 0$ we see that $R_1$, $R_2$, and $R_3$ will all be connected in parallel between the amplifier inverting input terminal and ground. As a result, the closed-loop gain with respect to $V_4$ can be written as $V_O/V_4 = 1 + R_F/(R_1\|R_2\|R_3)$. The net output voltage resulting from the combined action of all four input voltages can now be expressed as

$$V_O = -V_1\frac{R_F}{R_1} - V_2\frac{R_F}{R_2} - V_3\frac{R_F}{R_3} + V_4\left(1 + \frac{R_F}{R_1\|R_2\|R_3}\right) \tag{5.6}$$

## 5.4 GAIN ERROR AND GAIN STABILITY

Looking at the equation for the output voltage for the case of the op amp with finite open-loop gain, it is apparent that the output voltage and therefore the closed-loop gain will be a function of the open-loop gain. It should also be apparent that as the open-loop gain gets very large (with respect to $1 + Z_2/Z_1$), the closed-loop gain will become less and less dependent on the open-loop gain, and will approach the value obtained for the "infinite gain approximation."

Considering, for example, the case in which $v_2 = 0$, the equation for $v_o$ will be

$$v_o = v_1\frac{1 + Z_2/Z_1}{1 + (1/A_{OL})(1 + Z_2/Z_1)}$$

so that the closed-loop gain will be

$$A_{CL} = \frac{v_o}{v_1} = \frac{1 + Z_2/Z_1}{1 + (1/A_{OL})(1 + Z_2/Z_1)}$$

As the open-loop gain $A_{OL}$ becomes very large and approaches infinity, the value of the closed-loop gain $A_{CL}$ will approach a limiting value to be denoted by $A_{CL}(\infty)$. The value of $A_{CL}(\infty)$ for the case under consideration will be $A_{CL}(\infty) = 1 + Z_2/Z_1$. Using this relationship, we can now rewrite the expression for $A_{CL}$ as

$$A_{CL} = \frac{A_{CL}(\infty)}{1 + A_{CL}(\infty)/A_{OL}}$$

From this equation we note that for small values of $A_{OL}$ such that $A_{OL} \ll A_{CL}(\infty)$, we will have that $A_{CL} \simeq A_{OL}$. When $A_{OL} = A_{CL}(\infty)$, then $A_{CL} = \frac{1}{2}A_{CL}(\infty)$. For large values of open-loop gain such that $A_{OL} \gg A_{CL}(\infty)$, which is the usual case of interest, $A_{CL}$ will approach $A_{CL}(\infty)$ and the equation for $A_{CL}$ can be written in approximate form as

$$A_{CL} = \frac{A_{CL}(\infty)}{1 + A_{CL}(\infty)/A_{OL}} \simeq A_{CL}(\infty)\left[1 - \frac{A_{CL}(\infty)}{A_{OL}}\right] \simeq A_{CL}(\infty)(1 - \epsilon) \tag{5.7}$$

where $\epsilon$ is the *gain error*. The gain error is defined as the *fractional change* in the closed-loop gain as the open-loop gain varies from some specified finite value up to infinity. The gain error can thus be expressed as

$$\text{gain error} = \epsilon = \frac{A_{CL}(\infty) - A_{CL}}{A_{CL}(\infty)} \tag{5.8}$$

From the equation given above for $A_{CL}$ it can readily be verified that $\epsilon \simeq A_{CL}(\infty)/A_{OL}$. We see that large values of $A_{OL}$ with respect to the closed-loop gain will lead to small values for the gain error.

**Example.** To consider an example, let us take the case of an op amp with $Z_2 = 99\ \text{k}\Omega$, $Z_1 = 1.0\ \text{k}\Omega$, and an open-loop gain of $A_{OL} = 10^5$ (min.). The gain error for this case will be approximately given by

$$\epsilon \simeq \frac{A_{CL}(\infty)}{A_{OL}} = 100/10^5 = 0.001 \text{ or } 0.10\%$$

We can therefore say that the maximum variation in the closed-loop gain as the open-loop gain varies from a minimum value of $10^5$ up to infinity will be only 0.10%. Thus the closed-loop gain is stabilized against changes in the open-loop gain, and a large change in the open-loop gain will produce only a very small variation in the closed-loop gain.

To look at this question of gain stability from a slightly different viewpoint, let us start off again with the expression for $A_{CL}$ as given by

$$A_{CL} = \frac{A_{CL}(\infty)}{1 + A_{CL}(\infty)/A_{OL}} \tag{5.9}$$

and take the derivative of $A_{CL}$ with respect to $A_{OL}$. Doing this gives

$$\frac{dA_{CL}}{dA_{OL}} = \frac{-A_{CL}(\infty)}{[1 + A_{CL}(\infty)/A_{OL}]^2}\frac{-A_{CL}(\infty)}{A_{OL}^2} = \frac{A_{CL}^2}{A_{OL}^2} \tag{5.10}$$

This can be rewritten in the form

$$\boxed{\frac{dA_{CL}}{A_{CL}} = \frac{A_{CL}}{A_{OL}}\frac{dA_{OL}}{A_{OL}}} \tag{5.11}$$

From this we see that the *fractional* (or percentage) change in the closed-loop gain, $dA_{CL}/A_{CL}$, that results from a fractional change in the open-loop gain, $dA_{OL}/A_{OL}$, will be equal to the fractional (or percentage) change in the open-loop gain multiplied by the factor $A_{CL}/A_{OL}$.

Going back now to the example just considered for which $A_{CL} = 100$ and $A_{OL} = 10^5$, we have that $dA_{CL}/A_{CL} = (100/10^5)(dA_{OL}/A_{OL}) = 0.001(dA_{OL}/A_{OL})$.

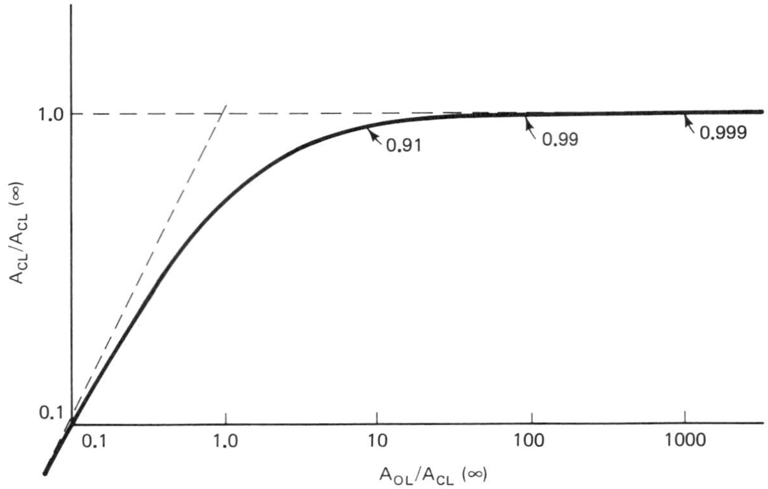


**Figure 5.12** Variation of closed-loop gain with open-loop gain.

Thus if the open-loop gain changes by 10%, the corresponding change in the closed-loop gain will be only 0.01%.

From the previous discussion we have seen that the gain error can be reduced and the gain stability improved by making the open-loop gain large compared to the closed-loop gain. Indeed, the factor by which the gain error or the gain stability is improved is the same factor by which the gain is sacrificed in going from open-loop conditions to closed-loop conditions.

In Figure 5.12 a doubly normalized graph of the closed-loop gain as a function of the open-loop gain is presented. Note that as $A_{OL}$ increases, $A_{CL}$ becomes increasingly *less* dependent on the open-loop gain. When $A_{OL} = 10\ A_{CL}(\infty)$ the closed-loop gain will be $0.91A_{CL}(\infty)$. For $A_{OL} = 100A_{CL}(\infty)$ the closed-loop gain will be $0.99A_{CL}(\infty)$, and for $A_{OL} = 1000A_{CL}(\infty)$ the closed-loop gain will be $0.999A_{CL}(\infty)$. We see therefore that as $A_{OL}$ increases, the closed-loop gain will asymptotically approach $A_{CL}(\infty)$, and that the rate of change of $A_{CL}$ with respect to $A_{OL}$ will decrease rapidly. At the other extreme, as $A_{OL}$ becomes small compared to $A_{CL}(\infty)$, we see that the closed-loop gain will asymptotically approach $A_{OL}$.

## 5.5 FREQUENCY RESPONSE

An op amp is a multistage electronic amplifier and the variation of the open-loop gain with frequency can generally be expressed in the form

$$A_{OL}(f) = \frac{A_{OL}(0)}{(1 + jf/f_1)(1 + jf/f_2)(1 + jf/f_3) \cdots} \tag{5.12}$$

where $A_{OL}(0)$ is the zero-frequency value of the open-loop gain and the breakpoint frequencies are in the sequence $f_1 < f_2 < f_3 < \cdots$. For most op amps the first breakpoint frequency, $f_1$, will be very small (~10 Hz) compared to $f_2$ (~1 to 3 MHz) and the other breakpoint frequencies. The frequency range that will be of greatest interest will usually be that for which $f^2 \gg f_1^2$ and $f^2 \ll f_2^2, f_3^2$, and so on. For this frequency range the expression for $A_{OL}$ can be written as $A_{OL} \simeq A_{OL}(0)/j(f/f_1) = A_{OL}(0)f_1/jf$. The frequency condition for this approximation will generally be satisfied for frequencies such that $f$ is at least one-half of a decade ($\sqrt{10}$:1 or approximately 3:1) away from $f_1$ and $f_2$ such that $3f_1 \lesssim f \lesssim f_2/3$.

The *unity-gain frequency* $f_u$ is defined as the frequency at which the open-loop gain as given in the approximate expression above has dropped to unity. Therefore, we have that at $f = f_u$, $A_{OL}(0)f_1/f_u = 1$, so that

$$\boxed{f_u = A_{OL}(0)f_1} \tag{5.13}$$

The open-loop gain can therefore be expressed in terms of the unity-gain frequency as

$$\boxed{A_{OL} \simeq \frac{f_u}{jf}} \tag{5.14}$$

In Figure 5.13 the open-loop gain versus frequency characteristics (Bode plot) are shown. The frequency response of the op amp is flat down to zero frequency (d-c). At $f = f_1$ (the first breakpoint frequency), the open-loop gain will be 3 dB down from the zero-frequency value of $A_{OL}(0)$, so that the open-loop 3-dB bandwidth will be $BW_{OL} = f_1$.

The closed-loop gain for the simple noninverting amplifier case considered before is given by

$$A_{CL} = \frac{1 + R_2/R_1}{1 + (1/A_{OL})(1 + R_2/R_1)} \tag{5.15}$$

Assuming that the zero-frequency open-loop gain $A_{OL}(0)$ is large enough, which will virtually always be the case, the zero-frequency value of the closed-loop gain $A_{CL}(0)$ will be given by $A_{CL}(0) = 1 + R_2/R_1$. As a result, the closed-loop gain as a function of frequency can now be expressed as

$$A_{CL}(f) = \frac{A_{CL}(0)}{1 + A_{CL}(0)/A_{OL}} \tag{5.16}$$

If we now introduce the approximate expression for $A_{OL} = f_u/jf$, we obtain

$$A_{CL}(f) = \frac{A_{CL}(0)}{1 + jfA_{CL}(0)/f_u} \tag{5.17}$$

We see that when $f = 0$, the closed-loop gain is indeed $A_{CL}(0)$ and as $f$ increases, $A_{CL}(f)$ decreases monotonically, as expected.

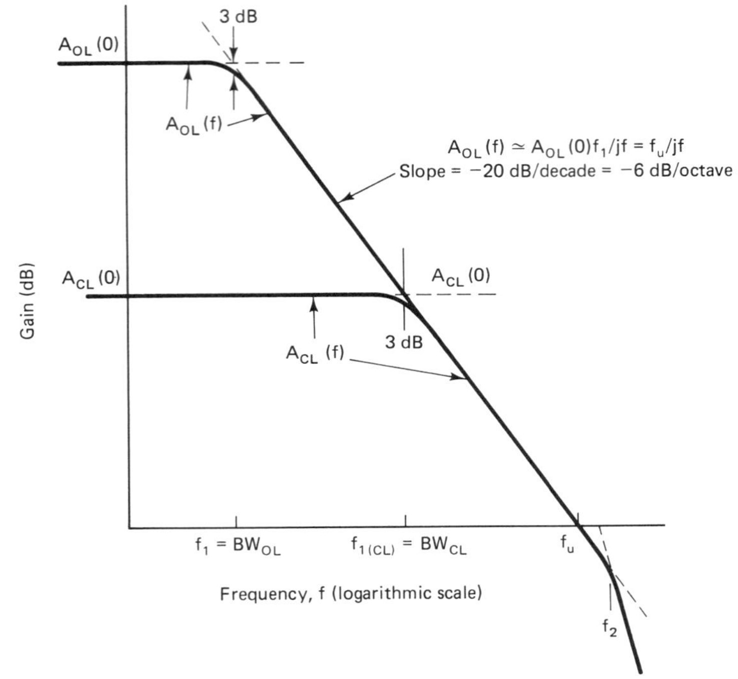


**Figure 5.13** Operational-amplifier gain versus frequency characteristics (Bode plot).

When $fA_{CL}(0)/f_u = 1$, the expression for $A_{CL}(f)$ becomes

$$A_{CL}(f) = \frac{A_{CL}(0)}{1 + j1} = \frac{A_{CL}(0)}{\sqrt{2}\,\angle -45^\circ} \tag{5.18}$$

Thus $A_{CL}$ will be down by a factor of $\sqrt{2}$ or 3 dB from the zero frequency value at the frequency given by

$$\boxed{f = \frac{f_u}{A_{CL}(0)}} \tag{5.19}$$

so that this corresponds to the *half-power* or *3-dB frequency* for the closed-loop gain and will be denoted as $f_{1(CL)}$.

The expression above for the variation of closed-loop gain with frequency can be rewritten as

$$A_{CL}(f) = \frac{A_{CL}(0)}{1 + jA_{CL}(0)/|A_{OL}|} \tag{5.20}$$

The *magnitude* of the closed-loop gain will be

$$A_{CL}(f) = \frac{A_{CL}(0)}{\sqrt{1 + [A_{CL}(0)/A_{OL}]^2}} \tag{5.21}$$

From this last equation we see that at the frequency at which the open-loop gain has decreased to the point where it is equal to $A_{CL}(0)$, the closed-loop gain will be down by a factor of $\sqrt{2}$ or 3 dB below the zero-frequency value of $A_{CL}(0)$. This frequency will therefore correspond to the closed-loop half-power or 3-dB frequency. Looking at the Bode plot of Figure 5.13, we see that the closed-loop 3 dB frequency $f_{1(CL)}$ will thus occur at the point of intersection of the open-loop gain curve, $A_{OL}(f)$, with the horizontal line corresponding to $A_{CL}(0)$.

The 3-dB bandwidth of a system is the range of frequencies over which the gain remains within 3 dB of the maximum value. Since op amps are direct-coupled or d-c amplifiers with no coupling capacitors between stages and no bypass capacitors, the open-loop frequency response will remain flat down to zero frequency (d-c). The range of frequencies over which the gain stays within 3 dB of the maximum value therefore will extend from 0 up to the 3 dB (or half-power) frequency. Therefore, the 3-dB bandwidth will be equal to the 3-dB frequency $f_1$ and thus $BW_{OL} = f_1$, where $BW_{OL}$ is the *open-loop 3-dB bandwidth*.

Since the open-loop gain remains flat down to zero frequency, the closed-loop gain will similarly be flat down to zero frequency. Therefore, the *closed-loop 3-dB bandwidth*, $BW_{CL}$, will be equal to the closed-loop 3-dB frequency, $f_{1(CL)}$, so that $BW_{CL} = f_{1(CL)}$. Since $BW_{CL} = f_{1(CL)} = f_u/A_{CL(0)}$, the closed-loop gain–bandwidth product will be equal to the unity-gain frequency as given by

$$\boxed{A_{CL}(0) \times BW_{CL} = f_u} \tag{5.22}$$

For $f_u$ we have also had the relationship that

$$\boxed{f_u = A_{OL}(0) f_1 = A_{OL}(0) \times BW_{OL}} \tag{5.23}$$

so that we see that the open-loop gain–bandwidth product will also be equal to the unity-gain frequency. For either gain–bandwidth product case given above, the gain is the zero-frequency (d-c) gain and the bandwidth is the 3-dB bandwidth. Thus we see that for both the closed-loop and open-loop situations the gain–bandwidth product will be a constant and equal to the unity-gain frequency.

From the gain–bandwidth product relationship, we see that by sacrificing gain in going from the open-loop situation to the closed-loop case there will be a corresponding improvement in the bandwidth. Indeed, by whatever factor the gain is sacrificed by using negative feedback, the bandwidth will be improved by the same factor.

If we take the ratio of the closed-loop bandwidth to the open-loop bandwidth, we obtain

$$\frac{BW_{CL}}{BW_{OL}} = \frac{A_{OL}(0)}{A_{CL}(0)} \tag{5.24}$$

We thus see that the factor by which the bandwidth is improved in going from open-loop conditions to closed-loop conditions is exactly the factor by which the gain is sacrificed in going from the open-loop gain to the closed-loop gain.

To consider some simple examples of bandwidth improvement, let us assume a representative value for $f_u$ of 1.0 MHz. For a closed-loop gain of $A_{CL}(0) = 1000$ (or 60 dB) the closed-loop bandwidth will be 1.0 MHz/1000 = 1.0 kHz. For $A_{CL}(0) = 100$ the bandwidth will be 10 kHz, and for $A_{CL}(0) = 10$ it will increase to 100 kHz. Finally, for the unity-gain voltage follower case the bandwidth will be equal to $f_u$ or 1.0 MHz. In contrast to these bandwidth values, if the open-loop gain for $A_{CL}(0)$ is $10^5$ (or 100 dB) the open-loop bandwidth will be only 10 Hz!

For the bandwidth relationships above it is to be understood that $f_u$ is the *extrapolated unity-gain frequency* as obtained from the extrapolation of the $A_{OL} = f_u/jf$ response line down to the 0-dB (unity-gain) level. In most cases this will correspond to the actual frequency at which the gain is unity. However, in the cases in which $f_2$ is less than $f_u$, the value of $f_u$ can be substantially greater than the value of frequency at which the gain has dropped to unity.

For the bandwidth relationships given above to be valid, it is necessary that the second op-amp breakpoint frequency $f_2$ be substantially above the closed-loop bandwidth $BW_{CL}$. If this is not the case, the closed-loop bandwidth can be considerably less than that predicted by the equations above.

For frequencies that are substantially above $f_{1(CL)} = BW_{CL}$, such that the open-loop gain has dropped considerably below $A_{CL}(0)$, we see from the equation for $A_{CL}(f)$ that the $A_{CL}(f)$ curve will asymptotically approach the $A_{OL}$ curve as shown in the Bode plot. The closed-loop gain starts off at $A_{CL}(0)$ at low frequencies, is down 3 dB at $f_{1(CL)} = BW_{CL} = f_u/A_{CL}(0)$, and then will asymptotically approach the $A_{OL}$ curve, dropping at a rate of 20 dB/decade.

## 5.6 TIME-DOMAIN RESPONSE

To evaluate the frequency response of a system a sinusoidal excitation or signal is applied to the system and the response or output of the system is measured as a function of the frequency of the signal source. A graph of the input-to-output gain or transfer ratio is then usually drawn. For the *time-domain* or *transient* response a step-function excitation is applied and the graph or display of the output response as a function of time is obtained. The time-domain response of most systems is usually most conveniently described by the *rise time*, which is the time required for the output response to rise from a specified lower value to some other specified upper value. The most commonly used lower and upper values are 10% and 90% of the maximum or ultimate response level, respectively. Unless otherwise indicated or implied, the rise time, $t_{\text{rise}}$, will be understood to be the 10 to 90% rise time. This is illustrated in Figure 5.14 for the case of a simple system that is characterized by a single time constant.

Although a step-function excitation and the corresponding response is the simplest to evaluate mathematically, from a practical standpoint the most convenient type of excitation is often a repetitive rectangular pulse waveform of suitable pulse duration and pulse period. This type of signal can be obtained from a pulse generator or square-wave generator and the input and output waveforms can be viewed on an oscilloscope.

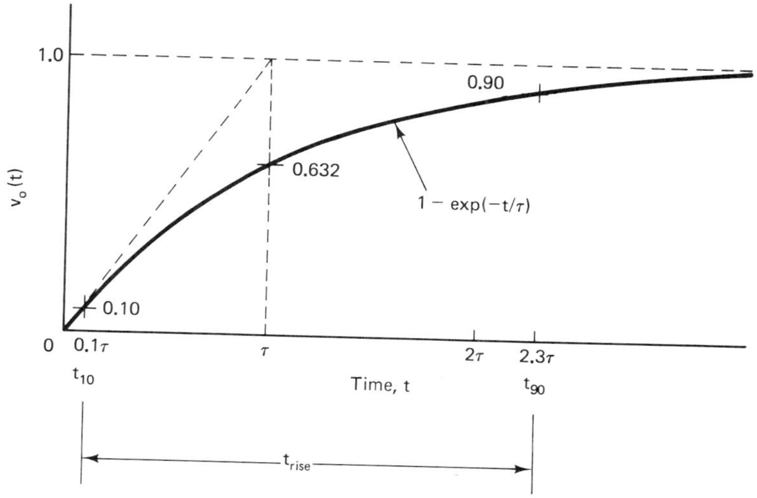


**Figure 5.14** Time-domain response characteristics (normalized).

For a system that has a gain versus frequency response characterized by an equation of the form

$$A(f) = \frac{A(0)}{1 + jf/f_1} \tag{5.25}$$

where $f_1$ is the half-power frequency or 3-dB bandwidth of the system, the corresponding response of the system to a unit-step-function excitation will be given by $v_o(t) = A(0)[1 - \exp(-t/\tau)]$, where $v_o(t)$ is the system response as a function of time and $\tau$ is the system time constant. The time constant of the system is related to the 3-dB frequency by $\tau = 1/\omega_1 = 1/(2\pi f_1)$.

For the 10 to 90% rise time we have that at the 10% response point $0.1 = 1 - \exp(-t_{10}/\tau)$, so that $t_{10} = 0.105\tau$. Similarly, at the 90% response point $0.9 = 1 - \exp(-t_{90}/\tau)$, so that $t_{90} = 2.303\tau$. We therefore will have that $t_{rise} = t_{90} - t_{10} = 2.2\tau$. Since $\tau = 1/(2\pi f_1)$, we can express the rise time in terms of the bandwidth as $t_{rise} = 2.2\tau = 2.2/(2\pi f_1) = 0.35/f_1 = 0.35/BW$. Thus the rise time can be obtained from the 3-dB bandwidth, and vice versa. This relationship between rise time and bandwidth can most conveniently be described in terms of the rise-time bandwidth product as given by

$$\boxed{t_{rise} \times BW = 0.35} \tag{5.26}$$

The closed-loop frequency response characteristics for most op-amp systems can adequately be described by an equation of the type given above with a single breakpoint frequency $f_1$, which will be the closed-loop bandwidth of the system. As a result, for these cases the simple rise time–bandwidth product relationship given above can be used. For example, if the closed-loop bandwidth is 10 kHz, the corresponding rise time will be $t_{\text{rise}} = 0.35/10 \text{ kHz} = 35\ \mu\text{s}$.

The relationship between the frequency response equation and the time-domain response equation will now be derived. If a system has a frequency response given by $A(\omega) = A(0)/(1 + j\omega/\omega_1)$, the corresponding expression in terms of the Laplace transform variable $s$ will be $A(s) = A(0)/(1 + s/\omega_1)$. The response of the system to a unit-step-function excitation $v_i(s) = 1/s$ will be

$$v_o(s) = A(s)v_i(s) = \frac{A(0)}{s(1 + s/\omega_1)} \tag{5.27}$$

In order to obtain the response in the time domain, we will now take the inverse Laplace transform of the foregoing expression for $v_o(s)$ to obtain $v_o(t)$ as given by

$$v_o(t) = A(0)[1 - \exp(\omega_1 t)] = A(0)\left[1 - \exp\left(\frac{-t}{\tau}\right)\right] \tag{5.28}$$

where $\tau = 1/\omega_1 = 1/2\pi f_1)$, thus demonstrating the relationship between the time constant and the 3-dB frequency $f_1$.

## 5.7 OFFSET VOLTAGE

For the ideal op amp we have that $v_o = A_{OL}v_i$, so that $v_o = 0$ when $v_i = 0$. In any real op amp, however, there are various component mismatches and other circuit imbalances within the op amp so that with zero input voltage ($v_i = 0$) the output voltage will not be zero. To produce a condition of zero output voltage, a small input voltage must be applied, this input voltage being equal to the *input offset voltage*, $V_{OS}$. The transfer function for the real op amp must now be expressed as $v_o = A_{OL}(v_i - V_{OS})$.

The offset voltage $V_{OS}$ is a small d-c voltage generally in the range of about 1 mV, although some op amps may have maximum offset voltage specifications as large as 5 or 10 mV. Other "precision op amps" may have maximum $V_{OS}$ specifications as low as in the range 10 to 100 $\mu$V. If we consider a large number of op amps of the same type, the distribution of offset voltage values will be a symmetrical Gaussian or "bell-shaped" curve with a mean value of zero and truncated at the maximum $V_{OS}$ values specified by the manufacturer. An individual op amp is thus just as likely to have a $V_{OS}$ of one polarity as it is to have one of the opposite polarity. The specification sheet for the op amp will list a guaranteed maximum value for the offset voltage under certain specified conditions, and sometimes a typical value will also be listed.

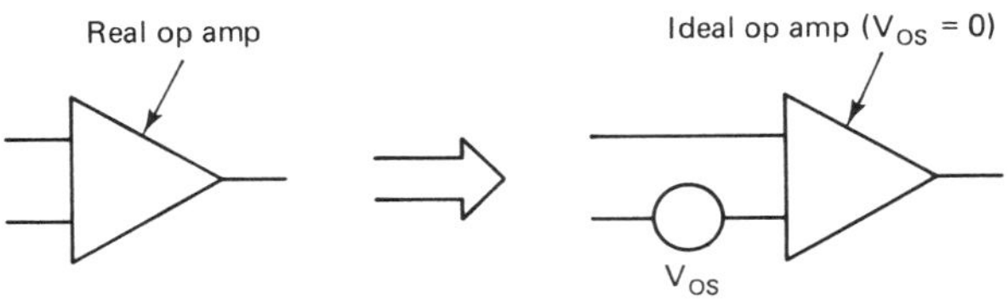


**Figure 5.15** Offset voltage representation.

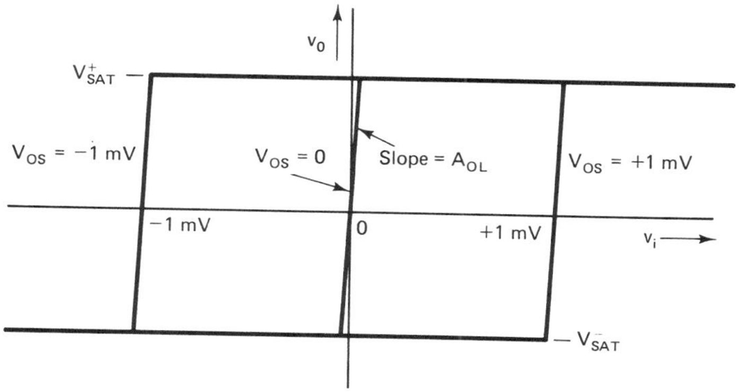


**Figure 5.16** Open-loop transfer characteristics showing effect of $V_{OS}$.

For the purposes of analysis the effect of the offset voltage for a real op amp can be considered by replacing the real op amp by an "ideal" op am (i.e., $V_{OS} = 0$) and by placing a d-c voltage source of strength $V_{OS}$ in series with one of the input terminals of the op amp, as shown in Figure 5.15. It is usually most convenient to insert the $V_{OS}$ "source" in series with the noninverting input terminal.

In Figure 5.16 the open-loop input–output transfer characteristics are presented for three different cases of $V_{OS}$. The magnitude of $V_{OS}$ will usually be such that under open-loop conditions the output voltage of the op amp will be saturated in either the high ($V_{SAT}^+$) or low ($V_{SAT}^-$) states, even when no input voltage is applied. In Figure 5.17 the transfer characteristics under closed-loop conditions are presented. As a result of the greatly expanded input voltage range, the output voltage can be kept well away from saturation. Thus again we see the advantage of closed-loop operation of the op amp.

To illustrate the effect of the offset voltage on the output voltage for the closed-loop op-amp system, let us consider the simple circuit of Figure 5.18. To determine the effect of $V_{OS}$ we will now consider the op amp to be ideal and insert a voltage source of strength $V_{OS}$ in series with the noninverting input terminal. The resulting output voltage will be $v_o = V_{OS}A_{CL} = V_{OS}(1 + R_2/R_1)$. In general we can say

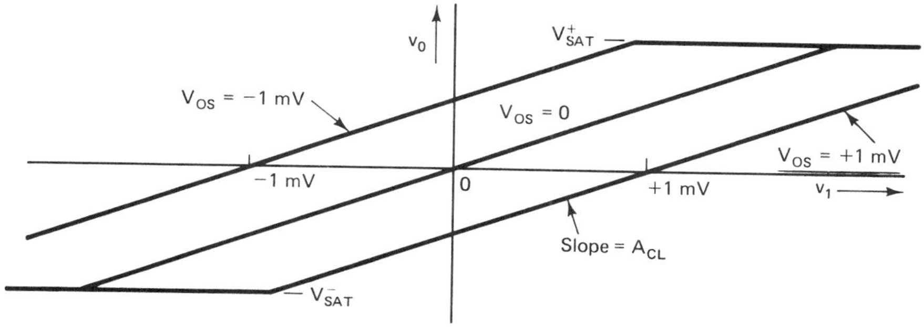


**Figure 5.17** Closed-loop transfer characteristics showing effect of $V_{OS}$.

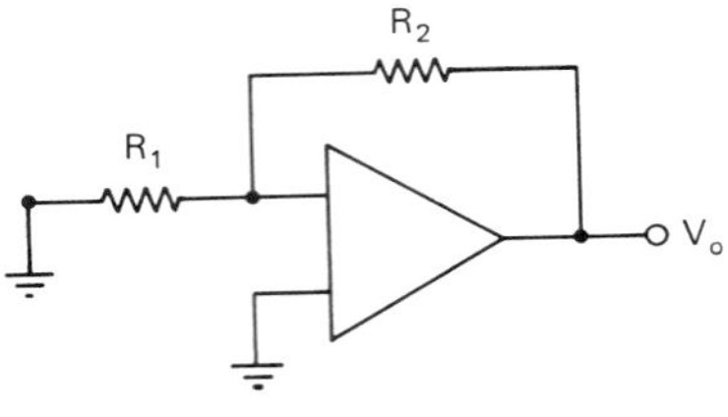


**Figure 5.18** Circuit for calculation of $V_o$ due to $V_{OS}$.

that the output voltage component that results from the input offset voltage will be equal to the offset voltage $V_{OS}$ multiplied by the closed-loop gain $A_{CL}(0)$, taken with respect to the noninverting input terminal.

### 5.7.1 Offset Voltage Compensation

For many applications, especially those in which the input signal level is large compared to the offset voltage $V_{OS}$, the effect of $V_{OS}$ will present no serious problem. Also, precision op amps are available with maximum input offset voltage values down in the range 10 to 100 $\mu$V. Furthermore, since $V_{OS}$ is a d-c voltage, in many cases

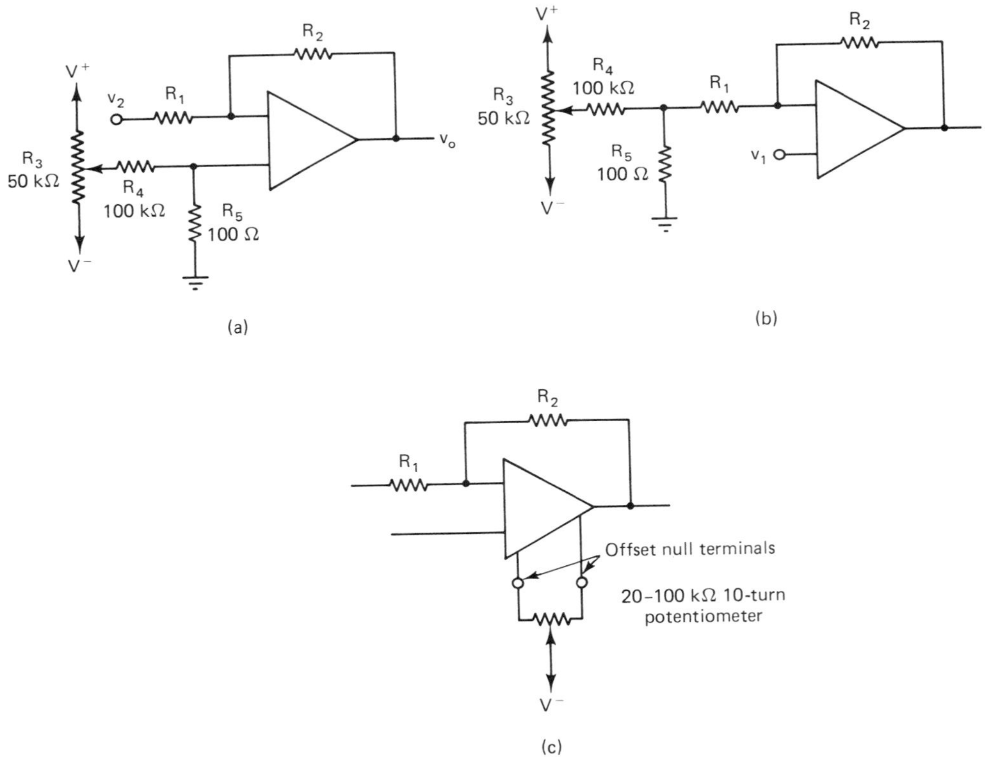


**Figure 5.19** (a) and (b) Circuits for offset voltage compensation; (c) op amp with offset null terminals.

the effect of $V_{OS}$ can be removed by the use of capacitative or other coupling such that only the a-c component of the output voltage is obtained. Nevertheless, there are situations where it is desirable to compensate for or cancel out the offset voltage. We will now examine some methods for offset voltage compensation.

In Figure 5.19 a simple means of $V_{OS}$ compensation is shown. By the use of potentiometer $R_3$ and the resistive voltage divider $R_4$ and $R_5$ a suitably small and adjustable voltage of either polarity can be applied to the noninverting input terminal of the op amp to cancel out $V_{OS}$. With the circuit as shown, the range of cancellation voltages will extend from $-15$ to $+15$ mV, which should prove to be a sufficient voltage range to cover almost all $V_{OS}$ situations. Indeed, it may be desirable to reduce this voltage range by making $R_5$ smaller, such as 50 Ω or even 20 Ω, so as to make the offset voltage compensation adjustment more precise. In Figure 5.19b a similar compensation circuit is shown, but this time the compensation voltage is applied to the inverting input terminal.

Some op amps provide specific offset voltage compensation (or "nulling") terminals as shown in Figure 5.19c. A potentiometer, generally in the range of 20 to 100 kΩ, is connected between the op-amp $V_{OS}$ compensation (or "$V_{OS}$ null") terminals. The wiper arm of the potentiometer is connected to the negative supply, $V^-$.

For any of the compensation techniques described, the compensation is usually carried out with no signal applied to the input terminals and a sensitive d-c voltmeter connected to the output. The potentiometer is then adjusted to the point where the output voltage becomes zero (i.e., is "nulled out").

### 5.7.2 Offset Voltage Temperature Coefficient

The input offset voltage will be a function of device temperature. The rate of change of the offset voltage with temperature is denoted by

$$\text{temperature coefficient of } V_{OS} = TC_{V_{OS}} = \frac{dV_{OS}}{dT} \tag{5.29}$$

For op amps with bipolar transistor input stages the $TC_{V_{OS}}$ will be related to the offset voltage itself by the following approximate relationship: $TC_{V_{OS}} \simeq V_{OS}/T$, where $T$ is the absolute temperature (~300 K in most cases). Thus, for $V_{OS} = 1$ mV, a temperature coefficient of about $TC_{V_{OS}} = 1$ mV/300 K $= 1000$ μV/300 K $= 3$ μV/K $= 3$ μV/°C would be expected. Thus a temperature change of 10°C will result in a drift of the input offset voltage of about 30 μV.

As a result of the variation of the input offset voltage with temperature, the $V_{OS}$ compensation techniques described earlier will be completely effective at only one temperature. As the temperature of the device drifts in either direction from the temperature at which the compensation was initially made, the effect of the offset voltage will again be felt. Nevertheless, the net effect of the offset voltage will generally be substantially reduced by the compensation techniques. For example, if the offeset voltage is 1 mV and is nulled out by a compensation circuit at a given temperature, a subsequent temperature drift of 10°C may result in an effective uncompensated offset voltage of around 30 μV. This is, however, some 30 times smaller than the initial $V_{OS}$, so that the offset voltage compensation is indeed worthwhile.

## 5.8 INPUT BIAS CURRENT

For any operational amplifier to operate properly, a current must be allowed to flow into (or out of) the two input terminals. This current is the *input bias current* and is shown in Figure 5.20. The two currents $I_{B_1}$ and $I_{B_2}$ in Figure 5.20 will not be exactly equal to each other, but will generally be within about 10% of each other. The *input bias current* $I_{\mathrm{BIAS}}$ or $I_B$ by strict definition is the average of the two input currents as given by

$$\boxed{I_B = \frac{I_{B_1} + I_{B_2}}{2}} \tag{5.30}$$

The difference between these two currents is called the *input offset current* $I_{OS}$, given by

$$\boxed{I_{OS} = I_{B_1} - I_{B_2}} \tag{5.31}$$

The algebraic sign of the offset current is usually of no importance, with equal probabilities for either algebraic sign.

For some op amps $I_B$ will be of positive polarity (i.e., flowing into the op amp), whereas for other op amps it will be of negative polarity (flowing out of the op amp). For op amps that use bipolar transistor input stages, the input bias current can range from 10 $\mu$A down to as little as a few nanoamperes. For op amps with field-effect transistor input stages, the input bias current will be very small, even down to the range of a few picoamperes for some op amps. On the manufacturer's specification sheet for a particular op amp the usual specification for the bias current will be the guaranteed maximum value for $I_B$ under certain specified operating conditions. Sometimes a typical value for $I_B$ is also indicated. Similarly, for the input offset current $I_{OS}$ a guaranteed maximum value is specified, with a typical value sometimes being given as well.

To determine the effect of the input bias current on the output voltage of an op amp, let us consider the basic circuit of Figure 5.21. We will evaluate $V_o$ using

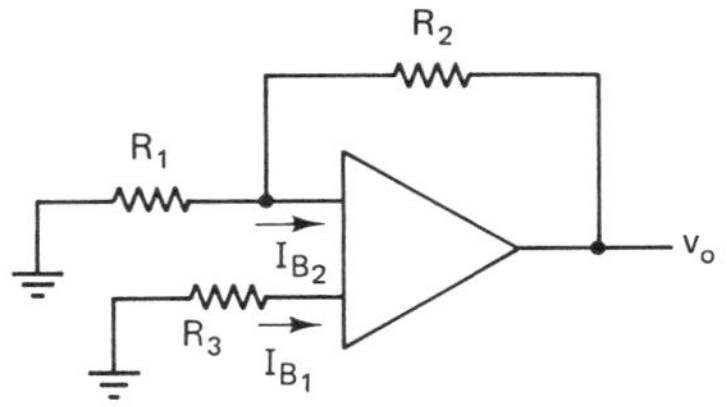


Figure 5.20 Input bias current.

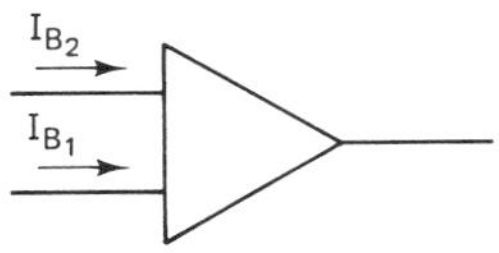


Figure 5.21 Circuit for calculation of effects of input bias current.

the superposition theorem, in that we will consider the effect on $V_o$ of each current, $I_{B_1}$ and $I_{B_2}$, acting separately and then taking the algebraic summation of the two results to obtain the net output voltage $V_o$. The current $I_{B_1}$ in flowing through $R_3$ will produce a voltage drop $-I_{B_1}R_3$ that will appear at the noninverting input terminal of the op amp. This voltage will be multiplied by the closed-loop gain of $1 + R_2/R_1$ to appear at the output as a voltage $V_o = (-I_{B_1}R_3)(1 + R_2/R_1)$. The component of output voltage that results from $I_{B_2}$ acting alone (i.e., $I_{B_1}$ is now considered to be zero) can be determined by first noting that the voltage at the noninverting input terminal will now be zero, and hence the inverting input terminal will also be essentially at ground potential (i.e., a "virtual ground"). The voltage across $R_1$ will therefore be essentially zero, so that there will be no current through $R_1$. All of current $I_{B_2}$ will therefore be flowing through $R_2$, producing a voltage drop $I_{B_2}R_2$. Since the voltage at the inverting input terminal is essentially zero and the voltage drop across $R_2$ is $I_{B_2}R_2$, the output voltage due to $I_{B_2}$ will be $V_o = I_{B_2}R_2$. The *net output voltage* due to both $I_{B_1}$ and $I_{B_2}$ will therefore be

$$\boxed{V_o = I_{B_2}R_2 - I_{B_1}R_3\left(1 + \frac{R_2}{R_1}\right)} \tag{5.32}$$

Since $I_{B_1}$ and $I_{B_2}$ will generally be close in value, we see that if the coefficients of the $I_{B_1}$ and $I_{B_2}$ terms are equal, there will be near cancellation of the effects of the bias current. For this to happen we will require that $R_2 = R_3(1 + R_2/R_1)$, so that $R_1R_2/(R_1 + R_2) = R_3$. Since $R_1R_2/(R_1 + R_2)$ is the parallel combination of $R_1$ and $R_2$, we see that this requirement means that $R_3$ should be equal to the parallel combination of $R_1$ and $R_2$. Under these conditions the output voltage due to the bias currents will be given by

$$\boxed{V_o = R_2(I_{B_2} - I_{B_1}) = -R_2I_{OS}} \tag{5.33}$$

Since the offset current is usually a small fraction of the bias current, this can represent a substantial improvement, so it is very often desirable to make $R_3$ equal to the parallel combination of $R_1$ and $R_2$.

The bias current will usually pose a problem only for very large values of $R_1$, $R_2$, and $R_3$. If this is the case, the situation can most easily be remedied by using an op amp with a very low value of input bias current, such as various types of FET-input op amps. Such op amps are available with $I_B$ values down as low as 10 pA. It is also possible to introduce various compensation circuits, including some similar to the offset voltage compensation circuits.

One basic problem with any bias current compensation circuit is that the bias current will be a function of temperature. The rate of change of bias current with temperature is the *temperature coefficient of the bias current* as given by $TC_{I_B} = dI_B/dT$. Similarly, the offset current will also exhibit a temperature coefficient as expressed by $TC_{I_{OS}} = dI_{OS}/dT$.

## 5.9 COMMON-MODE GAIN

An op amp is ideally sensitive only to the difference-mode signal $v_i$ applied to its input terminals, and is completely unresponsive to any common-mode input voltage. In the ideal op-amp case we have that $v_o = A_{OL}v_i$, where $v_i$ is the difference-mode input voltage, $v_i = v_x - v_y$, as shown in Figure 5.22. The open-loop gain is the same as the difference-mode gain $A_{DM}$, since both are taken with respect to the difference-mode signal. Any real op amp, however, will exhibit some small response to the common-mode component of the input voltages as given by $v_{cm} = (v_x + v_y)/2$. The output voltage resulting from this common-mode input voltage will be given by $v_o = A_{CM}v_{cm}$, where $A_{CM}$ is the common-mode voltage gain. The common-mode gain will, however, be very much smaller than the difference-mode gain. The ratio of the difference-mode gain, $A_{DM}$ or $A_{OL}$ to the common-mode gain, $A_{CM}$, is the *common-mode rejection ratio* (CMRR) and is usually expressed in decibels (dB). CMRR values in the range of 80 dB ($10^4$) to 120 dB ($10^6$) are typical for op amps.

To evaluate the effect of the common-mode gain in a closed-loop op-amp situation, let us consider the circuit of Figure 5.23. For the infinite open-loop gain ($A_{OL}$) case we have seen that this is a simple difference amplifier with $v_o = (R_2/R_1)(v_1 - v_2)$, so that this circuit will be sensitive only to the difference-mode component of the input signal, and will be completely insensitive to any common-mode component. We will now analyze this circuit for the case in which both the open-loop difference-mode gain $A_{DM}$ (or $A_{OL}$) and the common-mode gain $A_{CM}$ are finite (i.e., nonzero) in value.

From the circuit of Figure 5.23 we will have the following equations:

$$(1)\ v_o = A_{DM}(v_x - v_y) + \frac{A_{CM}(v_x + v_y)}{2} \tag{5.34}$$

$$(2)\ v_x = \frac{R_2 v_1}{(R_1 + R_2)}$$

$$(3)\ v_y = \frac{v_2 R_2}{R_1 + R_2} + \frac{v_o R_1}{R_1 + R_2} \tag{5.35}$$

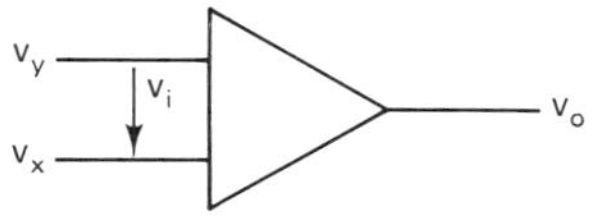


**Figure 5.22** Difference-mode input voltage.

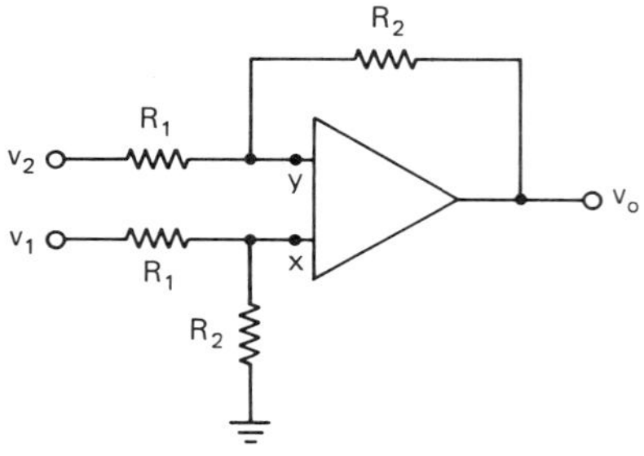


**Figure 5.23** Difference-amplifier circuit for the analysis of the effect of the common-mode gain.

Upon substituting equations (2) and (3) into equation (1), we obtain

$$v_o = A_{DM}\left[\frac{(v_1 - v_2)R_2}{R_1 + R_2} - \frac{v_o R_1}{R_1 + R_2}\right] + A_{CM}\left[\frac{v_1 + v_2}{2}\frac{R_2}{R_1 + R_2} + \frac{v_o R_1}{2(R_1 + R_2)}\right] \tag{5.36}$$

Collecting terms in $v_o$ on the left-hand side gives

$$v_o\left[1 + \frac{A_{DM}R_1}{R_1 + R_2} - \frac{A_{CM}R_1}{2(R_1 + R_2)}\right] = A_{DM}(v_1 - v_2)\frac{R_2}{R_1 + R_2} + A_{CM}\frac{v_1 + v_2}{2}\frac{R_2}{R_1 + R_2} \tag{5.37}$$

Multiplying both sides by the factor $(R_1 + R_2)/R_1$ gives

$$v_o\left(\frac{R_1 + R_2}{R_1} + A_{DM} + \frac{A_{CM}}{2}\right) = A_{DM}(v_1 - v_2)\frac{R_2}{R_1} + A_{CM}\frac{v_1 + v_2}{2}\frac{R_2}{R_1} \tag{5.38}$$

Solving for $v_o$ now yields

$$v_o = \frac{(R_2/R_1)[A_{DM}(v_1 - v_2) + A_{CM}(v_1 + v_2)/2]}{A_{DM} + (R_1 + R_2)/R_1 - A_{CM}/2} \tag{5.39}$$

Now dividing through numerator and denominator by $A_{DM}$ gives

$$v_o = \frac{(R_2/R_1)[(v_1 - v_2) + (A_{CM}/A_{DM})(v_1 + v_2)/2]}{1 + [(R_1 + R_2)/A_{DM}R_1] - A_{CM}/2A_{DM}} \tag{5.40}$$

Note that if $A_{OL}$ (i.e., $A_{DM}$) goes to infinity (the "infinite-gain case") this equation for $v_o$ reduces to simply $v_o = (R_2/R_1)(v_1 - v_2)$, as obtained previously. For the usual case of a very large open-loop gain such that $A_{OL} = A_{DM} \gg (R_1 + R_2)/R_1$ and $A_{DM} \gg A_{CM}$, the equation for $v_o$ can be expressed approximately as

$$v_o = \frac{R_2}{R_1}\left[(v_1 - v_2) + \frac{(A_{CM}/A_{DM})(v_1 + v_2)}{2}\right] \tag{5.41}$$

We see from this result that the common-mode component of the input voltage will have some influence on the output, but the effect will be reduced by the factor of $A_{DM}/A_{CM}$ (i.e., the common-mode rejection ratio) with respect to the difference-mode component. For most cases the CMRR will be such that the effect of the common-mode signal component will be very small indeed. For example, for a CMRR of 100 dB the gain accorded the common-mode signal component will be 100,000 times smaller than the gain received by the difference-mode signal.

## 5.10 INPUT IMPEDANCE

The ideal op amp has an infinite input impedance $Z_i$ such that the op amp can be driven by a signal source of any impedance level without any signal loss due to a high value of the source impedance. Any real op amp has a finite input impedance $Z_i$, which in some cases can seriously affect the operation of the circuit.

For the analysis of the effect of $Z_i$ the equivalent-circuit representation of Figure 5.24a will be used. The input impedance $Z_i$ can be represented as a combination of a *difference-mode input impedance* $Z_{DM}$, and a *common-mode input impedance* $Z_{CM}$, both of which are dynamic or a-c impedances.

We will evaluate the effect of the difference-mode impedance $Z_{DM}$ under closed-loop conditions with reference to the circuit of Figure 5.24b. The a-c current $i_s$ flowing into the noninverting input terminal will be given by

$$i_s = \frac{v_i}{Z_{DM}} = \frac{v_o/A_{OL}}{Z_{DM}} = \frac{v_s A_{CL}/A_{OL}}{Z_{DM}}$$

Solving this for the *closed-loop input impedance* $Z_{i(CL)}$, we have

$$\boxed{Z_{i(CL)} = \frac{v_s}{i_s} = \frac{A_{OL}}{A_{CL}} Z_{DM}} \tag{5.42}$$

Thus the closed-loop input impedance resulting from $Z_{DM}$ is increased by a factor of $(A_{OL}/A_{CL})$ over the open-loop value. This can represent a very substantial improvement. We note again in this case as well that the factor of improvement is exactly that factor by which the gain is sacrificed in going from open-loop conditions to closed-loop conditions.

The open-loop value of $Z_{DM}$ for op amps will generally be in the range of from 100 kΩ up to as large as many gigaohms ($10^9$ Ω) for the case of FET-input op amps. These large $Z_{DM}$ values when coupled with the very large $A_{OL}/A_{CL}$ ratios available will result in extremely large values for $Z_{i(CL)}$ such that the difference-mode input impedance will very rarely produce any significant loading effect on the signal source, and can therefore be neglected from further consideration.

We will now consider the effect of the common-mode input impedance $Z_{CM}$, and we will make use of Figure 5.24a. For large values of the open-loop gain the voltage $v_i$ across $Z_{DM}$ will be very small, such that the current through $Z_{DM}$ will be negligible. Therefore, the signal current $i_s$ will flow principally through $Z_{CM}$ and will be given by $i_s = v_s/Z_{CM}$. The closed-loop input impedance will therefore be given by $Z_{i(CL)} = v_s/i_s = Z_{CM}$.

The value of $Z_{CM}$ for op amps will generally be in the range of several megohms up to 100 GΩ ($10^{11}$ Ω) for FET op amps in parallel with a capacitance of the order of 3 to 10 pF.

The net closed-loop input impedance considering the effects of both $Z_{DM}$ and $Z_{CM}$ will be the *parallel combination* of $Z_{CM}$ and $(A_{OL}/A_{CL})Z_{DM}$. Although $Z_{CM}$ is much larger than $Z_{DM}$, the $A_{OL}/A_{CL}$ factor will usually make the $(A_{OL}/A_{CL})Z_{DM}$ impedance very much greater than $Z_{CM}$, so that the net closed-loop input impedance

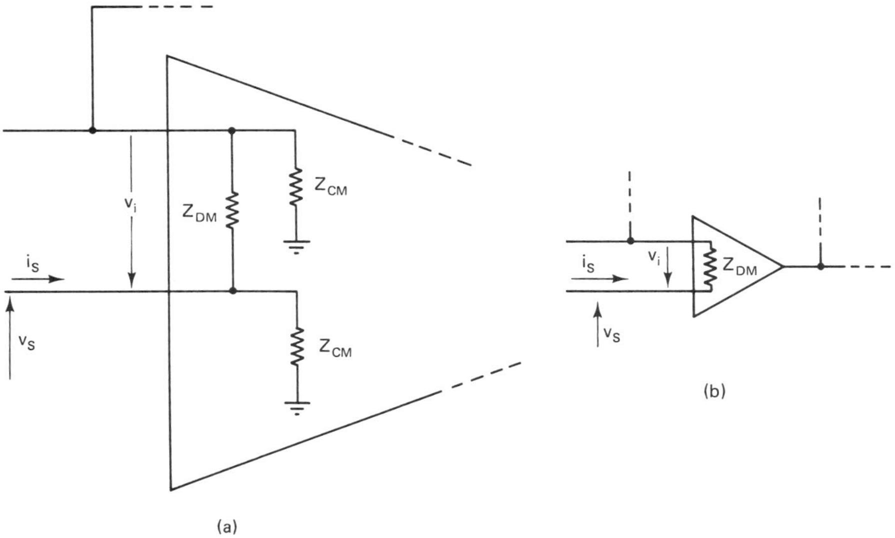


**Figure 5.24** (a) Equivalent-circuit representation of input impedance; (b) evaluation of effect of $Z_{DM}$.

will indeed be approximately equal to just $Z_{CM}$ alone as expressed by $Z_{i(CL)} \simeq Z_{CM}$.

To consider a very simple example of the effect of the op-amp input impedance on the circuit performance, let us refer to the voltage-follower circuit of Figure 5.25. Let us assume that the op-amp open-loop gain is very large and that the common-mode input impedance $Z_{CM}$ is 1 GΩ, in parallel with 10 pF. The source resistance is chosen to be 50 MΩ for this example, so that at low frequencies the 50-MΩ source resistance and the 1000-MΩ closed-loop input impedance of the op amp will constitute a simple resistive voltage divider with a voltage-division ratio of 1000/(1000 + 50) = 1000/1050 = 0.952 MΩ. We therefore will have that the gain of this system at low frequencies will be $v_o/v_s = 0.952$ MΩ.

At higher frequencies the op-amp input impedance will look primarily capacitative, so that at 10 kHz, for example, we have that

$$Z_{i(CL)} = Z_i = \frac{1}{j\omega C_i} = \frac{1}{j(2\pi \times 10 \text{ kHz}) \times 10 \text{ pF}} = -j1.59 \text{ M}\Omega \tag{5.43}$$

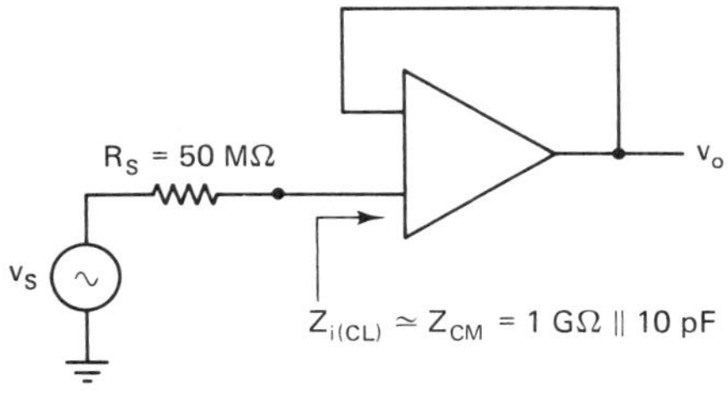


**Figure 5.25** Input impedance example.

The system gain at 10 kHz will therefore be given by

$$\frac{v_o}{v_s} = \frac{-j1.59\ \text{M}\Omega}{50\ \text{M}\Omega - j1.59\ \text{M}\Omega} = -j\left(\frac{1.59}{50}\right) = -j0.0318\ = 0.0318\ \underline{/-90^\circ} \qquad (5.44)$$

This represents a very severe signal attenuation resulting from the combination of the op-amp input capacitance $C_i$ and the high value of source resistance $R_S$. Indeed, $R_S$ and $C_i$ can be considered to form a simple $RC$ low-pass filter with a time constant of $\tau = (50h \parallel 1000)\text{M}\Omega \times 10\ \text{pF} \simeq 500\ \mu\text{s}$, and a breakpoint frequency or 3-dB bandwidth of only $BW \simeq 1/(2\pi 500\ \mu\text{s}) = 318\ \text{Hz}$.

## 5.11 OUTPUT IMPEDANCE

The ideal op amp acts as a voltage source of strength $A_{OL}v_i$ with a zero output impedance. For this case the output voltage $v_o$ will not be a function of the load impedance $Z_L$ that is driven by the op amp. Any real op amp has a finite *output impedance* $Z_O$, such that the output voltage and therefore the closed-loop gain will be a function of the load impedance.

To analyze the effect of the finite output impedance $Z_O$ on the circuit performance of an op amp the equivalent circuit representation of Figure 5.26 will be used. The op-amp output impedance $Z_o$ and the load impedance $Z_L$ will constitute a voltage divider that results in a voltage-division ratio of $Z_L/(Z_o + Z_L)$, so that as a result the output voltage will be given by $v_o = A_{OL}v_i \times Z_L/(Z_L + Z_O) = A_{OL}'v_i$, where $A_{OL}' = A_{OL}Z_L/(Z_L + Z_O)$. The output voltage can now be expressed in terms of the signal voltage $v_s$ by

$$v_o = \frac{1 + R_2/R_1}{1 + (1 + R_2/R_1)/A_{OL}'}\, v_s \qquad (5.45)$$

If we say that $1 + R_2/R_1 = A_{CL}$, then we will have that

$$v_o = \frac{A_{CL}v_s}{1 + A_{CL}/A_{OL}'} = \frac{A_{CL}v_s}{1 + [A_{CL}(Z_L + Z_O)/A_{OL}Z_L]}$$

$$= \frac{A_{CL}Z_L v_s}{Z_L + (A_{CL}/A_{OL})(Z_L + Z_O)} \qquad (5.46)$$

This can be rewritten as

$$v_o = A_{CL}v_s \frac{Z_L}{Z_L + (A_{CL}/A_{OL})Z_L + (A_{CL}/A_{OL})Z_O} \qquad (5.47)$$

Noting that $A_{CL}/A_{OL} \ll 1$, this can be written approximately as

$$v_o \simeq A_{CL}v_s \frac{Z_L}{Z_L + (A_{CL}/A_{OL})Z_O} = A_{CL}v_s \frac{Z_L}{Z_L + Z_{O(CL)}} \qquad (5.48)$$

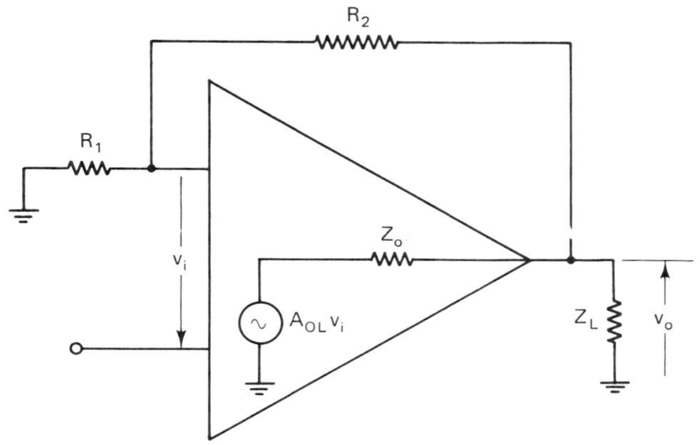


**Figure 5.26** Equivalent circuit for the evaluation of the closed-loop output impedance.

where

$$Z_{O(CL)} = \frac{A_{CL}}{A_{OL}} Z_O \tag{5.49}$$

is the *closed-loop output impedance*. From this last result we see that under closed-loop conditions the effective voltage-division ratio for $v_o$ due to the voltage-divider action of $Z_L$ and $Z_O$ will be $Z_L/(Z_L + Z_{O(CL)})$, so that the effective output impedance is $Z_{O(CL)}$.

The open-loop output impedance of op amps will generally be of the order of 10 to 100 Ω. Since $A_{CL}$ will usually be very much smaller than $A_{OL}$, we see that the closed-loop output impedance can be made to be very small indeed, with values down in the milliohm range being readily obtainable. We must again take note of the fact that the factor of improvement, $A_{CL}/A_{OL}$, in the output impedance is exactly the same factor by which the gain is sacrificed in going from open-loop to closed-loop conditions.

## 5.12 POWER DISSIPATION AND CURRENT LIMIT

For any electronic device there is a *maximum power dissipation*, $P_{d\,(\mathrm{MAX})}$, limitation. If the power dissipated in the device goes beyond the $P_{d\,(\mathrm{MAX})}$ value, the consequent temperature rise in the device may be such that the device temperature will exceed the maximum allowable temperature rating $T_{J\,(\mathrm{MAX})}$ and the device will be subject to permanent irreversible damage.

For op amps packaged in the plastic dual-in-line package (the "DIP" package), the $P_{d\,(\mathrm{MAX})}$ rating will be about 500 to 750 mW. For other packages the $P_{d\,(\mathrm{MAX})}$ rating can be as high as around 1 W. For some ICs with heat sinks attached to

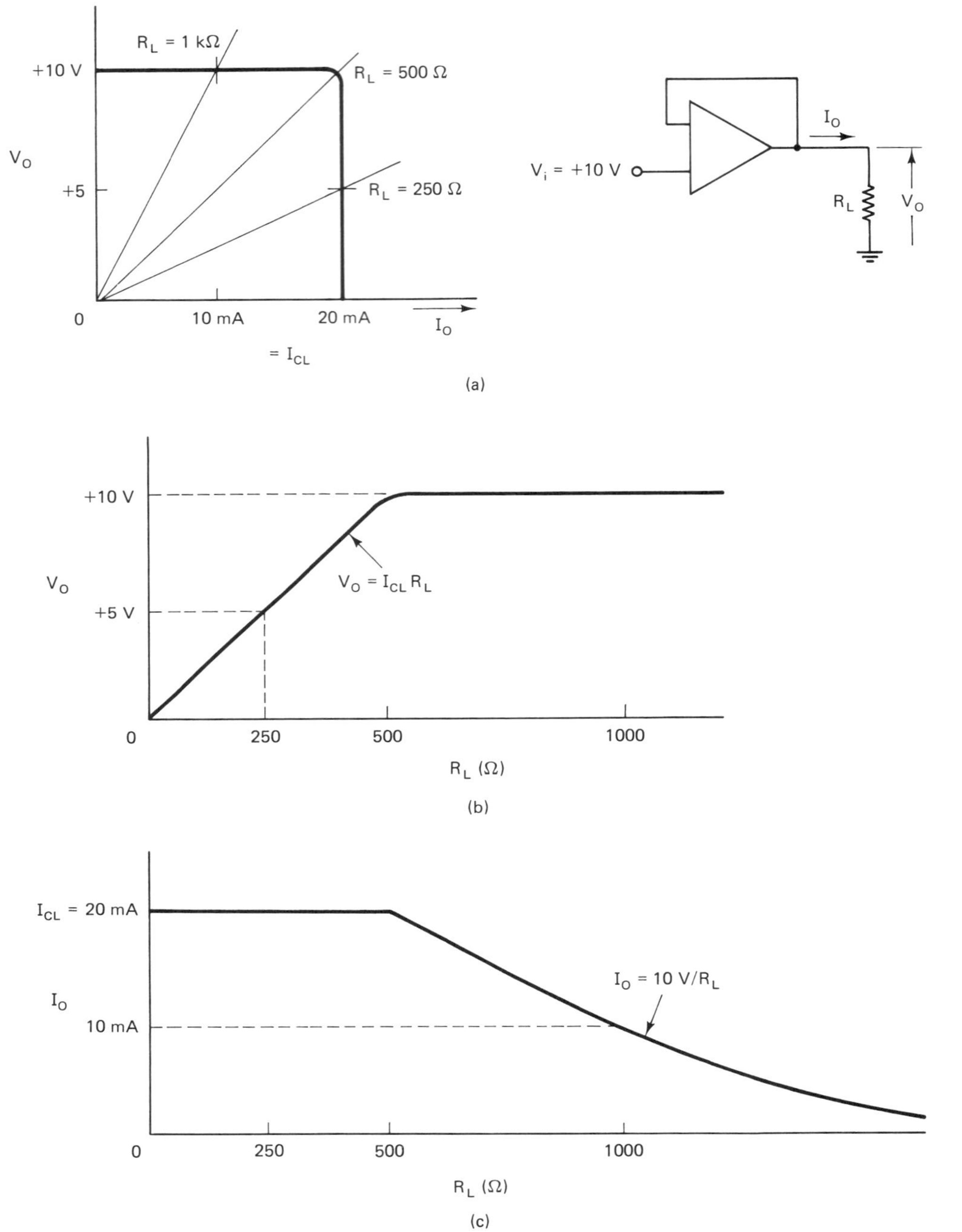


**Figure 5.27** Operational amplifier output voltage and current characteristics: (a) $V_O$ versus $I_O$ output characteristic; (b) $V_O$ versus $R_L$; (c) $I_O$ versus $R_L$.

provide more efficient heat flow away from the device by either conduction or convection, maximum power dissipation ratings of up to 10 W are possible.

Most op amps have some internal circuitry to limit the output current to a safe maximum value such that the power dissipated in the device will not exceed the maximum power dissipation rating, even under a "worst case" situation. Let us consider the op-amp circuit of Figure 5.27a. The worst-case situation from the standpoint of op-amp power dissipation will occur when the supply voltage is at its maximum rated value and the load resistance $R_L$ is zero. In this case the full supply voltage of $V^+$ (or $V^-$) will appear across the output stage of the op amp. The power dissipation of the op amp will be $P_d = V^+ I_O + P_{d(Q)}$, where $P_{d(Q)}$ is the *quiescent power dissipation* of the op amp. This is the power dissipation of the op amp when it is not supplying any current to the load. Normally, this quiescent power dissipation $P_{d(Q)}$ is very small, often less than 1 mW, so it will not be an important factor in the determination of the maximum allowable output current.

The maximum supply voltage rating of most op amps is around $\pm 18$ V, so we will choose this value. If the $P_{d(\text{MAX})}$ rating of the op amp is a representative value of 500 mW, the output current of the op amp should be limited to a value of

$$I_{o(\text{MAX})} = I_{CL} = \frac{P_{d(\text{MAX})}}{V^+} = \frac{500 \text{ mW}}{18 \text{ V}} = 28 \text{ mA} \tag{5.50}$$

A reasonable choice for the current limit $I_{CL}$ would therefore be in the range 20 to 25 mA, which allows for a suitable safety factor, and indeed this is the typical range of current limit values for most op amps.

In Figure 5.27 the output voltage and current characteristics are present for an op amp with internal current-limiting circuitry. A current limit value of 20 mA is chosen and the output voltage is assumed to be +10 V when the op amp is not operating in the current-limited region. In Figure 5.27a the $V_o$ versus $I_o$ output characteristic is shown. Note that for load resistance values greater than 500 Ω the output current will be less than 20 mA and the op amp will not be operating in the current-limited region. The output voltage will now be essentially independent of $R_L$. For $R_L$ less than 500 Ω, however, the op amp will be driven into the current-limited mode of operation and the output voltage will decrease linearly with $R_L$, whereas the output current remains relatively constant at 20 mA. This behavior is also evident from inspection of Figure 5.27b and c.

Most op amps are designed to be operated with split power supplies and will have internal current limiting in both directions so that the maximum current that the op amp can source will be limited to $I_{CL}^+$, and the maximum current that the op amp will be allowed to sink will be $I_{CL}^-$. Usually, the two currents will be approximately equal and will be in the range 20 to 25 mA.

## 5.13 EFFECTS OF FEEDBACK ON DISTORTION

The closed-loop gain of an amplifier is given by $A_{CL} = A_{OL}/(1 + FA_{OL})$ in terms of the open-loop gain $A_{OL}$ and the feedback factor $F$. From this we see that for a large loop gain such that $FA_{OL} \gg 1$, the closed-loop gain will be given approximately by $A_{CL} \simeq 1/F$. Since the feedback factor $F$ is usually determined by a resistive voltage divider, it will be a constant and virtually independent of the input signal level, whereas the open-loop gain will reflect the nonlinearities of the various stages of the amplifier, especially that of the output stage, where the signal level is the largest. Therefore, since $A_{CL} \simeq 1/F$, we see that the effect of feedback will be to "linearize" the amplifier and thus to cause the output to be a more exact replica of the input signal. In other words, the amplifier distortion will be greatly reduced.

The fractional variation in the closed-loop gain due to a fractional change in the open-loop gain can be obtained from the foregoing expression for $A_{CL}$ and expressed in the form

$$\frac{dA_{CL}}{A_{CL}} = \frac{dA_{OL}}{A_{OL}} \frac{1}{1 + FA_{OL}} = \frac{dA_{OL}}{A_{OL}} \frac{A_{CL}}{A_{OL}} \tag{5.51}$$

Thus any instantaneous variations in the open-loop gain resulting from the nonlinearity of the open-loop $v_o$ versus $v_i$ transfer characteristics will result in a considerably smaller fractional change in the closed-loop gain. We see that, in effect, the nonlinearity or distortion of the amplifier has been reduced by the factor $A_{CL}/A_{OL}$. There is, however, a corresponding loss in the gain, but this can easily be made up by increasing the gain of the stages preceding this amplifier. Since this increased gain occurs where the signal level is considerably lower than in the output stage, the problem of distortion is considerably reduced.

For a representative example, let us consider an amplifier with $A_{OL} = 1000$ and $A_{CL} = 50$. The distortion of the output signal under open-loop conditions with an output voltage of 10 V peak-to-peak will be assumed to be 10%. Under closed-loop conditions the distortion will be reduced to only 10% × (50/1000) = 0.5%, with the same output voltage swing. The input signal level of $v_s = v_o/A_{CL} = 10$ V/50 = 0.2 V (peak-to-peak) will be sufficiently small such that there will be no significant distortion introduced by any preamplifier stages that may be required.

To look at this from a slightly different approach, let us write the expression for the $v_o$ versus $v_i$ transfer characteristics under open-loop conditions in the form of $v_o = a_1 v_i + a_2 v_i^2 + a_3 v_i^3 + \cdots$. The $a_1$ term represents the linear response of the amplifier, and all of the other terms represent the amplifier nonlinearity. The open-loop voltage gain will therefore be given by

$$A_{OL} = \frac{v_o}{v_i} = a_1 + a_2 v_i + a_3 v_i^2 + \cdots \tag{5.52}$$

We see from this that the open-loop gain will be a function of the signal level, so that we will indeed have a nonlinear response characteristic and the distortion that will result therefrom. For the closed-loop gain we have the expression

$$A_{CL} = \frac{A_{OL}}{1 + FA_{OL}} = \frac{1/F}{1 + (1/FA_{OL})} \simeq \frac{1}{F}\left(1 - \frac{1}{FA_{OL}}\right)$$

since $FA_{OL}$ is large compared to unity. If we insert the foregoing expression for $A_{OL}$ into this last equation, we obtain

$$A_{CL} \simeq \frac{1}{F}\left(1 - \frac{1/F}{a_1 + a_2 v_i + a_3 v_i^2 + \cdots}\right)$$

$$\simeq \frac{1}{F}\left\{1 - \frac{1/F}{a_1[1 + (a_2/a_1)\, v_i + a_3/a_1\, v_i^2 + \cdots]}\right\} \tag{5.53}$$

Since the $a_2/a_1$, $a_3/a_1$, and so on, terms will be small compared to unity, we can rewrite this last expression as

$$A_{CL} \simeq \frac{1}{F}\left[1 - \frac{1}{a_1 F}\left(1 - \frac{a_2}{a_1}\, v_i - \frac{a_3}{a_1}\, v_i^2 - \cdots\right)\right] \tag{5.54}$$

Since $A_{OL} \simeq a_1$ and $A_{CL} \simeq 1/F$, we can write an equation for the output voltage $v_o$ as

$$v_o \simeq A_{CL} v_i \left[1 - \frac{A_{CL}}{A_{OL}}\left(1 - \frac{a_2}{a_1} v_i - \frac{a_3}{a_1} v_i^2 - \cdots\right)\right] \tag{5.55}$$

From this last expression we see that the nonlinearities of the $v_o$ versus $v_i$ transfer characteristic will have been reduced by the $A_{CL}/A_{OL}$ factor, and if $A_{CL} \ll A_{OL}$, a very linear response characteristic can indeed be obtained.

## 5.14 POWER SUPPLY REJECTION RATIO

For the ideal op amp the output voltage $v_o$ is independent of the d-c supply voltage. For any real op amp, however, the supply voltage will influence the output voltage due to both the variation of the open-loop gain $A_{OL}$ with the supply voltage and a feedthrough of supply voltage fluctuations into the signal path of the circuit. The open-loop gain of an op amp will increase slowly with supply voltage, a typical variation being an increase in $A_{OL}$ by a 2:1 or 3:1 ratio for a supply voltage variation from a minimum value of $\pm 5$ V up to a maximum rated value of $\pm 18$ V. Although this is a significant variation, it must be remembered that the closed-loop gain will be relatively independent of the open-loop gain, so that the resulting effect on the closed-loop gain, and therefore on the output voltage, will usually be very small.

A principal cause of supply voltage fluctuations is the a-c ripple voltage from a d-c power supply that uses a rectifier to convert an a-c line voltage to d-c followed by a filter circuit. The most important ripple component will usually be at 120 Hz. This a-c ripple voltage can produce a small contribution to the output voltage. This effect of the power supply ripple voltage or other voltage fluctuations can be referred back to the input of the op amp and can be described in terms of an effective change in the offset voltage.

The *power supply rejection ratio* (PSRR) is defined as the ratio of change in the input offset voltage resulting from a change in the supply voltage to the change in the supply voltage, and is usually expressed in decibels (dB). The PSRR is usually very small, typically in the range of $-80$ dB ($10^{-4}$) to $-100$ dB ($10^{-5}$). For example, with a PSRR of 100 dB, a 1-V a-c ripple on the power supply voltage will be equivalent to an input voltage of only 0.01 mV.

## 5.15 NOISE

In any communications system unwanted and extraneous signals which interfere with the desired signal will set a lower limit to the strength of the desired signal that can reliably and accurately be received and understood. The interfering "signals" that are of a *random* nature and are not emitted by any other communications system are called "noise." It is the random, unpredictable nature of noise that makes it so difficult to deal with and eliminate. Interfering signals that have a reasonably fixed or predictable nature, such as having a constant frequency or band of frequencies, can be reduced by special signal processing techniques, such as the use of band-pass and band-stop filters. Although there are techniques for noise reduction, it can never be entirely eliminated. Furthermore, the signal processing that is required for noise reduction will generally involve a trade-off of some other aspect of overall system performance. A good example of this trade-off is the reduction in the system bandwidth in order to reduce the overall noise level. The reduced system bandwidth will correspondingly reduce the information transmission rate of which the system is capable.

In an electronic communications system noise is generated internally both within the transmitter and much more important, in the receiver, as a result of a number of random processes. These processes include the random thermal motion of electrons and holes and the random nature of the electron–hole generation and recombination processes in semiconductor devices. In addition, noise generated by external sources may enter the receiver together with the desired signal. Some examples of external noise sources are lightning discharges, electrical machinery, and the ignition systems of gasoline engines.

The most critical component of an electronic communications system from the standpoint of the ability of the system to detect weak signals is the first or input stage of the receiver. It is here that the signal is the weakest and therefore the most susceptible to interference by noise, either internally generated in the first stage or coming into the first stage together with the signal.

The ratio of the signal voltage (rms) to the noise voltage (rms) will be called the *signal-to-noise ratio* (SNR). At the receiver input there will be a certain SNR depending on the signal strength and the external noise level. The internally generated noise in the first stage will add to the noise already present, so that there will generally be a significant degradation of the SNR in the first stage. It is of course true that the signal is amplified by the first stage, but the noise is also amplified by the same factor, so that due to the additional noise that is generated in the first stage there will actually be some decrease or loss in the SNR.

Following the first stage there will be further amplification of both the signal and the noise by the other stages of the receiver. In each stage there will be additional noise added to the signal due to the internally generated noise of that stage. As a result, the SNR will continue to decrease as the signal proceeds through the various stages of the amplifier. Nevertheless, it is the first stage that is generally the most important by far in determining the SNR of the receiver. If the gain of the first stage is denoted by $A_1$, that of the second stage by $A_2$, and so on, the overall amplifier gain will be $A_T = A_1A_2A_3 \cdots$. If the incoming signal level is $S_i$ and the noise level is $N_i$, the SNR at the receiver input is $(\text{SNR})_i = S_i/N_i$. The noise added by the first stage will be $N_1$, so that the SNR following the first stage will be $A_1S_i/A_1(N_i + N_1) = S_i/(N_i + N_1)$.

At the output of the second stage the SNR will be given by

$$(\text{SNR})_2 = \frac{A_1A_2\,S_i}{A_1A_2(N_i + N_1) + A_2N_2} = \frac{S_i}{(N_i + N_1) + (N_2/A_1)} \tag{5.56}$$

where $N_2$ is the noise added by the second stage. Similarly, at the output of the third stage we will have

$$\begin{aligned}(\text{SNR})_3 &= \frac{A_1A_2A_3S_i}{A_1A_2A_3(N_i + N_1) + A_2A_3N_2 + A_3N_3} \\ &= \frac{S_i}{(N_i + N_1) + (N_2/A_1) + (N_3/A_1A_2)}\end{aligned} \tag{5.57}$$

We see from this that for reasonably large gains per stage, the SNR of the first stage will largely determine the overall SNR of the receiver.

The *sensitivity* of a system is a measure of the ability of the system to detect or recover weak signals. It is usually described in terms of the lowest input signal level that will result in an acceptable SNR. In many cases the minimum SNR needed to detect the presence of a signal is equal to unity, for which case the sensitivity is the signal level that is equal to the noise present at the first amplifier stage.

In the ideal situation the amplified signal that is produced by the receiver will be an exact replica of the original signal produced by the transmitter. This will never be the case in practice, due to the various nonlinearities in the system that will result in distortion and the presence of noise.

In general, in order to produce an acceptable signal level, an SNR of at least around 10 (20 dB) is necessary, and for some applications SNR values as high as 30 or 40 (30 to 36 dB) are required. For example, to produce a good-quality television picture free of any apparent noise, an SNR of about 30 (36 dB) is required. If the SNR drops below 10 (20 dB), the amount of noise (or "snow") in the picture becomes quite noticeable. For SNR values less than about 3 (10 dB) the noise will be very objectionable, and if the SNR drops below about 1 or 2 (0 to 6 dB), it will be very difficult to discern the picture at all.

There will be two basic sources of noise in an electronic amplifier system, *thermal noise* and *shot noise*. Thermal noise (or Johnson noise) is due to the random thermal motion of charge carriers (electrons and holes) in the various circuits components of the amplifier, principally the resistors and transistors. Shot noise is the result of random fluctuations in the flow of current through electronic devices. Both thermal noise and shot noise are random functions of time with a zero average value, and both will have a Gaussian probability distribution function.

In any electronic device or component there will be minute random voltage

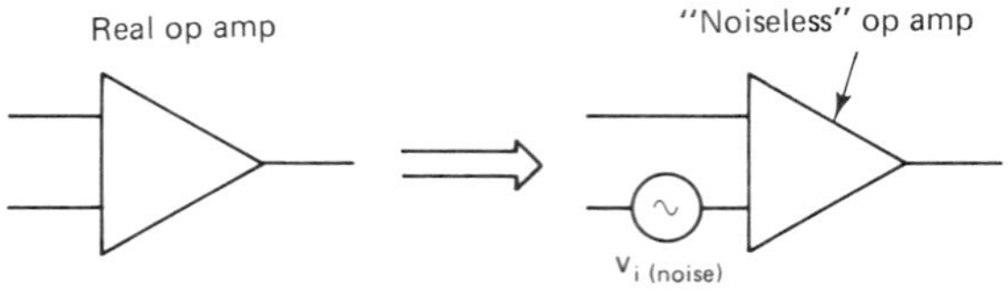


**Figure 5.28** Equivalent input noise voltage.

and current fluctuations known as "noise." An op amp is no exception to this. Electrical noise generated within the op amp, especially that produced by the transistors of the first or input stage of the op amp, will appear together with the desired signal as part of the output voltage. The noise generated within an op amp can most easily be described and its effect evaluated by replacing the real op amp by an ideal "noiseless" op amp with an *equivalent input noise voltage* source, $v_{i\,(\text{noise})}$, in series with one of the input terminals as shown in Figure 5.28. It is usually most convenient to place the equivalent noise source in series with the noninverting input terminal.

The noise voltage is a completely random function of time with a zero average or d-c value. The noise voltage can, however, be specified in terms of its rms ("root mean square") value over a given frequency range or bandwidth. The noise voltage will have a flat spectral distribution, except at low frequencies ($\lesssim$100 Hz), as shown in Figure 5.29. The rms voltage over a given bandwidth range will be proportional to the *square root* of the bandwidth.

The equivalent input noise voltage is usually expressed in terms of its *spectral density*, which is the rms noise voltage per unit bandwidth and is often given in such units as $\text{nV}/\sqrt{\text{Hz}}$. A typical value for the equivalent input noise voltage of an op amp is 20 $\text{nV}/\sqrt{\text{Hz}}$. The effective bandwidth to be used for noise calculations (i.e., "noise bandwidth") is related to the 3-dB bandwidth of the system under consideration by $(BW)_{\text{noise}} = (\pi/2)(\text{BW})_{3\text{dB}}$ for systems characterized by a single breakpoint frequency (i.e., a 20-dB/decade roll-off). For example, an op amp with an equivalent input noise voltage (spectral density) of 20 $\text{nV}/\sqrt{\text{Hz}}$ and a 10-kHz 3-dB bandwidth will have a total input noise voltage of $20\ \text{nV}/\sqrt{\text{Hz}} \times \sqrt{(\pi/2) \times 10\ \text{kHz}} = 2.5\ \mu\text{V}$ rms. Therefore, the input signal level that is needed in order to produce a *signal-to-noise ratio* of 10:1 will be 25 $\mu$V. A signal level of 2.5 $\mu$V will result in an SNR

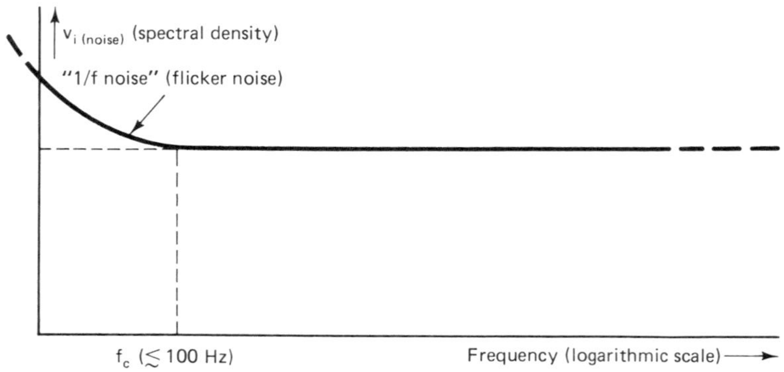


**Figure 5.29** Noise voltage spectral distribution.

of unity, which results in a barely detectable, extremely noisy signal. If the SNR drops below unity, the signal will become "buried" in the noise and in most cases will not be recoverable. The input noise voltage will therefore determine the *sensitivity* of the system for detecting low-level signals.

One other op-amp noise source of interest is due to the input bias current $I_B$. The input bias current is essentially a d-c current, but it will exhibit minute random fluctuations due to the particulate nature of electrical current flow. These minute fluctuations constitute a "noise current" called the *shot noise*, $i_{sn}$, which like the noise voltage has a zero average or d-c value, but can be described in terms of its rms value. It will have a reasonably flat spectral distribution, except at low frequencies ($\lesssim$100 Hz) and the rms value will be proportional to the square root of the bandwidth. The shot noise current is usually described in terms of its spectral density, which is the shot noise current per unit bandwidth, often specified in units of pA/$\sqrt{\text{Hz}}$ or sometimes fA/$\sqrt{\text{Hz}}$ (1 fA $= 10^{-15}$ A). The shot noise current (spectral density) is related to the bias current by the simple relationship

$$i_{sn} \text{ (rms, spectral density)} = \sqrt{2eI_B}$$

where $e$ is the electronic charge ($1.6 \times 10^{-19}$ C). For example, if $I_B = 100$ nA, the shot noise current (rms, spectral density) will be $i_{sn} = 0.18$ pA/$\sqrt{\text{Hz}}$.

One other noise source of importance in op-amp systems is the noise voltage developed in resistors, known as the *thermal noise voltage* because it is proportional to the square root of the temperature. This is a random voltage of zero average or d-c value and can be described by its rms value over a specified bandwidth. The rms thermal noise voltage is proportional to the square root of the bandwidth and is usually specified in terms of the spectral density, which is the thermal noise voltage per unit bandwidth. The rms thermal noise voltage spectral density is given by $v_{th} = \sqrt{4kTR}$, where $k$ is Boltzmann's constant ($1.38 \times 10^{-23}$/K), $T$ is the absolute temperature (K), and $R$ is the resistance value. At a temperature of 293K (20°C) this equation becomes $v_{th} = \sqrt{1.6 \times 10^{-20} R}$. For $R = 1$ M$\Omega$ this gives $v_{th} = 127$ nV/$\sqrt{\text{Hz}}$. The various (uncorrelated) noise sources in a system can be combined by taking the square root of the sum of the squares of the various noise voltages, wherever they appear in series.

To consider some simple examples of noise calculations in an op-amp system, we will use the circuit of Figure 5.30. We will consider an op amp with $v_{i\,(\text{noise})} = 20$ nV/$\sqrt{\text{Hz}}$ and $I_B = 100$ nA such that $i_{sn} = 0.18$ pA/$\sqrt{\text{Hz}}$. We will assume a

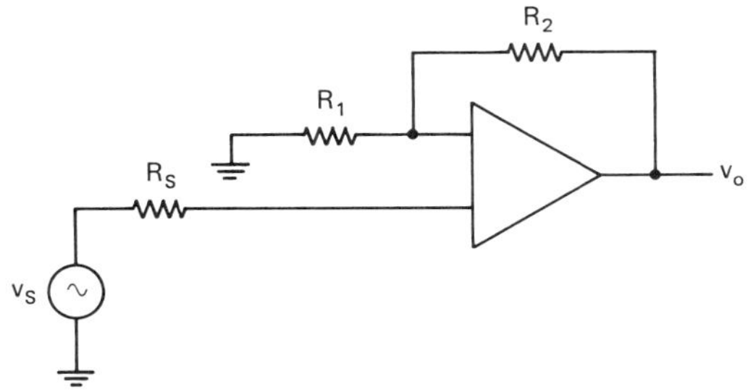


**Figure 5.30** Circuit for noise calculation example.

closed-loop bandwidth of 10 kHz, and will also assume for simplicity that $R_1$ and $R_2$ are sufficiently small ($\lesssim$10 kΩ) such that they will not contribute significantly to the overall noise.

When $R_S$ is small ($\lesssim$10 kΩ), the dominant noise source will be the equivalent input noise voltage of the op amp of 20 nV/$\sqrt{\text{Hz}}$ that will result in a net rms noise voltage of 2.5 μV for the 10-kHz 3-dB bandwidth.

If $R_S$ = 250 kΩ, it will produce a thermal noise voltage of 63 nV/$\sqrt{\text{Hz}}$, so that it now becomes the dominant noise source. The noise voltage resulting from the flow of the shot noise current through $R_S$ will be 0.18 pA/$\sqrt{\text{Hz}}$ × 250 kΩ = 45 nV/$\sqrt{\text{Hz}}$. The net noise voltage, considering the effects of all three noise sources, will now be 80 nV/$\sqrt{\text{Hz}}$, or 10 μV for the 10-kHz bandwidth.

When $R_S$ = 1 MΩ, the thermal noise voltage will now be up to $v_{th}$ = 127 nV/$\sqrt{\text{Hz}}$, and the flow of the shot noise current through $R_S$ will produce a voltage drop of $i_{sn}R_S$ = 0.18 pA/$\sqrt{\text{Hz}}$ × 1 mΩ = 180 nV/$\sqrt{\text{Hz}}$, so that the shot noise current is now the most important noise source. The net noise voltage due to all three sources will be 221 nV/$\sqrt{\text{Hz}}$, or 28 μV for the 10-kHz bandwidth.

We see from this example that when the resistance value is very large a low value for the shot noise current can be a very desirable feature. This, in turn, means that the input bias current should be very small. This requirement can be satisfied most readily by using a FET-input op amp for which $I_B$ values down as low as 10 pA can be obtained.

## 5.16 DEFINITIONS OF TERMS

Now that some of the basic characteristics of op amps have been discussed, a list of definitions of various terms relating to op amps is presented.

1. *Amplifier-to-amplifier coupling*: In some cases there is more than one op amp on the same IC chip, such as the case of the 747 device, which consists of two complete 741-type op amps on the same IC chip. Another example is a "quad" op amp such as the 124 type, in which there are four op amps on one chip.

   The amplifier-to-amplifier coupling is a measure of the degree of interaction between the op amps on a chip. It is usually expressed as the ratio of the change in the input offset voltage of one op amp as a result of the change in the output voltage of another op amp on the same chip. This ratio is usually expressed in dB, with the frequency or frequency range at which the measurement is made being specified.
2. *Bandwidth, 3-dB*: This is the frequency at which the voltage gain of an op amp has dropped by 3 dB from its zero-frequency value. Numerically, this corresponds to the voltage gain being 0.7071 (i.e., $1/\sqrt{2}$) times the zero-frequency value. The *open-loop* bandwidth is the frequency at which the open-loop gain of the op amp has decreased by 3 dB from the zero-frequency value of the open-loop gain, $A_{OL}(0)$. For internally compensated op amps this bandwidth is typically just of the order of 10 Hz. The *closed-loop* bandwidth, $BW_{CL}$, is

the frequency at which the closed-loop gain has dropped 3 dB from the zero-frequency value, $A_{CL}(0)$. The closed-loop bandwidth is generally much higher than the open-loop bandwidth, $BW_{OL}$.

3. *Common-mode input impedance*, $Z_{i(CM)}$: This is the ratio of the change in the common-mode input voltage to the change in the current at either input terminal.
4. *Common-mode rejection ratio* (CMRR): This is the ratio of the difference-mode voltage gain to the common-mode voltage gain, so that CMRR = $A_{DM}/A_{CM}$. This ratio is usually expressed in decibels, so that $\text{CMRR}_{\text{dB}} = 20 \log_{10} (A_{DM}/A_C)$.
5. *Common-mode voltage gain*, $A_{CM}$: This is the ratio of the change in the output voltage to the common-mode input voltage of the op amp. The common-mode input voltage is the average of the two input voltages of the op amp.

$$V_{DM} = V_A - V_B$$
$$V_{CM} = (V_A + V_B)/2$$
$$A_{DM} = A_{OL} = \Delta V_O/\Delta V_{DM}$$
$$A_{CM} = \Delta V_O/\Delta V_{CM}$$

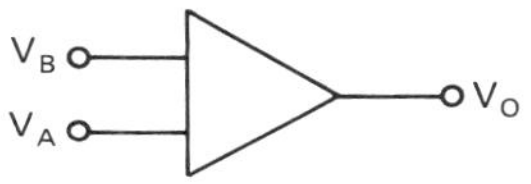


6. *Difference-mode voltage gain*, $A_{DM}$: This is the ratio of the change in the output voltage to the change in the voltage between the two input terminals of the op amp. The difference-mode voltage gain is essentially identical to the op-amp open-loop voltage gain, $A_{OL}$.
7. *Equivalent input noise current*: The input current (i.e., the bias current) of an op amp is a d-c current, but will not be absolutely constant. There will be very minute fluctuations of a random nature in the input currents. These minute random fluctuations can be considered to be a noise current that is superimposed on the d-c value of the input current, and can be represented by a current source external to the op amp. The input noise current is commonly expressed in terms of a spectral density, typically in units of $\text{pA}/\sqrt{\text{Hz}}$.
8. *Equivalent input noise voltage*: The various circuit components within an op amp will generate electrical noise. This noise will be amplified together with the desired signal in the op amp and appear at the output. The effect of the internally generated noise in an op amp can be expressed in terms of an equivalent noise voltage applied across the input terminals of the op amp. The *equivalent input noise voltage*, $v_{i(\text{noise})}$, will be a random voltage of zero d-c value, but it will have some rms value. The noise voltage will generally be proportional to the square root of the amplifier (closed-loop) bandwidth, so that the frequency range involved must be included as part of the specification of $v_{i(\text{noise})}$. Very often the "spectral density" of the noise voltage is specified. This is the noise voltage expressed on a per unit bandwidth (i.e., 1 Hz) basis, with the usual units being expressed as $\text{nV}/\sqrt{\text{Hz}}$.
9. *Full-power bandwidth* (*FPBW*): The full-power bandwidth of an op amp is the frequency at which the closed-loop voltage gain has decreased by 3 dB below the zero-frequency value of the closed-loop gain under large-signal conditions such that the amplifier response is slew-rate limited.

10. *Harmonic distortion*: As a result of nonlinearities in the op amp, the output voltage waveform will not be an exact replica of the input voltage. A quantitative index of the amount of distortion produced by an op amp under specified conditions can be obtained under conditions of a sinusoidal input signal by obtaining the ratios of the amplitudes of the various harmonic terms in the output voltage to the fundamental component of the output voltage. Such ratios are usually expressed on a percentage basis, such as percent second harmonic distortion, and so on. An index of the total amount of distortion can be obtained from the following equation:

$$\%\ \text{THD} = \frac{\sqrt{(V_2^2 + V_3^2 + V_4^2 + \cdots)} \times 100\%}{V_1}$$

where % THD is the percent *total harmonic distortion,* and $V_1$, $V_2$, $V_3$, and so on, are the amplitudes of the fundamental ($V_1$), second harmonic ($V_2$), third harmonic ($V_3$), and so on, of the output voltage of the op amp.

11. *Input bias current,* $I_B$ or $I_{\text{BIAS}}$: For an op amp to operate properly, there must be some quiescent (d-c) current flowing into (or out of) the two input terminals. The average of the two input currents is called the input bias current, $I_B$ or $I_{\text{BIAS}}$. For bipolar input op amps, $I_B$ is generally in the range of 1 $\mu$A to 1 nA. For JFET op amps it will be of the order of 1 to 10 pA, and for MOSFET input op amps it can be even smaller.

12. *Input bias current temperature coefficient,* $TCI_B$: This is the rate of change of the input bias current with temperature as given by $TCI_B = dI_B/dT$.

13. *Input impedance,* $Z_i$: The *difference-mode input impedance* is the ratio of the change in the input voltage to the change of the input current on either input with the other input terminal grounded (a-c). This is the a-c (dynamic) input impedance looking into either input terminal with the other input terminal grounded.

14. *Input offset current,* $I_{OS}$ or $I_{\text{OFFSET}}$: This is a d-c current that is the difference between the two input currents. The value of $I_{OS}$ is generally about 10 to 20% of $I_B$.

15. *Input offset voltage,* $V_{OS}$ or $V_{\text{OFFSET}}$: This is the voltage that must be applied between the two input terminals of the op amp to obtain zero output voltage. A primary source of $V_{OS}$ is the mismatch between the transistors making up the input stage difference amplifier. For op amps with a bipolar transistor (NPN or PNP) input stage, $V_{OS}$ will generally be of the order of 1 mV, although for some op amps it can be down as low as about 10 to 100 $\mu$V. For FET input stage op amps the $V_{OS}$ values will generally be somewhat larger, typically around 5 mV.

16. *Input offset voltage temperature coefficient,* $TC_{V_{OS}}$: This is the rate of change of $V_{OS}$ with temperature. For bipolar transistor input stage op amps, $TC_{V_{OS}}$ will be given approximately by $TCV_{OS} \simeq V_{OS}/T$, so that if $V_{OS} = 1.0$ mV = 1000 $\mu$V, $TC_{V_{OS}} \simeq 3\ \mu$V/°C.

17. *Input voltage range*: This is the range of voltages applied to the input terminals

of the op amp for which the device will operate within specifications. This is sometimes expressed in terms of the *common-mode input voltage range* and the *difference-mode voltage range*. The common-mode voltage range is the range of input voltages that can be applied to either or both input terminals and still have the op amp operate properly. The difference-mode voltage range is the maximum voltage that can be applied between the two input terminals and still allow for proper operation of the op amp.

18. *Large-signal voltage gain*: This is the ratio of the output voltage swing to the change in the voltage applied between the two input terminals required to change the output voltage from zero to some specified value. This is a dimensionless quantity and is sometimes expressed in "units" of V/mV, and is essentially the open-loop gain of the op amp, $A_{OL}(0)$.
19. *Long-term stability*: This is the drift of the input offset voltage as a function of time, and is typically expressed in units of $\mu$V/month.
20. *Operating temperature range*: This is the range of temperatures over which the op amp will operate within the stated specifications. Many manufacturers specify and test op amps for the following three temperature ranges:

| | |
|---|---|
| Commercial: | 0 to +70°C |
| Industrial: | −25 to +85°C |
| Military: | −55 to +125°C |

21. *Output impedance*, $Z_O$: This is the ratio of the change in the output voltage to the change in the output current.
22. *Output short-circuit current*: This is the maximum current that can be supplied by an op amp to a load. For most op amps this maximum current is set by an internal current-limiting circuit so as to protect the device against excessive power dissipation. Generally, there will be two somewhat different values of maximum output current for an op amp, one value corresponding to the device supplying current (i.e., sourcing current) to a load, and the other corresponding to current coming into the op amp (i.e., sinking current) from the load.
23. *Output voltage swing*: This is the peak output voltage swing, referred to zero, that can be obtained without clipping. This will generally be limited by the supply voltage such that the output voltage can not get to less than about 1 or 2 V of either the positive or the negative supply voltage.
24. *Power supply rejection ratio* (PSRR): This is the ratio of the change in the input offset voltage to the change in the power supply voltage producing this change. The PSRR is usually expressed in decibels (dB).
25. *Settling time*: This is the time interval between the application of a step-function input voltage and the time when the output voltage has settled to a value that is within a specified range (error band) of the final output voltage.
26. *Slew rate*: This is the maximum rate of change of the output voltage with time when a large-amplitude step-function voltage is applied to the input terminals. The slew rate of an op amp is approximately inversely proportional to the size of the compensation capacitor, $C_{COMP}$, that is used for feedback stability. The slew rate is usually expressed in units of volts/$\mu$s.

27. *Supply current*: This is the quiescent current required to operate the op amp supplied by the power supply. This quiescent current is measured under conditions of no load (i.e., no output current) and the output voltage at (or near) zero.
28. *Thermal feedback coefficient*: The temperature of an IC chip will be a function of the IC power dissipation, and in turn the input offset voltage will be a function of the chip temperature. The ratio of the change in the input offset voltage to the change in the power dissipation is called the thermal feedback coefficient. Typical units for this quantity are nV/mW or $\mu$V/mW.
29. *Transient response*: This term refers to the closed-loop step-function response of the amplifier under specified small-signal conditions. The transient response is usually expressed in terms of the 10 to 90% rise time.
30. *Unity-gain frequency*, $f_u$: This is the frequency at which the open-loop voltage gain, $A_{OL}$, of the op amp has decreased to unity. The value of $f_u$ is typically in the range of about 1 to 10 MHz.

Some of the preceding definitions dealing with large-signal frequency-domain and time-domain response characteristics such as the slewing rate and the full-power bandwidth (FPBW), as well as some other definitions, will be considered in more detail later when the internal circuitry of op amps is investigated.

## 5.17 COMPARISON OF IDEAL AND ACTUAL OPERATIONAL-AMPLIFIER RESPONSE CHARACTERISTICS

The basic operation of op amps has now been discussed and some of the nonideal characteristics have been investigated. It is now of interest to summarize the operation of op amps in terms of the difference between the behavior of the ideal op amp and nonideal or real op amps.

### 5.17.1 Characteristics of the Ideal Operational Amplifier

For the ideal op amp:

$$\boxed{V_O = A_{OL}(V_A - V_B)} \qquad \text{where } A_{OL} \text{ (open-loop voltage gain)} = \textit{constant}$$

*This equation implies the following*:

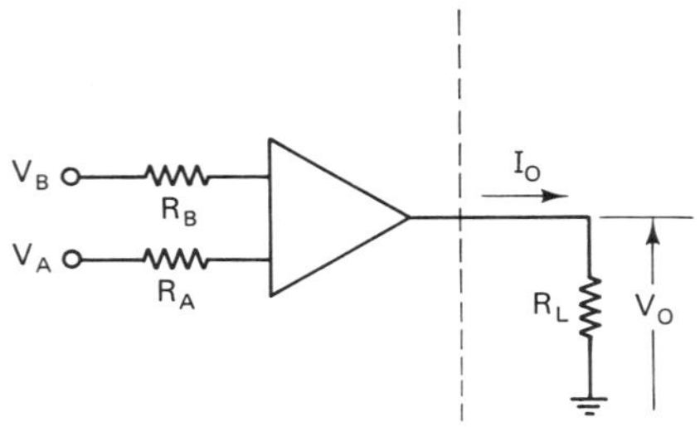

1. $V_O = 0$ when $V_A = V_B = 0$. Therefore:
   (a) There is no input offset voltage (i.e., $V_{OS} = 0$).
   (b) The equivalent input noise voltage, $v_{i(\text{noise})}$, is zero, meaning that there is no internally generated electrical noise in the op amp.

2. $V_O = 0$ when $V_A = V_B \neq 0$. Therefore, the common-mode voltage gain, $A_{CM} = 0$. As a result, the common-mode rejection ratio (CMRR) will be infinite.

3. $V_O$ is independent of $R_A$ and $R_B$. Therefore:
   (a) There is no voltage drop across $R_A$ and $R_B$ due to the input bias current so that $I_{\text{BIAS}} = 0$ (and therefore $I_{OS} = 0$), so that even if $R_A \neq R_B$, there will be no difference-mode input voltage resulting from the bias currents.
   (b) There is no voltage drop across either $R_A$ or $R_B$ due to a flow of current that would result from a finite input impedance, so that $Z_i$ is infinite.

4. The output voltage is independent of the load resistance, $R_L$, and thus of the output current, $I_O$. Therefore, the op-amp output impedance is *zero*.

5. The voltage gain, $A_{OL}$, is independent of $V_A$, $V_B$, and $V_O$. Therefore:
   (a) The input voltage range is unlimited.
   (b) The output voltage swing is unlimited.
   (c) Since $A_{OL}$ is a constant, the $V_O$ will be an exact replica of $V_A - V_B$, so that there is no distortion. The op amp therefore acts as an exactly linear device.

6. A change in the supply voltage will not change the output voltage, so that the power supply rejection ratio (PSRR) is infinite, and $A_{OL}$ is independent of the supply voltage.

7. $V_O(t) = A_{OL}[V_A(t) - V_B(t)]$, meaning that the output voltage will follow exactly the variations in $V_A$ and $V_B$ as a function of time. Therefore, the rise, fall, delay, and settling times are all zero. For the ideal op amp, the output voltage will be an exact and instantaneous replica of the input voltage.

8. $dV_O/dt = A_{OL}(dV_A/dt - dV_B/dt)$ for any combination of $dV_A/dt$, $dV_B/dt$, and $dV_O/dt$. Therefore, the amplifier never becomes slew-rate limited; that is, the slew rate is infinite.

9. The voltage gain, $A_{OL}$, is independent of frequency. Therefore, the bandwidth is infinite, both for small-signal conditions (i.e., $BW_{OL}$) and for large-signal conditions (i.e., FPBW).

10. $V_O$ is independent of $R_L$, so there is no output short-circuit current limit.

11. $V_O$ is independent of temperature, so that:
   (a) $TCV_{\text{OS}} = 0$.
   (b) $TCI_{\text{B}} = 0$.
   (c) The value of $A_{OL}$ is independent of temperature.

12. The input noise current is zero, so that there will be no noise voltage drop developed across $R_A$ and $R_B$ as a result.

## 5.18 OPERATIONAL-AMPLIFIER STABILITY

Let us consider the closed-loop op-amp system shown in Figure 5.31 together with the following definitions:

$A_{OL}$ = open-loop gain

$F$ = feedback factor = fraction of the output voltage that is fed back to the input

$FA_{OL}$ = loop gain = signal gain around the feedback loop

$A_{CL}$ = closed-loop gain

For the system of Figure 5.31a the relationship between the output voltage $v_o$ and the input or signal voltage $v_s$ can be written as $v_o = A_{OL}v_s - FA_{OL}v_o$, so that $v_o(1 + FA_{OL}) = A_{OL}v_s$, and thus the closed-loop gain will be given by

$$A_{CL} = \frac{v_o}{v_s} = \frac{A_{OL}}{1 + FA_{OL}} = \frac{\text{open-loop gain}}{1 + \text{loop gain}} \tag{5.58}$$

If we divide both the numerator and the denominator by the loop gain $FA_{OL}$ we will obtain

$$A_{CL} = \frac{1/F}{1 + (1/FA_{OL})} \tag{5.59}$$

If the loop gain is large compared to unity ($FA_{OL} \gg 1$), the closed-loop gain will be approximately given by $A_{CL} \simeq 1/F$. We see that the closed-loop gain becomes approximately equal to the reciprocal of the feedback factor $F$, and thus will be dependent essentially only on the parameters of the feedback network and will be relatively independent of the amplifier open-loop gain.

At the other extreme we note that as the loop gain becomes small compared to unity ($FA_{OL} \ll 1$), the closed-loop gain will approach the open-loop gain. We also note that as long as the loop gain is algebraically positive, the closed-loop gain will be less than the open-loop gain. For large values of loop gain the closed-loop gain will be much smaller than the open-loop gain.

The equation for the closed-loop gain for an op amp with negative feedback is given by $A_{CL} = A_{CL}/(1 + FA_{OL})$ and we see that the closed-loop gain will be less than the open-loop gain. The op amp is a multistage electronic amplifier, and as the frequency increases the input-to-output phase shift $\phi$ will increase monotonically with frequency. At some frequency, to be called $f_{180°}$, the phase shift will have reached 180°. At this frequency of $f_{180°}$ what was the inverting input at low frequencies

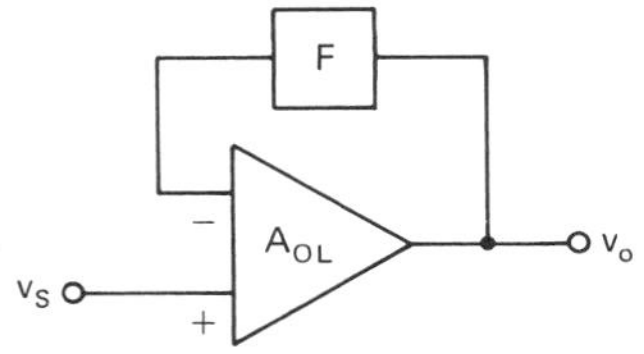


**Figure 5.31** Operational-amplifier closed-loop configuration.

becomes, in effect, the noninverting input terminal, and vice versa for the other input terminal. Thus what was *negative feedback* at low frequencies becomes actually *positive feedback* at $f = f_{180°}$. At $f = f_{180°}$ the loop gain becomes algebraically negative and the closed-loop gain expression can be written as

$$A_{CL} = \frac{A_{OL}}{1 - |FA_{OL}|}$$

As long as the magnitude of the loop gain is less than unity (i.e., $|FA_{OL}| < 1$) the system will be *stable* with a finite closed-loop gain. However, if the magnitude of the loop gain at $f = f_{180°}$ is equal to, or greater than unity (i.e., $FA_{OL} \geq 1$ at $f = f_{180°}$) the closed-loop gain will go to infinity, and the system will be said to be *unstable* (or *oscillatory*).

If the system is unstable with an infinite closed-loop gain, the system can produce an output voltage with zero signal input. The voltage fed back from the output to the input via the feedback network $F$ is sufficient to drive the system, so that the system is supplying its own input voltage by means of the positive feedback loop that exists at $f = f_{180°}$. The output voltage waveform will generally be a sine wave or a square wave with a frequency of $f = f_{180°}$. The op amp is now operating as an *oscillator* rather than an *amplifier* and will be relatively unresponsive to any signal input.

The low-frequency value of the closed-loop gain is related to the feedback factor $F$ by $A_{CL}(0) \simeq 1/F$. Therefore, the stability condition that $|FA_{OL}| < 1$ at $f = f_{180°}$ can be expressed as $|A_{OL}|/A_{CL}(0) < 1$ at $f = f_{180°}$, or as follows:

$$\boxed{\text{For stability: } A_{CL}(0) > |A_{OL}| \qquad \text{at } f = f_{180°}} \tag{5.60}$$

Since the open-loop gain will decrease *monotonically* with increasing frequency, if the open-loop gain becomes equal to $A_{CL}(0)$ at a frequency that is less than $f_{180°}$, then at $f = f_{180°}$ the open-loop gain will surely be less than $A_{CL}(0)$. Therefore, the stability condition can be reexpressed as

$$\boxed{\text{For stability: } |A_{OL}| = A_{CL}(0) \qquad \text{at } f < f_{180°}} \tag{5.61}$$

That is, *the frequency at which $A_{OL}$ becomes equal to $A_{CL}(0)$ should be less than $f_{180°}$.*

The foregoing stability conditions hold true only for the situations in which there is no significant phase shift contributed by the feedback network $F$. For the usual case of a feedback network containing only resistive elements, this requirement will be satisfied.

We will now consider another derivation for the closed-loop gain equation that will give us further insight into the question of stability. Let us consider again the system of Figure 5.31. An input signal $v_s$ will produce an output voltage $v_o = A_{OL}v_s$. A fraction $F$ of this output voltage is fed back to the input, and after passing through the op amp with gain $-A_{OL}$ it appears at the output as $(A_{OL}v_s)(-FA_{OL})$. A fraction $F$ of this output voltage component is then fed back to the input, and after passing through the op amp with gain $-A_{OL}$ appears at the output as $(A_{OL}v_s)(-FA_{OL})(-FA_{OL}) = (A_{OL}v_s)(-FA_{OL})^2$. This process keeps repeating indefinitely and a *net* output voltage is produced as given by

$$\begin{aligned} v_o &= (A_{OL}v_s) + (A_{OL}v_s)(-FA_{OL}) + (A_{OL}v_s)(-FA_{OL})^2 + \cdots \\ &= (A_{OL}v_s)[1 + (-FA_{OL}) + (-FA_{OL})^2 + (-FA_{OL})^3 + \cdots] \\ &= (A_{OL}v_s)(1 + x + x^2 + x^3 + \cdots) \end{aligned} \tag{5.62}$$

where $x = -FA_{OL}$.

As long as $x < 1$, this infinite power series will be convergent, and can be written in closed form as

$$v_o = (A_{OL}v_s)\frac{1}{1-x} = \frac{A_{OL}}{1 + FA_{OL}} v_s$$

so that

$$A_{CL} = \frac{A_{OL}}{1 + FA_{OL}} \tag{5.63}$$

and $A_{CL}$ is finite, so that the system is stable. This last expression for $A_{CL}$ can be recognized as being identical to the one obtained previously. If $x \geq 1\underline{/0°}$, the series will not be convergent, so that the closed-loop will go to infinity. Thus for stability we must have that $x = -FA_{OL} < 1\underline{/0°}$. Since $F$ is a positive real quantity and $A_{OL}$ becomes negative (i.e., $\underline{/A_{OL}} = 180°$) at $f = f_{180°}$, this stability condition can be written as

$$\text{For stability: } F|A_{OL}| < 1 \qquad \text{at } f = f_{180°} \tag{5.64}$$

Since $A_{CL}(0) = 1/F$ it can also be expressed as $A_{CL}(0) > |A_{OL}|$ at $f = f_{180°}$. This is the same stability condition as that obtained in the previous analysis.

### 5.18.1 Determination of $f_{180°}$

From the preceding discussions it is apparent that the value of $f_{180°}$ and the open-loop gain at $f = f_{180°}$ will be key factors in the stability of op-amp and other feedback systems. We will now develop an approximate analytical expression for $f_{180°}$ and find the corresponding value of the open-loop gain. We will consider the open-loop frequency response of the op amp to be of the form $A_{OL}(f) = A_{OL}(0)/(1 + jf/f_1)(1 + jf/f_2)(1 + jf/f_3) \cdots$, where the breakpoints are in the sequence $f_1 < f_2 < f_3 < \cdots$ and $f_1$ is considerably smaller than all the rest, so that $f_1^2 \ll f_2^2$.

At $f = f_1$, the phase angle of $A_{OL}$ will be approximately $-45°$. For frequencies in the range given by $f_1^2 < f^2 < f_2^2$ the phase angle will be approximately $-90°$. The phase angle will be in the neighborhood of $-135°$ at $f = f_2$, although it may be somewhat larger if $f_3$ is close to $f_2$. For the remainder of this discussion let us assume that $f_4$ is well beyond $f_3$. If this is the case, the phase angle of $A_{OL}$ will reach $-180°$ somewhere between $f_2$ and $f_3$. We will now determine the frequency at which the phase angle will be 180° (i.e., $f_{180°}$).

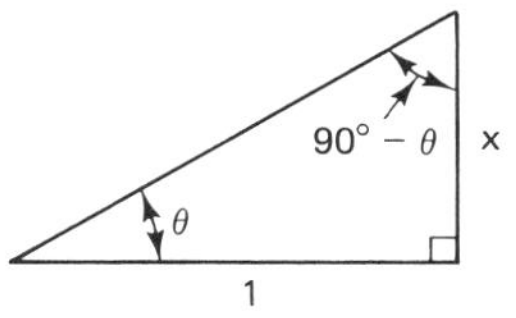


**Figure 5.32** Triangle for determination of $f_{180^\circ}$.

The phase angle of $A_{OL}$ will be given by

$$-\underline{/A_{OL}} = \underline{/1 + jf/f_1} + \underline{/1 + jf/f_2} + \underline{/1 + jf/f_3} + \cdots$$
$$= \tan^{-1}\frac{f}{f_1} + \tan^{-1}\frac{f}{f_2} + \tan^{-1}\frac{f}{f_3} + \cdots \tag{5.65}$$

For frequencies such that $f^2 > f_1^2$, $\tan^{-1}(f/f_1) \simeq 90°$. Furthermore, assuming that $f_4^2 > f_3^2$, the phase angle contributions of $f_4$ and all of the other higher breakpoint frequencies will be small. Therefore, for $-\underline{/A_{OL}} = 180°$ we have

$$180° \simeq 90° + \tan^{-1}\frac{f}{f_2} + \tan^{-1}\frac{f}{f_3}$$

so that

$$\tan^{-1}\frac{f}{f_2} + \tan^{-1}\frac{f}{f_3} \simeq 90° \tag{5.66}$$

Looking at the triangle shown in Figure 5.32, we see that since $\tan\theta = x$, $\tan(90° - \theta) = 1/x$, so that if we take the inverse tangent functions we obtain $\theta = \tan^{-1}x$ and $90° - \theta = \tan^{-1}(1/x)$. If we now add these two inverse tangent functions together, we obtain $\tan^{-1}x + \tan^{-1}(1/x) = 90°$. Therefore, for $\tan^{-1}(f/f_2) + \tan^{-1}(f/f_3) = 90°$, we must have that if $f/f_2 = x$, then $f/f_3 = 1/x$ and thus $f/f_2 = f_3/f$, giving $f^2 = f_2 f_3$. Thus $f \simeq \sqrt{f_2 f_3}$ as the frequency at which the total phase angle of the open-loop gain is equal to 180°.

It has been determined that the frequency at which the phase angle of the open-loop gain is 180° is given approximately by $f = \sqrt{f_2 f_3}$, subject to the conditions that $f_2$ and $f_3$ be well separated from both $f_1$ and $f_4$. The condition with respect to $f_1$ is usually very well satisfied. The condition with respect to $f_4$ and the rest of the higher breakpoint frequencies is sometimes not well satisfied. For those cases, then, $f_{180^\circ}$ will occur at a frequency somewhat lower than $\sqrt{f_2 f_3}$, but usually not less than $f_2$. We note that $f_{180^\circ} = \sqrt{f_2 f_3}$ means that $f_{180^\circ}$ will be the *geometric mean* of $f_2$ and $f_3$, and on a logarithmic frequency scale $f_{180^\circ}$ will in terms of actual distance be exactly *halfway* between $f_2$ and $f_3$.

We will now determine the value of the open-loop gain at $f = f_{180^\circ} = \sqrt{f_2 f_3}$. If we substitute $f = \sqrt{f_2 f_3}$ into the expression for $A_{OL}$, we obtain

$$A_{OL}(f = f_{180^\circ}) \simeq \frac{A_{OL}(0)}{(1 + j\sqrt{f_2 f_3}/f_1)(1 + j\sqrt{f_2 f_3}/f_2)(1 + j\sqrt{f_2 f_3}/f_3)\cdots} \tag{5.67}$$

If we now drop the higher breakpoint frequency terms (i.e., $f_4$, $f_5$, etc.) as being small, and take the product of the second and third terms in the denominator, we obtain

$$A_{OL}(f_{180^\circ}) \simeq \frac{A_{OL}(0)}{(j\sqrt{f_2 f_3}/f_1)(1 - 1 + j\sqrt{f_3/f_2} + j\sqrt{f_2/f_3})}$$
$$\simeq \frac{A_{OL}(0)}{-(\sqrt{f_2 f_3}/f_1)(\sqrt{f_3/f_2} + \sqrt{f_2/f_3})} \tag{5.68}$$
$$\simeq \frac{A_{OL}(0)\ \underline{/180^\circ}}{(f_2 + f_3)/f_1} = \frac{A_{OL}(0)f_1\ \underline{/180^\circ}}{f_2 + f_3}$$

If we use the definition $A_{OL}(0)f_1 = f_u =$ (extrapolated) unity-gain frequency, we can write the expression for $A_{OL}(f_{180^\circ})$ as

$$A_{OL}(f_{180^\circ}) \simeq \frac{f_u}{f_2 + f_3} \underline{/180^\circ} \tag{5.69}$$

Note that the phase angle of the open-loop gain does indeed turn out to be 180°, as expected. We see that at $f = f_{180^\circ}$, the magnitude of the open-loop gain is smaller than the zero-frequency value by a factor $f_1/(f_2 + f_3)$. In terms of the unity-gain frequency, the magnitude of the open-loop gain at $f = f_{180^\circ}$ can be expressed very simply as $A_{OL}(f_{180^\circ}) \simeq f_u/(f_2 + f_3)$.

The stability condition can now be expressed as

$$\text{For stability: at } f = f_{180^\circ},\ |FA_{OL}| < 1 \tag{5.70}$$

and since $F = 1/A_{CL}(0)$, this becomes $A_{CL}(0) > A_{OL}(f_{180^\circ})$. Since $A_{OL}(f_{180^\circ}) = A_{OL}(0)f_1/(f_2 + f_3) = f_u/(f_2 + f_3)$, this will give

$$\text{For stability:} \qquad A_{CL}(0) > \frac{A_{OL}(0)f_1}{f_2 + f_3}$$
$$\text{or in terms of } f_u,\ A_{CL}(0) > \frac{f_u}{f_2 + f_3} \tag{5.71}$$

### 5.18.2 Phase Margin

For many practical amplifier designs it is desirable to allow for some safety margin in the design for stable operation. The term *phase margin* will be defined as follows:

$$\text{phase margin} = 180^\circ - \underline{/FA_{OL}(f)}$$

where $f$ is the frequency at which the magnitude of the loop gain is unity. A value of phase margin that is frequently chosen is 45°. With this choice of phase margin, the angle of the loop gain at the frequency at which $|FA_{OL}| = 1$ will be 135°, and will thus be 45° away from the critical phase angle of 180°.

If the breakpoint frequencies are well separated such that $f_1^2 \ll f_2^2 \ll f_3^2$, then the frequency at which the phase angle of the amplifier open-loop gain is 135° will be approximately $f_2$. At this frequency we have a contribution of 90° from the $(1 + jf/f_1)$ term and 45° from the $(1 + jf/f_2)$ term, so that the total phase angle will be 135° (assuming that the phase-angle contribution from the other terms is small).

At $f = f_2$, the value of the open-loop gain will be given by

$$A_{OL}(f_2) = \frac{A_{OL}(0)}{(1 + jf_2/f_1)(1 + jf_2/f_2)} = \frac{A_{OL}(0)}{j(f_2/f_1)(1 + j1)} = \frac{A_{OL}(0)}{(f_2/f_1)\sqrt{2}\,\underline{/135^\circ}}$$
$$= \frac{A_{OL}(0)f_1}{(\sqrt{2}\,f_2)\underline{/-135^\circ}} \tag{5.72}$$

Since $f_u = A_{OL}(0)f_1$, this last equation can be rewritten as

$$\boxed{A_{OL}(f_2) = \frac{f_u}{\sqrt{2}\,f_2}} \tag{5.73}$$

For a phase margin of at least 45°, we must satisfy the condition $|FA_{OL}| \leqslant 1$ at $f_{135^\circ}$. Since $F = 1/A_{CL}(0)$, this can be rewritten as $A_{CL}(0) \geqslant |A_{OL}(f_{135^\circ})|$. If we now use the expression obtained above for $A_{OL}(f_{135^\circ})$, we have

$$\boxed{A_{CL}(0) \geqslant \frac{A_{OL}f_1}{\sqrt{2}\,f_2} = \frac{f_u}{\sqrt{2}\,f_2} \qquad \text{for a phase margin} \geqslant 45^\circ} \tag{5.74}$$

The preceding expressions for $A_{OL}(f_{135^\circ})$ and the corresponding phase margin condition was based on the assumption that the second breakpoint frequency ($f_2$) was well separated from its closest neighbors (i.e., $f_1$ and $f_3$). The condition with respect to $f_1$ is usually very well satisfied, but there can be cases in which $f_3$ may be close to $f_2$. If this is the case, then $f_{135^\circ}$ can be less than $f_2$. Let us consider, as an example, the *worst-case* condition, in which $f_3 = f_2$. For this case, at $f = f_{135^\circ}$ we must have that $\tan^{-1}(f/f_1) + \tan^{-1}(f/f_2) + \tan^{-1}(f/f_3) = 135^\circ$. Since $f_2 \gg f_1$, $\tan^{-1}(f/f_1) \simeq 90^\circ$, and noting that for this case $f_2 = f_3$, we now have $2\tan^{-1}(f/f_2) = 45^\circ$, so that $\tan^{-1}(f/f_2) = 22.5^\circ$, giving $f_{135^\circ} = f_2 \tan(22.5^\circ) = 0.4142f_2$. Therefore, we see that in this worst-case condition $f_{135^\circ}$ is a little less than one-half of $f_2$.

The value of the open-loop gain under these conditions (at $f = f_{135^\circ}$) will be given by

$$A_{OL}(f_{135^\circ}) = \frac{A_{OL}(0)}{(j0.4142f_2/f_1)(1 + j0.4142f_2/f_2)^2} = \frac{A_{OL}(0)f_1}{0.4853f_2\,\underline{/135^\circ}}$$
$$= \frac{A_{OL}(0)f_1}{(0.4853f_2)\underline{/-135^\circ}} = \frac{f_u}{(0.4853f_2)\underline{/-135^\circ}} \tag{5.75}$$

As a result, for a 45° phase margin the condition on the closed-loop gain will be $A_{CL}(0) \geqslant f_u/0.4853f_2$ for a phase margin $= 45^\circ$.

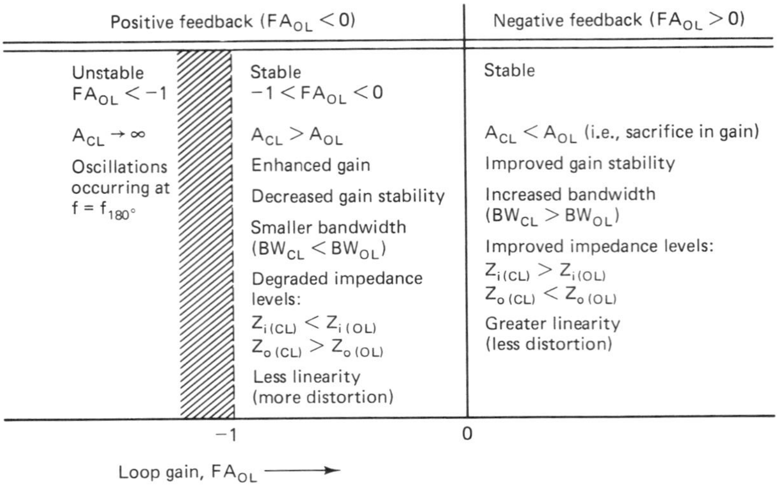

Positive feedback ($FA_{OL} < 0$)
Negative feedback ($FA_{OL} > 0$)
Unstable
$FA_{OL} < -1$
$A_{CL} \to \infty$
Oscillations occurring at $f = f_{180°}$
Stable
$-1 < FA_{OL} < 0$
$A_{CL} > A_{OL}$
Enhanced gain
Decreased gain stability
Smaller bandwidth ($BW_{CL} < BW_{OL}$)
Degraded impedance levels:
$Z_{i(CL)} < Z_{i(OL)}$
$Z_{o(CL)} > Z_{o(OL)}$
Less linearity (more distortion)
Stable
$A_{CL} < A_{OL}$ (i.e., sacrifice in gain)
Improved gain stability
Increased bandwidth ($BW_{CL} > BW_{OL}$)
Improved impedance levels:
$Z_{i(CL)} > Z_{i(OL)}$
$Z_{o(CL)} < Z_{o(OL)}$
Greater linearity (less distortion)
−1
0
Loop gain, $FA_{OL}$


**Figure 5.33** Characteristics of closed-loop (feedback) systems.

In Figure 5.33 the important characteristics of negative and positive feedback are summarized. In Figure 5.34 a representative open-loop op-amp Bode plot is shown. The critical frequencies of $f_{180°}$ and $f_{135°}$ are indicated as well as the open-loop gains at these frequencies. If the closed-loop gain $A_{CL}(0)$ falls below the $A_{CL}(f_{180°})$ line, the op amp will be operating in the unstable region. For stability it is necessary that $A_{CL}(0) > A_{OL}(f_{180°})$. If $A_{CL}(0) > A_{CL}(f_{135°})$, not only will stability be ensured, but the phase margin will be greater than 45°. If the unity-gain frequency $f_u$ is less than $f_2$, the op amp will be stable with a 45° phase margin for all closed-loop gain situations with resistive feedback elements, even down to the worst-case condition of unity feedback ($F = 1$), which corresponds to a closed-loop gain of unity. This situation in which $f_u < f_2$ will generally be the case for internally compensated op amps.

### 5.18.3 Operational-Amplifier Compensation

The open-loop gain of op amps at low frequencies [i.e., $A_{OL}(0)$] is usually extremely large ($\gtrsim$ 100,000). As a result, for almost all op amp circuits it will be the case that $A_{CL}(0) \ll A_{OL}(0)$. If we now take a look at the stability conditions, we see that it will be necessary that the first breakpoint frequency ($f_1$) be much smaller than the second breakpoint frequency ($f_2$). As the closed-loop gain decreases, so must the value of $f_1$. For example, if $A_{OL}(0) = 100{,}000$ (100 dB), $f_2 = 1.0$ MHz, and $f_3 = 4.0$ MHz, for stable operation with a closed-loop gain, $A_{CL}(0) = 100$, the value of $f_1$ must be no greater than 5.0 kHz. If stable operation with closed-loop gain values as small as 1.0 is desired (i.e., a unity-gain voltage follower), the value of $f_1$ must be restricted to a maximum of 50 Hz. If it is desired to operate the voltage follower with a 45° phase margin, then $f_1$ must be no greater than 14 Hz.

To achieve such low values for the first breakpoint frequency, it is necessary to either use an *internally compensated* op amp, or else modify the frequency response

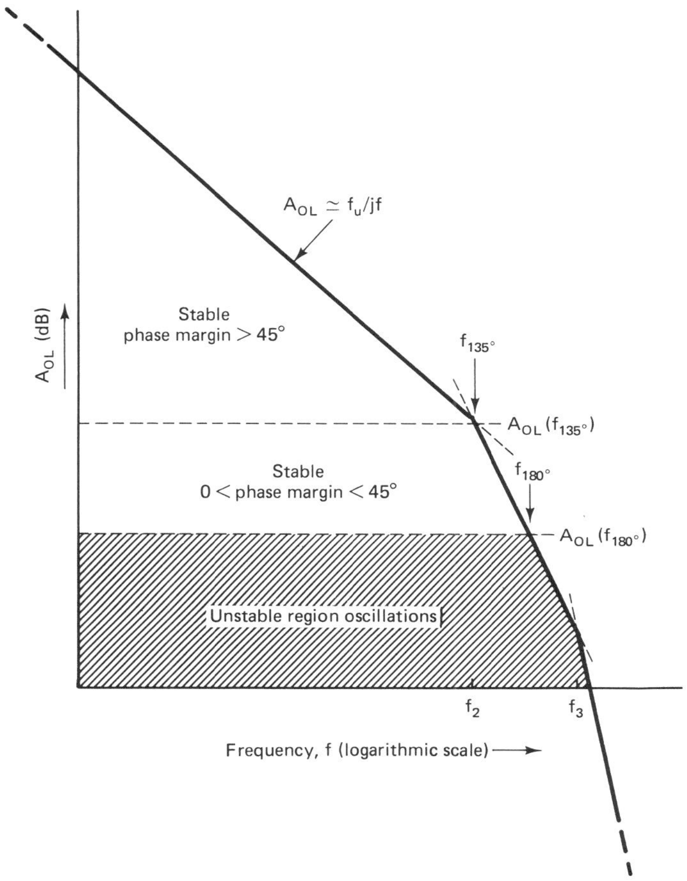


**Figure 5.34** Open-loop frequency response (Bode plot).

of the op amp by adding sufficient capacitance across the *compensation terminals* of the op amp. In either case, a "roll-off" network is produced that will introduce a factor of the form $1/(1 + jf/f_1)$ into the open-loop gain versus frequency characteristics of the op amp. The roll-off network can be considered to be basically similar to a simple $RC$ low-pass network that has a transfer characteristic expressed by

$$\frac{v_o}{v_i} = \frac{1/j\omega C}{R + (1/j\omega C)} = \frac{1}{1 + j\omega RC}$$
$$= \frac{1}{1 + j(f/f_1)} \qquad \text{where } f_1 = \frac{1}{2\pi RC} \tag{5.76}$$

**Internally compensated operational amplifiers.** In internally compensated op amps the value of the first breakpoint frequency, $f_1$, is usually chosen such that the op amp will be stable, with a phase margin of at least 45° for all closed-loop conditions which involve feedback networks containing only resistive elements. In particular, for the worst-case condition of the unity-gain voltage follower for which $A_{CL}(0) = 1$, a phase margin of at least 45° is obtained. As a result, for any other closed-loop configuration (with resistive elements in the feedback loop) the phase margin will be greater than 45°.

For the internally compensated op amps the required value of $f_1$ will usually be in the range 3 to 30 Hz, often around 10 Hz. Since this is so much lower than the other breakpoint frequencies, the open-loop 3-dB bandwidth will be essentially determined by just $f_1$, and therefore we will have that $BW_{OL} \simeq f_1$.

The compensation in the internally compensated op amps is achieved by the addition of a small (30 to 50 pF) capacitor to the silicon chip during the manufacturing process. This capacitor added to the appropriate points in the circuit will shift the breakpoint frequency $f_1$ all the way down to about 10 Hz.

**Operational amplifiers that are not internally compensated.** Internally compensated op amps have the advantage that they are simple to use since they will be stable (with a 45° phase margin) under any feedback condition that involves just resistive elements in the feedback network. They do have the disadvantage that they will result in a lower closed-loop bandwidth for all closed-loop gains above unity than would otherwise be the case. This is because the value of $f_1$ (and hence the closed-loop bandwidth) is chosen on the basis of producing a 45° phase margin under unity-gain conditions. This value of $f_1$ therefore does not correspond to the value needed for other closed-loop gain conditions.

For an op amp that is not internally compensated, the compensation is usually achieved by adding a small capacitor (usually in the range 5 to 50 pF) to specified compensation terminals. The net result, in terms of the op-amp circuitry, is the same as for the internally compensated op amp except that now the compensation capacitor is an external element and can thus be chosen for optimum circuit performance for any given closed-loop gain configuration. For a unity-gain voltage follower there would be essentially no difference between the bandwidth obtained with the op amp that is not internally compensated compared to the device that is internally compensated, and thus the internally compensated op amp would be the one to choose. For closed-loop gains greater than unity, it may be possible to obtain closed-loop bandwidths that are substantially higher by using a non-internally compensated op amp together with the correct choice of compensation capacitor. In some cases the closed-loop bandwidth obtained with the non-internally compensated op amp can be as much as 5 or even 10 times higher than the values that would be obtained by using an internally compensated op amp.

### 5.18.4 Gain Peaking

At the critical frequency of $f = f_{180°}$ what was negative feedback at low frequencies has turned around to become positive feedback. If the loop gain, $FA_{OL}$, is equal to or greater than unity in magnitude at this critical frequency, the op amp will be able to supply its own input and the circuit will break into oscillations. We have already examined the conditions necessary for amplifier stability. However, even if the circuit is indeed stable, the positive-feedback condition that exists in the region of $f_{180^\circ}$ can cause the closed-loop gain in this frequency range to be higher than the low-frequency value of the closed-loop gain $A_{CL}(0)$. If this is the case, a condition of *gain peaking* is said to exist. In this section we examine gain peaking by obtaining an expression for the quantity $A_{CL}(f_{180^\circ})/A_{CL}(0)$. If this quantity is greater than unity, there will be gain peaking. From the expression for $A_{CL}(f_{180^\circ})/A_{CL}(0)$ the required conditions for no gain peaking will be obtained.

We have that $A_{CL} = A_{OL}/(1 + FA_{OL})$. At $f = 0$ this becomes $A_{CL}(0) = A_{OL}(0)/[1 + FA_{OL}(0)] \simeq 1/F$, since $FA_{OL}(0) \gg 1$, so that $F \simeq 1/A_{CL}(0)$. We can therefore express $A_{CL}$ as $A_{CL} = A_{OL}/[1 + A_{OL}/A_{CL}(0)]$, and in normalized fashion as

$$\frac{A_{CL}}{A_{CL}(0)} = \frac{A_{OL}/A_{CL}(0)}{1 + A_{OL}/A_{CL}(0)} = \frac{A_{OL}}{A_{CL}(0) + A_{OL}} \tag{5.77}$$

At $f = f_{180^\circ}$, we have that

$$A_{OL}(f_{180^\circ}) = \frac{f_u}{f_2 + f_3} \angle 180^\circ = \frac{-f_u}{f_2 + f_3} \tag{5.78}$$

so that for the normalized closed-loop gain we have

$$\frac{A_{CL}(f_{180^\circ})}{A_{CL}(0)} = \frac{-f_u/(f_2 + f_3)}{A_{CL}(0) - f_u/(f_2 + f_3)} = \frac{-1}{[A_{CL}(0)(f_2 + f_3)/f_u] - 1} \tag{5.79}$$

For stability, the closed-loop gain at $f = f_{180^\circ}$ must remain finite, so that we require that $A_{CL}(0) > f_u/(f_2 + f_3)$. This is the same stability requirement that was given earlier.

In order not to have any gain peaking, we must require that $A_{CL}(f_{180^\circ})/A_{CL}(0) \leqslant 1$. For this to be the case, then $A_{CL}(0)(f_2 + f_3)/f_u - 1 \geqslant 1$, so that $A_{CL}(0)(f_2 + f_3)/f_u \geqslant 2$, giving the requirement that $A_{CL}(0) \geqslant 2f_u/(f_2 + f_3)$. Note that the lower limit on the closed-loop gain, $A_{CL}(0)$, obtained here for no gain peaking is exactly *twice* the minimum allowable $A_{CL}(0)$ value for stability. In terms of the $f_u$ values for a given value of $A_{CL}(0)$, the $f_u$ value for no gain peaking is one-half of the maximum allowable $f_u$ value for stability.

**Examples of gain peaking.** Let us consider the worst-case situation of $A_{CL}(0) = 1$ (i.e., a unity-gain voltage follower circuit). Let $f_2 = 1.0$ MHz and $f_3 = 2.0$ MHz, so that $f_2 + f_3 = 3.0$ MHz. For stability, the requirement therefore is that $f_u < 3.0$ MHz.

*Case 1*: $f_u = 2.5$ MHz. For this case we have that $A_{CL}(f_{180^\circ}) = 1/[(3\text{ MHz}/2.5\text{ MHz}) - 1] = 5.0$. Thus we have a 400% gain peaking factor!

*Case 2*: $f_u = 2.0$ MHz. For this case, $A_{CL}(f_{180^\circ}) = 1/[(3\text{ MHz}/2.0\text{ MHz}) - 1] = 2.0$, so that we now have 100% gain peaking.

*Case 3*: $f_u = 1.75$ MHz. For this case, $A_{CL}(f_{180^\circ}) = 1/[(3\text{ MHz}/1.75\text{ MHz}) - 1] = 1.40$, so that there is now a 40% gain peaking.

*Case 4*: $f_u = (f_2 + f_3)/2 = 1.5$ MHz. For this case $A_{CL}(f_{180^\circ}) = 1/[(3\text{ MHz}/$

1.5 MHz) − 1] = 1.00, so that there is now no gain peaking. This case represents the borderline situation between the conditions of gain peaking and no gain peaking.

*Case 5:* $f_u = 1.0$ MHz. Now we have that $A_{CL}(f_{180°}) = 1/[(3 \text{ MHz}/1 \text{ MHz}) - 1] = 0.50$, so that not only is there no gain peaking, but the gain at $f = f_{180°}$ is down substantially below $A_{CL}(0)$.

The condition for a 45° phase margin has been given as $A_{CL}(0) \geqq f_u/\sqrt{2}\, f_2$, for the case of $f_3^2 \gg f_2^2$, and in the worst-case situation of $f_3 = f_2$ the requirement becomes $A_{CL}(0) = f_u/(0.485 f_2)$. In terms of $f_u$ we have $f_u \leqq \sqrt{2}\, f_2 A_{CL}(0)$ for the case in which $f_3$ is well separated from $f_2$, and $f_u \leqslant 0.485 f_2 A_{CL}(0)$ for the worst-case situation of $f_3 = f_2$. Therefore, for the example above, we see that for a 45° phase margin, an $f_u$ value of about 1.0 MHz maximum would be required. We note that with this value of $f_u$ there will be no gain peaking, and in general, there will be no gain peaking if the phase margin is at least 45°.

We have examined the "gain peaking" situation at the critical frequency of $f = f_{180°}$. From that analysis we see that to prevent gain peaking the requirement is that $A_{CL}(0)(f_2 + f_3)/f_u \geqslant 2$, so that $f_u \leq A_{CL}(0)(f_2 + f_3)/2$. For the worst-case situation of $f_2 = f_3$, this will become $f_u \leq A_{CL}(0) f_2$. At the other extreme, however, when $f_3$ is remote from $f_2$ there may still be gain peaking even if the conditions above are satisfied, although the gain peaking will not occur at $f_{180°}$.

To examine this situation, let us assume that

$$A_{OL} = \frac{A_{OL}(0)}{(jf/f_1)(1 + jf/f_2)} = \frac{f_u}{jf(1 + jf/f_2)} \tag{5.80}$$

Since

$$A_{CL} = \frac{A_{OL}}{(1 + FA_{OL})} = \frac{1/F}{1 + (1/FA_{OL})} = \frac{A_{CL}(0)}{1 + A_{CL}(0)/A_{OL}}$$

we have that

$$\begin{aligned}\frac{A_{CL}}{A_{CL}(0)} &= \frac{1}{1 + A_{CL}(0)/A_{OL}} = \frac{1}{1 + [A_{CL}(0)/f_u]jf[1 + j(f/f_2)]} \\ &= \frac{1}{1 + [A_{CL}(0)/f_u][jf - (f^2/f_2)]} \\ &= \frac{1}{1 - (f^2/f_2 f_u)A_{CL}(0) + jfA_{CL}(0)/f_u}\end{aligned} \tag{5.81}$$

When $f = \sqrt{f_2 f_u/A_{CL}(0)}$, the normalized closed-loop gain will be given by

$$\frac{A_{CL}}{A_{CL}(0)} = \frac{1}{j\sqrt{[f_u f_2/A_{CL}(0)][A_{CL}(0)/f_u]}} = \frac{1}{j\sqrt{(f_2/f_u)A_{CL}(0)}} = -j\sqrt{\frac{f_u}{f_2 A_{CL}(0)}} \tag{5.82}$$

Therefore, for no gain peaking we require that $f_u < f_2 A_{CL}(0)$. When this condition is compared to the previous condition pertaining to gain peaking at $f_{180°}$, we see that in general if $f_u < f_2 A_{CL}(0)$ there will be no gain peaking.

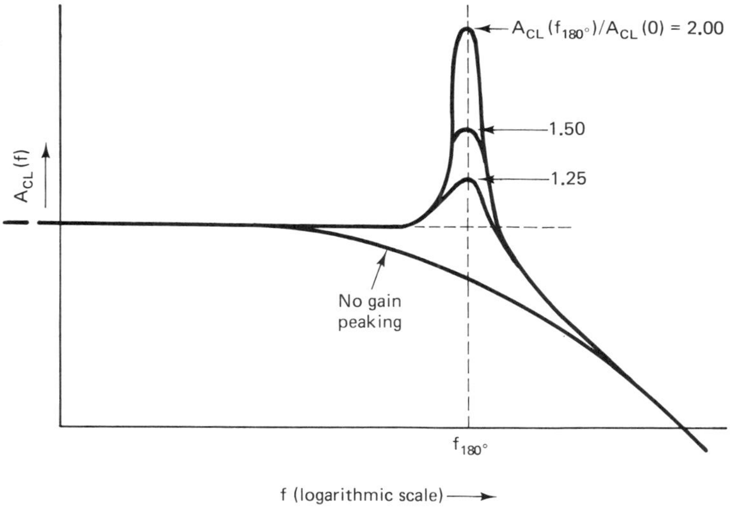


**Figure 5.35** Gain peaking due to inadequate phase margin.

In Figure 5.35 the closed-loop frequency response characteristics of an op amp are shown with various amounts of gain peaking. In Figure 5.36 the op amp response to a rectangular pulse input voltage is shown. Gain peaking in the frequency-domain response has as its counterpart *overshoot* in the time-domain response. In addition to the overshoot, there is often a damped oscillatory type of transient response known as *ringing*.

## 5.18.5 Effect of Load Capacitance and Input Capacitance on Stability

Internally compensated operational amplifiers are normally stable with a phase margin of at least 45° for all closed-loop situations involving just resistive elements in the feedback loop. A large load capacitance $C_L$ can, however, have a destabilizing effect on the op amp. The combination of the load capacitance and the open-loop output

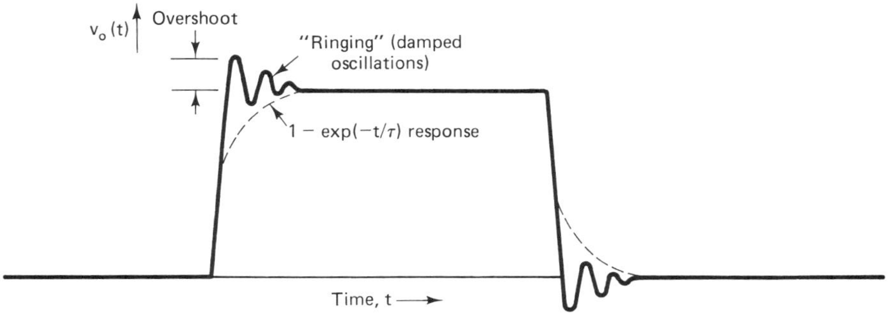


**Figure 5.36** Time-domain response showing overshoot and ringing.

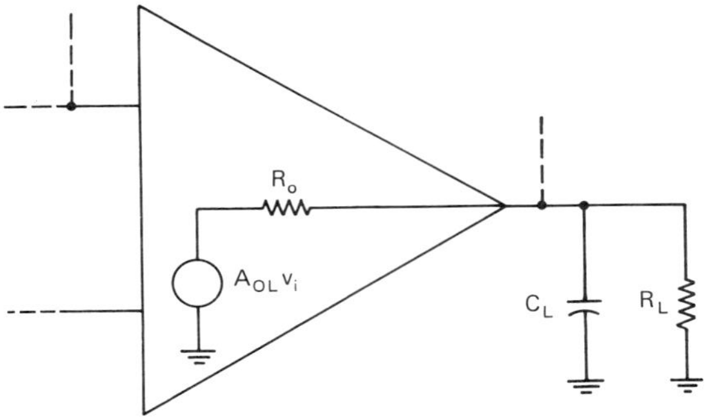


**Figure 5.37** Effect of load capacitance on amplifier stability.

resistance $R_o$ of the op amp will result in a low-pass or "lag" network as shown in Figure 5.37.

This lag network will have a transfer function given by

$$T(f) = \frac{1}{1 + j\omega\tau_L} = \frac{1}{1 + jf/f_L} \tag{5.83}$$

where $\tau_L = (R_o \| R_L)C_L$ and $f_L = 1/(2\pi\tau_L)$. This will add a new breakpoint frequency to the loop gain, which can now be written as

$$\text{loop gain} = FA_{OL} = \frac{FA_{OL}(0)}{(1 + jf/f_1)(1 + jf/f_2) \cdots (1 + jf/f_L)}$$

As long as $f_L$ is substantially larger than $f_2$ it will have no serious effect on the operation and stability of the circuit. However, if $C_L$ is such that $f_L$ drops below $f_2$, the sequence of frequencies in the stability equations must be reordered, with $f_L$ replacing $f_2$, $f_2$ replacing $f_3$, and so on. The condition for a 45° phase margin in this case will become $A_{CL}(0) > f_u/\sqrt{2}\, f_L$. Thus as $C_L$ increases and $f_L$ correspondingly decreases, this condition may no longer be satisfied. Indeed, with a large enough value of $C_L$ the op amp may become unstable and break into oscillations.

As an example of the effect of load capacitance, let us consider an op amp with $f_u = 1.0$ MHz and an open-loop output resistance of $R_o = 100\ \Omega$. Taking the case of a unity-gain voltage-follower circuit, as $f_L$ drops below $f_2$ the requirement for stability with a 45° phase margin becomes

$$A_{CL}(0) = 1.0 > \frac{f_u}{\sqrt{2}\, f_L} = \frac{1.0\ \text{MHz}}{\sqrt{2}\, f_L} = \frac{707\ \text{kHz}}{f_L} \tag{5.84}$$

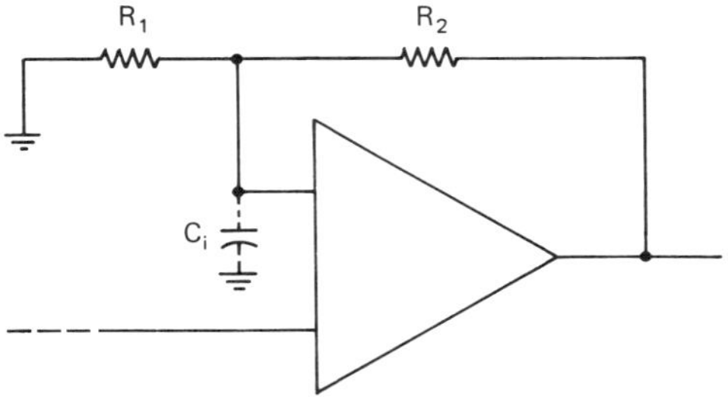


**Figure 5.38** Effect of input capacitance on amplifier stability.

Therefore, $f_L = 1/[2\pi(R_o \| R_L)C_L]$ should be greater than 707 kHz. Taking the worst-case condition of $R_L \gg R_O$, the corresponding requirement on the load capacitance is that $C_L < 2.25$ nF.

The combination of large-value resistors in the feedback circuit and the input capacitance $C_i$ of the op amp can also act to destabilize the op amp. The combination of the feedback resistors $R_1$ and $R_2$ and the input capacitance $C_i$ results in a low-pass or lag network, as shown in Figure 5.38. The transfer function of this lag network will be given by

$$T(f) = \frac{R_1/(R_1 + R_2)}{1 + j\omega\tau_i} = \frac{R_1/(R_1 + R_2)}{1 + j(f/f_i)} \tag{5.85}$$

where the time constant $\tau_i$ is given by $\tau_i = (R_1 \| R_2)C_i$ and $f_i = 1/(2\pi\tau_i)$.

This transfer function will be part of the loop gain and will thus add a new breakpoint frequency to the loop gain, which can now be written as

$$\text{loop gain} = FA_{OL} = \frac{R_1/(R_1 + R_2)A_{OL}(0)}{(1 + jf/f_1)(1 + jf/f_2) \cdot \cdot \cdot (1 + jf/f_i)} \tag{5.86}$$

The situation is now similar to that encountered with respect to the effect of the load capacitance. As long as $f_i$ is substantially above $f_2$, it will have no significant effect on the op-amp stability. If $f_i$ drops below $f_2$, however, it must now be considered and the sequence of breakpoint frequencies in the stability equations must be reordered, with $f_i$ replacing $f_2$, $f_2$ replacing $f_3$, and so on.

As an example of the effect of the input capacitance on stability, let us take an op amp with $f_u = 1.0$ MHz and $C_i = 5.0$ pF. For the case of a voltage-follower circuit with $A_{CL}(0) = 1$ the condition for stability with a 45° phase margin becomes $A_{CL}(0) = 1 > f_u/\sqrt{2}\, f_i = 1.0\ \text{MHz}/\sqrt{2}\, f_i = 707\ \text{kHz}/f_i$. The corresponding requirement in terms of $R_1$ and $R_2$ will be that $f_i = 1/[2\pi(R_1 \| R_2)C_i] > 707$ kHz, so that $(R_1 \| R_2) < 45\ \text{k}\Omega$.

The effect of $C_i$ can be compensated for by using a small feedback capacitor $C_2$ in parallel with the $R_2$ feedback resistor, as shown in Figure 5.39. The feedback factor for this circuit can now be written as $F = Z_1/(Z_1 + Z_2) = 1/[1 + (Z_2/Z_1)]$, where $Z_1 = R_1/(1 + j\omega\tau_1)$ and $Z_2 = R_2/(1 + j\omega\tau_2)$ with $\tau_1 = R_1C_i$ and $\tau_2 = R_2C_2$. Since the impedance ratio is $Z_2/Z_1 = (R_2/R_1) \times (1 + j\omega\tau_1)/(1 + j\omega\tau_2)$, we see that if the two time constants are equal such that $\tau_1 = \tau_2$, the impedance ratio and therefore the feedback factor will become independent of frequency, and

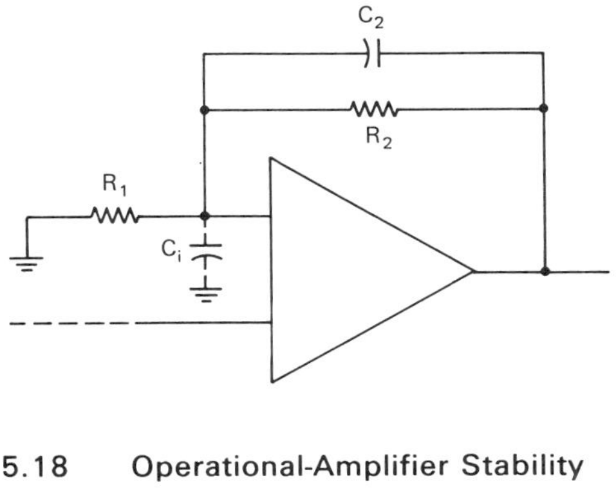


**Figure 5.39** Compensation for input capacitance.

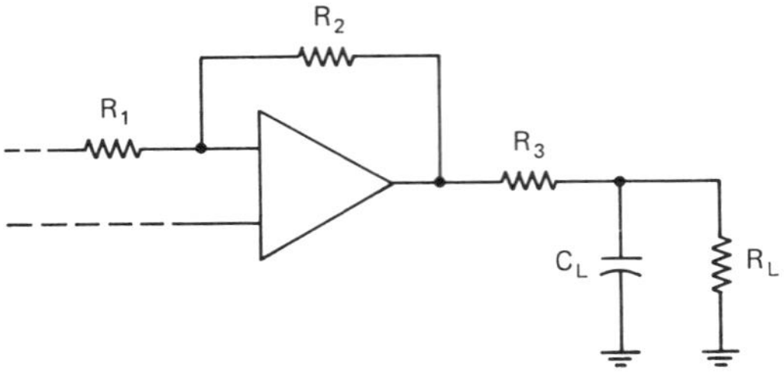


**Figure 5.40** Use of resistor to isolate load capacitance.

the effect of $C_i$ will therefore be canceled out. The basic requirement on $C_2$ can therefore be written as $R_1C_i = R_2C_2$.

The same basic technique described can be used in some cases to compensate for the effects of the load capacitance. If $R_2 \gg R_1$, the required condition for this compensation is that $\tau_2 = \tau_L$, where $\tau_2 = R_2C_2$ and $\tau_L = R_oC_L$.

Another technique that can be used to minimize the destabilization effects of large load capacitances is to use a small resistor $R_3$ between the op amp and the load, as shown in Figure 5.40. This provides some degree of isolation between the op amp and $C_L$, and generally, resistance values of the order of 100 Ω will be sufficient. The use of this isolating resistor can ensure stability of the op amp, but it carries with it the disadvantages of reducing the output voltage, increasing the output impedance of the circuit, and reducing the overall bandwidth.

## 5.19 OPERATIONAL-AMPLIFIER APPLICATIONS

A number of representative operational-amplifier applications are shown in the problems at the end of this chapter. Some brief remarks concerning these applications are given below. *Note:* The figures referenced in this section can be found at the end of this chapter in the Problem section.

5.1. *Difference amplifier* (see Figure P5.1): In this circuit the output voltage is proportional to the difference of the two input voltages. In the ideal situation any common-mode component of the two input voltages will be totally rejected. In the actual situation, because of the finite common-mode gain of the operational amplifier and the mismatches in the resistance ratios there will be some small common-mode response. Notice also that the input resistance with respect to the $V_A$ and $V_B$ inputs will not be the same, being $R_A$ for the $V_A$ input and $R_A + R_F$ for the $V_B$ input.

5.2. *Summing amplifier* (see Figure P5.2): A number of input voltages can be arithmetically combined with various weighting factors.

5.3. *Current-to-voltage converter—transimpedance amplifier* (see Figure P5.3): This curcuit produces an output voltage that is directly proportional to the input current. Note that the voltage drop at the input $V_i$ will be $V_i = -V_O/A_{OL} = I_iR_F/A_{OL}$, so that the effective input resistance looking into this circuit will be $R_i = V_i/I_i = R_F/A_{OL}$. Therefore, even if $R_F$ is relatively large the effective input resistance $R_i$ will be very small, due to the very large open-loop gain of the amplifier. This very small input resistance will be of great advantage since it will not produce a significant effect on the circuit in which the current is to be measured. Note also that the output voltage will be substantially independent of the load that is driven by this circuit.

5.4. *Voltage-to-current converter—voltage-controlled current source* (see Figure P5.4): The current through the load resistance will be independent of the value of the load resistance, but will be directly proportional to the input voltage, so that this circuit will act as a voltage-controlled constant current source. Note, however, that neither end of $R_L$ can be grounded.

5.5. *Low-pass active filter—integrator* (see Figure P5.5): This circuit is a simple single-pole low-pass active filter with a 3-dB bandwidth given by $BW = 1/(2\pi R_F C_F)$. At low frequencies the gain asymptotically approaches the zero-frequency gain value of $A_{CL}(0) = -R_F/R_A$. At high frequencies the gain will asymptotically approach a response given by $A_{CL}(\omega) = -1/j\omega R_A C_F$ and thus will drop off at a rate of $-20$ dB/decade or $-6$ dB/octave.

Looking at this in the time domain, this circuit can be used as an integrator. In the limiting case in which $R_F$ goes to infinity, we will have $V_O = -(1/C_F) \int^t i\, dt = -(1/C_F) \int^t (V_S/R_A)\, dt = -(1/R_A C_F) \int^t V_S\, dt$, so that this circuit will indeed act as an ideal integrator. Since the d-c gain under these conditions will be $A_{OL}$, the output voltage will be very strongly affected by the op-amp $V_{OS}$ and will either be $A_{OL} V_{OS}$ or more likely will be saturated near the positive or negative supply voltages. To prevent this from happening, a feedback resistor $R_F$ of suitable size is usually included in the circuit, although it makes the circuit a less than ideal integrator.

5.6. *High-pass active filter—differentiator* (see Figure P5.6): This circuit is a simple single-pole high-pass active filter with a breakpoint frequency of $f_{bp} = 1/(2\pi R_A C_A)$. The gain will asymptotically approach $-R_F/R_A$ at high frequencies, and will asymptotically go to zero at low frequencies.

Looking at the operation of this circuit in the time domain, and letting $R_A$ go to zero, we will have that $i_s = C_A(dv_s/dt)$, so that $v_o = -i_s R_F = -R_F C_A\, (dv_s/dt)$. We see that this circuit will act as an ideal differentiator. With $R_A = 0$, however, the gain at high frequencies will be relatively large. This large high-frequency gain can cause problems due to circuit noise and other factors, so that a resistor $R_A$ of some suitable size is often included in the circuit, although this will result in a less than ideal differentiator characteristic.

5.7. *Precision detector or rectifier* (see Figure P5.7): The input voltage $V_i$, that is required to turn the diode on will be $V_i = V_D/A_{OL} \simeq 0.5\ \text{V}/A_{OL}$, so that the effect of the diode forward voltage drop will be essentially nullified. As a result, this circuit can detect (or rectify) very low-level signals.

5.8. *Precision full-wave rectifier* (see Figure P5.8): Note again the effect of the op amp in essentially nullifying the forward voltage drop of the diode. The first and third op amps are voltage followers for source and load isolation, respectively.

5.9. *Precision peak detector* (see Figure P5.9): Again note the action of the op amp in reducing the effect of the diode forward voltage drop. The second op amp is used for load isolation such that the load resistance will not affect the discharge rate of the capacitor.

5.10. *Logarithmic converter* (see Figure P5.10): This circuit uses nonlinear elements (transistors) in the feedback loop to obtain a nonlinear (logarithmic) transfer characteristic. The gain of the difference amplifier ($A_3$) is set so as to obtain a logarithmic conversion scale factor of 1.0 V/decade.

5.11. *Exponential amplifier—antilogarithmic converter* (see Figure P5.11): This circuit produces an output voltage that is an exponential function of the input voltage. Note that the overall transfer characteristic of this circuit represents the inverse function of the feedback device.

5.12. *Current integrator—charge amplifier* (see Figure P5.12): This is another integrator circuit in which the output voltage will be directly proportional to the net flow of charge $Q$ into the circuit. Note that the voltage drop across the input terminals will be very small. Note also that the $V_{OS}$ of the op amp can cause problems so that it may be desirable to connect a feedback resistor $R_F$ across $C_F$, although this would make this circuit a less than ideal current integrator.

5.13. *Schmitt trigger* (see Figure P5.13): This circuit uses positive feedback and will operate in the switching mode with the output voltage being in either of two states, $V_O^+$ and $V_O^-$. For most op amps the high state $V_O^+$ will be near the positive supply $V^+$ and the low output state $V_O^-$ will be close to the negative supply voltage $V^-$. The voltage applied to the noninverting input terminal will be a function of the fixed reference voltage $V_{\text{REF}}$ and the output voltage $V_O$. As a result, the input threshold voltage for switching will be a function of the output voltage, and this results in the hysteresis loop of the input–output transfer characteristic.

5.14. *Positive voltage regulator* (see Figure P5.14): In this circuit the output voltage is limited to values less than $V_Z$. The voltage follower is used to keep load resistance or current variations from affecting the output voltage (i.e., to give good load regulation).

5.15. *Positive voltage regulator with* $V_O$ *greater than* $V_Z$ (see Figure P5.15): In this circuit the output voltage can be greater than $V_Z$, but cannot be any less than $V_Z$. Note that the current through the zener diode will be given by $I_Z = I_{R3} = V_O R_2/(R_1 + R_2)$, so that it will be essentially independent of the supply voltage. As a result, the output voltage will be relatively insensitive to changes in the supply voltage (i.e., good line regulation) as long as the supply voltage is above the minimum value required for the operation of this circuit.

5.16. *High current voltage regulator with current limiting* (see Figure P5.16): The maximum current output of this circuit is not limited by the maximum current capability of the op amp, but is increased by the net current gain of the $Q_1$-$Q_2$ Darlington circuit. Transistor $Q_3$ and resistor $R_{CL}$ are used for current limiting in order to limit $I_L$ to a safe value to prevent excessive power dissipation in $Q_2$.

5.17. *Constant-current sink* (see Figure P5.17): The output current $I_O$ will be approximately independent of the output voltage $V_O = V_{C2}$ as long as both $Q_1$ and $Q_2$ remain in the active region. This requires that $V_{CE2} > 0.9\text{ V} + V_{\text{REF}}$, and less than the collector-to-emitter breakdown voltage. Note that $I_O + I_{B1} - I_{\text{BIAS}} = I_{R1} = (V_{\text{REF}} + V_{OS})/R_1$, so that for precision operation down to low current levels a small $V_{OS}$ and $I_{\text{BIAS}}$ is desirable. The high net current gain of the Darlington pair will make $I_{B1}$ very small compared to $I_O$, which is also a desirable condition.

5.18. *Precision constant-current source for low current levels* (see Figure P5.18): In this circuit $I_O + I_G - I_{\text{BIAS}} = (V_{\text{REF}} + V_{OS})/R_1$, where $I_G$ is the gate current of the JFET. Again we see that for precision operation down to very low current levels the op-amp offset voltage and bias current should be very small. The use of a JFET is of advantage in this circuit due to the very small gate current $I_G$, which will usually be less than 1 nA.

5.19. *Operational amplifier with electronic gain control* (see Figure P5.19): The drain-to-source resistance of a JFET when $V_{DS} \lesssim V_P/3$ will be a function of the gate-to-source voltage $V_{GS}$ and is given approximately by $r_{ds} = r_{ds(\text{ON})}/(1 - \sqrt{V_{GS}/V_P})$, where $r_{ds(\text{ON})}$ is the drain-to-source resistance when $V_{GS} = 0$. This voltage-variable resistance characteristic of the JFET can be used for the electronic gain control of an op-amp circuit. Note, however, that the closed-loop gain will be a nonlinear function of the gain control voltage, $-V_{\text{BIAS}}$.

5.20. *Operational amplifier with electronic gain control* (see Figure P5.20): Again in this circuit the voltage-variable resistance characteristics of JFETs are used to control the closed-loop gain of an op-amp circuit. Note that in this circuit, however, the gain will be a linear function of the control voltage.

5.21. *Tracking voltage regulator* (see Figure P5.21): This circuit produces two voltages of equal magnitude proportional to the input voltage $V_{\text{REF}}$, but of opposite algebraic sign. The two transistors are used just to increase the output current capability of the circuit.

5.22. *Precision phase splitter with high input impedance and low output impedance* (see Figure P5.22): This circuit produces two output voltages of equal magnitude, but of opposite algebraic sign.

5.23. *Instrumentation amplifier with high input impedance and low output impedance* (see Figure P5.23): This is basically a difference amplifier, but with a very high input impedance for both inputs.

5.24. *Instrumentation amplifier with high input impedance and low output impedance* (see Figure P5.24): This again is a difference amplifier with a very high input impedance for both input channels and a very low output impedance.

5.25. *Exponential converter—antilogarithmic amplifier* (see Figure P5.25): This circuit produces an output voltage that increases at an exponential rate with increasing input voltage. Note that the input voltage $V_i$ can be of either positive or negative polarity, but the output voltage will be of positive polarity in either case.

5.26. *Circuit for raising a variable to a power with logarithmic techniques* (see Figure P5.26): This circuit uses a combination of a logarithmic converter and an antilogarithmic converter to produce the input–output relationship $V_O = (V_i)^{R_F/R_L}$. Note that the exponent $R_F/R_L$ is not restricted to integer values and it can be less than unity or greater than unity as required.

5.27. *Amplifier with exponential gain control* (see Figure P5.27): The exponential gain control characteristics of this circuit makes it possible to obtain a very large variation in the closed-loop gain with only a relatively small variation in the gain control (or AGC) voltage.

5.28. *Function generator* (see Figure P5.28): Note that the functional relation obtained from this circuit is the inverse of the functional relationship of the device in the feedback loop.

5.29. *Dump-and-integrate circuit* (see Figure P5.29): In the dump-and-integrate circuit of Figure P5.29, transistor $Q_1$ is turned on prior to the integration time, to discharge capacitor $C_1$ and set the output voltage to zero. Then the transistor is turned off. The output voltage will then be equal to the integral of the input voltage as given by

$$V_O = -\int_0^t \frac{1}{R_1 C_1} V_i \, dt$$

5.30. *Precision voltage-controlled limiting ("clipping" or "bounding") circuit* (see Figure P5.30): This is a noninverting unity gain amplifier circuit that clips the output voltage at $V_{\mathbf{REF}}$ such that $V_O = V_S$ for $V_S < V_{\mathbf{REF}}$ and $V_O = V_{\mathbf{REF}}$ for $V_S > V_{\mathbf{REF}}$ where $V_{\mathbf{REF}}$ can be of either polarity. The forward voltage drop of the diode is, in effect, divided by the amplifier open-loop gain such that it will have very little effect on the operation of this circuit. If diode $D_1$ is reversed, the transfer relationship will be $V_O = V_S$ for $V_S > V_{\mathbf{REF}}$ and $V_O = V_{\mathbf{REF}}$ for $V_S < V_{\mathbf{REF}}$.

5.31. *Precision voltage-controlled clamping circuit* (see Figure P5.31): This circuit produces an output voltage that has the same a-c variation as the input voltage, but the d-c level is shifted such that the output voltage will not drop below $V_{\mathbf{REF}}$. The voltage $V_{\mathbf{REF}}$ can be of either polarity. If diode $D_1$ is reversed, the output voltage will again have the same a-c variation as the input, but the shift in the d-c level is such that the output voltage will not rise above $V_{\mathbf{REF}}$.

The diode forward voltage drop is, in effect, divided by the open-loop gain of A2 such that it will have a negligible effect on the performance of the circuit in most cases.

5.32. *Voltage-controlled gain polarity switching circuit* (see Figure P5.32): This circuit is an inverting amplifier with a closed-loop gain given by $A_{CL} = -(1 + R_2/R_1)$ when transister $Q_1$ is turned on ($V_{\text{control}} = 0$), and a noninverting amplifier with $A_{CL} = 1 + R_2/R_1$ with $Q_1$ is turned off ($V_{\text{control}} < V_P$). The magnitudes of the two voltage gains will be approximately equal subject to the condition that $r_{ds} \ll R_4$.

5.33. *Voltage-controlled impedance multiplier* (see Figure P5.33): The feedback impedance $Z_F$ of this circuit is transformed to appear as an input impedance of value $Z_i = Z_F/(1 + A)$ and correspondingly the input admittance is given by $Y_i = (1 + A)Y_F$. If the feedback impedance is a capacitor of value $C_F$ the input capacitance of this circuit will be $C_i = C_F\ (1 + A)$ so that this circuit can act as a capacitance multiplier. If the gain of amplifer $A2$ is varied by the gain control voltage, the input capacitance will be a voltage-variable capacitance.

5.34. *Inductance simulator* (see Figure P5.34): The input impedance of this circuit is proportional to the reciprocal of the feedback impedance $Z_F$, and will be given by $Z_i = R_2R_3/Z_F$. If $Z_F$ is a capacitor of value $C_F$ the input impedance will be given by $Z_i = j\omega R_2R_3C_F = j\omega L_{eq}$ so that the input impedance appears as an inductance of value $L_{eq} = R_2R_3C_F$.

5.35. *Analog signal multiplexing circuit* (see Figure P5.35): For this circuit, $\phi_1$ through $\phi_N$ are non-overlapping active low clock pulses such that transistors $Q_1$ through $Q_N$ are "on" except when the clock pulses applied to one of the transistors is active which will turn that transistor "off." This will then permit the signal transmission from the selected input to the output. The resistance $R_1$ should be chosen such that $R_1 \gg r_{ds\,(ON)}$.

One variation of this circuit is to use a series-shunt switch arrangement with additional FETs in series with $R$ and driven with active high-clock pulses such that when a shunt switch transistor is "on" the corresponding series switch transistor will be "off," and vice-versa.

5.36. *Symmetrical bipolar limiter* (see Figure P5.36): This circuit provides symmetrical limiting of the output voltage. This is done without excessive current drain from the op-amp output due to the limiting diodes being in the feedback loop rather than between the output terminal and ground.

5.37. *Constant-amplitude phase shifter* (see Figure P5.37): This circuit has a closed-loop voltage gain that has a magnitude of unity, independent of frequency. The phase shift, however, will change with frequency, ranging from 0° at d-c to a value approaching −180° at high frequencies. This circuit can be used as a time-delay circuit, with the time delay $T_d$ being given by $T_d = \phi/\omega$, where $\phi$ is the phase shift.

5.38. *Bridge amplifier* (see Figure P5.38): This circuit is useful for thermometry. If a piezoresistive element is used in the bridge, this circuit can be used as a pressure or strain transducer.

5.39. *Active RC bandpass amplifier* (see Figure P5.39): This circuit can be used as a bandpass amplifier to amplify a narrow band of frequencies (the passband) and to reject frequencies that are outside the passband. This circuit is especially advantageous for low-frequency applications since only $RC$ elements and no inductors are required.

5.40. *Square-wave oscillator* (see Figure P5.40): This simple circuit produces a square wave with a peak-to-peak amplitude that is almost equal to the net supply voltage ($V^+ - V^-$) and with a frequency determined by the $R_1C_1$ time constant. Note the combined use of positive and negative feedback in this circuit.

5.41. *Precision triangle wave generator* (see Figure P5.41): This circuit produces a triangular waveform that has very straight sides. Op amp $A_1$ is a Schmitt trigger, $A_2$ is a an integrator, and $A_3$ serves as an inverting amplifier.

5.42. *Sample-and-hold circuit* (see Figure P5.42): This is a simple sample-and-hold circuit that can sample the input voltage over a short time interval, and then hold that sampled value over an extended period of time. Note the use of the two voltage followers for source and load isolation.

5.43. *Howland current source* (see Figure P5.43): This is a voltage-controlled constant current source that produces an output current of $I_L = (V_1 - V_2)/R_1$. The output current can be of either polarity and one side of the load resistance that is driven by this current source can be grounded.

5.44. *Circuit to produce an output voltage that increases linearly with temperature* (see Figure P5.44): This is an interesting circuit that produces an output voltage that is directly proportional to the absolute temperature.

5.45. *Voltage regulator* (see Figure P5.45): This is a simple voltage regulator circuit for $V_O > V_{REF}$. By adding a transistor or a Darlington transistor pair on the output side of the op amp the current output capability of this circuit can be greatly increased.

5.46. *Voltage regulator with current foldback* (see Figure P5.46): This somewhat more complex voltage regulator circuit has a "current foldback" current-limiting characteristic to increase the output current available to drive the load, yet still provide full protection to the output transistors.

5.47. *Analog multiplier* (see Figure P5.47): With a combination of three op amps and a set of four matched transistors a circuit is obtained that can be used as a multiplier, a divider, or as a square-rooting circuit. Note that this circuit can be connected as a four-quadrant multiplier in which input voltages of either polarity can be accepted. This circuit is also available as a monolithic IC.

5.48. *Phase shift oscillator* (see Figure P5.48): This is a feedback oscillator. For oscillations to occur the net gain around the feedback loop must be greater than unity at the frequency at which the net phase shift around the feedback loop is zero. The frequency of oscillation will be given by $f_{osc} = 1/(2\pi\sqrt{3}\, R_1C_1)$ and the gain condition will be satisfied if $R_3/R_2 > 8$. If the net gain around the feedback loop is adjusted to a value that is just slightly above unity, the output voltage will be a relatively undistorted sinusoidal waveform.

This circuit will also work if voltage-followers $A_2$ and $A_3$ are omitted. The frequency of oscillation will then be given by $f_{osc} = 1/(2\pi\sqrt{6}\, R_1C_1)$ and the gain condition will require that $R_3/R_2 > 29$.

5.49. *Wien bridge oscillator* (see Figure P5.49): This circuit has both a positive and negative feedback loop. For oscillations to occur, the net feedback must be positive with a phase angle of zero. This gives the required condition that $(R_3/R_4) > (R_1/R_2) + (C_1/C_2)$ and the corresponding frequency of oscillation will be $f_{osc} = 1/(2\pi\sqrt{R_1R_2C_1C_2})$. If the $R_3/R_4$ ratio is adjusted to a value that is just slightly above that required for oscillations, then a relatively undistorted sinusoidal waveform will be obtained.

5.50. *Optoelectronic sensor* (see Figure P5.50): This simple light sensing circuit consists of a photodiode and a current-to-voltage converter op-amp circuit. The photocurrent produced by the photodiode will be a linear function of the light intensity so that the resulting output voltage of this circuit will also be a linear function of the light intensity. The photodiode bias voltage could be set to zero, but a reverse bias voltage across the photodiode will have the advantage of reducing the junction capacitance and thereby decreasing the response time of this circuit.

5.51. *Circuit for evaluating the power supply rejection ratio* (*PSRR*) (see Figure P5.51): The coupling capacitor $C_C$ is used to block the output voltage component that is the result of the d-c input offset voltage of the op amp. As a result, only the a-c output voltage that results from the power supply a-c ripple will be measured.

5.52. *Circuit for obtaining the open-loop gain* (see Figure P5.52): Note that although it is the open-loop gain of the op amp that is to be measured, this circuit is nevertheless operated as a closed-loop system. This is to prevent the output voltage from being driven into saturation near either supply voltage due to the effect of the input offset voltage $V_{OS}$.

5.53. *Evaluation of the equivalent input noise voltage* (see Figure P5.53): This op amp is operated with a high closed-loop gain in order to greatly amplify the input noise voltage to make it easier to measure.

5.54. *Common-mode input impedance* (see Figure P5.54): This simple voltage-follower circuit illustrates the effect of the common-mode input resistance and the input capacitance on the overall circuit performance and can be used for the measurement of these two parameters.

5.56. *Common-mode gain* (see Figure P5.56): If the op amp has a nonzero common-mode gain, or if the resistance ratios are not exactly equal, this circuit will not act as an ideal difference amplifier, but there will be some response to a common-mode input voltage.

## 5.20 ACTIVE FILTERS

An important category of op-amp applications is that of active filters. In the following discussion various low-pass, high-pass, bandpass, and band-reject active filters will be presented together with the appropriate design equations.

The ideal frequency selective filter is a device or system that has an input-to-output transfer characteristic that is constant over the specified passband and provides zero signal transmission in the stop bands. In Figure 5.41 the transfer characteristics of some ideal and actual low-pass, high-pass, bandpass, and band-reject filters are shown.

Passive filters use only passive elements, resistors, capacitors, and inductors. An active filter uses one or more active devices, usually an operational amplifier, in the filter circuit. Some advantages of active filters over their passive counterparts are:

1. *Gain*: with active filters transfer functions with maximum values greater than unity are possible.

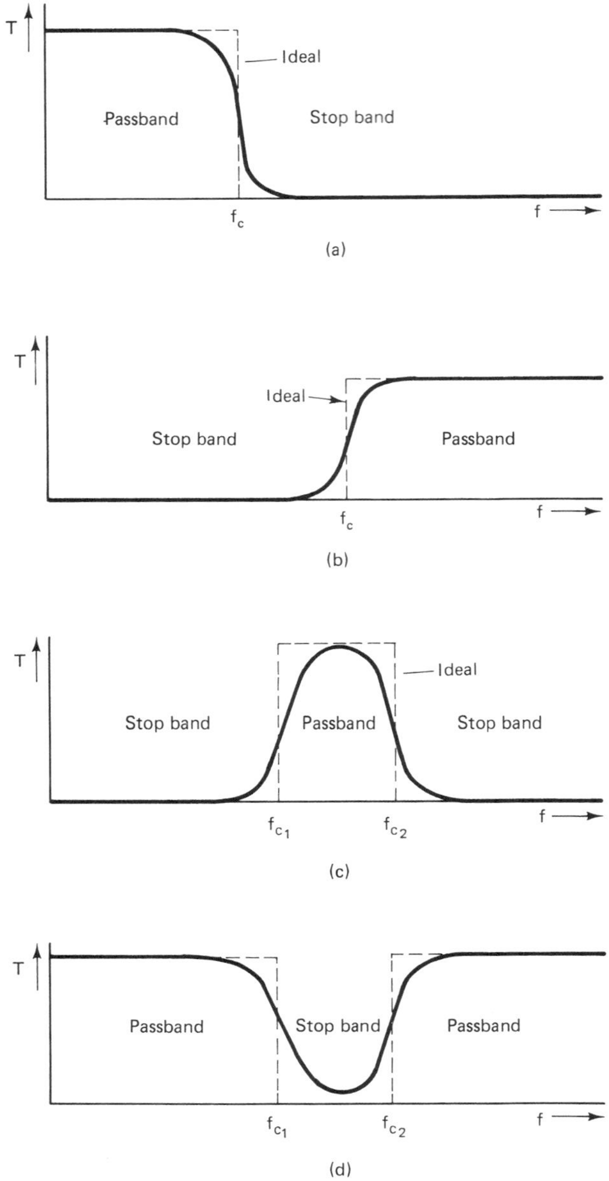


**Figure 5.41** Frequency selective filter characteristics: (a) low-pass filter; (b) high-pass filter; (c) bandpass filter; (d) band-reject filter.

2. *Minimal loading effects*: the transfer characteristics of an active filter can be substantially independent of the load driven by the filter or of the source that drives the filter.
3. *Inductorless filters*: with active filters only resistive and capacitative elements are required; no inductors are needed. This can be an especially useful feature

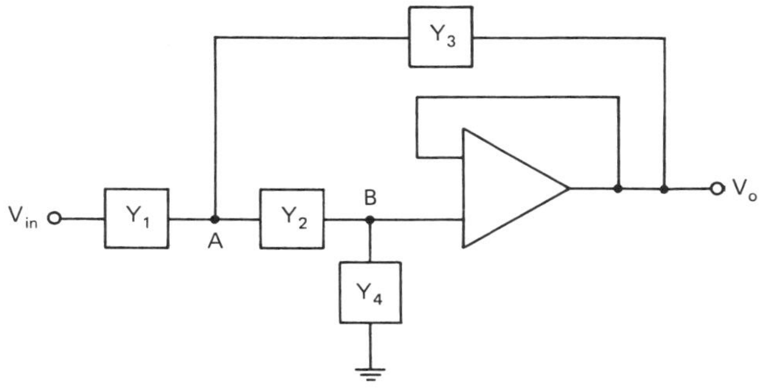


**Figure 5.42** General two-pole active filter circuit.

for operation at relatively low frequencies (<10 kHz) where otherwise large inductors would be required.

### 5.20.1 Analysis of General Two-Pole Active Filter Circuit

Let us consider the circuit of Figure 5.42. If we assume an ideal infinite-gain op amp the output voltage $V_o$ will be equal to the voltage at node $B$, so that $V_o = V_B$. The node voltage equation for node $A$ will be $V_{in}Y_1 + V_OY_3 + V_oY_2 = V_A(Y_1 + Y_2 + Y_3)$ and for node $B$ we will have that $V_AY_2 = V_O(Y_2 + Y_4)$. Solving this last node voltage equation for $V_A$ gives $V_A = V_O(Y_2 + Y_4)/Y_2$. Substituting this back into the equation for node $A$ gives

$$V_{in}Y_1 + V_o(Y_2 + Y_2) = \frac{V_o(Y_2 + Y_4)(Y_1 + Y_2 + Y_3)}{Y_2} \tag{5.87}$$

Multiplying this equation through by $Y_2$ yields

$$V_{in}Y_1Y_2 + V_o(Y_2^2 + Y_2Y_3) = V_o(Y_1Y_2 + Y_2^2 + Y_2Y_3 + Y_1Y_4 + Y_2Y_4 + Y_3Y_4) \tag{5.88}$$

and upon cancellation of like terms this reduces to

$$V_{in}Y_1Y_2 = V_o(Y_1Y_2 + Y_1Y_4 + Y_2Y_4 + Y_3Y_4) \tag{5.89}$$

Solving for the filter transfer function $T = V_o/V_{in}$ gives

$$T = \frac{V_o}{V_{in}} = \frac{Y_1Y_2}{Y_1Y_2 + Y_1Y_4 + Y_2Y_4 + Y_3Y_4} = \frac{Y_1Y_2}{Y_1Y_2 + Y_4(Y_1 + Y_2 + Y_3)} \tag{5.90}$$

**Two-pole low-pass active filter.** We will now consider the special case of a two-pole low-pass active filter for which $Y_1 = G_1$, $Y_2 = G_2$, $Y_3 = sC_3$, and $Y_4 = sC_4$, as shown in Figure 5.43. The transfer function for this case will be

$$T = \frac{G_1G_2}{G_1G_2 + sC_4(G_1 + G_2 + sC_3)} \tag{5.91}$$

Note that at zero frequency where $s = j\omega = 0$ we will have that $T = 1$, and as

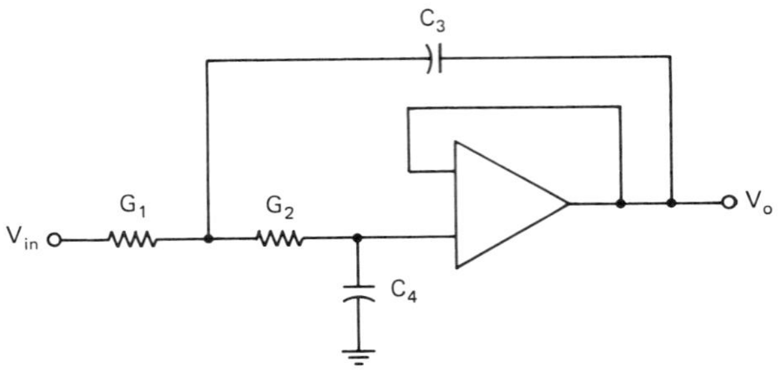


**Figure 5.43** Two-pole active low-pass filter.

the frequency ($s = j\omega$) goes to infinity $T$ will go to zero, so that this will indeed act as a low-pass filter.

For simplicity let us now further specify that $G_1 = G_2 = G = 1/R$, so that the transfer function can now be written as

$$T = \frac{G^2}{G^2 + sC_4(2G + sC_3)} = \frac{1}{1 + (sC_4/G)(2 + sC_2/G)}$$
$$= \frac{1}{1 + sRC_4(2 + sRC_3)} = \frac{1}{1 + s\tau_4(2 + s\tau_3)} \tag{5.92}$$

where $\tau_4 = RC_4$ and $\tau_3 = RC_3$. To examine the response in the frequency domain, we will let $s = j\omega$ and obtain

$$T = \frac{1}{1 + j\omega\tau_4(2 + j\omega\tau_3)} = \frac{1}{1 - \omega^2\tau_3\tau_4 + j(2\omega\tau_4)} \tag{5.93}$$

The square of the magnitude of the filter transfer function will be

$$|T|^2 = \frac{1}{1 - 2\omega^2\tau_3\tau_4 + \omega^4\tau_3^2\tau_4^2 + 4\omega^2\tau_4^2}$$
$$= \frac{1}{1 + \omega^2(4\tau_4^2 - 2\tau_3\tau_4) + \omega^4\tau_3^2\tau_4^2} \tag{5.94}$$

If we want a "maximally flat" filter frequency response that falls off monotonically with frequency, we will first set $d|T|^2/d\omega = 0$ and then solve for the condition of a zero slope occurring only at $\omega = 0$. Doing this gives $2\omega(4\tau_4^2 - 2\tau_3\tau_4) + 4\omega^3(\tau_3^2\tau_4^2) = 0$, so that $2\tau_4^2 - \tau_3\tau_4 + \omega^2\tau_3^2\tau_4^2 = 0$. For the zero slope to occur only at $\omega = 0$ we must therefore require that $2\tau_4 = \tau_3$ and thus

$$\boxed{C_3 = 2C_4} \tag{5.95}$$

Under these conditions the square of the transfer function magnitude becomes

$$|T|^2 = \frac{1}{1 + 4\omega^4\tau_4^4} = \frac{1}{1 + 4(\omega\tau_4)^4} \tag{5.96}$$

The 3-dB point will occur when $|T|^2 = \frac{1}{2}$, so that $4(\omega\tau_4)^4 = 1$ and thus $\omega\tau_4 = 1/\sqrt{2}$ and the *3-dB bandwidth* or *cutoff frequency* will be given by

$$\omega_{3\text{dB}} = \frac{1}{\sqrt{2}\,\tau_4} = \frac{0.7071}{RC_4} \tag{5.97}$$

This type of filter with a "maximally flat" transfer characteristic in the passband is called a Butterworth filter, and in Figure 5.44a this two-pole Butterworth active filter circuit is shown. In Figure 5.45a Bode plot of the frequency response of this filter is given. Note that the high-frequency response drops off as $1/\omega^2$ or at a 40-dB/decade (12-dB/octave) rate. This is to be compared to the case of a simple single-pole $RC$ low-pass network for which the high-frequency response in the stop band drops off as $1/\omega$ or at a 20-dB/decade (6-dB/octave) rate. A corresponding two-pole high-pass Butterworth active filter is shown in Figure 5.44b. The 3-dB frequency for both the low-pass and high-pass filters will be given by $f_{3\text{dB}} = 1/(2\pi RC)$, with the passband being in the region $f < f_{3\text{dB}}$ for the low-pass case and $f > f_{3\text{dB}}$ in the high-pass case.

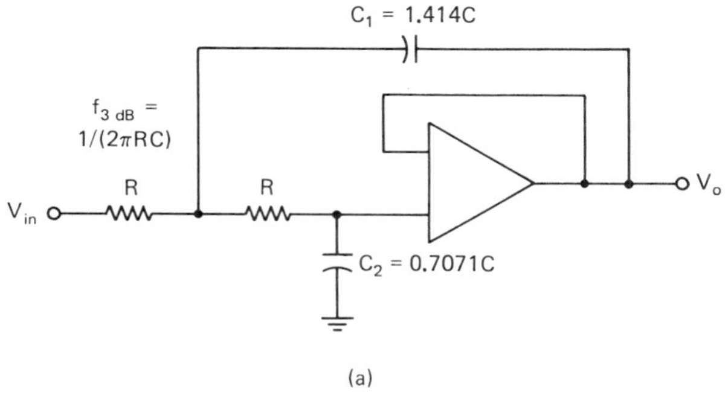


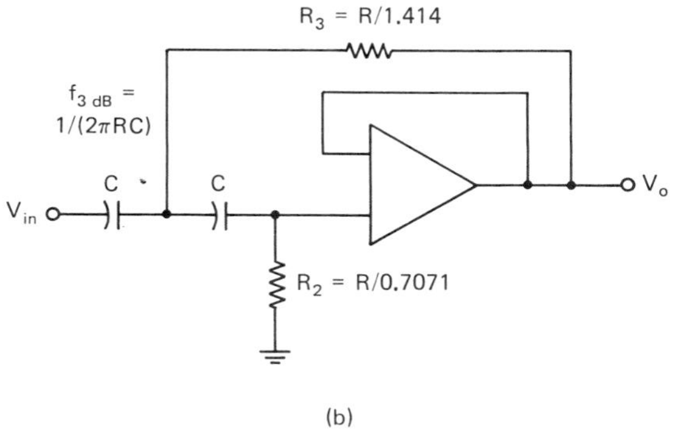


**Figure 5.44** (a) Two-pole Butterworth low-pass active filter; (b) two-pole Butterworth high-pass active filter.

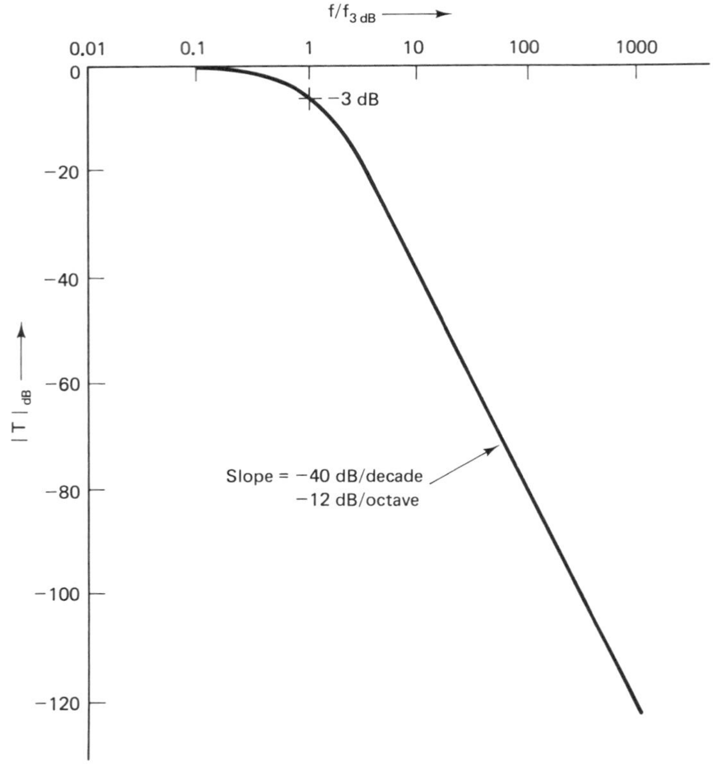


**Figure 5.45** Frequency response of two-pole Butterworth filters. (From L. G. Cowles, *A Source Book of Modern Transistor Circuits*, Prentice-Hall, 1976.)

## 5.20.2 Higher-Order Butterworth Active Filters

An $N$-pole active low-pass filter will have a high-frequency asymptotic response that falls off at a rate of $N \times 20$ dB/decade or $N \times 6$ dB/octave, as shown in Figure 5.46a. Correspondingly, an $N$-pole high-pass filter will have a low-frequency stop-band response that increases at a rate of $N \times 20$ dB/decade (or $N \times 6$ dB/octave), as shown in Figure 5.46b.

In Figure 5.47 a number of low-pass and high-pass Butterworth active filter circuits are shown, ranging from the simple case of a single-pole ($N = 1$) filter to the more complex five-pole ($N = 5$) filter. For every case the 3-dB frequency will be given by $f_{3dB} = 1/(2\ \pi RC)$.

Active filter circuits with an even greater number of poles are possible. In general, an $N$-pole active filter where $N$ is odd will have one three-pole section and $(N - 3)/2$ two-pole sections. If $N$ is even, the filter will have $N/2$ two-pole sections. In

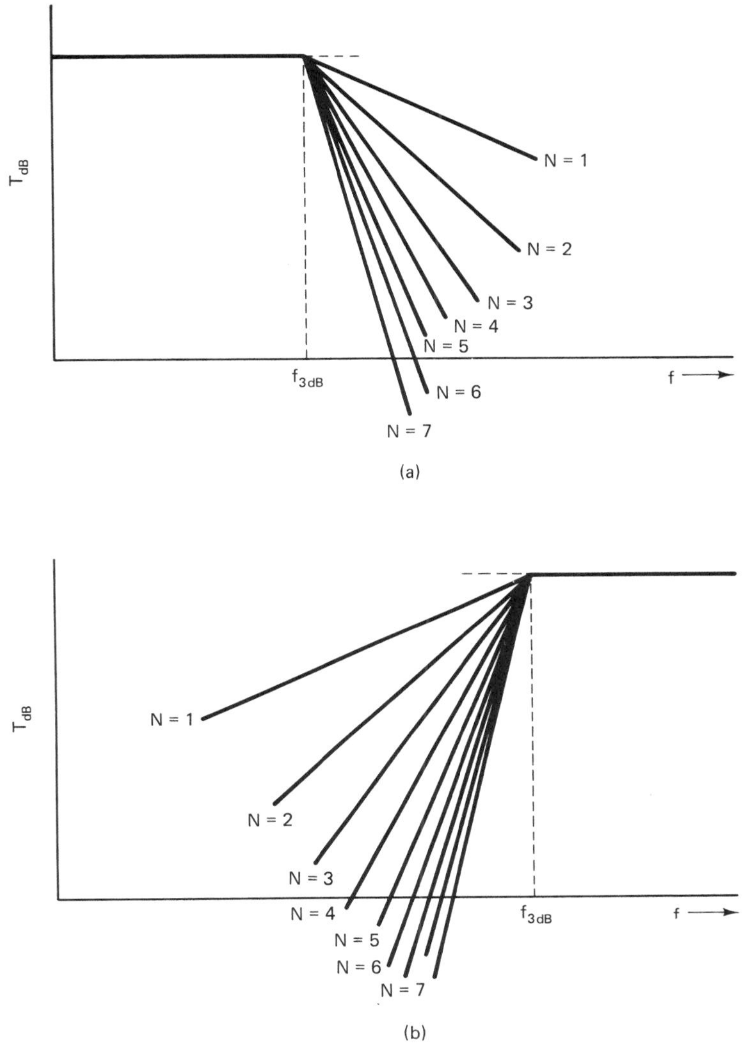


**Figure 5.46** Frequency response characteristics of $N$-pole filters: (a) low-pass filters; (b) high-pass filters.

Table 5.1 capacitance and resistance values are given for Butterworth active filters with $N$-values ranging from 2 to 10. For a two-pole section the circuit configuration is as shown in Figure 5.44 and for a three-pole section the circuit arrangement is as shown in Figure 5.47c for the low-pass case and Figure 5.47d for a high-pass filter section. For all cases the 3-dB frequency will be given by $f_{3dB} = 1/(2\pi RC)$.

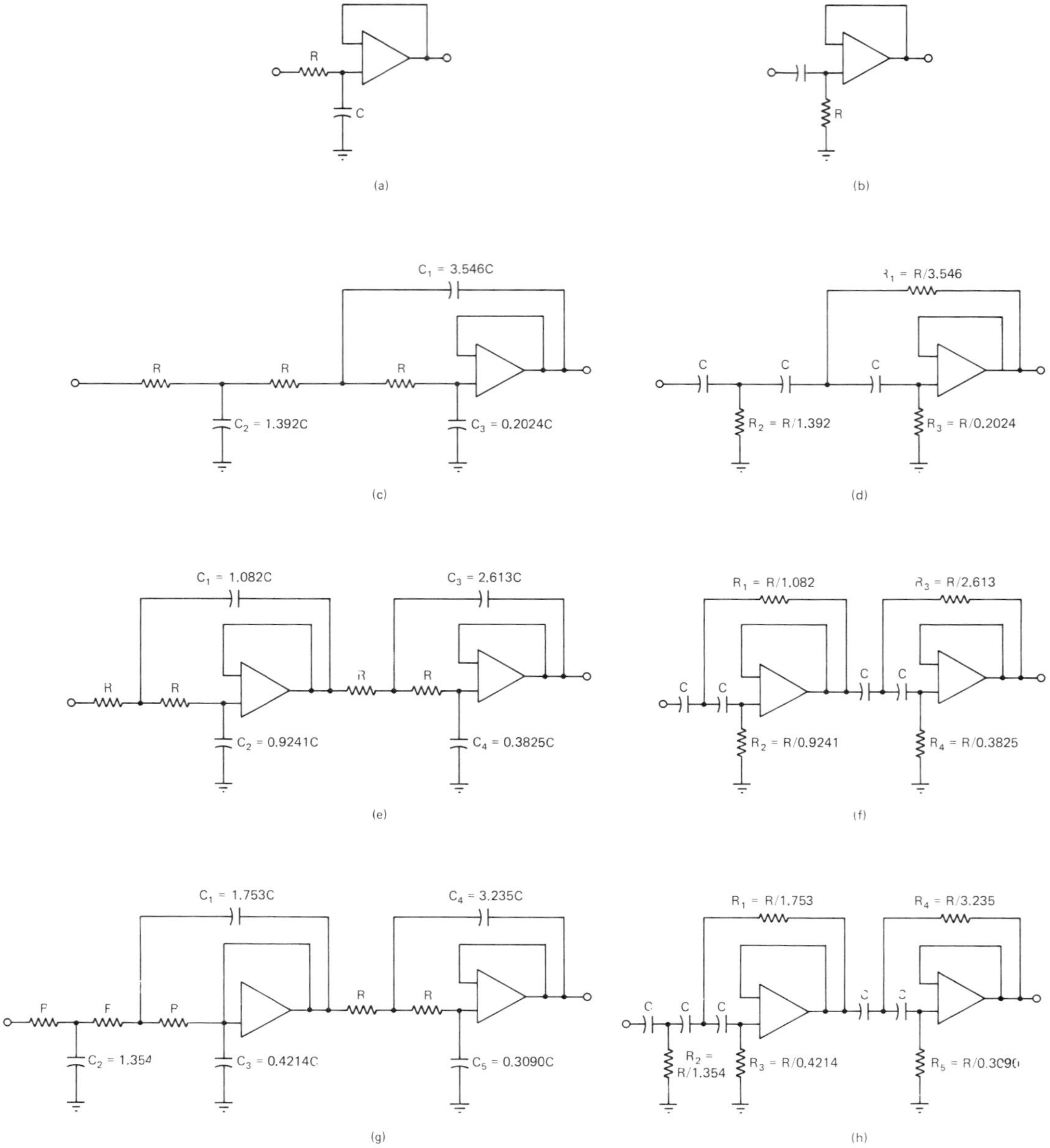


**Figure 5.47** Active Butterworth filter circuits: (a) single-pole low-pass filter; (b) single-pole high-pass filter; (c) three-pole low-pass filter; (d) three-pole high-pass filter; (e) four-pole low-pass filter; (f) four-pole high-pass filter; (g) five-pole low-pass filter; (h) five-pole high-pass filter.

**TABLE 5.1** COMPONENT VALUES FOR BUTTERWORTH ACTIVE FILTERS

| Number of poles | $C_1/C$ or $R/R_1$ | $C_2/C$ or $R/R_2$ | $C_3/C$ or $R/R_3$ |
|---|---|---|---|
| 2 | 1.414 | 0.7071 | |
| 3 | 3.546 | 1.392 | 0.2024 |
| 4 | 1.082 | 0.9241 | |
| | 2.613 | 0.3825 | |
| 5 | 1.753 | 1.354 | 0.4214 |
| | 3.235 | 0.3090 | |
| 6 | 1.035 | 0.9660 | |
| | 1.414 | 0.7071 | |
| | 3.863 | 0.2588 | |
| 7 | 1.531 | 1.336 | 0.4885 |
| | 1.604 | 0.6235 | |
| | 4.493 | 0.2225 | |
| 8 | 1.020 | 0.9809 | |
| | 1.202 | 0.8313 | |
| | 1.800 | 0.5557 | |
| | 5.125 | 0.1950 | |
| 9 | 1.455 | 1.327 | 0.5170 |
| | 1.305 | 0.7661 | |
| | 2.000 | 0.5000 | |
| | 5.758 | 0.1736 | |
| 10 | 1.012 | 0.9874 | |
| | 1.122 | 0.8908 | |
| | 1.414 | 0.7071 | |
| | 2.202 | 0.4540 | |
| | 6.390 | 0.1563 | |

### 5.20.3 Bandpass Active Filters

Let us now consider the operational amplifier circuit of Figure 5.48. We have that

$$v_{o2} = -\frac{v_o(1/sC)}{R_2} = -\frac{v_o}{sCR_2} \quad \text{and} \quad v_{o3} = -v_{o2} = \frac{v_o}{sCR_2}$$

At node $(x)$ the node voltage equation will be

$$\frac{v_{\text{in}}}{R_4} + \frac{v_o}{R_1} + sCv_o + \frac{v_o}{sCR_2R_3} = 0 \tag{5.98}$$

so that

$$\frac{v_{\text{in}}}{R_4} = -v_o\left(\frac{1}{R_1} + sC + \frac{1}{sCR_2R_3}\right) \tag{5.99}$$

The voltage gain of this circuit will therefore be

$$A_V = \frac{V_o}{V_{\text{in}}} = \frac{-1/R_4}{(1/R_1) + sC + 1/(SCR_2R_3} \tag{5.100}$$

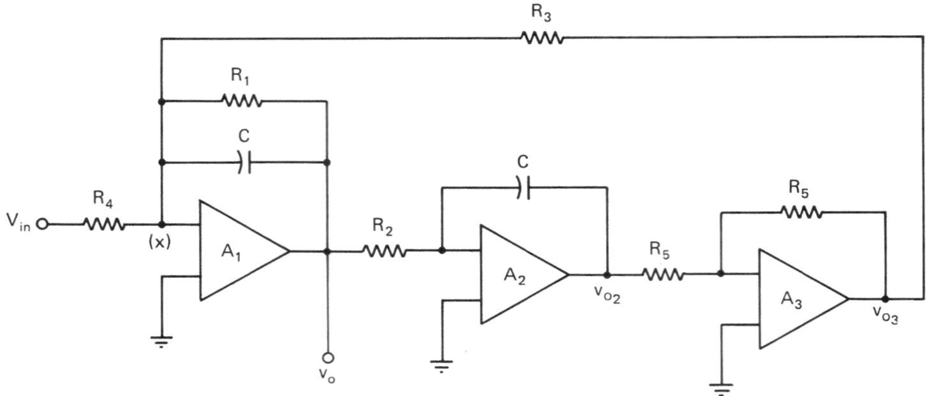


**Figure 5.48** Bandpass amplifier.

In the frequency domain $s = j\omega$ and this equation becomes

$$A_V(\omega) = \frac{-1/R_4}{(1/R_1) + j\omega C + 1/j\omega C R_2 R_3} \tag{5.101}$$

Let us now consider the *RLC* parallel resonant circuit of Figure 5.49. For this circuit we have that

$$A_V(\omega) = \frac{V_o}{V_{\text{in}}} = -\frac{G_f}{Y} = \frac{-G_f}{(1/R_1) + j\omega C + 1/j\omega L} \tag{5.102}$$

If we let $G_f = 1/R_4$ and $L = CR_2R_3$, this circuit will be equivalent to the bandpass amplifier circuit considered above. The maximum gain or gain at resonance will be

$$\boxed{A_{V\text{MAX}} = A_V(\omega_o) = -G_f R_1 = -\frac{R_1}{R_4}} \tag{5.103}$$

The frequency at which the maximum gain occurs or the "resonant frequency" will be given by

$$\omega_o^2 = \frac{1}{LC} = \frac{1}{C^2 R_2 R_3}$$

so that

$$\boxed{\omega_o = \frac{1}{C\sqrt{R_2 R_3}}} \tag{5.104}$$

The $Q$-factor at resonance ($\omega_o$) will be given by

$$Q_o = \frac{R_1}{\omega_o L} = \omega_o C R_1 = \frac{CR_1}{C\sqrt{R_2R_3}} = \frac{R_1}{\sqrt{R_2R_3}}$$

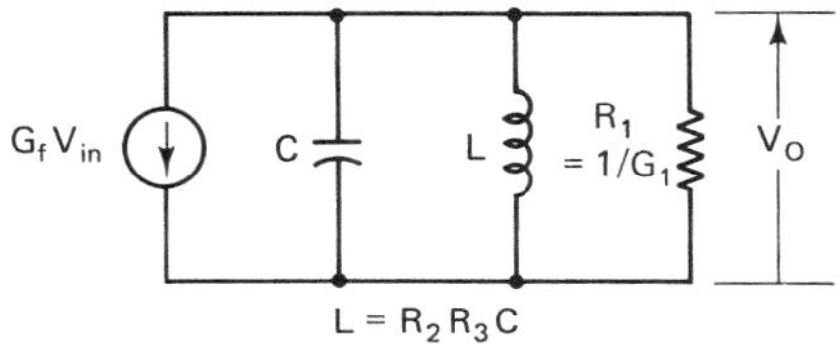


**Figure 5.49** Equivalent representation of bandpass amplifier.

The 3-dB bandwidth will be

$$BW = \frac{f_o}{Q_o} = \frac{\omega_o}{2\pi Q_o} = \frac{\omega_o}{2\pi \omega_o C R_1} = \frac{1}{2\pi C R_1} \tag{5.105}$$

**Design example for bandpass amplifier.** For a design example of this bandpass active filter, let us assume the following requirements: $f_o = 1000$ Hz, $BW = 10$ Hz, and $A_{V(\text{MAX})} = -50$. Note that since there are five components ($R_1$, $R_2$, $R_3$, $R_4$, and $C$) to be determined and three design parameters, two of the component values can be chosen arbitrarily. Let us choose $C = 1.0$ $\mu$F and we will also set $R_2 = R_3$.

For the 3-dB bandwidth we have that $BW = 1/(2\pi R_1 C)$, so that solving for $R_1$ gives $R_1 = 15.916$ k$\Omega$. Since $A_{V(\text{MAX})} = A_V(\omega_o) = -R_1/R_4 = -50$, we obtain that $R_4 = 318$ $\Omega$. For the resonant radian frequency we have that $\omega_o = 1/C\sqrt{R_2 R_3} = 1/CR_2$ since $R_2 = R_3$. Solving this for $R_2$ gives $R_2 = R_3 = 159.2$ $\Omega$. We see that reasonable component values are obtained. If a capacitance value of $C = 0.1$ $\mu$F had been chosen, all of the resistance values above will be increased by a factor of 10, which will still represent acceptable values.

**Bandpass amplifier using just one operational amplifier.** For a second example of a bandpass active filter, let us consider the general operational amplifier circuit of Figure 5.50. If we write the node voltage equations for this circuit, we obtain the result given by

$$A_{CL} = \frac{V_o}{V_{\text{in}}} = \frac{-Y_1 Y_2}{Y_2 Y_3 + Y_5(Y_1 + Y_2 + Y_3 + Y_4)} \tag{5.106}$$

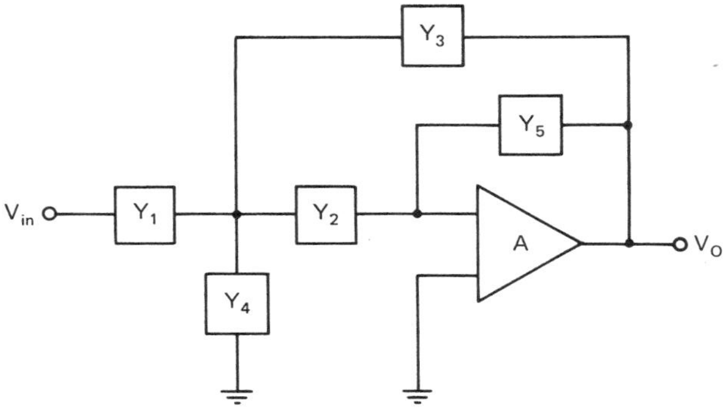


**Figure 5.50** Active filter circuit.

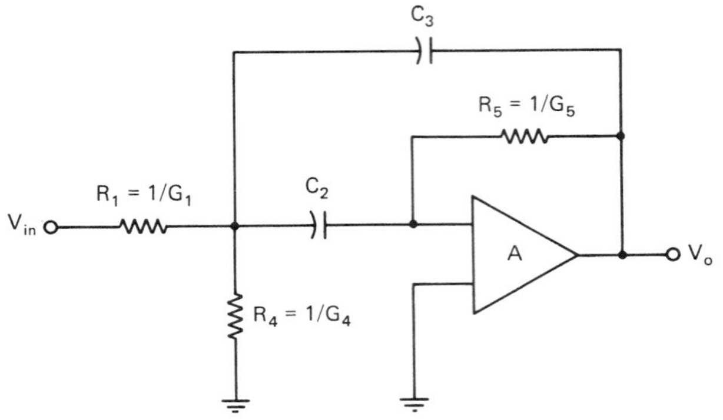


**Figure 5.51** Bandpass active filter.

For this circuit to be a bandpass active filter (Figure 5.51), we will let $Y_1 = G_1$, $Y_2 = sC_2$, $Y_3 = sC_3$, $Y_4 = G_4$, and $Y_5 = G_5$. With these circuit parameters the equation for the gain becomes

$$A_{CL} = \frac{-sG_1C_2}{s^2C_2C_3 + G_5(G_1 + sC_2 + sC_3 + G_4)} \tag{5.107}$$

After dividing through by $sC_2$, this equation can be rewritten as

$$A_{CL} = \frac{-G_1}{sC_3 + (G_1 + G_4)G_5/sC_2 + G_5(C_2 + C_3)/C_2} \tag{5.108}$$

We will now compare the gain equation obtained for this bandpass active filter to that obtained for the parallel resonant $RLC$ circuit of Figure 5.52. For that circuit we have that

$$A_V = \frac{V_O}{V_{\text{in}}} = -\frac{G_1}{Y} = \frac{-G_1}{sC_3 + 1/sL + G} \tag{5.109}$$

The two circuits will be equivalent if we let

$$L = \frac{C_2}{G_5(G_1 + G_4)} \qquad \text{and} \qquad G = \frac{G_5(C_2 + C_3)}{C_2}$$

The maximum gain or gain at "resonance" will be given by

$$\boxed{A_{CL\,(\text{MAX})} = A_{CL}(\omega_o) = -\frac{G_1}{G} = -\frac{(G_1/G_5)C_2}{C_2 + C_3} = -\frac{(R_5/R_1)C_2}{C_2 + C_3}} \tag{5.110}$$

The "resonant frequency" will be given by

$$\boxed{\omega_o^2 = \frac{1}{LC_3} = \frac{G_5(G_1 + G_4)}{C_2C_3}} \tag{5.111}$$

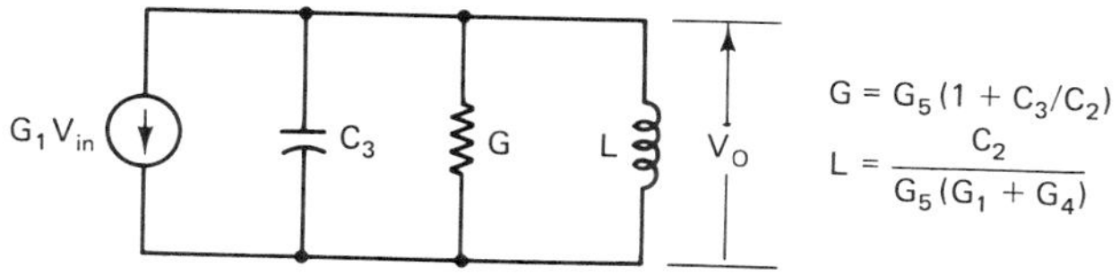


**Figure 5.52** Equivalent representation of bandpass active filter.

The $Q$-factor at resonance will be

$$Q_o = \frac{R}{\omega_o L} = R\omega_o C_3 = \frac{\omega_o C_3}{G} = \frac{\omega_o C_3 C_2}{(C_2 + C_3)G_5}$$

The 3-dB bandwidth will be given by

$$BW = \frac{f_o}{Q_o} = \frac{\omega_o}{2\pi Q_o} = \frac{\omega_o}{2\pi R\omega_o C_3} = \frac{1}{2\pi R C_3} = \frac{G}{2\pi C_3} = \frac{G_5(C_2 + C_3)}{2\pi C_2 C_3} \quad (5.112)$$

Note that we have five component values ($G_1$, $C_2$, $C_3$, $G_4$, and $G_5$) and three design parameters ($A_{CL\,(\mathrm{MAX})}$, $f_o$, and $BW$). If, for convenience, we set $C_2 = C_3 = C$, the equations above become

$$A_{CL\,(\mathrm{MAX})} = A_{CL}(\omega_O) = -\frac{R_5}{2R_1} \quad (5.113)$$

$$\omega_o = \frac{\sqrt{G_5(G_1 + G_4)}}{C} \quad (5.114)$$

$$BW = \frac{G_5}{\pi C} = \frac{1}{\pi R_5 C} \quad (5.115)$$

Note that there are now three design parameters and four circuit component values, so that one additional circuit parameter can still be specified arbitrarily.

### 5.20.4 Band-Reject Active Filters

A good example of a band-reject (or band-stop) active filter is the "boot-strapped twin-T active filter" shown in Figure 5.53. This filter produces the frequency response characteristics shown in Figure 5.54. The notch frequency $f_o$ will be given by

$$f_o = \frac{1}{2\pi RC} \quad (5.116)$$

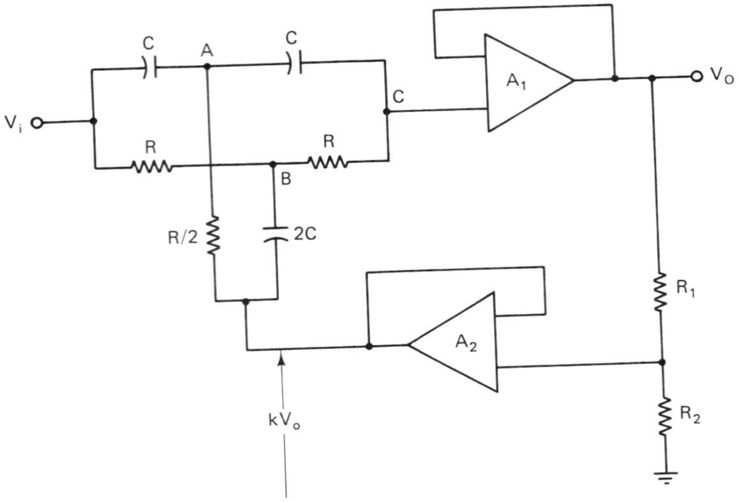


**Figure 5.53** Boot-strapped twin-T active band-reject filter.

and the $Q$-factor will be $Q_o = 1/4(1 - k)$, where $k = R_2/(R_1 + R_2)$. The 3-dB bandwidth will therefore be given by

$$\boxed{BW = \frac{f_o}{Q_o} = 4f_o(1 - k)} \qquad (5.117)$$

As $k$ approaches unity the bandwidth of this filter circuit will become very small, although it will never go to zero because of resistor and capacitor mismatches. For the case of $k = 1$ the second op amp ($A_2$) and resistors $R_1$ and $R_2$ can be eliminated and a direct feedback connection can be made from the output of $A_1$ back to the twin-T network.

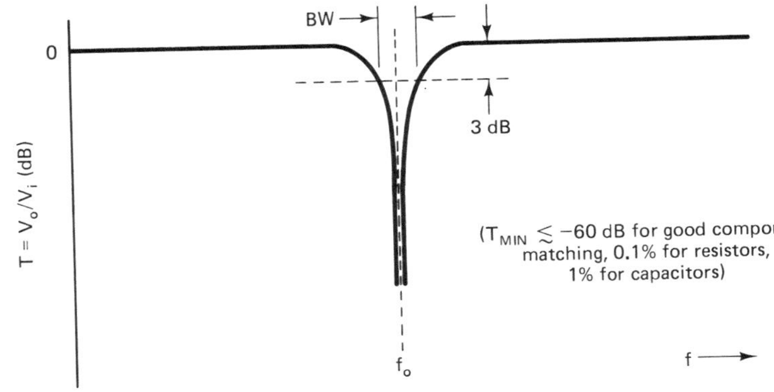


**Figure 5.54** Frequency response characteristics of twin-T active band-reject (notch) filter.

For the analysis of this active filter circuit we will first write the node voltage equations for nodes $A$, $B$, and $C$, which gives

$$\begin{aligned} &\text{Node A:} \quad sCV_i + sCV_o + 2kGV_o = 2(sC + G)V_A \\ &\text{Node B:} \quad GV_i + GV_o + 2ksCV_o = 2(G + sC)V_B \\ &\text{Node C:} \quad sCV_A + GV_B = (G + sC)V_o \end{aligned} \tag{5.118}$$

From these three node voltage equations the transfer function for this active filter can be obtained as

$$T = \frac{V_o}{V_i} = \frac{G^2 + s^2C^2}{G^2 + s^2C^2 + 4(1-k)sCG} = \frac{s^2 + (G/C)^2}{s^2 + (G/C)^2 + 4(1-k)s(G/C)} \tag{5.119}$$

For the frequency-domain response we will let $s = j\omega$ and we will also define $\omega_o$ as $\omega_o = G/C = 1/RC$. The transfer function can now be written as

$$T(\omega) = \frac{\omega^2 - \omega_o^2}{\omega^2 - \omega_o^2 - 4(1-k)j\omega\omega_o} \tag{5.120}$$

From this last equation we see that at $\omega = \omega_o$ the transfer function will be zero. As $\omega$ goes to zero the transfer function will approach unity, and as $\omega$ goes to infinity the transfer function will also asymptotically approach unity. In actual practice the high-frequency response will, of course, be limited by the frequency response of the operational amplifier.

At the 3-dB points we have that $T = 1/\sqrt{2}$, so that $\pm 4(1-k)\omega\omega_o = \omega^2 - \omega_o^2$ and thus $\omega^2 \pm 4(1-k)\omega\omega_o - \omega_o^2 = 0$. This quadratic equation can be rewritten as $(\omega/\omega_o)^2 \pm 4(1-k)(\omega/\omega_o) - 1 = 0$ and in terms of the frequency (Hz) this becomes $(f/f_o)^2 \pm 4(1-k)(f/f_o) - 1 = 0$. Solving this equation for the upper and lower 3-dB frequencies, we obtain

$$f_u = f_o\left[\sqrt{1 + 4(1-k)^2} + 2(1-k)\right] \quad \text{and} \quad f_L = f_o\left[\sqrt{1 + 4(1-k)^2} - 2(1-k)\right] \tag{5.121}$$

The 3-dB bandwidth will be

$$\boxed{BW = f_u - f_L = 4(1-k)f_o} \tag{5.122}$$

and the corresponding $Q$-factor will be $Q_o = f_o/BW = 1/4(1-k)$.

We see that as $k$ approaches unity the $Q$-factor becomes very large and the bandwidth approaches zero. In practice the $Q$-factor and the minimum bandwidth will be limited by mismatches between the resistors and capacitors and also by the frequency and phase shift response of the operational amplifiers.

One major applications area of the band-reject or "notch" filter is the elimination of the power-line frequency interference (60 Hz) in systems. The very narrow bandwidth and the very large attenuation at the notch frequency is useful for this type of application.

## 5.21 SWITCHED-CAPACITOR FILTERS

Active *RC* filters using ICs have the advantages of not requiring inductors and of offering easy implementation of various high-performance low-pass, high-pass, band-pass, and band-stop filters. The resistor and capacitor values needed for these filters are generally much too large for fabrication on a monolithic IC chip. Large-value resistors ($\gtrsim$ 10 k$\Omega$) take up an excessive amount of chip area and there are also problems with respect to the large parasitic capacitance, the absolute-value tolerance,

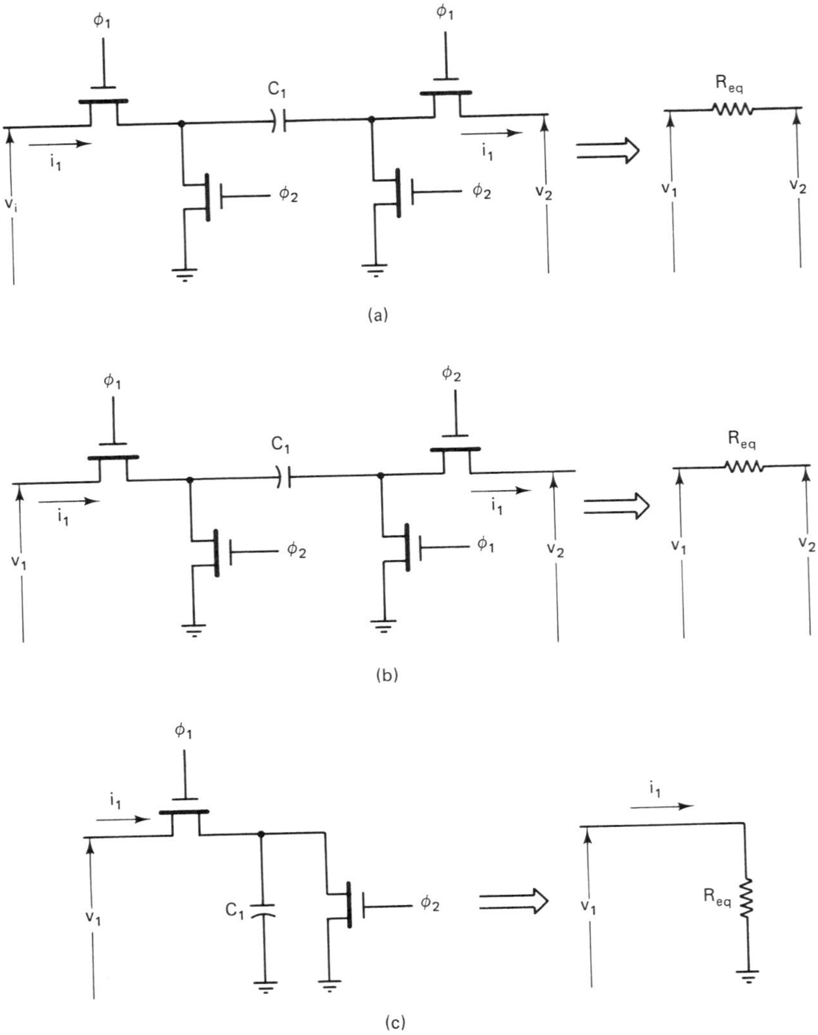


**Figure 5.55** Switched capacitor circuits: (a) noninverting switched capacitor circuit; (b) inverting switched capacitor circuit; (c) shunt capacitor circuit.

and the temperature coefficient. The upper limit for capacitors on a monolithic IC chip is generally around 100 pF because of the area requirements. There are also problems with respect to the absolute-value tolerance and the temperature coefficient. Therefore, the conventional active filters are either in the form of a hybrid IC using resistor and capacitor chips, or else use the combination of monolithic IC operational amplifiers together with discrete resistors and capacitors.

The switched-capacitor filter offers an attractive alternative to the conventional *RC* active filter. The large resistor values required are easily simulated by the combination of small-value capacitors ($\sim$1 to 10 pF) and MOS switching transistors. The equivalent resistance values that can be obtained such that the filter capacitance values will be small enough to be easily incorporated on a monolithic IC chip. As a result, the entire active filter circuit can be in the form of a monolithic integrated circuit.

Let us consider the switched-capacitor circuits of Figure 5.55. A two-phase clock will provide the complementary but nonoverlapping $\phi_1$ and $\phi_2$ clock pulses. The clock frequency will be assumed to be large compared to the signal frequency.

We will first consider the circuit shown in Figure 5.55a. When $\phi_1$ is high and $\phi_2$ is low, capacitor $C_1$ will charge up to a voltage of $v_1 - v_2$. In doing this the charge supplied to the capacitor and therefore passing through the circuit will be $Q_1 = C_1(v_1 - v_2)$. When $\phi_1$ goes low and $\phi_2$ goes high, $C_1$ will discharge back to zero. The current flow through capacitor $i_1$ will be equal to the rate at which charge is transferred through the circuit via $C_1$ and will therefore be given by

$$i_1 = \frac{Q_1}{T_c} = \frac{C_1(v_1 - v_2)}{T_c} = f_c C_1(v_1 - v_2) = \frac{(v_1 - v_2)}{R_{eq}} \tag{5.123}$$

where $T_c$ is the clock period, $f_c = 1/T_c$ is the clock frequency, and $R_{eq}$ is the equivalent resistance as given by $R_{eq} = T_c/C_1 = 1/f_c C_1$.

From this last result we see that a large resistance can be simulated by the use of a small IC capacitor and a suitable clock frequency. For example, an equivalent resistance of 1.0 MΩ can be obtained with a 10-pF MOS capacitor and a clock frequency of 100 kHz.

Another, closely related switched capacitor circuit is shown in Figure 5.55b. When $\phi_1$ is high, capacitor $C_1$ will charge up to $v_1$, corresponding to a charge transfer of $Q_1 = C_1 v_1$. When $\phi_1$ goes low and $\phi_2$ goes high, the capacitor will charge up in the opposite direction to $v_2$. The charge transfer from the output of the circuit back into $C_1$ will be $Q_1' = C_1(v_2 - v_1)$. The current flow $i_1$ will be the time average of this charge transfer and will be given by

$$i_1 = -\frac{Q_1'}{T_c} = -\frac{C_1(v_1 - v_2)}{T_c} = -f_c C_1(v_1 - v_2) = -\frac{v_1 - v_2}{R_{eq}} \tag{5.124}$$

The two circuits just considered used a series switched capacitor. In Figure 5.55c a shunt switched-capacitor circuit is shown. This circuit can be considered to be a special case of the circuit of Figure 5.55a, in which $v_2 = 0$. For this circuit capacitor $C_1$ charges up to $Q_1 = C_1 v_1$ when $\phi_1$ is high. Then when $\phi_1$ goes low and $\phi_2$ goes high $C_1$ will discharge to zero through the $\phi_2$ transistor. Current $i_1$ will therefore be given by

$$i_1 = \frac{Q_1}{T_c} = \frac{C_1 v_1}{T_c} = f_c C_1 v_1 = \frac{v_1}{R_{eq}} \qquad \text{where } R_{eq} = \frac{1}{f_c C_1} \tag{5.125}$$

Various types of low-pass, high-pass, bandpass, and band-stop active filter circuits can be implemented using switched capacitors to replace the resistors so that an "all-capacitor" filter circuit can be obtained. The various time constants that will govern the response characteristics of the filter circuits will be of the form given by

$$\tau = R_{eq} C_2 = \frac{C_2/C_1}{f_c} \tag{5.126}$$

We see that the filter time constants can easily and accurately be controlled by the clock frequency, so that various types of programmable and tracking filters can be obtained. The filter characteristics can be electronically varied over a wide range by control of the clock frequency.

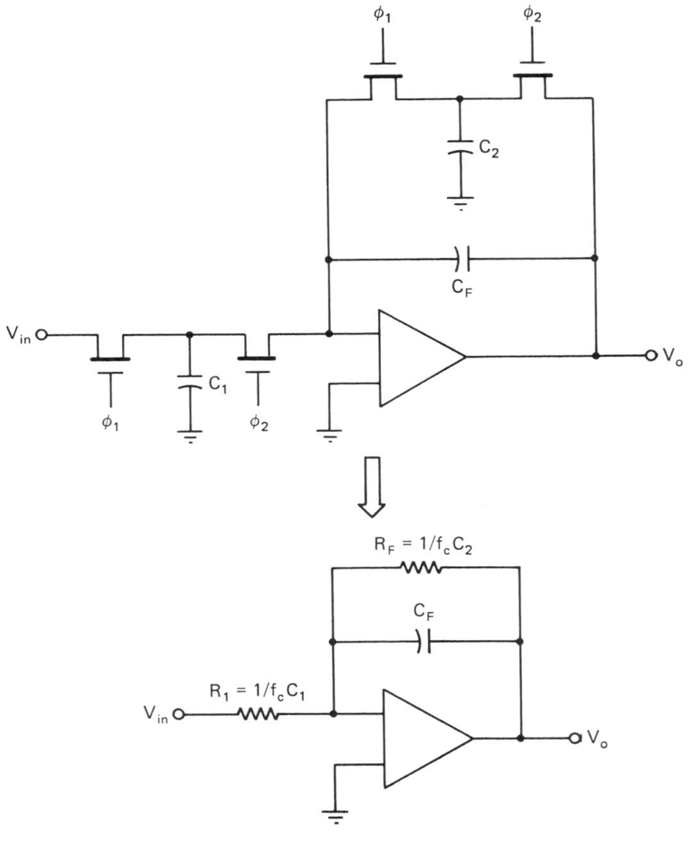


**Figure 5.56** Switched-capacitor low-pass filter.

For small-area MOS capacitors of around 1 pF or less a ratio accuracy in the range 1 to 2% can be expected, and for larger-area MOS capacitors of around 15 pF or greater the ratio accuracy improves to about 0.1%. As a result of this capacitance-ratio accuracy that is obtainable and the close control of the clock frequency, very precise control of the filter characteristics can be obtained.

A simple example of a switched-capacitor active filter is the low-pass filter of Figure 5.56. For this filter the transfer function will be given by

$$T(f) = \frac{V_o}{V_{\text{in}}} = \frac{R_F/R_1}{1 + j\omega R_F C_F} = \frac{R_F/R_1}{1 + j(f/f_L)} \tag{5.127}$$

where $R_F = 1/f_c C_2$, $R_1 = 1/f_c C_1$, and $f_L = 1/(2\pi R_F C_F) = f_c C_2/(2\pi C_F)$. The transfer function can thus be rewritten as

$$T(f) = \frac{C_2}{C_1} \frac{1}{1 + j(f/f_L)} \tag{5.128}$$

We see that the filter characteristics will be a function only of the capacitance ratio and the clock frequency.

## PROBLEMS

For these problems assume ideal, infinite-gain operational amplifiers, unless otherwise indicated or implied.

**5.1.** (*Difference amplifier*) Referring to Figure P5.1, show that $V_O = (R_F/R_A)(V_B - V_A)$.

**5.2.** (*Summing amplifier*) Referring to Figure P5.2, show that $V_O = -5V_1 - 5V_2 - 10V_3 - 20\ V_4 + 40V_5$.

**5.3.** (*Current-to-voltage converter—transimpedance amplifier*) Refer to Figure P5.3.

**(a)** Show that

$$V_O = \frac{-I_i R_F}{1 + (1/A_{OL})} \simeq -I_i R_F \qquad \text{for } A_{OL} \gg 1$$

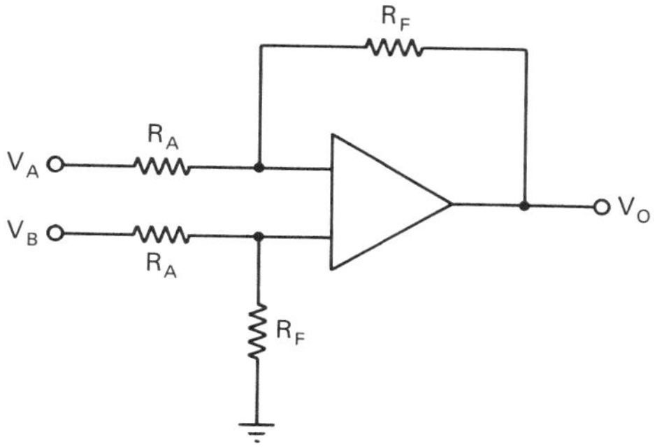


**Figure P5.1**

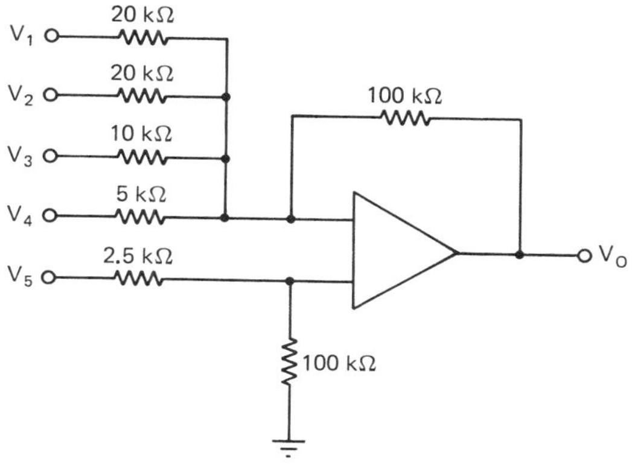


**Figure P5.2**

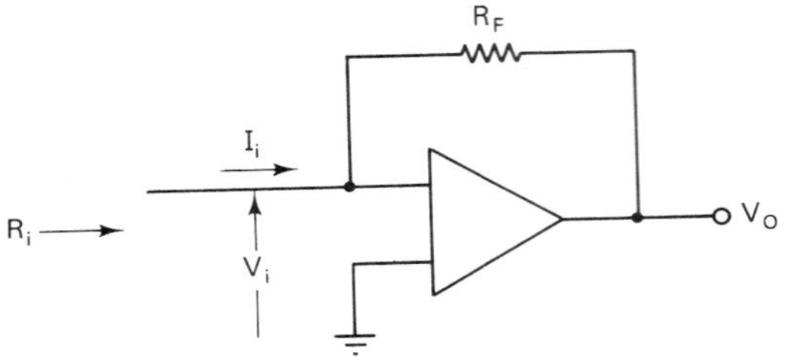


Figure P5.3

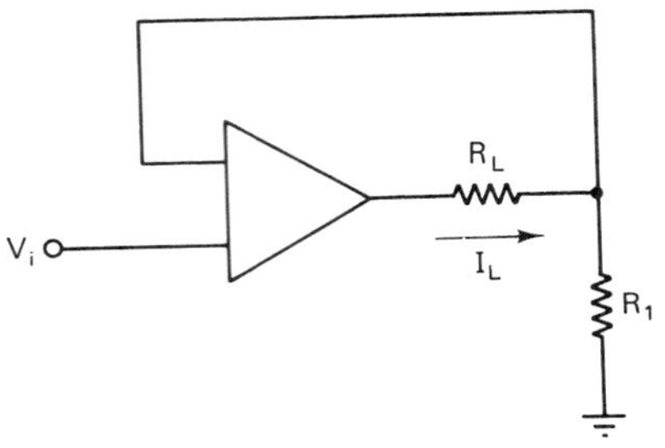


Figure P5.4

**(b)** Show that the input resistance will be given by

$$R_i = \frac{R_F}{1 + A_{OL}} \simeq \frac{R_F}{A_{OL}} \qquad \text{for } A_{OL} \gg 1$$

**5.4.** (*Voltage-to-current converter—voltage-controlled current source*) Referring to Figure P5.4, show that

$$I_L = \frac{V_i}{R_1[1 + (1/A_{OL})(1 + R_L/R_1)]}$$

$$\simeq \frac{V_i}{R_1} \quad \text{if } A_{OL} \gg 1 + \frac{R_L}{R_1}$$

**5.5** (*Low-pass active filter—integrator*) Refer to Figure P5.5.

**(a)** Show that for sinusoidal excitation,

$$A_{CL} = \frac{V_O}{V_S} = -\frac{R_F}{R_A} \frac{1}{1 + j\omega R_F C_F}$$

**(b)** Show that if $V_S$ is a step-function input of amplitude $V_S$, $V_O$ will be given by

$$V_O = -V_S \frac{R_F}{R_A}\left[1 - \exp\left(-\frac{t}{R_F C_F}\right)\right]$$

and that for $t \lesssim 0.1 R_F C_F$, $V_O \simeq -V_S(t/R_A C_F)$, so therefore for $t \lesssim 0.1 R_F C_F$ the output voltage is approximately the integral of the input voltage.

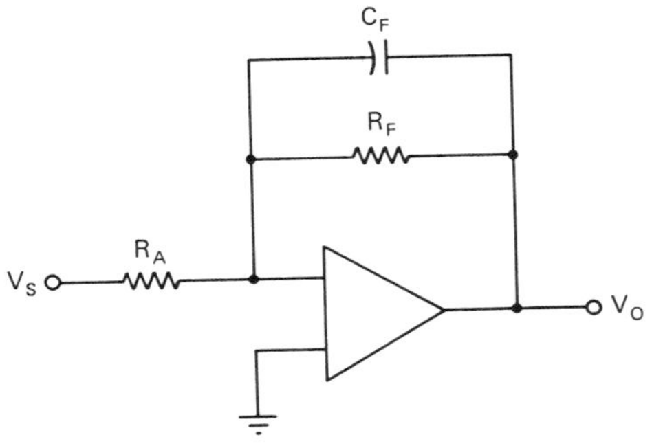


Figure P5.5

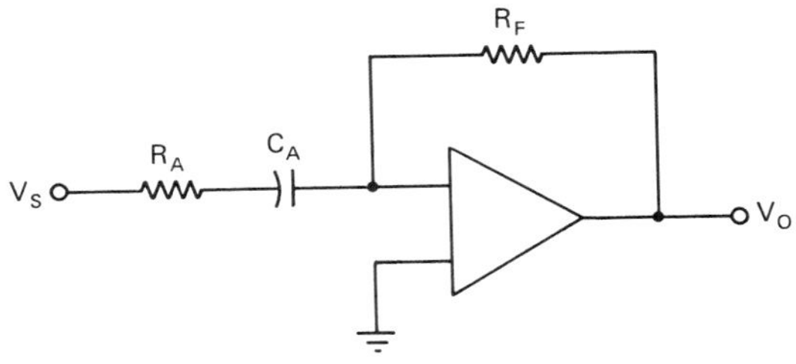


Figure P5.6

**5.6.** (*High-pass active filter—differentiator*) Refer to Figure P5.6.

**(a)** Show that for sinusoidal excitation:

$$A_{CL} = \frac{-j\omega C_A R_F}{1 + j\omega R_A C_A}$$

**(b)** Show that if $V_S$ is a step-function input, the output voltage will be given by

$$V_O = -V_S \frac{R_F}{R_A} \exp\left(-\frac{t}{R_A C_A}\right)$$ and therefore for $t \lesssim 0.1 R_A C_A$, $V_O \simeq -V_s(R_F/R_A)$.

**5.7.** (*Precision detector or rectifier*) Referring to Figure P5.7, show that $V_O = V_i$ for $V_i > 0$ (actually $> V_D/A_{OL}$, where $V_D$ is the forward voltage drop of the diode), and $V_o = 0$ for $V_i < 0$.

**5.8.** (*Precision full-wave rectifier*) Referring to Figure P5.8, show that $V_O = |V_i|$.

**5.9.** (*Precision peak detector*) Referring to Figure P5.9, show that $V_o$ will be the positive peak value of $V_i$. What is the function of the second op amp?

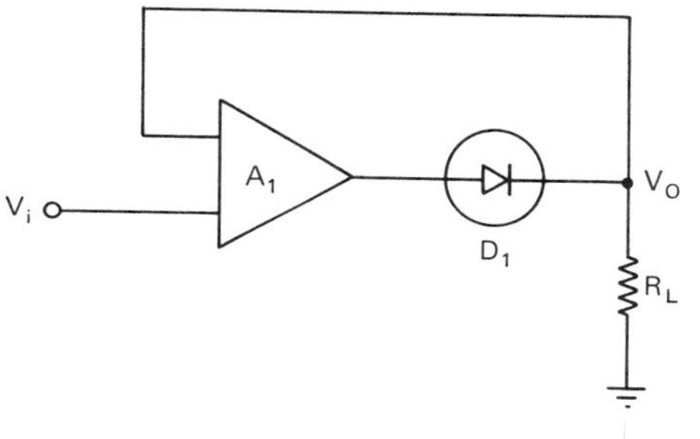


**Figure P5.7**

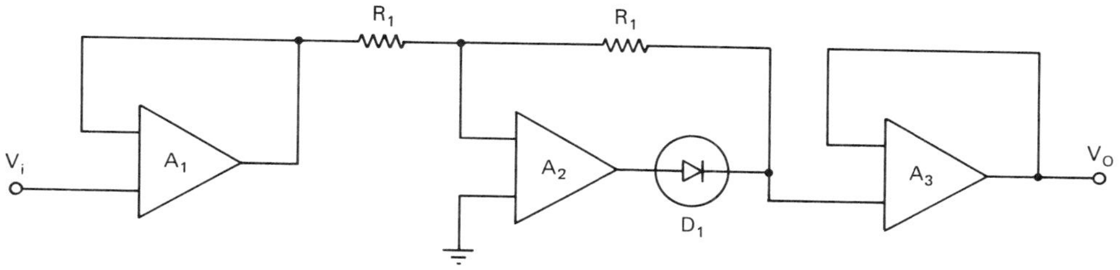


**Figure P5.8**

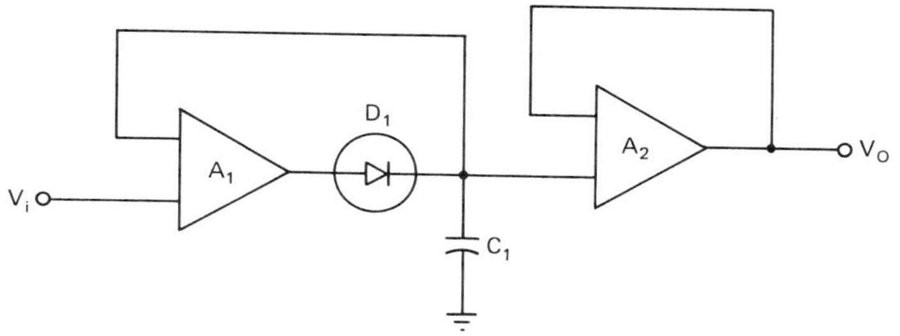


**Figure P5.9**

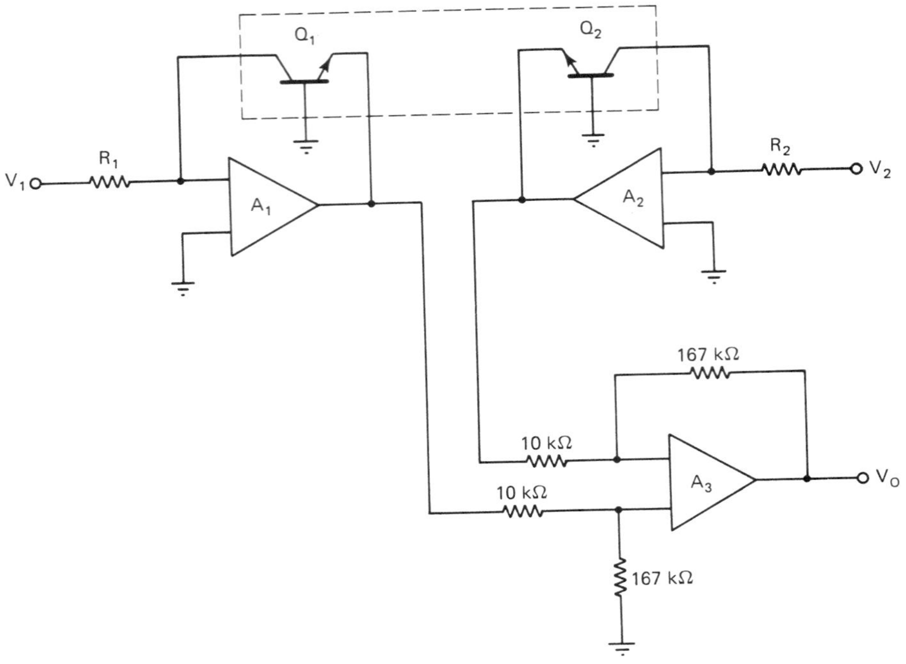


**Figure P5.10**

**5.10.** (*Logarithmic converter*) Referring to Figure P5.10, show that (at room temperature)

$$V_O = (1.0 \text{ V}) \log_{10} \frac{V_2 R_1}{V_1 R_2}$$

Assume that $Q_1$ and $Q_2$ are matched transistors.

**5.11.** (*Exponential amplifier—antilogarithmic converter*) Referring to Figure P5.11, show that $V_O = V_R 10^{-(v_i R_2 / K R_1)}$, where $K$ is a constant having units of volts.

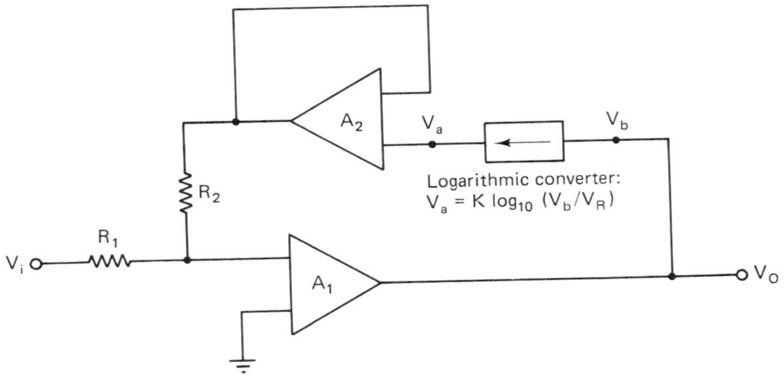


**Figure P5.11**

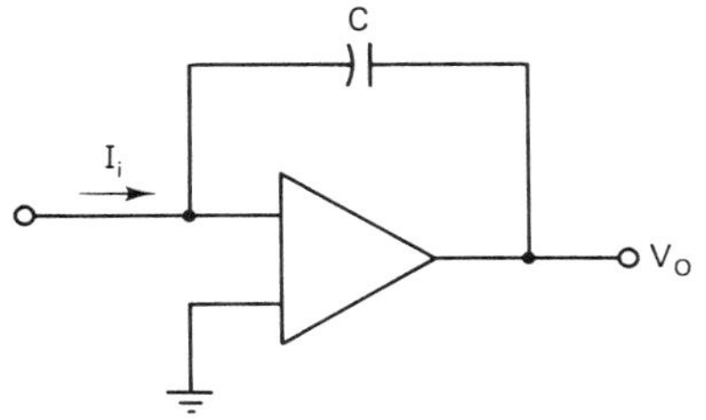


Figure P5.12

**5.12.** (*Current integrator—charge amplifier*) Referring to Figure P5.12, show that

$$V_O = -(1/C)\int_{-\infty}^{t} I_i dt = -Q_i/C$$

**5.13.** (*Schmitt trigger*) Referring to Figure P5.13, show that the transfer characteristics given are obtained for the circuit shown where $V_O^+$ and $V_O^-$ are the saturation output voltage levels of the amplifier.

**5.14.** (*Positive voltage regulator*) Referring to Figure P5.14, show that $V_O = V_Z R_2/(R_1 + R_2)$, where $V_Z$ is the breakdown voltage of the zener diode.

**5.15.** (*Positive voltage regulator with* $V_O$ *greater than* $V_Z$) Referring to Figure P5.15, show that $V_O = V_Z(1 + R_2/R_1)$.

**5.16.** (*High-current voltage regulator with current limiting*) Refer to Figure P5.16.

**(a)** Show that $V_L = V_{REF}(1 + R_1/R_2)$.

**(b)** Show that $V_{L(MAX)} = V^+ - 0.9$ V.

**(c)** Show that $I_{L(MAX)} \simeq 600$ mV$/R_{CL}$.

**(d)** If $V^+ = 20$ V (max.) and $P_{d(MAX)}$ for $Q_2$ is 50 W, find the appropriate value for $I_{L(MAX)}$. (*Ans.*: 2.5 A)

**5.17.** (*Constant-current sink*) Refer to Figure P5.17.

**(a)** Show that $I_o = V_{REF}/R_1$ as long as $V_C$ is greater than $+0.9$ V $+ V_{REF}$.

**(b)** Show an equivalent *constant-current source*.

**5.18.** (*Precision constant-current sink for low current levels*) Referring to Figure P5.18, show that $I_o = V_{REF}/R_1$ as long as $V_D > V_{REF} + V_p$ where $V_p$ is the JFET pinch-off voltage.

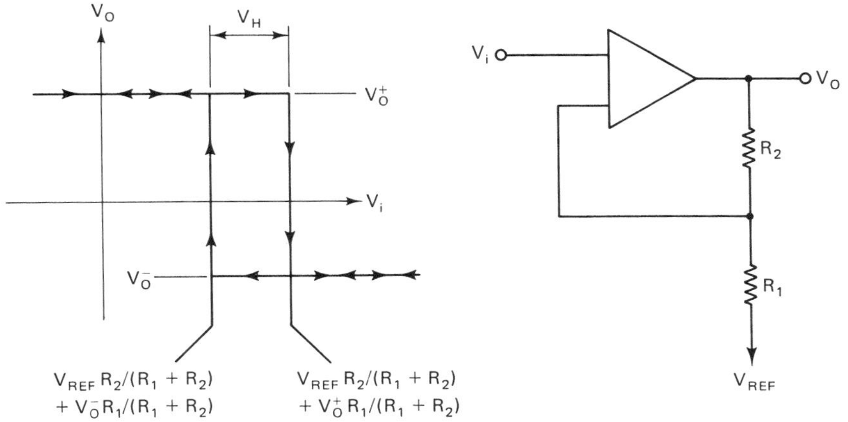


Figure P5.13

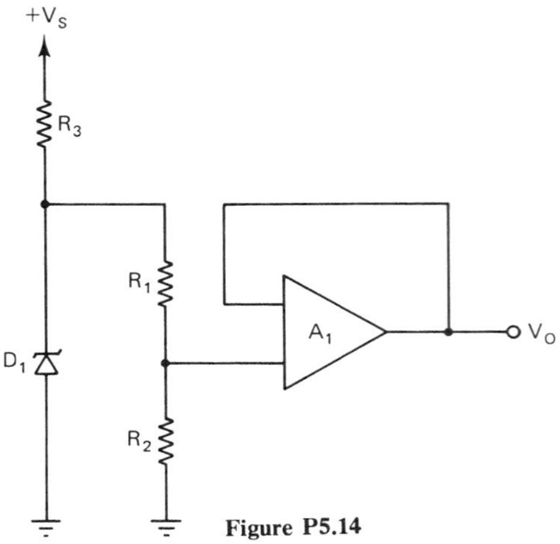


**Figure P5.14**

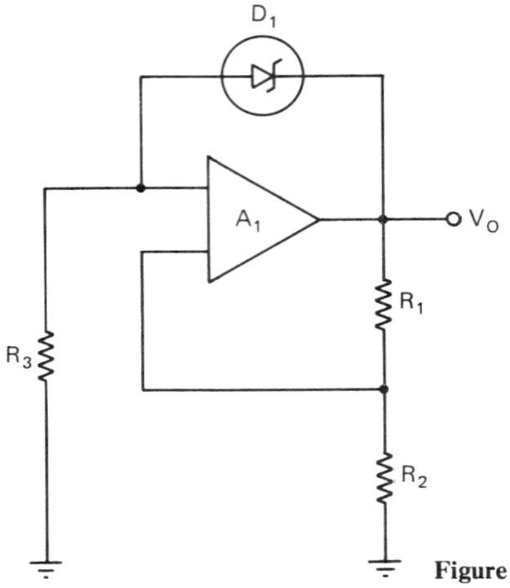


**Figure P5.15**

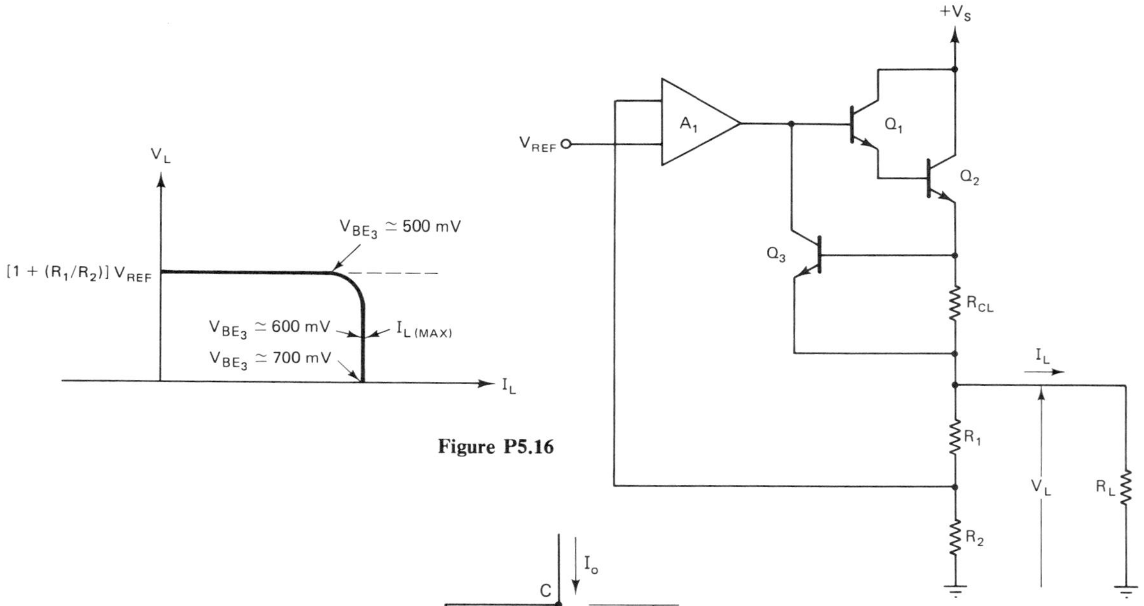


**Figure P5.16**

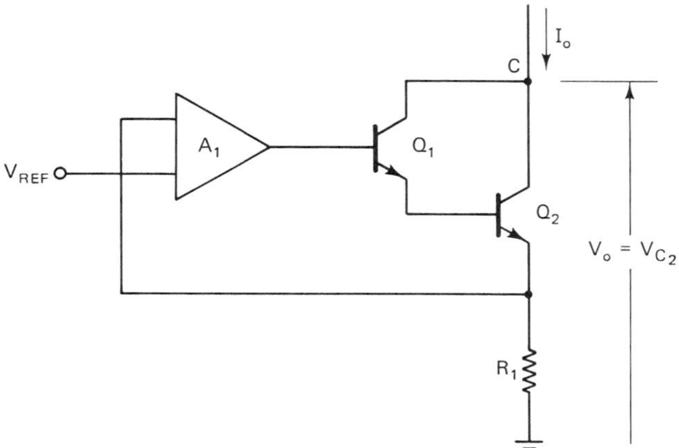


**Figure P5.17**

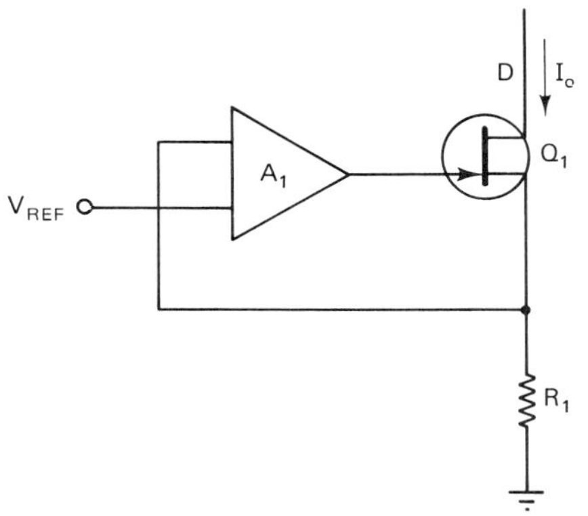


**Figure P5.18**

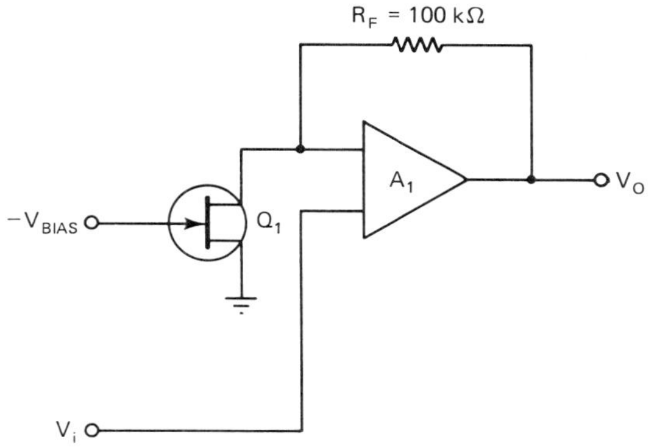


**Figure P5.19**

Show an equivalent constant-current source. Why is this circuit characterized as a *precision* current sink for low current levels?

**5.19.** (*Op amp with electronic gain control*) Refer to Figure P5.19. Given: $r_{ds(ON)} = 1000\ \Omega$ for $Q_1$.

**(a)** Show that for small $V_i$ (i.e., $V_i$ small compared to the JFET pinch-off voltage)

$$A_{CL} = \frac{V_O}{V_i} = 1 + \left(\frac{R_F}{r_{ds\,(\mathrm{ON})}}\right)\left(1 - \sqrt{\frac{V_{\mathrm{BIAS}}}{V_P}}\right) \qquad \text{for } V_{\mathrm{BIAS}} < V_P$$

**(b)** Find $A_{CL\,(\mathrm{MAX})}$ and $A_{CL\,(\mathrm{MIN})}$. (*Ans*.: 101, 1)

**5.20.** (*Op amp with electronic gain control*) Refer to Figure P5.20.

**(a)** If both $V_R$ and $V_S$ are small compared to the JFET pinch-off voltage, show that the drain-to-source resistance of both FETs will be the same and be equal to $r_{ds} = (V_R/V_C)R_1$ (for $V_C > 0$).

**(b)** Show that under the conditions above the voltage gain, $V_O/V_S$, will be given by $V_O/V_S = 1 + (R_2/R_1)(V_C/V_R)$.

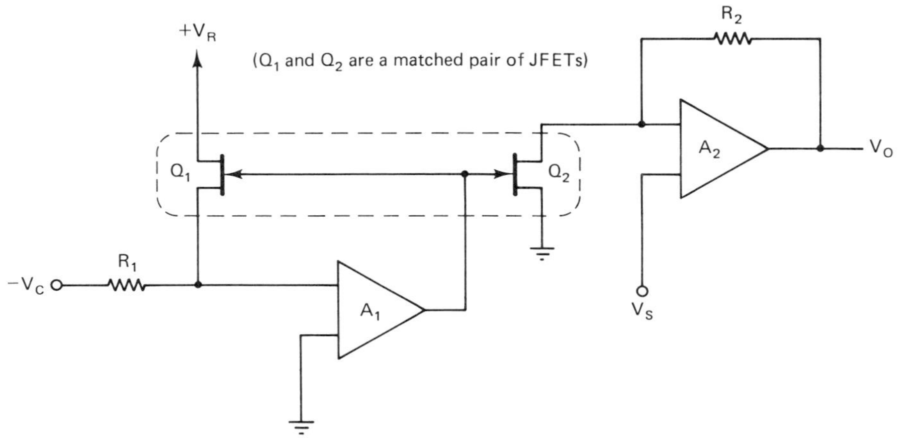


**Figure P5.20**

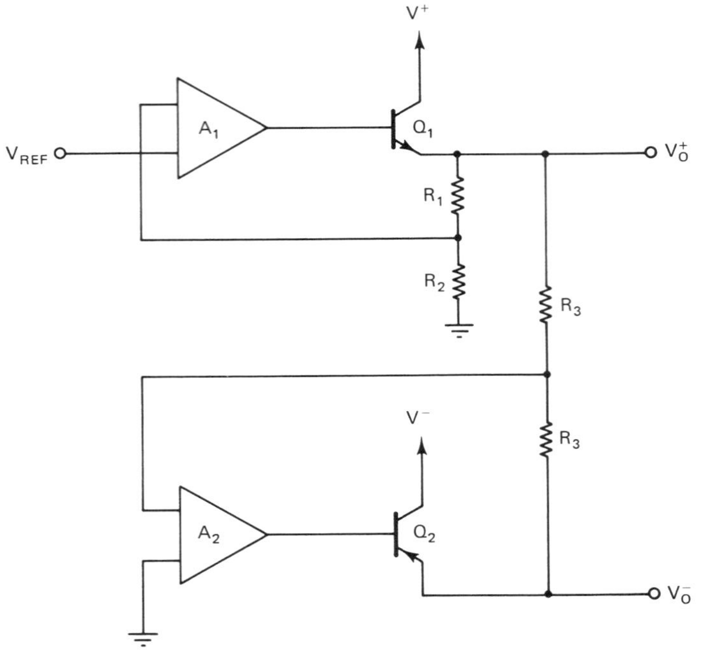


**Figure P5.21**

**(c)** If $V_{\text{pinch-off}} = 10$ V (min.) and $r_{ds\,(\text{ON})} = 100\ \Omega$ (max.), find $R_2$ for a gain range of 1 to 500 (min.). [*Ans.*: $R_2 = 50$ kΩ (min.)]

**(d)** If $V_R = +2.0$ V and $V_C = 10$ V (max.), find the required value of $R_1$ such that $r_{ds}$ can be driven down to its minimum value of $r_{ds\,(\text{ON})}$. (*Ans.*: $R_1 = 500\ \Omega$)

**5.21.** (*Tracking voltage regulator*) Referring to Figure P5.21, show that $V_O^+ = V_{\text{REF}}(1 + R_1/R_2)$ and that $V_O^- = -V_O^+$, so that this voltage regulator will produce two voltages,

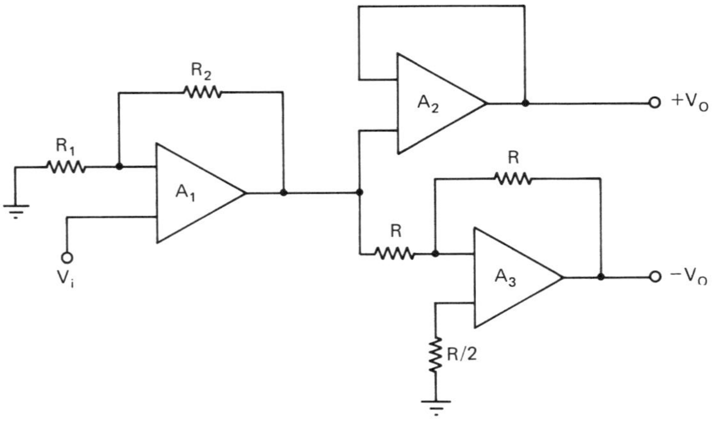


**Figure P5.22**

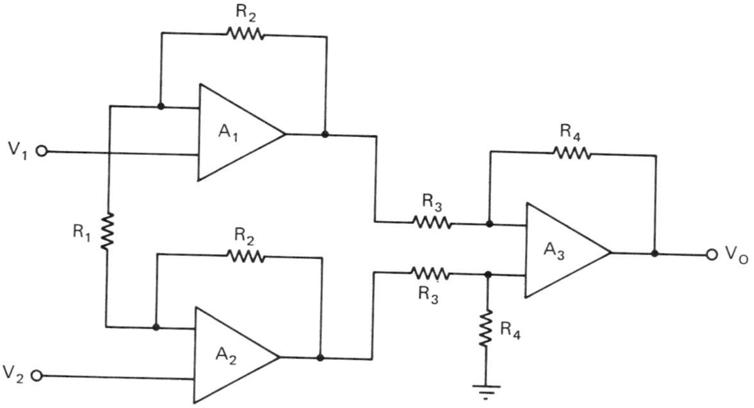


Figure P5.23

equal in magnitude but opposite in algebraic sign. (Note that this will be true even if $V_S^+$ and $V_S^-$ are not equal in magnitude.)

**5.22.** (*Precision phase splitter with high input impedance and low output impedance*) Referring to Figure P5.22, show that $|V_O|/V_i = 1 + R_2/R_1$.

**5.23.** (*Instrumentation amplifier with high input impedance and low output impedance*) Referring to Figure P5.23, show that $V_O = (R_4/R_3)(1 + 2R_2/R_1)(V_2 - V_1)$.

**5.24.** (*Instrumentation amplifier with high input impedance and low output impedance*) Referring to Figure P5.24, show that $V_O = (R_2/R_1)(V_2 - V_1)$.

**5.25.** (*Exponential converter—antilogarithmic converter*) Refer to Figure P5.25.

**(a)** Show that $V_O = R_3 I_R \exp\,[-V_i R_2/V_T(R_1 + R_2)]$.

**(b)** If $I_R = 10\ \mu\text{A}$, $R_3 = 100\ \text{k}\Omega$, $R_1 = 160\ \text{k}\Omega$, and $R_2 = 10\ \text{k}\Omega$, show that $V_O = (1.0\ \text{V})10^{-(V_i/1.0\ \text{V})}$, so that $V_O$ changes by a factor of 10 for every 1.0 V change in $V_i$, and is 1.0 V when $V_i = 0$.

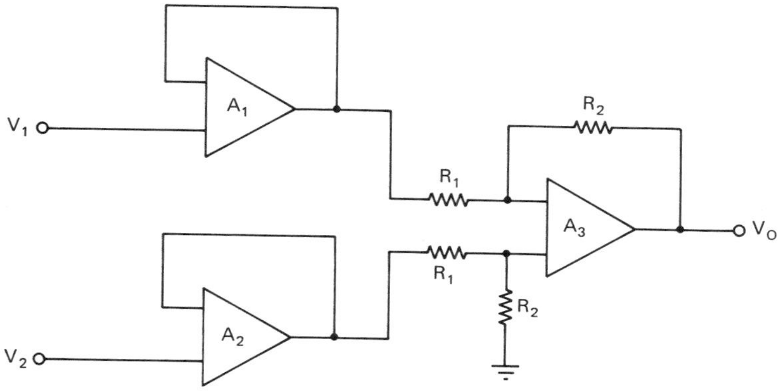


Figure P5.24

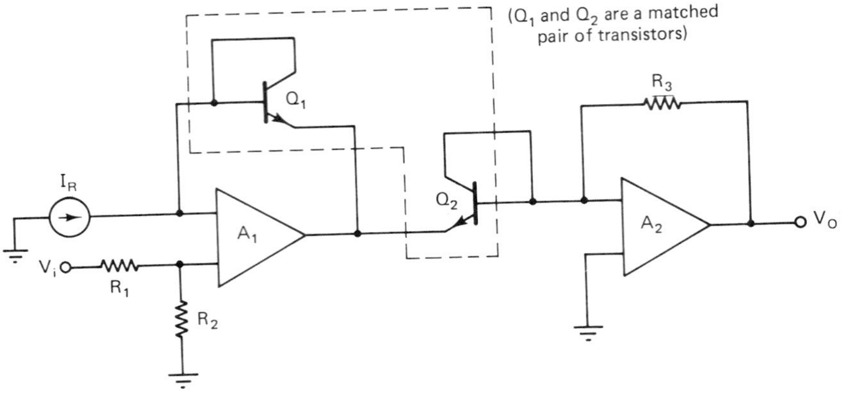


Figure P5.25

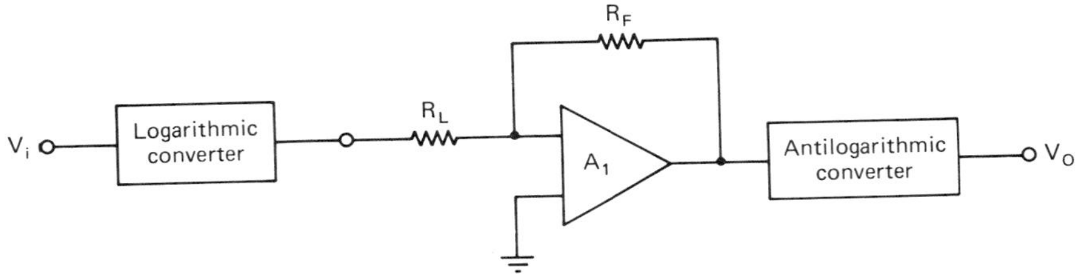


Figure P5.26

**5.26.** (*Circuit for raising a variable to a power with logarithmic techniques*) Refer to Figure P5.26. Given:

For the logarithmic converter: $V_O = (1.0\text{ V})\log_{10}(V_i/1.0\text{ V})$.
For the antilogarithmic converter: $V_O = (1.0\text{ V})10^{-(V_i/1.0\text{ V})}$.

Show that $V_O = (V_i)^{R_F/R_L}$.

**5.27.** (*Amplifier with exponential gain control*) Refer to Figure P5.27. Given: $Q_1$ and $Q_2$ are matched transistors.

**(a)** Show that $V_O = V_i(R_2/R_1)\exp(V_R/V_T)$, so that the voltage gain is exponentially dependent on $V_R$.

**(b)** Find the range in voltage gain available with $R_1 = R_2 = 10\text{ M}\Omega$ for $V_R$ ranging from 0 up to 200 mV. (*Ans.*: Voltage gain will vary from 1.0 to 2980.)

**5.28.** (*Function generator*) Referring to Figure P5.28, show that if $f(v) = av^n$, $V_O$ will be given by $V_O = (-R_2V_i/aR_1)^{1/n}$.

**5.29.** (*Dump-and-integrate circuit*) Refer to Figure P5.29. In the dump-and-integrate circuit transistor $Q_1$ is turned on prior to the integration time to discharge capacitor $C_1$ and set the output voltage to zero. Then the transistor is turned off.

**(a)** Show that

$$V_O = -\int_0^t \frac{1}{R_1C_1} V_i\,dt$$

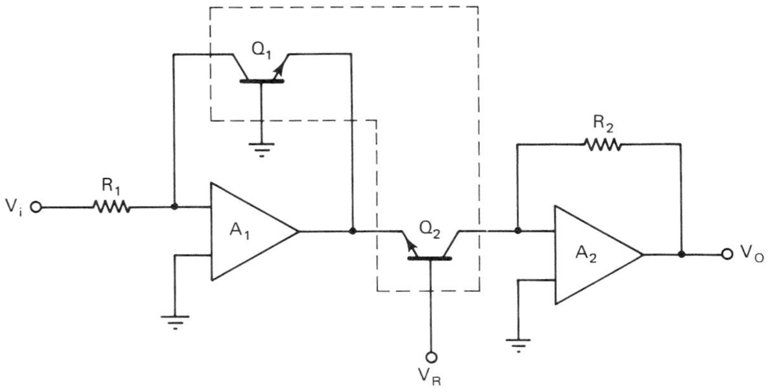


**Figure P5.27**

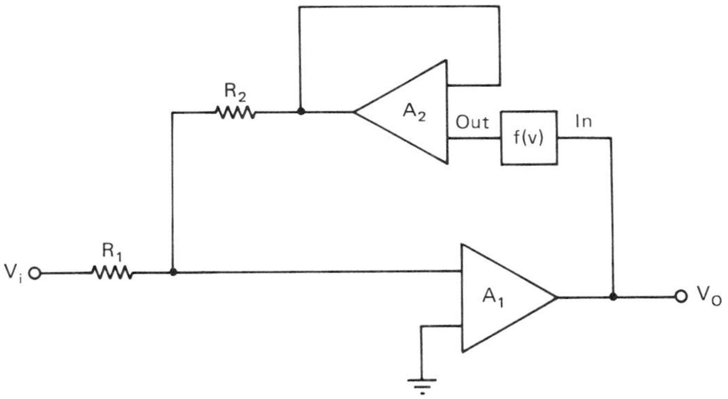


**Figure P5.28**

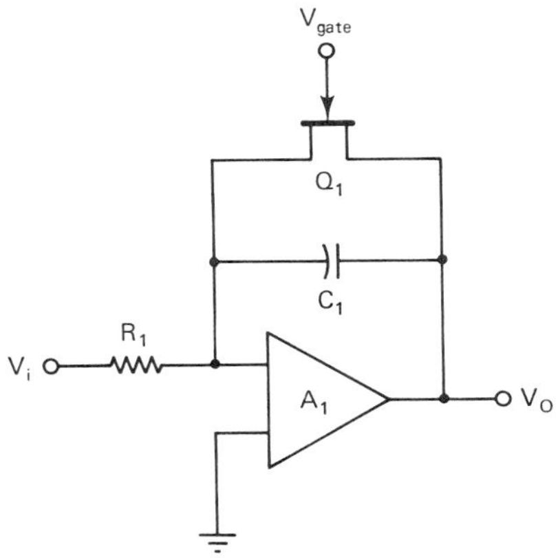


**Figure P5.29**

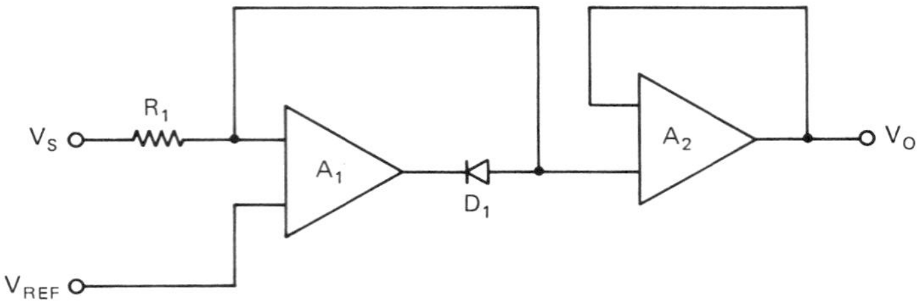


Figure P5.30

**(b)** If $V_1$ is +10 V and $R_1 = 1$ kΩ, find $C_1$ such that the output voltage at the end of a 1.0 ms integration period is +10 V. (*Ans.*: 1.0 μF)

**(c)** Describe the effect that the finite open-loop gain, the input offset voltage, and the input bias current have on the accuracy of this integrator circuit.

**5.30.** (*Precision voltage-controlled limiting ("clipping" or "bounding") circuit*) Refer to Figure P5.30.

**(a)** Draw the $V_O$ versus $V_S$ transfer characteristic and show that (1) $V_O = V_S$ for $V_S \leq V_{REF}$ and (2) $V_O = V_{REF}$ for $V_S \geq V_{REF}$.

**(b)** Repeat part (a) for the case in which diode $D_1$ is turned around.

**(c)** Draw a bilateral limiter circuit with clipping levels at $V_{REF1}$ and $V_{REF2}$. (*Hint*: Consider using two cascaded clipping circuits.)

**(d)** Why is this circuit called a "precision" limiting or clipping circuit? What effect will the forward voltage drop of diode $D_1$ have on the limiting characteristic?

**5.31.** (*Precision voltage-controlled clamping circuit*) Refer to Figure P5.31.

**(a)** Show that the output voltage $V_O$ will have the same a-c variation as the input voltage $V_S$, but that the d-c level will be shifted such that the output voltage will not drop below $V_{REF}$.

**(b)** If $V_S(t) = 10 \sin(\omega t)$ V and $V_{REF} = +5.0$ V, find $V_O(t)$. [*Ans.*: $V_O(t) = 10 \sin(\omega t) + 15$ V]

**(c)** Repeat part (b) for $V_{REF} = 0$. [*Ans.*: $V_O(t) = 10 \sin(\omega t) + 10$ V]

**(d)** Repeat part (b) for $V_{REF} = -5.0$ V. [*Ans.*: $V_O(t) = 10 \sin(\omega t) + 5$ V]

**(e)** If diode $D_1$ is turned around (reversed polarity) show that the output voltage $V_O$ will again have the same a-c variation as the input voltage, but that now the d-c level will be shifted such that the output voltage will not rise above $V_{REF}$.

**(f)** Repeat part (b), but with the diode turned around. [*Ans.*: $V_O(t) = 10 \sin(\omega t) - 5$ V]

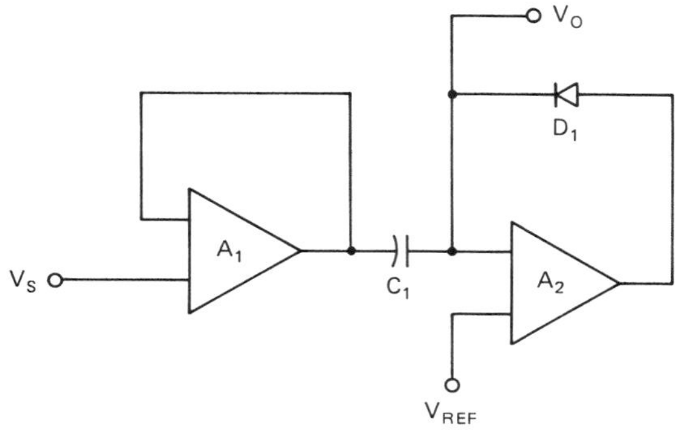


Figure P5.31

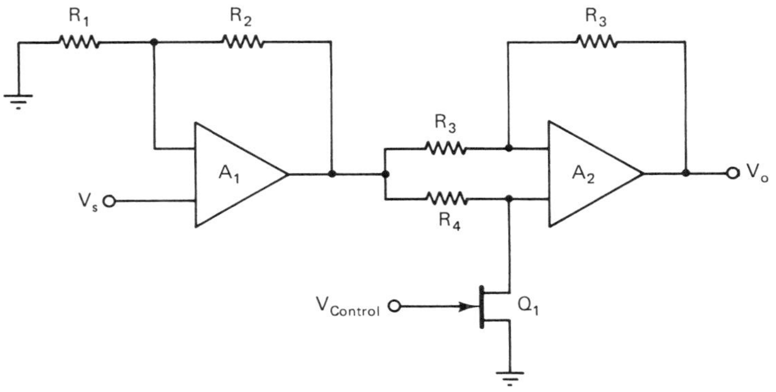


Figure P5.32

**(g)** Why is this called a "precision" clamping circuit? What effect does the forward voltage drop of the diode have on the performance of this circuit? Can the clamping level voltage $V_{REF}$ be of either positive or negative polarity?

**5.32.** (*Voltage-controlled gain polarity switching circuit*) Refer to Figure P5.32.

**(a)** Assuming that the drain-to-source resistance of $Q_1$, $r_{ds\,(ON)}$, is very small compared to $R_4$ when $Q_1$ is turned on ($V_{control} = 0$), show that the closed-loop gain of this circuit will be given by $A_{CL} = V_o/V_s = -(1 + R_2/R_1)$ *when* $Q_1$ *is turned on* and $A_{CL} = V_o/V_s = 1 + R_2/R_1$ *when* $Q_1$ *is turned off.*

**(b)** If $r_{ds\,(ON)} = 100\ \Omega$ (max) for $Q_1$, find $R_4$ such that the absolute values of the closed-loop gains given above will differ by no more than 1%. (*Ans.*: $R_4 = 20\ k\Omega$)

**(c)** Discuss the relative advantages and disadvantages, if any, of replacing $Q_1$ by a bipolar transistor.

**5.33.** (*Voltage-controlled impedance multiplier*) Refer to Figure P5.33.

**(a)** Given that the gain of $A_2$ is $-A$ show that the input impedance of this circuit will be given by

$$Z_i = \frac{v_i}{i_i} = \frac{Z_F}{1 + A}$$

and that $Y_i = (1 + A)\ Y_F$.

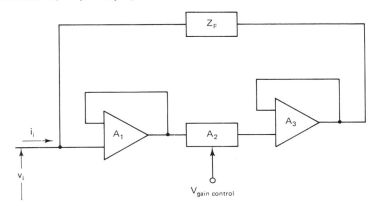

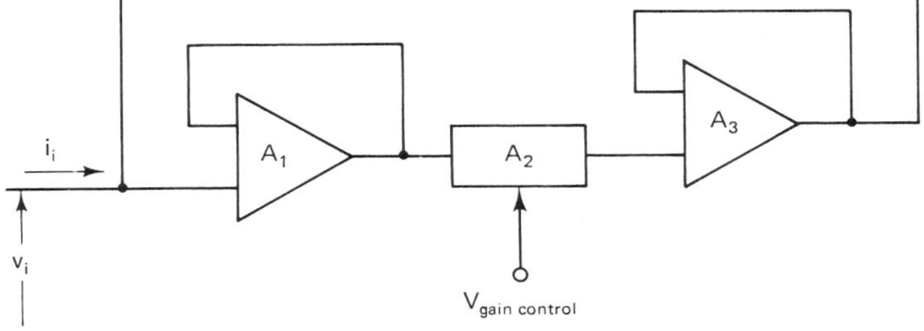


Figure P5.33

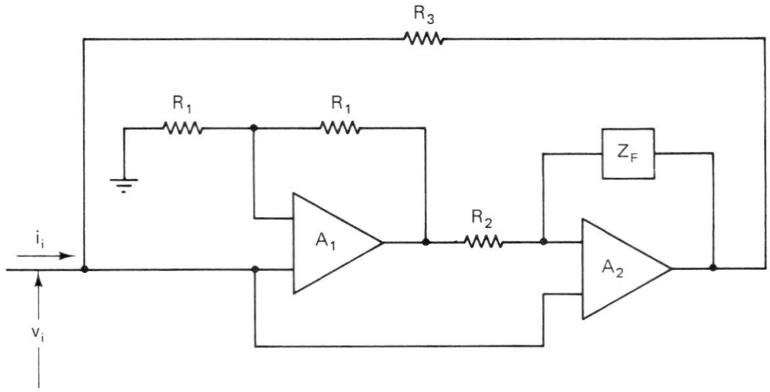


**Figure P5.34**

**(b)** If the feedback impedance is a capacitor of value $C_F$ show that the input capacitance of this circuit will be $C_i = C_F(1 + A)$.

**(c)** How is the impedance transformation produced by this circuit related to the Miller Effect?

**5.34.** (*Inductance simulator*) Refer to Figure P5.34.

**(a)** Show that the input impedance of this circuit will be given by

$$Z_i = v_i/i_i = R_2R_3/Z_F = R_2R_3\ Y_F$$

so that the input impedance is proportional to the reciprocal of the feedback impedance $Z_F$.

**(b)** If $Z_F$ is a capacitor of value $C_F$ show that the input impedance will be given by $Z_i = j\omega R_2R_3C_F = j\omega L_{eq}$, so that the input impedance of this circuit appears as an inductance of value $L_{eq} = R_2R_3C_F$.

**(c)** For the circuit of this problem the closed-loop voltage gain of $A_1$ is 2. Repeat part (a) for the more general case in which the closed-loop gain of $A_1$ is $K$. [*Ans.*: $Z_i = R_2\ R_3/(K - 1)Z_F$]

**(d)** The input impedance of this circuit is to appear to be an inductance of 1.0 mH. If $R_1 = R_2 = R_3 = 1000\ \Omega$, find the value of $C_F$ that is required. (*Ans.*: 1.0 nF)

**5.35.** (*Analog signal multiplexing circuit*) Refer to Figure P5.35.

**(a)** Given that the JFET pinch-off voltage is −5 V and the $r_{ds\,(ON)}$ is much less than $R_1$, and that $\phi_1$ through $\phi_N$ are non-overlapping active low (0 to −10 V) clock pulses, show that during the time when $\phi_i$ is active the output voltage will be given by $V_O = -(R_3/R_2)V_{Si}$, where $\phi_i$ represents any of the clock pulses between 1 and $N$.

**(b)** If $r_{ds\,(ON)} = 100\ \Omega$ (max), find $R_1$ such that the "channel crosstalk" will not exceed 1%. (*Ans.*: $R_1 \geqq 10\ k\Omega$)

**(c)** Describe the function of the voltage followers, $A_1$ through $A_N$.

**5.36.** (*Symmetrical bipolar limiter*) Refer to Figure P5.36. Given: $V_Z = 9.4$ V for $D_1$ and $D_2$.

**(a)** Draw and dimension the $V_O$ versus $V_i$ transfer characteristics.

**(b)** Repeat part (a) for the case of $V_Z = 9.4$ V for $D_1$ and 4.4 V for $D_2$.

**5.37.** (*Constant-amplitude phase shifter*) Refer to Figure P5.37.

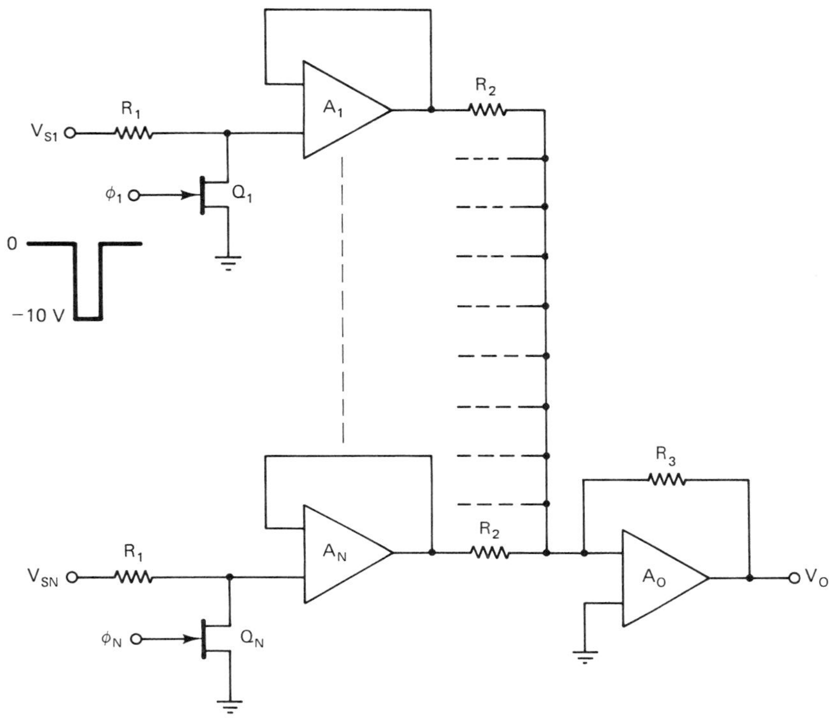


**Figure P5.35**

**(a)** Assuming a high-open-loop gain, ideal op amp, show that the closed-loop voltage gain will be given by

$$A_{CL} = \frac{v_o}{v_i} = 1.0 \underline{/-2 \tan^{-1}(\omega C_2/G_2)}$$

$$= 1.0 \underline{/-2 \tan^{-1}(\omega R_2 C_2)}$$

**(b)** Find $A_{CL}$ for: $R_1 = 2.0$ kΩ, $R_2 = 1.0$ kΩ, $C_2 = 10$ nF, and $f = 10$ kHz. (*Ans.*: $A_{CL} = 1.0 \underline{/-64.3°}$)

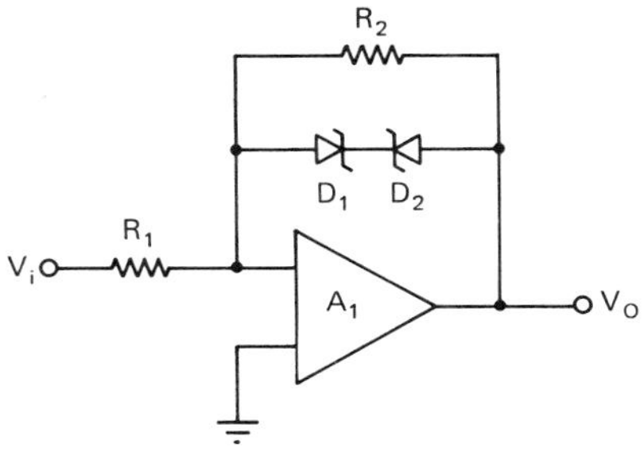


**Figure P5.36**

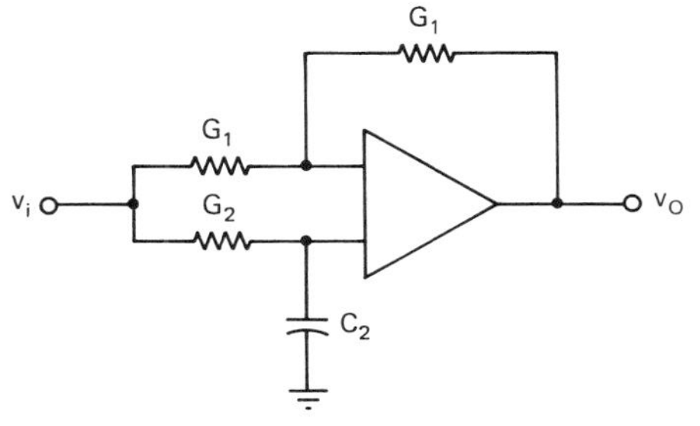


**Figure P5.37**

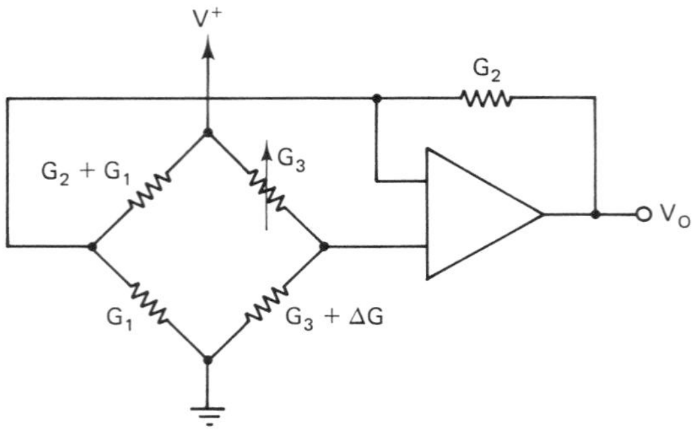


**Figure P5.38**

**5.38.** (*Bridge amplifier*) Refer to Figure P5.38.

**(a)** Show that

$$V_O = \left(1 + \frac{G_1}{G_2}\right) \frac{-\Delta G}{2G_3 + \Delta G} V^+$$

(*Hint*: Use nodal analysis.)

**(b)** We will now consider an application of the bridge amplifier. The bridge amplifier will be used to measure small changes in temperature. Given: $R_1 = 1.0$ kΩ, $R_2 = 9.0$ kΩ; $G_3 + \Delta G$ is a resistor with a temperature coefficient of 0.20%/°C; $G_3$ is a resistor similar to $G_3 + \Delta G$, but kept at a constant temperature and adjusted (with the aid of a trimming potentiometer) so that $V_O = 0$ at $T = 25$°C; and $V^+ = 10$ V. Find $V_O$ as a function of temperature. [*Ans*.: $V_O = (0.10\ \text{V/°C})(T - 25\text{°C})$]

**5.39.** (*Active RC bandpass amplifier*) Refer to Figure P5.39.

**(a)** Show that the closed-loop gain will be given by

$$A_{CL} = \frac{V_O}{V_i} = \frac{-y_1 y_3}{(y_1 + y_2 + y_3 + y_4)y_5 + y_3 y_4}$$

**(b)** If $y_1 = G_1$, $y_2 = G_2$, $y_3 = j\omega C_3$, $y_4 = j\omega C_4$, $y_5 = G_5$, obtain an equation for $A_{CL}$. [*Ans*.:

$$A_{CL} = \frac{-j\omega C_3 G_1}{G_5(G_1 + G_2) - \omega^2 C_3 C_4 + j\omega G_5(C_3 + C_4)}\Bigg]$$

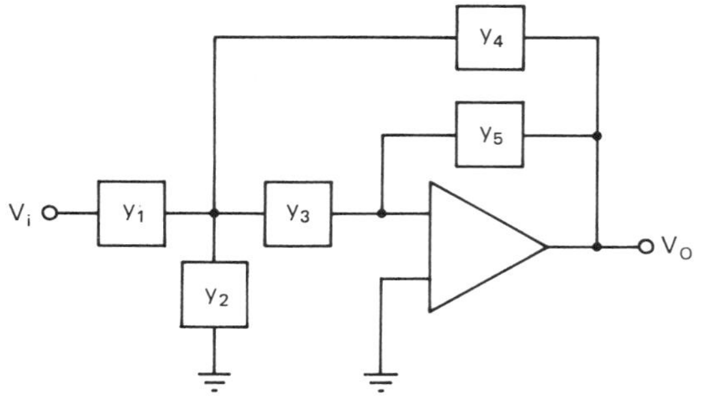


**Figure P5.39**

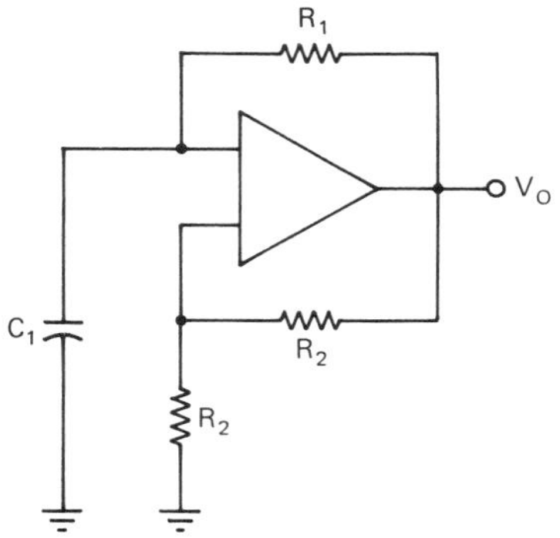


**Figure P5.40**

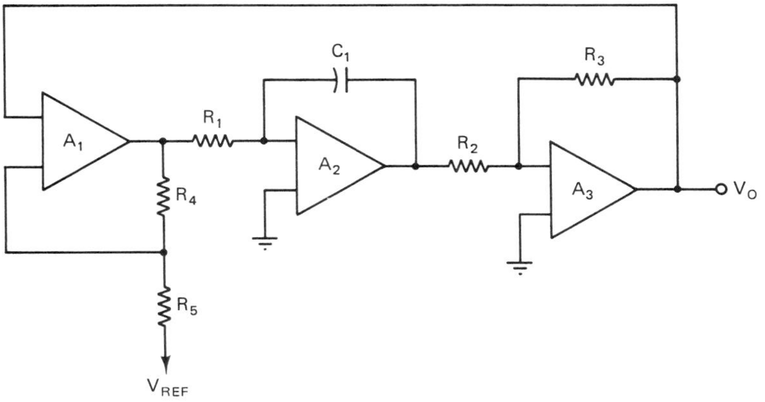


**Figure P5.41**

**(c)** Find the frequency at which the gain will be a maximum and the value of the gain at that frequency. [*Ans.*: At $\omega_o = \sqrt{G_5(G_1 + G_2)/(C_3C_4)}$ the gain will have a maximum value of

$$A_{CL\,(\mathrm{MAX})} = -\left(\frac{C_3}{C_3 + C_4}\right)\frac{G_1}{G_5} = -\left(\frac{C_3}{C_3 + C_4}\right)\frac{R_5}{R_1}\Bigg]$$

**(d)** Given that the 3-dB bandwidth (in *rad/s*) will be $BW = (C_3 + C_4)/(C_3C_4R_5)$ find $R_1$, $R_2$, and $R_5$ if $f_o = 100$ Hz, $BW = 10$ Hz, $A_{CL\,(\mathrm{MAX})} = -10$, and $C_3 = C_4 = 1.0\ \mu$F. (*Ans.*: $R_1 = 1.59$ k$\Omega$, $R_2 = 83.8\ \Omega$, $R_5 = 31.8$ k$\Omega$)

**5.40.** (*Square-wave oscillator*) Refer to Figure P5.40. Given: The positive and negative saturation voltages of the op amp are equal.

**(a)** Show that the frequency of oscillation will be given by

$$f_{\mathrm{osc}} = \frac{1}{2R_1C_1 \ln 3}$$

**(b)** Draw the waveform of $V_O$ and $V_{C1}$.

**5.41.** (*Precision linear triangle wave generator*) Refer to Figure P5.41. Given: The output voltage of $A_1$ will saturate at $V_H$ in the high state and $V_L$ in the low state, and $A_1$ can be either a voltage comparator or an operational amplifier.

**(a)** Show that the output voltage $V_O$ will be a linear triangular waveform with a peak-to-peak voltage of $V_{O\,(\mathrm{P-P})} = (V_H - V_L)R_5/(R_4 + R_5)$.

**(b)** Show that the period T of the triangular waveform will be given by

$$T = (R_1C_1)(R_2/R_3)\left[\frac{V_{O\,(\mathrm{P-P})}}{V_H} + \frac{V_{O\,(\mathrm{P-P})}}{V_L}\right]$$

$$= (R_1C_1)\left[\frac{R_2}{R_3}\right]\left[\frac{R_5}{R_4 + R_5}\right]\left[\frac{V_H}{-V_L} + \frac{-V_L}{V_H}\right]$$

**(c)** If $V_{\mathrm{REF}} = 0$, $V_H = +10$ V, $V_L = -10$ V, $R_4 = R_5$, $R_2 = R_3$, $R_1 = 10$ k$\Omega$, and $C_1 = 10$ nF, (1) draw and dimension the output voltage waveform, and (2) find the frequency of oscillation. (*Ans.*: $f_{\mathrm{osc}} = 10$ kHz)

**(d)** Describe the effect that voltage $V_{\mathrm{REF}}$ will have on the output voltage.

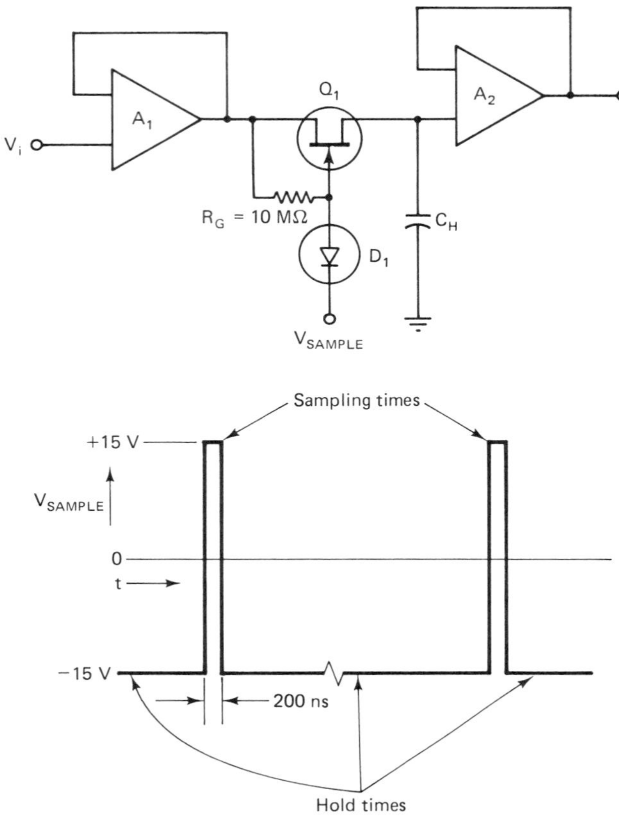


**Figure P5.42**

**(e)** Why is this called a "precision linear triangular wave generator"? What factors will limit the maximum frequency of oscillation and the linearity of the waveform?

**5.42.** (*Sample-and-hold circuit*) Refer to Figure P5.42. Given for the JFET: $r_{ds\,(ON)} = 300$ Ω (max.), $V_P = -3.0$ V (max.), and $I_{DS\,(OFF)} = 500$ pA (max.) Given for the op amp: $I_{BIAS} = 500$ pA (max.), sampling rate = 100 kHz, sampling time = 200 ns, and $V_{supply} = \pm 15$ V.

**(a)** Find the value of the capacitor $C_H$ such that (1) the output voltage, $V_O$, reaches at least 98% of $V_i$ during the *sampling time*, and (2) the output voltage decreases by no more than 0.1 mV during the hold interval. (*Ans.*: $C$ must be between the limits 100 to 170 pF. A good choice for $C_H$ might be around 130 pF.)

**(b)** What is the function of **(1)** $R_G$, **(2)** the diode, **(3)** the first op amp, and **(4)** the second op amp?

**5.43.** (*Howland current source circuit*) Refer to Figure P5.43.

**(a)** Show that the current $I_L$ will be given by $I_L = (V_1 - V_2)/R_1$ and will therefore be independent of $R_L$, so that this circuit acts as a constant current source.

**(b)** Given: $R_1 = 1$ kΩ, $R_2 = 250$ Ω, $V_2 = 0$, and the maximum operational amplifier output voltage swing is from −10 to +10 V. The voltage compliance range is the

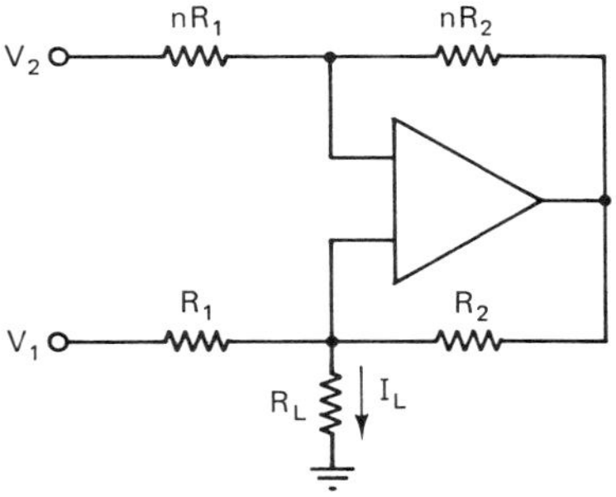


Figure P5.43

range of load voltages over which this circuit will act as a constant-current source. Find the voltage compliance range. (*Ans*.: $-8.0$ to $+8.0$ V)

**(c)** If the maximum input voltage range for $V_1$ is $-10$ to $+10$ V, find the maximum range of the output current $I_L$. Use the conditions of part (b). (*Ans*.: $-10$ to $+10$ mA)

**5.44.** (*Circuit to produce an output voltage that increases linearly with temperature*) Referring to Figure P5.44, show that $V_O$ will be given by $V_O = [(R_3/R_2)(k/q)(\ln 2)]\ T = (1.0\ \text{mV/°C})T$, so that $V_O$ will increase linearly with temperature with a coefficient of 1.0 mV/K = 1 mV/°C and $T$ is the absolute temperature (K).

**5.45.** (*Voltage regulator*) Refer to Figure P5.45.

**(a)** If $R_1 + R_2 = 10\ \text{k}\Omega$, find $R_1$ and $R_2$ for $V_O = +10.0$ V. (*Ans*.: $R_1 = 8.2\ \text{k}\Omega$, $R_2 = 1.8\ \text{k}\Omega$)

**(b)** If $A_{OL}(0) = 10\ \text{k}\Omega$, $f_u = 1.0$ MHz, and $R_{O(OL)} = 50\ \Omega$, find $Z_{O(CL)}$ at **(1)** d-c and **(2)** 10 kHz. (*Ans*.: **(1)** $28 + j0\ \text{m}\Omega$, **(2)** $0 + j2.8\ \Omega$)

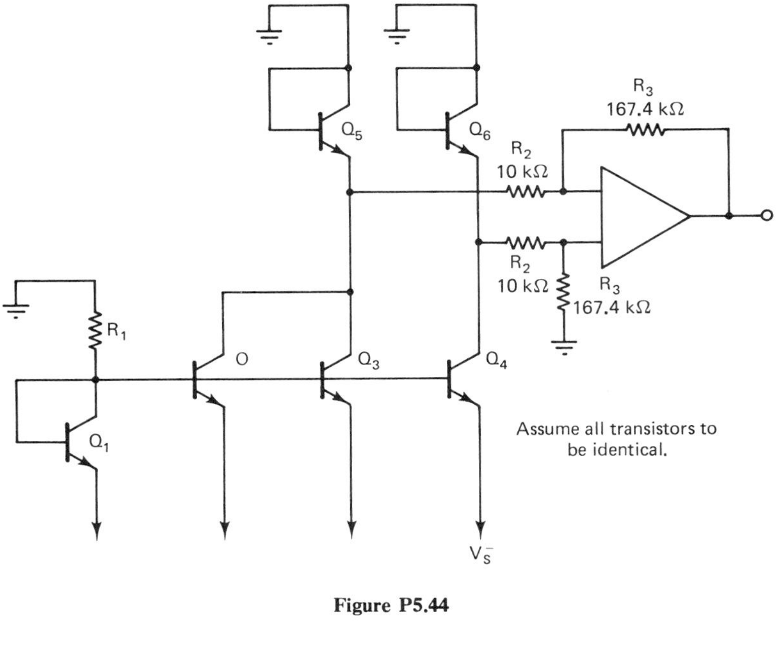


Figure P5.44

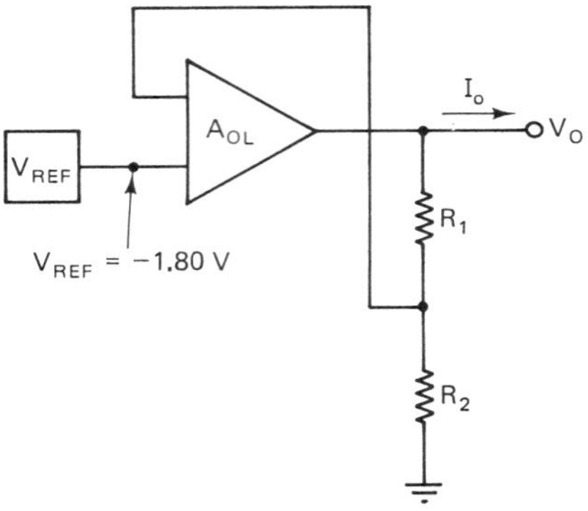


**Figure P5.45**

**(c)** If the reference voltage has a temperature coefficient given by $TC_{V(\mathrm{REF})} = 10\ \mu\mathrm{V}/°\mathrm{C}$, find the temperature coefficient of the regulated output voltage. (*Ans.*: $TC_{V_O} = 55.6,\ \mu\mathrm{V}/°\mathrm{C}$)

**(d)** Find the *load regulation* (change in $V_o$) as the output current $I_o$ goes from a no-load value of $I_{o(\mathrm{NL})} = 0$ to a full-load value of $I_{o(\mathrm{FL})} = 10$ mA. Express the result in terms of $\Delta V_o$ and the percentage change in $V_o$. (*Ans.*: $-280\ \mu$V or $-0.0028\%$)

**5.46.** (*Voltage regulator with current foldback*) Refer to Figure P5.46.

**(a)** Show that current limiting will occur when

$$\frac{I_O R_{CL} R_2}{R_1 + R_2} - \frac{V_O R_1}{R_1 + R_2} \simeq 0.6 \text{ V}$$

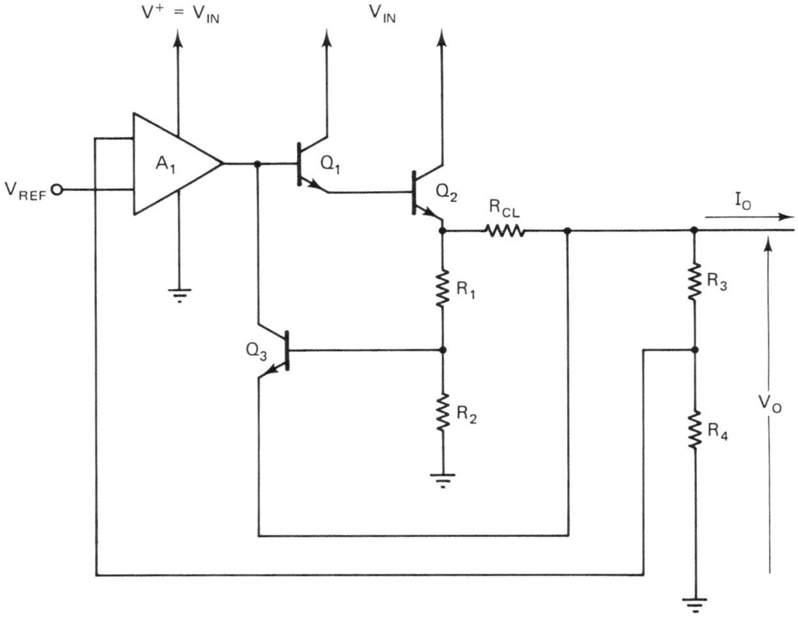


**Figure P5.46**

(b) Given that $P_{D(\text{MAX})}$ of $Q_2$ is 5.0 W, $V_{IN} = +15$ V. Find the value of $I_{O(\text{MAX})}$ for $V_O = +10$ V, and for $V_O = 0$. (*Ans.*: 1.0 A, 0.33 A)

(c) Given that $R_1 + R_2 = 1.0$ kΩ, find $R_1$, $R_2$, and $R_{CL}$ such that the current limiting corresponding to the conditions above will be obtained. (*Ans.*: $R_1 = 120$ Ω, $R_2 = 880$ Ω, $R_{CL} = 2.05$ Ω)

(d) Draw the $V_O$ versus $I_O$ characteristics of the voltage regulator corresponding to the conditions above.

(e) If $R_3 + R_4 = 2.0$ kΩ and $V_{REF} = 1.80$ V, find $R_3$ and $R_4$ for a regulated output voltage of 10 V. (*Ans.*: $R_3 = 1640$ Ω, $R_4 = 360$ Ω)

**5.47.** (*Analog multiplier*)

(a) For the circuit shown in Figure P5.47a, assume all transistors to have identical characteristics and all op amps to be ideal with very high open-loop gains. Show that the following relationship will be obtained, with all currents being positive:

$$I_1 I_2 = I_3 I_4$$

This circuit is available in monolithic IC form as the RC4200 (Raytheon).

(b) (*Four-quadrant multiplier*) For the circuit shown in Figure P5.47b, show that the following relationship is obtained.

$$V_o = \frac{V_X V_Y}{V_R} \frac{R_2 R_0}{R_1^2}$$

If $R_1 = R_2$ $R_0 = 20$ kΩ and $V_R = +10$ V, find $V_O$. (*Ans.*: $V_O = V_X V_Y / 10$ V)

If all four currents ($I_1$ through $I_4$) must be positive and limited to a maximum value of 1.0 mA for the accurate operation of this circuit, what will be the corresponding restrictions on $V_X$ and $V_Y$? (*Ans.*: $-10\text{ V} < V_X < +10\text{ V}$, $-10\text{ V} < V_Y < +10\text{ V}$)

Why is this circuit called a four-quadrant multiplier?

(c) (*One-quadrant analog divider*) For the circuit shown in Figure P5.47c, with $V_X$, $V_Z$, and $V_R$ being of positive polarity, show that the following relationship is obtained:

$$V_O = \frac{V_X}{V_Z} V_R \frac{R_4 R_0}{R_1 R_2}$$

Why is this circuit called a one-quadrant divider?

(d) (*Square-rooting circuit*) To produce a square-rooting circuit, $V_Z$ in the divider circuit (Figure P5.47c) is connected to $V_O$ such that $V_Z = V_O$. Show that $V_O$ will now become proportional to the square root of $V_X$ as given by

$$V_O = \sqrt{\frac{V_X V_R R_0 R_4}{R_1 R_2}}$$

If all resistors are equal and $V_R = +10$ V, find $V_O$. (*Ans.*: $V_O = \sqrt{10 V_X}$)

**5.48.** (*Phase-shift oscillator*) Refer to Figure P5.48. For oscillators to occur the net gain around the feedback loop must be greater than unity and the net phase shift around the feedback loop must be zero.

(a) Show that the frequency of oscillation will be given by

$$f_{osc} = \frac{1}{2\pi\sqrt{3}\, R_1 C_1}$$

and the condition for oscillation to occur will be that $R_3/R_2 > 8$.

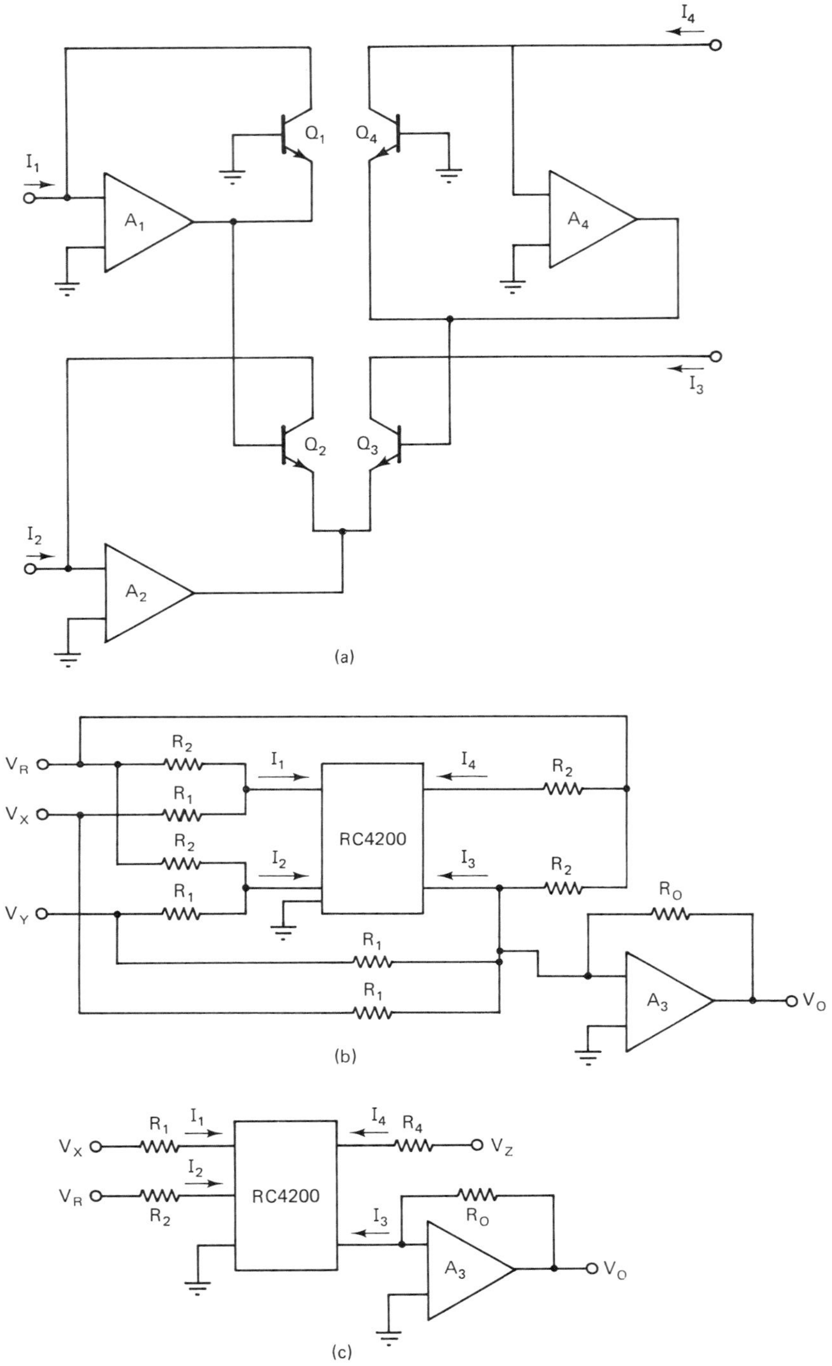

I4
I1
Q1
Q4
A1
A4
I3
Q2
Q3
I2
A2
(a)
VR
VX
VY
R2
R1
R2
R1
I1
I2
I4
I3
RC4200
R2
R2
RO
R1
R1
A3
VO
(b)
VX
VR
R1
R2
I1
I2
RC4200
I4
R4
VZ
I3
RO
A3
VO
(c)


**Figure P5.47**

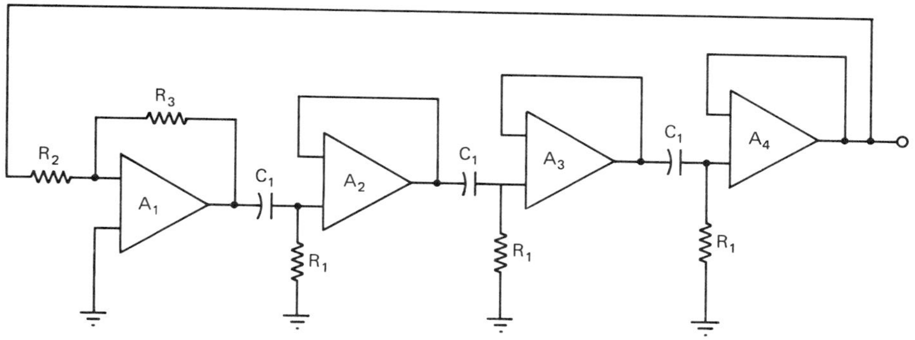


**Figure P5.48**

**(b)** If $R_1 = R_2 = 10\ \text{k}\Omega$ and $C_1 = 1.0$ nF, find the frequency of oscillation and the value of $R_3$ required for oscillations to occur. (*Ans*.: 9.19 kHz, 80 kΩ)

**(c)** If the net voltage gain around the feedback loop is adjusted to a value just slightly greater than unity, the output voltage will be a relatively undistorted sinusoidal waveform. Design a circuit that will automatically do this.

**5.49.** (*Wien bridge oscillator*) Refer to Figure P5.49. For this circuit to oscillate, the *net* feedback must be positive and have a phase angle of zero.

**(a)** Show that the frequency of oscillation will be given by

$$f_{osc} = \frac{1}{2\pi\sqrt{R_1 R_2 C_1 C_2}}$$

and the condition for oscillation will be

$$\frac{R_3}{R_4} > \frac{R_1}{R_2} + \frac{C_2}{C_1}$$

**(b)** If $R_1 = R_2 = R_4 = 10\ \text{k}\Omega$ and $C_1 = C_2 = 1.0$ nF, find the frequency of oscillation and the value of $R_3$ required for oscillations to occur. (*Ans*.: 15.9 kHz, 20 kΩ)

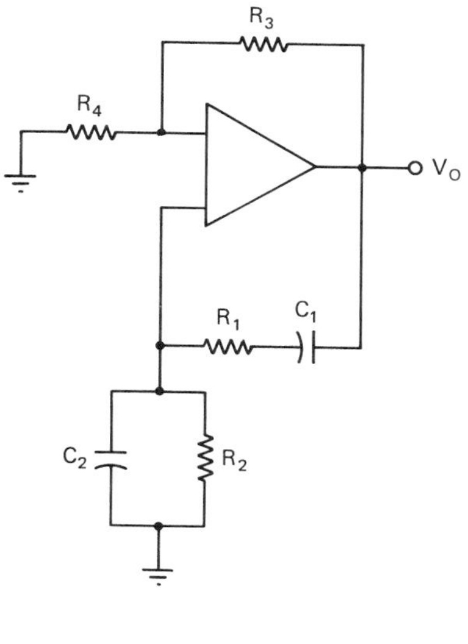


**Figure P5.49**

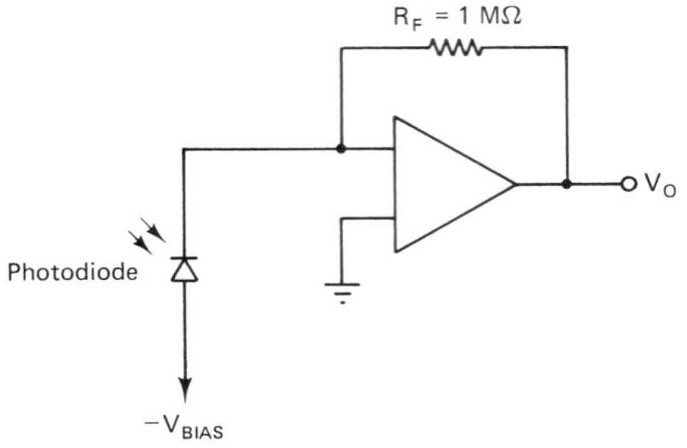


Figure P5.50

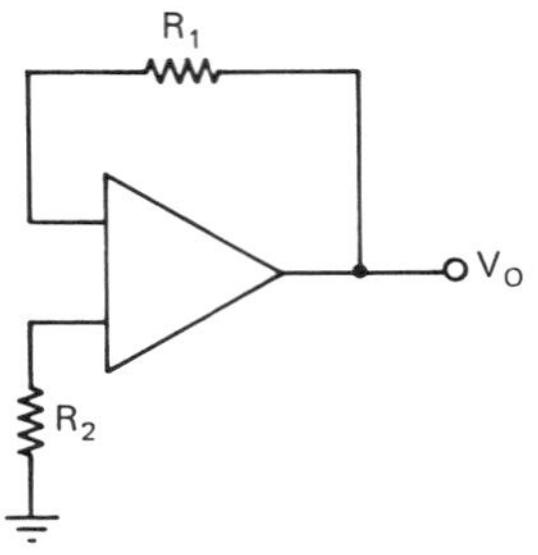


Figure P5.51

**5.50.** (*Optoelectronic sensor*) Given: The photodiode in Figure P5.50 has an active area of 10 mm² and a current responsivity (ratio of output current to incident optical power) of 0.5 A/W. Find the output voltage $V_o$ for an incident optical power density of 100 nW/cm². (*Ans*.: 10 mV)

**5.51.** (*Circuit for obtaining* $V_{OS}$, $I_B$, *and* $I_{OS}$) Refer to Figure P5.51. Given:

$$V_O = +2.2 \text{ mV when } R_1 = R_2 = 0$$

$$V_O = +20 \text{ mV when } R_1 = R_2 = 100 \text{ M}\Omega$$

$$V_O = -120 \text{ mV when } R_1 = 0 \text{ and } R_2 = 100 \text{ M}\Omega$$

Find:

**(a)** $V_{OS}$. (*Ans*.: 2.2 mV)
**(b)** $I_{BIAS}$. (*Ans*.: 1.2 nA)
**(c)** $I_{OS}$. (*Ans*.: 178 pA)

**5.52.** (*Effect of supply voltage on output voltage swing*) Find the output voltage $v_o$ for the following circuits.
**(a)** (*Ans*.: $v_o$ will be near +11 or −11 V)
**(b)** (*Ans*.: $v_o$ will be near +11 V if $I_B$ is positive, and near −11 V if $I_B$ is negative)
**(c)** (*Ans*.: $v_o$ will be approximately a square wave with an amplitude of ±11 V)

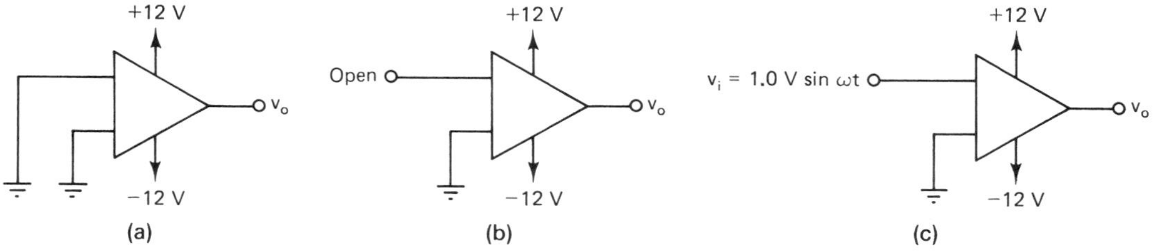


Figure P5.52

**5.53.** (*Effect of current limiting on output voltage*) Refer to Figure P5.53. Given: $I_{CL} = \pm 20$ mA. Find $V_O$. [*Ans*.: $v_o$ will be a 10-V amplitude sine wave clipped at the ±2.0-V level, and will therefore look approximately like a square wave of ±2.0-V amplitude]

**5.54.** (*Circuit for evaluating PSRR*) Refer to Figure P5.54. Given: $V_O$ = 2.0 mV (pms) at 1 kHz. Find PSRR (in dB). (*Ans*.: 94 dB)

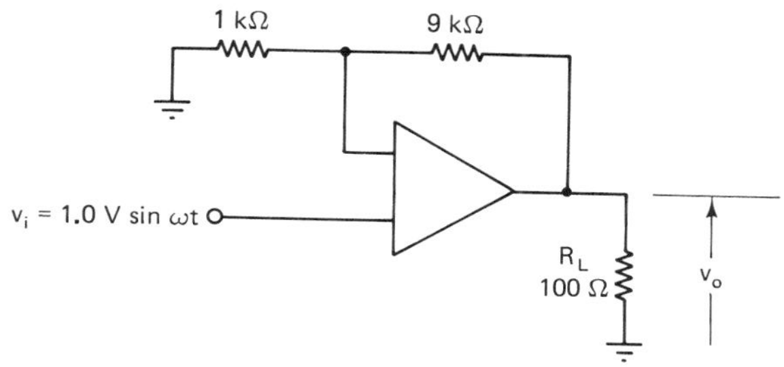


**Figure P5.53**

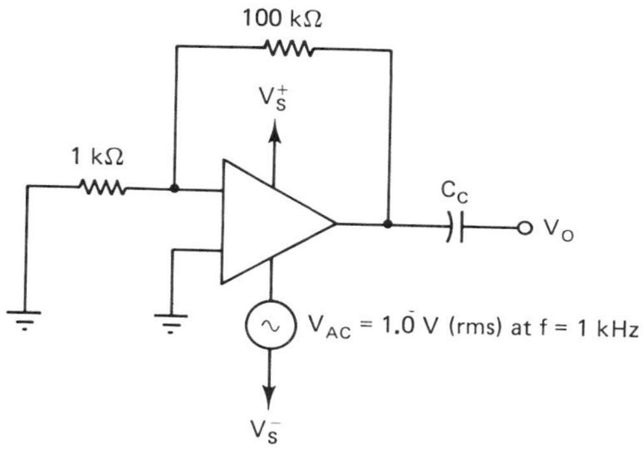


**Figure P5.54**

**5.55.** (*Circuit for obtaining open-loop gain*) Refer to Figure P5.55. Given: $V_O = 5.0$ V (a-c) and $V_1 = 20$ mV (a-c). Find the open-loop gain of the op amp. (*Ans.*: $A_{OL} = 250{,}000$ or 108 dB)

**5.56.** (*Evaluation of equivalent input noise voltage*) Refer to Figure P5.56. Given: $V_{O\,(\text{noise})} =$ 2.5 mV (rms). Bandwidth (3 dB) = 10 kHz.

**(a)** Find the amplifier equivalent input noise voltage. (*Ans.*: 2.5 μV)

**(b)** Find the amplifier equivalent input noise voltage spectral density. [*Ans.*: 20 nV/(Hz)$^{1/2}$]

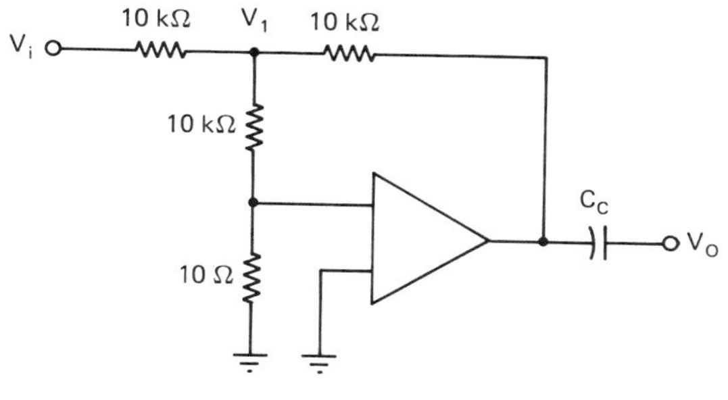


**Figure P5.55**

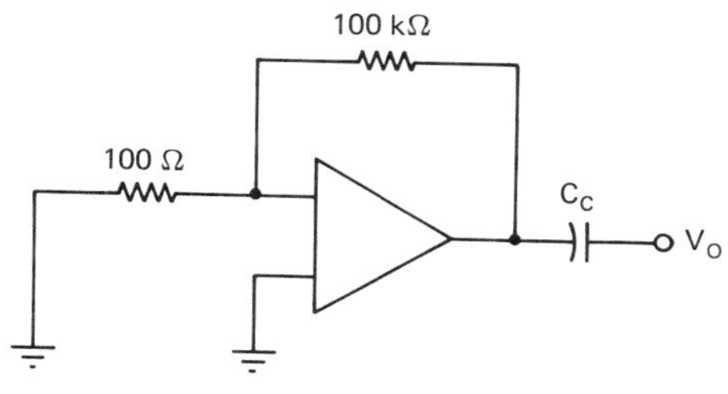


**Figure P5.56**

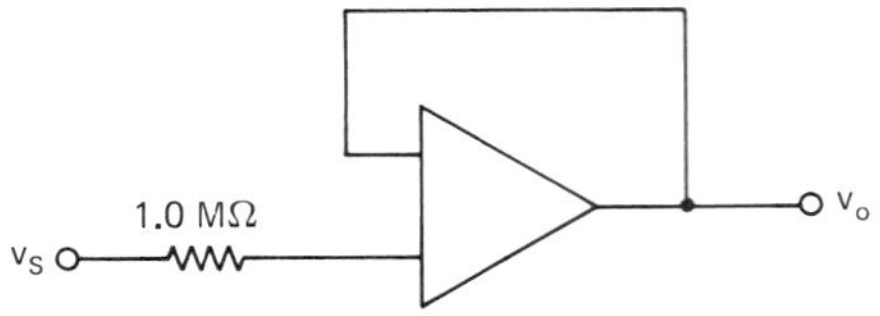


**Figure P5.57**

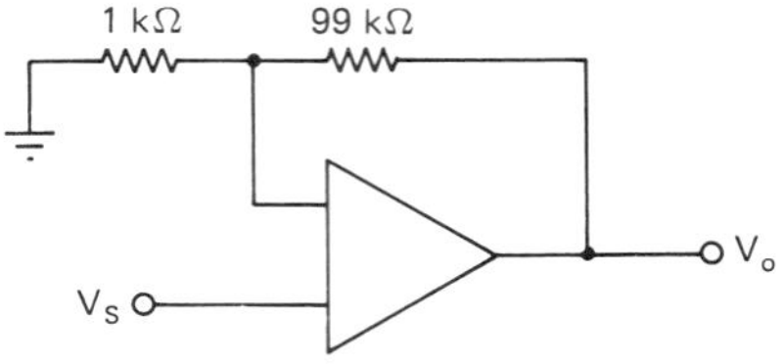


**Figure P5.58**

**5.57.** (*Common-mode input impedance*) Refer to Figure P5.57. Given: $f_u = 1.0$ MHz, $A_{OL}(0) =$ 106 dB (min.), $z_{i\,(CM)} = [1.0\ \text{G}\Omega \parallel 4.0\ \text{pF}]$. Find:

**(a)** $A_{CL}(0)$. (*Ans*.: 0.999)

**(b)** 3-dB bandwidth. (*Ans*.: 40 kHz)

**(c)** 0.5-dB bandwidth. (*Ans*.: 14 kHz)

**5.58.** (*Closed-loop gain versus frequency*) Refer to Figure P5.58. Given: $f_u = 1.0$ MHz, $A_{OL}(0) =$ 106 dB (min.). Find:

**(a)** Open-loop 3-dB bandwidth. [*Ans*.: 5.0 Hz (max.)]

**(b)** $A_{CL}$ at 1.0 kHz. (*Ans*.: 99.5 $\angle -5.7°$)

**(c)** $A_{CL}$ at 10 kHz. (*Ans*.: 71 $\angle -45°$)

**(d)** $A_{CL}$ at 100 kHz. (*Ans*.: 9.95 $\angle -84.3°$)

**5.59.** (*Common-mode gain*) Refer to Figure P5.59.

**(a)** Find $v_o$ if CMRR = 100 dB at 1 kHz. (*Ans*.: $v_o = 100\ \mu$V at 1 kHz)

**(b)** Find $v_o$ resulting from a ±1% mismatch between the two $R_2$ resistors. (*Ans*.: ±9 mV)

**(c)** Find $v_o$ resulting from a ±1% mismatch between the two $R_1$ resistors. (*Ans*.: ±9 mV)

**5.60.** (*Equivalent input noise voltage*) Refer to Figure P5.60. Given: $v_{o\,(\text{noise})} = 0.20$ mV (rms)

**(a)** Find the equivalent input noise voltage, $v_{i\,(\text{noise})}$. [*Ans*.: 2.0 $\mu$V (rms)]

**(b)** If the amplifier has a unity-gain frequency of 1.0 MHz, find the equivalent input noise voltage spectral density. [*Ans*.: 16 nV/(Hz)$^{1/2}$] (*Hint*: Remember the $\pi/2$ conversion factor to go from 3-dB bandwidth to equivalent noise bandwidth.)

**5.61.** (*Amplifier-to amplifier coupling*) Referring to Figure P5.61, find the amplifier-to-amplifier coupling (in dB). (*Ans*.: −94 dB)

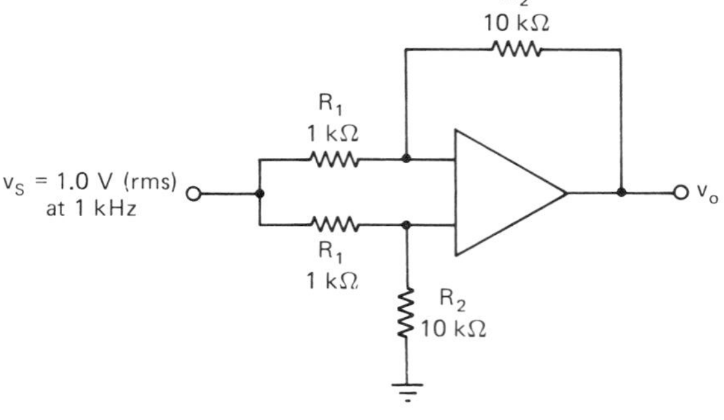


**Figure P5.59**

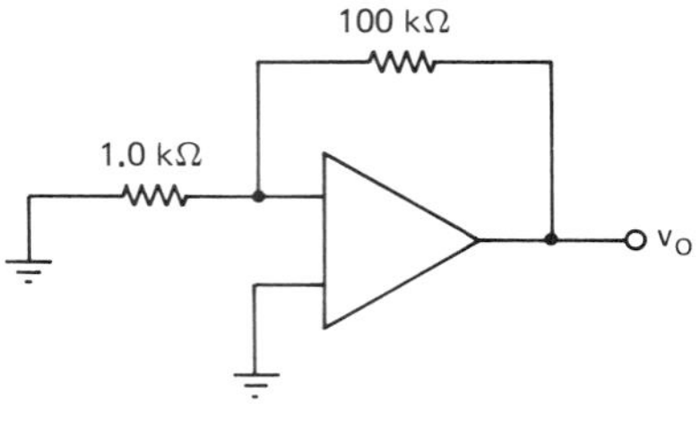


**Figure P5.60**

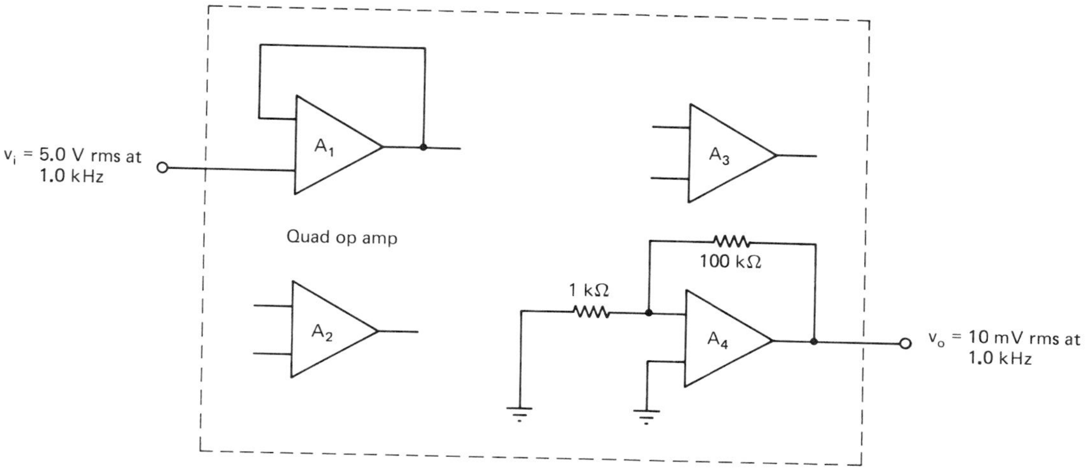


**Figure P5.61**

**5.62.** (*Power dissipation and temperature rise*) Refer to Figure 5.62. Given: Op amp with $T_{J(\text{MAX})} = 125°\text{C}$, thermal impedance $= \Theta = 100°\text{C/W}$, $I_{CL} = \pm 50$ mA, $TC_{V_{OS}} = 5$ $\mu$V/°C, and quiescent supply current $= I_{S(Q)} = 1.0$ mA. Find:

**(a)** Maximum power dissipation, $P_{D(\text{MAX})}$ (for $T_{\text{ambient}} = 25°\text{C}$). [*Ans*.: $P_{D(\text{MAX})} = 1.0$ W]

**(b)** Quiescent power dissipation, $P_{D(Q)}$. [*Ans*.: $P_{D(Q)} = 36$ mW]

**(c)** Maximum allowable current limit for full protection of op amp if maximum rated supply voltage is ±18 V. (*Ans*.: $I_{CL} = \pm 55$ mA)

**(d)** Power delivered to load, $P_O$. (*Ans*.: $P_O = 400$ mW)

**(e)** Op-amp power dissipation. (*Ans*.: $P_D = 356$ mW)

**(f)** Temperature rise of op-amp chip above ambient temperature, $\Delta T_J$. (*Ans*.: $\Delta T_J = 35.6°\text{C}$)

**(g)** Change in the offset voltage result from the temperature rise above. (*Ans*.: $\Delta V_{OS} = 178$ $\mu$V)

**5.63.** (*Bandwidth and rise-time calculations*) Refer to Figure P5.63. Given: $f_u = 1.0$ MHz and $f_2 > f_u$. Find:

**(a)** 3-dB bandwidth. (*Ans*.: 19.6 kHz)

**(b)** Rise time. (*Ans*.: 17.9 $\mu$s)

**5.64.** (*Compensation capacitance calculations*) Refer to Figure P5.64. Given: When $C_{\text{COMP}} = 40$ pF, $A_{OL} = 0$ dB at $f = f_2 = 1.0$ MHz. Assume that $f_u \propto 1/C_{\text{COMP}}$.

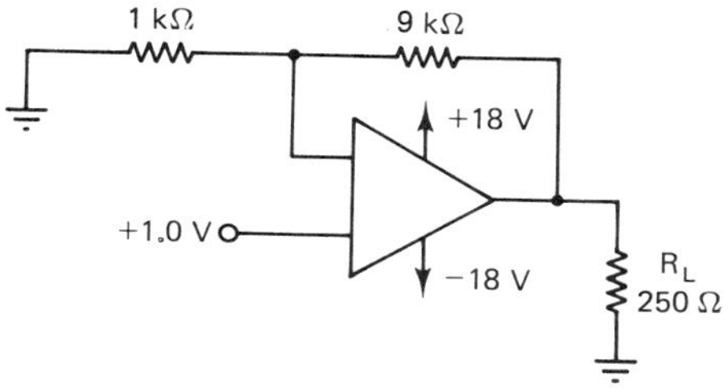


**Figure P5.62**

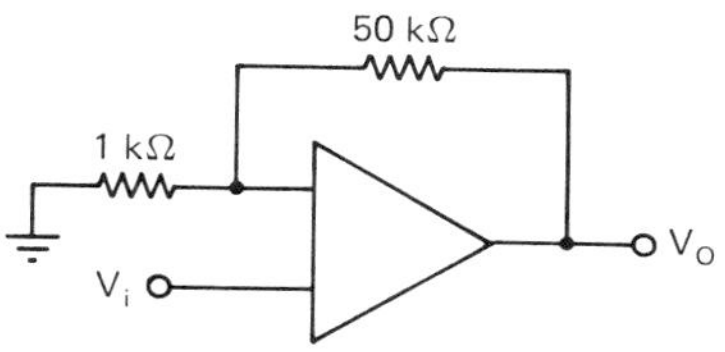


**Figure P5.63**

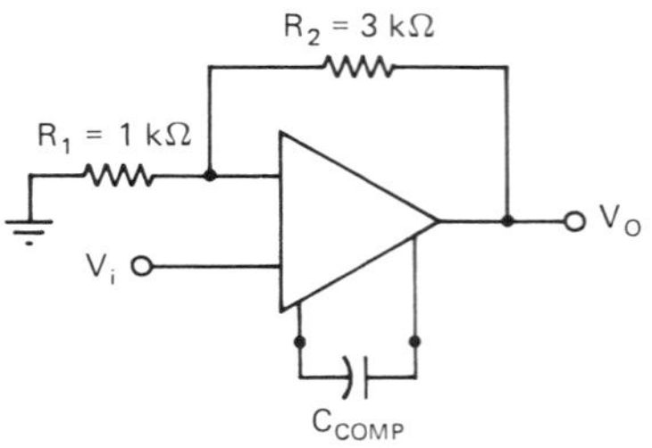


**Figure P5.64**

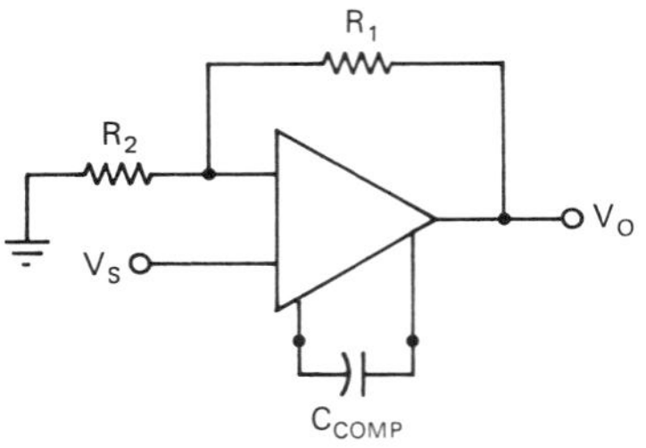


**Figure P5.65**

**(a)** Find the 3-dB bandwidth and the rise time (for small-signal conditions). (*Ans.*: 35 kHz, 1.0 $\mu$s)

**(b)** Find the smallest value of $C_{COMP}$ that can be used in order to maintain a phase margin of at least 45°. (*Ans.*: 10 pF)

**(c)** Find the corresponding bandwidth and rise time when this value of $C_{COMP}$ is used. (*Ans.*: $BW = 1.4$ MHz, $t_{rise} = 248$ ns)

**5.65.** (*Compensation capacitance calculations*) Refer to Figure P5.65. Given: $f_u = 1.0$ MHz when $C_{COMP} = 30$ pF, and $f_u = 21$ MHz when $C_{COMP} = 0$ (extrapolated values) and $f_2 = 1.0$ MHz. Assume that $f_u \propto 1/(C_{COMP} + C_{parasitic})$. Find:

**(a)** $f_u$ for $C_{COMP} = 3.0$ pF. (*Ans.*: 7.0 MHz)

**(b)** The 3-dB bandwidth if $R_1 = 39$ kΩ, $R_2 = 1.0$ kΩ, and $C_{COMP} = 30$ pF. (*Ans.*: 25 kHz)

**(c)** Repeat part (b) for $C_{COMP} = 3.0$ pF. (*Ans.*: 175 kHz)

**(d)** If $R_1 = 30$ kΩ and $R_2 = 10$ kΩ, find $C_{COMP}$ for a 45° phase margin. What will be the corresponding bandwidth? [*Ans.*: $C_{COMP} = 4.07$ pF (min.), $BW \simeq 1.0$ MHz]

**5.66.** (*Op amp stability*) Refer to Figure P5.66. Given: $f_1 = 10$ Hz; $f_2 = 1.0$ MHz, $f_3 = 4.0$ MHz, $C_{COMP} = 50$ pF, $f_u$ (extrapolated) $= 1.0$ MHz.

**(a)** Show that $V_O = 1.0 \text{ V} \log_{10} (V_S/V_{REF})$, for $V_S$ having a positive polarity.

**(b)** Show that the feedback factor, $F$, under small-signal a-c conditions will be $F = V_S/V_T$ for the $A_1$ circuit, and $F = V_{REF}/V_T$ for the $A_2$ circuit.

**(c)** For what values of $V_S$ and $V_{REF}$ will this circuit be stable? (*Ans.*: up to +125 mV for both $V_S$ and $V_{REF}$)

**(d)** If it is required that this logarithmic amplifier be stable for values of $V_S$ and $V_{REF}$ up to 1.0 V, what is the required value for $f_1$? If the 50-pF compensation capacitor gives an $f_1$ value of 10 Hz, what value of $C_{COMP}$ is now required? (*Ans.*: $f_1 = 1.25$ Hz, $C_{COMP} = 400$ pF)

**(e)** Repeat part (d) for input voltages going up as high as 50 V. (*Ans.*: $f_1 = 0.025$ Hz, $C_{COMP} = 20$ nF)

**(f)** Find the small-signal 3-dB bandwidths for the following values of $V_S$ assuming that a 20-nF compensating capacitor is used: 100 mV, 1.0 V, and 5.0 V. (*Ans.*: 10 kHz, 100 kHz, and 500 kHz)

**(g)** Given that $I_{CURRENT\ LIMIT} = 25$ mA, $I_{BIAS} = 10$ nA, and $I_{OS} = 1.0$ nA, find the acceptable range of values for $R_1$ that will limit the error, referred to the input voltage ($V_S$), to less than 1% for a $V_S$ range of 10 mV to 100 V (four decades). (*Ans.*: $R_I$ must be between 4 and 100 kΩ)

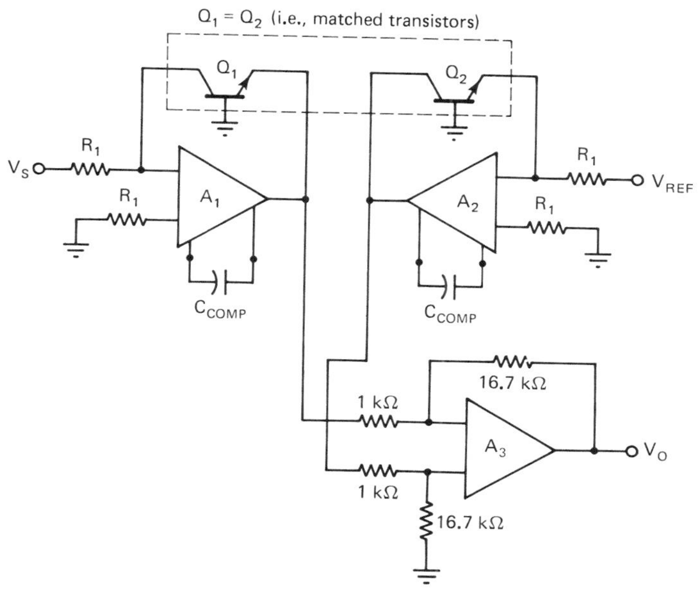


**Figure P5.66**

**(h)** If transistors $Q_1$ and $Q_2$ are changed from a common-base connection to a diode-connected transistor configuration, what effect will this have on the stability of the circuit and the logarithmic conversion accuracy? Does the use of diode-connected transistors offer any advantages over the use of a matched pair of diodes? Explain.

**5.67.** An op amp has $f_u = 1.5$ MHz and $f_2 = 6.0$ MHz. Find the closed-loop bandwidth for $A_{CL}(0) = 50$. (*Ans.*: 30 kHz)

**5.68.** Find the small-signal rise time for Problem 5.56. (*Ans.*: 11.7 $\mu$s)

**5.69.** If $f_u = 1.5$ MHz and $A_{OL}(0) = 106$ dB, find $A_{OL}$ (magnitude and angle) at $f = 25$ kHz. (*Ans.*: 60 $\underline{/-89.98°}$)

**5.70.** Find the open-loop bandwidth for the Problem 5.58. (*Ans.*: 7.5 Hz)

**5.71.** An op amp has a PSRR of $-90$ dB. The d-c power supply has a 10-mV a-c ripple and the op amp closed-loop gain is 40 dB. Find the a-c output voltage due to the power supply ripple. (*Ans.*: 31.6 $\mu$V)

**5.72.** An op amp has $A_{OL}(0) = 106$ dB, $f_1 = 10$ Hz, $f_2 = 500$ kHz, and $f_3 = 2.0$ MHz. Find the minimum closed-loop gain for a 45° phase margin. (*Ans.*: 2.828)

**5.73.** An op amp has $f_u = 1.0$ MHz, $A_{CL}(0) = 100$, and an output noise voltage of $v_{o\,(\text{noise})} =$ 0.20 mV (rms). Find the equivalent input noise voltage (spectral density). (*Ans.*: 16 $\text{nV}/\sqrt{\text{Hz}}$)

**5.74.** An op amp voltage follower is driven from a 1.0-MΩ source. The op amp has $A_{OL}(0) =$ 100 dB, $f_u = 2.0$ MHz, and the common-mode input impedance is 100 MΩ in parallel with 5.0 pF. Find the 3-dB bandwidth for this system. (*Ans.*: 31.86 kHz)

## REFERENCES


BARNA, A., *Operational Amplifiers*, Wiley, 1971.

CARR, J. J., *Linear IC/Op Amp Handbook*, TAB Books, 1983.

CONNELLY, J. A., *Analog Integrated Circuits*, Wiley, 1975.

FITCHEN, F., *Electronic Integrated Circuits and Systems*, Van Nostrand Reinhold, 1970.

GLASER, A. B., and G. E. SUBAK-SHARPE, *Integrated Circuit Engineering*, Addison-Wesley, 1977.

GRAEME, J. G., *Applications of Operational Amplifiers*, McGraw-Hill, 1973.

GRAY, P. R., and R. G. MEYER, *Analysis and Design of Analog Integrated Circuits*, Wiley, 1984.

GRINICH, V. H., and H. G. JACKSON, *Introduction to Integrated Circuits*, McGraw-Hill, 1975.

HAMILTON, D. J., and W. G. HOWARD, *Basic Integrated Circuit Engineering*, McGraw-Hill, 1975.

HNATEK, E. R., *A User's Handbook of Integrated Circuits*, Wiley, 1973.

HUGHES, F. W., *Op Amp Handbook*, Prentice-Hall, 1981.

JACOB, J. M., *Applications and Design with Analog Integrated Circuits*, Reston, 1982.

LENK, J. D., *Handbook of Integrated Circuits*, Reston, 1978.

LENK, J. D., *Manual for Integrated Circuit Users*, Reston, 1973.

LENK, J. D., *Manual for Operational Amplifier Users*, Reston, 1976.

MANASSE, F. K., *Semiconductor Electronics Design*, Prentice-Hall, 1977.

MILLMAN, J., *Microelectronics*, McGraw-Hill, 1979.

ROBERGE, J. K., *Operational Amplifiers*, Wiley, 1975.

SEDRA, A. S., and K. C. SMITH, *Microelectronic Circuits*, Holt, Rinehart and Winston, 1982.

SEIPPEL, R. G., *Operational Amplifiers*, Reston, 1983.

UNITED TECHNICAL PUBLICATIONS, *Modern Applications of Integrated Circuits*, TAB Books, 1974.

WAIT, J. V., *Introduction to Operational Amplifier Theory and Applications*, McGraw-Hill, 1975.

WONG, Y. J., and W. E. OTT, *Function Circuits*, McGraw-Hill, 1976.


### Active Filters


BOWRON, P., *Active Filters for Communications and Instrumentation*, McGraw-Hill, 1979.

GARRETT, P. H., *Analog I/O Design*, Reston, 1981.

HAYKIN, S. S., *Synthesis of RC Active Filters*, McGraw-Hill, 1969.

HILBURN, J. L., and D. E. JOHNSON, *Manual of Active Filter Design*, McGraw-Hill, 1973.

HUELSMAN, L. P., and P. E. ALLEN, *Introduction to the Theory and Design of Active Filters*, McGraw-Hill, 1980.

JOHNSON, D. E., and J. L. HILBURN, *Rapid Practical Designs of Active Filters*, Wiley, 1975.

LAM, H. Y.-F., *Analog and Digital Filters*, Prentice-Hall, 1979.

TEDESCHI, F. P., *The Active Filter Handbook*, TAB Books, 1979.

WILLIAMS, A. B., *Active Filter Design*, Artech House, 1975.

# 6

# OPERATIONAL-AMPLIFIER CIRCUIT DESIGN

## 6.1 OPERATIONAL-AMPLIFIER CIRCUIT ANALYSIS

We will now start to examine the *internal* circuitry of op amps. Much of this circuitry will be closely related to that of other analog ICs. A basic block diagram representative of most op amps is shown in Figure 6.1. There are two gain stages followed by an emitter-follower output stage. The voltage gain of each of the two gain stages will generally be in the range 300 to 1000 and the emitter-follower output stage will have a gain of around unity, so that the overall op-amp low-frequency open-loop voltage gain $A_{CL}(0)$ will be of the order of $10^5$ to $10^6$ (100 to 120 dB).

The emitter-follower output stage provides impedance transformation so that relatively low values of load resistance can be driven. This stage is usually of a "push-pull" configuration such that the op amp can *source* (i.e., $i_o > 0$) as well as *sink* ($i_o < 0$) output currents.

As is the case with all ICs, all of the stages are direct-coupled, there being no coupling or bypass capacitors in the circuit. Much of the design of the circuit relies on the close matching of transistors and resistors that are possible with ICs. This close matching comes about because the components are very close to each other on the same small silicon chip and they have undergone identical manufacturing processing. In addition, since capacitors and large-value (above 50 kΩ) resistors will occupy large areas on IC chips and also exhibit large undesirable parasitic effects, these components are usually avoided and the emphasis in the circuit design is on maximizing the number of transistors (and diodes) as compared to resistors.

For feedback stability it is desirable to minimize the total open-loop phase shift in the amplifier. This will result in a larger value for $f_{180}$ and thus allow for a

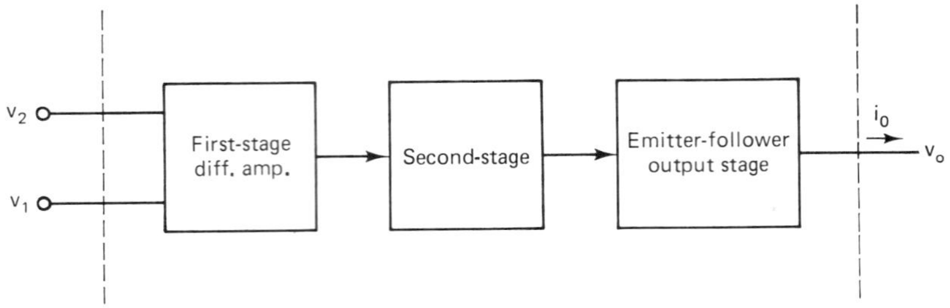


**Figure 6.1** Basic block diagram of an operational amplifier.

greater value for the unity-gain frequency $f_u$, which will mean a larger gain–bandwidth product. Since each gain stage will contribute to the total amplifier phase shift, it is therefore desirable to minimize the total number of stages while maintaining a large overall open-loop voltage gain. The earliest IC operational amplifiers had three gain stages, but almost all present-day op amps have just two gain stages, so that for a very large open-loop gain (100 kΩ to 1 MΩ) the voltage gain of each stage must be very large.

The principal exception to the general rule of two gain stages is op amps that use FETs. The FETs can be used just in the input stage or throughout the entire amplifier circuit, and these op amps offer the advantages of a very high input impedance and a very low input bias current. However, as a result of the much lower transfer conductance and therefore voltage gain available with FETs as compared to bipolar transistors, some FET op amps need three gain stages.

### 6.1.1 Operational-Amplifier Open-Loop Frequency Response

Now that the basic configuration of the op amp has been examined, the open-loop frequency response will be investigated. For this purpose the op-amp representation of Figure 6.2 will be used, in which the second gain stage and the emitter-follower output stage are represented as amplifiers with gains of $-A_2$ and $+1$, respectively. A feedback or compensation capacitor $C_c$ is shown connected across the second stage, although it could just as well be connected across both the second and output stages with essentially the same effect.

By inspection of this circuit we have that

$$i_c \simeq i_{o_1} = 2g_f v_i \tag{6.1}$$

and

$$\begin{aligned} i_c &= (v_o - v_{i_2}) j\omega C_c = \left(v_o + \frac{v_o}{A_2}\right) j\omega C_c \\ &= v_o\left(1 + \frac{1}{A_2}\right) j\omega C_c \end{aligned} \tag{6.2}$$

so that

$$v_o = \frac{i_c}{j\omega C_c(1 + 1/A_2)} = \frac{2g_f v_i}{j\omega C_c(1 + 1/A_2)} \tag{6.3}$$

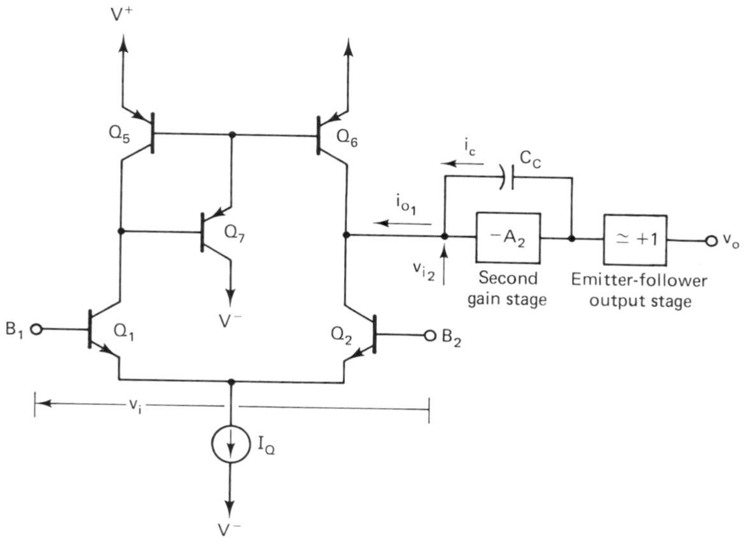


**Figure 6.2** Operational-amplifier circuit for analysis of frequency response.

If $A_2 \gg 1$ as will usually be the case, the preceding equation can be written as

$$v_o \simeq \frac{2g_f v_i}{j\omega C_c} \tag{6.4}$$

so that the equation for the op-amp open-loop gain can be written very simply as

$$\boxed{A_{OL} = \frac{v_o}{v_i} \simeq \frac{2g_f}{j\omega C_c}} \tag{6.5}$$

At the unity-gain frequency the magnitude of the open-loop gain is unity, so that $2g_f/\omega_u C_c = 1$ and thus

$$\boxed{\omega_u = \frac{2g_f}{C_c}} \tag{6.6}$$

where $\omega_u$ is the *unity-gain radian frequency*. The corresponding value of $f_u$ is

$$\boxed{f_u = \frac{\omega_u}{2\pi} = \frac{2g_f}{2\pi C_c}} \tag{6.7}$$

The capacitance $C_c$ will be principally the capacitance that is purposely added to the circuit to establish the required value for $f_u$, either as an on-chip capacitor in the case of internally compensated op amps, or as an external capacitor. There will also be some parasitic circuit capacitance of the order of a few picofarads such that $C_c$ should be considered to be given by $C_c = C_{\text{comp}} + C_p$, where $C_{\text{comp}}$ is the compensation capacitance added to the circuit and $C_p$ is the parasitic capacitance (~1 to 3 pF) already present in the circuit.

The equation for $A_{OL}$ given above is of course not valid at very low frequencies (i.e., below $f_1$) at which the gain approaches a limiting value of $A_{OL}(0)$. It is also not valid at frequencies above the op-amp second breakpoint frequency $f_2$, at which point the gain of the second stage ($-A_2$) is no longer large compared to unity. In a later section where an example of an op-amp circuit is discussed in detail, a more extensive analysis of the frequency response will be carried out and the effect of the second breakpoint frequency $f_2$ will be examined.

As an example of the relationship between $f_u$ and $C_c$, let us consider an op amp for which $I_Q = 20\ \mu\text{A}$ and specify that $f_u = 1.25$ MHz. For this value of $I_Q$, the differential-amplifier transfer conductance $g_f$ will be $g_f = I_Q/4V_T = 20\ \mu\text{A}/100\ \text{mV} = 200\ \mu\text{S}$. The required value for $C_c$ will be given by $f_u = 2g_f/2\pi C_c = 400\ \mu\text{S}/(2\pi C_c)$, so that solving for $C_c$ gives $C_c = 400\ \mu\text{S}/(2\pi \times 1.25\ \text{MHz}) = 51$ pF. Thus only a relatively small capacitance value is required, and this capacitance value is small enough such that it can be incorporated on the IC chip without occupying an excessive amount of area. Note that if a Darlington differential-amplifier configuration were used, the equation for $g_f$ above would have to be modified to become $g_f = I_Q/4(2V_T) = I_Q/8V_T$.

### 6.1.2 Slewing Rate

The *slewing rate* (*SR*) of an op amp is the maximum rate at which the output voltage can change with time, $(dv_o/dt)_{\text{MAX}}$. The output voltage of the op amp can be related to the current through $C_c$ as

$$i_c = C_c \frac{d(v_o - v_{i_2})}{dt} \simeq C_c \frac{dv_o}{dt} \tag{6.8}$$

if the gain of the second stage is large such that $v_{i_2} = v_o/A_2 \ll v_o$, as will usually be the case. Solving for $dv_o/dt$, we now have that

$$\frac{dv_o}{dt} \simeq \frac{i_c}{C_c} \simeq \frac{i_{o_1}}{C_c} \tag{6.9}$$

The output current of the first stage is limited by the $I_Q$ current source to a maximum limit of $i_{o_1(\text{MAX})} = +I_Q$ in one direction and $-I_Q$ in the other direction. As a result, the rate of change of output voltage will be limited to a maximum value given by

$$\text{Slewing rate (SR)} = \left(\frac{dv_o}{dt}\right)_{\text{MAX}} = \frac{i_{o_1(\text{MAX})}}{C_c} = \pm\frac{I_Q}{C_c} \tag{6.10}$$

There is a direct relationship between the slewing rate and the unity-gain frequency for the bipolar transistor case since $f_u = 2g_f/(2\pi C_c)$ and $g_f = I_Q/4V_T$. We have that

$$f_u = \frac{2g_f}{(2\pi C_c)} = \frac{2(I_Q/4V_T)}{2\pi C_c} = \frac{I_Q/C_c}{4\pi V_T} = \frac{SR}{4\pi V_T} \tag{6.11}$$

For the case of a Darlington differential amplifier $g_f = I_Q/8V_T$, so the corresponding relationship between $f_u$ and the slewing rate becomes $f_u = SR/8\pi V_T$.

To consider an example of a slewing rate calculation, let us choose $I_Q = 20$ $\mu$A and $f_u = 1.25$ MHz, as given above. The value of $C_c$ needed is 51 pF and the slewing rate will be

$$SR = I_Q/C_c = 20\ \mu\text{A}/51\ \text{pF} = 3.9 \times 10^5\ \text{V/s} = \underline{0.39\ \text{V}\mu\text{s}}$$

Most op amps will have slewing rates in the range 0.5 to 3 V/$\mu$s, and there are op amps available with slewing rates as high as 100 V/$\mu$s. In particular, op amps with FET input stages generally have very high slewing rates.

An op amp will become slewing-rate limited when the rate of change of the output voltage $dv_o/dt$ as predicted by the small-signal response equations exceeds the slewing rate. For a step-function type of input voltage we have that $v_o(t) = V_{O(\text{MAX})}\,[1 - \exp(-t/\tau)]$, so that

$$\left(\frac{dv_o}{dt}\right)_{\text{MAX}} = \left(\frac{dv_o}{dt}\right)_{t=0} = \frac{V_{O(\text{MAX})}}{\tau} \tag{6.12}$$

Thus if

$$\boxed{\frac{V_{O(\text{MAX})}}{\tau} > SR} \tag{6.13}$$

the op amp will be *slewing-rate limited* and the output voltage will increase linearly with time at the slewing rate. An example of this situation is shown in Figure 6.3.

The op-amp closed-loop bandwidth is related to the closed-loop time constant by $\tau = 1/(2\pi BW_{CL})$ so that $BW_{CL} = 1/(2\pi\tau)$. For the example of Figure 6.3 an op-amp slewing rate of 1 V/$\mu$s is chosen and a closed-loop bandwidth of $BW_{CL} = 1.0$ MHz is assumed, so that the corresponding time constant will be $\tau = 1/(2\pi BW_{CL}) = 0.16\ \mu$s. Thus for $V_{O(\text{MAX})} > SR \times \tau = 1\ \text{V}/\mu\text{s} \times 0.16\ \mu\text{s} = 0.16$ V, the op amp will be operating under slewing-rate-limited conditions.

In Figure 6.4 the "large-signal" or slewing-rate-limited response of an op amp to a rectangular pulse input voltage is shown. Note the trapezoidal shaped response characteristics. The time for the full-scale response (0 to $V_{O(\text{MAX})}$) will be approximately $V_{O(\text{MAX})}/SR$, so the 10 to 90% rise time will be approximately $t_{\text{rise}} = 0.8\,V_{O(\text{MAX})}/SR$.

The small-signal rise time for the example above will be $t_{\text{rise}} = 2.2\tau = 0.35/BW_{CL} = 0.35\ \mu$s. However, under "large-signal" slewing-rate-limited conditions the

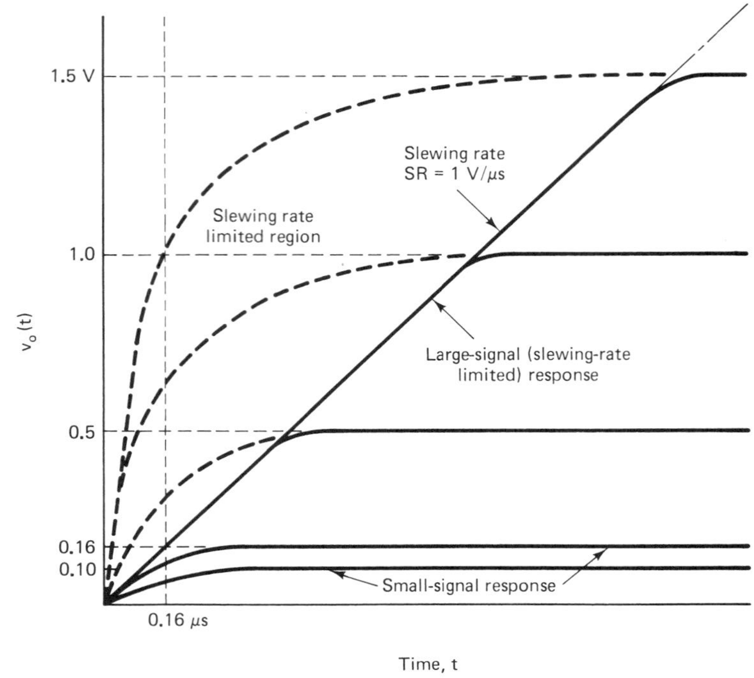


**Figure 6.3** Step-function response characteristics.

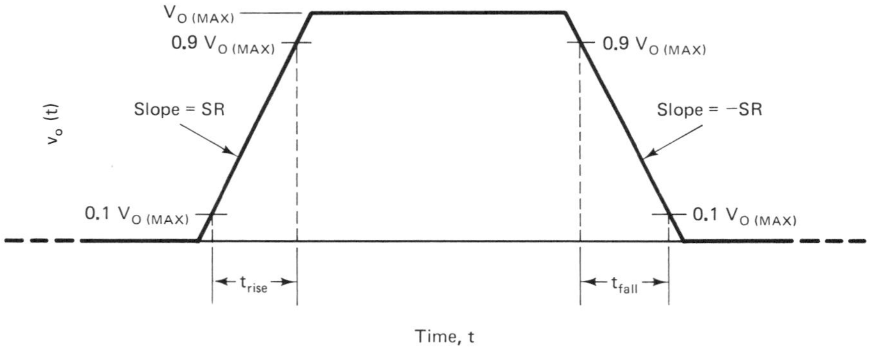


**Figure 6.4** Slewing-rate-limited response to rectangular pulse voltage input.

rise time will approach $t_{\mathbf{rise}} = 0.8\,V_{O\,(\mathbf{MAX})}/SR$. For a $V_{O\,(\mathbf{MAX})}$ of 10 V the rise time will be $t_{rise} = 0.8 \times 10$ V/1 V/μs = 8 μs, which is much longer than the small-signal rise time.

### 6.1.3 Full-Power Bandwidth

Let us now consider the application of a sinusoidal input voltage to an op amp. Under small-signal conditions such that the output is not slewing-rate limited, the output voltage can be expressed as $v_o(t) = V_{O\,(\mathbf{MAX})} \sin \omega t$. The maximum rate of change of output voltage with time will occur at the zero crossings and will be given by

$$\left(\frac{dv_o}{dt}\right)_{\text{MAX}} = \omega V_{O(\text{MAX})} (\cos \omega t)_{\text{MAX}} = \pm\, \omega V_{O(\text{MAX})} \tag{6.14}$$

If $\omega V_{O(\text{MAX})} > SR$, the op amp will be slewing-rate limited and the maximum slope of the output voltage will be limited to the slewing rate.

In Figure 6.5 some output voltage waveforms are shown for various values of $V_{O(\text{MAX})}$. When $\omega V_{O(\text{MAX})} > SR$ the op-amp output becomes slewing-rate limited and the output voltage waveform will start to become distorted from a sinusoidal shape to a waveform approaching a triangular wave with sides having slopes of $\pm SR$. The peak value of the output voltage will be limited to a value of $SR \times T/4 = SR/4f$, where $T = 1/f$ is the period of the waveform.

We see that the peak output voltage swing will be limited to a value of $\pm SR/4f$ under large-signal slewing-rate-limited conditions, and thus as the frequency increases the peak output voltage swing will decrease rapidly. In Figure 6.6 the variation of the peak output voltage swing as a function of frequency is shown. The *full-power bandwidth* (FPBW) is the frequency at which the op-amp output becomes slewing-rate limited and is obtainable from the condition that $\omega V_{O(\text{MAX})} = SR$, so that $\omega = SR/V_{O(\text{MAX})}$ and thus

$$\boxed{\text{FPBW} = \frac{\omega}{2\pi} = \frac{SR}{2\pi V_{O(\text{MAX})}}} \tag{6.15}$$

The full-power bandwidth can be considerably smaller than the bandwidth under small-signal non-slewing-rate-limiting conditions. For example, if $BW_{CL} = 1.0$ MHz, $SR = 1.0$ V/μs, and $V_{O(\text{MAX})} = 10$ V, we will have that

$$\text{FPBW} = \frac{SR}{2\pi V_{O(\text{MAX})}} = \frac{1.0\ \text{V}/\mu\text{s}}{2\pi 10\ \text{V}} = \underline{16\ \text{kHz}} \tag{6.16}$$

This value for the full-power bandwidth is to be compared with the small-signal bandwidth of 1.0 MHz, so that we see that the FPBW can be much smaller than

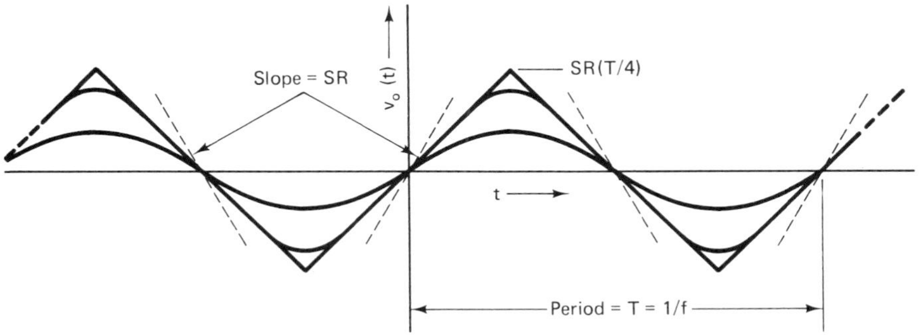


**Figure 6.5** Full-power bandwidth.

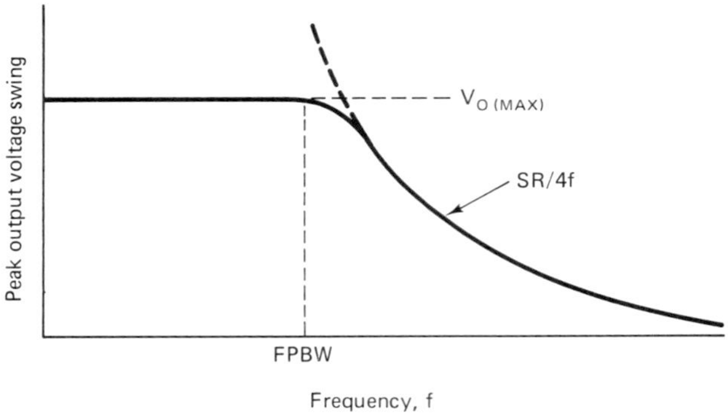


**Figure 6.6** Peak output voltage swing versus frequency.

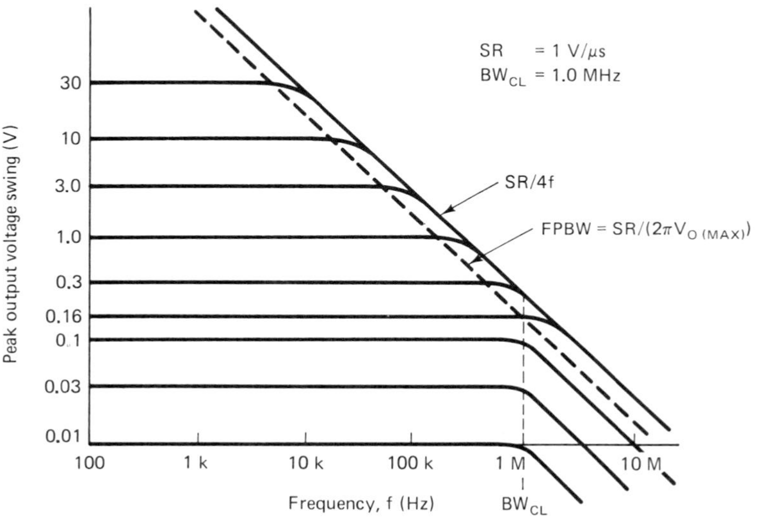


**Figure 6.7** Variation of full-power bandwidth with peak output-voltage swing.

the small-signal bandwidth. In Figure 6.7 the variation of the full-power bandwidth as a function of the maximum output-voltage swing $V_{O(\text{MAX})}$ is shown, with both variables being on logarithmic scales.

## 6.2 OPERATIONAL-AMPLIFIER CIRCUIT ANALYSIS EXAMPLE

Consider the operational-amplifier circuit shown in Figure 6.8. The circuit consists of two gain stages and a "push-pull" Darlington emitter-follower output stage. The first gain stage consists of the differential-amplifier transistor pair, $Q_1$ and $Q_2$, which is biased by the current source comprised of $Q_{12}$ and $Q_{13}$, and drives the current mirror active load comprised of $Q_3$ and $Q_4$, and the input transistor to the next stage, $Q_6$.

The second stage consists of the $Q_6$–$Q_7$ Darlington compound-transistor configu-

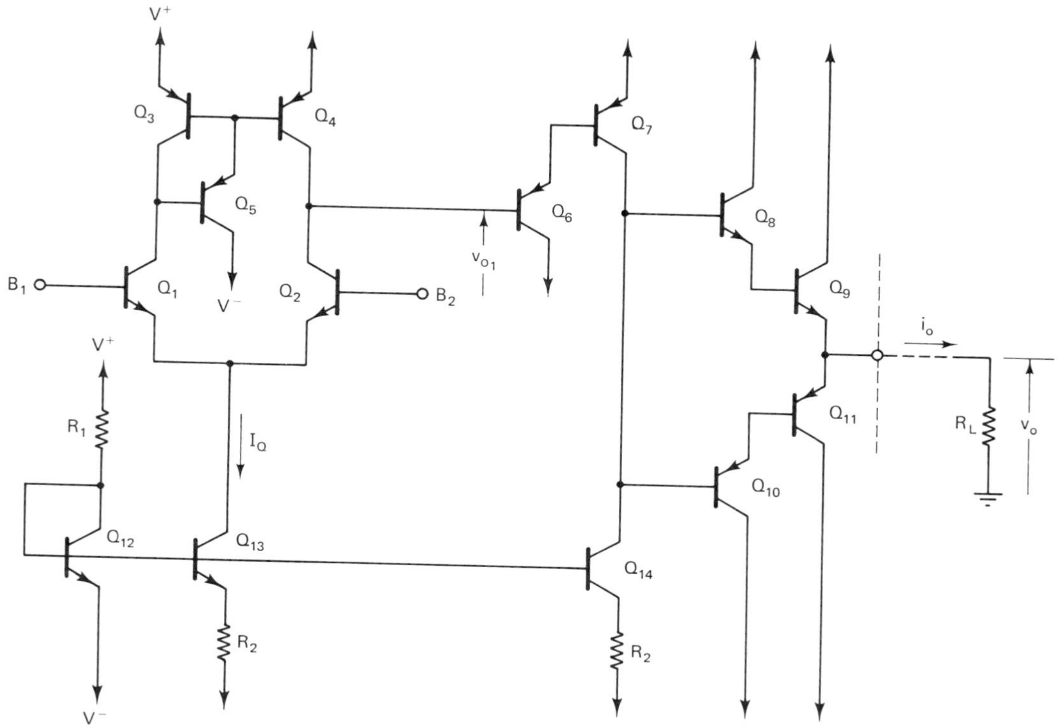


**Figure 6.8** Operational-amplifier circuit example.

ration, in which $Q_6$ is an emitter follower (i.e., common collector) and $Q_7$ is in the common-emitter configuration. This stage has as an active load current source transistor $Q_{14}$ (which in turn is biased by $Q_{12}$ and $R_1$) and drives the output stage transistors.

The output stage is a complementary push-pull emitter-follower configuration consisting of a Darlington NPN emitter-follower compound transistor $Q_8$ and $Q_9$, for sourcing currents into the load, and hence to produce the positive-going output voltage swing, and a Darlington PNP emitter-follower compound transistor, $Q_{10}$ and $Q_{11}$, for sinking currents from the load, and hence to produce the negative-going output voltage swing.

If we assume all transistors of the same type to be identical, then under quiescent conditions (i.e., $v_i = v_{B_1} - v_{B_2} = 0$) we will have that $I_{C_1} = I_{C_2}$ and $I_{C_3} = I_{C_4}$, so that by applying Kirchhoff's current law to the $C_1$–$C_3$–$B_5$ node and the $C_2$–$C_4$–$B_6$ node, we obtain that $I_{B_5} = I_{B_6}$. Since the transistor current gain values will be equal, we will have that $I_{E_6} = I_{E_5}$, so that $I_{B_7} = I_{B_3} + I_{B_4}$. Again noting the equality of the transistor current gains, we have that $I_{C_7} = I_{C_3} + I_{C_4}$. Since the base currents are small compared to the emitter and collector currents, we can say that $I_{C_3} + I_{C_4} = I_{C_1} + I_{C_2} = I_Q$, so that $I_{C_7} = I_Q$. Since $I_{C_{14}} = I_{C_{13}} = I_Q$, by applying Kirchhoff's current law to the $C_7$–$C_{14}$–$B_8$–$B_{10}$ node, we will have that $I_{B_8} + I_{B_{10}} = 0$. We now note that $Q_8$ and $Q_{10}$ cannot both be "on" at the same time, so that we must have that $I_{B_8} = 0$ and $I_{B_{10}} = 0$ under these (i.e., quiescent) conditions. As a result of these two base currents being zero, we will have that the two emitter currents, $I_{E_9}$ and $I_{E_{11}}$, will be zero, and thus $I_o = 0$. Therefore, $V_o = 0$ under quiescent conditions (i.e., $V_i = 0$). Thus with *zero input* ($V_i = 0$) we get *zero output* ($V_o = 0$) for this amplifier, which is precisely the condition that is desired.

### 6.2.1 Voltage Gain Analysis

The voltage gain of the first (i.e., differential amplifier) stage will be given by

$$A_{v_1} = \frac{v_{o_1}}{v_i} = \frac{2g_f}{g_{\text{total}}} \tag{6.17}$$

where $g_{\text{total}}$ is the total conductance from the $C_2$–$C_4$–$B_6$ node to ground and is given by $g_{\text{total}} = g_{i_6} + g_{o_2} + g_{o_4}$. The input conductance of $Q_6$ will be given by

$$g_{i_6} = \frac{I_{B_6}}{2nV_T} = \frac{I_Q}{2n\beta_6\beta_7 V_T} \tag{6.18}$$

The output conductances of $Q_4$ and $Q_2$ will be given by

$$g_{o_4} = g_{ce_4} = \frac{I_{C_4}}{V_{A\text{P}}} = \frac{I_Q}{2V_{A\text{P}}} \qquad \text{and} \qquad g_{o_2} = g_{e_2} = \frac{I_{C_2}}{V_{A\text{N}}} = \frac{I_Q}{2V_{A\text{N}}} \tag{6.19}$$

where $V_{A\text{P}}$ and $V_{A\text{N}}$ are the Early voltages of the PNP and NPN transistors, respectively (see Appendix B). The voltage gain of the first stage will thus be given by

$$\begin{aligned} A_{V_1} = \frac{2g_f}{g_{\text{total}}} &= \frac{2(I_Q/4V_T)}{(I_Q/2n\beta_6\beta_7 V_T) + (I_Q/2V_{A\text{P}}) + (I_Q/2V_{A\text{N}})} \\ &= \frac{1/V_T}{(1/n\beta_6\beta_7 V_T) + (1/V_{A\text{P}}) + (1/V_{A\text{N}})} \\ &= \frac{\beta_6\beta_7}{1/n + \beta_6\beta_7 V_T[(1/V_{A\text{P}}) + (1/V_{A\text{N}})]} \end{aligned} \tag{6.20}$$

The voltage gain of the second stage will be given by

$$A_{V_2} = \frac{v_o}{v_{i_6}} = \frac{v_o}{v_{o_1}} = -g_{f_{6\cdot 7}} R_{L_7} \tag{6.21}$$

The effective load driven by the $Q_6$–$Q_7$ compound transistor configuration will be the transformed value of the load resistance, $R_L$, which will be given by $R_{L_7} = \beta_{8,10}\beta_{9,11}R_L$. We are assuming here that the emitter-follower output stage has approximately unity gain, so that for the gain of the second stage we will have that

$$A_{V_2} = -(I_Q/2V_T)\beta_{8,10}\beta_{9,11}R_L \tag{6.22}$$

Thus the overall gain of the amplifier circuit will be given by

$$A_{OL}(0) = A_{V(\text{TOTAL})} = A_{V_1}A_{V_2} = \frac{-\beta_6\beta_7\beta_{8,10}\beta_{9,11}I_Q R_L/(2V_T)}{1/n + \beta_6\beta_7 V_T[(1/V_{A\text{P}}) + (1/V_{A\text{N}})]} \tag{6.23}$$

To now consider a numerical example, let us assume $\beta = 50$ (min.), $n = 1.5$ for all transistors, and $V_{AN} = V_{AP} = 200$ V. The $\beta$ product thus becomes $\beta_6\beta_7\beta_{8,10}\beta_{9,11} = (50)^4 = 6.25 \times 10^6$ (min.) and we also have that $\beta_6\beta_7 V_T = 62.5$ V. Therefore, the total voltage gain will be

$$A_{V(\text{TOTAL})} = \frac{6.25 \times 10^6 \text{ (min.)}(I_Q R_L/50 \text{ mV})}{1/n + 62.5[(1/200) + (1/200)]} = \frac{6.25 \times 10^6 \text{ (min.)}(I_Q R_L/50 \text{ mV})}{1.292} \tag{6.24}$$

If $R_L = 1.0$ kΩ and $I_Q = 20$ μA are chosen as typical values for this example, we have that $I_Q R_L/50$ mV $= 20$ mV/50 mV $= 0.4$, so that

$$A_{V(\text{total})} = 6.25 \times 10^6 \text{ (min.)} \times \frac{0.4}{1.292} = \underline{1.935 \times 10^6 \text{ (min.)}} \tag{6.25}$$

Note that this is the low-frequency gain, $A_{OL}(0)$, and that the gain will fall off at higher frequencies.

### 6.2.2 Operational-Amplifier Circuit with a Darlington Input Stage

In many cases it may be advantageous to use a Darlington differential-amplifier configuration as shown in Figure 6.9 to replace the differential amplifier that comprises just a pair of transistors.

If a Darlington differential amplifier is used in place of the simple two-transistor differential amplifier, the basic operation of the circuit will be the same as before. The two most important changes will be that the effective current gain (i.e., $\beta$) of the Darlington compound-transistor configuration will be the product of the individual transistor current gains (i.e., $\beta = \beta_A\beta_B$) of the two transistors that comprise the

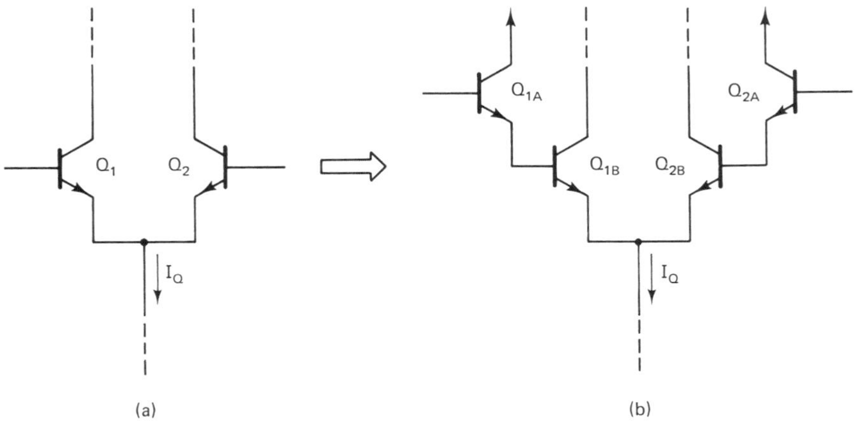


**Figure 6.9** Darlington differential amplifier: (a) simple differential amplifier; (b) Darlington differential amplifier.

compound configuration. This will give rise to a very large current gain value and thus correspondingly to a much reduced input bias current for the amplifier and a much larger input resistance.

In the equations for the differential amplifier, $V_T$ must be replaced by $2V_T$ wherever it appears, so that the dynamic forward transfer conductance of the differential amplifier becomes $g_f = I_Q/V_T = I_Q/200$ mV. Since the transfer conductance is reduced to one-half of its previous value the voltage gain will also be reduced to one-half of its previous value. In addition to these changes, there will also be some small changes in the input voltage range.

**Note on the input resistance of $Q_6$.** If we take into account the fact that the dependence of the base current of a transistor in the active mode on the base-to-emitter voltage is given by $I_B \propto \exp(V_{BE}/nV_T)$, where $n$ is a dimensionless constant with values between 1 and 2 and with a typical value of about 1.5 or 1.6, we will note that the input conductance will be given by $g_i = g_{be} = I_B/nV_T$. For the compound Darlington common-emitter configuration (such as $Q_6$ and $Q_7$) the input conductance will correspondingly be $g_{i_6} = I_B/2nV_T$.

### 6.2.3 Effect of the Transistor Current Gain ($\beta$) on the Voltage Gain

From inspection of the equation for the voltage gain we may note that the voltage gain is indeed highly dependent on the transistor $\beta$ values. Indeed, if all the transistor $\beta$ values turn out to be twice the minimum value, the total voltage gain will be 16 times larger than the voltage gain obtained when the $\beta$ values are all at the minimum values. Thus we may expect very great unit-to-unit variations in the open-loop gain of operational amplifiers. Operational amplifiers, are, however, usually operated in a closed-loop (i.e., negative feedback) configuration such that the closed-loop gain will be determined primarily by the feedback factor, and will be relatively independent of the open-loop gain, as long as the loop gain is much greater than unity. Thus, as long as the open-loop gain of the amplifier is sufficiently large so as to ensure that the loop gain will be much greater than unity, the closed-loop gain will be relatively independent of the open-loop gain, and thus relatively independent of the transistor current gain values.

### 6.2.4 Voltage Gain of the Darlington Emitter-Follower Output Stage

We will now consider the voltage gain of the Darlington emitter-follower push-pull output stage consisting of the NPN compound transistor, $Q_8$ and $Q_9$, and the PNP compound transistor, $Q_{10}$ and $Q_{11}$.

The voltage gain of a simple (i.e., single transistor) emitter-follower stage, from base (input) to emitter (output), is given by

$$A_V = \frac{g_m R_L}{1 + g_m R_L} = \frac{R_L}{(1/g_m) + R_L} = \frac{R_L}{(V_T/I) + R_L} \tag{6.26}$$

where $R_L$ is the net load driven by the emitter-follower transistor, and $I$ is the current through the transistor. For a Darlington (i.e., compound) emitter-follower configuration we simply replace $V_T$ by $2V_T$ in the expressions above to obtain

$$A_{V\,(\text{Darlington emitter-follower})} = \frac{R_L}{(2V_T/I) + R_L} = \frac{1}{1 + (2V_T/IR_L)} \tag{6.27}$$

Since the output voltage $V_o = IR_L$, this can be expressed as

$$A_{V\,(\text{DEF})} = \frac{1}{1 + (2V_T/V_o)} \tag{6.28}$$

From this last expression we see that as long as the output voltage satisfies the condition $V_o \gg 2V_T = 50$ mV, the output-stage voltage gain will be approximately (but always somewhat less than) unity.

We must note, however, that whenever the output voltage drops below $2V_T = 50$ mV, the output-stage voltage gain will decrease to values that are substantially less than unity. This reduced value of gain will produce a type of distortion known as *crossover distortion*. This is shown in Figure 6.10.

The circuit of Figure 6.11, showing the addition of transistor $Q_{15}$ and resistors $R_3$ and $R_4$, is one that can be used to minimize the crossover distortion. With this circuit there will be a small quiescent current flowing through the output-stage transistors even when $I_o = 0$, this being the result of the voltage drop produced by $Q_{15}$ in conjunction with $R_3$ and $R_4$. If we call this quiescent current (i.e., with $I_o = 0$)

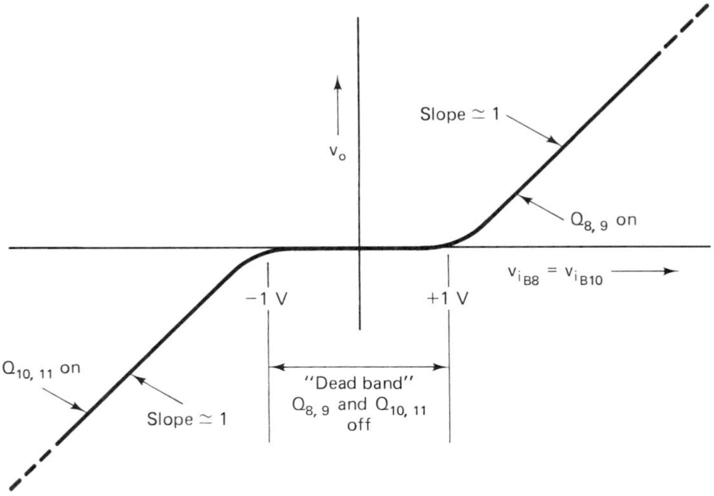


**Figure 6.10** Emitter-follower output-stage characteristics showing crossover distortion.

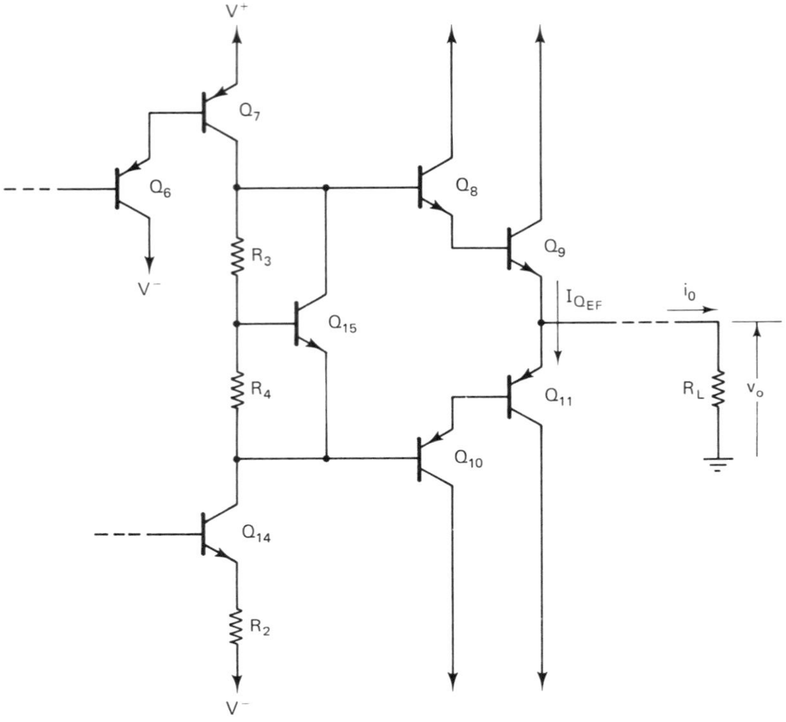


**Figure 6.11** Circuit for minimizing crossover distortion.

$I_{QEF}$, then the *minimum* voltage gain of the emitter-follower output stage will be given by

$$A_{V(\text{DEF}_{\text{minimum}})} = \frac{R_L}{R_L + (2V_T/I_{\text{minimum}})} = \frac{R_L}{R_L + (2V_T/I_{QEF})} \tag{6.29}$$

For example, if $I_{QEF} = 1.0$ mA, then $2V_T/I_{QEF} = 50 \text{ mV}/1.0 \text{ mA} = 50\ \Omega$, so that

$$A_{V(\text{DEF}_{\text{minimum}})} = \frac{R_L}{R_L + 50\Omega} \tag{6.30}$$

so that for $R_L = 1.0\ \text{k}\Omega$, we would have

$$A_{V(\text{DEF}_{\text{minimum}})} = \frac{1000}{1050} \simeq 0.95 \tag{6.31}$$

This would virtually eliminate the crossover distortion, as shown by Figure 6.12.

We should also note that op amps are usually operated in a closed-loop configuration, and that the closed-loop gain is usually relatively independent of the open-loop gain. The variation of closed-loop gain with respect to open-loop gain has been given by $dA_{CL}/A_{CL} = (A_{CL}/A_{OL})(dA_{OL}/A_{OL})$. Therefore, as long as $A_{CL} \ll A_{OL}$, a variation in $A_{OL}$ will have relatively little effect on $A_{CL}$. Looking at it from a slightly different viewpoint, we have that $A_{CL} = A_{OL}/(1 + FA_{OL})$, so that as long as the

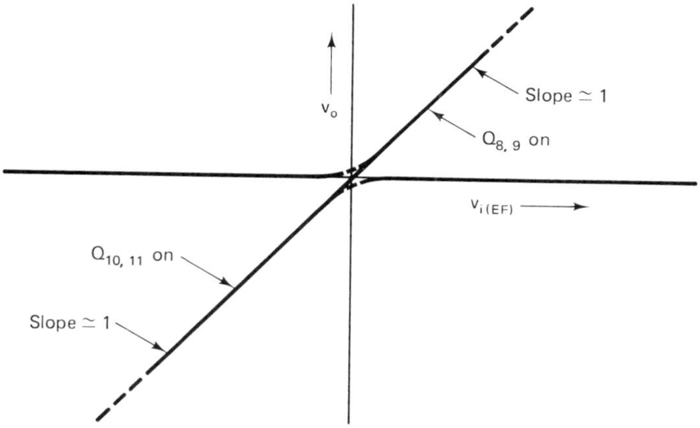


**Figure 6.12** Minimization of crossover distortion by the biasing of the output stage at the threshold of conduction.

*loop gain*, $FA_{OL}$, is large compared to unity, we will have that $A_{CL} \simeq 1/F$, and the closed-loop gain will be relatively independent of the open-loop gain. Thus as long as the open-loop gain is maintained above some suitable minimum value, we will have that $A_{CL} \simeq 1/F$, so that the nonlinear distortion in the open-loop gain characteristics will have relatively little impact on the closed-loop characteristics.

### 6.2.5 Current Limiting

In Figure 6.13 the output stage is modified to incorporate circuitry for bilateral output current limiting using $Q_{16}$ and $Q_{17}$ together with the two current-limiting resistors, $R_{CL\,1}$ and $R_{CL\,2}$. This circuit will limit the maximum current that the op amp can either source or sink to a safe maximum value, even under the worst-case condition of a short-circuit across the output ($v_o = 0$).

As long as the output current is such that the voltage drop across either current-limiting resistor is less than 0.5 V, both $Q_{16}$ and $Q_{17}$ will be *off*, and will therefore have no effect on the operation of the circuit. However, as the output current increases such that the voltage drop across either $R_{CL\,1}$ or $R_{CL\,2}$ rises above 0.5 V, then $Q_{16}$ or $Q_{17}$, respectively, will start to turn on. As either current limit transistor starts to go into conduction it will start to divert base drive away from the Darlington emitter-follower, $Q_8$–$Q_9$ or $Q_{10}$–$Q_{11}$. As the output current increases even more, the increased voltage drop across $R_{CL}$ will cause the current diverted through $Q_{16}$ or $Q_{17}$ to increase at an exponential rate and will ultimately limit the maximum output current to a value given by

$$\boxed{I_{O\,(\mathrm{MAX})} = I_{CL} = \frac{V_{BE\,(\mathrm{ON})}}{R_{CL}} \simeq \frac{0.6 \text{ to } 0.7\text{ V}}{R_{CL}}} \tag{6.32}$$

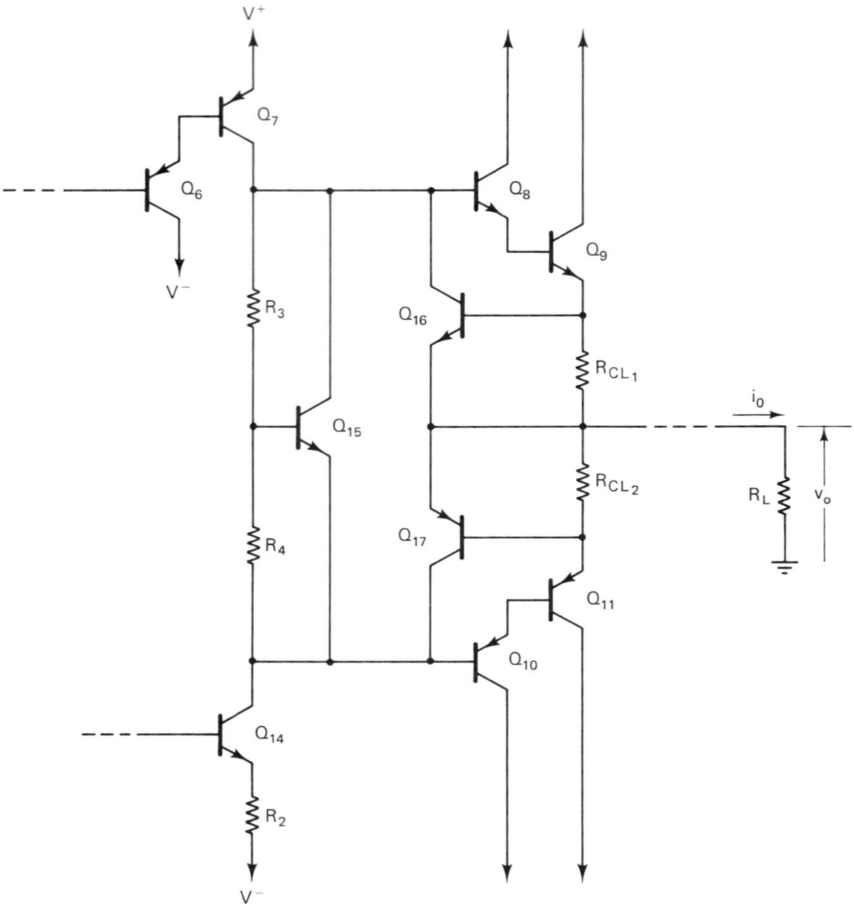


**Figure 6.13** Bilateral current-limiting circuit.

This will be *bilateral* current limiting with $Q_{16}$ and $R_{CL_1}$ acting to limit the maximum current that the op amp can source to a safe maximum value, and $Q_{17}$ and $R_{CL_2}$ acting to limit the maximum current that the op amp can sink. If, for example, a value of $I_{O(\text{MAX})} = I_{CL} = \pm 20$ mA is needed, the suitable choice for $R_{CL}$ will be given by

$$R_{CL} = V_{BE(\text{ON})}/I_{CL} \simeq 600 \text{ to } 700 \text{ mV}/20 \text{ mA} \simeq \underline{30 \text{ to } 35\ \Omega}$$

so that we see that $R_{CL}$ will be a relatively small resistor.

### 6.2.6 Effect of Supply Voltage on Voltage Gain

If we look at the equation for the overall gain of the op amp we do not see any explicit dependence of the voltage gain on the supply voltage. We must note, however, that the current $I_Q$, upon which the gain is directly dependent, is related to the supply voltage via an interrelationship with $I_{12}$ (the current through $Q_{12}$).

Since $I_Q = I_{13} = I_{14} = I_{12} \exp(-I_Q R_2/V_T)$, we will have that

$$\frac{dI_Q}{dV_S} = \frac{dI_{12}}{dV_s}\exp\left(-\frac{I_Q R_2}{V_T}\right) + \left[I_{12}\exp\left(-\frac{I_Q R_2}{V_T}\right)\right]\left[-\frac{R_2}{V_T}\frac{dI_Q}{dV_S}\right]$$
$$= \frac{dI_{12}}{dV_S}\frac{I_Q}{I_{12}} + (I_Q)\left(-\frac{R_2}{V_T}\frac{dI_Q}{dV_S}\right) \tag{6.33}$$

so that by collecting terms in $I_Q$ we have

$$\frac{1}{I_Q}\frac{dI_Q}{dV_S}\left(1 + \frac{I_Q R_2}{V_T}\right) = \frac{1}{I_{12}}\frac{dI_{12}}{dV_s} \tag{6.34}$$

Since $I_{12} = (2V_S - V_{BE})/R_1 \simeq 2V_S/R_1$, we obtain

$$\frac{1}{I_{12}}\frac{dI_{12}}{dV_S} = \frac{R_1}{2V_S}\frac{2}{R_1} = \frac{1}{V_S} \tag{6.35}$$

so that we now have

$$\frac{1}{I_Q}\frac{dI_Q}{dV_S} = \frac{1/V_S}{1 + (I_Q R_2/V_T)} \tag{6.36}$$

Since $\exp(I_Q R_2/V_T) = I_{12}/I_Q$, this can be rewritten as

$$\frac{dI_Q}{I_Q} = \frac{dV_S}{V_S}\frac{1}{1 + \ln(I_{12}/I_Q)} \tag{6.37}$$

Since the voltage gain is directly proportional to $I_Q$, we have that the fractional change in the voltage gain will be equal to the fractional change in $I_Q$, so that $dI_Q/I_Q = dA_V/A_V$, and thus we finally obtain

$$\frac{dA_V}{A_V} = \frac{dI_Q}{I_Q} = \frac{dV_S}{V_S}\frac{1}{1 + \ln(I_{12}/I_Q)} \tag{6.38}$$

For example, if $I_{12} = 1.0\text{ mA} = 1000\ \mu\text{A}$ and $I_Q = 25\ \mu\text{A}$, then

$$\ln\frac{I_{12}}{I_Q} = \ln\frac{1000\ \mu\text{A}}{25\ \mu\text{A}} = \ln 40 = 3.69 \tag{6.39}$$

so that

$$\frac{dA_V}{A_V} = \frac{dI_Q}{I_Q} = \frac{dV_S}{V_S}\left(\frac{1}{4.69}\right) = 0.213\left(\frac{dV_S}{V_S}\right) \tag{6.40}$$

Thus if $V_S = \pm 15$ V, a 1.0-V change in $V_S$ will produce a fractional change in the open-loop voltage gain given by

$$\frac{dA_V}{A_V} = 0.213\left(\frac{1}{15}\right) = 0.0142 \tag{6.41}$$

which corresponds to a 1.42% change, which is quite small. This is the change in the *open-loop* gain. The change in the closed-loop gain will be much smaller, usually by a very large factor, remembering that $dA_{CL}/A_{CL} = (dA_{OL}/A_{OL})(A_{CL}/A_{OL})$.

**Minimum voltage to operate op amp.** The minimum supply voltage required to operate this op amp is governed by the basic requirement that all of the transistors in the circuit operate in the active mode. Therefore, we must require that the collector-to-emitter voltage be greater than the saturation value, $V_{CE(\mathrm{SAT})}$, and that the base-to-emitter voltage drop be the value needed to operate the transistor in the active region. If we choose $V_{BE} = 700$ mV and $V_{CE(\mathrm{SAT})} = 200$ mV as conservative values to use in the determination of the minimum required operating voltage, we have the following requirements to meet:

$$V_S^+ - V_S^- = V_{BE\,3,4} + V_{BE\,5} + V_{CE\,1,2} + V_{CE\,13} + V_{R\,2}$$
$$= (700 + 700 + 200 + 200 + 92)\text{ mV} = 1892\text{ mV} \simeq 2.0\text{ V} \tag{6.42}$$

as a necessary condition, and

$$V_S^+ - V_S^- = V_{CE\,7} + V_{BE\,8} + V_{BE\,9} + V_{EB\,11} + V_{EB\,10} + V_{CE\,14} + V_{R\,2}$$
$$= (200 + 700 + 700 + 700 + 700 + 200 + 92)\text{ mV} \tag{6.43}$$
$$= 3292\text{ mV} = \underline{3.3\text{ V}} \text{ as a necessary } \textit{and sufficient} \text{ condition}$$

Thus it appears that this op amp can be operated with a supply voltage, $V_S^+ - V_S^-$, as low as about 3.3 V.

**Voltage gain variation.** We will now investigate the variation in the open-loop gain of the op amp as the supply voltage varies from a minimum of 3.3 V (total supply voltage) to a maximum of 30 V (i.e., $V_S = \pm 15$ V), assuming that at $V_S = \pm 15$ V we have $I_{12} = 1.0$ mA and $I_Q = 25\ \mu$A, corresponding to $R_1 = 29.3$ k$\Omega$ and $R_2 = 3.69$ k$\Omega$. At $V_S = \pm 1.65$ V we have that

$$I_{12} = \frac{3.3 - 0.7}{29.3\text{ k}\Omega} = 0.08874\text{ mA} = 88.74\ \mu\text{A} \tag{6.44}$$

By using an iterative calculation we obtain that $I_Q$ will now be 13.0 $\mu$A. Thus $I_Q$ has dropped by a factor of $13/25 = \underline{0.52}$, and we would expect the open-loop voltage gain to be reduced correspondingly by the same factor. Thus, in going from a total supply voltage of 3.3 V up to a total supply voltage of 30 V, the open-loop voltage gain increases by almost a factor of 2. The corresponding change in the closed-loop gain will, however, be generally very much less.

Since the unity-gain frequency, $f_u$, is given by $\omega_u = 2g_f/C_f$ and since $g_f$ is directly proportional to $I_Q$, we should see a similar variation in the unity-gain frequency. Thus, if $f_u = 1.0$ MHz when we have a total supply voltage of 30 V across the amplifier, the value of $f_u$ will drop to about 520 kHz when we have a total supply voltage of 3.3 V across the amplifier. Since the slewing rate is given by $SR = I_Q/C_F$, we should expect to see a similar variation in the slewing rate.

### 6.2.7 Common-Mode Rejection Ratio

The common-mode dynamic transfer conductances of a differential amplifier are given by $g_{f\,1(CM)} = (I_1/I_Q)g_o$ and $g_{f\,2(CM)} = (I_2/I_Q)g_o$, where $g_o$ is the output conductance of the current source that is used to bias the differential amplifier. For the op-amp circuit under consideration, $g_o = g_{o\,13}$. Since $i_3 \simeq i_1$ and $i_4 = i_3$, we will have that $i_o = i_4 - i_2 \simeq i_1 - i_2$. Therefore, in response to a common-mode input voltage we will have that $i_1 = g_{f\,1(CM)} v_{iCM}$ and $i_2 = g_{f\,2(CM)} v_{iCM}$, so that

$$i_o \simeq i_1 - i_2 = [g_{f\,1(CM)} - g_{f\,2(CM)}] v_{iCM} = \left(\frac{I_1 - I_2}{I_Q}\right) g_{o\,13} v_{iCM} \tag{6.45}$$

The common-mode output of the differential-amplifier stage will therefore be given by

$$v_{o\,1(CM)} = \frac{i_{oCM}}{g_{\text{total}}} = \frac{I_1 - I_2}{I_Q} \frac{g_{o\,13}}{g_{\text{total}}} v_{iCM} \tag{6.46}$$

and the common-mode gain of the differential-amplifier stage will thus be

$$A_{V\,1(CM)} = \frac{v_{o\,1(CM)}}{v_{iCM}} = \frac{I_1 - I_2}{I_Q} \frac{g_{o\,13}}{g_{\text{total}}} \tag{6.47}$$

If we now note that the gain with respect to the common-mode signal will be the same as that accorded to the difference-mode signal beyond the first stage, we can say that the common-mode rejection ratio will be determined by the ratio of the difference-mode gain to the common-mode gain in the first stage. We will therefore have $A_{V_{DM}}/A_{V_{CM}} = A_{V\,1(DM)}/A_{V\,1(CM)}$. Since $A_{V\,1(DM)} = 2g_f/g_{\text{total}}$, we now have

$$\text{CMRR} = \frac{A_{V\,1(DM)}}{A_{V\,1(CM)}} = \frac{2g_f/g_{\text{total}}}{[(I_1 - I_2)/I_Q](g_{o\,13}/g_{\text{total}})} = \frac{2g_f}{[(I_1 - I_2)/I_Q] g_{o\,13}} \tag{6.48}$$

If we let $V_i'$ represent the sum of the difference-mode input voltage and the input offset voltage, we have $V_i' = V_i + V_{OS}$. The resulting collector currents will be $I_1 = I_Q + \Delta I$ and $I_2 = I_Q - \Delta I$, where $\Delta I \simeq g_f V_i'$. Therefore, $I_1 - I_2 \simeq 2g_f V_i'$. The common-mode rejection ratio can now be written as

$$\text{CMRR} \simeq \frac{2g_f}{(2g_f V_i'/I_Q) g_{o\,13}} = \frac{I_Q}{g_{o\,13} V_i'} \tag{6.49}$$

The output conductance of the current sink transistor $Q_{13}$ will be given by

$$g_{o\,13} = \frac{I_Q}{V_A[1 + \ln(I_{12}/I_Q)]} \tag{6.50}$$

where $V_A$ is the Early voltage. Substituting this into the CMRR expression gives

$$\boxed{\text{CMRR} = \frac{I_Q}{g_{o\,13} V_i'} = \frac{V_A}{V_i'}\left(1 + \ln\frac{I_{12}}{I_Q}\right)} \tag{6.51}$$

To consider a representative example, let us take $V_A = 200$ V, $I_{12} = 1.0$ mA, and $I_Q = 25\ \mu$A. For $V_i'$ we will use a unit value of 1.0 mV. For these values we obtain CMRR $\simeq$ (200 V/1.0 mV)(4.69) = 938,000, corresponding to 119 dB. The actual value of the CMRR may be substantially lower than this as a result of other imbalances in the circuit and as a result of capacitative feedthrough, especially at higher frequencies.

### 6.2.8 Power Supply Rejection Ratio

The power supply rejection ratio (PSRR) is the ratio of the change in the input offset voltage to the change in the power supply voltage that produces this change, and is usually expressed in decibels.

If we inspect the operational-amplifier circuit under consideration we see that a change in the supply voltage can produce a change in the offset voltage by several means. First, a change in the supply voltage $V_S$ will change the current through $Q_{12}$ and thereby change the quiescent current $I_Q$, which in turn will produce a change in the input offset voltage. This will turn out to be by far the most important contribution to the PSRR.

A second but much smaller contribution to the PSRR is the change in $I_Q$ due to the change in the voltage across the current source comprised of $Q_{13}$ and $R_2$. A third, even smaller contribution to the PSRR is the change in the current through $Q_7$ resulting from the change in the voltage across $Q_7$.

We will now investigate the first and most important contribution to the PSRR, the change in $I_{12}$ due to the change in the supply voltage, which in turn results in a change in $I_Q$. We have obtained earlier that

$$\frac{dI_Q}{I_Q} = \frac{dV_S}{V_S} \frac{1}{1 + \ln (I_{12}/I_Q)} \tag{6.52}$$

for the type of current source under consideration here.

The output current of the differential amplifier stage $I_O$ will be given by $I_O = I_4 - I_2 \simeq I_1 - I_2 = 2g_f V_i'$, where $V_i' = V_i + V_{OS}$. Since $g_f = I_Q/4V_T$, this can be rewritten as $I_O = (I_Q/2V_T)V_i'$. The required change in $V_i'$ to cancel out a change in $I_O$ produced by a change in $I_Q$ will be given by $dV_i'/V_i' = -dI_Q/I_Q$. We therefore have that

$$\frac{dV_i'}{V_i'} = -\frac{dI_Q}{I_Q} = -\frac{dV_S}{V_S} \frac{1}{1 + \ln (I_{12}/I_Q)} \tag{6.53}$$

The power supply rejection ratio $dV_{OS}/dV_S = dV_i'/dV_S$ will therefore be given by

$$\boxed{\text{PSRR} = \left| \frac{dV_{OS}}{dV_S} \right| = \frac{V_i'}{V_S} \frac{1}{1 + \ln (I_{12}/I_Q)}} \tag{6.54}$$

To consider a representative example, let us take $I_{12} = 1.0$ mA, $I_Q = 25$ $\mu$A, $V_S = \pm 15$ V, and for $V_i'$ we will use a unit value of 1.0 mV. With these values we obtain

$$\text{PSRR} = \frac{1.0 \text{ mV}}{15 \text{ V}} \frac{1}{1 + \ln 40} = \frac{1}{7 \times 10^4}$$

which corresponds to <u>−97 dB</u>.

The change in the voltage across $Q_{13}$ will also produce a change in $I_Q$ and thus contribute to the overall PSRR. To evaluate this we note that the change in $I_Q$ resulting from the change in the voltage across $Q_{13}$ and $R_2$ will be given by

$$dI_Q = g_{o\,13}\,dV_S = \frac{I_Q}{V_A[1+\ln(I_{12}/I_Q)]}\,dV_S$$

The change in $V_i'$ required to compensate for the change in $I_Q$ will, as before, be given by $dV_i'/V_i' = dI_Q/I_Q$, so that we obtain

$$\frac{dV_i'}{V_i'} = -\frac{dI_Q}{I_Q} = -\frac{dV_S}{V_A}\,\frac{1}{1+\ln(I_{12}/I_Q)} \tag{6.55}$$

Therefore, this contribution to the PSRR will be given by

$$\text{PSRR} = \left|\frac{dV_i'}{dV_S}\right| = \frac{V_i'}{V_A}\,\frac{1}{1+\ln(I_{12}/I_Q)} \tag{6.56}$$

If we now use the same values as used in the previous example together with $V_A =$ 200 V, we obtain

$$\text{PSRR} = \frac{1.0\text{ mV}}{200\text{ V}}\,\frac{1}{1+\ln 40} = \frac{1}{9.4\times 10^5} \tag{6.57}$$

We therefore see that this contribution to the overall PSRR will be very small, and the dominant contribution is still due to the change in $I_{12}$ resulting from the change in the supply voltage.

We will now consider the contribution of the change in the current through $Q_7$ to the PSRR. The change in the current through $Q_7$, $\Delta I_7$, that results from a change in the supply voltage, $\Delta V_S$, will be given by

$$\Delta I_7 = g_{o\,7}\,\Delta V_S = g_{ce\,7}\,\Delta V_S = \frac{I_7}{V_A}\,\Delta V_S = \frac{I_Q}{V_A}\,\Delta V_S \tag{6.58}$$

where $V_A$ is the Early voltage of $Q_7$. The corresponding change in $I_{B\,6}$ that is required to cancel out the effect of the change in the current through $Q_7$ produced by the change in the supply voltage will be given by

$$\Delta I_{B\,6} = \frac{\Delta I_7}{\beta_6\beta_7} \tag{6.59}$$

This change in $I_{B\,6}$ will in turn require a change in the input offset voltage given by the following relationship:

$$\Delta I_{B\,6} = 2g_f\,\Delta V_{OS} = \frac{I_Q}{2V_T}\,\Delta V_{OS} \tag{6.60}$$

As a result, we now obtain

$$\frac{I_Q}{2V_T}\,\Delta V_{OS} = \frac{\Delta I_7}{\beta_6\beta_7} = \frac{(I_Q/V_A)\,\Delta V_S}{\beta_6\beta_7} \tag{6.61}$$

Solving for the ratio $\Delta V_S/\Delta V_{OS}$, we obtain

$$\frac{1}{\text{PSRR}} = \frac{\Delta V_S}{\Delta V_{OS}} = \frac{V_A(\beta_6\beta_7)}{2V_T} = \frac{V_A(\beta_6\beta_7)}{50 \text{ mV}} \tag{6.62}$$

For example, if we again choose $V_A = 200$ V and let $\beta_6 = 50$ (min.) and $\beta_7 = 50$ (min.), we obtain

$$\begin{aligned}\frac{1}{\text{PSRR}} &= \frac{200 \text{ V}}{50 \text{ mV}}(50 \times 50)(\text{min.}) \\ &= 4 \times 10^3 \times 2.5 \times 10^3 = 1 \times 10^7\end{aligned} \tag{6.63}$$

We therefore see that this contribution to the net value of the PSRR will be very small indeed.

Finally, if we were to investigate the effect of the change in the voltage across $Q_{14}$ and $R_2$ on the input offset voltage, we have essentially the same result as that given with respect to $Q_7$.

In summary, it indeed does appear that the major contributor to the PSRR will be the result of the change in $I_{12}$, and therefore of $I_{13}$ ($= I_Q$) due to a change in the supply voltage.

### 6.2.9 Frequency Response Analysis

The op-amp equivalent circuit shown in Figure 6.14 will be used for the analysis of the frequency response of the circuit of Figure 6.8. The capacitance $C_{O_1}$ represents the summation of all of the capacitances between the $C_2$–$C_4$–$B_6$ node and ground. Capacitance $C_{O_2}$ is the sum of all of the capacitances between the $C_7$–$C_{14}$–$B_8$–$B_{10}$ node and ground.

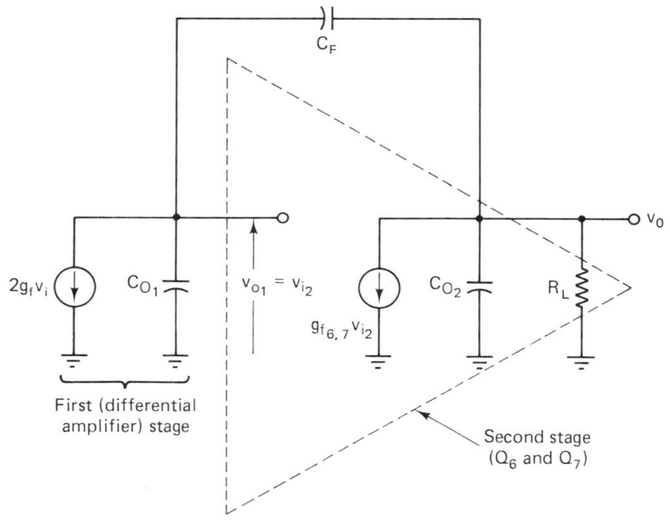


**Figure 6.14** Op-amp equivalent circuit for the analysis of frequency response.

If we write a node voltage equation for the input node of the second stage (i.e., the $C_2$–$C_4$–$B_6$ node), we obtain

$$(V_o - V_{i_2})(j\omega C_F) = 2g_f v_i + V_{i_2}(j\omega C_{0_1}) \tag{6.64}$$

Since $V_o = -A_{V_2} V_{i_2}$, we have that $V_{i_2} = -V_o/A_{V_2}$. If we now collect all terms of which $V_O$ is a factor on the left-hand side, we obtain

$$V_O\left[j\omega C_F\left(1 + \frac{1}{A_{V_2}}\right) + \frac{j\omega C_{0_1}}{A_{V_2}}\right] = 2g_f V_i \tag{6.65}$$

Therefore, the open-loop gain, $A_{OL} = V_O/V_i$, will be given by

$$\begin{aligned} A_{OL} = \frac{V_O}{V_i} &= \frac{2g_f}{j\omega C_F(1 + 1/A_{V_2}) + j\omega C_{0_1}/A_{V_2}} \\ &= \frac{2g_f/j\omega C_F}{1 + (1/A_{V_2}) + (C_{0_1}/C_F)/A_{V_2}} \end{aligned} \tag{6.66}$$

Since the (extrapolated) unity-gain frequency is given by the equation $2g_f/\omega_u C_F = 1$, we have that $\omega_u = 2g_f/C_F$, so that we can write the voltage-gain expressions as

$$A_{OL} = \frac{\omega_u/j\omega}{1 + [(1 + C_{0_1}/C_F)/A_{V_2}]} = \frac{f_u/jf}{1 + [(1 + C_{0_1}/C_F)/A_{V_2}]} \tag{6.67}$$

From this equation we see that as long as the voltage gain of the second stage, $A_{V_2}$, is large compared to unity, the frequency response of the second stage will not have an important effect on the overall frequency response. However, when this is no longer the case, the frequency response characteristics of the second gain stage must be taken into consideration. Since the voltage gain of the second gain stage at low frequencies is very large [i.e., $A_{V_2}(0) \gg 1$], typically of the order of 1000, we see that the frequency range wherein $A_{V_2} \lesssim 1 + (C_{0_1}/C_F)$ obtains will be considerably above the 3-dB (or breakpoint) frequency of the second stage. Indeed, in this frequency region we can express the voltage gain of the second stage as

$$\begin{aligned} A_{V_2} &= -g_{f_{6\cdot7}} Z_{L_7} = -\frac{g_{f_{6\cdot7}}}{Y_{L_7}} = \frac{-g_{f_{6\cdot7}}}{G_L + j\omega C_{0_2}} \\ &\simeq -\frac{g_{f_{6\cdot7}}}{j\omega C_{0_2}} \end{aligned} \tag{6.68}$$

where $g_{f_{6\cdot7}}$ is the dynamic transfer conductance of the $Q_6$–$Q_7$ compound transistor configuration as given by $g_{f_{6\cdot7}} = I_7/2V_T = I_Q/2V_T$, since the quiescent current of $Q_7$ is $I_Q$. If we define $f_2$ as the unity-gain frequency of the second stage, we have that $g_{f_{6\cdot7}}/\omega_2 C_{0_2} = 1$, so that $\omega_2 \simeq g_{f_{6\cdot7}}/C_{0_2}$, and we can therefore express the voltage gain of the second stage (in the frequency range of interest) as

$$A_{V_2} \simeq \frac{\omega_2}{j\omega} = \frac{f_2}{jf} \tag{6.69}$$

The overall gain of the op amp can now be expressed as

$$A_{OL} \simeq \frac{f_u/jf}{1+[(1+C_{O_1}/C_F)/f_2/jf]} = \frac{f_u/jf}{1+[jf(1+C_{O_1}/C_F)/f_2]} \tag{6.70}$$

Let us define $f_2'$ as $f_2' = f_2/(1 + C_{O_1}/C_F)$, so that we can now write

$$A_{OL} = \frac{f_u/jf}{1+j(f/f_2')} \tag{6.71}$$

For a 45° phase margin, we require that $|A_{OL}| \leqslant 1$ when $f = f_{135°}$. By inspection of the last equation for $A_{OL}$ we see that at $f = f_2'$ the phase angle of $A_{OL}$ will be $-135°$; therefore, $f_{135°} = f_2'$. The value of $A_{OL}$ at $f = f_{135°} = f_2'$ will be given by

$$|A_{OL}(f=f_2')| = \frac{f_u/f'_2}{|1+j(1)|} = \frac{f_u}{\sqrt{2}f_2'} \tag{6.72}$$

Thus, in order to satisfy the requirement that $|A_{OL}| \leqslant 1$ at $f = f_{135°}$ we have that $f_u/(\sqrt{2}f_2') \leqslant 1$, which gives $f_u \leqslant \sqrt{2}f_2'$, or correspondingly, $\omega_u \leqslant \sqrt{2}\omega_2'$.

We have already stated that $\omega_u = 2g_f/C_F$, and we now note that $g_f$ = dynamic transfer conductance of the first (i.e., differential amplifier) stage $= I_Q/4V_T$, so that $\omega_u = (I_Q/2V_T)/C_F$. For $\omega_2'$ we have

$$\omega_2' = \frac{\omega_2}{1+C_{O_1}/C_F} = \frac{g_{f_{6\cdot7}}/C_{O_2}}{1+C_{O_1}/C_F} \tag{6.73}$$

where $g_{f_{6\cdot7}} = I_7/2V_T = I_Q/2V_T$. If we now substitute these relationships into the equation relating $f_u$ to $f_2'$, we obtain

$$\frac{I_Q/2V_T}{C_F} \leqslant \frac{\sqrt{2}I_Q/2V_T}{C_{O_2}(1+C_{O_1}/C_F)} \tag{6.74}$$

After canceling out common factors, $I_Q$ and $2V_T$, this becomes

$$\frac{1}{C_F} \leqslant \frac{\sqrt{2}}{C_{O_2}(1+C_{O_1}/C_F)} \tag{6.75}$$

so that the required feedback capacitance is given by

$$C_F \geqslant \frac{C_{O_2}}{\sqrt{2}}\left(1+\frac{C_{O_1}}{C_F}\right) \tag{6.76}$$

The capacitance $C_{O_2}$ is the summation of all of the capacitances between the $C_7$–$C_{14}$–$B_8$–$B_{10}$ node and ground. Since $C_F$ is connected to this node, it will be part of $C_{O_2}$. We can thus express $C_{O_2}$ as $C_{O_2} = C'_{O_2} + C_F$, where $C'_{O_2}$ is the summation of all of the capacitances except $C_F$. The inequality for $C_F$ can now be written as

$$C_F \geqslant \frac{(C'_{O_2}+C_F)(1+C_{O_1}/C_F)}{\sqrt{2}} = \frac{C'_{O_2}+C_{O_1}+C'_{O_2}C_{O_1}/C_F+C_F}{\sqrt{2}} \tag{6.77}$$

By multiplying through this last equation by $2C_F$ and collecting terms we obtain a quadratic equation in $C_F$ as given by

$$C_F^2(\sqrt{2}-1)-(C_{O_1}+C'_{O_2})C_F - C_{O_1}C'_{O_2} = 0 \tag{6.78}$$

so that $C_F$ will be given by

$$C_F = \frac{C_{O1} + C'_{O2}}{2(\sqrt{2} - 1)} \left[ 1 + \sqrt{1 + \frac{4C_{O_1}C'_{O_2}(\sqrt{2} - 1)}{(C_{O_1} + C'_{O_2})^2}} \right] \tag{6.79}$$

For $C_{O_1}$ we have

$$C_{O_1} = C_{CB_2} + C_{CS_2} + C_{CB_4} + C_{CB_6} + C_{BE_6} \tag{6.80}$$

where $C_{CS}$ is the collector-to-substrate capacitance. For $C'_{O_2}$ we have

$$C'_{O_2} = C_{CB_7} + C_{CB_{14}} + C_{CS_{14}} + C_{CB_8} + C_{CS_8} + C_{CB_{10}} \tag{6.81}$$

Note that the base-to-emitter capacitances of $Q_8$ and $Q_{10}$ will not be involved since the voltage gain of the emitter-follower stage can be considered to be close to unity.

For a typical example, let us assume that for all transistors involved, $C_{CB}$ = 1.0 pF (typ.), 1.5 pF (max.); $C_{CS}$ = 1.0 pF (typ.), 1.5 pF (max.); and $C_{BE}$ = 10 pF (typ.), 15 pF (max.). With these capacitance values we obtain $C_{O_1}$ = 14 pF (typ.), 21 pF (max.), and $C'_{O_2}$ = 6 pF (typ.), 9 pF (max.). Solving for $C_F$ using the quadratic formula given above, we obtain $C_F$ = 52.2 pF (typ.), 78.3 pF (max). The corresponding unity-gain frequencies will be given by $f_u = 2g_f/2\pi C_F$. If $I_Q$ = 25 μA, then $g_f = I_Q/4V_T$ = 250 μs. For a conservative design we would use $C_F = C_{F(\max)}$ = 78.3 pF, which will give us a unity-gain frequency of $f_u$ = 1.02 MHz.

### 6.2.10 Equivalent Input Noise Voltage

Due to the finite charge on the electron, electrical current flow will not be absolutely constant, but will always exhibit minute random fluctuations called *shot noise*. The mean-squared shot noise current per unit bandwidth is given by the equation $\overline{i^2_{\text{noise}}} = 2qI$, where $q$ is the electronic charge of $1.6 \times 10^{-19}$ and $I$ is the average current.

As a result of the high voltage gain of the first (differential amplifier) stage, virtually all of the noise at the output of the amplifier will have its origin in the first stage. The mean-squared output noise current of the first stage will be contributed to by $Q_1$ and $Q_2$ of the differential amplifier, and $Q_3$ and $Q_4$ of the active load. The noise currents of these four transistors will be uncorrelated, so that the mean-squared noise currents will be directly additive. The mean-squared output noise current of the first stage will be given by $\overline{i^2_{o(\text{noise})}} = \overline{i^2_1} + \overline{i^2_2} + \overline{i^2_3} + \overline{i^2_4}$. For $\overline{i^2_1}$ we have that $\overline{i^2_1} = 2qI_{C_1} = 2q(I_Q/2) = qI_Q$, and we will obtain similar expressions for the other three noise currents. The output noise current will therefore be given by $\overline{i^2_{o(\text{noise})}} = 4qI_Q$ (A²/Hz).

The equivalent input noise voltage $v_{i(\text{noise})}$ is a noise voltage that when applied to the input of the amplifier will produce a noise current equal to $i_{o(\text{noise})}$. These two quantities will be related by the dynamic transfer conductance $g_f$ as given by $\overline{i^2_{o(\text{noise})}} = (2g_f)^2\, \overline{v^2_{i(\text{noise})}}$, where $g_f = I_Q/4V_T$ and the factor of 2 comes from the current doubling action of the current mirror-active load. Solving this equation for the *mean-squared equivalent input noise voltage* gives

$$\overline{v^2_{i(\text{noise})}} = \frac{\overline{i^2_{o(\text{noise})}}}{(2g_f)^2} = \frac{4qI_Q}{(I_Q/2V_T)^2} = 16qV_T\,\frac{V_T}{I_Q} \tag{6.82}$$

Since $V_T = kT/q$, $qV_T = kT$, so that

$$\boxed{\overline{v^2_{i\,(\text{noise})}} = 16kT\frac{V_T}{I_Q}} \tag{6.83}$$

At $T = 290$ K $= 17°$C, $4\ kT = 1.60 \times 10^{-20}$ J, so that $v_{i\,(\text{noise})}$ can be written as

$$\overline{v^2_{i\,(\text{noise})}} = 6.4 \times 10^{-20}\left(\frac{V_T}{I_Q}\right) \qquad \text{V}^2/\text{Hz} \tag{6.84}$$

In addition to the noise produced by the random fluctuations of the transistor currents, called the *shot noise*, there will also be a significant amount of noise due to the *thermal noise voltage* of the transistor base spreading resistances, $r_{bb'}$. For the two transistors of the differential amplifier the mean-squared thermal noise voltage due to the two base resistances will be $\overline{v^2_{th}} = 4kT(2r_{bb'})$. The *total* mean-squared equivalent input noise voltage of the amplifier will now be given by

$$\boxed{\overline{v^2_{i\,(\text{noise})}} = 16kT\frac{V_T}{I_Q} + 4kT(2r_{bb'}) = 8kT\left[2\left(\frac{V_T}{I_Q}\right) + r_{bb'}\right]} \tag{6.85}$$

If we use $I_Q = 25\ \mu$A and $r_{bb'} = 150\ \Omega$ as representative values, we obtain an equivalent mean-squared input noise voltage of

$$\begin{aligned} v^2_{i\,(\text{noise})} &= 3.20 \times 10^{-20}[2(25\text{ mV}/25\ \mu\text{A}) + 150]\Omega \\ &= 3.20 \times 10^{-20}\,(2000 + 150)\Omega \\ &= \underline{6.88 \times 10^{-17}\text{ V}^2/\text{Hz}} \end{aligned} \tag{6.86}$$

The rms noise voltage will be

$$v_{i\,(\text{noise})}\text{ (rms, spectral density)} = \underline{8.295\text{ nV}/\sqrt{\text{Hz}}} \tag{6.87}$$

The equivalent noise bandwidth is equal to the 3-dB bandwidth multiplied by $\pi/2$ for systems characterized by a single breakpoint frequency. Therefore, if for the example above the system 3-dB bandwidth is 10 kHz, the equivalent noise bandwidth will be 15.71 kHz. The total rms noise voltage over this system bandwidth will be given by

$$v_{i\,(\text{noise})}\text{ (rms)} = 8.295\text{ nV}/\sqrt{\text{Hz}} \times \sqrt{15.71\text{ kHz}} = \underline{1.04\ \mu\text{V}} \tag{6.88}$$

If the input stage is a Darlington configuration, the transfer conductance $g_f$ will be one-half of the value used in the preceding case. In addition, due to the use of four transistors in the differential amplifier input stage compared to two transistors in the preceding case, the thermal noise voltage will now result from a total resistance of $4r_{bb'}$. As a result, the mean-squared equivalent input noise voltage of a system using a Darlington differential-amplifier input stage will be given by

$$\overline{v^2_{i\,(\text{noise})}} = 16kT\left(\frac{2V_T}{I_Q} + r_{bb'}\right) \tag{6.89}$$

If we use the same values as in the preceding example—$I_Q = 25\ \mu A$ and $r_{bb'} = 150\ \Omega$—we now obtain

$$\overline{v^2_{i\,(\text{noise})}} = 6.40 \times 10^{-20}(2150\ \Omega) = 1.377 \times 10^{-16}\ \text{V}^2/\text{Hz} \tag{6.90}$$

The rms noise voltage will be

$$v_{i\,(\text{noise})}\,(\text{rms}) = \underline{11.7\ \text{nV}/\sqrt{\text{Hz}}}$$

Over a 10-kHz 3-dB system bandwidth the total noise voltage will be $\underline{1.47\ \mu V}$.

### 6.2.11 Input Noise Current

The input bias current of an amplifier will exhibit minute random fluctuations (shot noise) just as is the case for the collector currents. The rms input noise current (spectral density) will be given by $i_{i\,(\text{noise})}$ (rms) $= \sqrt{2\,qI_B}$, where $I_B$ is the input bias (or base) current.

If $I_Q = 25\ \mu A$ and $\beta = 50$ (min.), the input noise current for the two-transistor differential amplifier will be

$$i_{i\,(\text{noise})} = \sqrt{2qI_B} = \sqrt{\frac{2q(I_Q/2)}{\beta}} = 0.283\ \text{pA}/\sqrt{\text{Hz}} \tag{6.91}$$

For a Darlington differential amplifier with the same $I_Q$ and $\beta$ values the input noise current will be only $0.040\ \text{pA}/\sqrt{\text{Hz}}$ and is therefore much smaller.

Thus, although the input noise voltage of the Darlington circuit is $\sqrt{2}$ times as large as the two-transistor differential amplifier, the input noise current is much smaller. For situations involving very large values of source resistance the Darlington circuit may be the preferable choice from the standpoint of noise and the signal-to-noise ratio. For these situations the use of a JFET or MOSFET operational amplifier should also be considered as a possible choice.

## 6.3 FIELD-EFFECT-TRANSISTOR OPERATIONAL AMPLIFIERS

The op-amp circuits discussed thus far have been bipolar op amps in that they contain only bipolar transistors. Op amps using field-effect transistors (FETs) in the input stage can offer some very significant advantages over bipolar op amps, especially in such areas as input impedance, input bias and offset currents, and slewing rate.

Bipolar transistors have much higher transfer conductances than those of FETs. The transfer conductance of bipolar devices is often some 30 to 100 times larger than that of FETs operating at comparable current levels. As a result the voltage gain for bipolar transistors will be substantially higher than that obtained with FETs.

The input resistance seen looking into the base of a bipolar transistor is limited to only moderate values, generally in the kilohm range, by the fact that the base–emitter junction is *forward biased*. In the case of a *junction field-effect transistor* (JFET) the input terminal is the *gate*, which is one side of the *reverse-biased* gate–channel junction. As a result, extremely high values of input resistance can be obtained, generally well up into the gigaohm ($10^9\ \Omega$) range. The input current of the device is the *gate current* $I_G$ and is the reverse leakage current of the gate–channel junction. This current will usually be well below 1 nA ($10^{-9}$ A) and is often around 10 pA. In contrast to this, the input or base current of a bipolar transistor is related to the output or collector current by $I_B = I_C/\beta$ and will generally be somewhere in the microampere range.

In the case of the *metal-oxide-silicon FET* (MOSFET) the gate electrode is separated from the rest of the device by a thin insulating layer of silicon dioxide ($SiO_2$). The input resistance is much larger than in the case of the JFET, with values well up into the teraohm ($10^{12}$ Ω) range usually being obtained. Correspondingly, the gate current $I_G$ will be extremely small, often below 1 pA. Thus a differential amplifier input stage using FETs will have a very high input impedance and a very low input bias current.

A JFET differential amplifier is shown in Figure 6.15a and a MOSFET differential amplifier is presented in Figure 6.15b. The devices shown are N-channel FETs, although P-channel devices can also be used. Various types of loads can be used, including active loads with bipolar or with FET transistors. In the case of bipolar differential amplifiers the input bias current and the input resistance will both be directly dependent on the quiescent current $I_Q$ as given by

$$I_B = \frac{I_C}{\beta} = \frac{I_Q}{2\beta} \qquad \text{and} \qquad R_i = \frac{2V_T}{I_B} = \frac{2\beta V_T}{I_C} = \frac{4\beta V_T}{I_Q} \tag{6.92}$$

For field-effect transistors the gate current is not only very small, but it is also not directly dependent on the differential-amplifier quiescent current level so that relatively large values of $I_Q$ can be used. The input resistance will be very large and will also not be directly affected by $I_Q$. The large values of $I_Q$ possible in the case of FET differential amplifiers can make available very large slewing rates (~50 to 75 V/μs) and also relatively large values for the unity-gain frequency (~20 MHz).

We will now consider some examples of JFET and MOSFET op amps. Some

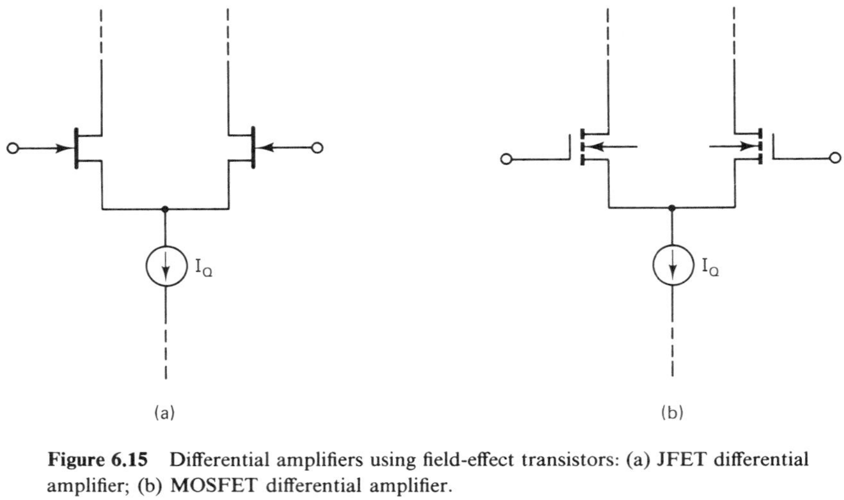


(a) (b)

**Figure 6.15** Differential amplifiers using field-effect transistors: (a) JFET differential amplifier; (b) MOSFET differential amplifier.

of these op amps use the FETs only in the input stage, but others will use FETs throughout the entire amplifier circuit.

### 6.3.1 JFET Operational Amplifiers

**Hybrid FET operational amplifiers: the LH0022/42/52.** A representative example of a FET input hybrid op amp is the LH0022/42/52 (National Semiconductor) series. This op amp uses a pair of JFET transistors only for the input-stage differential amplifier, and the rest of the circuit uses bipolar transistors. A simplified diagram of this op amp is shown in Figure 6.16.

The input-stage differential amplifier consists of $Q_1$ through $Q_4$, in which $Q_1$ and $Q_2$ are N-channel JFETs operating in a *common-drain* or *source-follower* configuration. These JFETs drive $Q_3$ and $Q_4$, which operate in the *common-base* configuration. Thus the differential amplifier uses a common-drain/common-source compound JFET/bipolar transistor configuration which provides the very high input resistance and very low input current characteristic of JFETs, while producing the high-voltage gain that is obtainable from bipolar transistors. This compound differential amplifier drives an active load consisting of $Q_5$ through $Q_7$, operating as a current mirror.

The rest of this op amp is of conventional design, with the second gain stage composed of $Q_{16}$ and $Q_{17}$ in a Darlington common-emitter configuration with a

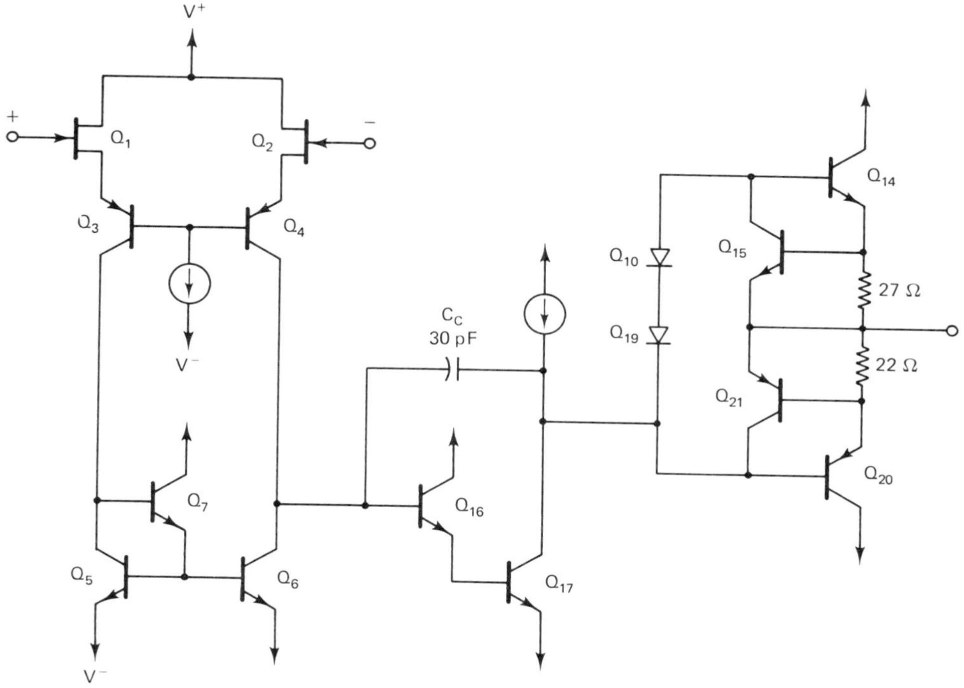


**Figure 6.16** Hybrid JFET-input operational amplifier: LH0022/42/52 series. (National Semiconductor).

**TABLE 6.1** LH0042C[a]

| Parameter | Min. | Typ. | Max. | Units |
|---|---|---|---|---|
| Offset voltage, $V_{OS}$ | | 6.0 | 20 | mV |
| Temperature coefficient of $V_{OS}$, $TCV_{OS}$ | | 10 | | $\mu$V/°C |
| Input bias current, $I_B$ | | 15 | 50 | pA |
| Input offset current, $I_{OS}$ | | 2 | 10 | pA |
| Difference-mode input resistance | | $10^{12}$ | | Ω |
| Common-mode input resistance | | $10^{12}$ | | Ω |
| Input capacitance, $C_i$ | | 4.0 | | pF |
| Open-loop voltage gain, $A_{OL}(O)$ ($R_L = 1\text{k}\Omega$) | 25 | 100 | | V/mV |
| CMRR | 70 | 80 | | dB |
| PSRR | 70 | 80 | | dB |
| Unity-gain frequency, $f_u$ | | 1.0 | | MHz |
| Slewing rate, $SR$ | 1.0 | 3.0 | | V/$\mu$s |

[a] $T = 25°\text{C}$, $V^+ = -15$, $V^- = -15$.

30-pF feedback or compensation capacitor. The output stage is a complementary push-pull emitter follower using $Q_{14}$ and $Q_{20}$. The voltage drop produced across $Q_{10}$ and $Q_{19}$ produces a "pre-bias" across $Q_{14}$ and $Q_{20}$ so as to bias these transistors slightly into conduction under quiescent condition, thereby minimizing crossover distortion. Transistors $Q_{15}$ and $Q_{21}$ together with the associated 27- and 20-Ω resistors are used for current limiting.

Some of the important specifications of the LH0042C are listed in Table 6.1. This device is the lowest-cost unit of these series and is specified for operation over the industrial temperature range −25 to +85°C. We take note in particular of the high value of input resistance of $10^{12}$ Ω (typ.) and the very small values of bias and offset currents of 15 pA and 2 pA (typ.), respectively, that are direct results of the use of the JFET input transistors.

**Another hybrid operational amplifier: the LH0062.** A second example of a hybrid FET op amp is the LH0062 (National Semiconductor). A simplified schematic diagram of this device is shown in Figure 6.17. The input differential-amplifier stage consists of a pair of N-channel JFETs $Q_1$ and $Q_2$ biased by a current source transistor $Q_3$ and driving a passive load consisting of resistors $R_1$ and $R_2$ (20 kΩ). The quiescent current of the differential amplifier is set at 600 mV/600 Ω = 1.0 mA by the combination of $D_1$ (1.2 V), $Q_3$, and $R_5$ (600 Ω), and this current will be relatively independent of the supply voltage.

The second gain stage is a bipolar differential amplifier using transistors $Q_4$ and $Q_5$ with the 6-kΩ emitter resistors, $R_3$ and $R_4$, setting the quiescent current level through each side at approximately a level given by (0.5 mA × 20 kΩ − 0.6 V)/6 kΩ = 1.6 mA. These emitter resistors also produce negative feedback within the second stage which acts to increase the input impedance and to extend the frequency response of this stage. The second stage drives a current mirror active load comprised of $Q_6$ and $Q_7$.

The output stage is a complementary push-pull emitter-follower circuit using

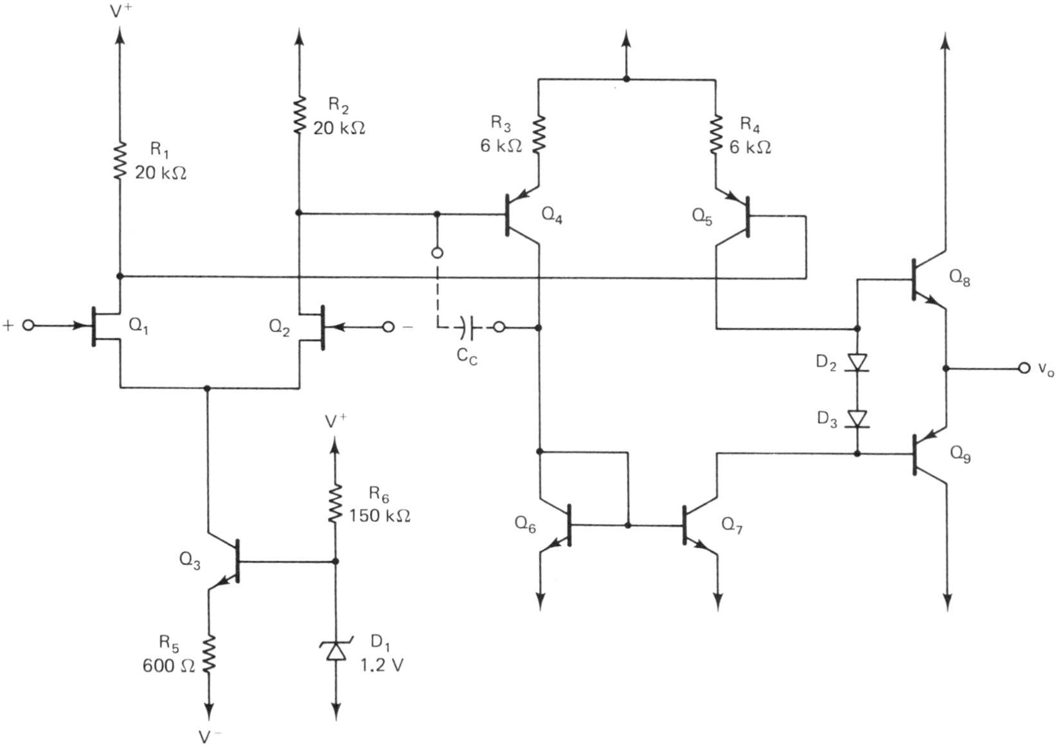


**Figure 6.17** Hybrid JFET-input operational amplifier: LH0062 (simplified schematic diagram). (National Semiconductor).

$Q_8$ and $Q_9$. The voltage drop that is produced across diodes $D_2$ and $D_3$ is used to pre-bias the output stage transistors so as to minimize crossover distortion.

This device is specified for operation over the industrial temperature range −25 to +85°C. Some of the important parameters of this device are listed in Table 6.2 for a supply voltage of ±15 V and at $T = 25°C$.

This op amp has the very high input impedance and very low input bias current that are characteristic of FET-input op amps. We note also that this op amp has a large value of unity-gain frequency $f_u$ [15 MHz (typ.)] and a very large slewing rate [50 (min.), 75 (typ.) V/μs]. This very large slewing rate is made possible by the large value of quiescent current ($I_Q = 1$ mA) of the JFET input stage. We note again that a large $I_Q$ can be used for JFET differential amplifiers without seriously affecting the input (gate) current or the input resistance. This is not the case with bipolar differential amplifiers, for which $I_Q$ must be kept small in order to have a low input bias current and a high input resistance.

**Monolithic FET-input operational amplifiers: the LF155/156/157.** As a third example of a JFET op amp we will consider the LF155/156/157 (National Semiconductor) series of devices. A simplified schematic diagram of

**TABLE 6.2** LH0062C

| Parameter | Min. | Typ. | Max. | Units |
|---|---|---|---|---|
| Offset voltage, $V_{OS}$ | | 10 | 15 | mV |
| Temperature coefficient of $V_{OS}$, $TCV_{OS}$ | | 10 | 35 | $\mu$V/°C |
| Input bias current, $I_B$ | | 10 | 65 | pA |
| Input offset current, $I_{OS}$ | | 1 | 5 | pA |
| Difference-mode input resistance | | $10^{12}$ | | Ω |
| Common-mode input resistance | | $10^{12}$ | | Ω |
| Input capacitance, $C_i$ | | 4 | | pF |
| Open-loop voltage gain, $A_{OL}(O)$ ($R_L = 2$ kΩ) | 25 | 160 | | V/mV |
| CMRR | 70 | 90 | | dB |
| PSRR | 70 | 90 | | dB |
| Unity-gain frequency, $f_u$ | | 15 | | MHz |
| Slewing rate, $SR$ | 50 | 75 | | V/$\mu$s |

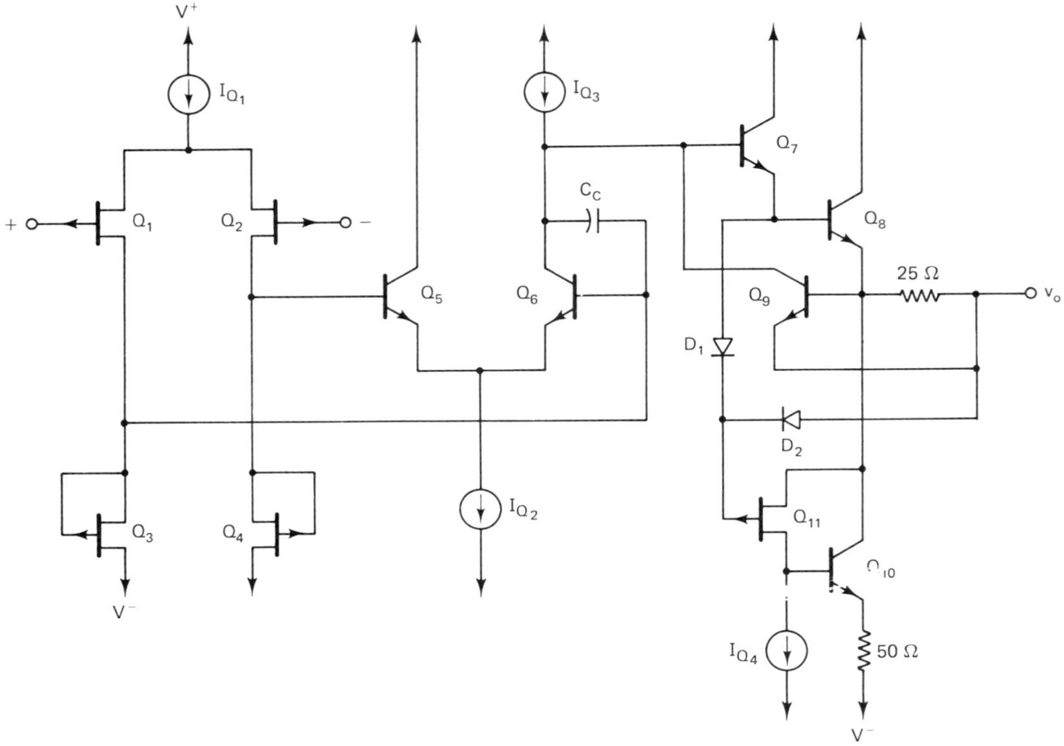


**Figure 6.18** Simplified schematic diagram of a monolithic JFET-input operational amplifier: LF155/156/157 series. (National Semiconductor).

this IC is shown in Figure 6.18. This is a monolithic IC with all transistors, both JFET and bipolar, on the same chip.

The input differential-amplifier stage consists of a pair of P-channel JFETs $Q_1$ and $Q_2$ biased by a current source $I_Q$. This differential amplifier drives a current source active load using $Q_3$ and $Q_4$, which are diode-connected JFETS ($V_{GS} = 0$) operating as current regulator diodes.

The second stage is a bipolar differential amplifier with current sink biasing and a current source active load. This stage drives a Darlington emitter-follower output stage with an internal current-limiting circuit. Transistor $Q_{10}$, together with $Q_{11}$, $D_1$, $D_2$, and the $I_{Q4}$ current source, are used for sinking currents from the load.

An on-chip feedback capacitor across the second stage provides frequency compensation. This capacitor is 10 pF for the LF155 and LF156 device series and provides for stabilization down to the unity-gain case with a unity-gain frequency of 2.5 MHz for the LF155 and 5 MHz for the LF156. The corresponding slewing rates are 5 V/$\mu$s for the LF155 and 12 V/$\mu$s for the LF156. For the LF157 the feedback capacitor is only 2 pF. As a result of this the op amp will be stable (with a 45° phase margin) only for closed-loop gains of 5 or greater. However, the unity-gain frequency (or gain–bandwidth product) will now be up to 20 MHz (typ.) and the slewing rate will be 50 V/$\mu$s (typ.).

The LF355/356/357 device series (National Semiconductor) is the same as the LF155/156/157 series except that the 300-series devices are specified over the commercial temperature range. Some of the important characteristics of the LF355/356/357 op amps are tabulated in Table 6.3 for a supply voltage of $\pm 15$ V and at $T = 25°\text{C}$.

**TABLE 6.3** LF355/356/357

| Parameter | Min. | Typ. | Max. | Units |
|---|---|---|---|---|
| Input offset voltage, $V_{OS}$ | | 3 | 10 | mV |
| Temperature coefficient of $V_{OS}$, $TCV_{OS}$ | | 5 | | $\mu$V/°C |
| Input bias current, $I_B$ | | 30 | 200 | pA |
| Input offset current, $I_{OS}$ | | 3 | 50 | pA |
| Input resistance | | $10^{12}$ | | Ω |
| Input capacitance, $C_i$ | | 3 | | pF |
| Open-loop voltage gain, $A_{OL}(O)$ ($R_L = 2$ kΩ) | 25 | 200 | | V/mV |
| CMRR | 80 | 100 | | dB |
| PSRR | 80 | 100 | | dB |
| Unity-gain frequency, $f_u$ (gain–bandwidth product) | | | | |
| LF355 | | 2.5 | | MHz |
| LF356 | | 5 | | MHz |
| LF357 | | 20 | | MHz |
| Slewing rate | | | | |
| LF355 | | 5 | | V/$\mu$s |
| LF356 | | 12 | | V/$\mu$s |
| LF357 ($A_V = 5$) | | 50 | | V/$\mu$s |

**Quad FET-input operational amplifiers: the LF347.** The LF347 (National Semiconductor) is an example of a monolithic JFET-input quad operational amplifier. There are four electrically independent op amps on the same silicon IC chip. The four op amps share only the positive and negative supply terminals. This device is available in a 14-pin DIP package. A simplified schematic diagram of this device is shown in Figure 6.19.

The input stage is a P-channel JFET differential amplifier that drives a bipolar transistor current mirror active load. The second stage is a common-emitter transistor with a current source active load. A 10-pF compensation capacitor $C_c$ is placed as a feedback capacitor across the second stage to ensure stability. The output stage is a complementary push-pull emitter-follower configuration. The voltage drop across the two diodes is used to bias the output stage for the minimization of crossover distortion.

The LF347 is specified over the commercial temperature range 0 to 70°C. Listed in Table 6.4 are some of the important characteristics of this device when operated with a supply voltage of ±15 V and at a temperature of 25°C.

We again note the very low input bias current, very large input resistance, and high slewing rate that are characteristic of FET-input op amps. The amplifier-to-amplifier coupling coefficient of −120 dB indicates a very high degree of electrical isolation between the four op amps that share the same IC chip.

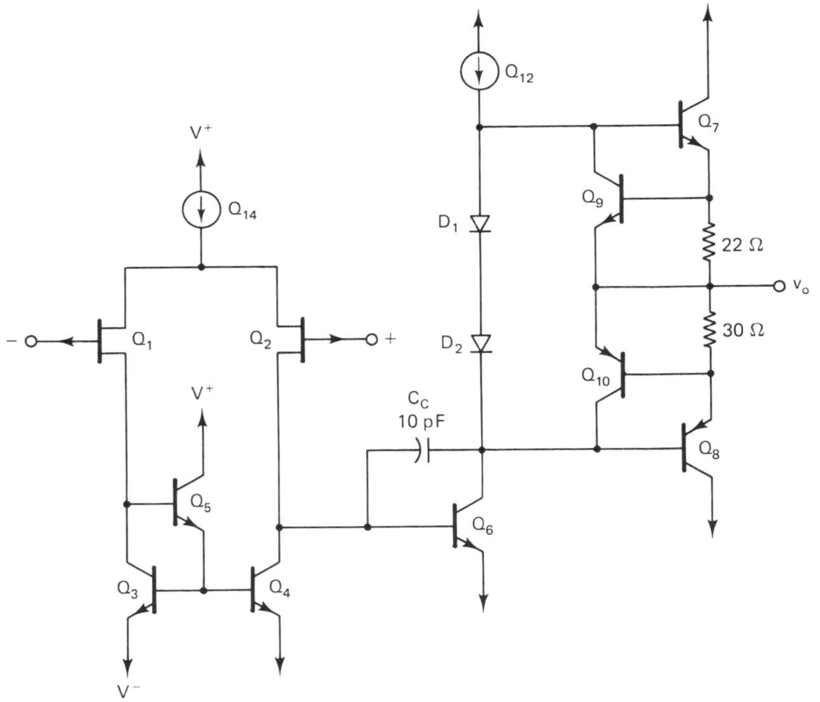


**Figure 6.19** Quad JFET-input operational amplifier: LF347. (National Semiconductor).

**TABLE 6.4** LF347

| Parameter | Min. | Typ. | Max. | Units |
|---|---|---|---|---|
| Input offset voltage, $V_{OS}$ | | 5 | 10 | mV |
| Temperature coefficient of $V_{OS}$, $TCV_{OS}$ | | 10 | | $\mu V/°C$ |
| Input bias current, $I_B$ | | 50 | 200 | pA |
| Input offset current, $I_{OS}$ | | 25 | 100 | pA |
| $TCV_{OS}$ | | 10 | | $\mu V/°C$ |
| Input resistance | | $10^{12}$ | | $\Omega$ |
| Open-loop voltage gain, $A_{OL}(O)$ ($R_L = 2\ k\Omega$) | 25 | 100 | | V/mV |
| CMRR | 70 | 100 | | dB |
| PSRR | 70 | 100 | | dB |
| Gain–bandwidth product, $f_u$ | | 4 | | MHz |
| Slewing rate, $SR$ | | 13 | | $V/\mu s$ |
| Amplifier-to-amplifier coupling | | −120 | | dB |
| Equivalent input noise voltage ($f = 1$ kHz) | | 16 | | $nV/\sqrt{Hz}$ |
| Input noise current ($f = 1$ kHz) | | 0.01 | | $pA/\sqrt{Hz}$ |

The LF353 is a monolithic dual op amp with circuitry and characteristics similar to the LF347 quad op amp, and the LF351 is a single op amp that is also similar to the LF347.

**Monolithic bipolar-JFET construction.** The monolithic FET-input op amps just considered have both bipolar transistors and JFETs on the same IC chip. In Figure 6.20 a cross-sectional view is presented showing an NPN transistor and a P-channel JFET of the type used for these ICs. The NPN transistor is of the standard IC construction.

The JFET uses the same processing steps as the bipolar transistor, including the P-type and N-type diffusions, except that there is a boron ion implantation to produce the thin channel region of the JFET. This implantation is a relatively low-dosage, low-energy implantation to produce a very thin channel region that will have a low pinch-off voltage $V_p$, and at the same time produce an acceptably high value for $I_{DSS}$. The uniformity of the implantation also results in well-matched JFETs.

A good example of what can be achieved with well-matched ion-implanted JFETs in the input stage is the OPA111 (Burr-Brown) with a $TCV_{OS}$ of only 1 $\mu V/°C$ and an input bias current of only 1 pA. This op amp has a dielectrically isolated JFET differential amplifier input stage that is biased by a 800 $\mu A$ current source. Each side of the differential amplifier drives a 4 k$\Omega$ resistive load. These load resistances are laser trimmed to null out the input offset voltage resulting in a maximum offset voltage of only 250 $\mu V$. Another important feature of this op-amp design is its low noise characteristics. The equivalent input noise voltage is only 6 $nV/\sqrt{Hz}$, and the low input bias current results in an input noise current of only 1.6 $fA/\sqrt{Hz}$ (0.0016 $pA/\sqrt{Hz}$). Other features of this op amp are an open-loop gain of 125 dB, a unity gain frequency of 2 MHz, and a slewing rate of 2 $V/\mu s$.

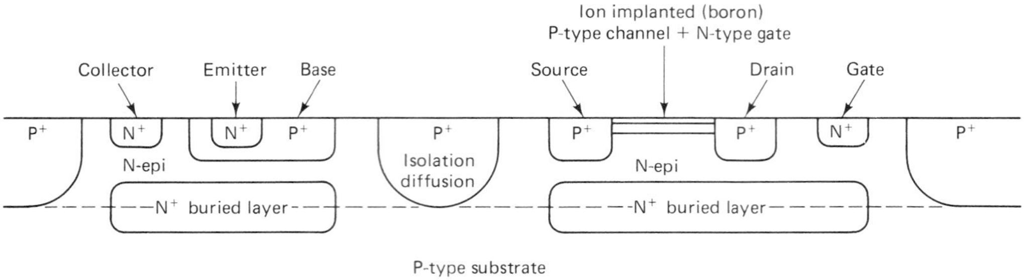


**Figure 6.20** Monolithic bipolar-JFET construction: cross-sectional view.

### 6.3.2 MOSFET Operational Amplifiers

**The CA3130.** An interesting example of a monolithic MOSFET operational amplifier is the CA3130 (RCA), which contains both MOSFETs and bipolar transistors on the same chip. A simplified schematic diagram of this device is shown in Figure 6.21a and a block diagram is given in Figure 6.21b.

The input stage is a differential amplifier consisting of a pair of P-channel enhancement-mode MOSFETs, $Q_6$ and $Q_7$. This differential amplifier is biased by a 200-$\mu$A MOSFET current source, and it drives a current mirror active load using $Q_9$ and $Q_{10}$. Resistors $R_5$ and $R_6$ (1 k$\Omega$) in the active load circuit are used together with an externally connected potentiometer (across pins 1 and 5) for offset voltage nulling. The quiescent voltage drop across these resistors is 100 $\mu$A $\times$ 1 k$\Omega$ = 100 mV, so that a $\pm$100-mV offset adjustment range is available. Diodes $D_5$ and $D_8$ are connected between the input terminals to protect the thin gate oxide of the input MOSFETs against excessive voltage spikes and static electricity discharge, which could cause breakdown of the oxide layer and result in irreversible damage to the transistors. The voltage gain of this first stage is only about 5, this low value being the result of the relatively low transfer conductance of the MOSFETs.

The second stage consists of a bipolar transistor connected as a common-emitter amplifier, with a 200-$\mu$A MOSFET current source serving as the active load. As a result of the high dynamic impedance provided by the current source active load and the very high input impedance seen looking into the MOSFET output stage, the voltage gain of the second stage is very large ($\sim$6000).

The output stage is a *complementary-symmetry* pair of MOSFETs (CMOS) with $Q_8$ being the P-channel device (PMOS) and $Q_{12}$ being the N-channel (NMOS) transistor. As the voltage at the collector of $Q_{11}$ goes up above the quiescent level the NMOS transistor $Q_{12}$ will turn on and the PMOS device $Q_8$ will turn off. This will drive the output voltage down toward the negative supply. Conversely, as the voltage at the collector of $Q_{11}$ goes down, the PMOS transistor $Q_8$ will be turned on and the NMOS transistor $Q_{12}$ will be turned off. This will pull the output voltage up toward the positive supply. The drain-to-source resistance of the CMOS output transistors is about 300 $\Omega$ when the transistor is turned on, so that for large-value load resistances ($\gtrsim$1 M$\Omega$) the output voltage can be driven to within a few millivolts of either supply voltage. As the load resistance drops down into the kilohm range,

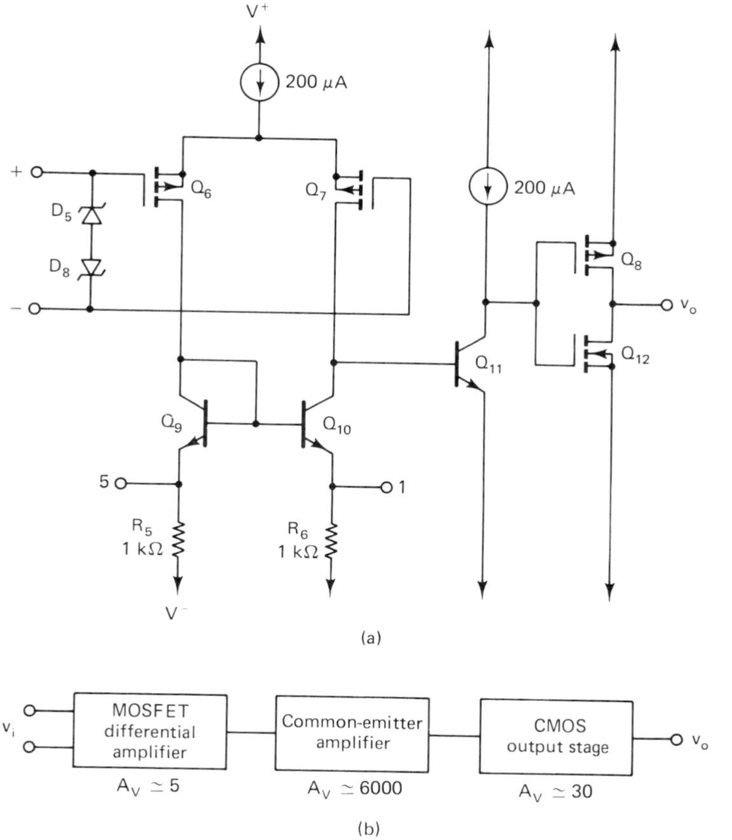


**Figure 6.21** MOSFET-input CMOS-output operational amplifier: CA3130 (RCA Corporation): (a) simplified schematic diagram; (b) block diagram.

the output voltage swing will, of course, get smaller. The voltage gain of this CMOS output stage will be approximately 30.

Some of the important specifications for the CA3130 are listed in Table 6.5 for operating conditions of a ±15-V supply and at a temperature of 25°C.

In Figure 6.22 a diagram is presented of the current source circuitry of the CA3130. Transistor $Q_1$ is a diode-connected MOSFET, which together with $Q_2$ constitutes a current mirror. The current through $Q_1$ is set at 100 μA by the combination of the voltage drops across the zener diode $Z_1$, diodes $D_1$ through $D_4$, and the 40-kΩ resistor $R_1$. Transistors $Q_2$, $Q_3$, $Q_4$, and $Q_5$ are scaled to have twice the channel width of $Q_1$, so that the current through these transistors will be twice that of $Q_1$, or 200 μA. Transistor pairs $Q_2$–$Q_4$ and $Q_3$–$Q_5$ are cascode-connected transistors, which results in a lower output conductance for the two current sources. This leads to a better common-mode rejection ratio for the differential amplifier stage and a higher voltage gain for the second stage.

For bipolar transistors, the base voltage must not be allowed to go beyond

**TABLE 6.5** CA3130

| Parameter | Min. | Typ. | Max. | Units |
|---|---|---|---|---|
| Input offset voltage, $V_{OS}$ | | 8 | 15 | mV |
| Input bias current, $I_B$ | | 5 | 50 | pA |
| Input offset current, $I_{OS}$ | | 0.5 | 30 | pA |
| Input resistance | | 1.5 | | TΩ |
| Input capacitance, $C_i$ | | 4.3 | | pF |
| Open-loop voltage gain, $A_{OL}(O)$ ($R_L = 2$ kΩ) | 50 | 320 | | V/mV |
| CMRR | 70 | 90 | | dB |
| PSRR | 70 | 90 | | dB |
| Unity-gain frequency, $f_u$ | | 15 | | MHz |
| Slewing rate, $SR$ | | 10 | | V/μs |

the collector voltage such that the collector–base voltage becomes forward biased by more than 0.5 V and the transistor goes into saturation. This restriction can severely limit the input voltage range of bipolar-input op amps such that operation with a split power supply is required. Unlike the case of the bipolar transistor, the gate of a MOSFET can go well beyond the drain voltage. As a result of this the input voltage range of the CA3130 extends down to 0.5 V below the negative supply

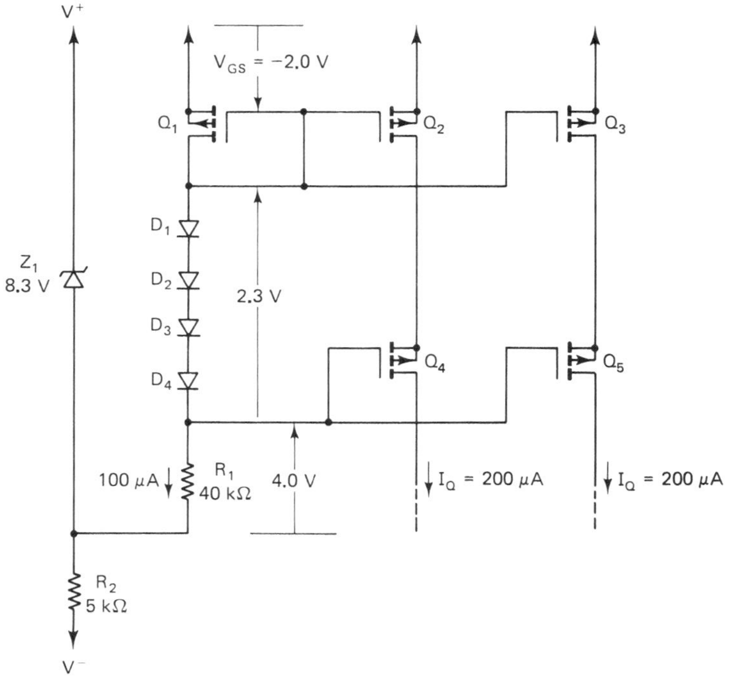


**Figure 6.22** Cascode MOSFET current source circuit for the CA3130 (RCA Corporation).

voltage. This fact makes possible the operation of this device with a single positive supply voltage.

**The CA3140.** A second example of a MOSFET-input op amp is the CA3140 (RCA). A simplified schematic diagram of this device is given in Figure 6.23. The input stage is a P-channel MOSFET differential amplifier, $Q_9$ and $Q_{10}$, biased by a 200-$\mu$A current source and driving a current mirror active load comprised of $Q_{11}$ and $Q_{12}$. Resistors $R_4$ and $R_5$ (500 $\Omega$) are used together with an externally connected potentiometer ($\sim$10 k$\Omega$) between pins 1 and 5 for offset-voltage nulling. The voltage drop across $R_4$ and $R_5$ of 100 $\mu$A $\times$ 500 $\Omega$ = 50 mV provides for a $\pm$50-mV offset adjustment range. Diodes $D_3$ and $D_4$ are used for the protection of the MOSFET gate oxides. The voltage gain of this first stage is approximately 10, this relatively low voltage gain again being the result of the low transfer conductance of the MOSFET transistors.

The second stage uses transistor $Q_{13}$ in a common-emitter configuration with a 200-$\mu$A current source active load. A 12-pF feedback capacitor across this stage is used for feedback stabilization. As a result of the high dynamic impedance presented by the active load and the high input impedance of the Darlington emitter-follower output stage, the voltage gain of the second stage will be about 10,000. Since the voltage gain of the output stage will be around unity, the overall amplifier open-loop gain, $A_{OL}(0)$, will be about 100,000, or 100 dB.

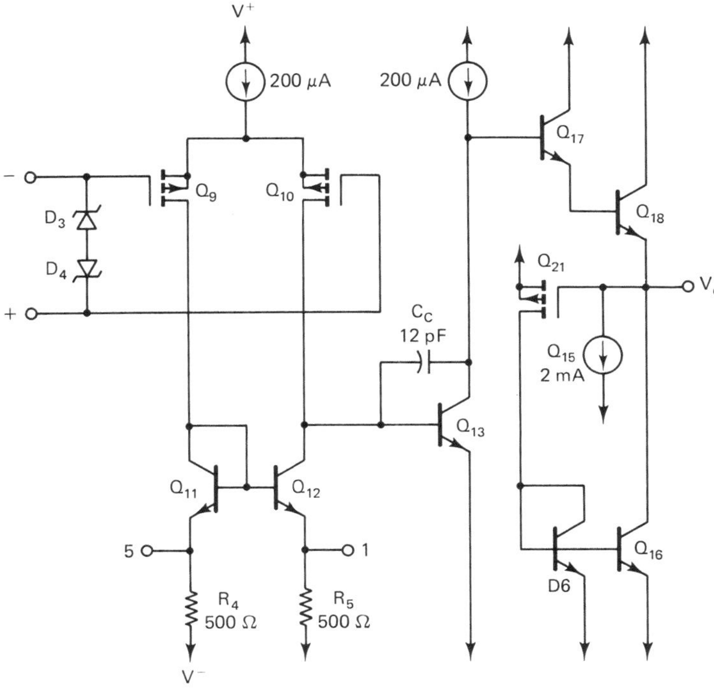


**Figure 6.23** MOSFET-input operational amplifier: CA3140 (RCA Corporation).

**TABLE 6.6** CA3140

| Parameter | Min. | Typ. | Max. | Units |
|---|---|---|---|---|
| Input offset voltage, $V_{OS}$ | | 8 | 15 | mV |
| Input bias current, $I_B$ | | 10 | 50 | pA |
| Input offset current, $I_{OS}$ | | 0.5 | 30 | pA |
| Input resistance | | 1.5 | | TΩ |
| Open-loop voltage gain, $A_{OL}(0)$ ($R_L = 2$ kΩ) | 20 | 100 | | V/mV |
| CMRR | 70 | 90 | | dB |
| PSRR | 76 | 80 | | dB |
| Unity-gain frequency, $f_u$ | | 4.5 | | MHz |
| Slewing rate, $SR$ | | 9 | | V/μs |

The output stage uses a Darlington NPN emitter follower $Q_{17}$–$Q_{18}$ for sourcing output currents, thus driving the output voltage up in the positive direction above the quiescent level. For sinking load currents, transistor $Q_{16}$ is used. As the current delivered by $Q_{18}$ drops below the 2-mA current level of the $Q_{15}$ current sink, the differences between these two currents will be the load current sunk by the op amp. This will cause the output voltage to drop below the quiescent level, which will cause the MOSFET $Q_{21}$ to turn on. As a result of $Q_{21}$ turning on, current will be supplied to the diode-connected transistor $D_6$, which together with $Q_{16}$, constitute a current mirror. Thus the current flowing through $Q_{21}$ and thence through $D_6$ will produce a similar current flow through $Q_{16}$, which can now sink the major part of the output current.

Some of the important specifications for the CA3140 are listed in Table 6.6 for the case of operation with a supply voltage of ±15 V and at $T = 25°C$.

**Quad CMOS operational amplifier.** An example of an all-MOSFET quad operational amplifier is the MC14573 (Motorola). A simplified circuit diagram is shown in Figure 6.24. Transistors $Q_1$ and $Q_2$ constitute a PMOS differential amplifier with NMOS transistors $Q_3$ and $Q_4$ serving as a current mirror active load. The differential amplifier is biased by the $Q_5$–$Q_6$ current mirror with a quiescent current $I_Q$ given approximately by $I_Q \simeq I_{SET} \simeq (V^+ - V^- - 1.0 \text{ V})/R_{SET}$, where $R_{SET}$ is an external programming resistor. The second and output stage uses $Q_7$ in a common-source configuration and $Q_8$ serves as a current source active load.

The op amp features an open-loop gain, $A_{OL}(0)$, of 90 dB, an amplifier-to-amplifier coupling (or channel separation) of −100 dB, an input bias current of 1 nA (max.) and an offset current of 200 pA (max.). When operated with $I_{SET} = 40\ \mu A$, the slewing rate will be 2.5 V/μs.

We will now consider a simple a-c analysis of this operational amplifier circuit. The low-frequency voltage gain of the first stage will be given by $A_{V1}(0) = 2g_f/g_{total} = 2g_f/(g_{o_2} + g_{o_4})$. The transfer conductance of the differential amplifier is given by $g_f = \sqrt{KI_Q/2} = (I_Q/2)/(V_{GS} - V_t)$ and the output conductances of $Q_2$

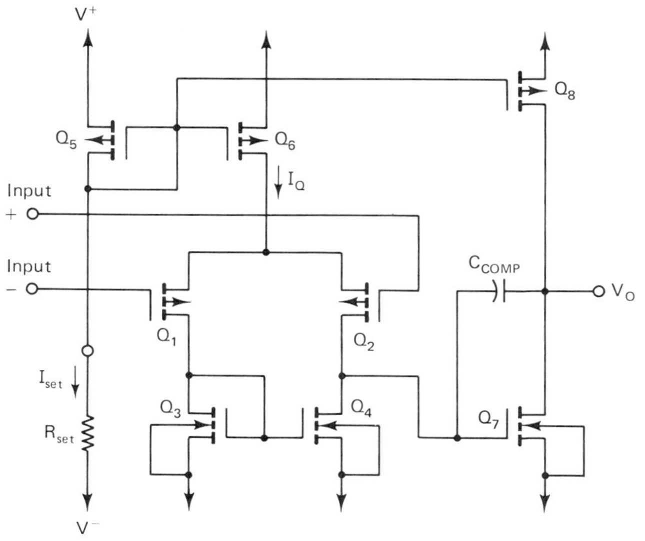


**Figure 6.24** CMOS operational amplifier: MC14573 (Motorola) simplified circuit diagram. (Courtesy Motorola Semiconductor Products Inc.)

and $Q_4$ will be $g_{o_2} = g_{ds_2} = (I_Q/2)/V_{A\,(\mathrm{PMOS})}$ and $g_{o_4} = g_{ds_4} = (I_Q/2)/V_{A\,(\mathrm{NMOS})}$, respectively. The voltage gain can therefore be written as

$$A_{V1}(0) = \frac{I_Q/(V_{GS} - V_t)}{(I_Q/2)\left[\dfrac{1}{V_{A\,(\mathrm{NMOS})}} + \dfrac{1}{V_{A\,(\mathrm{PMOS})}}\right]} = \frac{2/(V_{GS} - V_t)}{\dfrac{1}{V_{A\,(\mathrm{NMOS})}} + \dfrac{1}{V_{A\,(\mathrm{PMOS})}}} \tag{6.93}$$

For the MC14573 at $I_Q = I_{\mathrm{SET}} = 40\ \mu\mathrm{A}$, the value of $V_{GS}$ will be approximately 1 V. If we assume $V_t = 0.5$ V and $V_{A\,(\mathrm{NMOS})} = V_{A\,(\mathrm{PMOS})} = 75$ V as reasonable values we obtain a voltage gain of $A_{V1}(O) = 150$ for the first stage.

The voltage gain for the second stage will be given by

$$A_{V_2}(0) = g_{fs_7}/(g_{\mathrm{total}})_7 = g_{fs_7}/(g_{o_7} + g_{o_8} + g_L) \tag{6.94}$$

where $g_{fs_7}$ = transfer conductance of $Q_7 = 2K(V_{GS} - V_t) = 2I_{DS_7}/(V_{GS} - V_t) = 2I_Q/(V_{GS} - V_t)$. The conductance $g_{o_7}$ is the output conductance of $Q_7$ as given by $g_{o_7} = g_{ds_7} = I_{DS_7}/V_{A\,(\mathrm{NMOS})} = I_Q/V_{A\,(\mathrm{NMOS})}$, and $g_{o_8}$ is the output conductance of the current source active load transistor $Q_8$ as given by $g_{o_8} = g_{ds_8} = I_{DS_8}/V_{A\,(\mathrm{PMOS})} = I_Q/V_{A\,(\mathrm{PMOS})}$. The conductance $g_L$ represents conductance of the load driven by this operational amplifier. For the rest of the analysis we will assume $g_L = 0$. We now obtain for the voltage gain of the second stage the expression

$$A_{V_2}(0) = \frac{\dfrac{2I_Q}{V_{GS} - V_t}}{\dfrac{I_Q}{V_{A\,(\mathrm{NMOS})}} + \dfrac{I_Q}{V_{A\,(\mathrm{PMOS})}}} = \frac{\dfrac{1}{V_{GS} - V_t}}{\dfrac{1}{V_{A\,(\mathrm{NMOS})}}\ \dfrac{1}{V_{A\,(\mathrm{PMOS})}}} \tag{6.95}$$

If we use the same parameters as used for the first stage we now obtain a voltage gain of $A_{V_2}(0) = 150$.

The overall amplifier open-loop voltage gain $A_{OL}(0)$ will be $A_{OL}(0) = A_{V_1}(0) \times A_{V_2}(0) = 150 \times 150 = 22{,}500$ which corresponds to 87 dB. This is close to the manufacturer's specification of 90 dB (typ.). Thus we see that although low voltage gains are generally associated with MOSFET amplifiers, the use of active loads can provide for an acceptably large open-loop gain for this two-stage operational-amplifier circuit.

The unity gain frequency for this operational-amplifier circuit can be obtained from the equation $f_u = 2g_f/(2\pi C_C)$ where $g_f$ is the transfer conductance of the differential amplifier and $C_C$ is the compensation or feedback capacitor placed across the second stage. If we substitute in the equation for $g_f$ we obtain $f_u = I_Q/[2\pi C_C(V_{GS} - V_t)]$. If we assume $I_Q = I_{SET} = 50\ \mu A$ and use a value of 16 pF for $C_C$ we obtain a unity gain frequency of $f_u = 0.80$ MHz. The corresponding slewing rate will be given by slewing rate = SR = $I_Q/C_C = 40\ \mu A/16\ pF = 2.5\ V/\mu s$, which is in agreement with the manufacturer's specification.

## 6.4 NORTON OPERATIONAL AMPLIFIER

The operational amplifiers described thus far use a differential-amplifier input stage to produce an output voltage that is proportional to the difference-mode input voltage. In the Norton operational amplifier a current mirror circuit is used at the input to produce an output voltage for the amplifier that is proportional to the difference between the two input currents. This characteristic leads to many interesting applications and the circuit configuration of the Norton amplifier is such that it can easily be biased from a single d-c supply. In addition, a very large input voltage range can be obtained by the use of large-value resistors placed in series with the two amplifier input terminals.

The circuit symbol of the Norton operational amplifier is shown in Figure 6.25. In Figure 6.26 the circuit diagram of the LM3900 (National Semiconductor) Norton amplifier is shown. The LM3900 is actually a quad operational amplifier, and it can be operated with a single supply voltage as low as 4 V and as high as 36 V. It is internally compensated by means of capacitor $C_1$ to produce a unity gain frequency of 2.5 MHz.

In the LM3900 circuit, transistors $Q_1$ and $Q_2$ constitute a simple current mirror circuit, and if we neglect the base currents we will have that $I_{C_2} = I^+$. The base current of $Q_3$ will therefore be equal to the difference of the two input currents as given by $I_{B_3} = I^- - I_{C_2} = I^- - I^+$.

This amplifier has a single gain stage in which $Q_3$ is used as a common-emitter amplifier. Transistors $Q_5$ and $Q_6$ are a compound PNP–NPN pair of emitter followers for the impedance transformation of the load resistance that is driven by this amplifier.

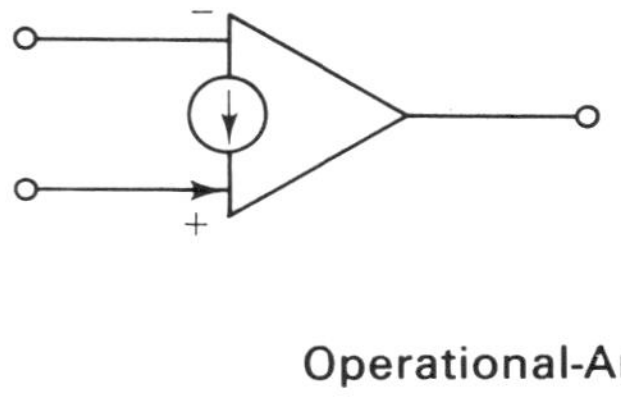


**Figure 6.25** Norton amplifier circuit symbol.

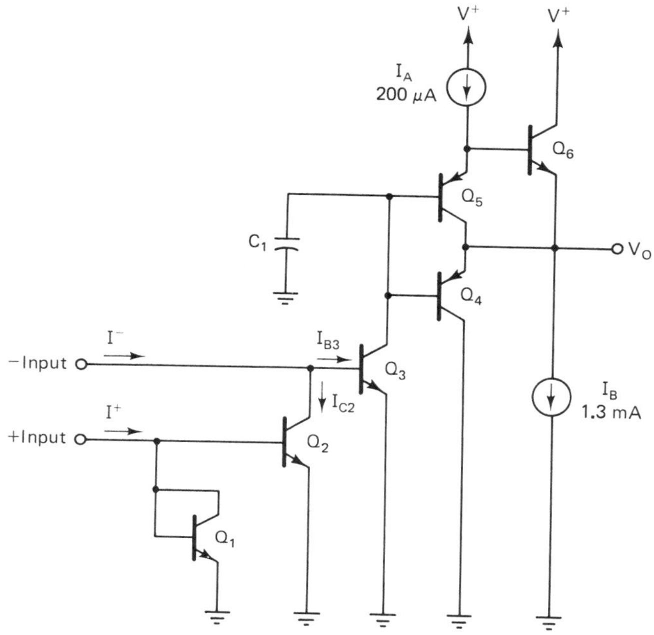


**Figure 6.26** Norton amplifier (LM3900) circuit diagram (National Semiconductor).

The large impedance transformation ratio that is produced by this compound emitter-follower stage will result in a very large voltage gain for the $Q_3$ amplifier stage.

Current sink $I_B$ (1.3 mA) is used to sink output currents, so that this circuit can both source as well as sink currents from the load. Under large-signal conditions transistor $Q_4$ will be turned on to provide an additional current sinking capability.

For a simple voltage gain analysis of this circuit we will assume a current gain of $\beta = 50$ (min.) for the NPN transistors. We will also assume that the voltage gain of the $Q_5$–$Q_6$ emitter-follower output stage is close to unity. The voltage gain provided by the $Q_3$ common-emitter gain stage will be $A_V = -g_{m\,3}R_L'$, where $R_L'$ is the transformed value of the load resistance $R_L$ as given by $R_L' = \beta_5\beta_6 R_L$. The transfer conductance of $Q_3$ will be given by

$$g_{m\,3} = \frac{I_{C\,3}}{V_T} = \frac{I_{B\,5}}{V_T} \simeq \frac{I_{E\,5}}{\beta_5 V_T} = \frac{200\ \mu\text{A}}{\beta_5 V_T} \tag{6.96}$$

Thus the voltage gain will be given by

$$\begin{aligned} A_V &= -\,g_{m\,3}R_L' = -\frac{200\ \mu\text{A}}{\beta_5 V_T}\beta_5\beta_6 R_L = \frac{200\ \mu\text{A}(\beta_6 R_L)}{25\ \text{mV}} \\ &= -\,8\ \text{mS} \times \beta_6 R_L \end{aligned} \tag{6.97}$$

For a load resistance of $R_L = 10$ kΩ this gives a voltage gain of $A_V = -8$ mS × 50 (min.) × 10 kΩ = 4000 (min.) = 70 dB (min., which is close to the manufacturer's specification.

### 6.4.1 Norton Amplifier Closed-Loop Analysis

For the general closed-loop gain analysis of the Norton operational amplifier, we will use the circuit of Figure 6.27 and reference will also be made to Figure 6.26. The voltage levels at the two input nodes of the amplifier will both be one base-to-emitter voltage drop ($V_{BE} \simeq 0.6$ V) above ground. As a result of the very high gain of the circuit the action of the feedback loop will be such that the two input currents will be essentially equal, so that $I^+ = I^-$. This is analogous to the case of the ordinary operational amplifier where the action of the feedback loop causes the voltages at the two input terminals to be essentially equal.

In general, the current through resistor $R_N$ will be given by $I_N = (V_N - V_{BE})/R_N$, where $N = 1, 2, 3, \ldots$ To simplify the writing of the equations we will define $V'_N$ as $V'_N = V_N - V_{BE}$. At node 1 we have

$$V'_1 G_1 + V'_3 G_3 + V'_5 G_5 = I^+ \tag{6.98}$$

and at node 2

$$V'_o G_F + V'_2 G_2 + V'_4 G_4 + V'_6 G_6 = I^- \tag{6.99}$$

Since $I^+ = I^-$, we obtain

$$V'_o G_F = V'_1 G_1 + V'_3 G_3 + V'_5 G_5 - (V'_2 G_2 + V'_4 G_4 + V'_6 G_6) \tag{6.100}$$

so that

$$V'_o = \frac{V'_1 G_1}{G_F} + \frac{V'_3 G_3}{G_F} + \frac{V'_5 G_5}{G_F} - \frac{V'_2 G_2}{G_F} - \frac{V'_4 G_4}{G_F} - \frac{V'_6 G_6}{G_F} \tag{6.101}$$

In terms of resistances this equation for $V'_o$ can be rewritten as

$$V'_o = \frac{V'_1 R_F}{R_1} + \frac{V'_3 R_F}{R_3} + \frac{V'_5 R_F}{R_5} - \frac{V'_2 R_F}{R_2} - \frac{V'_4 R_F}{R_4} - \frac{V'_6 R_F}{R_6} \tag{6.102}$$

The input-voltage range for any input signal that is active ranges down to a minimum value of $V_{BE} \simeq 0.6$ V. For the proper operation of the LM3900 Norton amplifier the input currents $I^+$ and $I^-$ should not go above 6 mA. As a result the

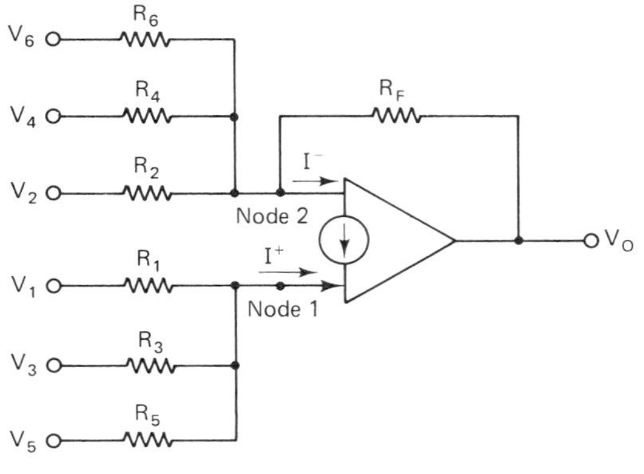


**Figure 6.27** Circuit for the closed-loop gain analysis of the Norton operational amplifier.

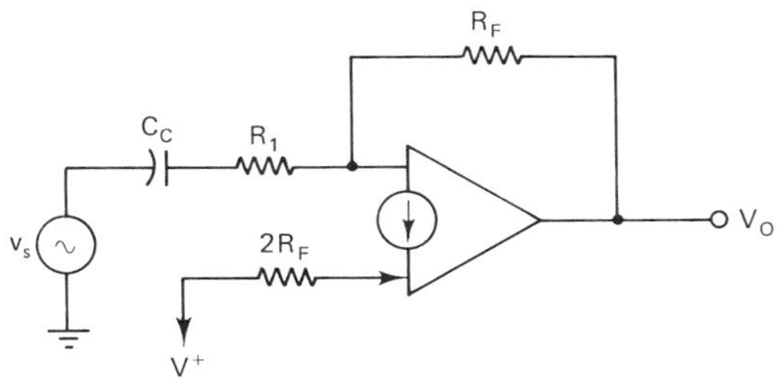


**Figure 6.28** Inverting a-c amplifier.

maximum input voltage will be limited solely by the requirement that neither input current exceed 6 mA. Therefore, with the choice of suitably large resistance values, very large input voltage levels can be accommodated by this circuit.

For a simple example of a Norton amplifier application, let us consider the inverting a-c amplifier shown in Figure 6.28. In this circuit the d-c or quiescent output voltage level will be set at

$$V_{O(\mathrm{DC})} = (V^+ - V_{BE})\frac{R_F}{2R_F} = \frac{V^+ - V_{BE}}{2} \simeq \frac{V^+}{2} \tag{6.103}$$

This setting of the quiescent output voltage to a point approximately midway between the positive supply and ground will allow for the maximum possible symmetrical output voltage swing. The a-c voltage gain will be $A_V = v_o/v_s = -R_F/R_1$. Values for $R_F$ in the range 100 kΩ to 10 MΩ will generally be satisfactory.

Another quad single-supply Norton operational amplifier is the MC3401 (Motorola). Its circuitry is similar to that of the LM3900 considered earlier. This amplifier has an open-loop voltage gain of 60 dB (min.) [66 dB (typ.)] and the 3-pF on-chip compensation capacitor results in a unity-gain frequency of 5.0 MHz. It can be operated from a single supply voltage over the range 5 to 18 V.

## 6.5 DUAL, QUAD, AND SINGLE SUPPLY OPERATIONAL AMPLIFIERS

There are many types of op amps available with various particular features such as low offset voltage, low input bias current, low d-c power supply current, wide bandwidth, high slewing rate, and low noise. The FET-input op amps discussed in previous sections offered some of these features.

Most op amps require both a positive and a negative d-c supply voltage for proper operation. There are some op amps, however, that can be operated from a single positive d-c supply voltage. Furthermore, the single d-c supply voltage can be as small as +5V. This can be a great convenience in many applications. Another category of op amps are the dual and the quad op amps. The dual op amp has two electrically independent op amps in one IC package, and the quad op amp offers the convenience of four independent op amps in the same package. In most cases, the op amps share the same silicon chip. Quad op amps are available in a 14-pin

DIP package. The input and output terminals of the four op amps are separate, but they do share common positive d-c supply and ground pins. Since the input, output, positive supply, and ground connections use up all of the available pins, such features as offset nulling pins and pins for an external compensation capacitor are not present.

Since the four op amps generally share a common silicon chip there will inevitably be some some amount of electrical coupling between the op amps. This *amplifier-to-amplifier coupling,* or channel separation, is very small and generally down in the range of about −120 dB.

An example of an IC that is a quad op amp and can also be used with just a single positive d-c supply voltage is the LM124/224/324 series (National Semiconductor). A schematic of this device is shown in the figure for problem 6.2 (Figure P6.2). This device can be operated from either single or dual supply voltages from as low as +3 V (or +1.5 and −1.5 V) to as high as +30 V (or +15 and −15 V). It is an internally compensated op amp with a unity gain frequency of 1 MHz, a slewing rate of 0.5 V/$\mu$s, and a d-c open-loop gain of 100 dB. The coupling between the four op amps on the chip is very small, the amplifier-to-amplifier coupling being down to about −120 dB.

Another interesting example of a quad op amp is the TL074/075 series (Texas Instruments) which comes in a 14-pin DIP package. Other devices in this same series are the TL072 which is a dual op amp in a mini-DIP 8-pin package, and the TL070 and TL071, also in 8-pin mini-DIP packages. These are not single supply op amps, but do offer the special feature of a JFET input stage. The JFET input stage uses a pair of P-channel JFETs that are biased by a current source and drive a current mirror active load. The second stage is a Darlington common-emitter configuration and has a current source active load. The output stage is a complementary push-pull emitter-follower with two diode-connected transistors used for the class AB biasing of the output stage to minimize the crossover distortion.

The JFET input stage results in a low input bias current of 30 pA and an offset current of 5 pA. The unity gain frequency is 3 MHz and the slewing rate is 13 V/$\mu$s. This device is also characterized as a low-noise op amp. This is due principally to the low input bias current which results in an input noise current of only 0.01 pA/$\sqrt{\text{Hz}}$. The equivalent input noise voltage is 18 nV/$\sqrt{\text{Hz}}$. The amplifier-to-amplifier coupling is rated at −120 dB for both the dual and the quad op amps.

An example of an all-bipolar transistor single supply quad op amp that offers a wide bandwidth and high slewing rate capability is the MC34074 (Motorola). This op amp has a unity gain bandwidth of 4.5 MHz and a slewing rate of 13 V/$\mu$s. It can operate with a single positive d-c supply voltage of as low as +3 V to up to a maximum of +44 V and comes in a 14-pin DIP package.

## 6.6 HIGH PERFORMANCE OPERATIONAL AMPLIFIERS

There are many op amps that exhibit various high performance characteristics such as very low input offset voltage (precision op amps), ultra-low input bias current, very wide bandwidth, very high slewing rate, very low quiescent current drain (micropower op amps), high voltage, and high output current. Examples of a few of these op amps will now be presented.

### 6.6.1 Precision Operational Amplifiers

A precision op amp is one that has a very low input offset voltage, generally less than 1 mV. Along with a low offset voltage, the temperature coefficient of the offset voltage is very small, generally less than 5 μV/°C. A good example of a precision op amp is the OP-07 (Precision Monolithics Inc.) which has an offset voltage of only 10 μV typical and 25 μV maximum. The temperature coefficient of the offset voltage is only 0.2 μV/°C typical and 0.6 μV/°C maximum, and the long-term drift of the offset voltage is specified as 0.2 μV/month typical and 1 μV/month maximum. This is a bipolar transistor op amp that uses laser-trimmed load resistances in the input differential amplifier stage to minimize the offset voltage. There are two offset null pins available for nulling out the offset voltage. A 20 kΩ potentiometer can be connected between the offset null pins with the wiper arm connected to the positive supply for the offset voltage cancellation.

Other examples of precision op amps are the LT1008C (Linear Technology Inc.), 3510 (Burr-Brown), LM108A/208A/308A series (National Semiconductor), LM11C (National Semiconductor), AD547 (Analog Devices), and the AD611KH (Analog Devices).

### 6.6.2 Ultra-Low Input Current Operational Amplifiers

Operational amplifiers with FET input stages will generally have input bias currents down in the range 10 to 100 pA. An op amp with an input current below 10 pA can be considered to have an ultra-low input bias current. A good example of such an op amp is the OPA104 series (Burr-Brown) with maximum input bias currents less than 1 pA at 25°C. The OPA104CM has an input bias current of a mere 75 fA (0.075 pA) maximum at 25°C. Associated with the ultra-low input bias currents is a very high input impedance. The difference-mode input impedance is $10^{14}$ Ω in parallel with a capacitance of 0.5 pF, and the common-mode input impedance is specified as $10^{15}$ Ω in parallel with a capacitance of 1.0 pF. This op amp also has a relatively low input offset voltage of 200 μV typical and 500 μV maximum with a temperature coefficient of 5 μV/°C typical and 10 μV/°C maximum.

### 6.6.3 Wide Bandwidth and Fast Slewing Operational Amplifiers

Most op amps have unity gain frequencies (or gain-bandwidth products) in the range 1 to 3 MHz with slewing rates in the range 0.5 to 5 V/μs. There are many op amps that can be characterized as very wide bandwidth and fast slewing op amps. A good example is the HA2539 (Harris Semiconductor) with a gain-bandwidth product of 600 MHz and a slewing rate of 600 V/μs. The NE5539 (Signetics) is a very wide bandwidth monolithic bipolar operational amplifier with a gain-bandwidth product of 1200 MHz at a gain of 7 and a slewing rate of 600 V/μs. The full-power bandwidth is 20 MHz with a peak-to-peak output of 3 V into a 150 Ω load. A very wide bandwidth hybrid op amp is the 3554 (Burr-Brown) with a gain-bandwidth product of 1700 MHz (at a gain of 1000) and a slewing rate of 1000 V/μs. The small-signal bandwidth is 22.5 MHz at a gain of 10; 7.25 MHz at a gain of 100; and 1.7 MHz at a gain of 1000. The full-power bandwidth is 19 MHz when driving a 100 Ω load with a 20 V peak-to-peak output voltage. Another wide bandwidth hybrid op amp is the LH0032 (National Semiconductor). This JFET-input op amp has a unity gain frequency of 70 MHz and a slewing rate of 500 V/μs along with an input impedance of $10^{12}$ Ω and an input bias current of just 10 pA.

A series of very wide bandwidth hybrid op amps are made by the Comlinear Corporation. The CLC200 features a 100 MHz 3 dB small-signal bandwidth for gains of 1 to 50. The rise time is 3.6 ns and the settling time for the output voltage to settle to within 0.02 % of the final value is 25 ns. The full-power bandwidth is 25 MHz for a 40 V peak-to-peak output voltage swing and 50 MHz for a 20 V peak-to-peak output, while delivering 100 mA to the load. The CLC210 combines a wide bandwidth with a large output voltage swing capability. It has a small-signal bandwidth of 70 MHz at a gain of 10, and it can supply 50 mA to a load with a 60 V peak-to-peak output voltage swing at a full-power bandwidth of 10 MHz. The CLC220 has a 200 MHz bandwidth with rise and fall times of only 1.6 ns and a 0.02% settling time of 12 ns. The slewing rate of this op amp is an impressive 8000 V/μs and it offers a full-power bandwidth of 50 MHz for a 40 V peak-to-peak output and 80 MHz for a 20 V peak-to-peak output.

### 6.6.4 High Voltage Operational Amplifiers

Most op amps have maximum total supply voltage ratings of no more than 36 V, and, correspondingly, a peak-to-peak output voltage swing that is limited to about 32 V. There are high voltage op amps available, however, with supply voltage ratings much greater than this. An example of a high voltage op amp is the 3584 (Burr-Brown) with a total supply voltage rating of 300 V and a maximum peak-to-peak output voltage swing of 290 V. This JFET input op amp offers a unity gain bandwidth of 7 MHz, a slewing rate of 150 V/μs, and a full-power bandwidth of 135 kHz.

## 6.7 UNITY-GAIN BUFFERS

A unity gain buffer is an amplifier with an internally set gain of unity and can be considered to perform basically similar to a voltage-follower op-amp circuit. The unity gain buffers can, however, offer substantial performance advantages over op-amp voltage-followers, especially in the areas of wide bandwidth and high slewing rate. Most unity gain buffers consist of a JFET source-follower input stage followed by one or two emitter-follower stages. An example of a unity gain buffer circuit is shown in Figure 6.29. The input stage is a JFET source-follower with a JFET current sink active load. This drives two cascaded emitter-follower stages, both of which use JFET current sink active loads. Diode-connected transistors, $Q_7$ and $Q_8$, to-

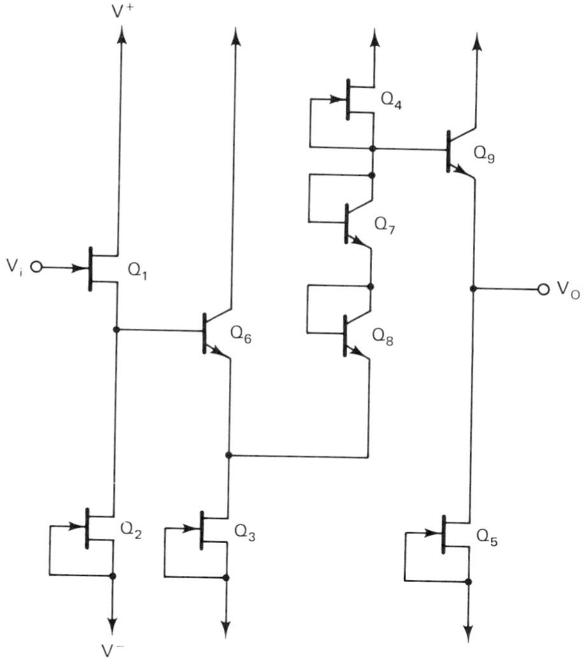


**Figure 6.29** Unity gain buffer

gether with the JFET current source transistor $Q_4$ are used for level shifting such that $V_o = 0$ when $V_i = 0$.

An example of a high performance unity gain buffer is the 3553 (Burr-Brown) which has a −3 dB bandwidth of 300 MHz and a slewing rate of 2000 V/$\mu$s, together with an input impedance of $10^{11}$ $\Omega$, an output current capability of 200 mA, and an output resistance of only 1 $\Omega$. The full-power bandwidth is 32 MHz for a 10 V peak output voltage swing. The voltage gain is 0.98 under no load conditions and 0.92 with a 50 $\Omega$ load. The output offset voltage is 50 mV (max.) with a temperature coefficient of 300 $\mu$V/°C. Another fast unity gain buffer is the LH0063 (National Semiconductor) with a bandwidth of 100 MHz and a slewing rate of 6000 V/$\mu$s with a 1000 $\Omega$ load. The input resistance is 10 $\Omega$, the output resistance is 6 $\Omega$, and it can supply 10 V to a 50 $\Omega$ load with a slewing rate of 2400 V/$\mu$s. The output offset voltage is 5 mV with a temperature coefficient of 50 $\mu$V/°C.

## PROBLEMS

**6.1.** (*Two-stage amplifier with active loads and complementary push-pull emitter-follower output stage*) Refer to Figure P6.1. Given: $V_{SUPPLY} = \pm 12$ V, $I_4 = 500$ $\mu$A, $I_Q = 10$ $\mu$A, $\beta_{NPN} = 100$ (min.), $\beta_{PNP} = 50$ (min.), and assume that $V_{BE} \simeq 0.6$ V wherever applicable.

**(a)** Find $R_3$ and $R_4$. (*Ans.*: $R_3 = 9.8$ k$\Omega$, $R_4 = 22.8$ k$\Omega$)

**(b)** What value should $R_{11}$ have in order that $V_o = 0$ when $V_i = 0$? (*Ans.*: 9.8 k$\Omega$)

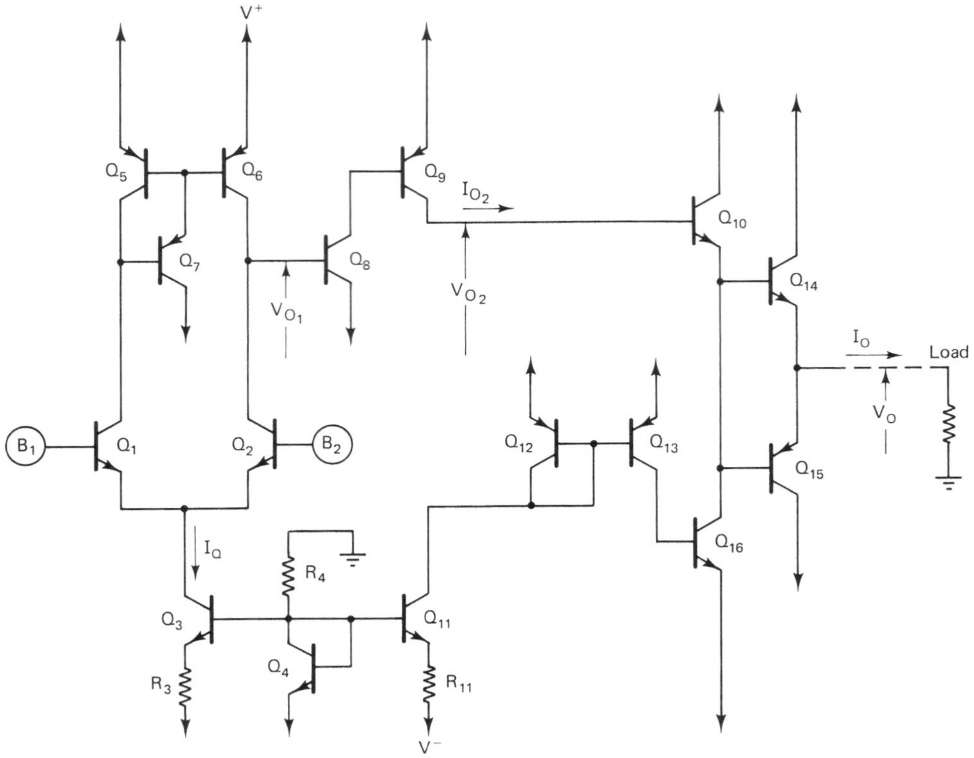


**Figure P6.1**

**(c)** Find $I_{\text{BIAS}}$. [*Ans.*: 50 nA (max.)]

**(d)** Find the input resistance, $R_i$. [*Ans.*: 1.0 MΩ (min.)]

**(e)** Find the input voltage range (common-mode). (*Ans.*: +11.3 to −11.1 V)

**(f)** If the emitter–base junction breakdown voltage, $BV_{EBO}$, is 7.0 V, what is the maximum allowable difference-mode voltage? (*Ans.*: ±7.6 V)

**(g)** Which input ($B_1$ or $B_2$) is the inverting input? Explain.

**6.2.** (*Operational-amplifier circuit—LM124 type*) Refer to Figure P6.2. Given: $V_S = +10$ V, $\beta_{\text{NPN}} = 100$ (min.), and $\beta_{\text{PNP}} = 50$ (min.).

**(a)** Find $R_6$, $R_5$, $R_{12}$, $R_{13}$, and $R_{10}$. (*Ans.*: $R_6 = 18.8$ kΩ, $R_5 = 18.4$ kΩ, $R_{12} =$ 30 kΩ, $R_{13} = 402$, $R_{10} = 30$ kΩ)

**(b)** Find $R_{SC}$ for a short-circuit current limit of 20 mA. (*Ans.*: $R_{SC} = 25$ Ω)

**(c)** Find $I_{\text{BIAS}}$. [*Ans.*: $I_{\text{BIAS}} = 1.2$ nA (max.)]

**(d)** Find the input resistance, $R_i$. [(*Ans.*: 83 MΩ (min.)]

**(e)** Find the unity-gain frequency, $f_u$. (*Ans.*: $f_u = 1.6$ MHz)

**(f)** Find the slew rate. (*Ans.*: $SR = 1.0$ V/μs)

**(g)** Find $A_{OL}$ at $f = 10$ kHz. (*Ans.*: $-j160$)

**(h)** Find the input (common-mode) voltage range. (*Ans.*: −0.5 to +8.5 V)

**(i)** Find the maximum output-voltage swing. (*Ans.*: +8.6 to 0 V)

**(j)** Find the quiescent power dissipation. (*Ans.*: 6.3 mW)

**(k)** For a closed-loop gain of 30, find the 3-dB bandwidth and the rise time. (*Ans.*: 53 kHz, 6.6 μs)

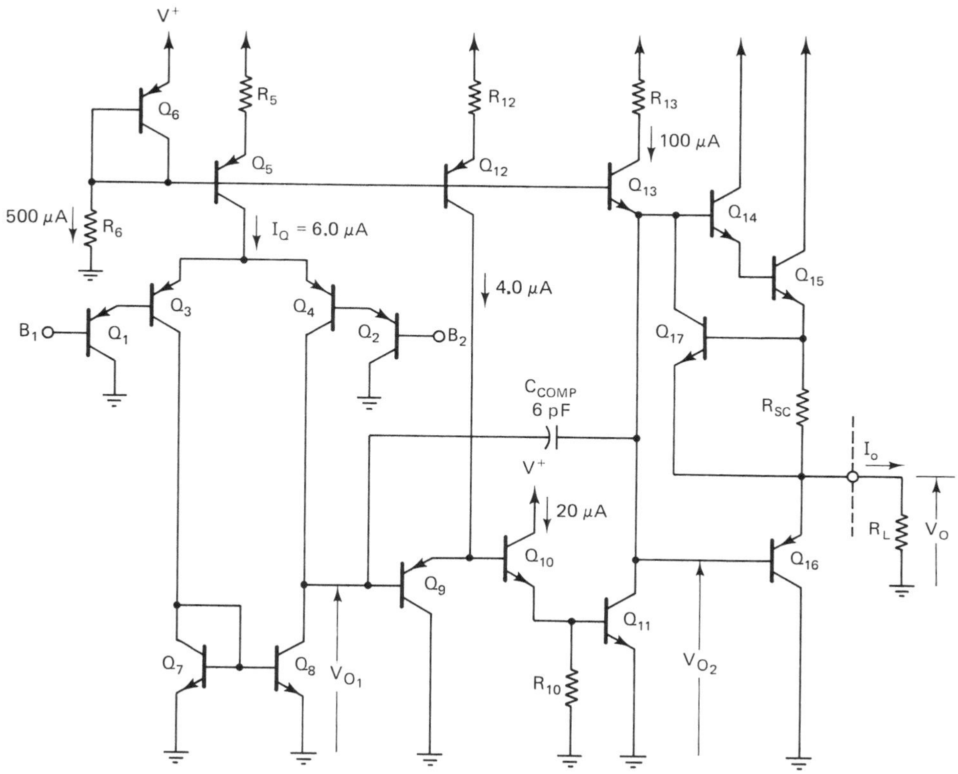


**Figure P6.2**

(l) For an output-voltage peak swing of 3.0 V, at what frequency will the op amp become slew-rate limited? (*Ans.*: 53 kHz)

**6.3.** (*Operational-amplifier circuit*) Refer to Figure P6.3. Given: Operational-amplifier circuit with $\beta = 50$ (min.) for all transistors, $V_S^+ = 15$ V, $V_S^- = -15$ V, $I_{12} = 1.0$ mA, $I_Q = 25\ \mu$A, $n = 1.5$ (factor in exponent for the base current), and $V_{A_{\text{NPN}}} = V_{A_{\text{PNP}}} = 250$ V (the Early voltage).

**(a)** Find $R_1$ and $R_2$. (*Ans.*: $R_1 = 29.3$ kΩ, $R_2 = 3.69$ kΩ)

**(b)** Find $g_f$ (differential amplifier). (*Ans.*: $g_f = 250\ \mu$S)

**(c)** Find the input-voltage range (common mode) and the maximum output-voltage swing. (*Ans.*: +14.1 to −14.0 V, +13.4 to −13.3 V)

**(d)** Find the input bias current and the input resistance (difference mode). [*Ans.*: 250 nA (max.), 200 kΩ (min.)]

**(e)** Find the quiescent current drain and the power dissipation (i.e., with $V_o = 0$). (*Ans.*: 1.05 mA, $P_{d(Q)} = 31.5$ mW)

**(f)** Find the input conductance of $Q_6$, the output conductance of $Q_4$ and $Q_2$, and the total conductance driven by the first stage. [*Ans.*: $g_{i_6} = 133$ nS (max.); $g_{o_4} = 50$ nS, $g_{o_2} = 50$ nS; $g_{\text{total}} = 233$ nS (max.)]

**(g)** Find the voltage gain of the first stage, $v_{o_1}/v_i$. [*Ans.*: $A_{v_1} = v_{o_1}/v_i = 2146$ (min.)]

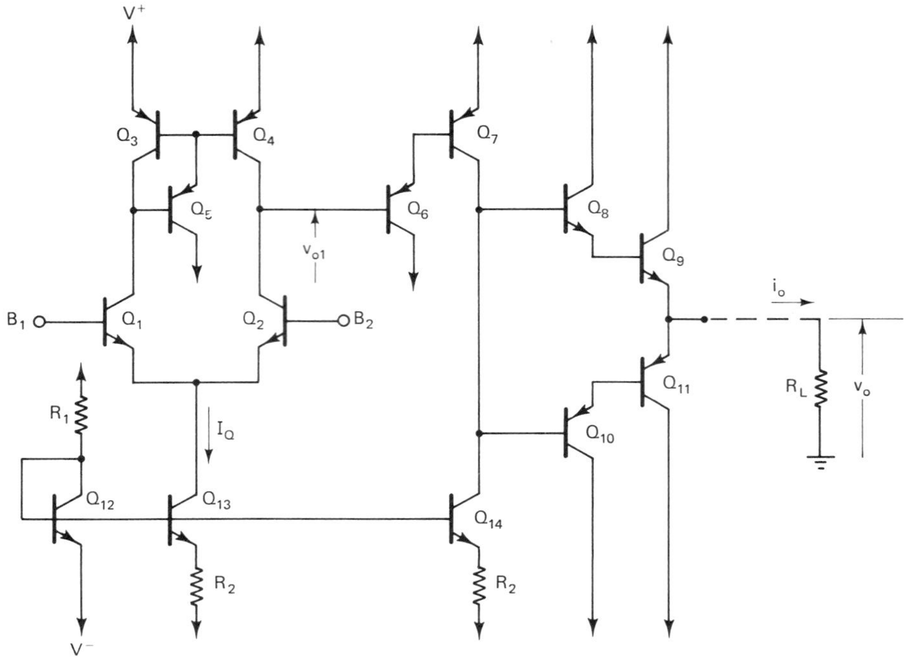


**Figure P6.3**

**(h)** If $R_L = 1.0\ \text{k}\Omega$, find the voltage gain of the second stage [assume that emitter-follower gain ($Q_8$ and $Q_9$ and $Q_{10}$ and $Q_{11}$) is approximately unity]. [*Ans.*: $A_{V_2} = v_o/v_{o_1} = 1250$ (min.)]

**(i)** Find the open-loop gain of the op amp, $A_{OL}(0)$. ($R_L = 1.0\ \text{k}\Omega$) Express the result numerically and in decibels. [*Ans.*: $A_{OL}(0) = 2.68 \times 10^6$ (min.), 128.6 dB (min.)]

**(j)** Show that the voltage gain of the emitter-follower output stage ($Q_{8,9}$ and $Q_{10,11}$) will be close to unity for output voltage swings of about 1 V or greater.

**(k)** Show that $B_1$ is the inverting input and $B_2$ is the noninverting input of the op amp.

**(l)** If a Darlington differential amplifier as shown in Figure 6.9 is used, describe how the answers above would change (qualitatively and quantitatively).

**(m)** If the output stage is modified by the addition of $R_3$, $R_4$, and $Q_{15}$ as shown in Figure 6.11, find $R_3$ and $R_4$ if $I_9 = I_{11} = I_Q$ (typ.) under quiescent conditions and $I_{R_3} = I_{R_4} = I_Q/2$. Assume that $V_{BE} = 600$ mV at $I_C = I_Q$ for the NPN and PNP transistors and $\beta = 100$ (typ.). (*Ans.*: $R_3 = 127\ \text{k}\Omega$, $R_4 = 46\ \text{k}\Omega$)

**(n)** If the output stage is further modified by the addition of $R_{CL}$, $Q_{16}$, and $Q_{17}$ as shown in Figure 6.13, find $R_{CL}$ for a current limit of 25 mA. (*Ans.*: $R_{CL} = 24\ \Omega$)

**(o)** If the maximum supply voltage is $\pm 13$ V find the corresponding maximum power dissipation if $I_{CL} = 25$ mA. (*Ans.*: $P_{d(\text{MAX})} = 495$ mW)

**(p)** Find the required value for the compensation capacitor (to be connected between $B_6$ and $C_7$) for a unity gain frequency of 1.0 MHz. (*Ans.*: $C_{\text{COMP}} = 80$ pF)

**(q)** Find the slewing rate and the full-power bandwidth (FPBW) for the above for a peak output swing of $\pm 10$V. (*Ans.*: 0.3125 V/$\mu$s, 4.97 kHz)

**(r)** Find the output conductance of the current source that biases the differential amplifier ($Q_{13}$), and the corresponding output resistance. (*Ans.*: $g_{o_{13}} = 22$ nS, $r_{o_{13}} = 44.1$ M$\Omega$)

**(s)** Find the common-mode transfer conductance, $g_{f_{CM}}$. (*Ans.*: $g_{f_{CM}} = 11$ nS)

**(t)** Find the common-mode gain, and the common-mode rejection ratio assuming that the offset voltage of the $Q_1$–$Q_2$ pair is 1.0 mV ($R_L = 1.0$ k$\Omega$). [*Ans.*: $A_{V_{CM}} = 2.3$ (min.), CMRR = 121 dB]

**(u)** The proper operation of the op-amp circuit is dependent on the matching of circuit components that is inherent in devices on IC chips. The components are made very close together on the same IC chip and have undergone exactly the same processing so that a very high degree of matching should be expected. If there were *exact* matching, the input offset voltage of this op amp would be zero. Find the contribution to the input offset voltage ($V_{OS}$) of the following component mismatches:

**(1)** $Q_1$ and $Q_2$: a 1-mV offset voltage. (*Ans.*: 1 mV)

**(2)** $Q_3$ and $Q_4$: a 1-mV offset voltage. (*Ans.*: 1 mV)

**(3)** $Q_5$ and $Q_6$: a 10% $\beta$ mismatch. (*Ans.*: 2 $\mu$V)

**(4)** $Q_7$ and $Q_{3,4}$: a 10% $\beta$ mismatch. (*Ans.*: 2 $\mu$V)

**(5)** A mismatch between the emitter resistors ($R_2$) of $Q_{13}$ and $Q_{14}$ sufficient to cause the current of $Q_{14}$ to differ from that of $Q_{13}$ by 10%. (*Ans.*: 2 $\mu$V)

What is the corresponding resistor mismatch ($R_2$)? (*Ans.*: 14%)

**(v)** In the voltage-follower configuration, the input impedance seen looking into the amplifier will be essentially the common-mode input impedance because the voltage that is present at the two input terminals will be essentially the same. Find the common-mode input impedance at zero frequency. [*Ans.*: $Z_{i_{CM}}(0) = R_{i_{CM}} = 4.55$ G$\Omega$ (min.)]

**(w)** If $C_{CB} = 1.0$ pF and $C_{CS} = 1.0$ pF, such that $C_i = 2.0$ pF, find $Z_{i_{CM}}$ at 1.0 kHz. (*Ans.*: $Z_{i_{CM}} = -j80$ M$\Omega$)

Find the breakpoint frequency for $Z_{i_{CM}}$. (*Ans.*: $f = 16.7$ Hz)

**(x)** Find the equivalent input noise voltage (spectral density). Use $r_{bb'} = 200\ \Omega$. Find the total rms noise over a 20-kHz 3-dB system bandwidth and the input signal level needed to produce a 10:1 signal-to-noise ratio. (*Ans.*: 8.3905 nV/$\sqrt{\text{Hz}}$, 1.49 $\mu$V, 14.9 $\mu$V)

**(y)** This op amp is driven from a signal source that has a 1000-$\Omega$ source resistance. Using the noise voltage obtained in part (x) together with the additional noise due to the 1000-$\Omega$ source resistance, find the input signal level required to produce a 10:1 signal-to-noise ratio with a 20-kHz 3-dB bandwidth. (*Ans.*: 16.5 $\mu$V)

What would be the result in the case of a 10-k$\Omega$ source resistance? (*Ans.*: 27.4 $\mu$V)

**(z)** The *noise figure* (NF) of an amplifier is the factor (expressed in decibels) by which the amplifier decreases the signal-to-noise ratio of the system below what would be obtained if the amplifier were noiseless. Find the noise figure for the two preceding problems. (*Ans.*: 0.89 dB, 5.28 dB)

**6.4.** (*JFET-input operational amplifier*) Given: An op amp has an input stage that is a JFET differential amplifier with a current mirror active load. The JFETs have $I_{DSS} = 400$ $\mu$A and $V_P = -4.0$ V. The differential amplifier is biased by a current sink of $I_Q = 200$ $\mu$A.

**(a)** Find the compensation capacitance needed for a unity-gain frequency of $f_u = 2.5$ MHz. (*Ans.*: 6.366 pF)

**(b)** Find the slewing rate ($SR$). (*Ans.*: 31.4 V/$\mu$s)

**(c)** Find the full-power bandwidth (FPBW) for a peak output voltage swing of ±10 V. (*Ans.*: 500 kHz)

**(d)** If the JFETs have a channel-length modulation coefficient of $1/V_A = 1/(50\text{ V})$, find the maximum possible gain that can be obtained from this JFET input stage (with a bipolar current mirror active load). Compare this result with what is generally available with a bipolar transistor input stage. (*Ans.*: 50)

**(e)** Operational amplifiers with JFET input stages often have much higher slewing rates than those of all-bipolar operational amplifiers. Explain.

**(f)** If the input bias current of the JFET input operational amplifier is 1.0 nA (max.) at 25°C and doubles for every 11°C temperature rise, what will $I_{BIAS}$ be at 100°C? [*Ans.*: 113 nA (max.)]

**(g)** Repeat part (f) for 125°C. [*Ans.*: 545 nA (max.)]

**(h)** The input bias current of a JFET-input op amp is very small at 25°C, but increases at an exponential rate with increasing temperature. In contrast, the input bias current of an all-bipolar amplifier is much higher at 25°C, and actually decreases very slowly with increasing temperature. Explain.

**6.5.** An op amp has $A_{OL}(0) = 100$ dB, $f_1 = 12$ Hz, $f_2 = 2.5$ MHz, and $f_3 = 5.0$ MHz.

**(a)** Find the unity-gain frequency $f_u$. (*Ans.*: 1.2 MHz)

**(b)** If the breakpoint frequencies given above are for $C_{COMP} = 40$ pF, find the corresponding values for a compensation capacitance of 200 pF. (*Ans.*: 2.4 Hz, 2.5 MHz, 5.0 MHz)

**6.6.** An op amp has a slewing rate of 2.0 V/μs. Find the full-power bandwidth (FPBW) for a peak output voltage swing of ±10 V. (*Ans.*: 31.8 kHz)

**6.7.** An op amp has a slewing rate of 2.0 V/μs. Find the rise time for an output voltage of 10 V amplitude resulting from a rectangular pulse input, if the op amp is slewing rate limited. (*Ans.*: 4.0 μs)

**6.8.** An op amp has a slewing rate of 1.0 V/μs when $C_{COMP} = 25$ pF. Find the slewing rate when $C_{COMP}$ is increased to 100 pF. (*Ans.*: 0.25 V/μs)

**6.9.** MOSFET-input op amps use an internal pair of gate protection diodes, connected back to back in series, across the input terminals of the circuit. What is the function of these diodes? Do they have any effect on the characteristics of the op amp? Explain.

**6.10.** Given: For the op-amp circuit for this problem (Figure P6.10), $R_1 \parallel R_2 = R_S$.

**(a)** Show that the signal-to-noise ratio (SNR) at the amplifier output will be given by

$$(\text{SNR})^2 = \left[\frac{v_{o\,(\text{signal})}}{v_{o\,(\text{noise})}}\right]^2 = \frac{v_S^2}{[(v_{i\,(\text{noise})})^2 + 2(i_{i\,(\text{noise})}R_S)^2 + 8kTR_S](\pi/2)BW}$$

where $v_{i\,(\text{noise})}$ and $i_{i\,(\text{noise})}$ are the equivalent input noise voltage and current of the op amp, respectively.

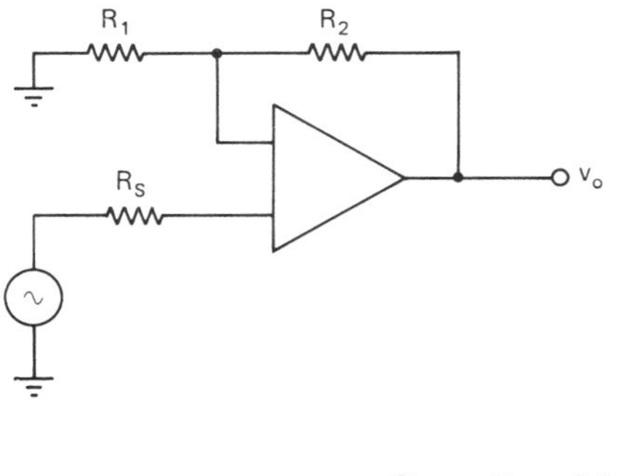


**Figure P6.10**

**(b)** If $v_{i\,(\text{noise})} = 20\ \text{nV}/\sqrt{\text{Hz}}$, $i_{i\,(\text{noise})} = 0.1\ \text{pA}/\sqrt{\text{Hz}}$, and $R_S = 100\ \Omega$, find $v_s$ for a 10:1 SNR. The 3-dB bandwidth is 10 kHz. (*Ans.*: 25.2 $\mu$V)

**(c)** Repeat for $R_S = 1000\ \Omega$. (*Ans.*: 26.0 $\mu$V)

**(d)** Repeat for $R_S = 10\ \text{k}\Omega$. (*Ans.*: 33.7 $\mu$V)

**(e)** Repeat for $R_S = 30\ \text{k}\Omega$. (*Ans.*: 46.5 $\mu$V)

**(f)** Repeat for $R_S = 100\ \text{k}\Omega$. (*Ans.*: 77.2 $\mu$V)

**6.11.** (*Unity gain buffer*) In Figure 6.29 a unity gain buffer (or voltage-follower) circuit is shown. This buffer is characterized by a very high input impedance, a low output impedance, and a very wide bandwidth along with a very high slewing rate. Assume that all transistors of the same type are identical, except that transistor $Q_3$ has twice the channel width as the other JFETs. Also assume that the base currents of the bipolar transistors are much smaller than the saturated drain currents of the JFETS.

**(a)** Show that when the input voltage $V_i$ is zero, the output voltage will be zero.

**(b)** Given that the current gain of the bipolar transistors is 200 and for the JFETs: $V_P = -2$ V, $V_A = 100$ V, and $I_{DSS} = 5.0$ mA, except for $Q_3$ for which $I_{DSS} = 10$ mA. Find the voltage gain of this buffer under no load conditions. (*Ans.*: 0.9891)

**(c)** Find the output impedance. (*Ans.*: 5.08 $\Omega$)

**(d)** Find the voltage gain of this buffer when driving a load resistance of 100 $\Omega$. (*Ans.*: 0.9413)

**(e)** Find the quiescent supply current. (*Ans.*: 20 mA)

**(f)** Given that: $C_{gs} = 3$ pF, $C_{gd} = 3$ pF, $C_{b'e} = 10$ pF, and $C_{cb'} = 2$ pF, find the 3 dB bandwidth. (*Ans.*: 159 MHz)

**(g)** Repeat parts (b) through (f) for $I_{DSS} = 1.0$ mA for all JFETs, except for $Q_3$ for which $I_{DSS} = 2.0$ mA. (*Ans.*: 0.9891, 25.4 $\Omega$, 0.7888, 4.0 mA, 31.8 MHz)

**(h)** Find the input capacitance and the 3 dB bandwidth of the system when driven from a 10 k$\Omega$ source. (*Ans.*: 3 pF, 5.3 MHz)

## REFERENCES


CONNELLY, J. A., *Analog Integrated Circuits*, Wiley, 1975.

COWLES, L. G., *Sourcebook of Modern Transistor Circuits*, Chap. 7, Prentice-Hall, 1976.

FITCHEN, F. C., *Electronic Integrated Circuits and Systems*, Van Nostrand Reinhold, 1970.

GLASER, A. B., and G. E. SUBAK-SHARPE, *Integrated Circuit Engineering*, Addison-Wesley, 1977.

GRAEME, J. G., G. E. TOBEY, and L. P. HUELSMAN, *Operational Amplifiers—Design and Applications*, McGraw-Hill, 1971.

GRAY, P. R., and R. G. MEYER, *Analysis and Design of Analog Integrated Circuits*, Wiley, 1984.

GREBENE, A. B., *Analog Integrated Circuit Design*, Van Nostrand Reinhold, 1972.

HAMILTON, D. J., and W. G. HOWARD, *Basic Integrated Circuit Engineering*, McGraw-Hill, 1975.

ROBERGE, J. K., *Operational Amplifiers*, Wiley, 1975.

SEDRA, A. S., and K. C. SMITH, *Microelectronic Circuits*, Holt, Rinehart and Winston, 1982.

SEVIN, L. J., JR., *Field-Effect Transistors*, McGraw-Hill, 1965.

WALLMARK, J. T., and H. JOHNSON, *Field-Effect Transistors*, Prentice-Hall, 1966.

# 7

# VOLTAGE COMPARATORS

A comparator is an integrated circuit as shown in Figure 7.1 that is used to compare two input voltages $V_1$ and $V_2$ and to produce an output voltage of $V_o = V_H$ if $V_1 > V_2$ and $V_o = V_L$ if $V_1 < V_2$, where $V_H$ and $V_L$ are two fixed output-voltage levels. In Figure 7.2 the $V_o$ versus $V_i = V_1 - V_2$ transfer characteristics of a comparator are shown. Depending on the particular type of comparator and the supply voltage levels, $V_H$ and $V_L$ may be of opposite polarity with $V_H$ positive and $V_L$ negative, or they both may be positive, or they both may be negative.

The open-loop gain of comparators are usually very large, although generally less than those of operational amplifiers. Low-frequency voltage gains in the range 3000 to 100,000 are typically the case for comparators. As a result, the required change of input voltage to produce a transition from one output state to the opposite state is very small, generally in the range of only 0.1 to 3 mV. As a consequence of this very small transition voltage range, the output of the comparator will virtually always be saturated at either $V_L$ or $V_H$.

A comparator can be thought of as being a simple one-bit analog-to-digital (A/D) converter, producing a digital "1" output (i.e., $V_O = V_H$) if the analog input voltage $V_1$ is above the reference level of $V_{REF} = V_2$. A digital "0" (i.e., $V_O = V_L$) will result if the input voltage level falls below the reference level.

In many respects comparators are very similar to op amps, and indeed op amps can be used as comparators. A comparator is, however, designed specifically to operate under open-loop conditions, basically as a switching device. An op amp, on the other hand, is almost always used in a closed-loop configuration and is usually operated as a linear amplifier.

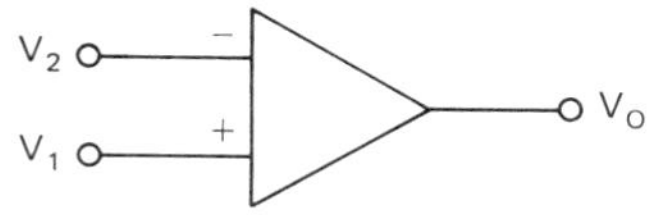


**Figure 7.1** Voltage comparator.

Being designed to be used in a closed-loop configuration, the frequency response characteristics of an op amp are generally tailored to ensure an adequate measure of stability against an oscillatory type of response. This will result in a sacrifice being made in the bandwidth, rise time, and slewing rate of the device. In contrast to this, since a comparator operates as an open-loop device, no sacrifices have to be made in the frequency response characteristics so that a very fast response time can be obtained.

An op amp is designed to produce a zero output voltage when the difference mode input signal is nominally zero. A comparator, in contrast, operates between two fixed output-voltage levels so that the output voltage for zero input voltage will generally be either $V_H$ or $V_L$, depending on the polarity of the input offset voltage.

The output voltage of an op amp will saturate at levels that are generally about 1 or 2 V away from the positive and negative power supply voltage levels. The comparator output is often designed to provide some degree of flexibility in fixing the high- and low-state output voltage levels, and for ease in interfacing with digital logic circuits.

There are many applications of comparators. These include pulse generators, square-wave and triangular-wave generators, pulse-width modulators, level detectors, zero-crossing detectors, pulse regenerators, line receivers, limit comparators, voltage-controlled oscillators, A/D converters, and time-delay generators.

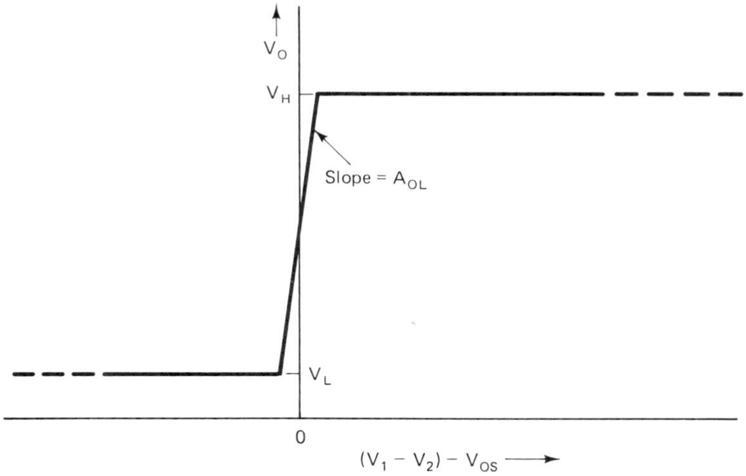


**Figure 7.2** Comparator transfer characteristic.

## 7.1 COMPARATOR CHARACTERISTICS

The most important characteristic of a comparator is generally the response time or propagation delay time. This is the time between the input-voltage transition and some specified point on the output-voltage transition, as shown in Figure 7.3. A commonly used point is the 50% point, although sometimes the 90% point is specified. For the response time measurement the noninverting input terminal is usually grounded (i.e., $V_{REF} = 0$). The input voltage is applied to the inverting input terminal and is such that a rapid transition is made from some level such as +100 mV or −100 mV to a voltage level of opposite polarity, the magnitude of which is the input overdrive voltage. Input overdrive voltages of 2, 5, 10, 20, and 100 mV are commonly used.

Response times of voltage comparators generally range from about 1 μs down to as fast as about 10 ns. The open-loop voltage gain of comparators is generally in the range 3000 to 100,000 so that the input voltage swing necessary to produce the output-voltage transition between the two saturated states will be in the range of about 0.1 to 3 mV.

The offset voltage $V_{OFFSET}$ or $V_{OS}$ of a comparator is another important specification since it results in a shift in the transition point for the input voltage from $V_{REF}$ to $V_{REF} + V_{OFFSET}$. The offset voltage for comparators will typically be in the range 1 to 10 mV.

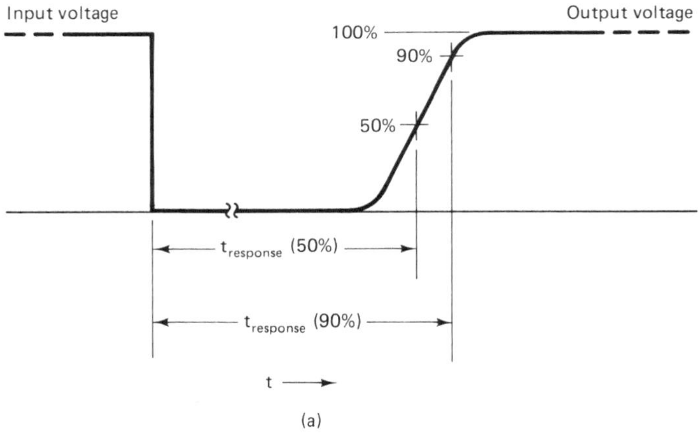


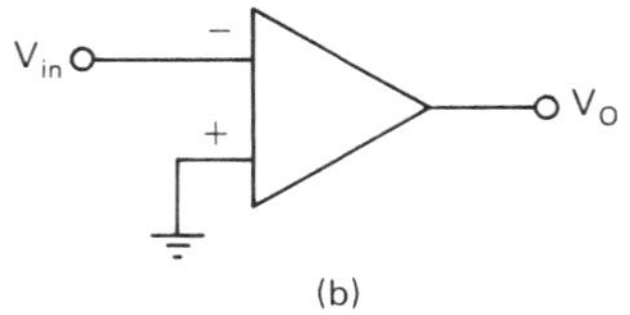


**Figure 7.3** Comparator response time: (a) input and output voltage waveforms; (b) test circuit.

## 7.2 COMPARATORS WITH POSITIVE FEEDBACK

Since comparators are not used as linear amplifiers, but rather as switching-mode devices, positive feedback can be used advantageously to increase the gain of the circuit and to produce some degree of hysteresis in the $V_O$ versus $V_i$ transfer characteristics. The increased gain will reduce the input voltage swing necessary to produce the output voltage transition to negligibly small values.

A comparator circuit with positive feedback is shown in Figure 7.4a, and the transfer characteristics are presented in Figure 7.4b. The input-voltage level at which the low-to-high output transition takes place is

$$V_{i(L-H)} = \frac{V_{\mathrm{REF}} R_1 + V_{\mathrm{L}} R_2}{R_1 + R_2} \tag{7.1}$$

and the input-voltage level at which the high-to-low transition occurs is

$$V_{i(H-L)} = \frac{V_{\mathrm{REF}} R_1 + V_H R_2}{R_1 + R_2} \tag{7.2}$$

The width of the hysteresis loop $V_W$ is

$$V_W = \frac{(V_H - V_L) R_2}{R_1 + R_2} \tag{7.3}$$

For many applications the presence of this hysteresis loop is desirable to produce a more definitive switching action of the comparator with less "contact bounce" or "chattering," as illustrated in Figure 7.4c. Note that in Figure 7.4d, which represents the case when hysteresis is present, if the width of the hysteresis loop $V_W$ is made larger than the peak-to-peak fluctuation in the input voltage $V_i$, a definitive switching action occurs without the intermittent switch "chattering" that would otherwise be present.

The use of positive feedback to produce a hysteresis loop and to increase the closed-loop gain is especially useful for input voltages that change very slowly near the reference voltage level. If the comparator is part of some larger feedback control system there could also be a problem with an oscillatory type of response if hysteresis were not used.

For a very simple example of this problem, consider the case of a comparator that is part of a thermostatic control system. The output of the comparator is used to control the heating element or system and the input voltage is proportional to the temperature. Starting with temperatures below the thermostat set point, the comparator input voltage will be less than the reference voltage and the comparator output will be in the high state, turning the heating system on. The temperature rises producing a corresponding increase in the comparator input voltage. In the absence of hysteresis, as soon as the input voltage exceeds the reference voltage, the comparator output will be switched to the low state, thereby turning the heating system off. The resulting decrease in the temperature will very soon cause the heater to be turned back on again. This will therefore result in a rapid off–on cycling of the system, continuing indefinitely at a rapid rate as the temperature fluctuates very

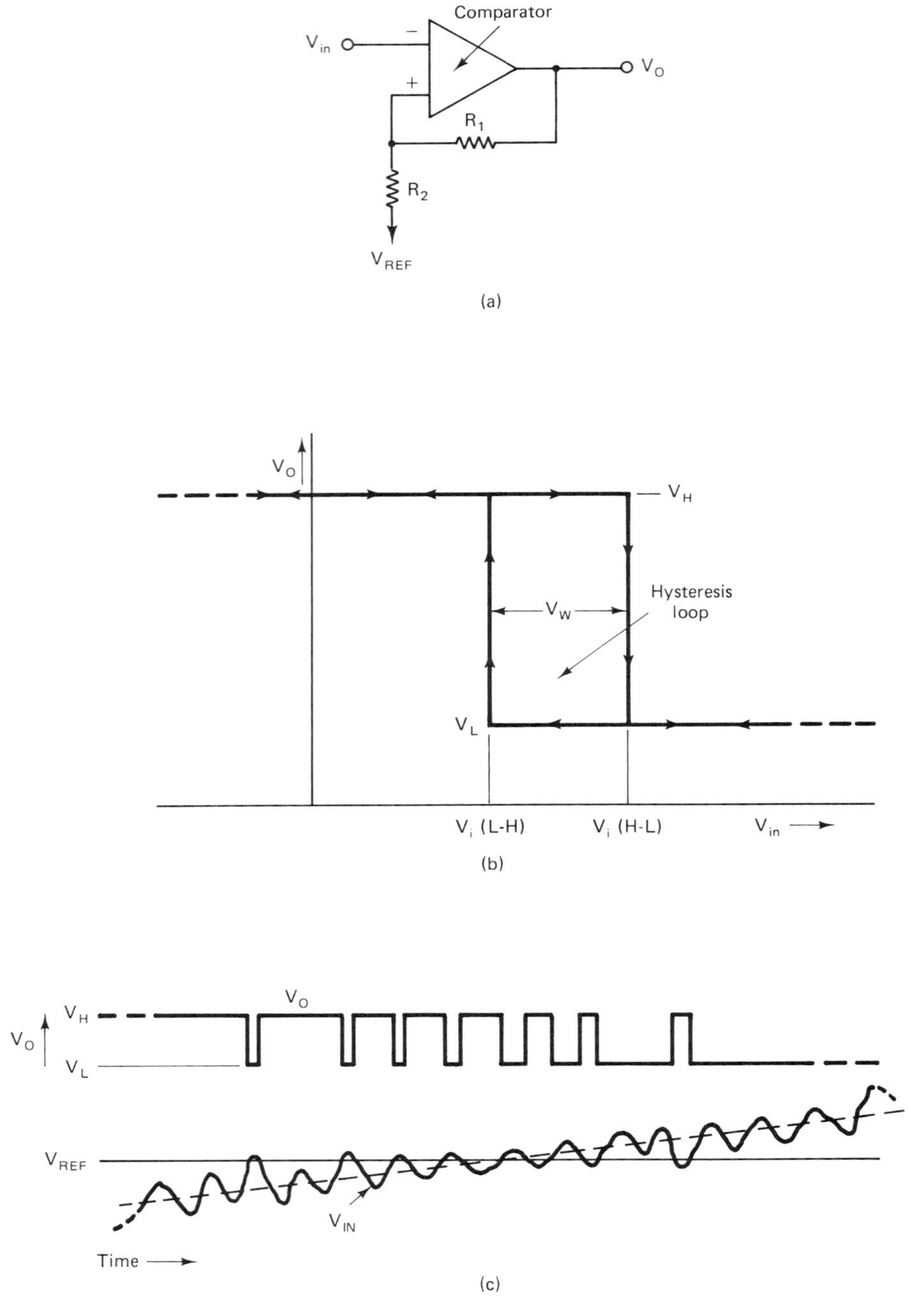


**Figure 7.4** Comparator with positive feedback: (a) circuit diagram; (b) transfer characteristics; (c) comparator switching characteristics without hysteresis (no positive feedback); (d) comparator switching characteristics with hysteresis (positive feedback present).

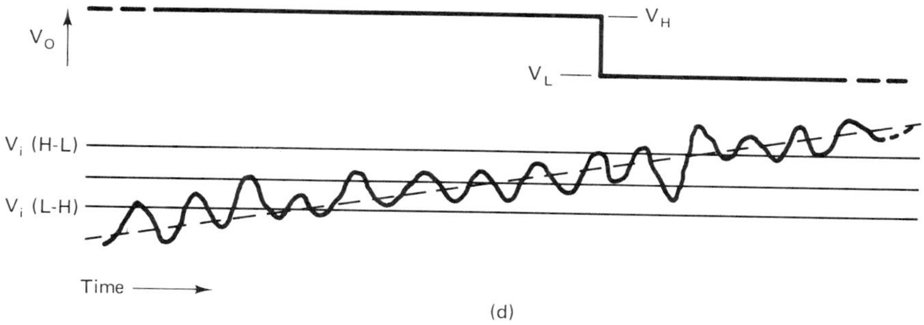


**Figure 7.4** *(cont.)*

slightly above and below the set point. This rapid "off–on" switching is generally very undesirable. It can be very greatly reduced, however, by the introduction of a hysteresis loop of suitable width.

## 7.3 COMPARATOR CIRCUITRY

The internal circuitry of comparators is in many respects very similar to that of op amps, especially with respect to the differential-amplifier input stages. The major differences occur in the output stages.

For an example of a fairly simple comparator circuit, we will examine the LM139/239/339 series (National Semiconductor). These are "quad" comparators, which means that there are four electrically independent comparators in one package (DIP or flat package).

A circuit diagram of this comparator is presented in Figure 7.5. It may be

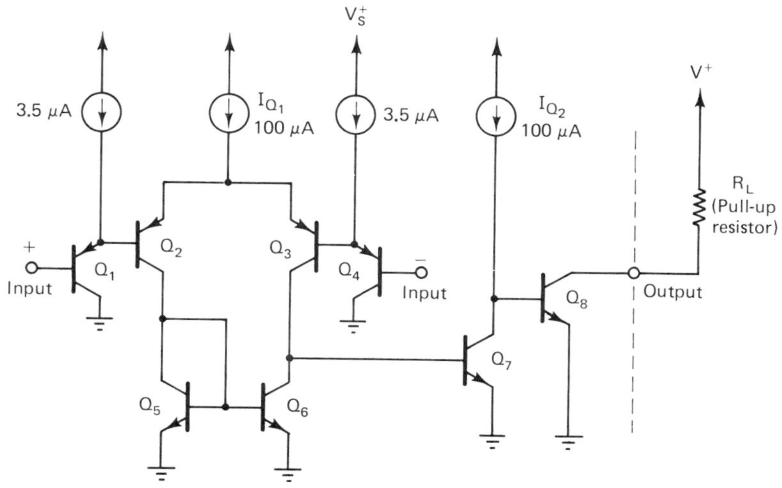


**Figure 7.5** Comparator circuit: LM139 type. (National Semiconductor).

noted that the input stage is very similar to that of the LM124/224/324 operational amplifiers (see Figure P6.2). Like these op amps, this comparator can be operated from a single positive supply voltage, although operation from a dual supply is also possible. This device can operate from supply voltages as low as 2.3 V to as much as 36 V (or ±18 V if operated from a dual supply).

### 7.3.1 Circuit Description: LM139

The differential-amplifier input stage consists of $Q_1$, $Q_2$, $Q_3$, and $Q_4$ connected as a Darlington (i.e., CC–CE) differential amplifier. The quiescent current of $Q_2$ and $Q_3$ is 50 $\mu$A each, and that of $Q_1$ and $Q_4$ will be 3.5 $\mu$A + (50 $\mu A/\beta_{PNP}$). Note that the 3.5-$\mu$A current sources will ensure that the quiescent currents of $Q_1$ and $Q_4$ will always be at least 3.5 $\mu$A. As a result the current gain ($\beta$) of these two transistors, as well as the dynamic transconductance ($g_m$), will always be maintained at an acceptably high value.

Transistors $Q_5$ and $Q_6$ constitute a current mirror active load for the input stage. The second gain stage is comprised of $Q_7$ in a common-emitter configuration with a current source (100 $\mu$A) active load. The third gain stage, which is also the output stage, consists of $Q_8$, which is also connected in a common-emitter configuration. Note that $Q_8$ has an "uncommitted" collector. In the actual operation of this comparator there will generally be a load resistor (also called a "pull-up" resistor) connected from the output (i.e., the collector of $Q_8$) to some positive supply voltage, which can be the comparator supply voltage, $V_S^+$, although this is not necessarily always the case.

This comparator is usually operated open-loop. Since the open-loop gain is very large, operation in this mode will usually mean that the second- and third-stage transistors will be driven back and forth between cutoff and saturation, passing very quickly through the active region.

If the input voltage is such that the −input is *higher* in voltage than the +input, this will cause the base current of $Q_7$ to be cut off, thus reducing the collector current of $Q_7$ to zero. This will cause all of the current $I_{Q2}$ (100 $\mu$A) to become the base drive of $Q_8$. The output transistor $Q_8$ will thus be turned on, and be in either the active region or the saturation region.

If the collector of $Q_8$ is connected to a positive supply voltage $V^+$ via a pull-up resistor, $R_L$, the saturation current will be

$$I_{C(SAT)} = \frac{V^+ - V_{CE(SAT)}}{R_L} \simeq \frac{V^+}{R_L} \tag{7.4}$$

If $\beta \times 100\ \mu A \geqslant I_{C(SAT)}$, $Q_8$ will be in the saturation region with $I_C = I_{C(SAT)}$. This will be the usual case. If, however, $\beta \times 100\ \mu A < I_{C(SAT)}$, then $Q_8$ will be in the active region with $I_C = \beta \times 100\ \mu A$. If $Q_8$ is driven into saturation, the output voltage will be given by $V_O = V_{CE_8} = V_{CE(SAT)}$. Therefore, the output voltage in the "low state" will be close to ground potential. If, however, $Q_8$ is not in saturation, the output voltage will be given by $V_O = V_{CE_8} = V^+ - I_C R_L = V^+ - (\beta \times 100\ \mu A)R_L$, and will thus be dependent on the current gain value of $Q_8$.

If the input voltage is such that the −input is *lower* in voltage than the +input, the current through $Q_2$, and therefore that through $Q_5$ and $Q_6$, will be lower than the current through $Q_3$. The difference between $I_{C_3}$ and $I_{C_6}$ ($\simeq I_{C_5} \simeq I_{C_2}$) will become the base current of $Q_7$. If the base current of $Q_7$ is such that $\beta_7 I_{B_7} > 100\ \mu A$, $Q_7$ will be in saturation. If $Q_7$ is in saturation, $Q_8$ will be cut off and the output voltage will become $V_O = V_{C_8} \simeq V^+$. The "high-state" output voltage will be essentially equal to $V^+$.

To switch the output voltage from the low state to the high state, the required change in the base current of $Q_7$ is thus from $I_{B_7} = 0$, corresponding to $Q_7$ cutoff and $Q_8$ in saturation, to $I_{B_7} = 100\ \mu A/\beta_7$, corresponding to $Q_7$ in saturation and $Q_8$ in cutoff. The change in $I_{B_7}$ in terms of the change in the input voltage, $\Delta V_i$, will be given by

$$\Delta I_{B_7} = \Delta I_{C_3} - \Delta I_{C_6} \simeq \Delta I_{C_3} - \Delta I_{C_5} \simeq \Delta I_{C_3} - \Delta I_{C_2} = 2g_f\, \Delta V_i \qquad (7.5)$$

where $g_f$ is the dynamic transfer conductance of the differential amplifier, which in this case will be given approximately be $g_f = I_Q/4V_T$. Note that the voltage gain of the $Q_1$ and $Q_4$ emitter-follower stages will be close to unity due to the 3.5-$\mu$A current sources that provide for the additional quiescent current for these transistors.

We can therefore now write

$$\Delta I_{B_7} = \frac{100\ \mu A}{\beta_7} = 2g_f\, \Delta V_i = \frac{I_Q}{2V_T}\, \Delta V_i \qquad (7.6)$$

so that

$$\Delta V_i = \frac{100\ \mu A/\beta_7}{I_Q/2V_T} = \frac{100\ \mu A/\beta_7}{100\ \mu A/50\ mV} = \frac{50\ mV}{\beta_7} \qquad (7.7)$$

This is the required change in the input voltage to switch the output from the low state ($V_O = V_{CE(SAT)} \simeq 0.1$ V) to the high state ($V_O \simeq V^+$).

If, for example, $\beta = 50$ (min.), 100 (typ.), we will have that $\Delta V_i = 50\ mV/\beta_7 = 1.0$ mV (max.), 0.5 mV (typ.).

The $V_O$ versus $V_i$ transfer characteristic for this comparator is shown in Figure 7.6. The offset voltage, $V_{OS}$, is defined as the voltage required on the input to produce an output voltage that is midway between the low and high states, or approximately $V^+/2$.

The condition on the load resistance required to cause $Q_8$ to be in saturation for the low state will be given by $\beta_8 \times 100\ \mu A \geq I_{C(SAT)} \simeq V^+/R_L$, so that $R_L = V^+/(\beta_8 \times 100\ \mu A)$. For example, if $\beta_8 = 50$ (min.), 100 (typ.) we will have that $R_L = V^+/\beta_8 \times 100\ \mu A = 1.0\ k\Omega$ (max.), 500 $\Omega$ (typ.) using $V^+ = 5.0$ V.

If instead of being connected to ground, the low side of the comparator circuit is connected to a negative supply voltage $V^-$, the low-state output voltage will be shifted downward by $V^-$, to $V^- + V_{CE(SAT)} \simeq V^-$. As a result, the low-state voltage will be approximately $V^-$ and the high-state voltage will be approximately $V^+$. If these two voltages are equal in magnitude, a symmetrical low-to-high transition will be obtained; that is, the low-state and high-state voltages will be equidistant from ground potential.

In Figure 7.7a curves are shown of the response times for negative (high-to-low) output voltage transitions, and in Figure 7.7b curves are presented for positive

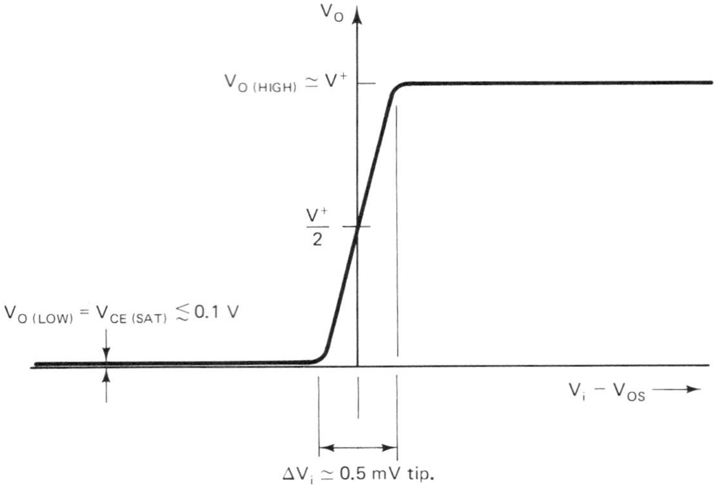


**Figure 7.6** Comparator transfer characteristic, $V_O$ versus $V_i$.

(low-to-high) transitions. The supply voltage is +5 V and the load (pull-up) resistor is 5.1 $k\Omega$. Curves are given for 5-, 20-, and 100-mV input-voltage overdrives.

The input voltage transition for a negative output voltage transition is from −100 mV to a level above ground ($V_{REF} = 0$) equal to the specified input-voltage overdrive. Thus for a 20-mV overdrive the input voltage swing will be from −100 to +20 mV. For the positive-going output voltage transition the input voltage swing will be from +100 mV to a level below the reference voltage (0 V) equal to the specified input overdrive.

Some amount of input overdrive is necessary to ensure that there will be an output voltage transition, and it should be at least equal to the maximum expected value of the input offset voltage [5 mV (max.) for the LM139 comparators]. Increasing the input overdrive will also reduce the response time, although increasing the over-

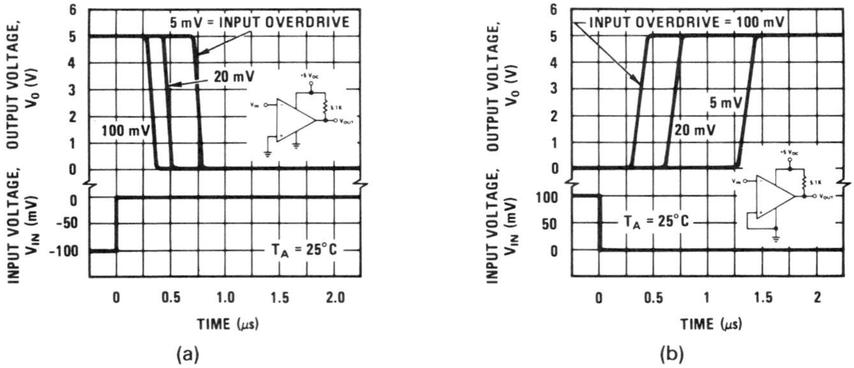


**Figure 7.7** Response times for various input overdrives for the LM139 voltage comparator (National Semiconductor): (a) negative transition; (b) positive transition.

drive above about 50 mV will result in very little improvement in the response time.

From the response time graphs for the LM139, we obtain the following set of response times (ns):

| Input overdrive (mV) | Negative output transition | Positive transition |
|---|---|---|
| 5 | 750 | 1375 |
| 20 | 500 | 700 |
| 100 | 300 | 375 |

Notice that the response can indeed be speeded up by increasing the input overdrive. It is also evident that the positive-going output transitions take appreciably longer than the negative-going transitions. This is the result of the fact that for the positive transition the output transistor $Q_8$ starts off in deep saturation and must then be driven out of saturation, across the active region, and then into cutoff. With a supply voltage of +5 V and a 5.1-kΩ pull-up resistor the current through $Q_8$ is limited to about 1.0 mA when in saturation, whereas $\beta I_B = 150 \times 100\ \mu\text{A} = 15$ mA, so that we see that $Q_8$ is indeed driven into deep saturation. When $Q_8$ is in deep saturation there is a considerable amount of stored charge in the transistor as a result of the large forward-bias voltage at the collector–base junction. The time required to remove this excess stored charge (i.e., the "storage time") is responsible for the longer transition time.

For the negative-going output-voltage transition it is $Q_7$ that must be driven out of saturation and into cutoff. However, the storage time for $Q_7$ will be appreciably less than that of $Q_8$ due to the fact that the saturation current of $Q_7$ is limited to 100 μA as compared to 1.0 mA for $Q_8$. As a result of the storage times of $Q_7$ and $Q_8$, the LM139 is not a very fast comparator.

In Figure 7.8 a graph of the "low-state" output voltage as a function of the output sink current is shown. We see that for currents less than 15 mA the output voltage will be small (less than 1 V). In this range the transistor is in saturation and we have that

$$V_{CE\,(\text{SAT})} = V_{CB} + V_{BE} + I_C\, r_{cc'} = -0.6\ \text{V} + 0.7\ \text{V} + I_C\, r_{cc'} = 0.1\ \text{V} + I_C\, r_{cc'}$$

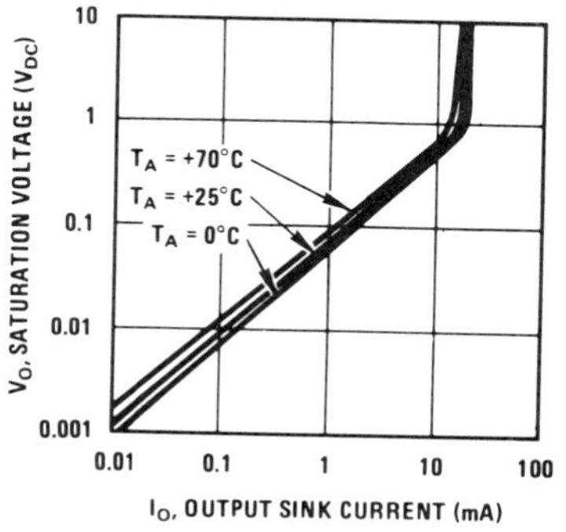


**Figure 7.8** Output saturation voltage for the LM139 (National Semiconductor).

where $r_{cc'}$ is the collector bulk series resistance ("saturation resistance"). At $I_C = 10$ mA, $V_O = V_{CE(SAT)} = 0.8$ V, so that $r_{cc'} = (0.8 - 0.1)$ V/10 mA $= 70\ \Omega$.

As the collector current ($I_C = I_0$) goes above 15 mA the output voltage increases very rapidly, for now the transistor is going out of the saturation region and into the active region. Since this transition occurs at $I_C = 15$ mA and the base current drive is $I_B = 100\ \mu$A, we can conclude that the current gain ($\beta$) of $Q_8$ is 150.

## 7.4 TRANSISTOR STORAGE TIME

We have seen that a major factor limiting the speed of a comparator is the transistor storage time. We will now look at the factors that determine the storage time and consider the techniques that can be used to reduce the storage time.

In the active mode of operation of a transistor the emitter–base junction is biased "on" and the collector–base junction is biased "off." The emitter emits minority carriers (electrons for an NPN transistor) into the base, whereupon they flow across the base and are collected by the collector. In this mode of operation the collector only collects electrons, and does not emit any electrons back into the base. This situation is illustrated in Figure 7.9a, and in Figure 7.9b the variation of the injected minority carrier density with distance is shown.

The charge stored in the base region due to the electrons in transit across the base will be given by $Q_{\text{base(active)}} = I_C\, t_{\text{transit}}$, where $t_{\text{transit}}$ is the transit time of the electrons across the base region from the emitter to the collector. This time is very short, typically of the order of only 30 to 100 ps. There will also be a very small amount of charge stored in the emitter region due to the injection of holes from the base into the emitter as a result of the forward-bias voltage across the emitter–base junction. This current injection will be very small compared to the flow of electrons in the other direction due to the very large doping asymmetry of the $N^+/P$ emitter–base junction. The very heavy doping of the emitter region will also result in a very short lifetime for the injected holes, so that as a net result the charge stored in the emitter region, $Q_E$, will be very small.

When the transistor is driven into saturation both junctions will be turned "on," so that the collector will not only collect electrons, but will also emit electrons back into the base. Most of these electrons will then travel across the base region and be collected by the emitter, which will collect most of the electrons emitted by the collector, but will still also be emitting electrons back into the base region. This situation is illustrated in Figure 7.10.

When the transistor is driven deeply into saturation, both the emitter–base and the collector–base junctions are strongly forward biased and there will be a high flow of electrons across the base region in both directions. As a result the base region is flooded with electrons and the stored charge in the base region is now very much greater than it was in the active mode of operation.

At the same time there will also be a very substantial injection of holes from the base region into the collector. The large magnitude of this hole injection is the result of the collector region being much less heavily doped than the base region. As a result, when the collector–base junction is forward biased by more than about

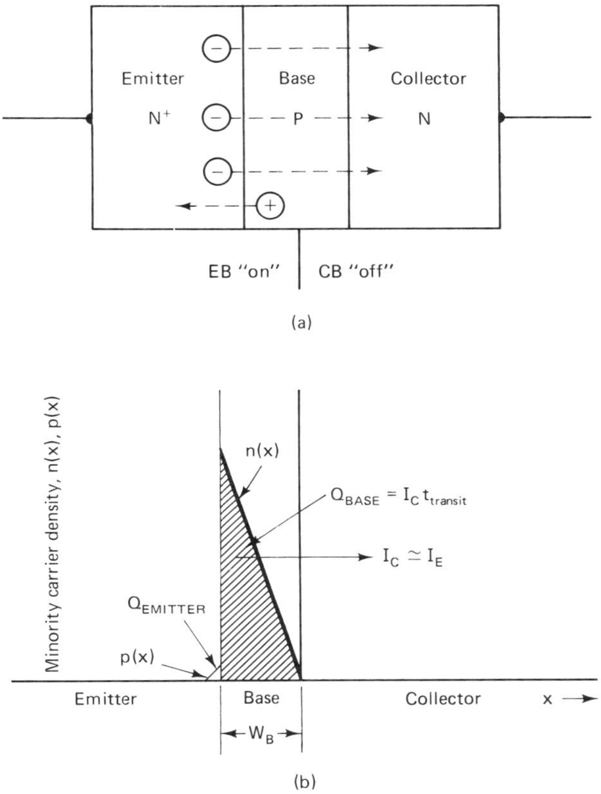


**Figure 7.9** Transistor operation in the active mode.

0.5 V there will be a large flow of electrons from the collector into the base and a large flow of holes from the base into the collector. This large injection of holes into the collector region and the relatively long lifetime of holes in this region $\tau_p$ will result in a large amount of stored charge.

The total stored charge in the transistor in the saturation mode will be $Q_{\text{stored(sat)}} = Q_{\text{base}} + Q_{\text{emitter}} + Q_{\text{collector}}$. We have already noted that $Q_{\text{emitter}}$ will be very small. As a result of the very narrow base width (~0.3 to 1.0 $\mu$m) and the resulting very small transit time we will generally have that the major component of the total stored charge will be $Q_{\text{collector}}$. This is the charge stored in the collector region due to the injection of holes from the base region. This stored charge will be related to the base current approximately by $Q_{\text{stored(sat)}} \simeq Q_{\text{collector}} = I_{B\,(\text{sat})}\tau_p$, where $\tau_p$ is the lifetime of holes in the collector region. This lifetime is typically in the range 1 to 10 $\mu$s, in contrast to the base transit time of the order of 30 to 100 ps. We therefore see that the stored charge in the transistor when operated in the saturation mode can be several orders of magnitude larger than when operated in the active mode.

When the transistor is now to be switched out of the saturation mode and across the active region into cutoff, the stored charge will produce a continued flow of current through the transistor for the period of time required to remove this

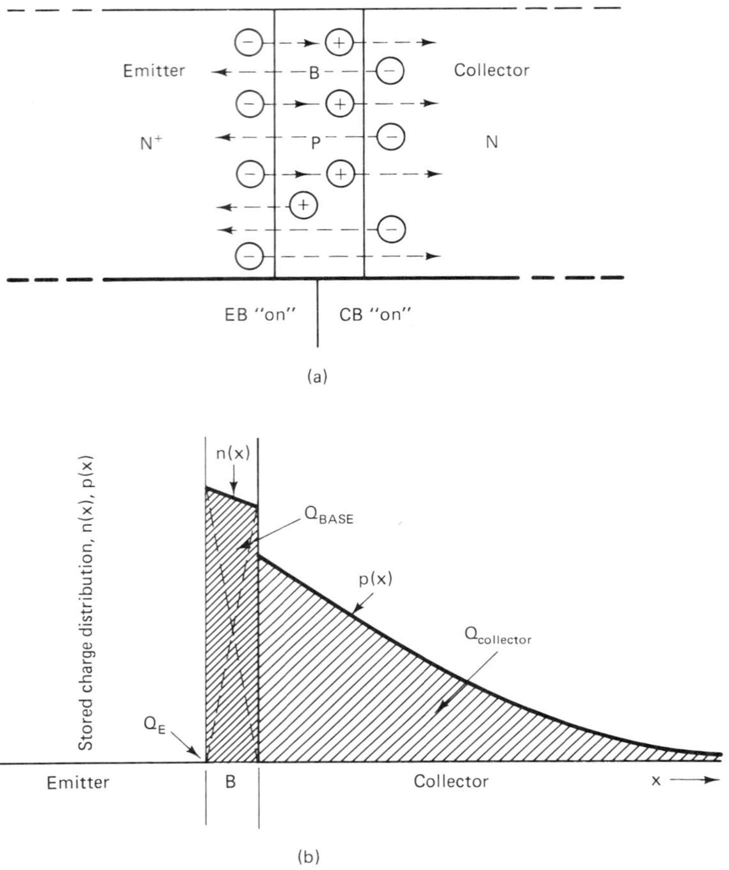


**Figure 7.10** Transistor operation in the saturation mode.

stored charge, as shown in Figure 7.11. This period of time of continued current flow is called the *storage time*, $t_s$. The stored charge is removed in two ways. One is by the reverse flow of base current $I_{B(R)}$, which directly removes the excess stored charge from the transistor. The other way is the recombination of the stored minority carriers with majority carriers. The rate at which the stored charge in the collector region recombines with electrons is $dQ/dt = Q_{\text{collector}}/\tau_p$. The total rate at which the stored charge decreases with time will be given by $dQ_{\text{stored}}/dt = -I_{B(R)} - Q_{\text{stored}}/\tau_p$, where $I_{B(R)}$ is the base current in the reverse (negative) direction and we have assumed that most of the stored charge is in the collector region.

If the base current during this switching transistor is limited to a very small value, or is zero, the decay of the stored charge will be given approximately by $dQ_{\text{stored}}/dt \simeq -Q_{\text{stored}}/\tau_p$, so that the decay of the stored charge as a function of time will be

$$Q_{\text{stored}}(t) = Q_{\text{stored}}(0) \exp\left(-\frac{t}{\tau_p}\right)$$

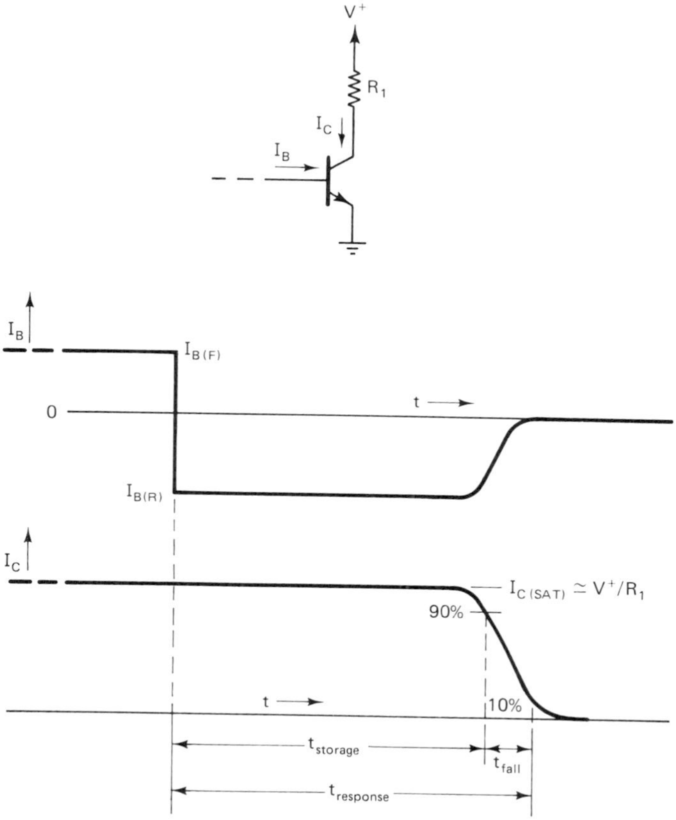


**Figure 7.11** Transistor turn-off transition.

where $Q_{\text{stored}}(0)$ is the stored charge at the beginning of the switching transition. From this we see that it will take a time of about 3 or $4\tau_p$ to dissipate the stored charge. This will result in a very long storage time, up to several microseconds.

If a substantial reverse base current is allowed to flow, the rate at which the stored charge is removed will be principally controlled by the reverse-base current $I_{B(R)}$. For this case we will have that $dQ_{\text{stored}}/dt \simeq -I_{B(R)}$, so that $Q_{\text{stored}}(t) \simeq Q_{\text{stored}}(0) - I_{B(R)}t$. The storage time is the time required to remove all of the stored charge so that it will be given by $t_{\text{storage}} \simeq Q_{\text{stored}}(0)/I_{B(R)}$. Since $Q_{\text{stored}}(0) \simeq Q_{\text{collector}}(0) = I_{B(\text{SAT})}\,\tau_p$, the storage time will become $t_{\text{storage}} \simeq [I_{B(\text{SAT})}/I_{B(\text{R})}]\tau_p$. Thus by forcing a large reverse base current to flow out of the transistor the storage time can be made to be very small. In many cases the storage time can be reduced to about 10 ns.

The fall-time portion of the total response time or switching time of the transistor is the time required to change the voltage across the emitter–base junction capacitance from $V_{BE(\text{sat})}$ to $V_{BE(\text{cutoff})}$. The emitter–base junction capacitance $C_{b'e}$ will generally be in the range 10 to 30 pF, and the voltage change necessary to drive the transistor from saturation to cutoff will be in the range of about 0.2 to 0.4 V. The fall time can be obtained from the relationship

$$\Delta V_{EE} = \frac{\Delta Q_{b'e}}{C_{b'e}} = \frac{I_{B(R)} t_f}{C_{b'e}}$$

so that (7.8)

$$t_f = \Delta V_{BE} \frac{C_{b'e}}{I_{B(R)}} = \frac{[V_{BE(\text{sat})} - V_{BE(\text{cutoff})}]\, C_{b'e}}{I_{B(R)}}$$

For example, if $V_{BE(\text{SAT})} = 0.75$ V, $V_{BE(\text{cutoff})} = 0.55$ V, $C_{b'e} = 20$ pF, and $I_{B(R)} = 1.0$ mA, the fall time will be $t_f = 0.2 \text{ V} \times 20 \text{ pF}/ 1.0 \text{ mA} = 4$ ns. We see that the reverse base current will be important in determining the fall time.

## 7.5 TECHNIQUES TO REDUCE RESPONSE TIME

There are several techniques that can be used to speed up the transition of the transistor from saturation to cutoff. These techniques are as follows.

1. Reduce the storage time by reducing the minority carrier lifetime.
2. Decrease the storage and fall times by increasing the reverse base current.
3. Reduce the charge stored in the base and collector regions by using a thinner and less heavily doped base region.
4. Use a Schottky-barrier diode across the collector–base junction to prevent the collector–base junction from turning on.

### 7.5.1 Minority Carrier Lifetime Reduction

The first technique of reducing the minority carrier lifetime involves adding a small amount of a "lifetime killer" impurity to the IC during the manufacturing process. This lifetime killer impurity acts to decrease the minority carrier lifetime and thereby reduces the stored charge, especially in the collector region. Gold is commonly used for this purpose, and it serves as a very effective recombination center in silicon in promoting or catalyzing the recombination of free electrons and holes. This increased recombination rate results in a reduced carrier lifetime. As a result, gold and other heavy metal impurities such as copper and iron are called "lifetime killers."

With the addition of small amounts of gold to silicon (about $10^{14}$ to $10^{15}$ $\text{cm}^{-3}$) the minority carrier lifetime can be reduced from the undoped level of around 1 to 10 $\mu$s down to around 10 ns. This will correspondingly result in a reduction of the storage time down to the range 10 ns.

The major disadvantage of the gold-doping process is that it is nonselective in that all of the devices on the IC chip will become gold-doped. The reduction in the minority carrier lifetime produced by the gold doping does have the beneficial effect of reducing the storage time, but there are some unwanted side effects as well. The reduction in the minority carrier lifetime will reduce the current gain $\beta$ of the transistors because of the increased recombination of minority carriers in transit across the base region. For the narrow-base NPN transistors with base transit times in the range 30 to 100 ps the reduction of the minority carrier lifetime to around 10 ns will not prove to be too severe an effect. We will still have that the minority carrier lifetime will be considerably greater than the transit time, so that most of the injected minority carriers will still make it safely across the base to the collector. However, the lateral and substrate PNP transistors will have base widths generally in the range 4 to 10 $\mu$m and therefore a much longer transit time, in the range 10 to 50 ns. As a result, a reduction in the minority carrier lifetime to about 10 ns can produce a very substantial reduction in the transistor current gain.

Another major disadvantage of gold doping is the increase in the reverse leakage current of the diodes and transistors. One component of the leakage current of a reverse-biased PN junction is the flow of the thermally generated minority carriers across the junction. The generation rate of these minority carriers is inversely proportional to the minority carrier lifetime, so that a decrease in the lifetime from about 10 $\mu$s to 10 ns can result in an increase in this leakage current component by a 1000:1 ratio.

### 7.5.2 Comparator Example: $\mu$A760

The transistor storage and fall times can be reduced substantially by increasing the reverse base current during the saturation to cutoff switching transition time. The storage time can also be minimized by limiting the amount of forward base drive that puts the transistor into saturation.

An example of a voltage comparator that uses these techniques together with the use of gold doping for minority carrier lifetime reduction is the $\mu$A760 (Fairchild). This device offers typical response times of 16 ns when driven with a 100-mV input step with a 5-mV overdrive.

A simplified circuit diagram of this comparator is shown in Figure 7.12. The input stage is an NPN differential amplifier comprised of $Q_1$ and $Q_2$ with current

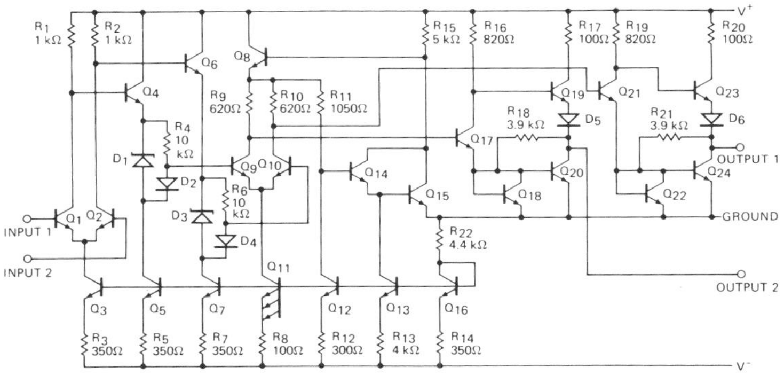


**Figure 7.12** High-speed voltage comparator circuit: $\mu$A760 (Fairchild Semiconductor).

sink biasing being provided by $Q_3$. With a typical supply voltage of $\pm 5.0$ V the quiescent current of this differential amplifier is set by $Q_3$ and $Q_{16}$ to

$$I_Q = I_3 = I_{16} = \frac{V^- - V_{BE}}{R_{14} + R_{22}} = \frac{(5.0 - 0.7)\text{ V}}{4.75\text{ k}\Omega} = 0.9\text{ mA}$$

Resistors $R_1$ and $R_2$ (1 k$\Omega$) provide a passive load for $Q_1$ and $Q_2$. The low value of these resistors helps to give this device a fast response time. A balanced output from the first stage is then fed to the differential amplifier second stage via emitter followers $Q_4$ and $Q_6$. These emitter-follower transistors, together with diodes $D_1$ through $D_4$ and the current sink transistors $Q_5$ and $Q_7$, provide d-c level shifting. The current provided by $Q_5$ and $Q_7$ is 0.9 mA. Since the voltage drop across the 10-k$\Omega$ resistors ($R_4$ and $R_6$) plus the diode drop of $D_2$ and $D_4$ is sufficient to drive the diodes $D_1$ and $D_3$ into breakdown ($V_Z \simeq 6.5$ V), the quiescent voltage at the bases of $Q_9$ and $Q_{10}$ will be approximately

$$\begin{aligned} V_{B9} = V_{B10} &= V^+ - I_1 R_1 - V_{BE4} - V_{Z1} + V_{D2} \\ &= 5.0 - (0.45\text{ mA})(1\text{ k}\Omega) - 0.7 - 6.5 - 0.7 = -1.95\text{ V} \end{aligned} \tag{7.9}$$

The second stage, comprised of $Q_9$ and $Q_{10}$, is biased $Q_{11}$ acting as a current sink and providing a quiescent current level of $I_{11} = 0.9\text{ mA} \times (350/100) = 3.15$ mA. The current through $Q_{12}$ will be $I_{12} = 0.9\text{ mA} \times (350/300) = 1.05$ mA. The quiescent voltage at the emitter of $Q_8$ will therefore be

$$V_{E8} = V_{BE15} + V_{BE14} + I_{12}R_{11} \simeq 0.6 + 0.7 + (1.05\text{ mA})(1.05\text{ k}\Omega) = +2.4\text{ V}$$

The quiescent voltage level at the collectors of $Q_9$ and $Q_{10}$ will then be

$$V_{C9} = V_{C10} = V_{E8} - I_9 R_9 \simeq 2.4 - (3.15\text{ mA}/2)(620\ \Omega) = 1.4\text{ V}$$

The balanced output of the second stage drives two identical output circuits. These output circuits provide the complementary TTL-compatible output voltages. A detailed view of the output circuit is shown in Figure 7.13. Note that the quiescent voltage level at the collectors of $Q_9$ and $Q_{10}$ of $+1.4$ V as obtained above is at the proper value to interface with the output circuit.

If the input signal is such that the current through $Q_9$ decreases substantially below the quiescent level, the excess of the current through $R_9$ becomes the forward base drive of $Q_{17}$ as indicated by $I_{B17(F)}$. This turns $Q_{17}$ on and also turns $Q_{18}$ on. Since $Q_{18}$ and $Q_{20}$ are connected as a current mirror, this will also result in $Q_{20}$ being turned on.

At the same time that $Q_{20}$ is being turned on, $Q_{19}$ will be rapidly driven out of saturation by the sinking of the reverse base current of $Q_{19}$, $I_{B19(R)}$, by $Q_{17}$.

The base drive of $Q_{17}$ will be sufficient to drive it into saturation. As a result, the voltage at the $C_{17}$–$B_{19}$ node will be approximately

$$V_{B19} = V_{C17} = V_{BE18} + V_{CE(\text{SAT})17} \simeq 0.9\text{ V}$$

Since a voltage of at least $V_{BE19} + V_{D5} \simeq 1.2$ V is required at the base of $Q_{19}$ in order to keep $Q_{19}$ on, we see that $Q_{19}$ will be turned off. The output will therefore be driven down to the "low" state with $Q_{20}$ sinking the load current.

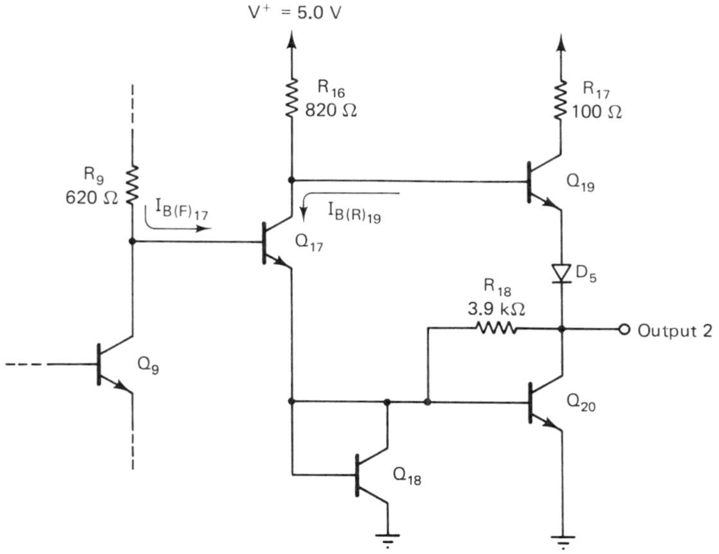


**Figure 7.13** μA760 output stage: high-to-low transition.

Note that although $Q_{20}$ may be driven into saturation, since $Q_{18}$ and $Q_{20}$ are connected in a current mirror configuration, the base drive of $Q_{20}$ is limited by $Q_{18}$ and keeps $Q_{20}$ from going very deeply into saturation.

For the transition from the low state to the high state, $Q_{17}$ must be rapidly brought out of saturation and into cutoff. This occurs when the input voltage is such that the current through $Q_9$ is greater than that through $R_9$, the difference between these two currents being the reverse base drive of $Q_{17}$, as indicated by $I_{B(R)17}$ in Figure 7.14. This reverse base current rapidly removes the stored charge in $Q_{17}$ and quickly brings it out of saturation and across the active region into cutoff. The rapidly rising voltage at the collector of $Q_{17}$ turns $Q_{19}$ and $D_5$ on. At the same time $Q_{18}$ and thus $Q_{20}$ also is being turned off. The rapid turn-off of $Q_{20}$ is aided by the flow of reverse base current $I_{B(R)20}$ through $R_8$ (3.9 kΩ).

With $Q_{20}$ off and $Q_{19}$ and $D_5$ on, the output voltage will now be in the high state at a voltage level of $V_H = V^+ - V_{BE_{19}} - V_{D_5} = 5.0 - 0.7 - 0.7 = 3.4$ V. Since $V_L = V_{CE\,(SAT)_{20}} \simeq 0.2 - 0.4$ V, we see that this comparator produces output voltages that are TTL-compatible.

The input voltage swing necessary to produce the output voltage transition is approximately 1 mV, so that the open-loop voltage gain is around 3000. This relatively small value of the open-loop gain as compared to the values commonly encountered for operational amplifiers is the result of purposely trading off a reduced gain for a faster response time.

The input offset voltage $V_{OS}$ is 1.0 mV (typ.), 6.0 mV (max.), so that we see that with the combination of the offset voltage and the input voltage swing of 1 mV that is required to produce the output transition, this device will typically switch at an input voltage level that is within about 2 mV of the reference voltage.

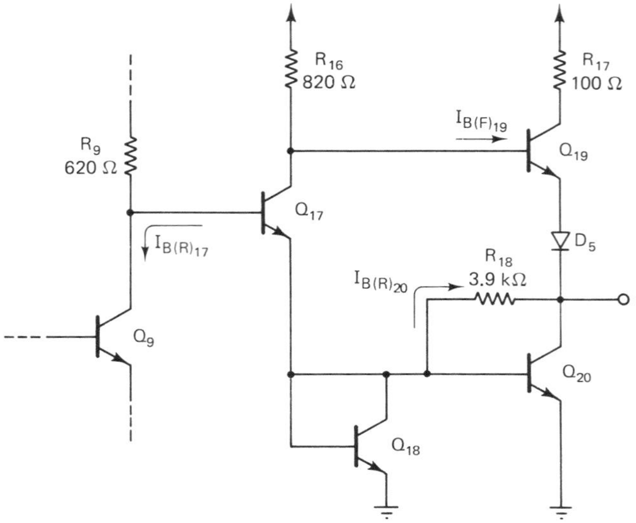


**Figure 7.14** $\mu$A760 output stage: low-to-high transition.

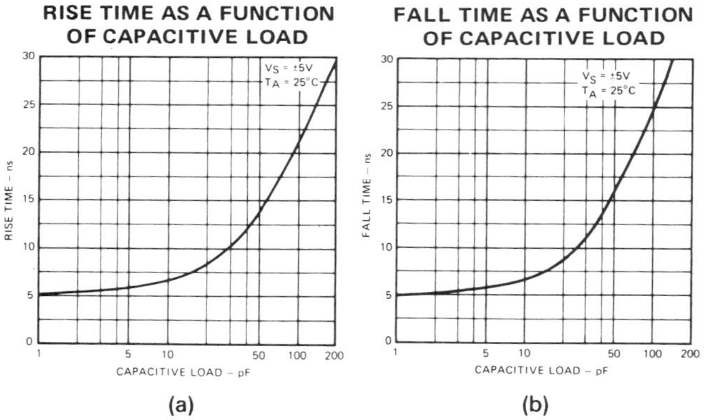


**Figure 7.15** Rise and fall times of the $\mu$A760 voltage comparator as a function of load capacitance (Fairchild): (a) rise time; (b) fall time.

This output circuit has an active "pull-up" and "pull-down" configuration, with $Q_{19}$ and $Q_{23}$ used for the active pull-up to source currents into the load, and $Q_{20}$ and $Q_{24}$ used for the pull-down to sink load currents. This type of output circuit will provide for the fast charging of load capacitances. Indeed, load capacitance of up to 10 pF will not produce any significant elongation of the response time. Large load capacitances, however, will start to produce a substantial lengthening of the response time, particularly with respect to the rise- and fall-time portions of the response time, as shown in Figure 7.15.

### 7.5.3 Comparator Example: LM160 and LM161

The use of a thinner, more lightly doped base region is an effective way of reducing the storage time. The thinner base width will result in a much smaller stored charge in the base region. The thinner and less heavily doped nature of the base region will also result in less injection of holes into the collector region, so that the stored charge in the collector region will be reduced. The only serious drawback of the use of the thin, lightly doped base region is the reduction in the collector-to-emitter breakdown voltage, $BV_{CEO}$. This reduction in the breakdown voltage is due to the occurrence of the collector-to-emitter "punch-through" or "reach-through" phenomenon. This happens when the width of the depletion region on the base side of the collector–base junction is such that it extends all the way across the base region, and reaches the emitter–base junction. The effective base width is reduced to zero and direct injection of electrons from the emitter to the collector can occur. The design of the IC and the specification of the maximum allowable applied voltages must be in conformance with the reduced breakdown voltage of the thin base transistors.

An example of comparators that use thin base width to achieve a fast response time is the LM160 and LM161 series (National Semiconductor). These comparators have propagation delay times of only 13 ns (typ.) for a 10-mV input overdrive. These comparators also have a storage time reduction resulting from the large reverse base current drawn from the saturated output transistors. In Figure 7.16 a circuit diagram of the LM161 comparator is shown. The input stage is an NPN differential amplifier with current sink biasing and low-value passive load resistors. This stage is coupled by emitter followers to a second differential-amplifier stage. The second stage has a current-to-voltage converter type of load.

The complementary output voltages obtained from the second stage drive two

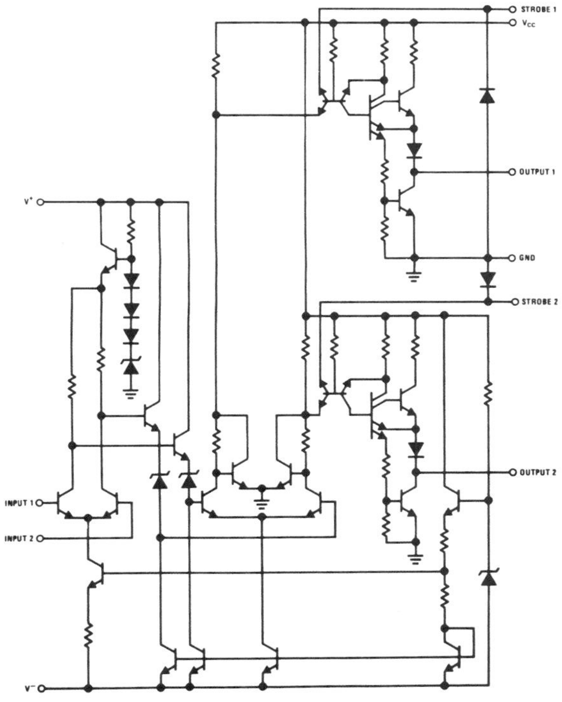


**Figure 7.16** Circuit diagram of the LM161 voltage comparator (National Semiconductor).

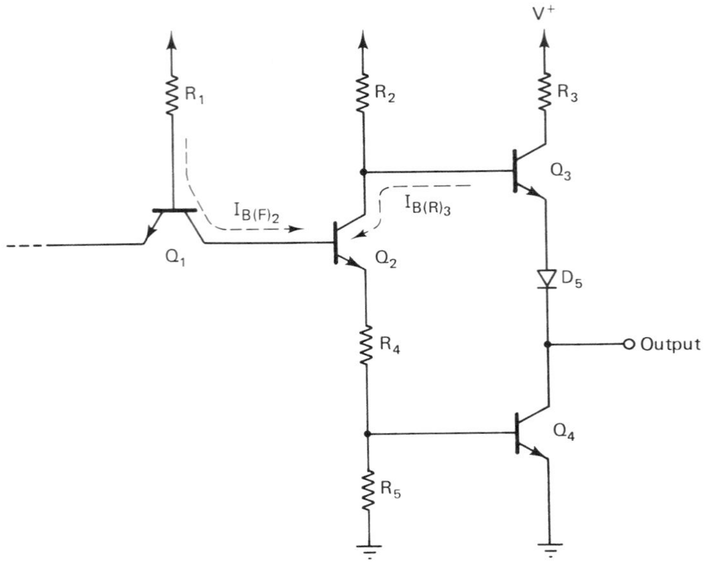


(a)

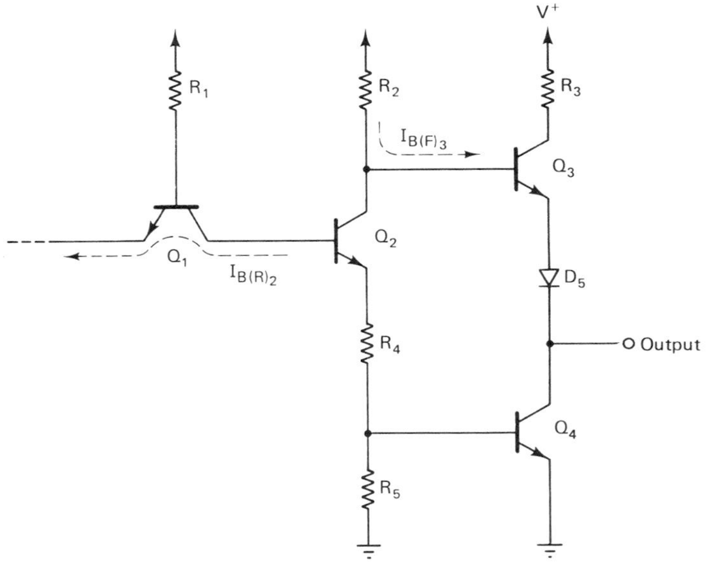


(b)

**Figure 7.17** LM161 output circuit: (a) high-to-low transition; (b) low-to-high transition.

identical output stages which are essentially high-speed TTL gates. These TTL gates are biased by a separate 5-V supply, so that the output voltage levels of the comparator will be at the standard TTL logic levels.

In Figure 7.17a a simplified diagram is shown of the TTL output stage. In this diagram the current directions are shown for the output high-to-low transition for which $Q_3$ is being brought out of saturation. Note that with the emitter of $Q_1$ high, the collector–base junction of $Q_1$ turns on. This results in the flow of current into the base of $Q_2$, which rapidly turns $Q_2$ on. As soon as $Q_2$ is brought into conduction, it sinks the reverse base current of $Q_3$ and rapidly brings $Q_3$ out of saturation. At the same time $Q_4$ is turned on, pulling the output down to the logic low level of $V_{CE(\text{SAT})_4} \simeq 0.1$ to 0.3 V, depending on the fan-out or the load being driven.

For the transition in the opposite direction the current directions are as shown in Figure 7.17b. For the output low-to-high transition the emitter of $Q_1$ goes low, putting $Q_1$ into the active region. The collector current of $Q_1$ now results in a reversal of the base current of $Q_2$, which rapidly pulls $Q_2$ out of saturation and into cutoff. As $Q_2$ goes into cutoff, $Q_4$ is turned off and $Q_3$ is turned on. This causes the output voltage to rise to the high logic level of $V^+ - V_{BE_3} - V_{D_5} \simeq 5 - 0.7 - 0.7 = 3.4$ V.

We therefore see that both $Q_2$ and $Q_3$ can rapidly be brought out of saturation by the current sinking action of the drive transistors, which for $Q_2$ is $Q_1$ and for $Q_3$ is $Q_2$. The reverse base current in either case will be much larger than the forward base current due to the current gain of the driving transistor.

### 7.5.4 Schottky-Clamped Transistors

A technique that is very effective in reducing the switching time of a transistor is to use a Schottky-barrier diode in parallel with the collector–base junction of the transistor. A "Schottky diode" or "Schottky-barrier diode" is a diode formed by a rectifying metal-semiconductor contact, as shown in Figure 7.18. The metals commonly used for the Schottky barrier include gold, aluminum, chromium, and platinum. The semiconductor is usually N-type silicon of moderate doping, generally in the

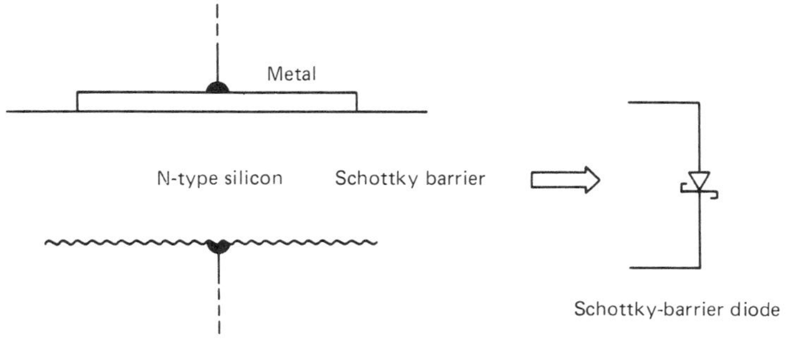


**Figure 7.18** Schottky diode.

range 0.1 to 10 Ω-cm. The metal is a thin film typically of 1 μm thickness deposited by a vacuum evaporation or sputtering process.

In Figure 7.19 the energy band (electric potential versus distance) diagram of a Schottky barrier formed between N-type silicon and a metal is shown. The height of the barrier (contact potential) $\phi_B$ at the metal/N-silicon junction is 0.56 V for chromium, 0.68 V for aluminum, 0.81 V for gold, and 0.90 V for platinum. This is the barrier for electron flow from the metal to the semiconductor and is high enough to prevent any significant flow of electrons in this direction.

The potential barrier for electron flow from the semiconductor to the metal is $\phi_B$ and this will typically be in the range of about 0.3 to 0.4 V. This is to be compared to the contact potential (or "built-in voltage") of a silicon PN junction, which is generally in the range 0.8 to 0.9 V. As a result of the much smaller barrier for electron flow in the Schottky diode, the forward-bias voltage that is required to produce a given level of forward current flow is about one-half of that for the PN junction diode.

In Figure 7.19b a diagram is shown for forward-bias conditions. Under forward-bias conditions the applied voltage is such that the metal (the anode) is made positive with respect to the semiconductor (the cathode). The voltage barrier for electron flow from the semiconductor to the metal is reduced from $\phi_B$ under zero-bias conditions to $\phi_B - V_F$, where $V_F$ is the applied forward-bias voltage. The forward current $I_F$ will vary exponentially with the forward-bias voltage as given by the diode equation

$$I_F = I_O \left[ \exp\left(\frac{V_F}{V_T}\right) - 1 \right] \simeq I_O \exp\left(\frac{V_F}{V_T}\right) \tag{7.10}$$

The forward-bias voltage level required to produce currents in the milliampere range is only about 0.3 V, compared to voltages in the range 0.6 to 0.7 V for the silicon PN junction. From the standpoint of the forward-bias voltage drop the silicon Schottky-barrier diode is similar to the germanium PN junction diode.

In Figure 7.19c the Schottky barrier is shown under reverse-bias conditions.

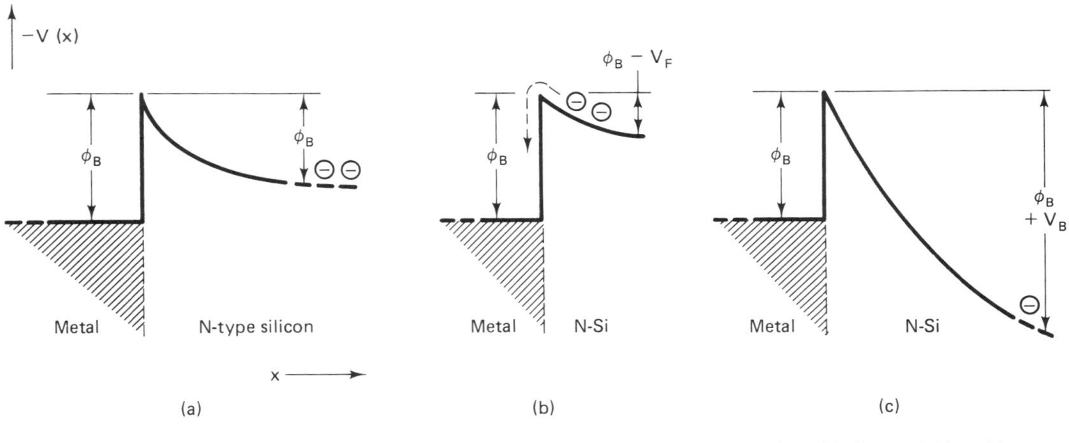


**Figure 7.19** Energy band diagram of Schottky barrier: (a) zero bias; (b) forward bias; (c) reverse bias.

Now the potential barrier for electron flow from the semiconductor to the metal has been increased to $\phi_B + V_R$, where $V_R$ is the reverse-bias voltage. This barrier is so high that there is negligible flow of electrons from the semiconductor to the metal. The high barrier $\phi_B$ for electron flow from the metal to the semiconductor will also prevent any significant flow of electrons from the metal to the semiconductor. Therefore, under these reverse-bias conditions the diode is "off." There will actually be a very small reverse current or "leakage" current, this being due principally to electrons flowing from the metal to the semiconductor by climbing over the $\phi_B$ potential barrier.

Under forward-bias conditions the current that flows in the device is due to the injection of electrons from the semiconductor into the metal. Since electrons are the majority carriers on both sides of the junction, as soon as an electron leaves the N-type semiconductor its place is taken by an electron entering the semiconductor from the external circuit. Similarly, as soon as an electron enters the metal from the semiconductor, another electron leaves the metal by going out into the external circuit. As a result, charge neutrality is preserved in both the semiconductor and the metal and there are no excess electrons anywhere.

Since the concentration of holes in the N-type silicon is extremely small and there are no holes in the metal, there will be no injection or flow of holes in either direction. As a result this is an entirely "majority carrier" device. There will be no accumulation or storage of excess charge carriers on either side of the junction. As a result of this absence of charge storage the Schottky-barrier diode proves to be a very fast device and exhibits no storage time.

In Figure 7.20 the combination of a transistor and a Schottky-barrier diode across the collector–base junction of the transistor is shown. Usually, the Schottky diode is fabricated as an integral part of the transistor structure, the result being known as a *Schottky transistor* or a *Schottky-clamped transistor*. The symbol for this device is shown in Figure 7.20. In Figure 7.21 there is a cross-sectional view of an IC NPN transistor with an integral Schottky diode. The Schottky diode is formed by the overlap of the base contact metallization over the collector region. It should be noted that a Schottky-barrier diode is not formed at the collector contact, for here the metallization is in contact with the heavily doped $N^+$ region. For $N^+$ or $P^+$ regions with doping levels of about $1 \times 10^{19}$ cm$^{-3}$ or greater, a rectifying Schottky barrier is not produced, but rather a nonrectifying or "ohmic" contact will result.

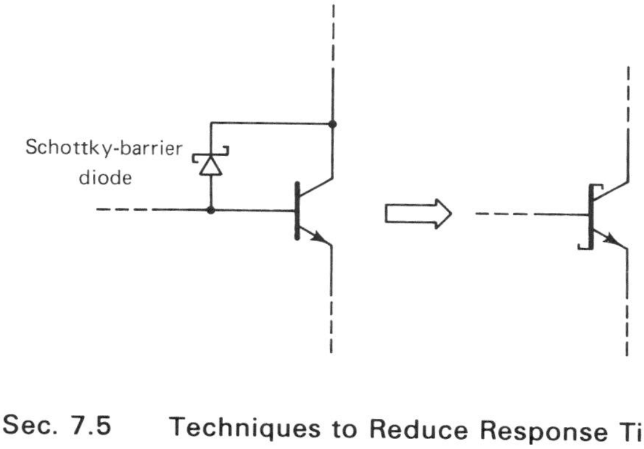


**Figure 7.20** Schottky-clamped transistor.

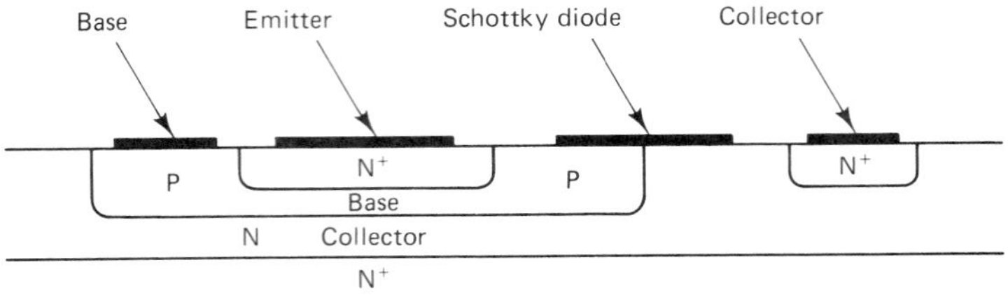


**Figure 7.21** Integrated Schottky diode–transistor combination.

When the collector–base junction is forward biased the Schottky diode turns on at about 0.2 to 0.3 V and will be in full conduction at about 0.3 to 0.4 V. The Schottky diode will bypass current around the collector–base junction of the transistor. The forward-bias voltage across the collector–base junction will thus be limited (or "clamped") to a maximum value of about 0.3 to 0.4 V by the Schottky diode. Since the threshold for conduction of a silicon PN junction is about 0.5 V, we see that the clamping action of the Schottky diode will prevent the collector–base junction from turning on.

As a result of the collector–base junction not being biased into conduction, there will be no injection of holes from the base into the collector, nor will there be any electrons emitted from the collector into the base. The only stored charge in the device will be the electrons in transit across the base region, as given by $Q_{\text{stored}} = Q_{\text{base}} = I_C t_{\text{transit}}$. Since $I_{B(R)} t_{\text{storage}} \simeq Q_{\text{base}}$, the storage time can be expressed approximately as

$$t_{\text{storage}} \simeq \frac{I_{C(\text{SAT})}}{I_{B(R)}} t_{\text{transit}} \tag{7.11}$$

where $I_{B(R)}$ is the reverse base current. Since the transit time is extremely short for the IC NPN transistors, typically in the range 30 to 100 ps, we see that storage times of less than 1 ns can be achieved.

The switching speed of the transistor will now be limited principally by the time required to change the voltage across the emitter–base junction capacitance, which is the fall time. Since

$$\Delta V_{BE} = \frac{\Delta Q_{BE}}{C_{BE}} \simeq \frac{I_{B(R)} t_{\text{fall}}}{C_{BE}}$$

we have that (7.12)

$$t_{\text{fall}} \simeq \frac{\Delta V_{BE} C_{BE}}{I_{B(R)}}$$

Since $V_{BE(\text{SAT})} \simeq 0.8$ V and $V_{BE(\text{cutoff})} \simeq 0.5$ V, we see that the change in the base-to-emitter voltage required to take the transistor from saturation to cutoff is only about 0.3 V. If $C_{BE} = 10$ pF and $I_{B(R)} = 10$ mA, this will give a fall time of $t_{\text{fall}} \simeq 0.3 \text{ V} \times 10 \text{ pF}/10 \text{ mA} = 3$ ns. We therefore see that with the Schottky-clamped transistor switching times down in the range of 3 ns can be obtained.

**Comparator using Schottky-clamped transistors.** An example of a comparator that uses Schottky-clamped transistors is the NE529 (Signetics). This device has the following propagation delay times for a 50-mV input overdrive:

$$\begin{aligned} t_{pd}(0) &= 10 \text{ ns (typ.), } 20 \text{ ns (max.)} \\ t_{pd}(1) &= 12 \text{ ns (typ.), } 20 \text{ ns (max.)} \end{aligned} \tag{7.13}$$

In Figures 7.22 and 7.23 the response-time characteristics are shown for various input overdrives.

In Figure 7.24 the circuit of this comparator is shown. The circuit consists of two cascaded NPN differential amplifiers which drive two complementary TTL gates. The input stage is the $Q_1$–$Q_2$ differential amplifier biased by the $Q_4$ current sink. Resistors $R_2$ and $R_3$ (1.5 kΩ each) serve as a low-resistance passive load for this stage to give a fast response time. The output voltage of the first stage is shifted downward in d-c level by the emitter followers $Q_4$ and $Q_5$ and the zener diodes $D_4$ and $D_5$. This level-shift circuit is biased by the current sink transistors $Q_6$ and $Q_7$. The total d-c shift is $V_{BE} + V_Z \simeq 7.0$ V.

The second differential-amplifier stage is $Q_{10}$ and $Q_{11}$ biased by the current sink transistor $Q_8$. The load for each half of the differential amplifier is a low-input-impedance current-to-voltage converter circuit consisting of $Q_{12}$ and $R_{11}$ (750 Ω) and $R_{13}$ (1 kΩ) on one side and $Q_{13}$, $R_{12}$, and $R_{14}$ on the other side.

The output voltage produced by the second stage is used to drive the TTL gates in a complementary fashion. The TTL gates have active pull-up ($Q_{17}$, $Q_{18}$ and $Q_{23}$, $Q_{24}$) and active pull-down ($Q_{19}$, $Q_{20}$ and $Q_{25}$, $Q_{26}$) configurations for fast charging and discharging of the load capacitance.

Notice that most of the transistors in the circuit are of the Schottky-clamped type. The principal exceptions are those transistors that are not subject to being driven into saturation, such as $Q_4$ and $Q_5$, $Q_{10}$ and $Q_{11}$, and $Q_{18}$. Also, the transistors used in the biasing circuitry do not need to be Schottky-clamped.

The TTL output stages can be biased separately from the rest of the circuit by supplying a d-c voltage of +5.0 V to the $V_2^+$ pin so as to give TTL-compatible logic outputs. The rest of the circuit can be biased with a substantially larger voltage,

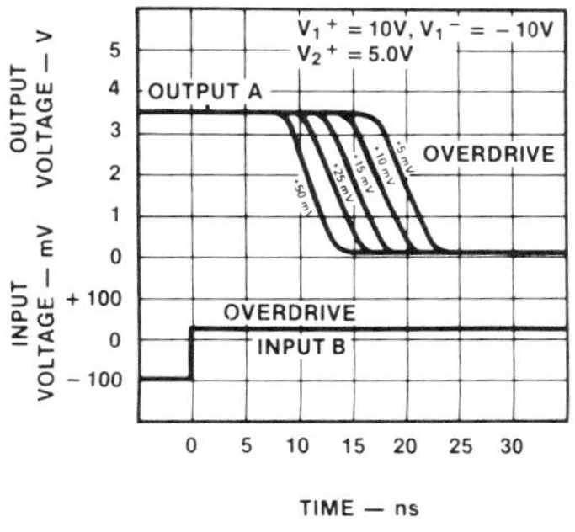


**Figure 7.22** NE529 response times for various input overdrives (Signetics).

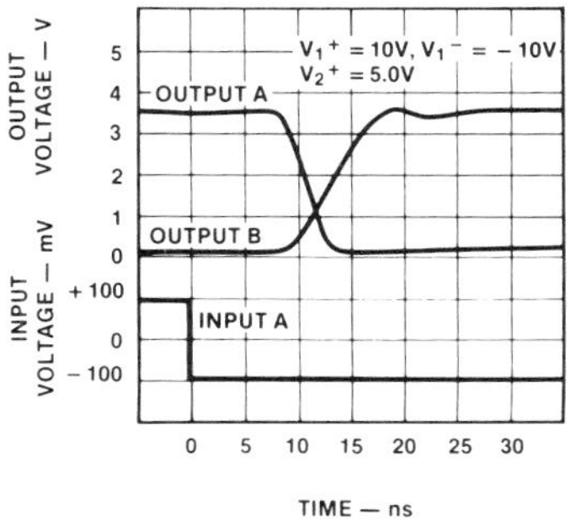


**Figure 7.23** NE529 output progation delays (Signetics).

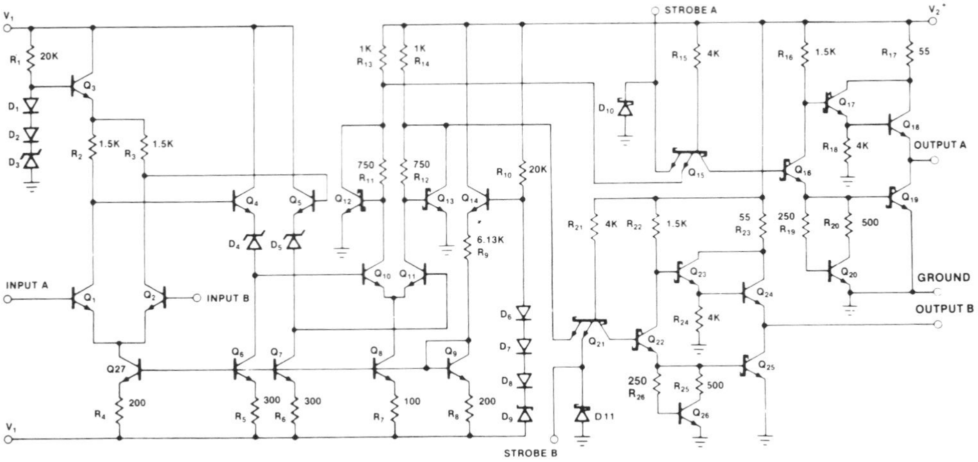


**Figure 7.24** Circuit diagram of the NE529 voltage comparator (Signetics).

which will typically be $V_1^+ = +10$ V and $V_1^- = -10$ V so as to produce a faster response time and to obtain a larger common-mode input-voltage range.

## 7.6 CMOS INVERTER VOLTAGE GAIN

A CMOS inverter as shown in Figure 7.25 can be used as the basis of a voltage comparator. For a CMOS inverter that is operating in the high-gain transition region both FETs will be on and operating in the active or saturated region. Each FET can be considered to operate as an active load for the other. For the two FETs we will have that

$$I_{DS} = K(V_{GS} - V_t)^2 \tag{7.14}$$

where the parameters $K$ and $V_t$ are the values that are appropriate for each transistor. The dynamic transfer conductance $g_{fs}$ will be given by

$$g_{fs} = \frac{dI_{DS}}{dV_{GS}} = 2K(V_{GS} - V_t) = \frac{2I_{DS}}{(V_{GS} - V_t)}$$

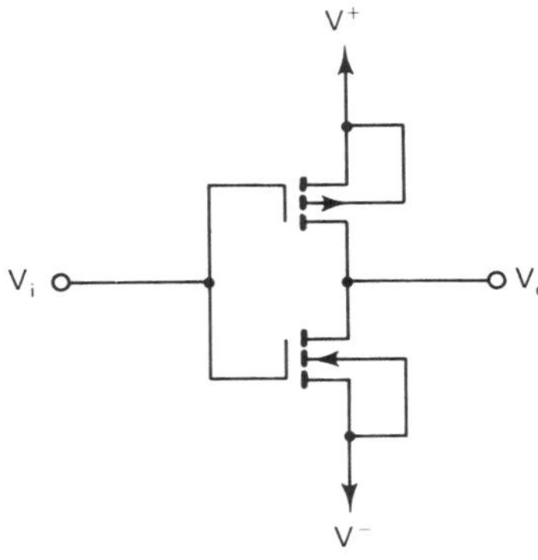


**Figure 7.25** CMOS inverter.

The active load resistance for each FET will be the dynamic drain-to-source resistance $r_{ds}$ of the other FET. This will be given approximately by $r_{ds} = V_A/I_{DS}$, where $1/V_A$ is the channel-length modulation coefficient and $V_A$ is typically in the range 30 to 300 V. Considering the combined effects of the two FETs and the active loads, the CMOS voltage gain in the transition region for the simple case of $K_N = -K_P$, $V_{t_N} = -V_{t_P}$, and $V_{AN} = -V_{AP}$ will be given by

$$A_V = 2g_{fs}\left(\frac{r_{ds}}{2}\right) = \frac{2I_{DS}}{V_{GS} - V_t}\frac{V_A}{I_{DS}} = \frac{2V_A}{V_{GS} - V_t} \tag{7.15}$$

For the case under consideration, a quiescent operating point in the middle of the transition region will be approximately halfway between the two supply voltages, $V^+$ and $V^-$, so that $V_{GS\,(\mathrm{NMOS})} = -V_{GS\,(\mathrm{PMOS})} = (V^+ - V^-)/2$. Therefore, the voltage gain can be written approximately as

$$A_V = \frac{2|V_A|}{(V^+ - V^-)/2 - |V_t|} \tag{7.16}$$

As an example, if $|V_t| = 2.0$ V and $|V_A| = 60$ V for both FETs, and the total supply voltage $(V^+ - V^-)$ is 10 V, we obtain for the voltage gain the result $A_V = 2 \times 30\ \mathrm{V}/(5 - 2)\ \mathrm{V} = 40$. Note that this is under no-load conditions, and the voltage gain can be considerably lower if a substantial load is driven. If the supply voltage is reduced to just 5 V the voltage gain in the transition region would reach a maximum of 240.

## 7.7 CMOS VOLTAGE COMPARATOR

A simple CMOS voltage comparator circuit is shown in Figure 7.26a, and the transfer curve is shown in Figure 7.26b. Transistors $Q_1$, $Q_2$, and $Q_3$ serve as analog switches, and the CMOS inverter can consist of one or several CMOS stages.

When $\phi_1$ is high, transistors $Q_1$ and $Q_2$ will be on and for the CMOS inverter we will have that $V_i = V_O = V_Q$. The voltage across capacitor $C_1$ will be $V_{\mathrm{REF}} - V_Q$. Then $\phi_1$ goes low and the MOSFET switches $Q_1$ and $Q_2$ turn off. When $\phi_2$ then goes high, transistor $Q_3$ turns on. The voltage across capacitor $C_1$ cannot change, however, since there will be no current flow through the capacitor when $Q_2$ is off. Since the voltage at the left side of the capacitor changes from $V_{\mathrm{REF}}$ to $V_A$, the voltage at the CMOS inverter input will correspondingly change by an amount $\Delta V_i = V_A - V_{\mathrm{REF}}$ from $V_Q$ to $V_Q + \Delta V_i$. As shown in Figure 7.26b, this can cause the output voltage $V_O$ to switch to the low state if $V_S > V_{\mathrm{REF}}$ (i.e., $\Delta V_i > 0$), or to the high state if $V_S < V_{\mathrm{REF}}$ (i.e., $\Delta V_i < 0$).

The entire voltage comparator circuit can be implemented with just MOS transistors. As a result, this comparator requires very little chip area, and can be especially useful for such applications as "flash" analog-to-digital converters that require very large numbers ($\gtrsim$100) of comparators on one chip. The very low power drain of the CMOS inverters will also be a very desirable feature for such applications.

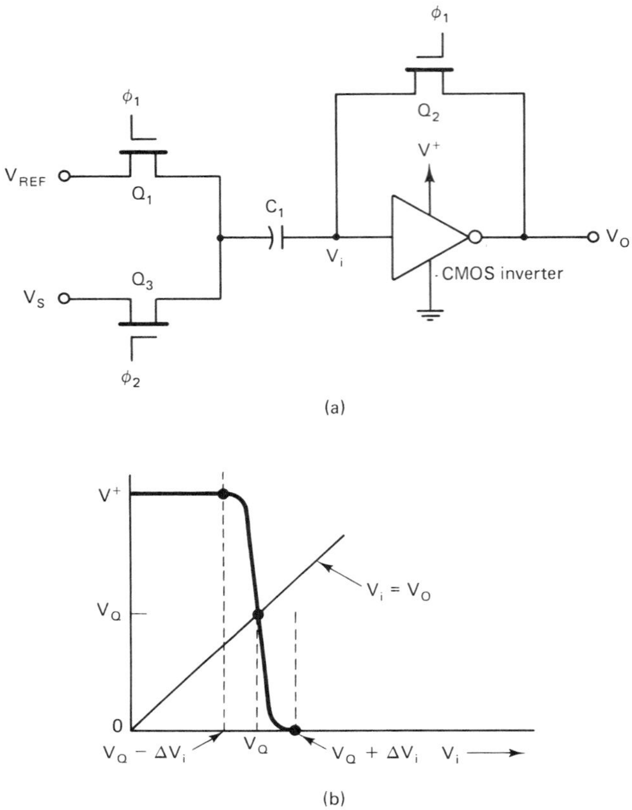


**Figure 7.26** CMOS voltage comparator: (a) CMOS comparator circuit; (b) CMOS transfer curve.

### 7.7.1 Quad CMOS Voltage Comparator

An example of a quad CMOS voltage comparator is the MC14574 (Motorola). A simplified circuit diagram is shown in Figure 7.27. Transistors $Q_1$ and $Q_2$ constitute a PMOS differential amplifier with $Q_3$ and $Q_4$ serving as a current mirror active load. The differential amplifier is biased by the $Q_5$–$Q_6$ current mirror with a quiescent current given approximately by

$$I_Q \simeq I_{\text{set}} \simeq \frac{V^+ - V^- - 1\text{ V}}{R_{\text{set}}}$$

where $R_{\text{set}}$ is an external programming resistor.

The second stage uses $Q_7$ as a common-source amplifier with $Q_8$ serving as a current source active load. The third stage uses the $Q_9$–$Q_{10}$ CMOS pair which drives the $Q_{11}$–$Q_{12}$ CMOS output stage.

When biased at $I_Q = I_{\text{set}} = 50\ \mu\text{A}$ this comparator offers an open-loop gain, $A_{OL}(0)$, of 96 dB, but the propagation delay time will be approximately 1 $\mu$s.

A second example of a quad CMOS voltage comparator is the TLC374 (Texas Instruments) which comes in a 14-pin DIP package. The differential-amplifier input

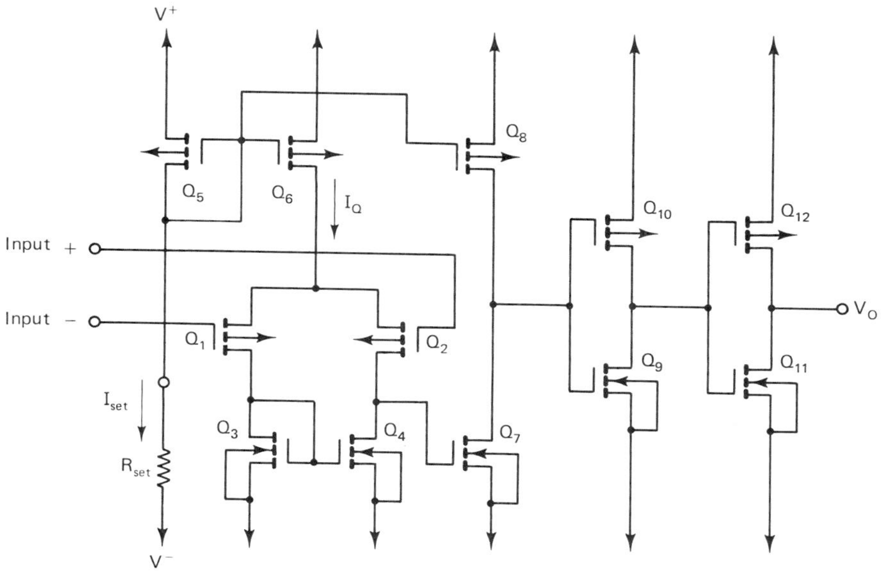


**Figure 7.27** CMOS voltage comparator: MC14574 simplified circuit diagram. (Courtesy of Motorola Semiconductor Inc.)

stage uses a PMOS source-follower/NMOS common-gate configuration with a PMOS current mirror active load. The second stage is a PMOS common-source amplifier with a current source active load. This is followed by three cascaded CMOS inverter stages. The output stage is an NMOS common-source amplifier with an uncommitted drain such that an external pull-up resistor must be used, as is the case of the LM139. MOSFET op amps and voltage comparators often exhibit large offset voltage temperature coefficients and long term drifts due to shifts in the threshold voltage caused by the migration of sodium ions through the gate oxide. This device uses a phosphorus-doped polycrystalline silicon gate technology which acts to immobilize sodium ions and as a result stabilize the MOSFET threshold voltage. For the TLC374, the offset voltage temperature coefficient is down to just 0.7 $\mu$V/°C and the long term drift is 0.1 $\mu$V/month (typ.). Along with the low input bias current of 20 pA and offset current of 1 pA, this device offers the special feature of single supply operation with d-c supply voltages as low as 2 V. The response time to a 100 mV step with a 5 mV overdrive is 650 ns, and for a TTL level input step the response time is down to 150 ns.

## PROBLEMS

**7.1.** (*Pulse-width modulator*) Given: Comparator with triangular waveform (see Figure P7.1a) applied to the inverting input and an analog signal applied to the noninverting input terminal.

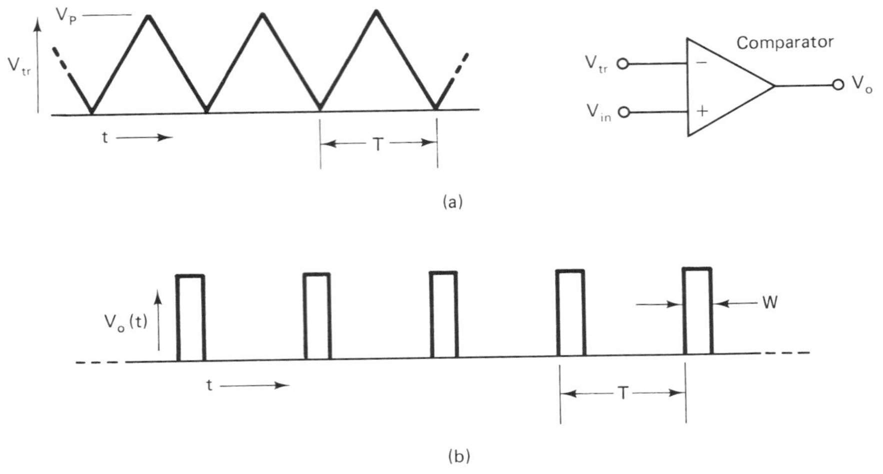


**Figure P7.1**

**(a)** Show that the output voltage waveform will be as shown in Figure P7.1b with a pulse width given by $W = (V_{\text{in}}/V_P)T$ for $0 \leq V_{\text{in}} \leq V_P$.

**(b)** If $V_P = +10$ V and the voltage-to-pulse width conversion accuracy is to be within $\pm 1\%$ over a 1000:1 input voltage range, find the maximum value for $V_{OS}$ if no offset voltage nulling is used. [*Ans*.: $\pm 0.1$ mV (max.)]

**(c)** If $V_{OS}$ is nulled out at 25°C and the operating temperature range is to be from 0 to +50°C, find the maximum acceptable value for $TC_{V_{OS}}$. (*Ans*.: $\pm 4$ $\mu$V/°C)

**(d)** If $TC_{V_{OS}}$ is related to $V_{OS}$ by $TC_{V_{OS}} = V_{OS}/T$, find the maximum acceptable value of $V_{OS}$ for the above. [*Ans*.: ($\pm 1.2$ mV (max.)]

**(e)** If $t_{pd\,(\text{L-H})} = 50$ ns and $t_{pd\,(\text{H-L})} = 40$ ns, find the maximum triangular wave frequency for the pulse width error due to the propagation delay times not to exceed 0.5% and the corresponding maximum signal frequency. (*Ans*.: 500 kHz, 250 kHz)

**7.2.** (*Square-wave generator*) Refer to Figure P7.2. The high-state and low-state output-voltage levels of the comparator are $V_{OH}$ and $V_{OL}$, respectively.

**(a)** Show that the frequency of oscillation $f_{\text{osc}}$ will be given by

$$f_{\text{osc}} = \frac{1}{2R_1C_1 \ln 3}$$

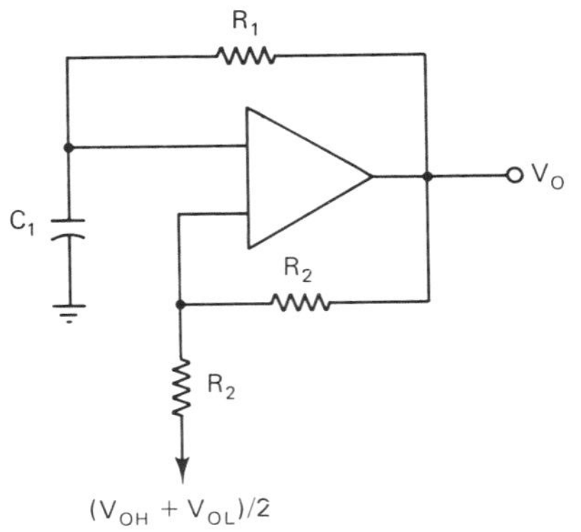


**Figure P7.2**

(b) Show that the output voltage of the comparator will be a square wave.

(c) If $R_1 = 1000\ \Omega$ and $C_1 = 1$ nF, find the frequency of oscillation. (*Ans.*: 455 kHz)

(d) What will limit the maximum frequency of oscillation available from this circuit?

**7.3.** (*Square-wave generator*) Refer to Figure P7.3. The high-state and low-state output-voltage levels of the comparator are $V_{OH}$ and $V_{OL}$, respectively.

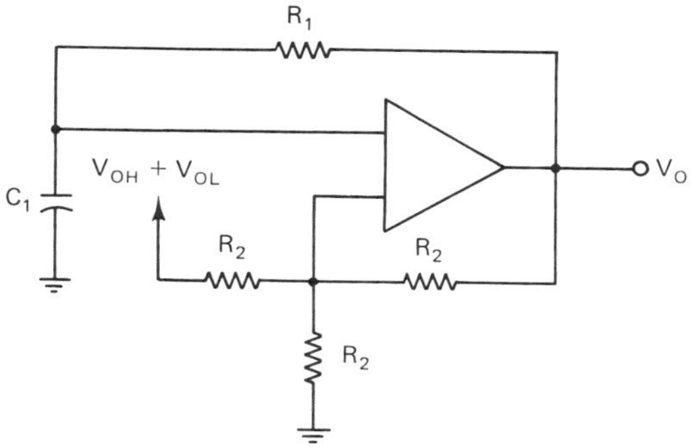


**Figure P7.3**

(a) Show that the frequency of oscillation $f_{osc}$ will be given by

$$f_{osc} = \frac{1}{2R_1C_1 \ln 2}$$

(b) Show that the output voltage of the comparator will be a square wave.

(c) If $R_1 = 1000\ \Omega$ and $C_1 = 1.0$ nF, find the frequency of oscillation. (*Ans.*: 721 kHz)

(d) What will limit the maximum frequency of oscillation available from this circuit?

**7.4.** (*Pulse generator*) Refer to Figure P7.4.

(a) Find: $t_1$. (*Ans.*: 6.9 $\mu$s)

(b) Find: $t_2$. (*Ans.*: 69 $\mu$s)

(c) Find the frequency of oscillation. (*Ans.*: 13.1 kHz)

(d) Sketch $V_O$ versus $t$ for the case in which diodes $D_1$ and $D_2$ are interchanged.

(e) By making the $RC$ time constants ($R_1C_1$ and $R_2C_1$) smaller, the frequency can be increased. What limits the maximum frequency of operation?

(f) What is the function of the pull-up resistor?

**7.5.** (*Time-delay generator*) Refer to Figure P7.5.

(a) Find the time delay of $V_{O_1}$, $V_{O_2}$, and $V_{O_3}$. (*Ans.*: 1.05 $\mu$s, 2.23 $\mu$s, 3.57 $\mu$s)

(b) Find the pulse lengths of $V_{O_1}$, $V_{O_2}$, and $V_{O_3}$. (*Ans.*: 8.95 $\mu$s, 7.77 $\mu$s, 6.43 $\mu$s)

**7.6.** (*Thermostatic control system*) Refer to Figure P7.6. Given: The comparator $C_1$ has $V_{O(\text{high})} = +10$ V (heater on) and $V_{O(\text{low})} = 0$ (heater off). The temperature sensor produces an output voltage given by $V_S = 10$ mV/°C × $T$, where $T$ is the absolute temperature (K). The temperature setpoint $T_{\text{SET}}$ is to be adjustable from 20 to 30°C. The heating system is to turn on when the temperature increases above $T_{\text{SET}} - 1$°C, and is to turn off when the temperature drops below $T_{\text{SET}} + 1$°C. The reference voltage is $V_{\text{REF}} = +1.00$ V.

(a) Find $R_2$, $R_{F_1}$, and $R_{F_2}$. (*Ans.*: 2.0 kΩ, 1.93 kΩ, 100 Ω)

(b) Repeat part (a) for the case in which the reference voltage $V_{\text{REF}}$ has a temperature coefficient of +1000 ppm/°C and is adjusted to +1.00 V at 20°C. (*Ans.*: 1.4 kΩ, 1.93 kΩ, 70 Ω)

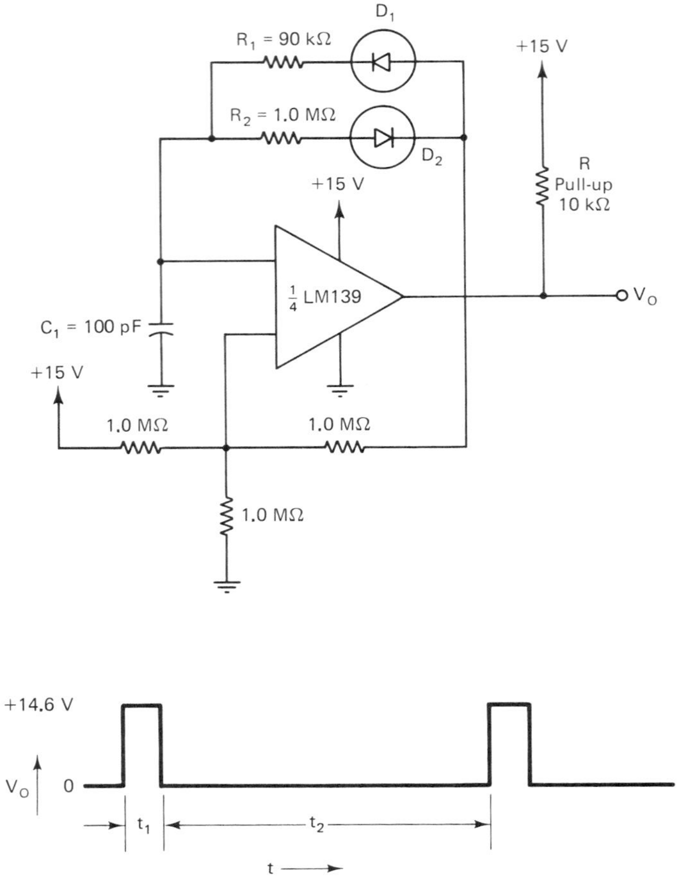


**Figure P7.4**

***7.7.** (*CMOS transfer characteristics*) The drain current for MOSFETs is given by $I_{DS} = k[2(V_{GS} - V_t)V_{DS} - V_{DS}^2]$ in the nonsaturated (triode) region for which $0 < |V_{DS}| < |V_{GS} - V_t|$, and $I_{DS} = k(V_{GS} - V_t)^2 \{1 + [V_{DS} - (V_{GS} - V_t)]/V_A\}$ in the saturated (active) region for which $|V_{DS}| > |V_{GS} - V_t|$ and $|V_{GS}| > |V_t|$, where $k$ and $V_A$ are constants, and $V_t$ is the threshold voltage. A CMOS inverter has a supply voltage of $V_{DD} = +10$ V and $V_{SS} = 0$. For the NMOS transistor $k_N = 0.1$ mA/V², $V_{t_N} = +2.0$ V, and $V_{A_N} = -V_{A_P} = 40$ V.

**(a)** Write a computer program to plot the transfer curve of the CMOS inverter as a function of the $k_P/k_N$ ratio and the PMOS threshold voltage $V_{t_P}$.

**(b)** Find the value of the input voltage $V_i$ at the midpoint of the transition region (i.e., $V_o = \frac{1}{2} V_{DD}$) for the following values of $k_P/k_N$ and $V_{t_P}$:

**(1)** $k_P/k_N = 1.0$, $V_{t_P} = -2.0$ V.

**(2)** $k_P/k_N = 0.4$, $V_{t_P} = -2.0$ V.

**(3)** $k_P/k_N = 1.0$, $V_{t_P} = -4.0$ V.

**(4)** $k_P/k_N = 2.0$, $V_{t_P} = -4.0$ V.

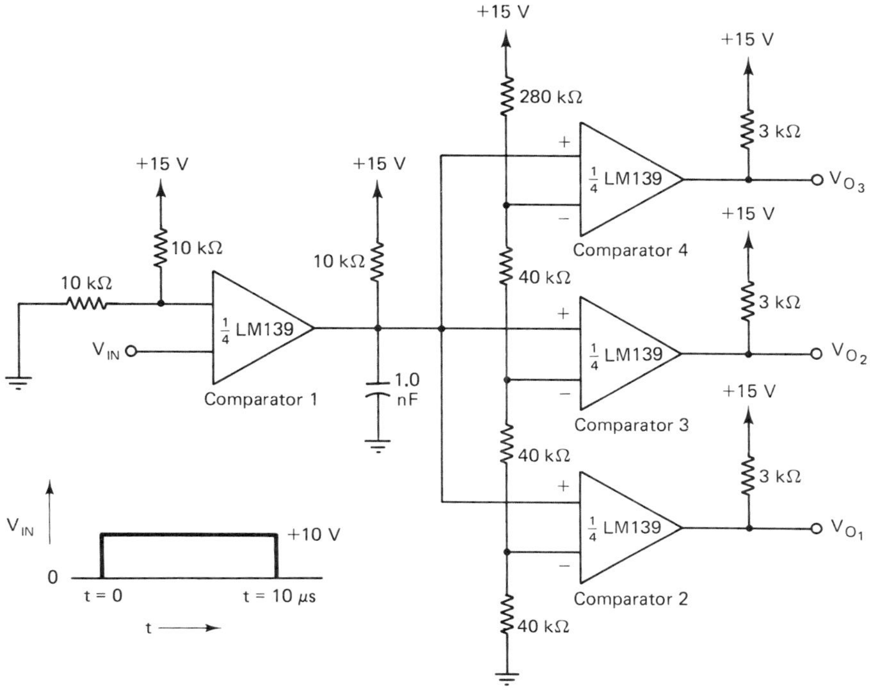


**Figure P7.5**

**(c)** Find the a-c small-signal voltage gain at the midpoint of the transition region for the curves obtained in part (b).

**(d)** Find the width of the transition region $\Delta V_i$ between the points where $V_O = 0.1\,V_{DD}$ and $V_O = 0.9\,V_{DD}$ for the curves obtained in part (b).

**(e)** A CMOS inverter is "self-biased" by a resistor connected between the output and input terminals such that $V_O = V_i$. Find the quiescent point ($Q$-point) for the curves obtained in part (b). Find the a-c small-signal voltage gain at these $Q$-points.

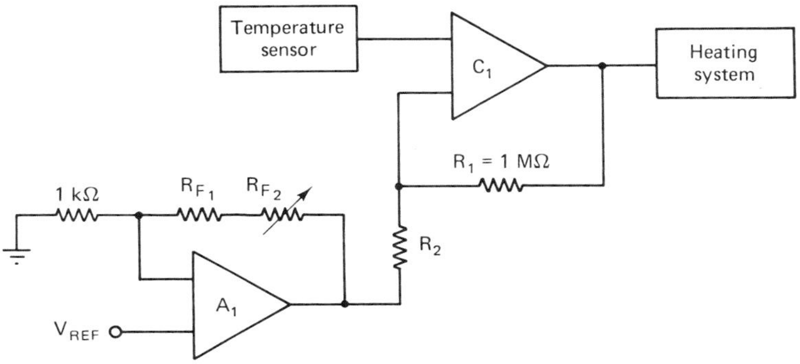


**Figure P7.6**

**7.8.** (*CMOS voltage comparator*) A voltage comparator as shown in Figure 7.26 uses a three-stage CMOS inverter. Each CMOS stage has a voltage gain of 20 in the transition region, and the total supply voltage is $V^+ = 15$ V.

**(a)** Find the approximate width of the transition region $\Delta V_i$ for the CMOS inverter. (*Ans*.: 1.9 mV)

**(b)** The leakage currents of MOSFET switches $Q_1$ and $Q_2$ are each 20 pA (min.) and the minimum clock frequency is to be 10 kHz. Find $C_1$ such that the voltage drift of $V_i$ when $\phi_2$ is high ($\phi_1$ low) will not exceed 1.0 mV. [*Ans*.: 2.0 pF (min.)]

**(c)** If $C_1 = 8.0$ pF and the net shunt capacitance at the input of the CMOS inverter is 4.0 pF, find $\Delta V_i$ in terms of $V_S - V_{REF}$ when $\phi_2$ is high. [*Ans*.: $\Delta V_i = \frac{2}{3}(V_S - V_{REF})$]

**(d)** If the inverter output resistance is 300 Ω, the MOSFET switch on resistance is 200 Ω, and the net capacitance driven by the inverter is 10 pF, find the time required for the CMOS inverter voltage to settle to within 10 mV of the quiescent level ($V_i = V_O = V_Q$) when $\phi_1$ is high. (*Ans*.: 31 ns)

**(e)** Find the maximum clock frequency subject to the conditions of part (d), and the requirement that the time that $\phi_2$ is high should not be less than the time that $\phi_1$ is high. (*Ans*.: 15 MHz)

**(f)** For both the NMOS and PMOS transistors $k = 10\ \mu A/V^2$ and $V_{t_N} = -V_{t_P} = 2.5$ V. Find $V_Q$, $I_{DS}$ at $V_i = V_Q$, and the average power dissipation at $f = f_{MAX}$. (*Ans*.: 7.5 V, 250 μA, 938 μW)

**(g)** Repeat part (d) for a total supply voltage of $V^+ = 10$ V. (*Ans*.: 5.0 V, 63μA, 156 μW)

**(h)** This CMOS voltage comparator is a sampled-data system. Sketch the output voltage waveform for **(1)** $V_S > V_{REF}$ and **(2)** $V_S < V_{REF}$.

## REFERENCES


CONNELLY, J. A., *Analog Integrated Circuits*, Wiley, 1975.

GRAEME, J. G., G. E. TOBEY, and L. P. HUELSMAN, *Operational Amplifiers—Design and Applications*, McGraw-Hill, 1971.

HAMILTON, D. J., and W. G. HOWARD, *Basic Integrated Circuit Engineering*, McGraw-Hill, 1975.

LENK, J. D., *Handbook of Integrated Circuits*, Reston, 1978.

LENK, J. D., *Manual for Integrated Circuit Users*, Reston, 1973.

MILLMAN, J., *Microelectronics*, McGraw-Hill, 1979.

MITRA, S. K., *An Introduction to Digital and Analog Integrated Circuits and Applications*, Harper & Row, 1980.

SEDRA, A. S., and K. C. SMITH, *Microelectronic Circuits*, Holt, Rinehart and Winston, 1982.

# 8

# SPECIAL-FUNCTION INTEGRATED CIRCUITS

The operational amplifier and the voltage comparator are widely used devices for a multitude of circuit and system functions. There are, however, many other types of analog integrated circuits that are somewhat more specialized in function and application. These integrated circuits are used in a wide variety of applications in the areas of communications, control, signal processing, optoelectronics, and digital systems and computer interfacing.

In this chapter a very brief discussion of a number of these special-function integrated circuits will be presented. A more detailed discussion of these integrated circuits is beyond the scope of this book, but is to be found in another book by the author, *Applications of Analog Integrated Circuits*. There are also many references listed at the end of this chapter.

## 8.1 VOLTAGE REGULATORS

A voltage regulator is an electronic device that supplies a constant voltage to a circuit or load. The output voltage of the voltage regulator is regulated by the internal circuitry of the regulator to be relatively independent of the current drawn by the load, the supply or line voltage, and the ambient temperature. A voltage regulator may be part of some larger electronic circuit, but is often a separate unit or module, usually in the form of an integrated circuit.

A basic block diagram of a voltage regulator in its simplest form is shown in Figure 8.1. It is comprised of three basic parts:

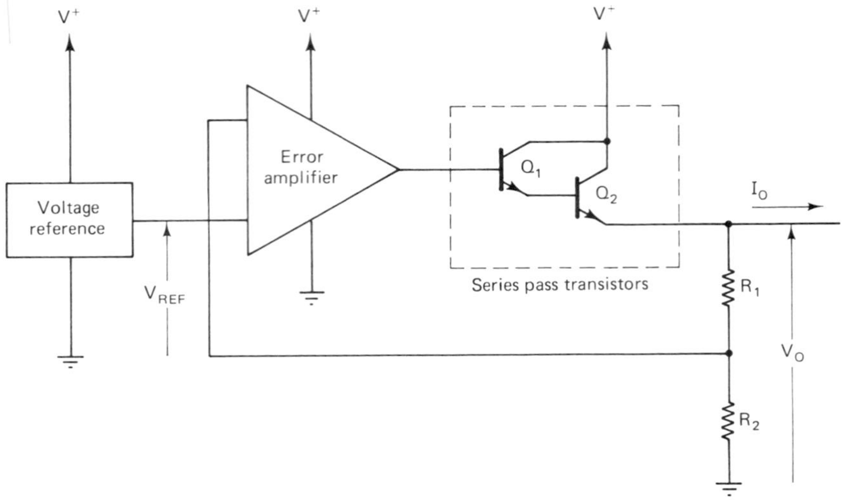


**Figure 8.1** Voltage regulator–basic block diagram.

1. A voltage reference circuit that produces a reference voltage that is independent of temperature and the supply voltage;
2. An amplifier to compare the reference voltage with the fraction of the output voltage that is fed back from the voltage regulator output to the inverting input terminal of the amplifier;
3. A series pass transistor or combination of transistors to provide an adequate level of output current to the load being driven.

The combination of the amplifier (often called an *error amplifier*) and the series pass transistors, together with the resistive voltage divider to tap off a portion of the output voltage, constitutes a feedback amplifier.

In the basic circuit of Figure 8.1 the closed-loop amplifier configuration acts to maintain the fraction of the output voltage fed back to the amplifier inverting input terminal equal to the reference voltage that is applied to the non-inverting input terminal. As a result we have that

$$V_O \frac{R_2}{R_1 + R_2} = V_{\mathbf{REF}}, \text{ so that } V_O = V_{\mathbf{REF}} \frac{R_1 + R_2}{R_2} = V_{\mathbf{REF}}\left(1 + \frac{R_1}{R_2}\right) \tag{8.1}$$

### 8.1.1 Output Resistance-Load Regulation

The ideal voltage regulator would be similar to an ideal voltage source in that the output voltage would be completely independent of the output current, or equivalently of the load impedance. In an actual voltage regulator, as is the case with an actual voltage source, there will be some variation of the output voltage with load or output current. The rate of change of the output voltage with output current has been defined as the output resistance as given by

$$\text{dynamic output resistance} = r_o = -\frac{\Delta V_O}{\Delta I_O}. \tag{8.2}$$

The negative sign in the preceding expression is used because for the reference direction of $I_O$ that has been chosen, an increase in $I_O$ will lead to a decrease in $V_O$. Therefore, with the negative sign in front of $\Delta V_O/\Delta I_O$, the output resistance will turn out to be an algebraically positive quantity.

For a closed-loop amplifier the following relationship exists between the closed-loop output impedance and the open-loop output impedance

$$z_{oCL} = z_{oOL}\,(A_{CL}/A_{OL}) \tag{8.3}$$

For the voltage regulator circuit shown in Figure 8.1 the amplifier input voltage is $V_{REF}$ and the output voltage is $V_O$ so that the closed-loop voltage gain will be given by $A_{CL} = V_O/V_{REF}$. Note that the series pass transistors (the Darlington configuration) is within the feedback loop and so can be considered to be part of the amplifier. Since the voltage gain of the Darlington emitter-follower stage will be close to unity, the open-loop gain, $A_{OL}$, will be essentially that of the error amplifier.

In addition to the basic elements of the voltage regulator that are shown in Figure 8.1, most voltage regulators also include circuitry to protect the voltage regulator from excessive power dissipation which can lead to overheating and permanent damage. For most voltage regulators this internal protection circuit takes the form of a current limiting circuit that senses the output current and acts to prevent the current from exceeding some predetermined safe limit. In some cases the output voltage or the input-to-output voltage differential is also sensed to produce a "foldback" type of current limiting characteristic which can permit a much larger output current limit than for the case of simple current limiting in which only the output current is monitored. In many IC voltage regulators a thermal shutdown circuit is also incorporated on the chip. This circuit senses the chip temperature. When the temperature exceeds some predetermined design limit, the circuit acts to shut down, or at least decrease the output current supplied by the regulator.

There are many types of IC voltage regulators with output voltages of both positive and negative polarities being available, including adjustable voltage regulators, fixed output voltage, three terminal regulators, and switching regulators. The adjustable voltage regulators use an external voltage divider to tap off part of the output voltage and feed it back to the error amplifier in the regulator where it is compared to an internally generated reference voltage. The output voltage can be adjusted from a value equal to the reference voltage (usually about 1.8 V) up to a maximum value that is about 2 V below the unregulated input voltage.

In the three terminal voltage regulator the output voltage is fixed at some value determined by the design of the regulator. The three terminals are for the unregulated input voltage, the regulated output voltage, and the common or ground terminal. The three terminal regulators are the simplest to use and although they offer only fixed output voltages, there are a wide variety of voltages available of both polarities. The output voltages of commercially available three terminal regulators are 5, 5.2,6, 8,10,12,15,18, and 24 V with output currents in the range 100 mA to 3 A. The line regulation for these regulators is generally in the range 0.005 to 0.02 %/V, the no-load to full-load output or load regulation in the range 0.1 to 1%, and a ripple rejection in the range 65 to 85 dB.

The output current capability of both types of regulators can be greatly increased by adding one or several external current boost transistors. The output current rating will then be principally determined by the power dissipation rating of the external current boost transistors.

A useful and efficient type of voltage regulator is the switching regulator. In the non-switching type of regulator, the series pass transistors operate continuously in the active region in which the combination of a large current flow and voltage drop produce a large power dissipation. In contrast to this, for the switching regulator, the series pass transistors are not operated in the active mode, but rather are switched back and forth between cutoff and saturation. The average or d-c output voltage level is controlled by the fraction of time these transistors are on. When operating in the cutoff region the power dissipation is negligible due to the very small amount of current through the transistors, and in the saturation mode the power dissipation is also very small due to the small voltage drop across the output transistor. As a result, efficiencies in the 90% range can be obtained as compared to values generally in the 50 to 70% range for the non-switching regulators.

## 8.2 IC POWER AMPLIFIERS

Most ICs have a power dissipation rating of 1 watt or less, and the maximum amount of a-c power that can be delivered to a load will generally be less than $\frac{1}{2}$ watt. There are, however, some ICs that are capable of delivering amounts of a-c power to a load ranging up to several watts, and with the addition of external power transistors, output power levels in excess of 50 watts can be obtained. The IC power amplifiers are generally fixed gain, single supply audio amplifiers. Some examples of these are the MC1306P (Motorola), the LM380, and the LM384 (both National Semiconductor). These IC audio power amplifiers have internal feedback loops to set the quiescent output voltage level at a value approximately equal to one-half of the d-c supply voltage. This permits the maximum symmetrical output voltage swing for the class AB push-pull emitter-follower output stage. The power dissipation rating of these ICs is very much dependent on the thermal impedance between case and ambient. This thermal impedance can be greatly reduced by the addition of a suitable heat sink to the IC with a consequent increase in the amount of a-c power that can be delivered to the load.

The power output capability of op amps can be greatly increased by using external "current boost" power transistors as shown in Figure 8.2. In Figure 8.2a transistors $Q_3$ and $Q_5$ are the current boost transistors and are driven as a complementary push-pull output stage. Transistors $Q_4$ and $Q_6$ together with the $R_{CL}$ resistors perform the current limiting function to protect $Q_3$ and $Q_5$ against excessive power dissipation. For even greater output currents, the Darlington configuration of Figure 8.2b can be used. In this circuit the $Q_3$-$Q_7$ combination is a direct-coupled PNP-NPN combination used for sourcing currents to the load, and in a similar fashion the $Q_5$-$Q_8$ PNP-NPN combination is used for sinking load currents.

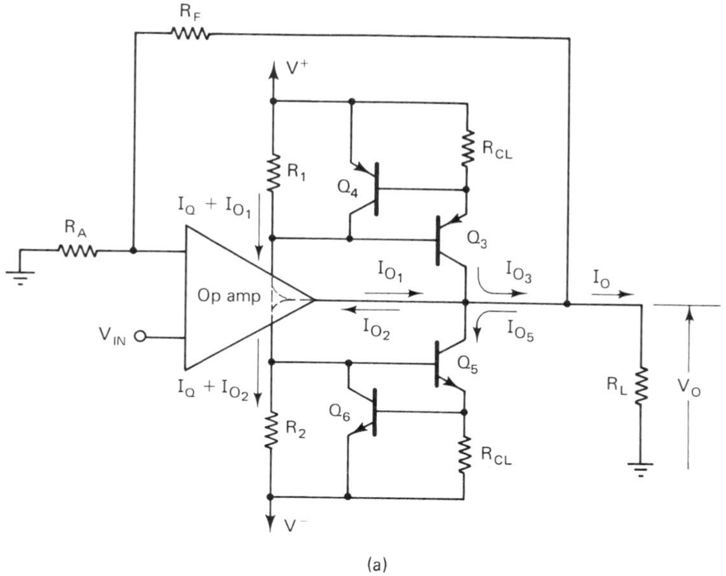


(a)

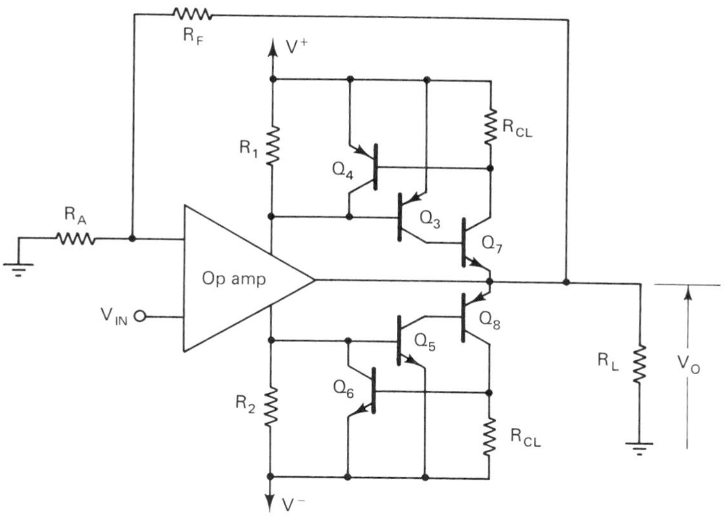


(b)

**Figure 8.2** Operational amplifier with external power transistors.

## 8.3 VIDEO AMPLIFIERS

For operational amplifiers the emphasis is on a high open-loop voltage gain, typically of the order of 100,000 (100 dB) to 1,000,000 (120 dB) at low frequencies. Operational amplifiers are almost always operated in a closed-loop configuration and the high open-loop gain values are needed in order to obtain a high loop gain. The high loop gain is desirable in order to minimize the gain error, to produce a large input impedance, and to obtain a very small output impedance.

In order to obtain the high voltage gain, sacrifices are made in the frequency response. Indeed, in order to stabilize the amplifier against an oscillatory response, the open-loop frequency response is usually made to start rolling off at a very low frequency, often at only around 10 Hz. The resulting unity gain frequencies are

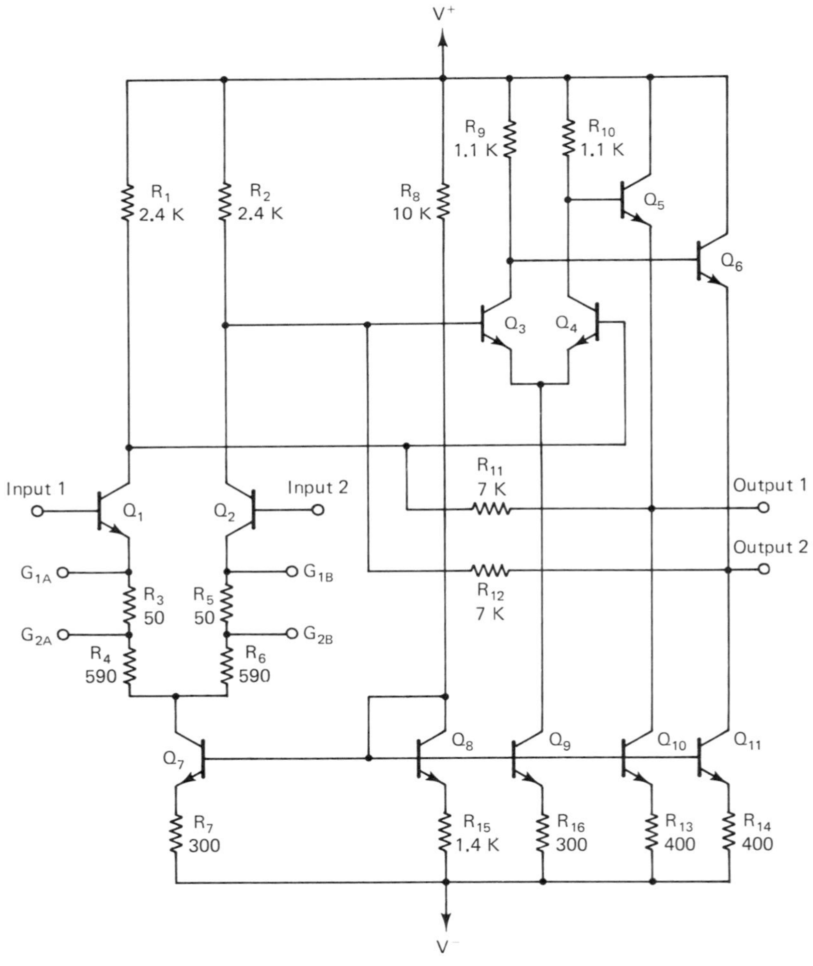


**Figure 8.3** Type 733 video amplifier.

generally in the 1 to 3 MHz range, with some operational amplifiers extending up to about 10 MHz.

Video or wideband amplifiers, are designed to give a relatively flat gain versus frequency response characteristics over the frequency range that is generally required to transmit video information. This frequency range is from low frequencies, generally down in the range of about 20 Hz, up to several MHz. For television applications bandwidths generally in the range of 4 to 6 MHz are required. For some other applications bandwidths as high as 50 MHz may be needed.

In contrast with this, the bandwidth required for audio amplifiers extend only over the frequency range corresponding to what the human ear is sensitive to, or from about 50 Hz to 15 kHz. For many applications, such as telephony or AM radio, a considerably narrower bandwidth is used, typically from 100 Hz to 5 kHz. In Figure 8.1 the frequency response characteristics of video and audio amplifiers are compared.

The principal technique that is used to obtain the large bandwidths that are required for video amplifiers is the trading off of gain for increased bandwidth. This trade-off is accomplished by use of reduced load resistance for the various gain stages of the amplifier and by the use of negative feedback. In many, if not most video amplifiers, both techniques of reduced load resistance and negative feedback are employed at the same time.

A popular type of IC video amplifier is the 733 type as shown in Figure 8.3.

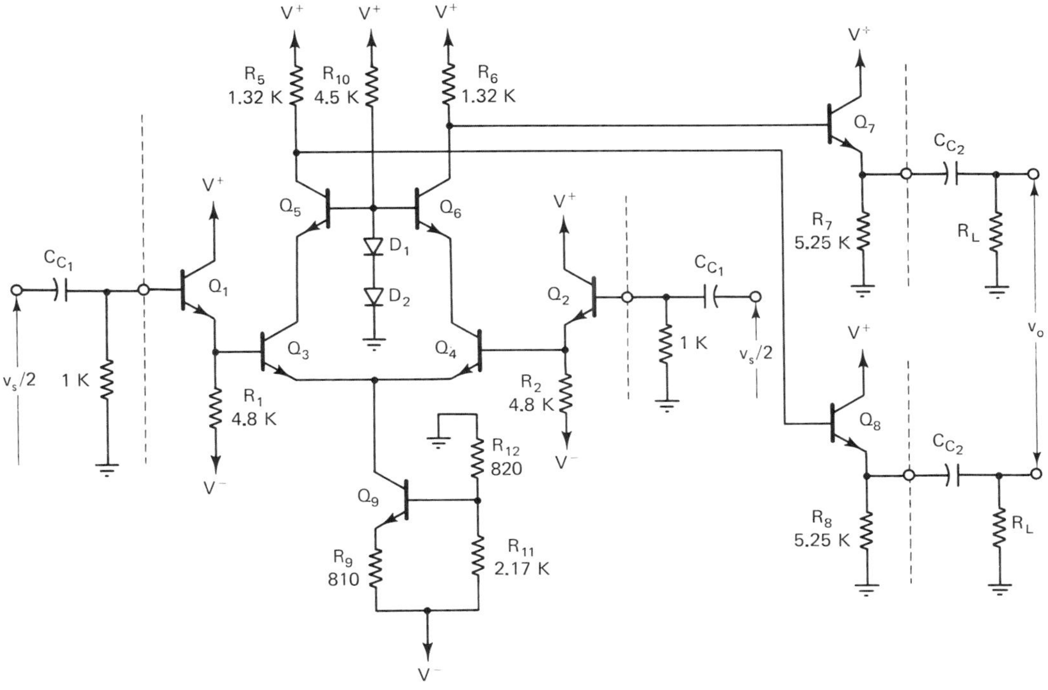


**Figure 8.4** Cascode video amplifier integrated circuit.

It consists of two cascaded NPN differential amplifiers and a balanced emitter-follower output stage. A wide bandwidth is achieved by the use of low value load resistances for both differential amplifier stages and the use of internal feedback loops. One feedback loop uses the emitter resistances of the first differential amplifier stage and the other uses the 7 kΩ feedback resistors around the second stage. This IC offers three fixed gains of 400, 100, and 10 and the corresponding bandwidths are 40 MHz, 90 MHz, and 120 MHz, respectively.

A second example of an IC video amplifier is the CA3140 (RCA). This IC uses a differential cascode amplifier configuration as shown in Figure 8.4. This device offers a bandwidth of 55 MHz at a fixed voltage gain of about 110.

Wide bandwidth op amps are also available as exemplified by the HA-2539 (Harris Semiconductor). This op amp has a gain-bandwidth product of 600 MHz, a slewing rate of 600 V/μs, and a full-power bandwidth of 9.5 MHz when driving a 1000 Ω load with a 20 V peak-to-peak output voltage swing.

## 8.4 DIGITAL-TO-ANALOG CONVERTERS

A digital-to-analog converter (DAC) is an integrated circuit that converts a digital input code to an analog output voltage or current that is the binary weighted equivalent of the digital input code. A simple example of an $N$-bit DAC is shown in Figure 8.5. The output voltage will be given by

$$\begin{aligned} V_O &= R_F\left(\frac{b_1 V_{\mathrm{REF}}}{R} + \frac{b_2 V_{\mathrm{REF}}}{2R} + \frac{b_3 V_{\mathrm{REF}}}{4R} + \cdots + \frac{b_N V_{\mathrm{REF}}}{2^{N-1}R}\right) \\ &= 2\left(\frac{R_F}{R}\right) V_{\mathrm{REF}}\left(\frac{b_1}{2} + \frac{b_2}{4} + \frac{b_3}{8} + \cdots + \frac{b_N}{2^N}\right) \end{aligned} \tag{8.4}$$

where $b_1$ is the most significant bit (MSB), $b_2$ is the next significant bit, . . . , and $b_N$ is the least significant bit (LSB).

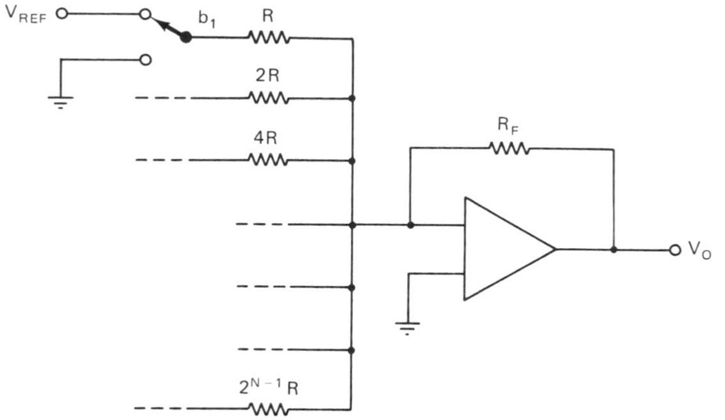


**Figure 8.5** $N$-bit digital-to-analog converter.

The output voltage corresponding to the LSB input (. . . 000001) will be $V_{O\,(LSB)} = 2(R_F/R)V_{REF}\,(1/2^N) = (R_F/R)V_{REF}\,(1/2^{N-1})$. The output voltage corresponding to a MSB digital input (10000. . .) will be $V_{O\,(MSB)} = V_{O\,(LSB)}2^{N-1}$.

The maximum output voltage will occur when all bits are "1" (1111. . .) and will be $V_{O\,(MAX)} = 2V_O(\text{MSB}) - V_{O\,(LSB)} = (2^N - 1)V_{O\,(LSB)} = 2V_{O\,(MSB)}(1 - 1/2^N)$. The *nominal full-scale output voltage* $V_{O\,(FS)}$ is defined as

$$V_{O\,(FS)} = 2V_{O\,(MSB)} = 2^N V_{O\,(LSB)} \tag{8.5}$$

and is thus greater than the maximum output voltage by an amount equal to $V_{O\,(LSB)}$.

Many DACs are set to have a full-scale output voltage of 10.000 V. For a 4-bit ($N = 4$) DAC, $V_{O\,(LSB)} = 10.00\text{ V}/2^N = 10.00/2^4 = 0.625$ V. The output voltage for a MSB input (1000) will be $V_{O\,(MSB)} = V_{O\,(FS)}/2 = 5.00$ V. The output voltage for a digital input of 1111 will be $V_{O\,(MAX)} = V_{O\,(FS)} - V_{O\,(LSB)} = 10.00 - 0.625 = 9.375$ V.

In Figure 8.6 the ideal transfer characteristic of a 4-bit DAC is shown. For this ideal DAC the envelope of the transfer characteristic is a straight line passing through the origin (i.e., $V_O = 0$ for a digital input code of 0000) and through the point $V_{O\,(MAX)} = V_{O\,(FS)} - V_{O\,(LSB)}$ for an input code of 1111. There are many types of DACs available, ranging from 8-bit devices up to 16-bit DACs, with conversion speeds from 5 μs to as fast as 35 ns.

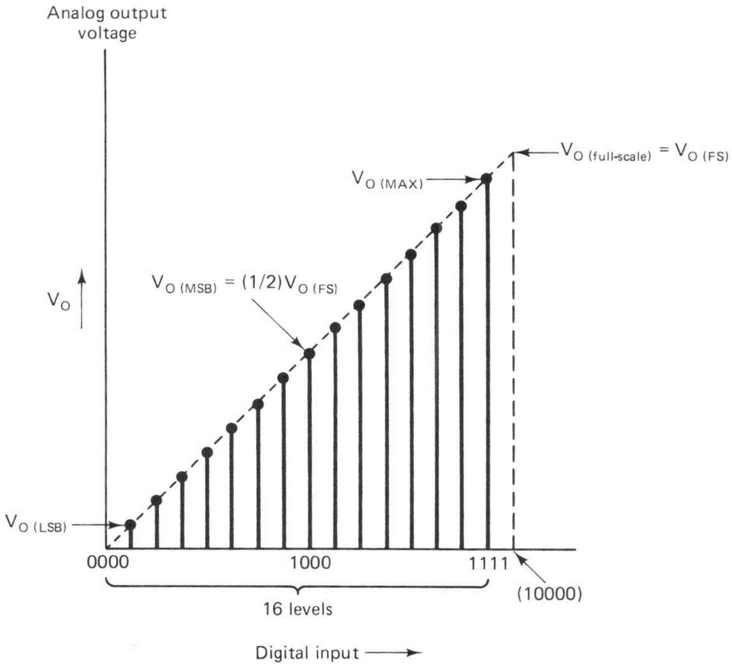


**Figure 8.6** Transfer characteristic of ideal 4-bit DAC.

## 8.5 ANALOG-TO-DIGITAL CONVERTERS

An analog-to-digital converter (ADC) converts an analog input voltage to a binary digital output code that corresponds on a stepwise basis to the analog input.

In Figure 8.7 the transfer characteristic of an ADC is shown. We note that since the digital output code can take on only certain discrete values, there will be an inherent error in the system called the *quantization error*. The quantization error will be defined as the difference between the analog voltage equivalent of the digital output code and the actual analog input voltage. A graph of the quantization error corresponding to the transfer characteristic of Figure 8.7 is presented in Figure 8.7. Note that the maximum value of the quantization error is 1 LSB.

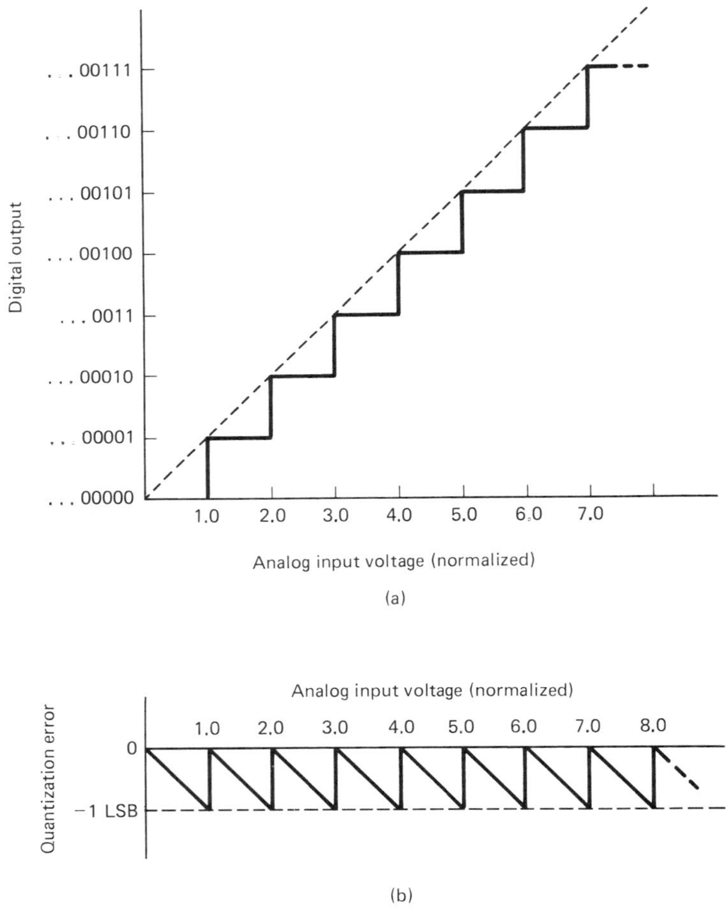


**Figure 8.7** ADC transfer characteristic and quantization error: (a) transfer characteristic; (b) quantization error.

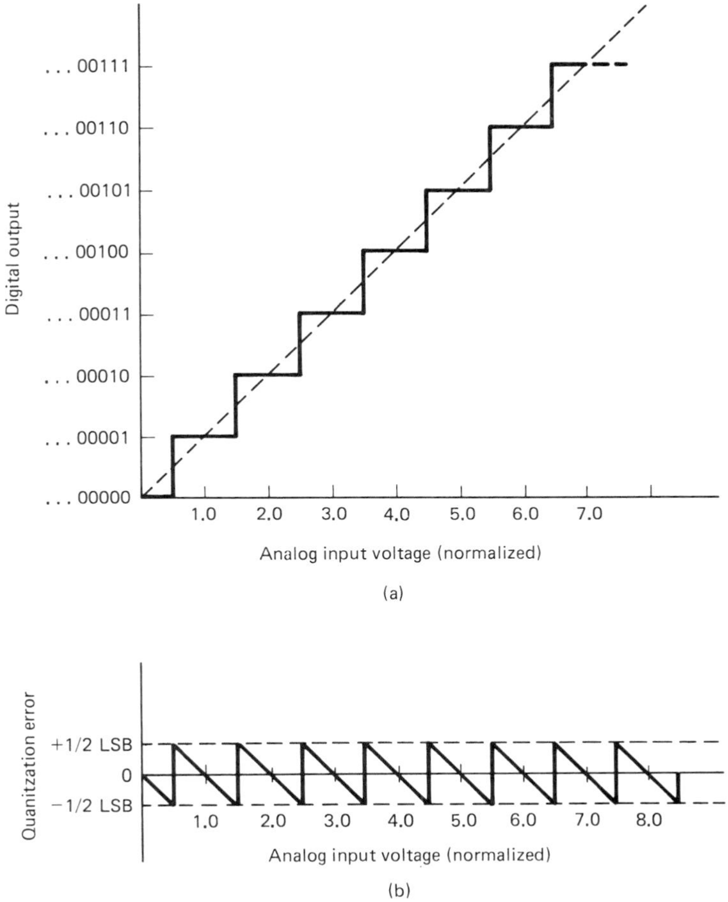


**Figure 8.8** ADC transfer characteristic and quantization error with $\frac{1}{2}$-LSB offset of analog input voltage: (a) transfer characteristic: (b) quantization error.

The maximum quantization error can be reduced to $\pm\frac{1}{2}$ LSB by adding a $-\frac{1}{2}$-LSB voltage offset to the analog signal, or equivalently adding $+\frac{1}{2}$ LSB to the digital reference voltage to which the analog signal is compared. The transfer characteristic obtained with this $\frac{1}{2}$-LSB offset is shown in Figure 8.8a and the corresponding quantization error is shown in Figure 8.8. Note that the maximum quantization error is the same for every range of input voltage and is equal to $\pm\frac{1}{2}$ LSB.

A simple example of the effect of $\frac{1}{2}$-LSB offset on the quantization error can be seen for the case of a 3-bit parallel comparator type of ADC. Let us assume a LSB voltage of 1.0 V and a full-scale input voltage of 8.0 V. Without the $\frac{1}{2}$-LSB offset the seven comparator reference voltages will be 1.0, 2.0, 3.0, 4.0, 5.0, 6.0, and 7.0 V. The digital output codes obtained will be as follows:

| Analog input voltage | Digital output | Analog equivalent of output |
|---|---|---|
| $0 < V_A < 1.0$ | 000 | 0 |
| $1.0 < V_A < 2.0$ | 001 | 1.0 |
| $2.0 < V_A < 3.0$ | 010 | 2.0 |
| $3.0 < V_A < 4.0$ | 011 | 3.0 |
| $4.0 < V_A < 5.0$ | 100 | 4.0 |
| $5.0 < V_A < 6.0$ | 101 | 5.0 |
| $6.0 < V_A < 7.0$ | 110 | 6.0 |
| $7.0 < V_A < 8.0$ | 111 | 7.0 |

The maximum digital output code will be 111 and the maximum quantization error will be 1 LSB in every 1.0 V (1 LSB) input voltage range. If the reference voltages are now shifted by $-\frac{1}{2}$ LSB to become 0.5, 1.5, 2.5, . . . , 6.5 V, the following transfer relationship will be obtained:

| Analog input voltage | Digital output | Analog equivalent of output |
|---|---|---|
| $0 < V_A < 0.5$ | 000 | 0 |
| $0.5 < V_A < 1.5$ | 001 | 1.0 |
| $1.5 < V_A < 2.5$ | 010 | 2.0 |
| $2.5 < V_A < 3.5$ | 011 | 3.0 |
| $3.5 < V_A < 4.5$ | 100 | 4.0 |
| $4.5 < V_A < 5.5$ | 101 | 5.0 |
| $5.5 < V_A < 6.5$ | 110 | 6.0 |
| $6.5 < V_A < 7.5$ | 111 | 7.0 |

The maximum quantization error is $\pm\frac{1}{2}$ LSB in every range.

### 8.5.1 Parallel Converter

The fastest type of analog-to-digital converter (ADC) is the *parallel comparator ADC*, also known as a *simultaneous* or *flash ADC*. An example of a 3-bit ADC of this type is shown in Figure 8.9. For this ADC seven comparators are required. In general, for an $N$-bit ADC of the parallel conversion type the number of comparators required will be $2^N - 1$. Although this type of ADC has the advantage of being a very fast ADC, the number of comparators required increases very rapidly with the number of bits $N$, and usually becomes an uneconomical approach for more than 3 or 4 bits.

The reference voltage divider sets up the following reference levels for the comparators: $V_{R1} = (R/2)/7R\,V_{\text{REF}} = \frac{1}{14}V_{\text{REF}}$, $V_{R2} = \frac{3}{14}V_{\text{REF}}$, $V_{R3} = \frac{5}{14}V_{\text{REF}}$, $V_{R4} = \frac{7}{14}V_{\text{REF}}$, $V_{R5} = \frac{9}{14}V_{\text{REF}}$, $V_{R6} = \frac{11}{14}V_{\text{REF}}$, and $V_{R7} = \frac{13}{14}V_{\text{REF}}$. The ADC transfer characteristic is as shown in Figure 8.8 with the normalized analog input voltage given by $V_A/(V_{\text{REF}}/7)$.

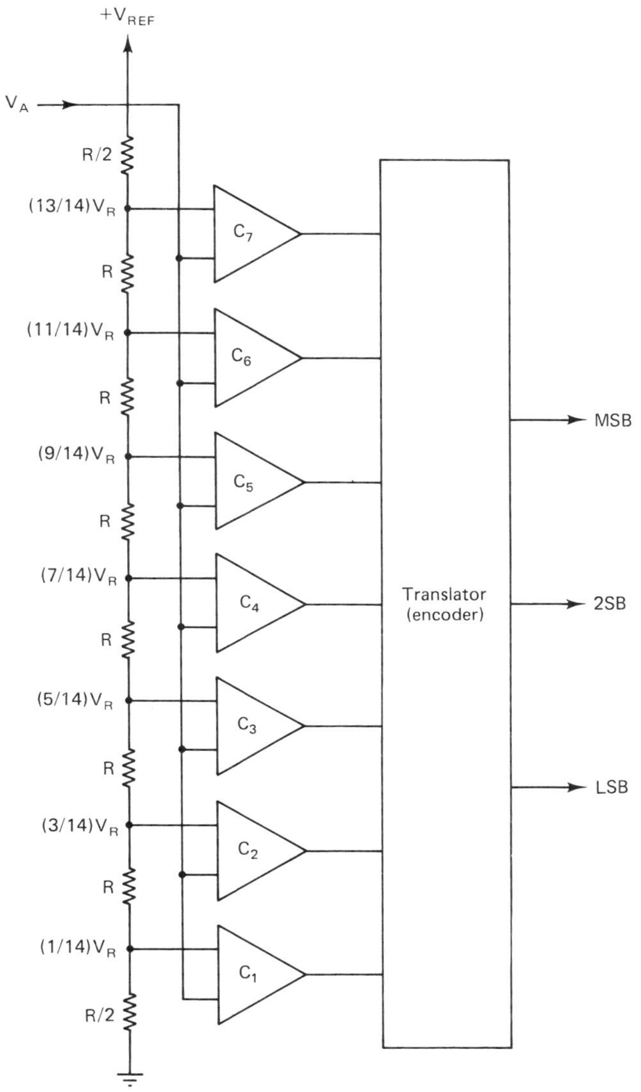


**Figure 8.9** Parallel comparator analog-to-digital converter.

The *quantization error* of an ADC is the range of analog input voltages that will produce a given digital output code, measured with respect to the midpoint value of the analog input. For the ADC of Figure 8.9 the choice of comparator reference voltage levels is such that the quantization error is uniformly distributed so that the quantization error is the same for every digital output level and is $\frac{1}{14}V_{\mathrm{REF}}$. Since the total span of the analog input voltage corresponding to a 1-LSB change in the digital output is $\frac{2}{14}V_{\mathrm{REF}}$, the quantization error of $\frac{1}{14}V_{\mathrm{REF}}$ can be expressed in terms of the digital output as $\pm\frac{1}{2}$ LSB.

Parallel comparator ADCs with conversion times down as low as 10 ns, which is equivalent to $10^8$ conversions per second, are available.

## 8.5.2 Counting Type

Many types of ADCs use DACs in feedback loop. A simple ADC that uses a DAC in its feedback loop is the counting type as shown in Figure 8.10. The digital output of a modulus $N$ binary counter goes to an $N$-bit digital-to-analog converter (DAC), which produces the staircase type of waveform shown in Figure 8.10b. The DAC output voltage goes up in increments of 1 $V_{LSB}$ up to a full-scale maximum of $2^N$ $V_{LSB}$. This voltage is compared to the analog input voltage by the comparator. As long as the DAC voltage stays below the analog input voltage $V_A$, the comparator

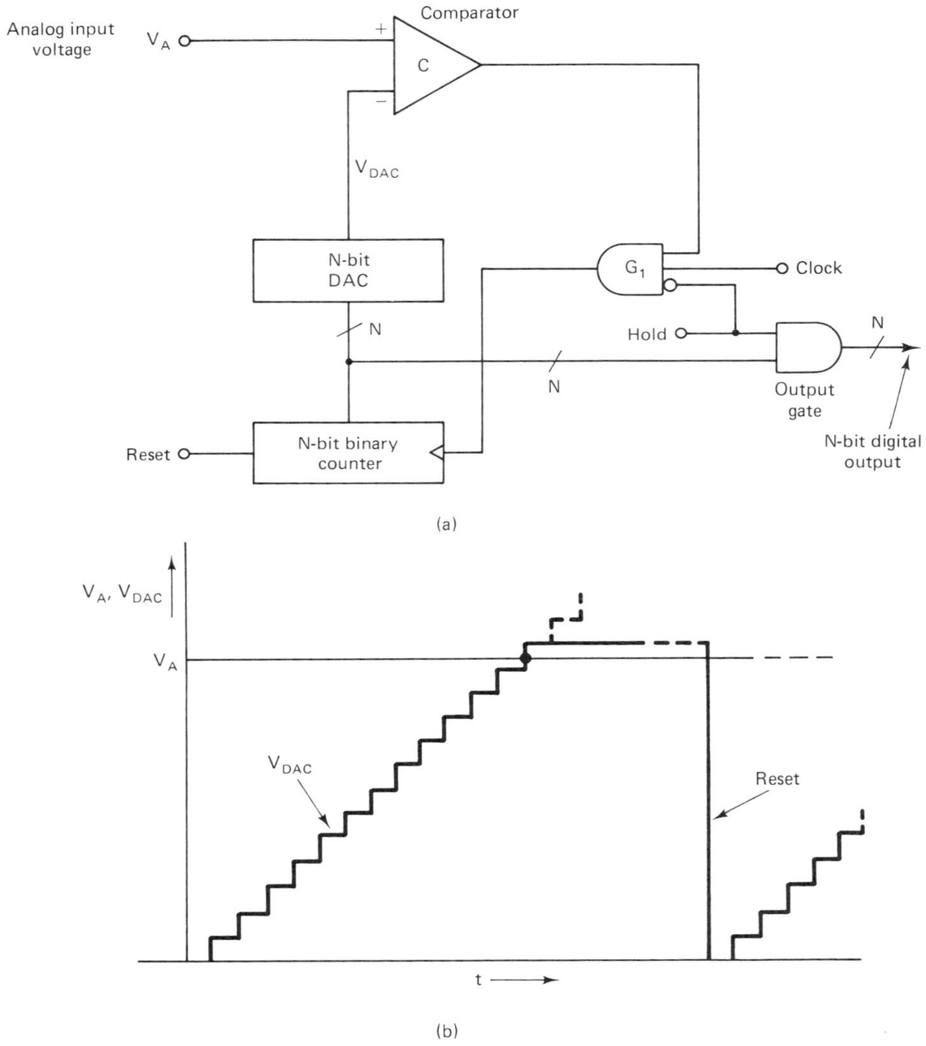


**Figure 8.10** Counting analog-to-digital converter: (a) block diagram of system; (b) DAC output waveform.

output voltage will stay in the *high* state. This will keep the clock gate $G_1$ enabled such that the clock pulses pass through to the counter so that the count will continue to increase and the DAC output voltage will continue with its incremental increases. As soon as the DAC voltage rises above the analog input voltage level the comparator output will go *low*. This will disable the clock gate $G_1$, so that the count will stop at this point.

Some time after this the HOLD voltage goes *high*, which enables the $N$-bit digital output gate so that the binary count of the counter now becomes available as the digital output of the ADC. The HOLD pulse then terminates and the RESET pulse then resets the counter back to zero and the counter starts to count up again, thus starting another conversion cycle.

For an $N$-bit counting converter a total conversion time of $2^N$ clock cycles is required for a full-scale input voltage. As a result, this type of ADC will be a relatively slow converter. For a 14-bit ADC of this type the conversion time will be $2^{14} T_{CLOCK} = 16{,}384 T_{CLOCK}$. If $T_{CLOCK} = 100$ ns is chosen as a representative value, the resulting conversion time will be 1.64 ms and the corresponding conversion rate will be 610 conversions per second.

The average conversion time can be reduced substantially if an up/down counter is substituted for the simple binary up counter and if the conversion cycles are terminated shortly after the counting stops. This type of counting ADC is called a *tracking* or *servo ADC*. In this type of converter the counter is not reset to zero at the beginning of every conversion cycle, but rather is given a command to either continue to count up or to count down from the previous count, this depending on whether the analog input voltage is above or below the DAC output voltage at the beginning of the conversion cycle, respectively.

For a slowly changing analog voltage the change in the count needed to match the DAC voltage with the analog input voltage will be small, so that the conversion time will be very short. For a full-scale analog voltage change occurring in a short period of time the required conversion time will become equal to that of the simple counting ADC.

The accuracy of the counting ADC is a function of the offset voltage and voltage gain of the comparator and of the accuracy of the DAC. It will often be the DAC that will be the limiting factor in the overall accuracy of the ADC.

This type of ADC is a feedback ADC because of the presence of a feedback loop around the comparator. This feedback loop consists of the clock gate $G_1$, the binary counter, and the DAC. Another type of feedback ADC that has a DAC in the feedback loop is the successive approximation converter (SAR) and it will be discussed next. The SAR converter offers the same basic accuracy as the counting ADC, but it offers a much shorter conversion time and is a very widely used ADC.

### 8.5.3 Successive Approximation Analog-to-Digital Converter

The *successive approximation register* (SAR) type of ADC is one of the most popular types of ADC because it offers the combination of high accuracy and high conversion speed. In Figure 8.11 a basic block diagram of a SAR ADC is shown. The successive

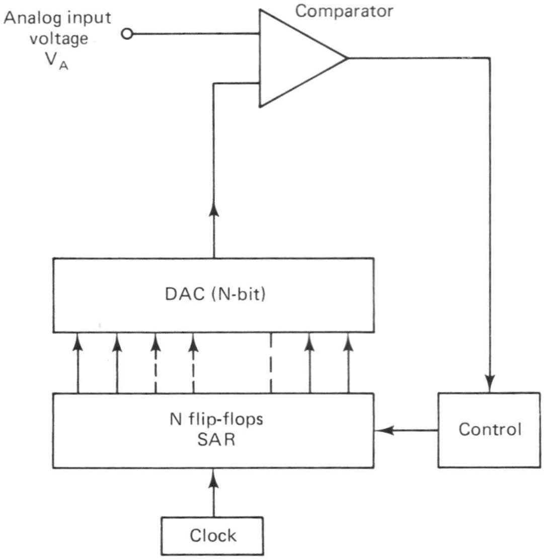


**Figure 8.11** Block diagram of a successive approximation register analog-to-digital converter (SAR ADC).

approximation register for an $N$-bit converter contains $N$ flip-flops (FFs) that are set to the high state ($b = 1$) one at a time to produce the bit inputs to the digital-to-analog (DAC). The output voltage of the DAC is compared to the analog input voltage.

The MSB FF is set to the high state ($b_1 = 1$) first so that the analog input voltage is first compared to a DAC output voltage that corresponds to the MSB voltage level of $V_{MSB} = V_{FS}/2$. If the analog voltage is greater than $V_{MSB}$, the MSB FF remains set ($b_1 = 1$) for the rest of the conversion period. If, however, the analog input voltage is below $V_{MSB}$, the MSB FF is reset to zero ($b_1 = 0$) and remains in this state for the rest of the conversion process.

The next significant bit (second bit) FF is set next ($b_2 = 1$) and the DAC output voltage now corresponds to $V_{FS}(b_1/2 + b_2/4)$ with $b_2 = 1$. If this voltage is less than the analog input voltage $V_A$, the second FF remains set at $b_2 = 1$. If this DAC voltage is more than the analog input voltage, the second FF is reset to give $b_2 = 0$. This process then continues for each successive FF for $N$ clock cycles to complete the conversion process. In each case if the analog input voltage is greater than the DAC voltage, the FFs retain their present state. If the analog voltage is below the DAC voltage, the last FF to be set ($b = 1$) is reset to zero ($b = 0$).

The DAC voltage and the corresponding digital code is thus a successive approximation to the analog input voltage, with each bit being tested at a time beginning with the most significant bit ($b_1$) and continuing one bit at a time down to the least significant bit ($b_N$) at the end of the conversion period.

Let us now consider the example of a 10-bit SAR ADC with a full-scale DAC output voltage of $V_{FS} = 10.24$ V so that $V_{LSB} = 10$ mV. We will also assume that a $\frac{1}{2}$-LSB offset voltage is added to the DAC output voltage in order to minimize the conversion quantization error.

For an analog input voltage of $V_A = 7.500$ V, the SAR bit outputs and the corresponding DAC output voltage values will be as given in Table 8.1. After 10 clock cycles the SAR conversion process is completed and the final status of the SAR FFs is 1011101101, which produces a DAC output voltage of 7.495 V. The resulting quantization error is 5 mV, which corresponds to $\frac{1}{2}$ LSB. If the $\frac{1}{2}$ LSB offset were not present, the DAC output voltage would be 7.490 V, which would result in a quantization error of 10 mV or 1 LSB.

For an analog input voltage of less than 5 mV the digital output will be 0000000000 and the DAC output voltage (including the $\frac{1}{2}$-LSB offset voltage) will be 5 mV. For an analog input voltage between 5 and 15 mV the digital output code will be 0000000001. The maximum DAC output voltage (including the $\frac{1}{2}$-LSB offset) corresponding to a digital output code of 1111111111 will be $V_{MAX} = 10.24 - 10 \text{ mV} + 5 \text{ mV} = 10.235$ V. The maximum analog voltage that can be converted with a quantization error not exceeding $\frac{1}{2}$ LSB (5 mV) is 10.240 V, which corresponds to the DAC full-scale output voltage.

The SAR ADC has the digital code output available in parallel form at the end of the conversion period. It is also possible to obtain a data output in serial form during the conversion period. As each data bit becomes valid at the end of the various clock cycle times during which the DAC output voltage is compared to the analog signal and the FFs are reset if necessary, the bits of the digital output code becomes available starting with the MSB and ending with the LSB at the end of the conversion process.

The total conversion time for an $N$-bit SAR ADC is approximately $(N + 2)T_{CLOCK}$, where $T_{CLOCK}$ is the clock period which is typically of the order of 1 $\mu$s. Therefore, for a 12-bit SAR ADC a total conversion time of around 14 $\mu$s would be required. This is to be compared to a conversion time of $2^N T_{CLOCK}$ for a digital ramp or counting type of ADC so that for a 12-bit digital ramp ADC with a 1-$\mu$s clock cycle time a total conversion time of around 4 ms will be required. Of course, a parallel comparator ADC can produce conversion times of well under 1 $\mu$s, but the hardware requirements for a 12-bit ADC (or even for an 8-bit ADC) would be quite excessive.

**TABLE 8.1.** SAR DAC EXAMPLE

| Clock cycle | Register FF states, $b_1$ (MSB)$\cdots b_N$(LSB) | DAC output voltage ($+\frac{1}{2}$ LSB offset) (V) | |
|---|---|---|---|
| 1 | 1000000000 | 5.125 | |
| 2 | 1100000000 | 7.685 | ($b_2$ FF now reset to $b_2 = 0$) |
| 3 | 1010000000 | 6.405 | |
| 4 | 1011000000 | 7.045 | |
| 5 | 1011100000 | 7.365 | |
| 6 | 1011110000 | 7.525 | ($b_6$ FF now reset to $b_6 = 0$) |
| 7 | 1011101000 | 7.445 | |
| 8 | 1011101100 | 7.485 | |
| 9 | 1011101110 | 7.525 | ($b_9$ FF now reset to $b_9 = 0$) |
| 10 | 1011101101 | 7.495 | |

## 8.6 BALANCED MODULATOR/DEMODULATOR

The balanced modulator/demodulator is a versatile type of integrated circuit that can be used as a nonlinear signal mixing device. It has two input ports, usually referred to as the carrier input port and the signal input port, and a single output port. When used as a balanced modulator, the signal supplied to the carrier input port will be amplitude modulated by the voltage applied to the signal input port to produce a double sideband/suppressed carrier (DSB/SC) output voltage. If the signal voltage is $v_s(t) = V_S \sin \omega_s t$ and the carrier voltage is $v_c(t) = V_C \sin \omega_c t$, the DSB/SC output voltage will be of the form $v_o(t) = V_o[\cos (\omega_c + \omega_s)t - \cos (\omega_c - \omega_s)t]$. The upper and lower sidebands of the AM signal will thus appear at the output at frequencies $\omega_c + \omega_s$ and $\omega_c - \omega_s$, respectively, but the carrier component at frequency $\omega_c$ is suppressed. A combination of two balanced modulators together with some phase-shifting circuits can be used to obtain a single sideband/suppressed carrier (SSB/SC) signal.

This integrated circuit can also be used as a demodulator for various types of double-sideband and single-sideband AM signals. The AM signal is applied to the signal input port and a voltage at the carrier frequency is supplied to the carrier input port. The output voltage, after passing through a low-pass filter, will be the demodulated signal which will be a replica of the original modulating voltage.

The balanced modulator/demodulator IC can also be used as a phase detector in which case the output voltage will be directly proportional to the phase difference between the voltages supplied to the carrier and signal input ports. A phase detector has many uses and is an essential element of phase-locked loops.

## 8.7 VOLTAGE-CONTROLLED OSCILLATOR

A voltage-controlled oscillator or VCO is an integrated circuit that produces an output voltage (usually a square wave) whose frequency is directly proportional to an input or control voltage. The frequency of a VCO can usually be expressed in the form $f_{VCO} = f_o + K_V V_C$, where $f_{VCO}$ is the VCO frequency, $f_o$ is the free-running frequency of the VCO, $K_V$ is the voltage-to-frequency transfer coefficient (Hz/V) of the VCO, and $V_C$ is the control voltage.

The VCO has many uses, and it serves as an essential element of phase-locked loops.

## 8.8 PHASE-LOCKED LOOPS

A *phase-locked loop* (PLL) is a feedback loop comprised of a phase detector, low-pass filter, amplifier, and voltage-controlled oscillator (VCO). The phase detector or phase comparator compares the phase angle of the signal to that of the VCO output voltage as shown in Figure 8.12 and produces an output voltage that is related to this phase angle difference.

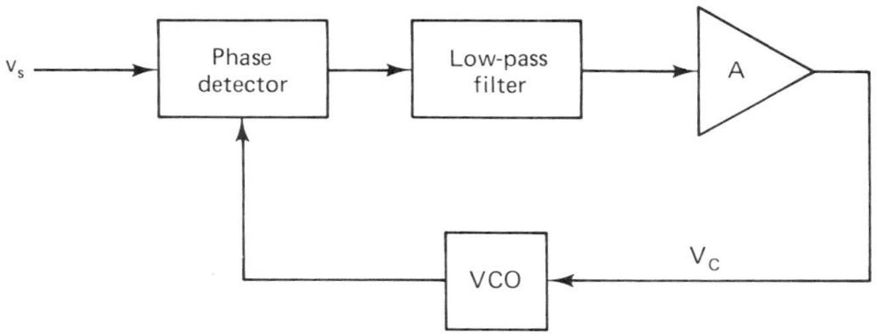


**Figure 8.12** Phase-locked loop.

If the phase difference between the signal and the VCO voltage is $\phi$ radians the output voltage of the phase detector will be given by

$$V_o = 2I_QR_L\left(\frac{2\phi}{\pi} - 1\right) = \frac{4I_QR_L}{\pi}\left(\phi - \frac{\pi}{2}\right) = K_\phi\left(\phi - \frac{\pi}{2}\right) \tag{8.6}$$

where $K_\phi$ is the *phase angle-to-voltage transfer coefficient* of the phase detector.

The output voltage of the phase detector is filtered by the low-pass filter (LPF) to remove the high-frequency components such as those at the signal and VCO frequencies and harmonics thereof. The output of the LPF is then amplified and applied as the input or control voltage $V_C$ of the VCO as given by $V_C = K_\phi A(\phi - \pi/2)$, where $A$ is the voltage gain of the amplifier.

This control voltage will result in a shift in the VCO frequency from its *free-running frequency* $f_o$ to a frequency $f$ given by $f = f_o + K_V V_C$, where $K_V$ is the *voltage-to-frequency transfer coefficient* of the VCO.

When the PLL is *locked in* to the signal frequency $f_s$ we have that $f = f_s = f_o + K_V V_C$. Since $V_C = (f_s - f_o)/K_V = K_\phi A(\phi - \pi/2)$ we will obtain that $\phi - \pi/2 = (f_s - f_o)/K_V K_\phi A$. Thus when the PLL is locked in to the signal there will be a phase angle difference $\phi$ established between the signal voltage and the VCO output voltage given by $\phi = (\pi/2) + [(f_s - f_o)/K_V K_\phi A]$ and the two frequencies will be in *exact synchronism*.

The maximum output-voltage magnitude available from the phase detector occurs for $\phi = \pi$ and for $\phi = 0$ radians and is $V_{O(\text{MAX})} = \pm 2I_QR_L = \pm K_\phi(\pi/2)$. The corresponding value of the maximum control voltage available to drive the VCO will be $V_{C(\text{MAX})} = \pm(\pi/2)K_\phi A$. The maximum VCO frequency swing that can be obtained will be $(f - f_o)_{\text{MAX}} = K_V V_{C(\text{MAX})} = \pm K_V K_\phi(\pi/2)A$. Therefore, the maximum range of signal frequencies over which the PLL can remain locked in on will be $f_s = f_o \pm K_V K_\phi(\pi/2)A = f_o \pm \Delta f_L$, where $2\Delta f_L$ will be the *lock-in frequency range* as given by

$$\boxed{\text{lock-in range:} = 2\Delta f_L = K_V K_\phi A \pi} \tag{8.7}$$

Note that the lock-in range will be symmetrically located with respect to the VCO free-running frequency $f_o$.

In Figure 8.13 a graph of the VCO control voltage $V_C$ versus the signal frequency

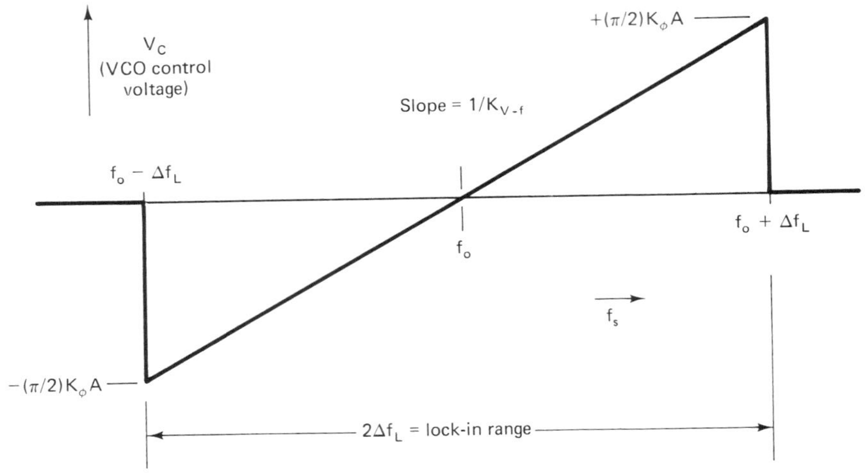


**Figure 8.13** PLL lock-in range.

$f_s$ is shown. Outside the lock-in range the VCO frequency can not be brought into synchronism with the signal frequency. The resulting phase angle difference will be $\phi = (\omega_s t + \theta_s) - (\omega_o t + \theta_o) = (\omega_s - \omega_o)t + (\theta_s - \theta_o)$ and will thus change rapidly with time. The rate of change of $\phi$ with time will be $d\phi/dt = \omega_s - \omega_o$. The phase detector output voltage will therefore vary rapidly with time and be heavily attenuated by the low-pass filter. As a result there will be very little voltage available to drive the VCO and the VCO frequency will revert back to essentially the free-running value of $f_o$. Thus we see that outside the lock-in range the VCO control voltage will essentially drop to zero.

When the VCO is locked in to the signal we have that $\phi = (\pi/2) + [(f_s - f_o)/K_\phi K_V A]$. Note that when $f_s = f_o$ the VCO voltage will be in *phase quadrature* (i.e., 90° phase difference) with respect to the signal voltage. As $f_s$ goes above $f_o$ the phase angle increases from 90° toward a maximum of 180° at the upper end of the lock-in range. As $f_s$ goes below $f_o$ the phase angle drops from 90° toward 0° at the lower end of the lock-in range.

### 8.8.1 Phase-Locked-Loop Applications

The phase-locked loop has many applications, especially in the area of communications circuits. The following is a brief description of some of the important applications of phase-locked loops.

**FM demodulator.** A phase-locked-loop FM demodulator circuit is shown in Figure 8.14. When the PLL is locked in on the FM signal we will have that the VCO frequency $f_{VCO} = f_o + K_V V_C$ will be equal to the instantaneous frequency of the FM signal $f_s$ so that $f_o + K_V V_C = f_s$. The VCO control voltage $V_C$ will therefore be given by $V_C = (f_s - f_o)/K_V$. If the instantaneous frequency of the FM signal is $f_s(t) = f_c + \Delta f \sin \omega_m t$, where $f_c$ is the (unmodulated) FM carrier frequency, $\Delta f$

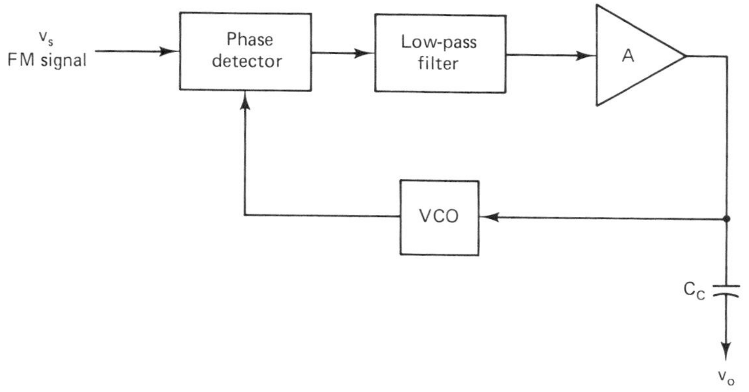


**Figure 8.14** PLL FM demodulator.

is the peak frequency deviation, and $\omega_m$ is the radian frequency of the modulating signal, we will have that

$$V_C(t) = \frac{f_s(t) - f_o}{K_V} = \frac{(f_c - f_o) + \Delta f \sin \omega_m t}{K_V} \tag{8.8}$$

The a-c component of $V_C(t)$ will be $v_c(t) = \Delta f \sin \omega_m t / K_V$, and this will be a true replica of the modulating voltage that is applied to the FM carrier at the transmitter. The VCO control voltage $V_C$ will be a linear function of the instantaneous frequency deviation of the FM signal, so that the FM signal will be demodulated with very little distortion.

**Frequency-shift keying demodulation.** Frequency-shift keying (FSK) is a type of frequency modulation in which the frequency of the FM signal is varied between two fixed levels. This can be considered to be a binary digital type of information transmission system in which one frequency $f_1$ represents a "0" and the other frequency $f_2$ represents a "1". The "0"-to-"1" frequency deviation $\Delta f$ will be $\Delta f = f_2 - f_1$.

A PLL FSK demodulator is shown in Figure 8.15. This circuit is similar to the FM demodulator, except for the addition of a comparator to produce a reconstructed digital output signal. If the PLL remains locked in to the FSK signal at both $f_1$ and $f_2$, the VCO control voltage levels will be $V_{C_1} = (f_1 - f_o)/K_V$ and $V_{C_2} = (f_2 - f_o)/K_V$, respectively. The difference between the two control voltage levels will be $\Delta V_C = (f_2 - f_1)/K_V$.

The reference voltage for the comparator is obtained by passing the VCO control voltage through a low-pass filter LPF-2. This filter has a very long time constant compared to the FSK pulse period such that an essentially d-c reference voltage that has a level midway between the "0" and "1" levels of the VCO control voltage is obtained. This reference voltage level is the optimum value to produce a digital output with a minimum bit error rate.

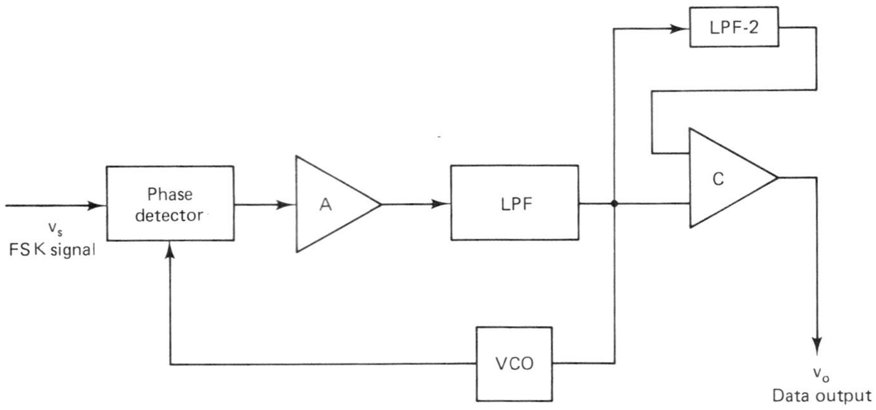


**Figure 8.15** Frequency-shift keying (FSK) demodulator.

**PLL frequency synthesizer.** The PLL can be used as the basis for frequency synthesizers that can produce a precise series of frequencies that are derived from a stable crystal-controlled oscillator. A basic diagram of a PLL frequency synthesizer is shown in Figure 8.16. The frequency of the crystal-controlled oscillator is divided by an integer factor $M$ by a counter circuit to produce a frequency $f_{osc}/M$, where $f_{osc}$ is the frequency of the crystal-controlled oscillator.

The VCO frequency $f_{VCO}$ is similarly divided by a counter circuit by an integer factor $N$ to become $f_{VCO}/N$. When the PLL is locked in on the divided down oscillator frequency we will have that $f_{osc}/M = f_{VCO}/N$ so that $f_{VCO} = (N/M)f_{osc}$.

The frequency counters or dividers can be programmed to produce a number of different frequency-division factors so that a large number of frequencies can be produced, all derived from the crystal-controlled oscillator.

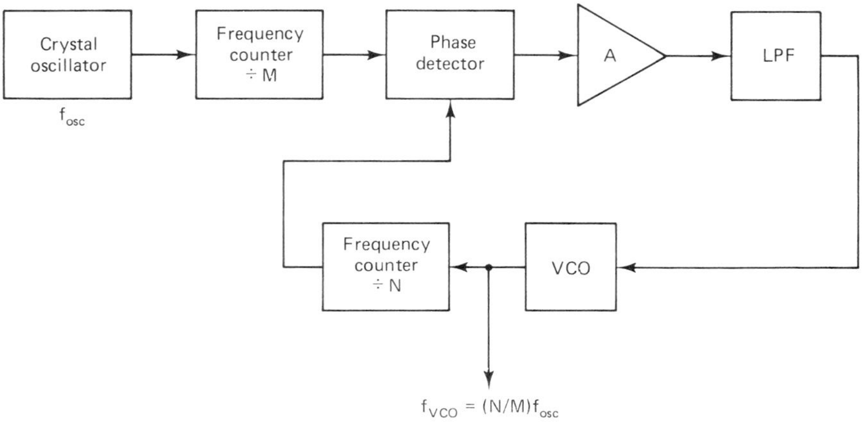


**Figure 8.16** PLL frequency synthesizer.

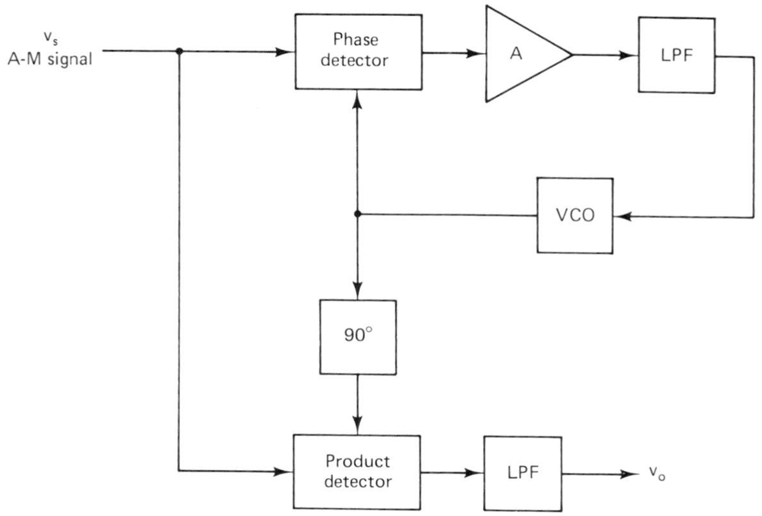


**Figure 8.17** PLL amplitude-modulation detector.

**AM detector using a phase-locked loop.** The use of a phase-locked loop for the demodulation of an AM signal is shown in Figure 8.17. This circuit is useful for the demodulation of AM signals that are transmitted with reduced carrier levels. The PLL locks in on the AM signal and regenerates the carrier, which is then supplied to a product detector, such as a balanced modulator/demodulator IC, for the demodulation of the AM signal.

**PLL motor speed control.** In the PLL circuit of Figure 8.18 an electric motor is part of the PLL feedback loop. The tachometer generates an output voltage waveform that has a frequency that is directly proportional to the motor speed. When this PLL is locked in on a control signal with frequency $f_s$, the motor speed in terms of the tachometer frequency $f_m$ will be exactly equal to the control signal frequency $f_s$.

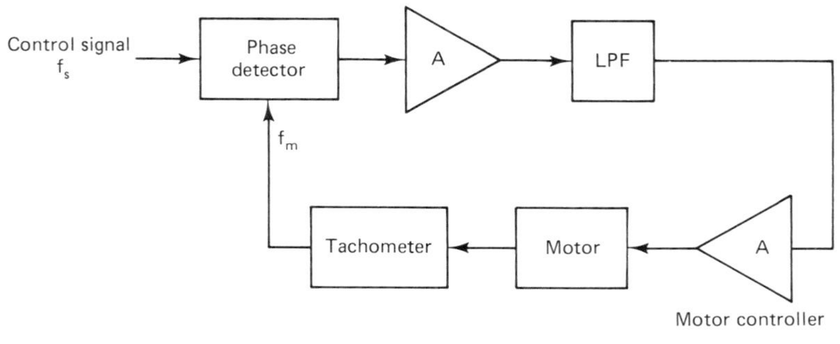


**Figure 8.18** Motor speed control.

## 8.9 MULTIPLIERS, DIVIDERS, AND FUNCTION GENERATORS

Many nonlinear functions can be implemented in integrated circuits making use of the exponential $I_C$ versus $V_{BE}$ transfer characteristic of bipolar transistors. A good example of this is the logarithmic-exponential multiplier-divider circuit of Figure 8.19. For this circuit the relationship between the currents is $I_1 I_2 = I_3 I_4$, subject to the restriction that all currents are algebraically positive. This circuit is available as a monolithic IC (Raytheon RC4200). The use of this circuit for a four-quadrant multiplier is shown in Figure 8.20 for which $V_O = (R_0 R_2 / R_1^2) V_X V_Y / V_R$. This circuit can also be used as a divider, as shown in Figure 8.21 for which

$$V_O = \frac{V_X}{V_Z} V_R \frac{R_0 R_4}{R_1 R_2} \tag{8.9}$$

A square-rooting circuit is shown in Figure 8.22, for which

$$V_O = \sqrt{V_X V_R \frac{R_0 R_4}{R_1 R_2}} \tag{8.10}$$

Many other functional relationships are possible using similar combinations of operational amplifiers and transistors, including logarithmic and exponential converters, sine and cosine function generators, arctangent function generators, and vector magnitude converters.

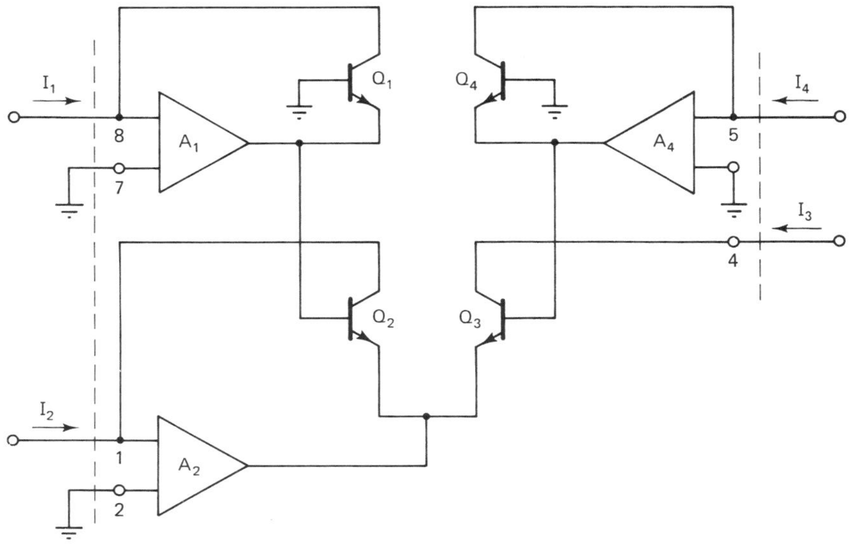


**Figure 8.19** Logarithmic-exponential multiplier-divider circuit.

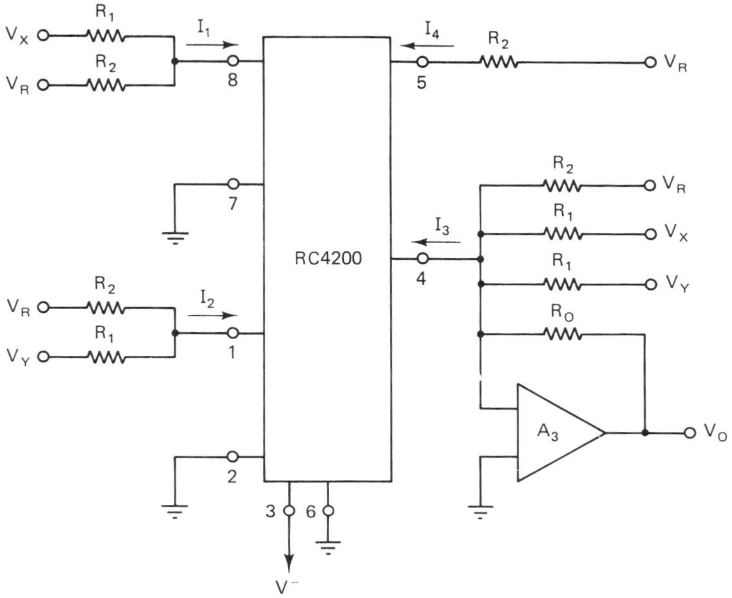


**Figure 8.20** Four-quadrant multiplier circuit.

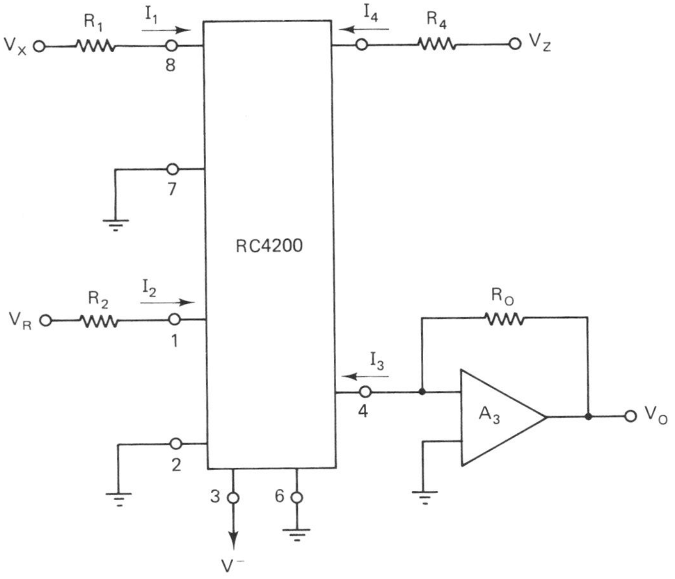


**Figure 8.21** One-quadrant divider.

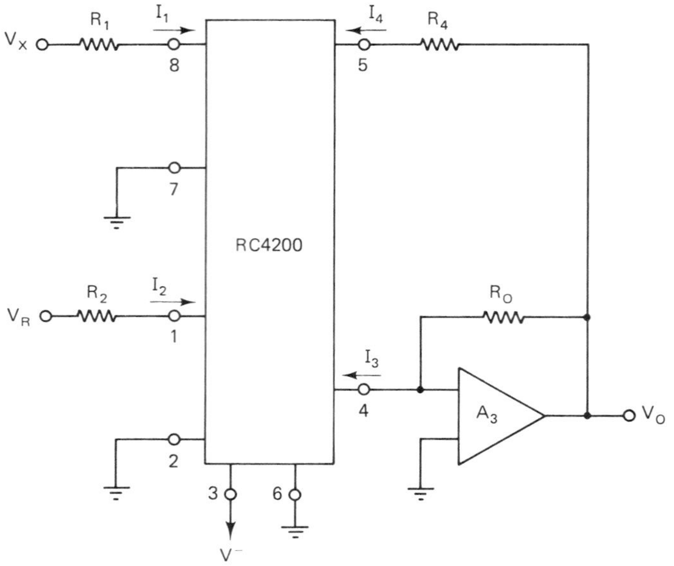


**Figure 8.22** Square-rooting circuit.

## 8.10 IC TEMPERATURE TRANSDUCERS

Semiconductor devices are very temperature sensitive. Usually, this temperature sensitivity represents a problem, but it can be used to advantage to make an integrated-circuit temperature sensor. Most IC temperature sensors make use of the temperature coefficient of the base-to-emitter voltage drop $V_{BE}$ of a transistor.

Monolithic IC temperature sensors are available that act as voltage sources and produce an output voltage that is directly proportional to the absolute temperature. Typical sensitivities are in the range 1 to 10 mV/K. There are also IC temperature sensors available that operate as a current source and produce a current that is directly proportional to the absolute temperature, with a typical sensitivity of 1 $\mu$A/K.

## 8.11 IC MAGNETIC FIELD SENSORS

Integrated circuits can be used to sense magnetic fields, making use of the Hall effect. In the Hall effect, the deflection of charge carriers due to a magnetic field will result in the generation of a voltage across a semiconductor sample that is in a direction perpendicular to the applied voltage. This transverse or Hall voltage is the result of the transverse force exerted on the free electrons and holes as they travel through the magnetic field and will be directly proportional to the strength of the magnetic field.

Integrated-circuit magnetic field sensors are available that incorporate the Hall effect sensor ("Hall plate") and an amplifier on a single silicon chip. Typical sensitivities are of the order of 1 mV/G. These sensors can be used for the direct measurement of magnetic field strengths, but most applications involve using the sensor for position or motion sensing.

## 8.12 IC PRESSURE TRANSDUCERS

The flow of electrons and holes in a semiconductor involves a complex interaction between the free charge carriers and the atoms of the crystal structure of the solid. A very slight change in the spacing or arrangement of the atoms can produce a very substantial change in the mobility of the charge carriers and therefore in the resistivity of the material. The change in the resistivity produced by a mechanical strain is called the *piezoresistive effect*.

Integrated-circuit pressure or strain transducers make use of diffused resistors in a Wheatstone bridge circuit such that as a result of the piezoresistive effect, the application of a mechanical stress will cause the bridge to become unbalanced. An amplifier can be included in the same IC so that a high-level, low-impedance output voltage that is proportional to the mechanical stress or pressure can be obtained.

## 8.13 ANALOG SWITCHES

Analog switches use transistors to serve as fast on–off switches. The ideal analog switch will have zero resistance in the "on" state and an infinite impedance in the "off" state. Although either bipolar or field-effect transistors can be used as the active element of an analog switch most analog switches use FETs. This is because of the inherent symmetry of FETs and the offset voltage that is present in bipolar transistors. While the offset voltage ($V_{CE}$ at $I_C = 0$) of bipolar transistors will generally be only a few millivolts, it can produce significant errors in the transmission of low-level analog signals. Another advantage of FETs over bipolar transistors for analog switch applications is the very high input impedance and very small gate current of the FETs compared to bipolar transistors.

Bipolar transistors can offer the advantage of relatively small "on" resistance, down to as low as 3 to 30 $\Omega$, but there are VMOS and DMOS transistors available with comparably low values of "on" resistance. Integrated-circuit analog switches generally incorporate the driving circuitry and several electrically independent JFET or MOSFET switching transistors in one package.

There are many applications of analog switches, including the following:

1. Sample-and-hold (S/H) circuits
2. Analog multiplexing and demultiplexing
3. Chopper stabilization of amplifiers
4. Digital-to-analog (D/A) conversion
5. "Integrate-and-dump" circuits
6. Programmable operational amplifier transfer characteristics, such as digital control of gain, frequency response, phase shift, and so on
7. Signal gating and squelch control

## 8.14 SAMPLE-AND-HOLD CIRCUITS

A major applications area for analog switches are in sample-and-hold (S/H) circuits. Sample-and-hold circuits are used to sample an analog input voltage for a very short time period, generally in the range 1 to 10 μs, and to hold the sampled voltage level for an extended period of time which can range from a few milliseconds to several seconds. The circuits are often used in analog-to-digital converters, especially the successive approximation type, to hold the analog input voltage level constant during the A/D conversion period. S/H circuits are also used in analog demultiplexing circuits, reset stabilized operational amplifier circuits, staircase generators, and other applications.

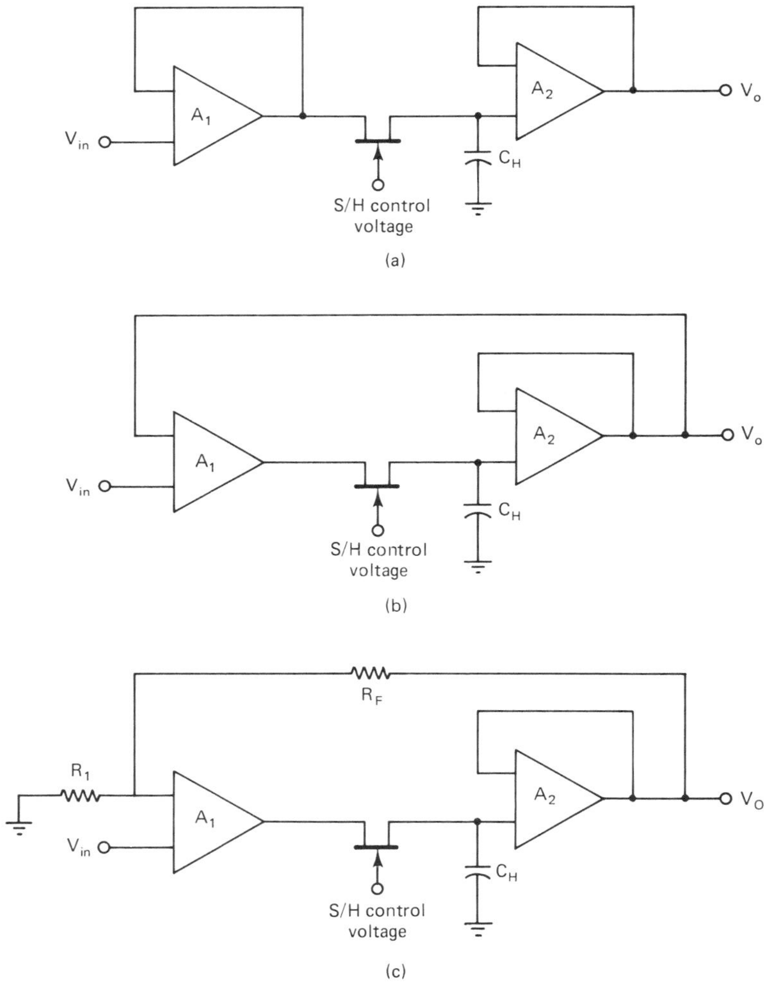


**Figure 8.23** (a) and (b) Some basic sample-and-hold circuits; (c) S/H circuit with voltage gain.

Three basic examples of sample-and-hold circuits are shown in Figure 8.23. In these circuits a JFET is used as an analog switch, although a MOSFET could also be used. During the sampling time the JFET switch is turned on, and in the case of Figure 8.23a and b the holding capacitor charges up to the level of the analog input voltage. At the end of this short sampling period the JFET switch is turned off. This isolates the holding capacitor $C_H$ from the input signal and the voltage across $C_H$, and therefore the output voltage $V_O$ will remain essentially at the value of the input voltage at the end of the sampling time. There will, however, be a small drop-off or droop of the capacitor voltage during the hold period due to the various leakage currents, including the $I_{DS(\mathrm{OFF})}$ of the JFET, the $I_{\mathrm{BIAS}}$ of the operational amplifier ($A_2$), and the internal leakage current of the holding capacitor $C_H$.

Sample-and-hold integrated circuits are available with all of the essential elements of a S/H circuit in one package, except for the holding capacitor $C_H$, which is usually an external component.

## 8.15 CHARGE-TRANSFER DEVICES

A charge-transfer device consists of an array of closely spaced charge storage elements in the form of MOS capacitors. The application of the clocking voltage will produce the transfer of the charge stored in one element of the array to the next element, and so on down the line. There are two basic types of charge-transfer devices (CTDs), the charge-coupled device (CCD) and the bucket-brigade device (BBD). Charge-transfer devices of both types are available with as many as 1000 charge storage elements.

Charge-transfer devices can be used for a wide variety of analog signal processing applications. The charge-transfer device for most of these applications is used as an analog signal delay line. The time delays available will range from as little as 25 μs (for a 512-stage device) to as much as 100 ms. These delay times can be very precisely controlled and adjusted by the clock frequency. As a result of the accurate control and variation of the delay time in a CTD it can offer substantial advantages over ultrasonic delay lines.

In the case of an ultrasonic delay line the analog signal is converted to a sound wave by a piezoelectric transducer, transmitted through a suitable medium such as a quartz rod, and then reconverted back to an electric signal by a second piezoelectric transducer. The resulting delay time is, however, not adjustable as it is in the case of a CTD.

Charge-transfer devices of both the CCD and BBD type can be used for many analog signal filtering applications, such as for comb filters, recursive filters, and transversal filters. Related applications involve analog signal correlators. Special audio effects can be obtained with CTDs, such as reverberation, echo, vibrato, tremulo, and chorus effects as well as speech compression and expansion and voice scrambling.

Charge-transfer devices can be used as analog serial memories or shift registers to provide temporary storage of an analog signal. This is useful for speed and time-base correction of audio and video tape recording systems. For videotape systems the CTD can be used for the temporary storage of one line of the video signal to provide a means of drop-out compensation. If there is a partial or complete loss of signal from one line of a taped TV signal, the preceding line can be substituted and the resulting effect on the TV picture will generally not be noticeable.

Charge-transfer devices can be used as serial input/parallel output and parallel input-serial output analog shift registers for signal multiplexing, demultiplexing, and scrambling.

Another applications area for CTDs is in ultrasonic systems. The electrical signal input to an array of ultrasonic transducers can be given a preprogrammed set of time delays by a number of CTDs placed in the signal paths. The ultrasonic transducers will become a phased array and the ultrasonic beam pattern, direction, and focusing can be electronically varied by control over the CTD time delays.

An important applications area for CCDs is in linear and area image sensing. In a linear image sensor photogenerated electrons that are collected by an array of small area light-sensing areas or photodiodes are transferred to an adjacent CCD array. The CCD array will operate as a parallel input/serial output analog shift register. The photogenerated electron packets after being transferred to the CCD in parallel can then be clocked out in serial fashion out of the CCD array. The output signal will then be a serial representation of the illumination pattern along the length of the array of light-sensing elements.

An area image sensor operates in a similar fashion. The photogenerated electrons that are produced in the parallel arrays of light-sensing elements are transferred to adjacent CCD arrays, which again serve as parallel input/serial output analog shift registers. The electron packets are then transferred out line by line from these interline CCD shift registers through another parallel input/serial output shift register.

## 8.16 OPTOELECTRONIC INTEGRATED CIRCUITS

The subject of optoelectronics is a vast one and a detailed presentation of this subject is well beyond the scope of this book. A summary of some of the important optoelectronic ICs and some related discrete devices will now be presented.

**Photodiodes and photodiode arrays.** A photodiode is a PN junction diode with most of the top surface of the chip not covered with metallization and in a package with a transparent window. This allows the incident light (photons) to penetrate into the semiconductor. The photons that are absorbed in the semiconductor produce photogenerated free electrons and holes which when collected by the PN junction can result in a greatly increased reverse current flow through the device. A *PIN photodiode* uses the $P^+NN^+$ (or $N^+PP^+$) diode structure, and when biased to produce full depletion of the lightly doped N (or P) epitaxial layer region has a very low device capacitance. This leads to a very high speed capability. An *avalanche photodiode* is a photodiode designed to be biased in the avalanche multiplication region, close to the avalanche breakdown voltage. The avalanche multiplication process can result in a very high internal current gain.

**Phototransistor.** A phototransistor is an integrated combination of a photodiode and a transistor on the same silicon chip. The photocurrent produced by the photodiode is multiplied by the current gain of the transistor. A *photodarlington* is an integrated combination of a photodiode and two transistors connected in a Darlington configuration. This results in a very high current gain for the photocurrent produced by the photodiode.

**Light-emitting diodes.** Light-emitting diodes (LEDs) are diodes constructed from various binary and ternary III–V compound semiconductors, such as GaAs, GaAlAs, and GaAsP. When forward biased there will be a flow of electrons and holes across the junction. After the electrons and holes flow across the junction they become minority carriers, and the subsequent recombination of these minority carriers with the resident majority carriers on both side of the junction will result in the emission of photons. Silicon cannot be used for an LED because the electron–hole recombination events are nonradiative in nature, producing heat rather than light.

**Optically coupled isolator.** An optically coupled isolator (OCI) is simple hybrid IC, consisting of an LED and a photodiode or phototransistor in one package. The light emitted by the LED is detected by the photodiode or phototransistor and this provides a unilateral optical coupling of a signal between input and output with a very high degree of electrical isolation.

**Slotted and reflective emitter-sensor modules.** These are special types of OCIs in which the emitter-to-sensor coupling can be interrupted in the case of the slotted module or established in the case of the reflective module. These devices are useful for a variety of measurement and control applications.

**Image sensors.** These ICs are linear or area (two-dimensional) arrays of photosensing elements on a common silicon substrate. The sensing devices can be photodiodes or MOS capacitor-potential well photoelements. The array can be electronically scanned using a digital or analog shift register to produce a video output signal. The analog shift register can be in the form of a bucket-brigade device (BBD) or a charge-coupled device (CCD). Linear arrays of up to 4096 elements and area arrays with as many as 185,000 sensing elements are available.

## 8.17 OTHER SPECIAL-FUNCTION INTEGRATED CIRCUITS

**FM sound circuit IC.** These monolithic ICs can serve as complete FM sound circuits for TV and other applications, combining the functions of an FM RF amplifier, limiter, demodulator, and audio amplifier. Examples are the MC1351 (Motorola) and the LM1351 (National Semiconductor).

**Frequency-to-voltage converter.** A frequency-to-voltage converter is a circuit that produces an output voltage that is linearly proportional to the frequency of the input signal. Examples of such ICs are the LM2907 and the LM2917 (National Semiconductor).

**Compandors.** Compandors are ICs with a nonlinear transfer characteristic that can be used to compress or expand the dynamic range of an analog signal. The signal compression is often performed at the transmitting end of the system and a complementary signal expansion is done at the receiving end in order to improve the signal-to-noise ratio of the received signal. Some examples of analog signal compandor ICs are the LM2704 through LM2707 series (National Semiconductor). These devices feature a complementary pair of a signal compressor and signal expandor in one IC.

**Electronic attenuators.** These ICs can be used to provide a variable gain or attenuation of an input signal. An example is the MC3340 (Motorola) in which by a variation of a d-c control voltage, the voltage gain can be varied from a maximum of +13 dB down to a minimum of −77 dB.

**Two-wire transmitter.** A two-wire transmitter is an IC useful for remote-sensing applications in which the transmitter d-c bias power is supplied over the same pair of wires as is used for the signal. For the signal transmission, the IC operates as a current source which greatly reduces the effects of the transmission line resistance and the interference that may be picked up along the transmission line. An example of this type of IC is the LH0045 (National Semiconductor).

**FM stereo demodulator.** An FM stereo demodulator IC contains all of the active circuitry needed to demodulate a stereo audio signal and produce the audio signals for the left and right speakers. Examples are the MC1310 (Motorola), LM1310, and LM1800 (National Semiconductor).

**AM radio system IC.** An AM radio system consisting of an RF amplifier, local oscillator, mixer, IF amplifier, AGC detecter, and zener regulator is available as a monolithic IC. An example of such a system is the LM3820 (National Semiconductor).

**Chroma demodulator.** ICs can be used to good advantage in the signal processing circuitry of TV receivers. Indeed, ICs find application in almost all areas of a TV receiver, except in the VHF/UHF tuner, the high-voltage circuit, and in the power supply circuit. An important application of ICs is in the color signal (chroma) processing circuitry—in particular, the chroma demodulator, which takes the $I$ and $Q$ color sidebands of a composite video signal and produces the separate red, green, and blue signals. Some examples of chroma demodulator ICs are the MC1324 (Motorola), LM746, LM1828, and LM1848 (National Semiconductor).

**Chroma subcarrier regenerator.** These ICs are basically phase-locked loops for the regeneration of the 3.58-MHz color subcarrier from the short "color burst" of 8 to 10 cycles of this signal that occurs during the blanking interval of every 63.5-μs horizontal line period. An example is the μA780 (Fairchild).

**Tuned amplifier ICs.** These ICs use externally connected *LC* tuned circuits to produce a frequency selective or bandpass amplifier, often with a relatively narrow bandwidth. They can be used as radio-frequency (RF) or intermediate-frequency (IF) amplifiers. A common application of these ICs is as a multistage "IF strip" for radio and TV receivers. Examples of these integrated circuit IF strips are the LM172 (National Semiconductor) for AM radios, LM3089 (National Semiconductor) for FM radios, and the LM3071 (National Semiconductor) for the TV chrominance signal.

## REFERENCES

### Digital-to-Analog and Analog-to-Digital Converters


Analog Devices, Inc., *Analog-Digital Conversion Notes*, Analog Devices, Inc., P.O. Box 796, Norwood, MA 02062.

CONNELLY, J. A., *Analog Integrated Circuits*, Wiley, 1975.

GARRETT, P. H., *Analog I/O Design*, Reston, 1981.

GRAEME, J. G., G. E. TOBEY, and L. P. HUELSMAN, *Operational Amplifiers—Design and Applications*, McGraw-Hill, 1971.

GREBENE, A. B., *Analog Integrated Circuit Design*, Van Nostrand Reinhold, 1972.

GRINICH, V. H., and H. G. JACKSON, *Introduction to Integrated Circuits*, McGraw-Hill, 1975.

HOESCHEL, D. F., *Analog-to-Digital/Digital-to-Analog Conversion Techniques*, Wiley, 1968.

MILLMAN, J., *Microelectronics*, McGraw-Hill, 1979.

MITRA, S. K., *An Introduction to Digital and Analog Integrated Circuits and Applications*, Harper & Row, 1980.

STOUT, D. F., *Handbook of Microcircuit Design and Applications*, McGraw-Hill, 1980.

TAUB, H., and D. SCHILLING, *Digital Integrated Electronics*, Chap. 14, McGraw-Hill, 1977.

WAIT, J. V., *Introduction to Operational Amplifier Theory and Applications*, McGraw-Hill, 1975.


### Balanced Modulator/Demodulator


GRAEME, J. G., G. E. TOBEY, and L. P. HUELSMAN, *Operational Amplifiers—Design and Applications*, McGraw-Hill, 1971.

GREBENE, A. B., *Analog Integrated Circuit Design*, Van Nostrand Reinhold, 1972.

KRAUSS, H. L., C. W. BOSTIAN, and F. H. RAAB, *Solid State Radio Engineering*, Wiley, 1980.

LENK, J. D., *Handbook of Integrated Circuits*, Reston, 1978.

LENK, J. D., *Manual for Integrated Circuit Users*, Reston, 1973.

MITRA, S. K., *An Introduction to Digital and Analog Integrated Circuits and Applications*, Harper & Row, 1980.

### Voltage-Controlled Oscillator

GRAEME, J. G., G. E. TOBEY, and L. P. HUELSMAN, *Operational Amplifiers—Design and Applications*, McGraw-Hill, 1971.

MILLMAN, J., *Microelectronics*, McGraw-Hill, 1979.

STOUT, D. F., *Handbook of Microcircuit Design and Applications*, McGraw-Hill, 1980.

United Technical Publications, *Modern Applications of Integrated Circuits*, TAB Books, 1974.

### Phase-Locked Loops

BERLIN, H. M., *Design of Phase-Locked Loop Circuits with Experiments*, Howard M. Sams, 1978.

BLANCHARD, A., *Phase-Locked Loops*, Wiley, 1976.

CONNELLY, J. A., *Analog Integrated Circuits*, Wiley, 1975.

EGAN, W. F., *Frequency Synthesis by Phase-Lock*, Wiley, 1981.

GARDNER, F. M., *Phaselock Techniques*, Wiley, 1979.

GLASER, A. B., and G. E. SUBAK-SHARPE, *Integrated Circuit Engineering*, Addison-Wesley, 1977.

GREBENE, A. B., *Analog Integrated Circuit Design*, Van Nostrand Reinhold, 1972.

KLAPPER, J., and J. T. FRANKLE, *Phase-Locked and Frequency Feedback Systems*, Academic Press, 1972.

KRAUSS, H. L., C. W. BOSTIAN, and F. H. RAAB, *Solid State Radio Engineering*, Wiley, 1980.

LENK, J. D., *Handbook of Integrated Circuits*, Reston, 1978.

STOUT, D. F., *Handbook of Microcircuit Design and Applications*, McGraw-Hill, 1980.

United Technical Publications, *Modern Applications of Integrated Circuits*, TAB Books, 1974.

### Multipliers, Dividers, and Function Converters

GLASER, A. B., and G. E. SUBAK-SHARPE, *Integrated Circuit Engineering*, Addison-Wesley, 1977.

GREBENE, A. B., *Analog Integrated Circuit Design*, Van Nostrand Reinhold, 1972.

LENK, J. D., *Handbook of Integrated Circuits*, Reston, 1978.

MITRA, S. K., *An Introduction to Digital and Analog Integrated Circuits and Applications*, Harper & Row, 1980.

STOUT, D. F., *Handbook of Microcircuit Design and Applications*, McGraw-Hill, 1980.

United Technical Publications, *Modern Applications of Integrated Circuits*, TAB Books, 1974.

WONG, Y. J., and W. E. OTT, *Function Circuits*, McGraw-Hill, 1976.

### IC Temperature, Pressure, and Magnetic Field Sensors

CHIEN, C. L., and C. R. WESTGATE, *The Hall Effect and Its Applications*, Plenum Press, 1980.

GARRETT, P. H., *Analog I/O Design*, Reston, 1981.

PUTLEY, E. H., *The Hall Effect and Related Phenomena*, Butterworth, 1960.

SACHSE, H., *Semiconducting Temperature Sensors and Their Application*, Wiley, 1975.

SURINA, T., and C. HERRICK, *Semiconductor Electronics*, Holt, Rinehart, and Winston, 1964.

TIMKO, M. P., "A Two-Terminal Temperature Transducer," *IEEE Journal of Solid State Circuits*, Vol. SC-11, No. 6, pp. 784–788, December 1976.

United Technical Publications, *Modern Applications of Integrated Circuits*, TAB Books, 1974.

WOLF, H. F., *Semiconductors*, Chaps. 1 and 3, Wiley, 1971.

### Analog Switches and Sample-and-Hold Circuits

CONNELLY, J. A., *Analog Integrated Circuits*, Wiley, 1975.

LENK, J. D., *Handbook of Electronic Circuit Design,* Prentice-Hall, 1976.

LENK, J. D., *Manual for M.O.S. Users*, Reston, 1975.

MIRTES, B., *D-C Amplifiers*, Chap. 6, Butterworth, 1969.

MITRA, S. K., *An Introduction to Digital and Analog Integrated Circuits and Applications*, Harper & Row, 1980.

OXNER, E. S., *Power FETs and Their Applications*, Chap. 9, Prentice-Hall, 1982.

STOUT, D. F., *Handbook of Microcircuit Design and Applications*, McGraw-Hill, 1980.

TAUB, H., and D. SCHILLING, *Digital Integrated Circuits*, Chap. 13, McGraw-Hill, 1977.

UNITED TECHNICAL PUBLICATIONS, *Modern Applications of Integrated Circuits*, TAB Books, 1974.

### Charge-Transfer Devices and Analog Delay Lines

BEYMAN, J. D. E., and D. R. LAMB, *Charge-Coupled Devices and Their Applications*, McGraw-Hill, 1980.

EINSPRUCH, N. G., *VLSI Electronics: Microstructure Science*, Vol. 5, Academic Press, 1982.

GLASER, A. B., and G. E. SUBAK-SHARPE, *Integrated Circuit Engineering*, Addison-Wesley, 1977.

HOBSON, G. S., *Charge-Transfer Devices*, Wiley, 1978.

MELEN, R., and D. BUSS, *Charge-Coupled Devices: Technology and Applications*, IEEE Press, 1977.

MILLMAN, J., *Microelectronics*, McGraw-Hill, 1979.

SEQUIN, C. H., and M. F. TOMPSETT, *Charge Transfer Devices*, Academic Press, 1975.

YANG, E. S., *Fundamentals of Semiconductor Devices*, Chap. 11, McGraw-Hill, 1978.

### Optoelectronics

BAR-LEV, A., *Semiconductors and Electronic Devices*, Prentice-Hall, 1984.

BARNOSKI, M. K., *Fundamentals of Optical Fiber Communications*, Academic Press, 1981.

CHAPELL, A., *Optoelectronics*, McGraw-Hill, 1978.

ELION, G., *Fiber Optics in Communications Systems*, Marcel Dekker, 1978.

ELION, G. R., and H. A. ELION, *Electro-optics Handbook*, Marcel Dekker, 1979.

GAGLIAIDI, R. H., and S. KARP, *Optical Communications*, Wiley, 1975.

HERMAN, M. A., *Semiconductor Optoelectronics*, Wiley, 1980.

HOWES, M. J., and D. MORGAN, *Optical Fiber Communications*, Wiley, 1980.

KAO, C. K., *Optical Fiber Systems: Technology, Design, and Applications*, McGraw-Hill, 1982.

KUECKEN, J. A., *Fiberoptics*, TAB Books, 1980.

LENK, J. D., *Handbook of Electronic Circuit Designs*, Prentice-Hall, 1976.

MELEN, R., and D. BUSS, *Charge-Coupled Devices: Technology and Applications*, IEEE Press, 1977.

MILLER, S. E., and A. G. CHYNOWETH, *Optical Fiber Telecommunications*, Academic Press, 1979.

MIMS, F. M., *Light-Beam Communications*, Howard M. Sams, 1975.

PANKOVE, J. I., *Optical Processes in Semiconductors*, Prentice-Hall, 1971.

SEIPPEL, R. G., *Optoelectronics*, Reston, 1981.

SZE, S. M., *Physics of Semiconductor Devices*, Wiley, 1969.

WOLF, H. F., *Handbook of Fiber Optics: Theory and Applications*, Garland, 1979.

### Other Special-Function Integrated Circuits

FITCHEN, F., *Electronic Integrated Circuits and Systems*, Van Nostrand Reinhold, 1970.

GREBENE, A. B., *Analog Integrated Circuit Design*, Van Nostrand Reinhold, 1972.

LENK, J. D., *Manual for Integrated Circuit Users*, Reston, 1973.

MANDL, M., *Solid State Circuit Design Users Manual*, Chap. 6, Reston, 1977.

# Appendix A

# PHYSICAL CONSTANTS, CONVERSION FACTORS, AND PARAMETERS

| | | |
|---|---|---|
| Angstrom unit | Å | 1 Å $= 10^{-4}$ $\mu$m $= 10^{-8}$ cm $= 10^{-10}$ m |
| Boltzmann constant | $k$ | $k = 1.380 \times 10^{-23}$ J/K $= 8.62 \times 10^{-5}$ eV/K |
| Electron charge | $q$ | $q = 1.602 \times 10^{-19}$ C |
| Micron | $\mu$m | 1 $\mu$m $= 10^{-6}$ m $= 10^{-4}$ cm $=$ 10,000 Å |
| Mil | | 1 mil $=$ 0.001 in. $=$ 25.4 $\mu$m |
| Nanometer | nm | 1 nm $= 10^{-9}$ m $= 10^{-3}$ $\mu$m $=$ 10 Å |
| Permittivity of free space | $\epsilon_0$ | $\epsilon_0 = 8.854 \times 10^{-14}$ F/cm |
| Permeability of free space | $\mu_0$ | $\mu_0 = 4\pi \times 10^{-9}$ H/cm |
| Planck's constant | $h$ | $h = 6.625 \times 10^{-34}$ J-s |
| Thermal voltage | $V_T$ | $V_T = kT/q = 25.85$ mV at 300K (27°C) |
| Velocity of light in free space | $c$ | $c = 2.998 \times 10^{10}$ cm/s |

| | Silicon | Germanium | GaAs | $SiO_2$ |
|---|---|---|---|---|
| Dielectric constant, $\epsilon_r$ (relative permittivity) | 11.8 | 16.0 | 10.9 | 3.8 |
| Energy gap, $E_G$ (eV at 300 K) | 1.11 | 0.67 | 1.43 | |
| Electron drift mobility, $\mu_n$ ($cm^2/V$-$s$ at 300 K) | 1350 | 3900 | 8500 | |
| Hole drift mobility, $\mu_p$ ($cm^2/V$-$s$ at 300 K) | 500 | 1900 | 400 | |
| Intrinsic carrier concentration, $n_i$ ($cm^{-3}$ at 300 K) | $1.4 \times 10^{10}$ | $2.1 \times 10^{12}$ | $1.1 \times 10^{7}$ | |
| Electron saturation velocity, $v_{sat}$ (cm/s) | $0.8 \times 10^{7}$ | $0.6 \times 10^{7}$ | $2 \times 10^{7}$ | |
| Saturation electric field, $E_{sat}$ (V/cm) | $2 \times 10^{4}$ | $3 \times 10^{3}$ | $2 \times 10^{3}$ | |

# Appendix B

# TRANSISTOR CONDUCTANCES

## B.1 BIPOLAR TRANSISTOR CONDUCTANCES

For a bipolar transistor operating in the active mode (emitter–base junction "on" and collector–base junction "off") we have the following basic relationships for the transistor currents:

1. $I_C = I_{TO} \exp (V_{BE}/V_T)$
2. $I_E = -I_C$
3. $I_B = I_{BO} \exp (V_{BE}/nV_T)$, where $n$ is a dimensionless constant between 1 and 2 and usually near 1.5, and
4. By definition $I_B = I_C/\beta$, where $\beta = h_{FE}$ is the transistor current gain

The dynamic transfer and input conductances can readily be obtained from these simple exponential relationships, giving:

1. *Dynamic forward transfer conductance*: $g_m = dI_C/dV_{BE} = I_C/V_T$
2. *Dynamic emitter input conductance*: $g_{eb} = dI_E/dV_{EB} \simeq dI_C/dV_{BE} = g_m$
3. *Dynamic base input conductance*: $g_{be} = dI_B/dV_{BE} = I_B/nV_T$

The equation for $I_C$ given above shows that there is an exponential dependence of the collector current on the base-to-emitter voltage $V_{BE}$, but there is no explicit dependence of $I_C$ on the collector voltage. However, as will be seen, as a result of

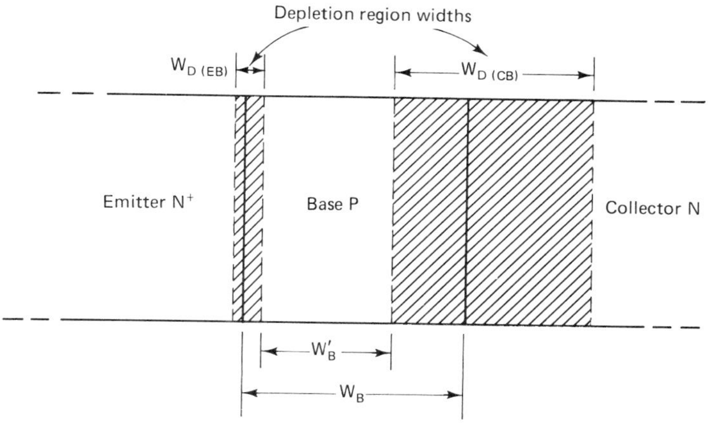


**Figure B.1** Base-width modulation.

the *base-width modulation* or *Early effect* there will indeed be a slight dependence of $I_C$ on the collector voltage.

The preexponential constant $I_{TO}$ is inversely proportional to the *effective* or *electrical* base width $W_B'$ of the transistor as given by $I_{TO} \propto 1/W_B'$. In Figure B.1 both the actual base width $W_B$ and the effective base width $W_B'$ are shown. The difference between these two is due to the incursion of the emitter–base and collector–base depletion regions into the base region so that the effective base width can become substantially smaller than the actual base width. The collector–base junction depletion region width will increase with increasing collector voltage, so that as the collector voltage increases the effective base width $W_B'$ will decrease, and as a result $I_{TO}$ will increase. This in turn will cause $I_C$ to increase. It is by this chain of events that $I_C$ becomes a function of the collector voltage, showing a slight increase as the collector voltage increases.

The *dynamic collector-to-emitter conductance* $g_{ce}$ will be defined as $g_{ce} = dI_C/dV_{CE}$ and we can write that

$$\begin{aligned} g_{ce} &= \frac{dI_C}{dV_{CE}} = \frac{dI_C}{dI_{TO}} \frac{dI_{TO}}{dW_B'} \frac{dW_B'}{dV_{CB}} \frac{dV_{CB}}{dV_{CE}} \\ &\simeq \exp\left(\frac{V_{BE}}{V_T}\right)\left(-\frac{I_{TO}}{W_B'}\right)\frac{dW_B'}{dV_{CB}} \times 1 \end{aligned} \tag{B.1}$$

We will now define the *base-width modulation coefficient* or *Early voltage* $V_A$ by the equation

$$\frac{1}{V_A} = -\frac{1}{W_B'}\frac{dW_B'}{dV_{CB}} \tag{B.2}$$

Using $V_A$ in the foregoing equation for $g_{ce}$ gives

$$\boxed{g_{ce} = I_{TO}\exp\left(\frac{V_{BE}}{V_T}\right)\frac{1}{V_A} = \frac{I_C}{V_A}} \tag{B.3}$$

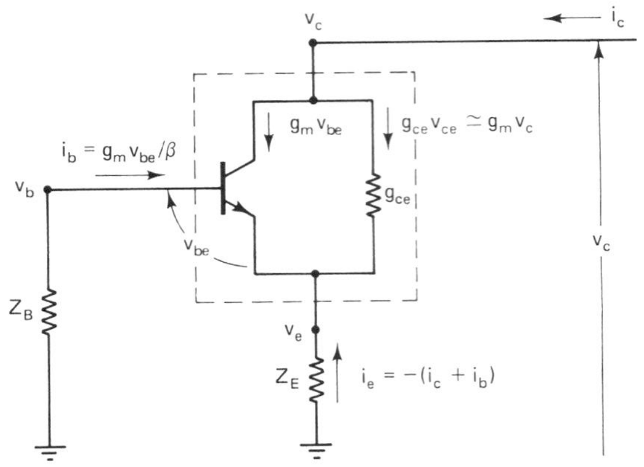


**Figure B.2** Circuit for the calculation of collector conductance.

The value of the Early voltage $V_A$ for most transistors will be in the range 100 to 300 V. For example, if $V_A = 200$ V, at a current of $I_C = 1.0$ mA the conductance $g_{ce}$ will be given by $g_{ce} = I_C/V_A = 1.0$ mA/200 V $= 1000$ $\mu$A/200 V $= 5$ $\mu$S. The reciprocal quantity of $r_{ce} = 1/g_{ce}$ is the *dynamic collector-to-emitter resistance*, and for this example will be $r_{ce} = 1/g_{ce} = V_A/I_C = 200$ V/1.0 mA $= 200$ k$\Omega$.

The dynamic conductance looking into the collector of a transistor $g_c$ will be a function of $g_{ce}$ and the impedances in series with the emitter and base of the transistor. For the analysis of the *dynamic collector conductance* $g_c$ we will use the circuit of Figure B.2. The transistor is represented as an ideal transistor (i.e., $g_{ce} = 0$) for which $i_c = g_m v_{be}$, in parallel with a conductance of value $g_{ce}$ to represent the collector-to-emitter conductance that is actually present in the transistor. The total a-c collector current of the transistor thus becomes $i_c = g_m v_{be} + g_{ce} v_{ce} \simeq g_m v_{be} + g_{ce} v_c$, since $v_c \simeq v_{ce}$.

Since $v_{be} = v_b - v_e$, $v_b = -i_b Z_B$, and $v_e = -i_e Z_E = (i_b + i_c) Z_E$, we have that $v_{be} = -i_b Z_B - (i_b + i_c) Z_E = -i_c Z_E - i_b (Z_E + Z_B)$. Since $i_b = g_m v_{be}/\beta$, we will have that $v_{be} = -i_c Z_E - (g_m v_{be}/\beta)(Z_E + Z_B)$. If we now solve this last expression for $v_{be}$ we obtain $v_{be}[1 + (g_m/\beta)(Z_E + Z_B)] = -i_c Z_E$, so that $v_{be} = -i_c Z_E/[1 + (g_m/\beta)(Z_E + Z_B)]$. If we now substitute this expression for $v_{be}$ into the equation for $i_c$ given earlier, we obtain

$$i_c = g_m v_{be} + g_{ce} v_c = -\frac{g_m (i_c Z_E)}{1 + (g_m/\beta)(Z_E + Z_B)} + g_{ce} v_c \tag{B.4}$$

Combining all terms involving $i_c$ on the left-hand side now gives

$$i_c \left[ 1 + \frac{g_m Z_E}{1 + (g_m/\beta)(Z_E + Z_B)} \right] = g_{ce} v_c \tag{B.5}$$

Now solving for the ratio $g_c + i_c/v_c$, we obtain

$$g_c = \frac{i_c}{v_c} = \frac{g_{ce}}{1 + g_m Z_E/[1 + g_m (Z_E + Z_B)/\beta]} = \frac{g_{ce}[1 + g_m (Z_E + Z_B)/\beta]}{1 + g_m Z_E + g_m (Z_E + Z_B)/\beta} \tag{B.6}$$

Since $g_m = I_C/V_T$ and $g_{ce} = I_C/V_A$, the expression for $g_c$ can be written as

$$\boxed{g_c = \frac{I_C}{V_A} \frac{1 + (I_C/V_T)(Z_E + Z_B)/\beta}{1 + (I_C/V_T)[Z_E + (Z_E + Z_B)/\beta]}} \tag{B.7}$$

From this last expression we note that if $Z_E = 0$, then $g_c = I_C/V_A = g_{ce}$. This result is independent of the value of $Z_B$. At the other extreme, for large values of $Z_E$ such that $(I_C/V_T)Z_E = g_m Z_E \gg \beta$ the collector conductance will approach a limiting value given by $g_c = I_C/\beta V_A$ and the corresponding dynamic collector resistance will become $r_c = \beta V_A/I_C$. This condition can result in a very small conductance. For example, if $V_A = 200$ V and $I_C = 1.0$ mA as in the preceding example, and if $\beta = 100$, then $g_c = I_C/\beta V_A = 1 \text{ mA}/100 \times 200 \text{ V} = \underline{50 \text{ nS}}$, and the corresponding resistance will be $r_c = 1/g_c = \underline{20 \text{ M}\Omega}$.

In Figure B.3 the common-emitter output or collector characteristics are shown, with curves of $I_C$ versus $V_{CE}$ for various constant values of base current $I_B$. The slope of the various $I_C$ versus $V_{CE}$ curves is the dynamic collector-to-emitter conductance $g_{ce}$ as given by $g_{ce} = dI_C/dV_{CE}$ and is equal to $g_{ce} = I_C/V_A$. If the curves are extrapolated back from the active region to the voltage axis intercept, they will all intersect the voltage axis at about the same place, this being at $-V_A$.

Finally, it is of interest to compare the effectiveness of the base-to-emitter voltage $V_{BE}$ to the collector voltage ($V_{CE}$ or $V_{CB}$) in controlling the collector current. We have that $i_c = g_m v_{be} + g_{ce} v_{ce}$. Since both conductances are proportional to the quiescent collector current as given by $g_m = I_C/V_T$ and $g_{ce} = I_C/V_A$, the ratio of $g_m$ to $g_{ce}$ will be independent of $I_C$ and will be $g_m/g_{ce} = V_A/V_T$. For $V_A \sim 200$ V and $V_T = 25$ mV this gives $g_m/g_{ce} = V_A/V_T \sim 200 \text{ V}/25 \text{ mV} = \underline{8000}$. Thus the input voltage $V_{BE}$ is very much more effective in controlling the collector current than is the output voltage ($V_{CE}$). It is due to this that very large voltage gains are possible with just a single transistor stage, especially when active loads are used.

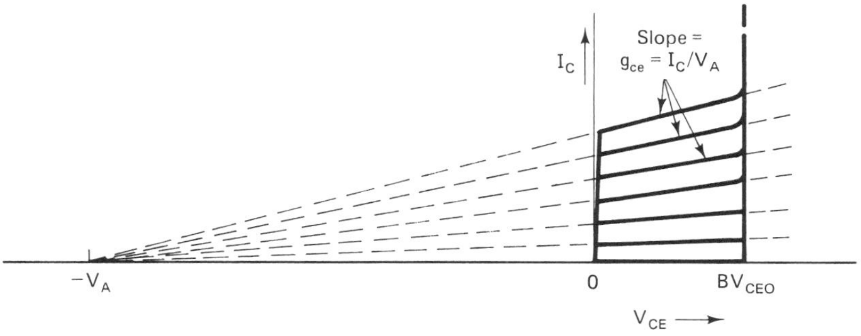


**Figure B.3** Transistor output (collector) characteristics.

## B.2 FIELD-EFFECT TRANSISTOR TRANSFER CONDUCTANCES

For a JFET in the saturated ("active") region for which $|V_{DS} - V_{GS}| > |V_P|$ the transfer relationship between the drain-to-source current $I_{DS}$ and the gate-to-source voltage $V_{GS}$ is given approximately by $I_{DS} = I_{DSS}(1 - V_{GS}/V_P)^2$ for $0 \leq |V_{GS}| \leq |V_P|$, and where $V_P$ is the pinch-off voltage and $I_{DSS}$ is the value of $I_{DS}$ when $V_{GS} = 0$. The dynamic transfer conductance, $g_{fs}$ or $g_m$, is given by

$$g_{fs} \equiv g_m = dI_{DS}/dV_{GS} = 2I_{DSS}/-V_P(1 - V_{GS}/V_P) \tag{B.8}$$

This can be rewritten as

$$\boxed{g_{fs} \equiv g_m = (2I_{DSS}/-V_P)\sqrt{I_{DS}/I_{DSS}} = 2\frac{\sqrt{I_{DSS}I_{DS}}}{-V_p} = g_{fsO}\sqrt{I_{DS}/I_{DSS}}} \tag{B.9}$$

where $g_{fsO} = g_{mO} = 2I_{DSS}/-V_P$. All of the preceding equations are applicable to both N-channel as well as P-channel JFETs, and the transfer conductances will be positive quantities in both cases.

For a MOSFET in the saturated ("active") region for which $|V_{DS}| > |V_{GS} - V_t|$ and $|V_{GS}| > |V_t|$ the transfer relationship between $I_{DS}$ and $V_{GS}$ will be given approximately by $I_{DS} = K(V_{GS} - V_t)^2$ where $K$ is a constant having units of $A/V^2$ and $V_t$ is the threshold voltage for channel formation (not to be confused with the thermal voltage). The dynamic transfer conductance, $g_{fs}$ or $g_m$, is given by

$$\boxed{g_{fs} \equiv g_m = dI_{DS}/dV_{GS} = 2K(V_{GS} - V_t) = 2I_{DS}/(V_{GS} - V_t) = 2\sqrt{K\,I_{DS}}} \tag{B.10}$$

Again, all of the preceding equations are equally applicable to both N-channel and P-channel MOSFETs and the transfer conductances will be positive in both cases.

The parameter $K$ for the MOSFET is given by

$$K = \tfrac{1}{2}\,\mu\,C_{ox}\,(W/L) = \tfrac{1}{2}\,\mu\,(\epsilon_{ox}/\mathrm{t}_{ox})(W/L) \tag{B.11}$$

where $\mu$ = mobility of the charge carriers in the surface inversion layer,
$C_{ox} = \epsilon_{ox}/t_{ox}$ = capacitance per unit area of the MOS capacitor formed by the gate, the gate oxide, and the surface inversion layer,
$\epsilon_{ox}$ = permittivity of the gate oxide ($SiO_2$) $\simeq 3.1 \times 10^{-13}$ F/cm,
$t_{ox}$ = thickness of the gate oxide,
$W/L$ = channel width to length ratio.

The charge carrier mobility in the surface inversion layer channel will be substantially less than the bulk mobility due to the scattering of the charge carriers in the very thin inversion layer at the $Si/SiO_2$ interface. The channel mobility will generally be in the range of about one-half of the corresponding bulk mobility so that for an NMOS transistor $\mu_{\text{channel}} \approx \mu_n/2 \approx 500$ cm²/volt-second, and for PMOS $\mu_{\text{channel}} \approx \mu_p/2 \approx 200$ cm²/volt-second.

For a representative example let us consider an NMOS transistor for which

$t_{ox}$ = 1000 Å = 0.1 μm and we will assume a channel mobility of 500 cm²/volt-second. For the parameter $K$ we obtain

$$K = 500 \text{ cm}^2/\text{volt-second} \times (3.1 \times 10^{-13} \text{ F/cm} /2 \times 10^{-5} \text{ cm})(W/L)$$

$$= 775 \times 10^{-8} (W/L) \text{ A/V}^2 = 7.75 (W/L) \ \mu\text{A/V}^2$$

If $L$ = 10 μm and $W$ = 100 μm then $K$ = 77.5 μA/V². At $V_{GS} - V_t$ = +5.0 V the drain current will be $I_{DS}$ = 77.5 μA/V² × (5 $V$)² = 1.94 mA. The transfer conductance at this current level will be $g_{fs} = 2I_{DS}/(V_{GS} - V_t)$ = 2 × 1.94 mA/ 5.0 V = 0.776 mS. For a PMOS transistor of similar construction the value of $K$ will be 31 μA/$V^2$. At $V_{GS} - V_t$ = 5.0 V we obtain $I_{DS}$ = 776 μA and $g_{fs}$ = 310 μS.

## B.3 FIELD-EFFECT TRANSISTOR OUTPUT CONDUCTANCE

The variation of the drain current $I_{DS}$ of a field-effect transistor, JFET or MOSFET, with respect to the drain-to-source voltage $V_{DS}$ in the saturated (active) region is due to the *channel length modulation effect* that is similar to the base width modulation effect or Early effect in a bipolar transistor. The dynamic drain-to-source conductance $g_{ds}$ of a FET in the saturated region can be expressed as

$$\boxed{g_{ds} = dI_{DS}/dV_{DS} = I_{DS}/V_A} \tag{B.12}$$

where the parameter $1/V_A$ is the channel length modulation coefficient. The parameter $V_A$ will have units of volts and is analogous to the Early voltage of a bipolar transistor. It will generally be in the range of 20 to 200 V for most FETs.

We will now have a simple analysis of the channel length modulation coefficient for MOSFETs. Similar results will apply for JFETs. The drain current of a MOSFET in the saturated region is given by $I_{DS} = K(V_{GS} - V_t)^2$ where $K = \frac{1}{2}\mu C_{ox} (W/L)$. The rate of change of the drain current with respect to the drain voltage is due to the variation in the channel length $L$ as a result of the expansion of the drain-body junction depletion region as shown in Figure B.4, and will be given by

$$g_{ds} = dI_{DS}/dV_{DS} = dI_{DS}/dL \times dL/dV_{DS} = -(I_{DS}/L)\ dL/dV_{DS}$$

$$= \mathrm{I}_{DS}\left(-\frac{1}{\mathrm{L}}\frac{dL}{dV_{DS}}\right) \tag{B.13}$$

We will define the channel length modulation coefficient as

$$\textit{channel length modulation coefficient} = 1/V_A = -(1/L)\ dL/dV_{DS} \tag{B.14}$$

The effective or electrical channel length $L$ will be given by $L = L_0 - \Delta L$. $L_0$ is the channel length when $V_{DS} = 0$ and $\Delta L$ is the depletion layer width at the drain-body junction. This depletion layer width will be given by $\Delta L = \sqrt{2\epsilon V_J/gN} = \sqrt{2\epsilon(V_{DS} + \phi)/qN}$ where $N$ is the body region doping (assuming heavily doped N⁺ or P⁺ source and drain regions) and $\phi$ is the contact potential of the junction and

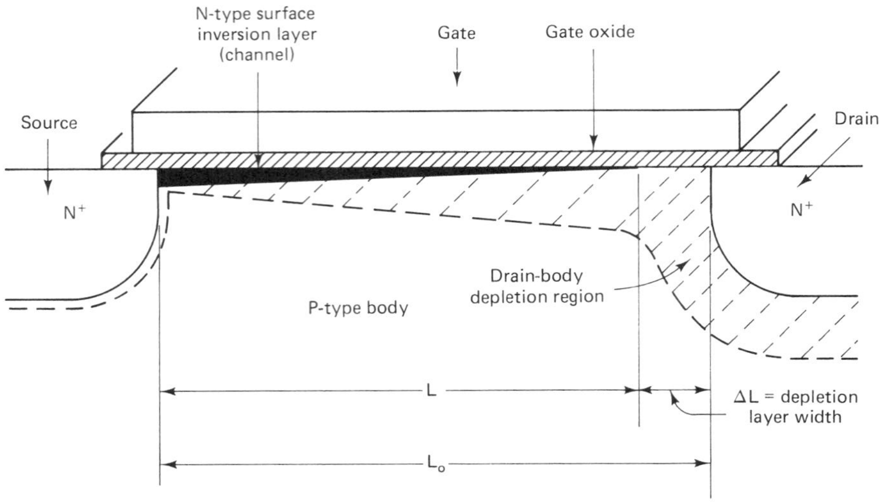


**Figure B.4** Channel length modulation effect.

is typically about 0.8 V. For $dL/dV_{DS}$ we now have $dL/dV_{DS} = dL/d(\Delta L) \times d(\Delta L)/dV_{DS} = -1 \times \frac{1}{2}\Delta L/(V_{DS} + \phi)$ so that

$$\frac{1}{V_A} = -\frac{1}{L}\frac{dL}{dV_{DS}} = \frac{\Delta L/L}{2(V_{DS} + \phi)}$$

and thus

$$\boxed{V_A = 2(V_{DS} + \phi)(L/\Delta L) = 2(V_{DS} + \phi)(L_o/\Delta L - 1)} \qquad \text{(B.15)}$$

In Table B.1, some representative values of $V_A$ are given based on this equation and for the case of $V_{DS} = 10$ V, $\phi = 0.8$ V, and for channel lengths of 5.0, 7.5, and 10 $\mu$m.

From Table B.1 we note that $V_A$ increases substantially with increases in the body doping level. However, increases in the body doping level will also result in

**TABLE B.1** $V_A$ **VERSUS BODY DOPING** $N$

$V_{DS} = 10$ V  $\phi = 0.8$ V

| | Body resistivity (ohm-cm) | | | $V_A$(volts) for $L =$ | | |
|---|---|---|---|---|---|---|
| $N$(cm$^{-3}$) | NMOS | PMOS | $\Delta L$ ($\mu$m) | 5 $\mu$m | 7.5 $\mu$m | 10 $\mu$m |
| $1 \times 10^{15}$ | 13 | 5.0 | 3.75 | 7.2 | 21.6 | 36.0 |
| $3 \times 10^{15}$ | 4.5 | 1.7 | 2.17 | 28.3 | 53.2 | 78.1 |
| $1 \times 10^{16}$ | 1.5 | 0.6 | 1.19 | 69.4 | 115 | 160 |
| $3 \times 10^{16}$ | 0.65 | 0.23 | 0.68 | 136 | 215 | 294 |

increases in the threshold voltage $V_t$ which in turn will result in lower values for the transfer conductance. We also see the important role of the channel length $L_o$ in that $V_A$ increases rapidly with channel length. However, $K$ and therefore $g_{fs}$ as well as the source-to-drain transit time will be inversely proportional to the channel length.

Short-channel MOSFETs will generally exhibit very low values of $V_A$ and also have relatively low values of the drain-to-source breakdown voltage $BV_{DS}$ due to the drain-to-source punch-through. This can be partially compensated for by using a larger body doping. A very effective technique however is the use of an $N/N^+$ drain region as is the case in the DMOS and VMOS transistors.

In the N-channel DMOS and VMOS devices the doping level of the N-type drain region adjacent to the P-type body is much lower than the body region doping. This is to be contrasted with the conventional N-channel MOSFET in which the $N^+$ drain region is much more heavily doped than the P-type body. As a result of the lightly doped N-type drain region in DMOS and VMOS transistors, the drain region will absorb most of the depletion region thickness so that the incursion of the depletion region into the P-type body, and therefore the decrease of the effective channel length $L$, will be relatively small. The result of this will be a much larger value of $V_A$ and drain-to-source breakdown voltage than would otherwise be the case for a short-channel device. Short-channel DMOS and VMOS transistors with channel lengths of less than 1 $\mu$m will have very large values for the $K$-parameter and therefore for the transfer conductance $g_{fs}$. The heavily doped $N^+$ portion of the drain region will insure a low drain series resistance. The short channel length will also lead to a very good frequency response due to the short source-to-drain transit time. At the same time these devices will exhibit large values for $V_A$ and therefore a relatively small value for $g_{ds}$, as well as a high value for the drain-to-source breakdown voltage.

### B.3.1 FET Output Conductance with Impedance in Series with the Source

Let us now consider the circuit of Figure B.5a in which there is an impedance $Z_S$ in series with the source. The FET can be either a JFET or a MOSFET, and in the equivalent circuit representation of Figure B.5b the drain to source conductance $g_{ds}$ of the FET is represented as a conductance external to the transistor.

The total FET a-c small-signal current $i_d$ will be given by $i_d = g_{fs}v_{gs} + g_{ds}v_{ds}$. Since the condition that $r_{ds} = 1/g_{ds} \gg Z_S$ will almost always hold true, we will have that $v_{ds} \simeq v_d$ so that $i_d$ can be written as $i_d \simeq g_{fs}v_{gs} + g_{ds}v_d$. For $v_{gs}$ we have that $v_{gs} = -i_dZ_S$ so that $i_d \simeq -g_{fs}Z_Si_d + g_{ds}v_d$. Solving this last equation for $i_d$ gives $i_d(1 + g_{fs}Z_S) \simeq g_{ds}v_d$ and thus $i_d \simeq g_{ds}/(1 + g_{fs}Z_S)\, v_d$.

The output or drain dynamic conductance of the FET circuit will be given by $g_o = i_d/v_d$. Solving for $g_o$ from the preceding equations gives

$$g_o \simeq \frac{g_{ds}}{1 + g_{fs}Z_S} \tag{B.16}$$

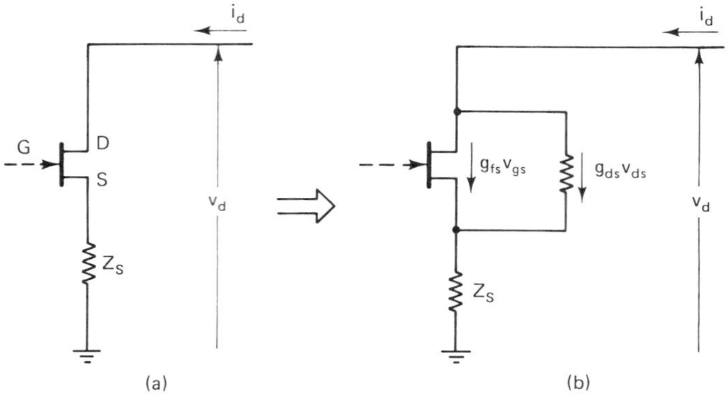


**Figure B.5** Field-effect-transistor output conductance.

## B.4 SUBSTRATE BIAS EFFECT (BODY EFFECT)

The MOSFET is actually a four electrode device with the gate, source, drain, and body (or substrate) electrodes. In most applications the body (or substrate) of the MOSFET is internally connected to the source so that the MOSFET appears as a three electrode or triode device. In many IC applications, however, this is not possible since the MOSFET body is the same as the IC substrate that is common to many transistors on the chip.

A voltage difference between the body and the source of a MOSFET can affect the net charge in the source-to-drain channel and the threshold voltage of the device. The body-to-source voltage $V_{BS}$ is, however, usually much less influential than the gate-to-source voltage $V_{GS}$ in controlling the channel conductance. This is due to the much greater thickness of the body-channel junction depletion region as compared to the thickness of the gate oxide. The relative influence of the body voltage compared to that of the gate voltage over the net charge in the channel will be proportional to the ratio of the body-channel junction capacitance $C_J$ to the gate-to-channel (or oxide) capacitance $C_{ox}$. The effect of the body voltage on the MOSFET characteristics can most conveniently be described in terms of a shift in the threshold voltage $\Delta V_t$ that results from a change in the body-to-source voltage $\Delta V_{BS}$.

The shift in the threshold voltage produced by a change in the body voltage will be given by

$$K_{BS} = \frac{-\Delta V_t}{\Delta V_{BS}} = \frac{C_J}{C_{ox}} = \frac{\sqrt{\dfrac{q\epsilon_{si}N_B}{2|V_{BS}+\phi|}}}{\epsilon_{ox}/t_{ox}} \tag{B.17}$$

where $N_B$ is the substrate (or body) doping and $\phi$ is the junction contact potential (~0.8 V). To obtain a normalized value of the $K_{BS}$ coefficient we will use $N_B = 1 \times 10^{16}\text{cm}^{-3}$, $V_{BS} + \phi = 1.0$ V, and $t_{ox} = 1000$ Å for which we obtain the following result

$$K_{BS} = \frac{\sqrt{\dfrac{1.602 \times 10^{-19}C \times 1.04043 \times 10^{-12}\text{F/cm} \times 10^{16}\text{cm}^{-3}}{2 \times 1.0 \text{ V}}}}{\dfrac{3.8 \times 10^{-13}\text{F/cm}}{1 \times 10^{-5}\text{cm}}} \tag{B.18}$$

$$= \underline{0.860}$$

For a general relationship for $K_{BS}$ we have

$$K_{BS} = 0.860 \frac{t_{ox}}{1000 \text{ Å}} \sqrt{\frac{N_B}{10^{16}\text{cm}^{-3}} \times \frac{1.0 \text{ V}}{V_{BS} + \phi}} \tag{B.19}$$

For a representative example we will let $t_{ox}$ = 800 Å, $N_B = 1 \times 10^{15}\text{cm}^{-3}$, and $V_{BS} + \phi = 10$ V, for which we obtain $K_{BS} = 0.0688$ so we see that the body voltage can produce a significant effect on the MOSFET characteristics, although it is considerably less influential than the gate voltage.

The influence of the body voltage on the drain current can be expressed in terms of a body-to-drain current dynamic transfer conductance, $g_{fb} = dI_{DS}/dV_{BS}$. This transfer conductance can be obtained from the relationship

$$g_{fb} = \frac{dI_{DS}}{dV_{BS}} = \frac{dI_{DS}}{dV_t} \times \frac{dV_t}{dV_{BS}} = g_{fs}K_{BS} \tag{B.20}$$

where $g_{fs}$ is the gate-to-source dynamic transfer conductance.

A principal consequence of the body effect will be an often substantial increase in the dynamic drain-to-source conductance of the MOSFET. This can result in a reduced voltage gain for MOSFET amplifiers using active loads. Another consequence can be an increased dynamic conductance for MOSFET current source circuits.

To evaluate the contribution of the body effect to the dynamic output conductance of a MOSFET let us consider the circuit of Figure B.6. In this circuit an N-channel depletion mode MOSFET is used as a current source of strength $I_O = I_{DSS}$. The dynamic output conductance $g_o$ will be given by

$$\begin{aligned} g_o &= -\frac{dI_O}{dV_O} = -\frac{dI_O}{dV_{DS}} \times \frac{dV_{DS}}{dV_O} - \frac{dI_O}{dV_{BS}} \times \frac{dV_{BS}}{dV_O} \\ &= -(g_{ds} \times -1) - (g_{fs}K_{BS} \times -1) \\ &= g_{ds} + g_{fs}K_{BS} \\ &= \frac{I_{DSS}}{V_A} + \frac{2I_{DSS}}{-V_P} K_{BS} \end{aligned} \tag{B.21}$$

The fractional change in $I_O$ per volt change in $V_O$ will therefore be given by

$$\frac{-1}{\text{I}_\text{O}} \frac{\text{dI}_\text{O}}{\text{dV}_\text{O}} = \frac{1}{\text{V}_\text{A}} + \frac{2\text{K}_\text{BS}}{-\text{V}_\text{P}} \tag{B.22}$$

Note that in the circuit of Figure B.6 the body of the MOSFET is not connected to the source so that a body-to-source voltage of $V_{BS} = -V_O$ will be present. This will result in a substantial increase in the output conductance. If the circuit construc-

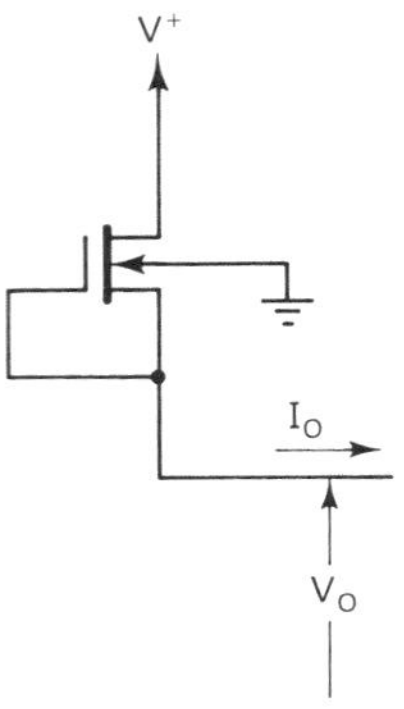


**Figure B.6** Contribution of the body effect to the dynamic output conductance.

tion were to allow the body to be connected to the source then we would have $V_{BS} = 0$ and the output conductance would reduce to just $g_o = g_{ds} = I_{DSS}/V_A$.

## B.5 COMPARISON OF BIPOLAR TRANSISTOR AND FET MODES OF OPERATION

It is of interest to compare the various regions or modes of operation of bipolar and field-effect transistors.

In the cutoff region for a bipolar transistor both junctions (EB and CB) are off and all transistor currents are reduced to zero, except for the small amount of junction leakage current. This corresponds to the cutoff region of a JFET (or a depletion mode MOSFET) in which the gate-to-source voltage is beyond the pinch-off value which fully depletes the channel all the way between the source and drain regions. The channel is thus pinched off at both the source and drain ends and the transistor currents are reduced to zero, except for the very small leakage currents. For the enhancement mode MOSFET the corresponding situation is that of the gate-to-source voltage being below the threshold value such that no conducting channel is formed between the source and drain regions.

In the active region for a bipolar transistor the emitter-base junction is on and the collector-base junction is off. For the case of an NPN transistor this corresponds to the situation where the emitter is emitting electrons into the base region, and these electrons after traversing the base region, are then collected by the collector. In this mode of operation the collector current will be relatively independent of the collector voltage, but exponentially dependent on the base-to-emitter voltage. For field-effect transistors (JFETs and MOSFETs) this corresponds to the situation of the channel being open at the source end, but being pinched off for a very short distance at the drain end. This produces the approximate saturation of the drain current such that the drain current increases very slowly with increasing drain voltage (but is very much dependent on the gate voltage). For this reason this active region for FETs is often called the "saturated region" (not to be confused with the saturation region of bipolar transistors). The FETs operation in the active ("saturated") region is obtained when the drain-to-gate voltage is such that the channel will be pinched off at the drain end, but at the same time the gate-to-source voltage will result in an open channel at the source end.

The saturation region for bipolar transistors occurs when both junctions (EB and CB) are on. In this region, the collector current is a strong function of the collector voltage, and the collector-to-emitter voltage drop across the transistor is small, generally less than 0.2 V. For field-effect transistors this corresponds to the channel being open (or "on") at both the source and drain ends, and as a result the drain current will be very much dependent on the drain voltage. This mode or region of operation for FETs is often called the "triode" or "nonsaturated" region. For small values of drain voltage the drain current will be an approximately linear function of the drain voltage, but as the drain voltage increases the channel starts to narrow at the drain end which produces a bending over of the drain current versus drain voltage curve. Finally, as the drain voltage increases further, the channel becomes closed ("pinched-off") at the drain end and the transistor then enters the "saturated" (i.e. active) region where the drain current becomes relatively independent of drain voltage.

# Appendix C

# TRANSISTOR SMALL-SIGNAL CIRCUIT MODELS

Although both bipolar and field-effect transistors are basically nonlinear devices, under a-c small-signal conditions these devices can be represented in the form of a linear equivalent-circuit model. In Figure C.1 a general transistor amplifier stage is represented, and in Figures C.2 and C.3 are some approximate small-signal a-c models for various bipolar and field-effect transistor configurations. These are simplified models but will be suitable for virtually all practical transistor amplifier applications.

The following is a summary of some of the basic parameters and the midfrequency voltage gain equations for the various transistor circuits. For the parameter values given, $V_T = kT/q$ = thermal voltage (~25 mV) and $V_A$ is the Early voltage or the reciprocal of the base-width modulation coefficient for bipolar transistors, and for field-effect transistors it is the reciprocal of the channel-length modulation coefficient. For both types of transistors $V_A$ is typically in the range 30 to 300 V. The dimensionless factor $n$ will be between 1 and 2 and usually will be close to 1.5. For the voltage-gain expressions, $A_{V(\mathrm{MID})}$ is the midfrequency a-c voltage gain as given by $A_{V(\mathrm{MID})} = v_o/v_i$, and $R_{L(\mathrm{NET})}$ is the net load resistance driven by the transistor. The input and output resistances, $r_i$ and $r_o$, are the midfrequency values. At higher

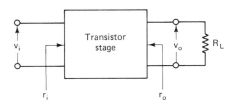

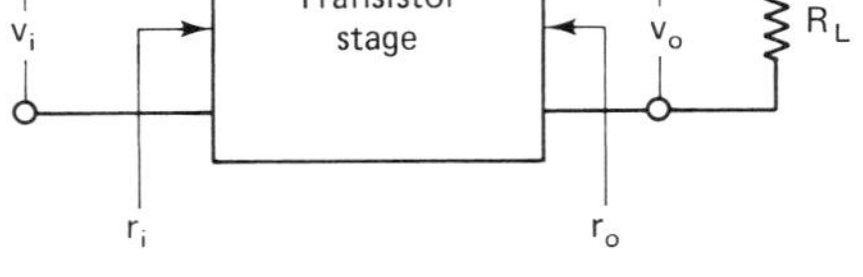


**Figure C.1** Transistor amplifier stage.

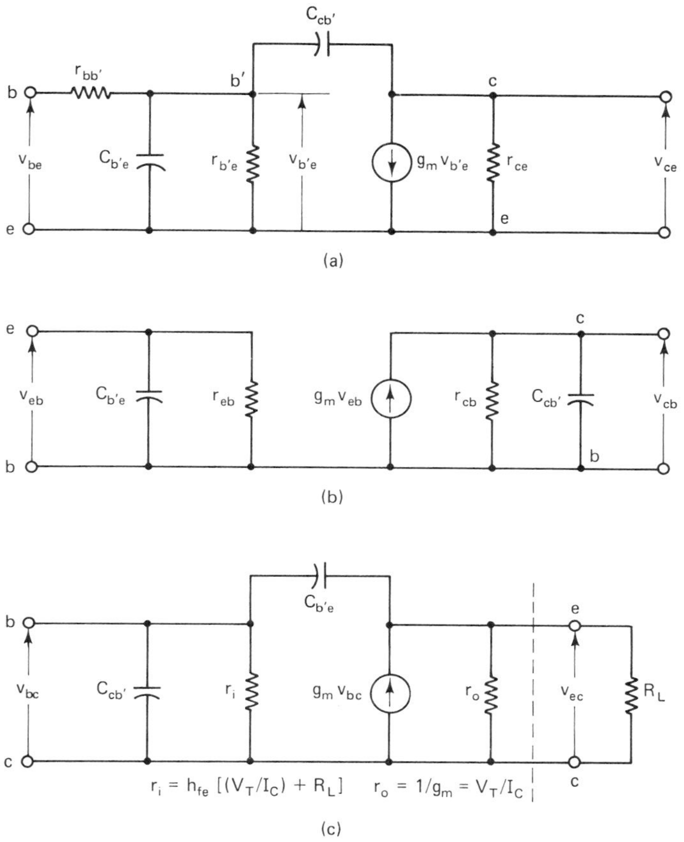


**Figure C.2** Transistor small-signal equivalent-circuit models for bipolar transistors: (a) common-emitter; (b) common-base; (c) common-collector (emitter-follower).

frequencies the capacitative reactances must be considered. For the bipolar transistor circuits some of the relationships with the hybrid or $h$ parameters are shown. The parameter $h_{fe}$ is the common-emitter (short-circuit) a-c current gain, sometimes called $\beta$, $h_{fb}$ is the common-base current gain, and $-h_{fb} = \alpha \simeq 1$.

## Common-Emitter (Figure C.2a)

1. $r_i = h_{ie} = r_{be} = r'_{bb} + r'_{be}$, where $r'_{bb}$ is the base spreading resistance and $r'_{be} = nV_T/|I_B|$
2. $g_m = h_{fe}/h_{ie} = |I_C|/V_T$ = dynamic transfer conductance
3. $r_o = r_{ce} = 1/h_{oe} = V_A/|I_C|$
4. $A_{V(\text{MID})} \simeq -g_m R_{L(\text{NET})} = -h_{fe} R_{L(\text{NET})}/h_{ie}$

## Common-Base (Figure C.2b)

1. $r_i = h_{ib} = V_T/|I_E| \simeq V_T/|I_C| = 1/g_m$
2. $g_m = h_{fe}/h_{ie} = -h_{fb}/h_{ib} = |I_C|/V_T$
3. $r_o = r_{cb} = 1/h_{ob} = h_{fe}/h_{oe} = h_{fe}(V_A/|I_C|)$
4. $A_{V(\mathrm{MID})} = g_m R_{L(\mathrm{NET})} = -h_{fb} R_{L(\mathrm{NET})}/h_{ib}$

## Common-Collector (Emitter-Follower) (Figure C.2c)

1. $r_i = h_{fe}[(V_T/|I_C|) + R_L] = h_{fe}[(1/g_m) + R_L]$
2. $g_m = |I_C|/V_T$
3. $r_o = 1/g_m = V_T/|I_C|$
4. $A_{V(\mathrm{MID})} = g_m[r_o \| R_L] = \dfrac{g_m}{g_m + 1/R_L} = \dfrac{g_m R_L}{1 + g_m R_L} = \dfrac{R_L}{(1/g_m) + R_L}$

$$= \frac{R_L}{R_L + V_T/|I_C|}$$

(Note that for the usual case of $R_L \gg V_T/I_C$, $A_{V(\mathrm{MID})} \simeq 1$.)

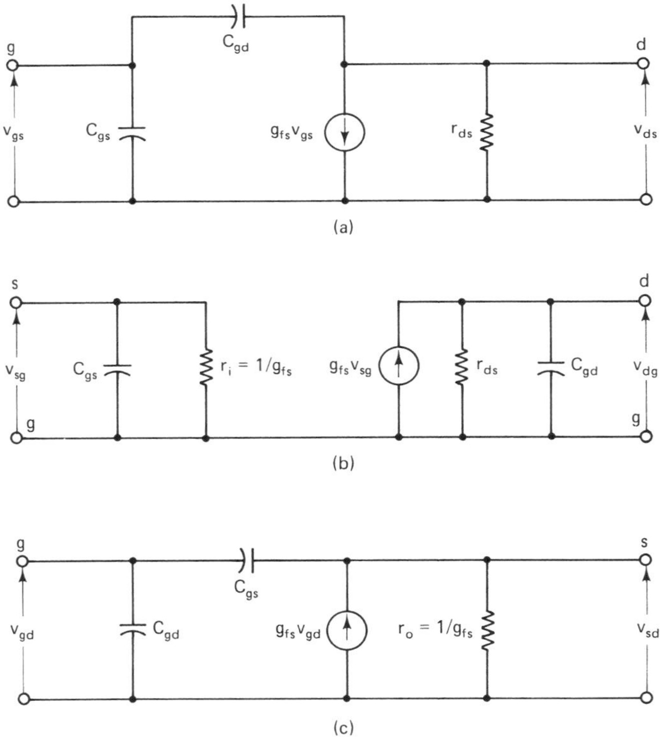


**Figure C.3** Transistor small-signal equivalent-circuit models for field-effect transistors: (a) common-source; (b) common-gate; (c) common-drain (source-follower).

### Common-Source (Figure C.3a)

1. $r_o = r_{ds} = V_A / |I_{DS}|$
2. $A_{V\,(\mathrm{MID})} = -g_{fs} R_{L\,(\mathrm{NET})}$

### Common-Gate (Figure C.3b)

1. $r_i = 1/g_{fs}$
2. $r_o = r_{ds} = V_A / |I_{DS}|$
3. $A_{V\,(\mathrm{MID})} = g_{fs} R_{L\,(\mathrm{NET})}$

### Common-Drain (Source-Follower) (Figure C.3c)

1. $r_o = 1/g_{fs}$
2. $A_{V\,(\mathrm{MID})} = g_{fs}\,[r_o \| R_L] = \dfrac{g_{fs}}{g_{fs} + 1/R_L} = \dfrac{g_{fs} R_L}{1 + g_{fs} R_L}$

# INDEX

## A



## B



## C

## D

## K



## L



## M



## N



## O



## P